PETER GOLDBLATT & JOHN MANNING

GLADIOLUS
in Southern Africa

Watercolours by Fay Anderson & Auriol Batten

Line drawings by John Manning

FERNWOOD
PRESS

FERNWOOD
PRESS

P.O. BOX 15344
VLAEBERG 8018

REGISTRATION NO. 90/04463/07

FIRST PUBLISHED 1998

TEXT © PETER GOLDBLATT AND JOHN MANNING 1998
PAINTINGS © FAY ANDERSON (PLATES 6, 16, 20, 21, 30, 31, 33, 34,
36–42, 44–46, 48–50, 52–62, 65–77, 79–88, 90, 91,
93–98, 100, 102–106, 109–137, 139–144)
© AURIOL BATTEN (FRONTISPIECE,
PLATES 1–5, 7–15, 17–19, 101, 107, 108, 138)

EDITED BY LENI MARTIN
DESIGNED BY JOANNE SIMPSON
DUSTJACKET DESIGN BY JOANNE SIMPSON
DTP BY EPS&M
MAPS BY RED ROOF DESIGN
PRODUCTION CONTROL BY ABDUL LATIEF (BUNNY) GALLIE

REPRODUCTION BY POSITIVE IMAGE CC, CAPE TOWN
REPRODUCTION CO-ORDINATOR ANDREW DE KOCK
PRINTED AND BOUND BY TIEN WAH PRESS (PTE) LTD, SINGAPORE
COLLECTORS' EDITION BOUND BY PETER CARSTENS, JOHANNESBURG

ISBN 1 874950 32 6 STANDARD EDITION
ISBN 1 874950 33 4 COLLECTORS' EDITION

CONTENTS

FOREWORD

*I*t is indeed a great pleasure for me to contribute this Foreword to a beautifully produced monograph on *Gladiolus* in southern Africa.

Botanists are stern critics of plant illustration, demanding scientific accuracy and, what is less often achieved, a satisfying representation of those less easily defined characteristics which give each species its individual 'stamp' and personality. In both respects, the splendid plates and drawings in this monograph, depicting 145 of the 163 southern African species, should please the most critical eye. They are technically accurate, scientifically exact and aesthetically so pleasing that I feel they must be preserved for the benefit of future beholders of the 'gallery of botanical art'.

I am sure that all those involved in this publication can claim with justification that this will be one of the most significant botanical monographs to appear in recent times. It is the result of many years of study in the field and in herbaria around the world. Thorough research, aided by modern technology, makes this the most complete revision of the genus to date. Together with senior author Peter Goldblatt's recent revisions of *Gladiolus* in Madagascar (1989) and in tropical Africa (1996), this book represents the highest level of knowledge about *Gladiolus*.

Moreover, the authors' extensive field work has yielded new information about its ecology and pollination biology, enabling them to present a broad treatment of the genus unrivalled in African flora in its scope and detail. Of particular merit, I think, is that the authors have not been shy about explaining why they classify as they do, and why they have sometimes disagreed with previous authors on the subject, mainly in the light of new information.

It is clear that southern Africa is a major centre of diversity and speciation for *Gladiolus*, and it is fitting that this work is the product of co-operation between four South Africans who have all won international acclaim in their respective fields of botanical research and botanical art.

The authors have gone to great lengths to collect as many species as possible for illustration. Of the 19 species not painted, nine are represented in the form of superbly precise line drawings by John Manning – a true botanist and artist.

Although *Gladiolus* species have been recorded in southern Africa since Simon van der Stel's day, we are still learning about the genus more than three hundred years later. Even in this latest revision, a number of new species have been described. When one considers this, it is not surprising that the fynbos biome – the home of the vast majority of southern African *Gladiolus* species – is regarded so highly in global terms. For this reason alone, as I have written so often, it is essential to conserve what little is left for future generations to appreciate.

Although it is aimed primarily at botanists, this monograph will also, I am certain, appeal to both nature-lovers and art connoisseurs. I commend wholeheartedly the contributions of all the authors – Fay Anderson, Auriol Batten, Peter Goldblatt and John Manning – to this marvellous creation. Artistry such as theirs, in paintings, drawings and words, deserves to find expression in a tangible form such as this book, and I wish that every beautiful plant could be made immortal in this way. Only in an archive like this can our plants find a measure of security in a world which sadly appreciates less and less 'such unprofitable gaiety and splendour'!

To the publisher, Pieter Struik of Fernwood Press – a hard taskmaster – I have the utmost admiration and respect for the success you have achieved during your career. May there be many more books to follow and may you continue to promote these archives for future generations. We should all be proud, and I am well satisfied that something is being done to preserve what little is left by those who are concerned. Let us all support this book and those to come! ❧

STEVE BALES

Group Art Custodian
First National Bank Group
Johannesburg, April 1998

PREFACE

*A*lthough the genus *Gladiolus* has been known in southern Africa for more than three hundred years and has been comprehensively revised twice in the last century, it is only in this book that an attempt is made to present a natural classification of the genus, and to include within it all species that share an immediate common ancestor with *Gladiolus*. This may sound strange, but the last revision, written largely by G.J. Lewis (1909–1967) and completed by Amelia Obermeyer (Mrs A.A. Mauve) and Captain T.T. Barnard, presented an avowedly artificial infrageneric classification. This treatment, as well as the account of *Gladiolus* in *Flora Capensis* by J.G. Baker, published in 1896, did not include many of the species that are now understood to be highly specialized species of *Gladiolus* with unusual flowers adapted for pollination by either sunbirds or long-tongued insects, including moths and nemestrinid flies. The bird-pollinated species of *Gladiolus* have long-tubed and bright red flowers, sometimes with enlarged upper and reduced lower tepals and were placed by Baker either in the genus *Homoglossum* or in *Antholyza*, the latter also including species now referred to *Babiana* and *Tritoniopsis*. *Homoglossum* was upheld by Lewis et al., and believed to include, in addition, *H. aureum* (now *G. aureus*), a species that was excluded from *Homoglossum* by M.P. de Vos who revised that genus in 1976. Tropical African bird-pollinated species were variously placed in *Homoglossum*, *Petamenes* or *Oenostachys*. Some southern African species of *Gladiolus* with long perianth tubes were assigned by Baker to a genus *Acidanthera*, the species of which shared a similar pale-coloured flower, and one species now regarded as a true *Gladiolus* was included in *Geissorhiza* because it had a fully actinomorphic flower.

What then holds *Gladiolus* together? This is always a difficult question to answer because systematic botany today demands that genera are not only monophyletic (the constituent species have a common ancestor also in that genus), but include all the descendants of that common ancestor. This means in practice that however bizarrely specialized a species may be, it must be included in *Gladiolus* if its closest relatives are also members of that genus. This is easy to say, but in practice how do we decide that, for example, the actinomorphic *G. stellatus* or the red-flowered, extraordinary *G. saccatus*, which has a floral spur and vestigial lower tepals, are members of the same genus. One of the most constant features of *Gladiolus* is the inflated capsule and the discoid, broadly winged seeds. A second feature is the ancestral basic chromosome number of $x = 15$, shared by nearly all species in the genus, including *G. saccatus* and *G. stellatus*. Reduction or even loss of the seed wing has occurred in *Gladiolus* at least twice, and changes in base number have occurred several

times, yet there seems no reasonable doubt in these examples that the plants are members of *Gladiolus* because of their other characters which include a characteristic flower shape, inclined and secund spike, and leaf features that are common to many species of the genus. In combination these features provide indirect evidence of the relationships of such species. Thus the final decision about whether a species belongs in *Gladiolus* can be the result of careful consideration of the phylogenetic relationships of the plant rather than the presence of a particular set of characters.

The decision to write this book was based foremost on the realization that the taxonomy of *Gladiolus* as presented in the 1972 revision by Lewis et al. had become outdated because of the publication of several new species since then, and by the addition of the species of *Anomalesia*, *Homoglossum* and *Oenostachys* to *Gladiolus*. This, combined with the fact that the 1972 revision was out of print, led to our embarking on the project. We were also completing a study of *Gladiolus* in tropical Africa, published in 1996 by Timber Press, and an extension of this work to embrace the species in southern Africa seemed logical. Having recently completed paintings of the woody Iridaceae for a book on the genera of the woody Iridaceae, published by Timber Press in 1993, Fay Anderson agreed to provide illustrations of southern African species of *Gladiolus*. Given the number of species in the genus and the logistics involved in studying and collecting *Gladiolus* species across so vast an area as southern Africa, we were fortunate enough to enlist the collaboration of Auriol Batten in East London.

We naively thought that *Gladiolus* was in the main well understood in southern Africa and that all of the basic nomenclature and the location and identification of type specimens had been completed by G.J. Lewis. Little did we realize that, as in most undertakings of this kind, surprises awaited us at almost every turn. We found, for example, that we could not accept the species concepts adopted by Lewis et al., even for some of the well-known plants such as the western South African *G. alatus* and *G. scullyi*, the widespread eastern southern African *G. woodii*, or the Mpumalanga *G. varius*. Field research and travel intended to collect species for illustration and description often resulted in the discovery of plants that did not conform to the current taxonomy. Sometimes these plants had been referred to existing species but were new, while in other cases they had been described, but were misunderstood and later synonymized. The total number of species of *Gladiolus* recognized by Lewis et al. in southern Africa was 103. We now believe a more realistic figure is 163. This includes the 13 species of *Homoglossum*, *Anomalesia* and *Oenostachys* added to the total species of *Gladiolus* in southern

Africa after the Lewis et al. revision, eight described after 1972, and the 15 new species that we have described since the beginning of the project, 12 of them in this monograph. Together with the 82 species of *Gladiolus* in tropical Africa, nine in Madagascar, and an estimated ten in Eurasia, there now appear to be more than 255 species in the genus, probably making it the largest genus in the Iridaceae. The other large genus in the family, *Iris*, has an estimated 225 species, but critical study may show that here too, the total number of species has been underestimated. It is clear from the foregoing that southern Africa is a major centre of diversity and speciation for *Gladiolus*, but the high species number for tropical Africa shows that radiation in the genus is by no means confined to the temperate south. The genus is in fact remarkable in being one of few African genera that has speciated significantly across the entire continent and in Madagascar.

Gladiolus is also a genus in which many of the characters on which species are based are floral. The shape of the flower, the relative lengths of the tepals, their orientation and colour are often the primary or the only features on which species are defined. Yet some or all of these characters preserve poorly in herbarium specimens, the basis for all work in plant systematics. As a result, although there are many thousands of preserved specimens available in collections throughout the world, field research was vital to achieving an understanding of species, their natural patterns of variation and their ecology. These limitations were overcome by extensive field work and the generous assistance of botanical colleagues and informed amateurs who, like us, were enchanted and excited by the genus.

Despite the economic importance of *Gladiolus* as an ornamental garden plant and in the cut-flower industry, the genus had never been studied in its entirety. The species of *Gladiolus* in southern Africa were relatively well understood, but the tropical African species had not been surveyed since J.G. Baker's treatment of the genus for the *Flora of Tropical Africa,* published in 1898, a time when the continent had barely begun to be explored botanically. With the publication of revisions of *Gladiolus* in Madagascar and in tropical Africa (Goldblatt, 1989, 1996), and the completion of this monograph for the southern African species, knowledge of the genus has reached a new level. In particular, our extensive field work has yielded new information on the ecology and pollination biology of the genus. This has enabled us to present a broad treatment of *Gladiolus* which we believe is unrivalled in the African flora in its scope and detail.

It has been a pleasure to work with Auriol Batten in the course of preparation of her watercolour illustrations. We thank

her for the hard work, long hours and wonderful results. We thank the following for generously sharing their knowledge in the field and assisting with other aspects of our research: Kevin Balkwill, Ashley Batten, Erich Bigalke, Estelle Brink, John and Sandy Burrows, Marge Courtenay-Latimer, Welland Cowley, Charles Craib, Zelda Crossman, Tony Dold, John Duckitt, Mark Duckitt, Graham Duncan, Desmond Elliott, Elsie Esterhuysen, Peter Fischer, Noel Gray, Robin Guy, Nel Hanekom, Olive Hilliard, Doreen Holliday, Colin King, Noelline Kroon, Sonette Krynauw, Bill Liltved, Peter Linder, Pixie Littlewort, Ben Loots, Mervyn Lotter, John MacGillivray, Neil and Neva MacGregor, William Massyn, Diana Matlock, Cameron and Rhoda McMaster, Henrietta Melck, Phyllis Nänni, Geoff Nichols, Jo Onderstall, Queenie Paine, Stefaan Pienaar, Darrel Plowes, Anne Rennie, Jill Renton, Charles Rostance, Pat Runnals, Averil Sonemann, Marina Swart, Allan Tait, Kobie Truter, Stella van Gass, Christa van Rooyen, Marjorie van Rooyen, Jan Vlok, Desmond Weeks, Ion Williams, Trevor Wolf and Mary Yates. Our several trips to Namaqualand were made more pleasant by the hospitality of Tant Kosie Roux of Springbok. Roy Gereau helped us with the translation of the diagnoses of new species into Latin, for which we are very grateful. We also thank especially Ingrid Nänni for unstinting help with pollination studies and for carefully reviewing the manuscript. The Indigenous Bulb Growers Association of South Africa (IBSA) is due a special acknowledgement for their generous assistance with growing rare species, for bringing others to our attention, and for allowing us access to their collections for study and illustration. In particular we thank Andries de Villiers, Johan Fourie, Alan Horstmann, Gordon Summerfield, Cameron and Margie Taswell-Yates and Paul von Stein. We also owe special thanks to Jan Vlok and Anne Lise Schutte for assisting us with species from the southern and eastern Cape when time did not permit travel there ourselves.

Lastly, we gratefully acknowledge our respective institutions, the Missouri Botanical Garden, St Louis, USA, and the National Botanical Institute, South Africa, for supporting our work, allowing us *carte blanche* to travel when and where it seemed necessary and, when needed, for providing financial support. Support for the preparation of the line drawings was provided by The Stanley Smith Horticultural Trust, and our research was funded in part by grants 5153–93 and 5408–95 from the National Geographic Society, USA, and by the National Botanical Institute, South Africa. Publication subsidies were provided by The Stanley Smith Horticultural Trust, First National Bank, Dr John Cook, the Ernest Oppenheimer Memorial Trust and the Mia Karsten Fund.

Gladiolus, the largest genus of the petaloid monocot plant family Iridaceae, is thought to comprise some 255 species. A member of the mainly African subfamily Ixioideae, it is far and away the largest genus in this, the largest subfamily of the Iridaceae with some 950 species in 27 genera. It is rivalled in size only by *Iris*, a member of subfamily Iridoideae, which may have close to 250 species. *Gladiolus* is also one of the world's most important horticultural plants, valued both as an ornamental garden subject and as a cut-flower crop. Yet despite its size and importance, the genus has until recently been surprisingly poorly understood in the wild. Such basic facts as the number of species that exist, their geographical ranges and their patterns of variation have not until now been known with any certainty. Not only has this information been lacking, but the generic definition and circumscription of *Gladiolus* have only recently been satisfactorily established. Beginning in 1988, the circumscription and systematics of the genus have been methodically investigated, first for Madagascar (Goldblatt, 1989) and then for tropical Africa (Goldblatt, 1993; 1996). Finally, the systematics of *Gladiolus* in southern Africa, undoubtedly the centre of the genus in terms of both species richness and overall diversity, is revised and presented here in monographic detail.

The definition of *Gladiolus* has now been firmly settled and the genus is seen as comprising not only a core of short-tubed species that have the classic flower form we recognize as that of *Gladiolus*, but also numerous others closely related to them. These include species with long-tubed, white flowers referred in the past to the genus *Acidanthera*. Likewise, now included in *Gladiolus* are the tropical African species with large, often reddish bracts and flowers and the lower tepals much reduced in size that were variously placed in *Petamenes* or *Oenostachys*. In southern Africa species with bright red flowers, either with a wide tube and segregated in the genus *Homoglossum*, or with an elongate dorsal tepal and vestigial lower ones and included in *Anomalesia* and *Kentrosiphon*, have also been transferred to *Gladiolus* (Goldblatt & De Vos, 1989). These several species are now understood to belong in *Gladiolus* in the phylogenetic sense, but with their flowers specialized for particular pollination systems, either sphinx moth (*Acidanthera*) or sunbird. Floral specialization for a particular pollination strategy alone is not sufficient grounds to warrant recognition of separate genera, and these genera have been merged with *Gladiolus*.

As currently understood then, *Gladiolus* is seen as including species with moderate-sized to large, usually zygomorphic flowers with unilateral, arcuate anthers and channelled style branches, variously expanded toward the apices. In addition to the floral characters, the genus is characterized by softly textured and relatively large, green floral bracts, flowers on a secund spike and inflated capsules containing numerous seeds with a broad membranous wing. An ancestral chromosome number in *Gladiolus* of $x = 15$ is also unusual in subfamily Ixioideae, where it is shared only with the monotypic African genus *Radinosiphon*. Exceptions to one or more of these features occur in several species and must be seen as isolated specializations within the genus and not as indications of taxonomic misplacement.

The geographic range of *Gladiolus* is Africa, Madagascar, Europe and the Middle East. Approximately ten of the 255 species in the genus occur north of the Sahara in Eurasia as far east as Afghanistan, and another eight occur in Madagascar. There are some 84 species in tropical Africa, where the genus is well understood as a result of research by the first author of this book (Goldblatt, 1996). In southern Africa – the area south of the Cunene–Okavango–Limpopo River axis – we now recognize 163 species. This part of Africa lies mostly south of the Tropic of Capricorn and includes the countries of Botswana, Lesotho, Namibia, South Africa and Swaziland, plus the southern corner of Mozambique. Species mostly have fairly narrow ranges and few occur in more than one of these main regions. Madagascar shares *G. dalenii* with tropical and southern Africa, while *G. crassifolius*, *G. magnificus*, *G. oatesii*, *G. permeabilis* and *G. sericeovillosus* are shared between southern Africa and southern tropical Africa.

Gladiolus has been fairly well studied in southern Africa and was revised as recently as 1972 by G.J. Lewis, A.A. Obermeyer and T.T. Barnard who admitted 103 species to the genus. At the time, however, four small genera, namely *Homoglossum* (sometimes also called *Petamenes*, comprising 12 species), *Anomalesia* (including *Kentrosiphon*, three species) and *Oenostachys* (about five species), were regarded as separate from *Gladiolus*, although they were generally acknowledged to be closely related to it. As a result of the union of these genera with *Gladiolus*, the latter acquired an additional 13 southern African species. Several new species, added to *Gladiolus* after the publication of the revision by Lewis et al., had brought the number to 125 (Hilliard & Burtt, 1979, 1983, 1986; Goldblatt, 1984a; Goldblatt & Vlok, 1989; Obermeyer, 1979, 1982). The total number of species now recognized, 163, is the result of the addition of these new and transferred species, a critical taxonomic re-evaluation of existing species, and the discovery of several new species made in the course of field work while studying *Gladiolus* for this monograph.

The reasons for the tremendous diversity and success of *Gladiolus* are complex. Obviously, several factors have contributed to this remarkable example of adaptive radiation across all of Africa, but especially in southern Africa with its preponderance of species. These include an unusual floral adaptability that over time has allowed species to become specialized for almost all the important floral pollinators in Africa, including long- and short-tongued bees, butterflies, long-tongued flies, sphinx and noctuid moths, sunbirds and, in one case, scarab beetles. Vegetative specialization has also been significant. Variation in leaf morphology, ranging from the broad, soft-textured leaves of some species to the tough, narrow leaves with thickened margins and midribs of others, has permitted radiation from moist, high-rainfall habitats to semi-arid sites. Various species in the genus are adapted to grow on most of the substrates encountered in southern Africa, including acidic sandstone, quartzitic and granitic soils, basic limestone and dolomite, toxic serpentine and shifting coastal dunes. Other significant adaptations are the ability to flower after the season's leaves have withered and before the next season's vegetative shoot has sprouted, and control of the flowering response to early, middle or late in the season. Flowering times in *Gladiolus* species, unlike in most petaloid monocots of southern Africa, thus extend almost throughout the year. This, like floral diversity, makes available more adaptive niches to the genus and in turn leads to greater species diversification. Finally, the specialized, wind-dispersed seeds have enabled the genus to spread widely and to colonize isolated and disjunct habitats in which populations may diverge and evolve into new species largely as a result of their geographic isolation.

HISTORY

Gladiolus has been known since classical times and has been a treasured wild flower in the Mediterranean and Middle East for thousands of years. The word *gladiolus* is Latin for 'little sword' and the genus was so known by the Romans. The name alludes to the flattened, sword-shaped leaves, actually a characteristic of the entire *Iris* family. In 1753 *Gladiolus* entered the era of modern botany with the introduction of the binomial system of botanical nomenclature. In that year *Species Plantarum*, the seminal work by the Swedish scientist Carl Linnaeus, was published, in which he laid the foundations of the modern system of naming plants. Six species of *Gladiolus* were included in this work: two Eurasian and four from the Cape of Good Hope. Just one of these last four, *G. angustus*, still belongs in the genus as it is understood today. The other three have long since been removed to different genera. By 1753, however, several more species of *Gladiolus* from southern Africa were already known in Europe, although their generic affiliation was not always clear. These were to be formally named according to the binomial system in the coming years.

Botanical exploration in southern Africa really began with Simon van der Stel, Governor of the Dutch colony at the Cape. Van der Stel led an expedition to the fabled copper mountains of Namaqualand in 1685, little more than 30 years after the establishment of a permanent European settlement at the Cape of Good Hope in 1652 (De Wet & Pheiffer, 1979). The expedition had scientific as well as economic aims. The plant and animal life encountered on the long journey was carefully documented and was illustrated by the artist, Claudius, who accompanied the expedition. Two species of *Gladiolus* were among the many plants he painted. One, called *G. esculentus*, is *G. speciosus* and the other, called *Aquilegia*, is *G. caryophyllaceus*. Van der Stel's account of the expedition, sometimes called the *Codex Witsenii* and including Claudius's natural history paintings, was never published. Several copies were, however, made and circulated in Europe. One of these came to the attention of the seventeenth-century English botanist, Leonard Plukenet, who published a copy of the illustration of '*Gladiolus esculentus*', calling it *G. viperatus*. This, the first published record of an African species of *Gladiolus*, appeared in Plukenet's *Phytogeographia* (1691). It was through this work that Linneaus first became aware of the presence of *Gladiolus* in southern Africa, and the illustration was cited in 1760 under his description of *G. alatus* which, incidentally, we now know is not the same species as Plukenet's *G. viperatus*.

As Linnaeus's fame grew after 1753 he was able to promote exploration of botanically unknown parts of the world, including the Cape of Good Hope. The many unusual plants that were encountered there included the first Bruniaceae, Proteaceae and Restionaceae seen by European scientists. These, as well as a quite remarkable range of geophytic plants, stimulated intense scientific interest, and gradually knowledge of the extraordinary diversity of Cape, and then of other southern African plants, increased. In 1772 Linnaeus's student, and later his successor, Carl Peter Thunberg arrived in Cape Town. He spent three years at the Cape, systematically exploring the local flora and making several major expeditions to the northern and eastern interior. This period was seminal for southern African botany. Thunberg collected hundreds of new species and sent thousands of specimens back to Europe. Some were described by Linnaeus's son and many more by Thunberg himself. The results were remarkable. Where a handful of species of Cape *Gladiolus* were known in 1770, the first *Flora Capensis*, published in various editions from 1807 to 1823 (Thunberg, 1823), included some 50 species (not all of which are retained in the genus), almost one-third of the total recognized today.

Botanical exploration continued to be encouraged in the late eighteenth and early nineteenth centuries as a result of the European fascination with exotic and new plants, especially those found at the Cape. First Dutch and French, then English patrons supported botanical exploration in an unprecedented manner so that by 1810 not only were the Cape species of *Gladiolus* being recorded and described, but many of them were being illustrated in wonderful detail at a time when botanical art was in its ascendency. Colour paintings published in sumptuous volumes and periodicals, notably Nicholas Jacquin's *Icones Plantarum Rariorum*, William Curtis's *Botanical Magazine* and Henry Andrews's *Botanist's Repository*, are a monument to these endeavours and remain a valuable botanical resource. The collectors of particular species that were discovered at this time included Franz Boos and Georg Schol, Francis Masson, James Niven, and William Roxburgh and his son John. Their contributions are discussed under the particular species they recorded.

The science of plant systematics was at this time in its infancy and generic concepts, in particular, were in flux. Initially *Gladiolus* included species of Iridaceae (the family name came later) with zygomorphic flowers in spikes and often with short tubes, but occasionally with elongate ones. Species with actinomorphic flowers in spikes were regarded as belonging to *Ixia*, while those with zygomorphic flowers of more complicated appearance, with strongly unequal tepals and often a bright red colour, were assigned to a third genus, *Antholyza*. Thus, when actinomorphic-flowered *Gladiolus quadrangulus* was first named in 1766 it was called *Ixia quadrangula* and later *I. linearis*, the identity of the earlier synonym being unknown. Surprisingly, Thunberg, who appears to have made the first collection of the actinomorphic-flowered southern Cape species, *Gladiolus stellatus*, realized that it was a *Gladiolus*. Flouting the generic conventions current in 1800, he called it *G. elongatus*. The name is unfortunately a homonym. When the species was rediscovered in 1912, the Cape Town botanist, H.M.L. (Louisa) Bolus, was less perceptive for she placed it in *Geissorhiza*, calling it *G. patersoniae*. Thunberg's generic placement of the species was reaffirmed by another Cape Town botanist, G.J. Lewis, when she transferred the species back to *Gladiolus*, where it required a new name and hence became *G. stellatus*.

Almost from the start, *Antholyza* was an inconsistently applied name and included an artificial assemblage of species. The first two species admitted to the genus were *A. cunonia* and *A. ringens* (Linnaeus, 1753). Twenty-five years later the German botanist, Joseph Gaertner, noted that the seeds of *A. cunonia* were broadly winged and exactly like those of *Gladiolus* and, in 1788, he transferred the species to that genus. The second species, *A. ringens*, is now recognized to be a species of *Babiana*, one of two in that genus with bright red flowers adapted for bird pollination.

As more African Iridaceae were discovered, *Gladiolus* became the repository for species of several other genera with zygomorphic flowers, including what we now recognize as *Babiana*, *Freesia* and *Tritoniopsis*. Likewise, *Ixia* and *Antholyza* became increasingly artificial assemblages. *Antholyza* was recognized throughout the nineteenth century when it included

species of what are now *Chasmanthe*, *Tritoniopsis* and *Watsonia*, as well as more species of *Gladiolus*. That this state of affairs was leading to a more and more unnatural system of classification was perceived by the English botanist, John Gawler. Gawler, who changed his last name to Bellenden Ker and is now known to botanists as Ker Gawler, restructured the genus *Ixia*, recognizing from its constituent species *Anomatheca* (now *Freesia*), *Babiana*, *Geissorhiza*, *Hesperantha*, *Sparaxis* and *Tritonia*. To these and other genera he added several anomalous species of *Gladiolus*, thus transferring, for example, *G. anceps* and *G. silenoides* to *Lapeirousia*, *G. villosus* to *Sparaxis*, and *G. viridis* to *Tritonia*. Most of them remain in these genera today.

Ker Gawler had no confidence in *Antholyza* at all and continued to recognize most of the several species of that genus as members of *Babiana* or *Gladiolus*, actually reserving *Antholyza* for species we now know as *Chasmanthe* and *Tritoniopsis*. Later botanists had a poorer understanding of Iridaceae and tended to include several species of *Gladiolus* in *Antholyza*, although admitting to it a range of plants of several other genera. *Gladiolus cunonius*, one of the two original species of Linnaeus's *Antholyza*, was placed in a new genus, *Cunonia*, by the English botanist, Philip Miller, in 1756. This generic name was ignored by Linnaeus, who continued to favour *Antholyza* and actually named another genus *Cunonia* in 1759. This is now the type genus of the dicot family Cunoniaceae.

When a second species closely related to *Gladiolus cunonius*, *G. splendens*, was flowered in Britain in 1826, the botanist Robert Sweet placed it in a new genus, *Anisanthus*. Clearly species of Iridaceae with unusually shaped, bright red flowers, especially those with elongated dorsal tepals and much reduced lower ones, were perceived as too unusual to be included in *Gladiolus*, even though at least some botanists realized that was where their relationships lay. The similarity between *Anisanthus splendens* and *Gladiolus cunonius* was obvious and the latter was transferred to *Anisanthus* by Sweet in 1830. *Anisanthus* was recognized by the German botanist, F.W. Klatt, who maintained a special interest in Iridaceae throughout his life. In 1863 Klatt added a third species to the genus, the Namaqualand and Namibian plant that is now *G. saccatus*.

Botanical exploration of southern Africa lost its momentum after the first third of the nineteenth century (Lewis et al., 1972), but not before exploration by William Burchell (1812–1815), C.F. Ecklon and C.L. Zeyher (1827–1843), and J.F. Drège (1826–1833) had brought to light many more new southern African plants. These collectors discovered the first species of *Gladiolus* from the interior and eastern parts of the subcontinent. *Gladiolus dalenii*, now known from across tropical Africa and Madagascar, was sent to Holland in 1825 and was named there in 1827. This species and *G. oppositiflorus*, named in 1837, were to become incorporated into living collections and proved vital in breeding the garden gladiolus hybrids of today (Barnard, 1972).

Later nineteenth-century exploration was largely the work of amateurs, sometimes at the behest of patrons in Great

Britain, and at other times in their employ. Thomas Cooper, who made many important collections – including *Gladiolus longicollis*, *G. saundersii* and *G. sericeovillosus* as well as the type collections of *G. crassifolius* and *G. papilio* – for British herbaria and gardens, was employed as a plant collector by W. Wilson Saunders, a wealthy patron of botany who kept a large garden at Reigate near London. *Gladiolus saundersii*, grown and flowered at Reigate, was named in Saunders' honour, and Cooper himself was celebrated in *G. cooperi*, a species now included in *G. dalenii*. In eastern South Africa the farmer, trader and amateur naturalist, John Medley Wood, was appointed Curator of the Botanical Garden in Natal in 1882 and for the following 30 years he was to make a major contribution to the knowledge of the plants of this botanically rich part of southern Africa. Wood is celebrated in *G. woodii*, of which he appears to have made the first collection in 1880.

Progress in classification – at least as we see the process today leading to a more natural system – has always been erratic, and the aim was not always to achieve a natural system. Unusual species have often been accorded generic rank even when their relationships have been known to lie with particular species of other genera. We are uncertain of the reasons that the German botanist, Friedrich Hochstetter, had for describing the genus *Acidanthera* in 1844 for the Ethiopian *A. bicolor*, now *Gladiolus murielae*. A species very like it, *G. aequinoctialis* from West Africa, had already been described in *Gladiolus* (Herbert, 1842). *Acidanthera*, based on a white-flowered plant with a perianth tube 120–150 mm long, nevertheless continued to be recognized for more than a century. As species with similar flowers were added to *Acidanthera*, however, it became more and more artificial. Long-tubed white flowers, especially those with a strong fragrance, are adapted for pollination by moths and the flower type does not necessarily signal natural relationships. Most of the species that were ultimately included in *Acidanthera* by the English botanist and specialist in bulbous plants, J.G. Baker, in his accounts of *Gladiolus* for *Flora Capensis* (1896) and *Flora of Tropical Africa* (1898) are now known to have inflated capsules and broadly winged seeds exactly like species of *Gladiolus* and they belong in this genus. *Acidanthera rosea* and *A. tubulosa* are, however, simply long-tubed species of *Geissorhiza*; *A. capensis* and *A. pallida* are long-tubed species of *Tritonia*; *A. brevicaulis*, *A. huttonii* and *A. tysonii* belong in *Hesperantha*; and *A. flexuosa* is a species of *Tritoniopsis*. *Acidanthera* was only formally sunk in *Gladiolus* in 1973, by which time the genus had been reduced to seven species (Marais, 1973), all tropical African and most closely allied to *G. murielae*. They are now all assigned to section *Acidanthera* (Goldblatt, 1996).

Another mis-step on the way to a natural classification of *Gladiolus* was the recognition of *Petamenes* by Jane Loudon in 1841. The species assigned to this genus, *G. abbreviatus*, admittedly has very unusual flowers for *Gladiolus* (see **Figure 11G and Plate 144**). Loudon might be forgiven for recognizing a new genus, except that the species had already been described by Henry Andrews in 1801 as *G. abbreviatus*. She was actually

validating an unpublished genus of R.A. Salisbury, a botanist of very mixed repute. Salisbury, who had grown and studied bulbous plants for many years, was working on a new classification of Iridaceae at the time of his death in 1829. The part of his manuscript published posthumously in 1866 stops in the middle of his highly idiosyncratic account of *Gladiolus*. Instead of one genus, Salisbury recognized eight: *Gladiolus*, which he reserved for Eurasian species with winged seeds, *Petamenes* and six completely new genera.

This odd work would have been consigned to the realm of historical curiosity except that one of Salisbury's genera, *Homoglossum*, became widely used for red-flowered southern African, and even some tropical African species with long perianth tubes. The genus was adopted by Baker in 1877 in his sweeping review and classification of the Iridaceae, *Systema Iridacearum*. There, *Homoglossum* comprised three species of *Gladiolus* and two of *Tritoniopsis*, all of which had at some time in the past been included in *Antholyza*. Baker (1896) subsequently reversed his decision and his account of the Iridaceae in *Flora Capensis* included *Homoglossum*, Loudon's *Petamenes* and Sweet's *Anisanthus* in *Antholyza*. For all their shortcomings, Baker's revisions of African *Gladiolus* and related genera were perfectly acceptable at the time. His account of *Gladiolus* in southern Africa is now outdated, but was extremely valuable for bringing together in one publication all available knowledge of the genus. Without it further work would have been severely handicapped.

As it turned out, the *Flora Capensis* treatment of the southern African Iridaceae came at a particularly opportune time, forming a foundation for new work that was starting in South Africa, where botany was beginning to stand by itself. In the last decades of the nineteenth century exploration by John Medley Wood in Natal and Harry Bolus, Peter MacOwan and Rudolf Marloth in the Cape Province began to fill in ranges of known species and to bring to light several new ones. Both Marloth and Bolus collaborated with botanically minded amateurs, as a result drawing H.K. Andreae, Francis Guthrie and C.B. Fair, amongst others, into an active group of naturalists. Their work was complemented by that of the German botanist, Rudolph Schlechter, who spent just three years in southern Africa, travelling energetically the length and breadth of the subcontinent. In this short time Schlechter amassed more than 10 000 collections, all amply duplicated and deposited in both European and South African herbaria. Schlechter also found the time to describe many of the new species he encountered, including *Gladiolus taubertianus*.

Species of Iridaceae adapted for bird pollination continued to confuse botanists and in 1930 the odd genus *Oenostachys* was described by A.A. Bullock. *Oenostachys dichroa* is peculiar in having enlarged, bright red floral bracts which serve to provide the colour signal to birds. The flowers themselves, resembling those of a fairly long-tubed *Gladiolus*, are quite small and pale-coloured. At this time the British botanist, N.E. Brown, working at the Royal Botanic Gardens, Kew, realized that plants included in *Antholyza* by Baker were an unnatural assemblage of species with diverse affinities. His solution was, however, to order the several apparently natural clusters of species in new genera. He placed species unrelated to *Gladiolus* in the new genera, including *Anapalina* (now incorporated in *Tritoniopsis*), and *Chasmanthe* and *Curtonus* (now *Crocosmia*). For species allied to *Gladiolus* Brown maintained *Homoglossum* and *Oenostachys*, and recognized in addition *Anomalesia* for *Gladiolus cunonius* and *G. splendens*, two of the three species Sweet and Klatt had placed in *Anisanthus*. The third, *A. saccatus*, Brown removed to another new genus, *Kentrosiphon*, in which he recognized four species. Brown's motive for renaming *Anisanthus* was his belief that the name was a homonym for *Anisanthes* Schultes, dating from 1819, and hence illegitimate.

Chromosome cytology and morphological comparison of the species of *Anomalesia, Homoglossum, Kentrosiphon* and *Petamenes* convinced Goldblatt (1971) that all four were related to *Gladiolus*. Provisionally Goldblatt recognized the first two, but sunk *Kentrosiphon* in *Anomalesia* and *Petamenes* in *Homoglossum*. The latter continued to be recognized until 1989, even though it was already apparent in 1976, when the genus was revised by M.P. de Vos, that its members were simply species of *Gladiolus* adapted for pollination by birds. Finally, when these genera were examined more critically in 1989, Goldblatt & De Vos concluded that they were nested within the larger genus and should all be included in *Gladiolus*. The erstwhile species of *Homoglossum* are now placed in three different sections of *Gladiolus*, one of which is closely allied to the species of Brown's *Anomalesia*.

Louisa Bolus continued in the mould of Schlechter, Harry Bolus and Marloth, making some new collections and promoting understanding of the flora through both scientific and popular publications. She established links with amateurs across the country and encouraged people like Florence Paterson at Redhouse, near Port Elizabeth, Emily Ferguson at Riversdale, and C.L. Leipoldt and T.P. Stokoe in the western Cape to collect specimens for her. As a result many new species of *Gladiolus* were brought to light in the 1920s and 1930s, several of which she immediately described. Meanwhile Bolus was also responsible for training a new generation of botanists, including the energetic collector, Elsie Esterhuysen, and the perceptive systematist, G.J. Lewis. Beginning in 1939, Esterhuysen began her systematic botanical exploration of the Cape montane flora, and discovered several new species in a part of southern Africa that had seen botanical exploration for more than 200 years. Lewis's major contribution was largely synthetic, bringing order through her revisions of *Thereianthus, Babiana, Ixia* and *Tritoniopsis*. She located the type specimens of species and identified them, a huge task in itself, and also re-evaluated the generic taxonomy of Iridaceae.

Her untimely death in 1967 left a revision of *Gladiolus* in southern Africa not quite finished. The treatment was completed by the systematist, A.A. Obermeyer, assisted by T.T. Barnard, a *Gladiolus* grower and student of early botanical literature. The revision, finally published in 1972, brought together our knowledge of *Gladiolus* in more critical fashion than did Baker's

account, and presented detailed information about the relationships and geographical ranges of species. In this work southern African species of *Gladiolus* until then assigned to *Acidanthera* were correctly treated. There is no doubt that Lewis also understood the relationships of the species then included in *Homoglossum*, but she apparently did not intend to revise them or transfer them to *Gladiolus*. We suspect that she was likewise aware of the affinities of the species of N.E. Brown's *Anomalesia*.

The understanding of pollination biology and the significance of floral adaptations to particular pollinating agents was late in coming. In 1989 Goldblatt & De Vos finally made the decision to break with tradition and included *Homoglossum* and *Anomalesia* in *Gladiolus*. This action rendered the 1972 revision seriously outdated. The publication was also out of print and these two factors influenced our decision to produce a new revision of *Gladiolus*. This work was initially planned as an updated treatment of Lewis, Obermeyer and Barnard's revision of the genus. Expeditions to gather material for description and illustration gradually revealed shortcomings in the current taxonomy and our work changed direction, developing into an entirely new account of *Gladiolus*. It has uncovered a number of new species and led us to revise the limits of several more. We now recognize 163 species in the genus in southern Africa, including the three species of *Anomalesia* and 11 of *Homoglossum* recognized when we began the project in 1992, a notable increase over the 103 admitted by Lewis et al. (1972) in the genus. We have also for the first time developed a natural classification of *Gladiolus* in southern Africa (**see table overleaf**) in which we recognize seven sections and 27 series.

Despite our extensive research for this book, we have no illusions that the last word has been said about the systematics of *Gladiolus* in southern Africa. New species almost certainly remain to be found. Several species in this account were only discovered in the last 25 years: *G. delpierrei* in 1975, *G. deserticola* in 1977, *G. atropictus* in 1978, *G. roseovenosus* in 1982, *G. uitenhagensis* in 1983 and *G. pavonia* in 1987. Several more are poorly understood and leave open the possibility that the taxonomy may need to be refined in some instances. The biology of the genus has only just begun to be investigated and much of what is known is still based on very few observations. Finally, our understanding of phylogeny and species relationships is in its infancy. The infrageneric taxonomy we present here, although having been subjected to careful scrutiny and evaluation, is based solely on external features which we know are highly labile in the evolutionary sense. Molecular-based phylogeny using DNA sequencing will almost certainly lead to a refinement but, we hope, not to radical change to our current classification (**see table overleaf**). A biologically complex and fascinating genus with as much horticultural potential as *Gladiolus* begs further study, and we hope this volume will stimulate new work. Time is running out, however, for habitat loss is proceeding rapidly in southern Africa as the human population continues to grow. Several species, already reduced to small, vulnerable populations, are now at the edge of extinction. ✢

CLASSIFICATION OF SOUTHERN AFRICAN *GLADIOLUS*

The 163 species are arranged in seven sections and 27 series. Species outside southern Africa include ten in Eurasia (assigned to a section *Gladiolus*); nine in Madagascar, one of these shared with Africa; and 83 in tropical Africa plus Arabia (six shared with southern Africa). The current total species in the genus is 255.

1. Section *DENSIFLORUS*

Series *PALUDOSUS*
1. *G. paludosus*
2. *G. papilio*
(several more in tropical Africa, including *G. gregarius*, *G. harmsianus*, *G. microspicatus* and probably *G. zambesiacus*)

Series *DENSIFLORUS*
3. *G. crassifolius*
4. *G. hollandii*
5. *G. serpenticola*
6. *G. exiguus*
7. *G. densiflorus*
8. *G. ferrugineus*
9. *G. lithicola*
10. *G. varius*

Series *CALCARATUS*
11. *G. appendiculatus*
12. *G. calcaratus*
13. *G. macneilii*

Series *SCABRIDUS*
14. *G. ochroleucus*
15. *G. mortonius*
16. *G. microcarpus*
17. *G. scabridus*
18. *G. cataractarum*
19. *G. pavonia*
20. *G. brachyphyllus*

2. Section *OPHIOLYZA*

Series *OPPOSITIFLORUS*
21. *G. dolomiticus*
22. *G. pole-evansii*
23. *G. oppositiflorus*
24. *G. sericeovillosus*
25. *G. elliotii*

Series *ECKLONII*
26. *G. ecklonii*
27. *G. vinosomaculatus*
28. *G. rehmannii*

Series *OPHIOLYZA*
29. *G. antholyzoides*
30. *G. aurantiacus*
31. *G. dalenii*
32. *G. magnificus*
33. *G. flanaganii*
34. *G. saundersii*
35. *G. cruentus*
(tropical African species include *G. melleri*, *G. roseolus*, *G. velutinus* and several more)

3. Section *BLANDUS*

Series *PHOENIX*
36. *G. oreocharis*
37. *G. crispulatus*
38. *G. phoenix*

Series *SABULOSUS*
39. *G. gueinzii*

Series *BLANDUS*
40. *G. carneus*
41. *G. pappei*
42. *G. geardii*
43. *G. aquamontanus*
44. *G. undulatus*
45. *G. angustus*
46. *G. buckerveldii*
47. *G. bilineatus*
48. *G. insolens*
49. *G. cardinalis*
50. *G. sempervirens*
51. *G. stefaniae*
52. *G. carmineus*

Series *FLORIBUNDUS*
53. *G. rudis*
54. *G. grandiflorus*
55. *G. floribundus*
56. *G. miniatus*

4. Section *LINEARIFOLIUS*

Series *PUBIGERUS*
57. *G. woodii*
58. *G. malvinus*
59. *G. pardalinus*
60. *G. pubigerus*
61. *G. parvulus*
(tropical African species, apparently a clade, series *Intonsus*, include *G. curtifolius*, *G. fuscoviridis*, *G. huillensis*, *G. intonsus*, *G. laxiflorus* and *G. zimbabweensis*)

Series *LINEARIFOLIUS*
62. *G. hirsutus*
63. *G. caryophyllaceus*
64. *G. guthriei*
65. *G. emiliae*
66. *G. overbergensis*
67. *G. bonaspei*
68. *G. aureus*
69. *G. brevifolius*
70. *G. monticola*
71. *G. nerineoides*
72. *G. stokoei*

5. Section *HETEROCOLON*

Series *UNGUICULATUS*
73. *G. oatesii*
(*G. unguiculatus*, probably *G. gracillimus* and *G. pusillus*, and possibly even *G. atropurpureus* and *G. serapiiflorus* in tropical Africa)

Series *HETEROCOLON*
74. *G. rubellus*
75. *G. pretoriensis*
76. *G. filiformis*
(several more in tropical Africa, including the Congo [Zaïre] copperbelt endemics [Goldblatt, 1996], the Angolan *G. fenestratus*, and possibly *G. juncifolius* from Zimbabwe, which may alternatively belong in series *Vernus*)

Series *VERNUS*
77. *G. rufomarginatus*
78. *G. vernus*
79. *G. kamiesbergensis*
80. *G. marlothii*
81. *G. mostertiae*

6. Section *HEBEA*

Series *INVOLUTUS*
82. *G. leptosiphon*
83. *G. loteniensis*
84. *G. involutus*
85. *G. vandermerwei*
86. *G. cunonius*
87. *G. splendens*
88. *G. saccatus*

Series *PERMEABILIS*
89. *G. permeabilis*
90. *G. uitenhagensis*
91. *G. acuminatus*
92. *G. stellatus*
93. *G. wilsonii*
94. *G. inandensis*
95. *G. robertsoniae*

Series *DESERTICOLA*
96. *G. arcuatus*
97. *G. viridiflorus*
98. *G. deserticola*
99. *G. scullyi*
100. *G. venustus*
101. *G. salteri*
102. *G. lapeirousioides*

Series *HEBEA*
103. *G. orchidiflorus*
104. *G. watermeyeri*
105. *G. virescens*
106. *G. ceresianus*
107. *G. uysiae*
108. *G. equitans*
109. *G. speciosus*
110. *G. pulcherrimus*
111. *G. alatus*
112. *G. meliusculus*

7. Section *HOMOGLOSSUM*

Series *CARINATUS*
113. *G. atropictus*
114. *G. violaceolineatus*
115. *G. comptonii*
116. *G. roseovenosus*
117. *G. carinatus*
118. *G. griseus*
119. *G. quadrangulus*

Series *MUTABILIS*
120. *G. mutabilis*
121. *G. exilis*
122. *G. vaginatus*
123. *G. maculatus*
124. *G. albens*
125. *G. meridionalis*
126. *G. priorii*

Series *BREVITUBUS*
127. *G. brevitubus*

Series *GRACILIS*
128. *G. rogersii*
129. *G. bullatus*
130. *G. blommesteinii*
131. *G. virgatus*
132. *G. debilis*
133. *G. variegatus*
134. *G. vigilans*
135. *G. ornatus*
136. *G. inflexus*
137. *G. taubertianus*
138. *G. gracilis*
139. *G. caeruleus*
140. *G. recurvus*

Series *TERETIFOLIUS*
141. *G. inflatus*
142. *G. cylindraceus*
143. *G. nigromontanus*
144. *G. engysiphon*
145. *G. patersoniae*
146. *G. subcaeruleus*
147. *G. martleyi*
148. *G. jonquilliodorus*
149. *G. trichonemifolius*
150. *G. sufflavus*
151. *G. pritzelii*
152. *G. delpierrei*

Series *TRISTIS*
153. *G. hyalinus*
154. *G. liliaceus*
155. *G. tristis*
156. *G. longicollis*
157. *G. symonsii*

Series *HOMOGLOSSUM*
158. *G. watsonius*
159. *G. teretifolius*
160. *G. quadrangularis*
161. *G. huttonii*
162. *G. fourcadei*
163. *G. abbreviatus*

CORM

The storage and perennating organ in *Gladiolus*, as in many other genera of the Iridaceae, is a corm (**Figure 1A–C**). In *Gladiolus* the corm is derived from the swollen lower internodes of the underground portion of the stem and is more or less globose to depressed-globose. On sprouting, adventitious roots are produced from the lower half. As well as assimilatory roots, corms usually produce one or more thick, fleshy contractile roots (**Figure 1B**). These have the special function of pulling the newly formed corm deeper into the ground, thus ensuring that it remains buried in the soil at a level suitable for the plant. Corms of different species vary considerably in size, a feature often directly related to plant size. The largest corms, often 30–40 mm in diameter, are found in some species of the *G. dalenii* complex (section *Ophiolyza*). The smallest occur in the southern African section *Homoglossum* and may be no more than 10 mm in diameter (**Figure 2H, J**). Perhaps the most unusual corms are those of *G. jonquilliodorus* and *G. martleyi* (section *Homoglossum*). At 30–40 mm in diameter, they are very large for the section, and fleshy in texture (**Figure 2I**). The surface and internal tissue are usually white, as in most Ixioideae, but are orange to red in some species of the *G. dalenii* alliance, and yellow in *G. bilineatus*. Two species of more or less permanently damp habitats, *G. aquamontanus* and *G. sempervirens* (**Figure 3D**), have very small, virtually vestigial corms, a reduction of the ancestral condition associated with the habitat.

Corms are almost always covered by protective layers of distinctive appearance, commonly called tunics (**Figure 1A**). These layers are derived from the bases of the cataphylls and lower leaves and are usually characteristic of species groups (**Figure 2**). Most closely resembling the bases of dried leaves – and hence thought to be least specialized – are the coriaceous to firmly papery, brown tunics such as those found in most members of sections *Blandus*, *Densiflorus* and *Ophiolyza* (**Figure 2A, B**). With age the tunics decay into irregularly broken fragments that sometimes become more or less fibrous because the vascular tissue is more resistant to decay than the mesophyll. Such unspecialized corm tunics occur in most species of these sections. Members of section *Linearifolius* have tunics that are composed of flat, vertically oriented fibres, and they can often be placed to section on this basis (**Figure 2C**).

More specialized corm tunics characterize species of other sections. Tunics in section *Hebea* are varied, but the plesiomorphic state is probably best represented by the tunics of rather stiff wiry fibres found in series *Permeabilis*. Members of this group can always be recognized by their corm tunics. In series *Deserticola* the corm tunics consist of hard, woody layers that decay into somewhat irregularly gnarled fibres (**Figure 2E**). *Gladiolus arcuatus* and most species of series *Hebea* have rather soft, papery tunics that quickly decay, often leaving corms

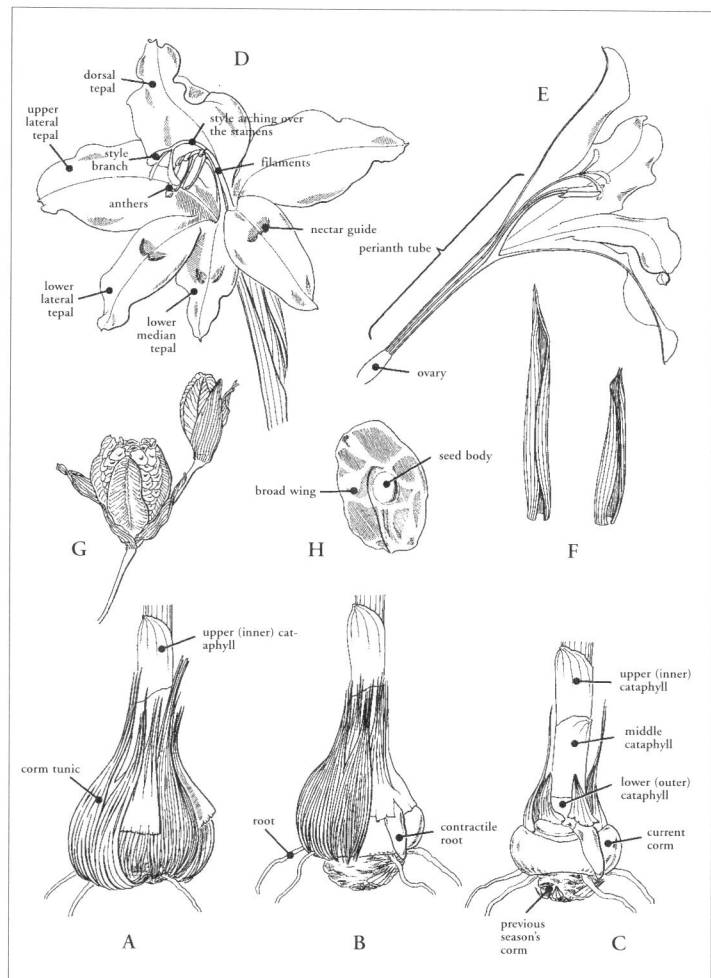

Figure 1 *Gladiolus carneus*, type species of section *Blandus*, illustrating the various reproductive and vegetative organs of the *Gladiolus* plant. A–C The base of the plant. D The whole flower. E A vertical section of the flower, showing the slender lower part of the perianth tube and the wider upper part or throat, with stamens attached at the top of the lower part of the tube. F Floral bracts. G An intact capsule and a dehiscing capsule with the seeds visible. H A single seed.

virtually naked (**Figure 2F**). *Gladiolus ceresianus* is exceptional in having almost woody tunic layers. In series *Involutus* tunics are often coriaceous to papery, as in *G. involutus*, rather than wiry but the latter type characterizes *G. leptosiphon* (**Figure 2D**) of the series. In section *Heterocolon* the tunics typically accumulate in a thick mass of coarse, shaggy fibres that extend upward around the base of the stem, a derived feature of the alliance.

Particularly specialized tunics are found in southwestern Cape species of section *Homoglossum* that grow on nutrient-poor sandstone soils. In this section the most frequent type of tunic is woody in texture, and consists of concentric layers which fragment below (or completely) with age into flat, vertical segments (**Figure 2G, H**). These may be truncate basally or acute, and then are often thickened into claw-like ridges.

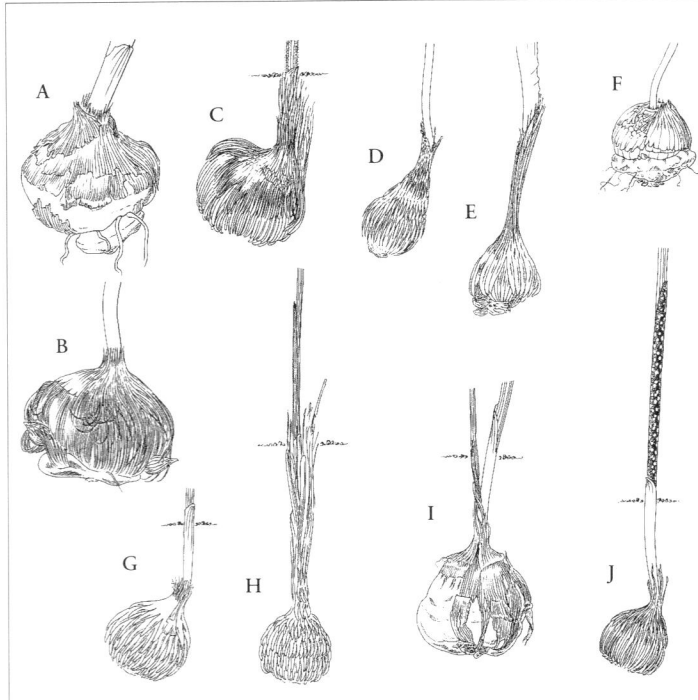

Figure 2 Corms of southern African *Gladiolus*.
In section *Ophiolyza* the soft, papery corm tunics of *G. aurantiacus* (**A**) and *G. dolomiticus* (**B**) are plesiomorphic for *Gladiolus*. The firm-papery tunics of species in section *Linearifolius*, e.g. *G. hirsutus* (**C**), decay with age into flat fibres. In section *Hebea* the plesiomorphic type is *G. lepto-siphon* (**D**) with wiry tunics; *G. scullyi* (**E**) has coarse, woody tunics; and *G. arcuatus* (**F**) has very softly membranous tunics. Hard, woody corm tunics are characteristic of section *Homoglossum*, e.g. in *G. gracilis* (**G**) and *G. inflatus* (**H**), the latter with the stem base enclosed in a thick, fibrous neck; in *G. jonquilliodorus* (**I**) the corm is moist and fleshy and the tunics consist of softly membranous layers; and in *G. carinatus* (**J**), which has characteristic mottled cataphylls, the fairly soft tunics decay into soft fibres. (approximately x ½)

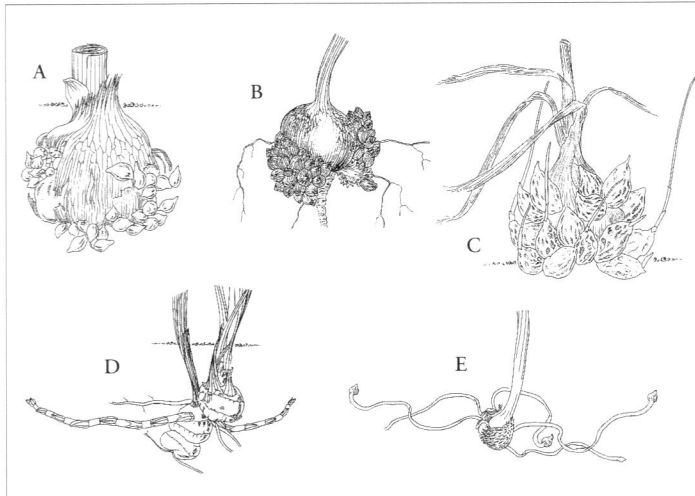

Figure 3 Vegetative reproduction in southern African *Gladiolus*.
In *G. saundersii* (**A**) cormlets are produced around the corm base on short, thread-like stolons. Cormlets are produced in flattened fascicles in *G. phoenix* (**B**), and in *G. gueinzii* (**C**) the distinctive large cormlets have thick, corky tunics. *Gladiolus sempervirens* (**D**) produces thick rhizomes from the base of the small, poorly developed corm. In *G. loteniensis* (**E**) the long stolons terminating in a fairly large cormlet are characteristic of several species of section *Hebea*. (approximately x ½)

Although such tunics are the most common kind in the section, several species have tunics that are less woody and sometimes even quite soft-textured, so that on decaying they become softly fibrous, as in *Gladiolus carinatus* (**Figure 2J**) and its allies. The unusually large, fleshy corms of *G. jonquilliodorus* (**Figure 2I**) and *G. martleyi*, mentioned above, have membranous, extremely soft-textured tunics quite unlike those of most related species.

The production of small cormlets on short, thread-like stolons is often characteristic of species or species groups (**Figure 3**). *Gladiolus carinatus* (section *Homoglossum*), *G. dalenii* and *G. saundersii* (section *Ophiolyza*), *G. orchidiflorus* (section *Hebea*) and *G. undulatus* (section *Blandus*) are especially notable for the production of masses of tiny cormlets around the base of the main corm (**Figure 3A**). *Gladiolus phoenix* (section *Blandus*) bears numerous cormlets on stiffly ascending, fasciated stolons (**Figure 3B**), a feature known elsewhere in the genus only in the distantly related central African *G. verdickii*. Cormlets are also typically produced in the axils of the lower leaves in *G. alatus* and *G. gueinzii*. Both species favour sandy soils and the lower part of the stem may be buried for some distance, permitting the development of several cormlets along the underground part. In *G. gueinzii* a second type of cormlet is produced when the corm is exposed by sand erosion. These cormlets are fairly large and have thick, corky tunics (**Figure 3C**) that float in water and are believed to play an important role in dispersal. *Gladiolus sempervirens* is remarkable in the genus in producing long rhizomes (**Figure 3D**) from the corm base, the apices of which develop directly into new plants. Most distinctive of the several means of vegetative reproduction is the formation of long stolons from the corm base, each terminating in a moderate-sized corm (**Figure 3E**). Several species of section *Hebea* related to *G. involutus* produce such stolons, including *G. cunonius*, *G. loteniensis*, *G. splendens* and *G. vandermerwei*. Other unrelated species of the section, *G. speciosus* and *G. uysiae* (series *Hebea*), as well as *G. pavonia* and *G. papilio* (section *Densiflorus*) produce similar stolons, apparently independently developed in each.

LEAVES AND LEAF-RELATED ORGANS

Cataphylls

Apart from developed foliage leaves, plants have three cataphylls that sheath the underground part of the shoot and always lack blades (**Figure 1A, C**). These foliar organs are the first ones produced by the shoot, and they grade from pale and membranous to firm in texture. The inner (uppermost) cataphyll generally extends a short distance above the ground and then becomes green or sometimes purplish. The distinction between the uppermost cataphyll and the lowermost leaf can be obscure when the foliage leaves have vestigial blades. The lower cataphylls are usually truncate, and the upper obtuse or subacute. Inserted on the lowermost nodes of the stem, the cataphylls sheath the part of the stem that later enlarges to become the new season's corm. The bases of the cataphylls give rise to the corm tunics during the course of the growing season, although they are initially soft, pale and more or less membranous. Light velvety pubescence is present on the upper cataphyll of several species, notably those of sections *Linearifolius* and *Ophiolyza*. Finely scabrid upper cataphylls are typical of species of series *Teretifolius* (section *Homoglossum*).

Distinctively mottled cataphylls, pale and dark green to purple, are characteristic of several species or lineages. In section *Homoglossum*, *Gladiolus carinatus*, *G. griseus* and the related *G. quadrangulus* (series *Carinatus*) are linked by this feature, as are *G. involutus* and a few closely related species of section *Hebea*, including the florally very different *G. vandermerwei* and *G. cunonius* and its allies. Several species of section *Blandus* have obscurely or very strongly mottled tunics, a feature well developed in *G. rudis* where the cataphyll is textured, with the white areas raised above the surface.

Foliage leaves

Varying in number from a maximum of eight to just one (**Figure 4**), foliage leaves are the main organs of photosynthesis. Leaves of *Gladiolus* and many other Iridaceae consist of an expanded portion which is isobilateral, and a basal part or sheath which encloses the stem. Outgroup comparison suggests that the production of several leaves is ancestral. The presence of long-bladed foliage leaves inserted on the stem below ground level (basal leaves) and smaller shorter-bladed leaves inserted on the stem above ground level (cauline leaves) is the ancestral (primitive) condition not only of *Gladiolus*, but of Ixioideae in general. Leaves are primitively crowded basally in a tight, distichous or fan-like arrangement (**Figure 4A**). The fan-like appearance is less obvious in many species of section *Hebea* because the leaves are often very narrow or few in number (**Figure 4B, C**). The fan-like arrangement is lost when the leaves are spaced well apart on the stem (**Figure 4D, E**), a condition called superposed. This is a specialized feature of section *Homoglossum*, but it is also found in a few isolated species of other sections.

Reduction in leaf number is a common trend in *Gladiolus*. Species of section *Homoglossum* have four leaves or fewer (**Figure 4D, E**), and within the section *G. carinatus*, *G. griseus* and *G. quadrangulus* (series *Carinatus*), *G. inflexus*, *G. ornatus*, *G. taubertianus* and usually *G. variegatus* (series *Gracilis*), *G. exilis* (series *Mutabilis*) and *G. nigromontanus* and *G. engysiphon* (series *Teretifolius*) consistently have three leaves, and *G. vaginatus* always just two. All the species of series *Tristis* and *Homoglossum* also have three leaves. Likewise leaf number has been progressively reduced in section *Linearifolius*, where *G. pubigerus* and *G. parvulus* (**Figure 4G**) have just two leaves. In the latter, the uppermost leaf is reduced to a tiny scale. Reduction in leaf number seems to have occurred repeatedly in different lineages within the genus, but it is, nevertheless, a useful guide both to species identification and to an understanding of relationships. We have found no convincing examples of phylogenetic increase in leaf number and we think the character is not readily reversed.

The degree of blade development is another important leaf feature. In some specialized species the leaf blades are substantially reduced, vestigial or completely lacking (**Figure 4F, G**). The condition is developed in a few species of most sections, and may be associated with the production of long-bladed foliage leaves at a different time of the year. The growth patterns that accompany leaf reduction are complex and are discussed more fully in the section on Ecology (page 24).

Additional leaf features of taxonomic importance include shape in both outline and transverse section, texture and pubescence. Sword-shaped to narrowly lanceolate and firm-textured

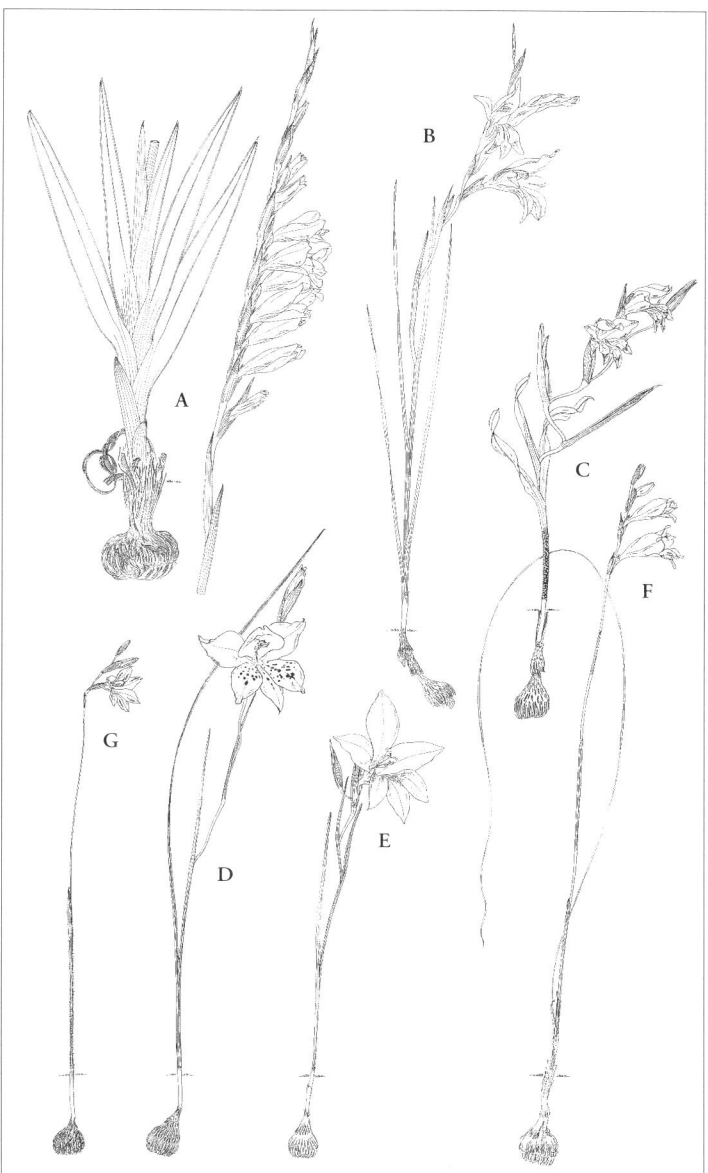

Figure 4 Habit and leaf arrangement in southern African *Gladiolus*. In section *Densiflorus G. exiguus* (**A**) has leaves arranged in a basal distichous fan and the spike inclined, with numerous flowers. In section *Hebea G. permeabilis* (**B**) and *G. viridiflorus* (**C**) have narrow leaves arranged in a loose fan typical of the section, and flexuous spikes. *Gladiolus variegatus* (**D**) and *G. trichonemifolius* (**E**) have flexuose stems with only three or four narrow leaves in superposed arrangement, and few-flowered flexuose spikes which are typical of section *Homoglossum*. Illustrating flowering spikes with reduced leaf blades, *G. subcaeruleus* (**F**) of section *Homoglossum* has a solitary long-bladed leaf, now dry, from the past season; and *G. parvulus* (**G**) of section *Linearifolius* has just two entirely sheathing leaves, a long basal one and a reduced scale-like one below the spike. (approximately x ¼)

(more or less coriaceous) leaves are primitive in Ixioideae and characterize most species of sections *Blandus*, *Densiflorus* and *Ophiolyza*, and many species of section *Hebea*, although several species of this last section have very narrow leaves. The leaves of all members of section *Homoglossum* are narrow, usually less than 2.5 mm wide (**Figure 4D, E**), and either linear or oval to round in transverse section. Typical leaf specializations in Ixioideae are the thickening of the midrib and the margins by the enlargement of the fibrous caps of the primary and marginal veins. These areas then usually appear semitransparent or hyaline in a living leaf, due to the absence of green tissue. When the leaf dries the difference in coloration often becomes more pronounced. In species such as *Gladiolus sericeovillosus* and *G. crassifolius* the margins and at least one pair of secondary veins also become thickened. In others, for example, *G. elliotii* and *G. densiflorus*, only the margins and midrib are strongly thickened and all the remaining veins are fine and closely set.

More elaborate development of the margins and midrib into raised ridges or wings extended at right angles to the leaf blade characterize several species of sections *Heterocolon* and *Homoglossum*. Extreme development of the midrib, for example in *Gladiolus tristis* and several other species of series *Homoglossum* and *Tristis*, has resulted in leaves that are symmetrically cross-shaped in transverse section (**Figure 5O**). Under particularly dry conditions the leaves of these same species become narrow in diameter and the marginal and midrib edges become more heavily thickened so the leaf assumes a terete appearance, now with four narrow, longitudinal grooves or sulci, sometimes so narrow as to be visible only under a hand lens (**Figure 5F, M**). Leaves of this type have evolved independently in series *Teretifolius* and *Tristis* of section *Homoglossum*, and in several species of section *Heterocolon*, including *G. pretoriensis*, its allies in the copperbelt of central Africa (Goldblatt, 1996), and *G. kamiesbergensis*. In *G. marlothii*, also section *Heterocolon*, and *G. cylindraceus* (**Figure 5N**), section *Homoglossum*, the leaves are cross-shaped rather than terete and bear a striking resemblance to those of *G. tristis*. Clearly a xeromorphic adaptation, leaf thickening has evolved repeatedly in *Gladiolus*. Similar leaf adaptations are present in the related *Geissorhiza* (Goldblatt, 1985) and occasionally in other genera, including *Hesperantha* (Goldblatt, 1984b), and are ancestral in *Romulea* and *Devia* (Goldblatt & Manning, 1990).

Leaves of several species of series *Gracilis* are unusual for section *Homoglossum* in being thick, even somewhat fleshy, so that neither the midrib nor the margins are visible (**Figure 5I**). In three species of the series, *Gladiolus caeruleus*, *G. gracilis* and *G. recurvus*, the leaf margins are extended into wings, rendering the leaf more or less H-shaped in transverse section (**Figure 5J**). *Gladiolus violaceolineatus* (series *Carinatus*) is unusual in having leaves more or less triangular in section, the midrib evidently enlarged only on one surface.

Leaf pubescence is relatively uncommon and is regarded as a synapomorphy for section *Linearifolius*. All the species of the section have the seedling leaves and usually the immature leaf blades, or at least the sheaths, covered with long hairs. A few

species of section *Densiflorus*, for example *Gladiolus mortonius*, sometimes have lightly pubescent leaves or leaf sheaths, and some species of section *Ophiolyza* may also have light pubescence, notably *G. dolomiticus*, *G. pole-evansii* and *G. sericeovillosus* (all series *Oppositiflorus*). *Gladiolus microcarpus* and *G. scabridus* (section *Densiflorus*) also have lightly pubescent to scabrid leaves, but the pubescence is weakly developed and usually more or less microscopic.

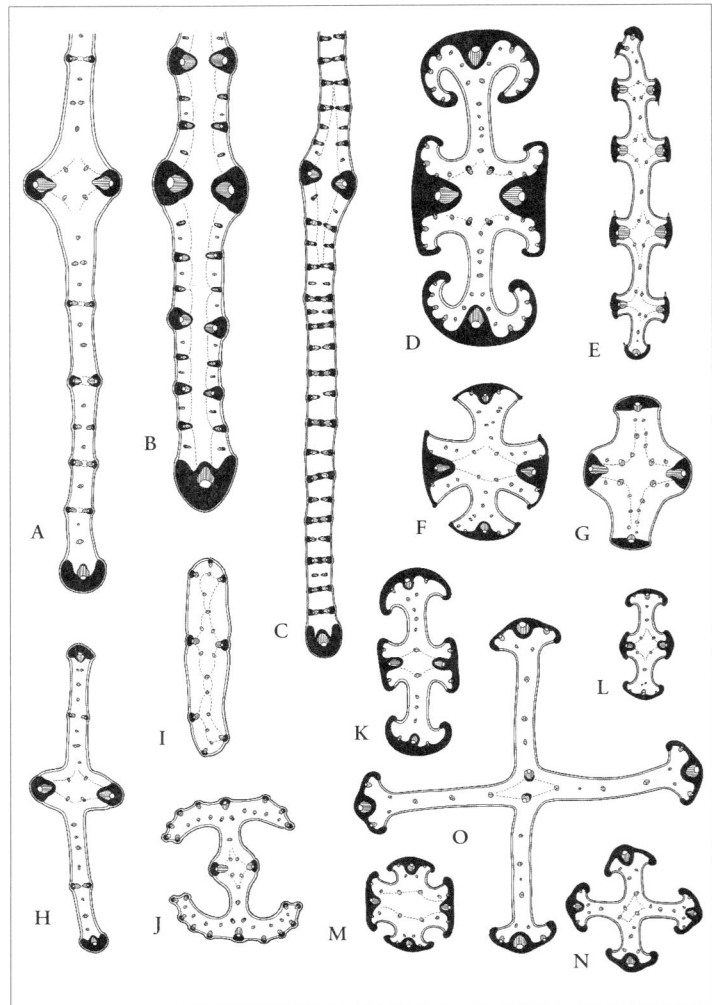

Figure 5 Leaf anatomy: transverse sections of leaves of southern African *Gladiolus,* showing the main leaf types in the genus.

Only a half leaf is shown for A, B and C (sclerechyma black). In section *Blandus* the leaf of *G. carneus* (**A**) is an unspecialized, presumably ancestral type. In section *Densiflorus* the leaf of *G. crassifolius* (**B**) has prominent secondary veins; and that of *G. densiflorus* (**C**) has numerous fine secondary veins. In section *Heterocolon* the leaf margins and midrib of *G. vernus* (**D**) are raised and thickened. In section *Hebea* *G. alatus* (**E**) has a characteristically ridged leaf surface; the leaf of *G. ceresianus* (**F**) is terete and four-grooved; and that of *G. acuminatus* (**G**) is whip-like. In section *Homoglossum* the leaf of *G. comptonii* (**H**) is the generalized type; *G. priorii* (**I**) has a thick blade lacking raised margins or midrib; *G. recurvus* (**J**) has an H-shaped leaf; the leaf of *G. bullatus* (**K**) has a heavily thickened midrib and less thickened margins; and that of *G. blommesteinii* (**L**) has both midrib and margins lightly thickened; in *G. trichonemifolius* (**M**) the terete blade with four grooves is typical of series *Teretifolius*; in *G. cylindraceus* (**N**) the cross-shaped blade is derived from the terete blade by the widening of the four grooves; and the leaf of *G. abbreviatus* (**O**) has the characteristic cross shape of series *Homoglossum*.

STEM

For convenience, we continue the practice of calling the flowering stalk the stem (Goldblatt, 1996 and elsewhere), although this is not strictly correct, the corm also being a part of the stem system. All species of *Gladiolus* have an aerial flowering stem bearing leaves or leaf-related organs along its length. The three lowermost leaves, or cataphylls, lack blades and sheathe the foliage leaves as they emerge from the ground.

Stems may be simple or branched, but in many species they are consistently unbranched (simple), a common specialization in *Gladiolus*. Species of sections *Linearifolius* and *Homoglossum* consistently lack branched stems and species of section *Ophiolyza*, including the *G. dalenii* complex (series *Ophiolyza*), seldom produce branches unless individuals are particularly robust.

The stem is terete in most species, but *Gladiolus alatus* and its immediate allies within series *Hebea* (*G. equitans*, *G. meliusculus*, *G. pulcherrimus* and *G. speciosus*) have compressed and angled to winged stems, a synapomorphy for this clade. Stem pubescence occurs only in *G. sericeovillosus* subsp. *sericeovillosus*, *G. dolomiticus* (section *Ophiolyza*) and *G. scabridus* (section *Densiflorus*). Stems are typically flexed outward and inclined above the sheath of the uppermost one or two leaves or just below the base of the spike. Stems that are erect throughout are usually associated with bird pollination and occur in many species of section *Ophiolyza* and in series *Homoglossum* of section *Homoglossum*, all members of which have flowers adapted for bird pollination. Stem thickness is usually related to plant size, and the diameter just below the base of the spike, given in all descriptions, is a useful measure. Many bird-pollinated species of whatever stature, however, also tend to have thick stems.

INFLORESCENCE

As in nearly all the genera of subfamily Ixioideae, the flowers of species of *Gladiolus* are sessile and arranged in a spike. Each flower is subtended by a pair of opposed bracts, a larger outer and smaller inner bract (bracteole) (**Figure 1F**). The spikes are typically secund and most often inclined at anthesis (**Figure 4**). Spirally arranged flowers occur in some species with actinomorphic flowers (apparently a derived condition in *Gladiolus*). Distichous and erect spikes characterize *G. sericeovillosus*, *G. elliotii* and *G. oppositiflorus*. The number of flowers per spike is particularly variable, even within populations, and depends largely on the health and age of the plants. A particular range is, however, characteristic of a species. This has resulted in the use of epithets such as '*multiflorus*' (a synonym of *G. gregarius*) and '*pauciflorus*' for particular species. Variability of flower number within species limits the usefulness of the character.

Bracts

The bracts are typically green and firm in texture in *Gladiolus* and are fairly constant in size for species and some species groups. The outer bracts are typically slightly longer than the inner, and the latter are usually apically forked for a short distance. Unusually attenuate bracts characterize *G. dolomiticus*. In most species of section *Ophiolyza* the bracts lightly clasp the stem in the lower half, and in *G. sericeovillosus* and related species of series *Oppositiflorus* the inner bract margins are united below, forming a tube around the flower bud and the perianth tube of the open flower. In species of dry habitats the bracts often become membranous or dry above, and in most members of section *Heterocolon* and a few of other sections, notably *G. ferrugineus* and *G. varius* (section *Densiflorus*), the bracts often become entirely dry and tawny-coloured as flowering progresses.

Perianth

More variable than any other organ in the genus, the perianth of *Gladiolus* closely reflects the adaptive radiation of the flower to a range of different pollinators and often does not reflect evolutionary relationships. A zygomorphic flower with unilateral, arcuate stamens and style (**Figure 1D, E**) is the most common condition in the genus, and most likely the basic or ancestral state. Scattered species across the genus have an actinomorphic perianth or completely actinomorphic flower. In the latter category are *G. quadrangulus* from the western Cape and *G. stellatus* from the southern Cape. Both have stellate, open flowers with prominently displayed, symmetrically arranged stamens around a central style. In some populations of *G. brevitubus*, *G. deserticola*, *G. gueinzii* and *G. trichonemifolius* the perianth is completely regular, and in the last-mentioned the stamens are also occasionally symmetrically disposed. *Gladiolus parvulus* and *G. pubigerus*, both species of eastern southern Africa, also have a completely regular perianth, except for the curved tube. Outside southern Africa *G. actinomorphanthus*, from southern Congo (Zaïre), and *G. bojeri*, from Madagascar, have actinomorphic flowers. In all these species we assume that the actinomorphic condition is secondary. The species to which each of these is most closely related all have zygomorphic flowers.

Outgroup comparison, the recommended method for determining ancestral traits, is not useful in the case of perianth symmetry in *Gladiolus* because the immediate generic relationships are uncertain. All that can be said is that it seems likely that floral zygomorphy is the plesiomorphic (ancestral) condition, as are relatively small flower size and a fairly short perianth tube. Comparison with other genera of Ixioideae in which a range of floral types occurs suggests that zygomorphy probably evolved as an integrated unit: a bilabiate perianth with a larger dorsal (upper or posterior) tepal inclined to arched over unilateral stamens, combined with closely aligned lower tepals marked in contrasting colours to form a nectar guide.

Perianth size, coloration and markings (nectar guides) are remarkably varied in *Gladiolus* and, associated with other floral features, constitute adaptations to particular pollinator groups or even to a single pollinator (**Figures 7–11**). Bee-pollinated species typically have the smallest flowers in the genus, usually 20–40 mm long or even smaller (**Figure 7**). Associated with this small flower, the lower tepals are united below for a short distance and narrowed in the lower half into claws in sections *Heterocolon*, *Hebea* and *Homoglossum*, and the nectar guides are often on the upper third of the tepals. Except for species with long, obviously modified perianth tubes, the tube is about half to two-thirds as long as the dorsal tepal, usually 10–20 mm long. The lower tepals are often channelled and usually more or less directed forward below, and sharply flexed downward distally. In profile, the lower tepals exceed the dorsal tepal, although they are seldom longer than the dorsal in linear measurement. The term 'prognathous' is sometimes applied to such flowers.

Bird- and butterfly-pollinated species have the largest flowers in the genus, usually 50–80 mm long, except for the long-tubed moth-pollinated species in which the flower may be up to 150 mm long. Associated with the larger flower, the lower three tepals are hardly, if at all, united to one another, are not, or only weakly, narrowed below, and are never claw-like. The nectar guides in these sections usually take the form of longitudinal blotches on the lower part of the tepals. Viewed in profile, the lower tepals often seem as long as or shorter than the upper. *Gladiolus dalenii*, probably the most important species in commercial *Gladiolus* breeding and one of the most well-known species, is typical of this type.

One repeated specialization is the development of a white or cream flower with an elongated perianth tube and a sweet or clove-like scent (**Figure 9**). *Gladiolus robertsoniae* and *G. acuminatus* (section *Hebea*) and *G. tristis* and its relative, *G. longicollis* (section *Homoglossum*), have just such flowers. These species are pollinated by a variety of night-flying moths. In tropical Africa this type of flower characterizes the species of section *Acidanthera*. The tube in this alliance is remarkably long, 100–120 mm. Other tropical African species with similar flowers include *G. curtifolius* and *G. harmsianus* (Goldblatt, 1996).

Another repeated floral specialization accompanies a shift to bird pollination. Most common in sections *Ophiolyza* and *Homoglossum* (**Figure 11**), this includes a bright red perianth colour and a cylindric perianth tube sharply divided into upper and lower parts. The tepals may be subequal, but more often the dorsal tepal is much enlarged, or the lower three tepals are much smaller than the upper. Prime examples of this flower type are found in the northeast African species *Gladiolus abyssinicus* and its close allies (Goldblatt, 1996). In southern Africa, *G. magnificus* has a similar type of flower, but this species is most likely independently derived from *G. dalenii*, itself pollinated by sunbirds although it lacks some of the adaptations usually associated with bird pollination. Species of *Gladiolus* with red flowers, elongate perianth tubes and lower tepals that are often somewhat to extremely reduced (**Figure 11H–K**) have in the past been regarded as belonging to separate genera, *Homoglossum* (= *Petamenes*) and *Anomalesia*. Some of the tropical African species with similar flowers have at times been included in *Homoglossum*. There seems no doubt that these bird-pollinated flowers have evolved independently at least twice in tropical Africa and six or seven times in southern Africa. Species with such flowers are best included in *Gladiolus* instead of being treated as one polyphyletic genus or two to four smaller genera, each derived from different ancestors within *Gladiolus*.

In species of section *Homoglossum* adapted for sunbird pollination the upper part of the floral tube is always as long as or longer than the narrow lower part. The tepals, however, range from subequal and large (*Gladiolus watsonius*) to unequal with the lower three somewhat smaller, and in *G. abbreviatus* the lower tepals are almost scale-like. Some species of section *Hebea* with flowers adapted for bird pollination have a somewhat different floral structure. Three species, *G. cunonius*, *G. saccatus* and *G. splendens*, have very short lower tepals which are united at the base and thus contribute to the tube length. The upper part of the tube in these species is short, rarely more than 10 mm long, and usually 3–5 mm long. The tube is also remarkable in having a cartilaginous tooth at the base of the abaxial filament that projects upward into the tube. This is presumably a strengthening device to prevent damage from the hard beaks of sunbirds. In *G. saccatus*, which has the shortest tube of the three species, the nectar capacity of the tube is increased by the formation of a sac-like spur, 2–6 mm long, that projects backwards from the tube base. The repeated development of different types of flowers adapted to different pollination systems demonstrates the plasticity of the *Gladiolus* genome and its ability to respond to changing adaptive pressures.

Fragrance

Fragrance is an important feature of the pollination strategy in many species of *Gladiolus*. All moth-pollinated species of the genus produce a strong, sweet scent, usually with an undertone of cloves. The scent may be present throughout the day and night, as in *G. emiliae* (section *Linearifolius*), *G. robertsoniae* (section *Hebea*) and *G. recurvus* (section *Homoglossum*). In the moth-pollinated species, *G. liliaceus*, *G. longicollis* and *G. tristis* of section *Homoglossum*, scent is produced only in the evenings. Scent production is accompanied by a shift in perianth colour from dull brown or red to pale mauve in *G. liliaceus*, a particularly unusual phenomenon.

Bee-pollinated species do not depend on fragrance to attract visitors to the same extent as do species pollinated by moths. Nevertheless, most bee-pollinated species of the predominantly winter-rainfall sections *Hebea* and *Homoglossum* produce strong, sweet odours, whereas bee-pollinated species of sections in the summer-rainfall region do not produce scent. Species of section *Linearifolius* which are pollinated by bees produce scent inconsistently. The winter-rainfall *Gladiolus brevifolius* and summer-rainfall *G. pubigerus* produce scent in some populations. Fragrance in section *Hebea* is characteristically very strong and is a combination of violet and freesia. In the field, flowers can often be located by scent before they are seen. Particularly strongly scented species are the cryptically coloured *G. orchidiflorus* and its allies, *G. ceresianus*, *G. uysiae* and *G. watermeyeri*. *Gladiolus permeabilis* subsp. *permeabilis* is usually scented and subsp. *edulis* frequently so, with most populations north of the Vaal River being unscented, whereas strongly scented populations occur in the Karoo, Free State and Namibia. In section

Homoglossum the southwestern Cape species *G. carinatus* and *G. gracilis* are well known for their highly fragrant flowers, but perhaps the finest fragrance is produced by *G. trichonemifolius*. In the nineteenth and early twentieth centuries the flowers of these species were treasured for the vase and a few stems will fill a house with their delightful fragrance.

The source of floral fragrance in *Gladiolus* has received little study. We suspect that the fragrance comes from the papillae on the lower tepals and the upper part of the perianth tube. These papillae are prominent in species that produce scent and can be stained with ruthenium red, a dye that is an indicator of lipid production. In sections *Homoglossum* and *Hebea* lines of papillae are most often present on the areas of pale or yellow pigmentation on the lower half of the lower tepals, but papillae are also present in the throat or at the base of the filaments, for example in *G. liliaceus* and *G. tristis*. Particularly prominent ridges of papillae along the sutures of the tepals in the upper perianth tube in series *Hebea* of section *Hebea* do not appear to be the source of fragrance in this alliance of strongly scented species.

Closing movements

Flowers of most *Gladiolus* species close partially in the late afternoon. This is accomplished by the movement of the tepals toward one another and the downward dipping of the dorsal tepal. Exceptions to this rule are species that are adapted for pollination by moths. These have pale-coloured flowers, often with brownish markings, and are often scented in the evenings. In *G. longicollis*, which is also thought to be pollinated by moths, the flowers close during the day and open only at sunset. Species with fairly complex flowers also may not close at all during the evening hours. These include *G. dalenii* and *G. ecklonii* (section *Ophiolyza*), the bird-pollinated species of sections *Hebea* and *Homoglossum* and all the species of series *Hebea*. *Gladiolus stellatus* is unusual in the genus in that the flowers are open for part of the day, from early in the morning until shortly after midday, when they close rapidly.

Androecium

As in all Ixioideae and most Iridaceae, there are three stamens in *Gladiolus* (**Figure 1D**). The filaments are inserted at the base of the upper part of the perianth tube, usually at the point where the tube widens (**Figure 1E**), but the distinction is not always evident in species that have slender perianth tubes. The stamens are unilateral and arcuate to horizontal in most species, but symmetrically arranged around the style in the few species with actinomorphic flowers. Filaments are normally exserted for some distance from the tube, although more or less included in the flower. However, in several species the filaments are included. This is often a useful diagnostic character, and an important distinction between *G. longicollis* and *G. tristis*, for example. Filaments are conspicuously hairy in several species of series *Hebea*, including *G. alatus*, *G. meliusculus* and *G. pulcherrimus*, and lightly pubescent below in *G. liliaceus* (section *Homoglossum*).

The anthers lie parallel to one another in species with zygomorphic flowers and are usually contiguous (**Figure 1E**). Variation in their size is sometimes a useful taxonomic character to separate allied species. Fairly large anthers are typical of sections *Blandus* and *Ophiolyza*, and smaller anthers of the smaller-flowered sections. Anthers are subbasifixed in most species. Three species of section *Hebea*, *Gladiolus cunonius*, *G. saccatus* and *G. splendens*, have the filaments apparently medially fixed and the anther thecae free below the filament insertion. Three species of section *Densiflorus*, *G. appendiculatus*, *G. calcaratus* and *G. macneilii*, have the anther bases drawn into sterile tails or spurs 2–4 mm long. The conspicuously long-apiculate anther apices of some tropical African species (Goldblatt, 1996), notably in section *Acidanthera*, do not occur in southern African species. Anther colour is sometimes useful taxonomically. Pale yellow to cream anthers and pollen are most common, but blue to violet anthers and pollen are typical of some species, for example, *G. undulatus* (section *Blandus*).

Pollen ultrastructure of *Gladiolus* species is broadly similar and does not differ from that of most other Ixioideae. Pollen grains are monosulcate and the exine is perforate-scabrate. The apertures of all species examined have a two-banded operculum of the type that is basic for the subfamily (Goldblatt et al., 1991, Goldblatt, 1996).

Gynoecium

Simple but apically expanded to spathulate style branches are apomorphic in *Gladiolus*. Most other genera of Ixioideae have either filiform style branches, or the branches are divided apically for some distance. Variation in style type is rare within genera of the subfamily. The only genus with style branches comparable to those of *Gladiolus* is *Babiana*, a largely southern African genus with one species on Socotra and one ranging across southern tropical Africa. Whether the style branches of *Gladiolus* and *Babiana* are structurally identical and homologous awaits investigation. In bud the margins of the style branches of *Gladiolus* are conduplicate, so that the stigmatic surfaces are covered. Typically the stigma margins unfold two days after anther dehiscence, exposing the minute stigmatic hairs. There are few useful taxonomic characters in either the ovary or style (**Figure 1D, E**). The ovary is ovoid to oblong and varies in size in relation to overall flower size. The style is slender and arches over the stamens, except in the few species with actinomorphic flowers when the style is erect and central. The point of division of the style into its three stigmatic branches relative to the anthers is often characteristic of a species, although the feature is somewhat variable. Comparatively short styles with short branches characterize *G. brevitubus*, *G. stellatus* and the coastal dune species, *G. gueinzii*, in which the style divides opposite the lower half of the anthers. Such short styles often occur in autogamous plants and *G. gueinzii*, at least, is known to be autogamous. More typically, the style divides near or just beyond the anther apices, and the branches are at least 3 mm long. *Gladiolus cunonius*, *G. splendens* and *G. saccatus* have unusually long style branches that divide just below the anther bases, and are expanded and stigmatic only at the apices.

FRUITS AND SEEDS

The *Gladiolus* fruit is a three-locular capsule and the genus can normally be recognized by the capsule's inflated, obovoid to almost ellipsoid shape (**Figure 1G, Figure 6A–J**). The capsules are also typically fairly coriaceous, although they become brittle when completely dry. At maturity the capsule valves spread on drying, often widely, but close in damp weather. Comparison with other genera of Ixioideae suggests that the obovoid capsule (**Figure 6A, B**) of section *Densiflorus* and most species of sections *Ophiolyza* and *Heterocolon* is the ancestral type. In most species of section *Linearifolius* in the winter-rainfall region the capsules are oblong to obovoid. More specialized narrowly oblong to ellipsoid capsules characterize most species of sections *Hebea* and *Homoglossum* (**Figure 6D, E, H–J**), and some species of series *Tristis*, particularly *G. longicollis* (**Figure 6J**) and *G. tristis*, have exceptionally long, narrow capsules. In an apparent reversal to an ancestral condition, *G. salteri*, *G. lapeirousioides* and *G. venustus*, specialized species of section *Hebea*, have

obovoid capsules with obtuse to emarginate apices (**Figure 6F**). In species of section *Homoglossum* the capsule valves typically curve outward when fully open, thereby releasing all their seeds.

Fairly large, light brown, broadly winged seeds are characteristic of *Gladiolus*. The seed body itself is globose, but part of the seed coat extends in a halo around the seed as a broad wing (**Figure 1H**). The seed surface is lightly reticulate and shows the outline of the epidermal cells. Seeds are oval to round and 8–12 x 5–7 mm in most large-flowered species, including sections *Ophiolyza* and *Blandus*, and series *Homoglossum* and *Tristis* of section *Homoglossum*, but they are substantially smaller in some species. In smaller-flowered species, notably in sections *Hebea* and *Heterocolon* and many species of section *Homoglossum*, the seeds, when winged, are 4–7 x 2–5 mm. In several species of section *Heterocolon*, including *G. pretoriensis* (**Figure 6M**) and *G. rufomarginatus* in southern Africa, the wing is vestigial or lacking, and it is reduced in several species of section *Densiflorus*, notably *G. ferrugineus* and *G. varius* (**Figure 6L**). In contrast to the usual oval to round shape, narrowly oblong seeds characterize *G. aquamontanus* and *G. buckerveldii* of section *Blandus* (**Figure 6Q**).

Seed number depends on capsule size, and to some extent on seed size. Species of series *Tristis* of section *Homoglossum* (**Figure 6I, J**) have particularly large capsules, mostly at least 30 mm long, and in *Gladiolus liliaceus*, as well as *G. watsonius* of series *Homoglossum*, the capsules are up to 50 mm long. Seed number is fairly high, rarely less than 20 per locule, and in species with large capsules such as *G. watsonius* there may be more than 65 seeds per locule, thus 200 or more seeds per capsule. At the opposite end of the scale, seed number in coastal populations of the Western Cape species, *G. speciosus* (section *Hebea*), is remarkably low. We have not seen capsules with more than three seeds per locule and some locules may contain no seeds at all. The seeds are among the largest in the genus, sometimes up to 12 mm long. In these populations the locule walls are unusually thick, although light and spongy. The dry, rather inflated capsules may act as sails, buoying the dry, dead stems and allowing the whole aerial part of the plant to be dispersed as a unit by wind, rather than the seeds alone. Inland populations of *G. speciosus* and its close relatives all have fairly large capsules with a normal complement of seeds.

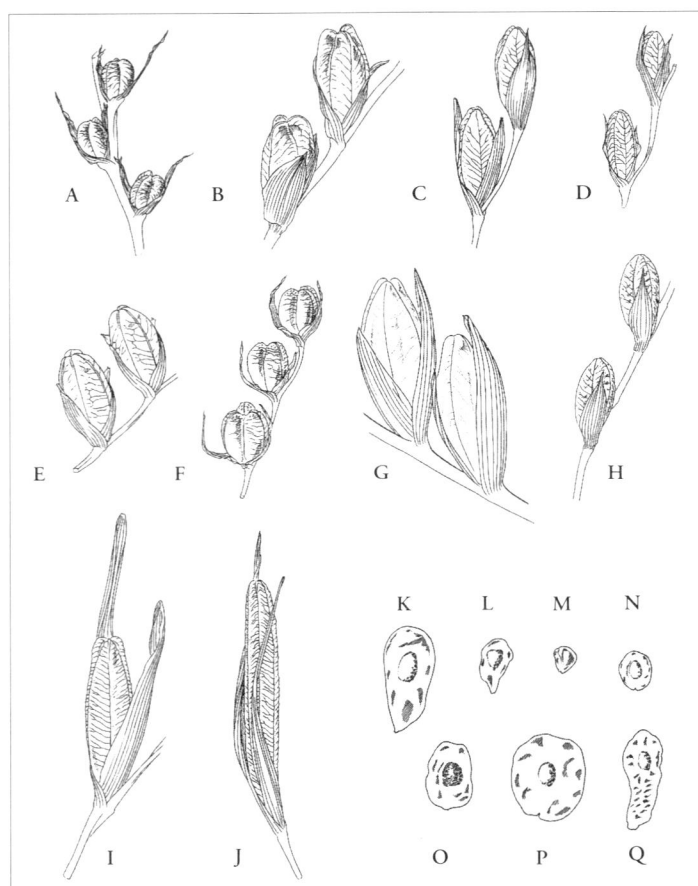

Figure 6 Capsules (**A–J**) and seeds (**K–Q**) in southern African *Gladiolus*. *Gladiolus crassifolius* (**A**) and *G. ochroleucus* (**B**) of section *Densiflorus*; *G. hirsutus* (**C**) of section *Linearifolius*; *G. permeabilis* (**D**), *G. orchidiflorus* (**E**), *G. salteri* (**F**) and *G. cunonius* (**G**) of section *Hebea*; *G. jonquilliodorus* (**H**), *G. liliaceus* (**I**) and *G. longicollis* (**J**) of section *Homoglossum*. (approximately x 2/3)
Gladiolus papilio (**K**) and *G. varius* (**L**) of section *Densiflorus*; *G. pretoriensis* (**M**) of section *Heterocolon*; *G. quadrangulus* (**N**) of section *Homoglossum*; *G. permeabilis* (**O**) and *G. virescens* (**P**) of section *Hebea*; and *G. aquamontanus* (**Q**) of section *Blandus*. (approximately x 2)

CHROMOSOME CYTOLOGY

Ever since the first chromosome counts (Vilmorin & Simonet, 1927; Ernst-Schwarzenbach, 1931; Bamford, 1935) were made in *Gladiolus*, $x = 15$ has been considered the likely ancestral base number for the genus (Goldblatt, 1971). Except for the occasional occurrence of B chromosomes, the several Eurasian, southern African and Madagascan species counted all have diploid numbers based on $x = 15$. Significant cytological variation includes polyploidy, which has been recorded, for example, in all the Eurasian species (Van Raamsdonk & De Vries, 1989) and in the widespread African and Madagascan *G. dalenii*, which has races with $n = 15$, 30, and 45 (Goldblatt et al., 1993). Other significant variation has been discovered in some small-flowered tropical African species which show patterns of descending dysploidy (also called aneuploidy). Dysploidy includes $2n = 22$ in *G. gregarius* and the allied *G. pseudospicatus* (related to southern African *G. crassifolius*). In section *Heterocolon*, the Congo (Zaïre) endemic, *G. actinomorphanthus*, which has $2n = 28$, and the widespread tropical African *G. unguiculatus*, which has $2n = 26$ (or $2n = 26+2B$) and $2n = 24$, are cytologically puzzling. Two more species, possibly also in section *Heterocolon* (the sectional classification of some small-flowered tropical African species is unsettled), *G. serapiiflorus* has $2n = 22$, and the widespread *G. atropurpureus* appears to have $x = 12$ ($2n = 24$ and 36).

Chromosomes of *Gladiolus* are relatively small, 0.7-2.9 micrometres, and few details of the karyotype apart from the diploid numbers are known. Karyotypes of most species comprise metacentric (symmetrical in arm ratio) to submetacentric chromosomes of fairly uniform size. All chromosomes have a large heterochromatic segment that stains darkly at early prophase. Counts are available for just 60 of the 163 species in southern Africa (**see table below**), of which some ten counts are new (Goldblatt & Takei, 1997). The pattern is surprisingly consistent. All species have numbers based on $x = 15$ and most are diploid, $2n = 30$. An identical base number and a similar chromosomal configuration in most species of *Gladiolus* explain why it is relatively easy to hybridize species, even from different sections, and why hybrids are so often fertile. Chromosome number is clearly of virtually no value in understanding the systematics and phylogeny of the genus in southern Africa.

Both polyploidy and dysploidy are rare in the southern African species. *Gladiolus dalenii* is polyploid in the subcontinent, $2n = 45$ or 60, and the only count for the southern Cape species, *G. leptosiphon*, is also polyploid, $2n = 60$. Polyploidy has evidently played a minimal role in the evolution of *Gladiolus* except in Eurasia and North Africa. Polyploidy elsewhere is rare; the reports of triploidy in the southern African *G. scullyi* (probably correctly *G. venustus*) and *G. orchidiflorus* (**see table below**) hardly suggest a need for additional study. Excluding these examples, only two southern African species have been reported as polyploid, *G. papilio* and *G. saundersii*, and both the counts are for so-called hybrids (Bamford, 1935). The significance of polyploidy in *G. dalenii* is difficult to gauge. It is one of the most successful species of the genus, and certainly the one with the widest geographical distribution (Goldblatt, 1989; 1996). It extends from the Eastern Cape, South Africa, to Senegal in West Africa and from Madagascar to Ethiopia and Yemen. Perhaps polyploidy has played a role in enabling the species to thrive under the wide variety of conditions that occur across its enormous geographic range. The role of humans in its dispersal may also be important. The species is cultivated in West Africa and, as it is an important medicinal plant, it may have been spread over long distances by human agency.

The presumed ancestral basic chromosome number in *Gladiolus*, $x = 15$, is shared in subfamily Ixioideae only with *Radinosiphon* (Goldblatt, 1971), and the two genera are evidently palaeotetraploid. This base number is apomorphic in Ixioideae and it is the only specialized feature shared by *Gladiolus* and *Radinosiphon*. Most other genera of Ixioideae have lower base numbers (Goldblatt, 1971), $x = 11$ and 10 being most frequent. Other high base numbers in Ixioideae are $x = 16$ in *Tritoniopsis*, 13 or 14 in *Romulea*, and 13 in *Geissorhiza* and *Hesperantha*, the two genera possibly most closely related to *Gladiolus* after *Radinosiphon*. ❧

Chromosome numbers in southern African species of *Gladiolus*

Data are taken from compilations by Goldblatt et al. (1993) and Goldblatt & Takei (1997).
Obviously erroneous counts have been ignored and the taxonomy of species counted has been corrected.
The identity of species counted has, however, not been verified, and in most cases this is not possible since voucher specimens do not exist. Doubtful counts are placed in parentheses.

Species	Diploid number	Species	Diploid number	Species	Diploid number	Species	Diploid number
G. abbreviatus	30	*G. floribundus*	30	*G. pappei*	30	*G. sericeovillosus*	
G. alatus	30	*G. gracilis*	30	*G. permeabilis*		subsp. *sericeovillosus*	30
G. angustus	30	*G. gueinzii*	30	subsp. *permeabilis*	30	subsp. *calvatus*	30
G. aquamontanus	30	*G. guthriei*	30	subsp. *edulis*	28, 30	*G. stefaniae*	30
G. bonaspei	30	*G. hirsutus*	30	*G. phoenix*	30	*G. teretifolius*	30
G. brevifolius	30	*G. huttonii*	30	*G. pole-evansii*	30	*G. trichonemifolius*	30
G. buckerveldii	30	*G. inandensis*	30	*G. pretoriensis*	30	*G. tristis*	30
G. cardinalis	30	*G. lapeirousioides*	30	*G. priorii*	30	*G. undulatus*	30
G. carmineus	30	*G. leptosiphon*	60	*G. pritzelii*	30	*G. uysiae*	30
G. carneus	30	*G. liliaceus*	30	*G. quadrangularis*	30	*G. varius*	30
G. caryophyllaceus	30	*G. longicollis*	30	*G. quadrangulus*	30	*G. watermeyeri*	30
G. ceresianus	30	*G. maculatus*	30	*G. recurvus*	30	*G. watsonius*	30
G. crassifolius	30	*G. martleyi*	30	*G. saccatus*	30	*G. wilsonii*	30
G. cunonius	30	*G. ochroleucus*	30	*G. saundersii*	30 (45)	*G. woodii*	30
G. dalenii	45, 60, 75, 90	*G. oppositiflorus*	30	*G. scabridus*	30		
G. debilis	30	*G. orchidiflorus*	(45)	*G. scullyi*	(45)		
G. equitans	30	*G. papilio*	(75)	*G. sempervirens*	30		

GROWTH STRATEGIES

Like nearly all geophytic plants, that is, those with underground storage and regenerative organs, most species of *Gladiolus* conform to the standard pattern of lying dormant underground during the dry season and producing leaves and then an aerial flowering stem during the course of the wet season. Flowering occurs after most vegetative growth has taken place and a new corm, the type of storage organ in *Gladiolus*, has begun to develop from the base of the stem. While capsules with their complement of seeds develop, the new corm is being provisioned under the ground. As the above-ground parts dry at the end of the wet season, the corm matures and becomes dormant. The dry flowering stem is shed and the corm enters a resting phase, the tiny subapical bud in the axil of a leaf base already formed and ready to sprout at the beginning of the next growing season.

In southern Africa with its two opposed growing seasons – a wet summer in the eastern part of the subcontinent and a wet winter in the west – growth cycles in the flora occur at quite different times of the year. In the winter-rainfall region, which comprises the southwestern and southern Cape, Namaqualand to the north and the western Karoo in the interior, winter rains usually begin in April or May. Most geophytes then commence their growth cycles. Flowering may begin as early as May or June in *Gladiolus griseus*, *G. maculatus*, *G. priorii* and *G. viridiflorus*, for example, but more often it commences in early August and lasts until October in the lowlands. At higher elevations in the mountains flowering finishes later, sometimes as late as December. Plants of wet montane habitats, such as *G. cardinalis*, *G. insolens* and *G. oreocharis*, are less immediately affected by the limitations of seasonal availability of water. They flower late in the season, in December and January. *Gladiolus sempervirens* is unusual for a species of the winter-rainfall region in that its flowering is delayed until March or April. It grows in the southeastern Cape in habitats that seldom, if ever, become dry, and the plants are evergreen.

In the summer-rainfall region the winters are dry and cold, except at the coast, and most species of *Gladiolus* sprout with the onset of spring rains in October and November, then flower in December or later. The summer rainfall season is longer than the winter rainfall season and flowering in different species may continue until April or May. Flowering in the season most favourable for growth means that species flower at the same time as most other plants in their communities and are thus exposed to high levels of competition for space, light and pollinators. Flowering during a season when competition is less intense confers certain advantages on plants and this strategy has been achieved in *Gladiolus* in diverse ways.

Flower production in autumn has evolved in the winter-rainfall region in several species of sections *Blandus*, *Homoglossum* and *Linearifolius*. Shortage of water is a limitation to

plant growth at this time of year, but has been overcome by the size of the leaves being reduced, often to small sheaths. Thus less water is lost through transpiration. With the onset of the rains, species adopt one of two strategies. In one of these, a fully developed foliage leaf is produced on a separate shoot. The leaf continues to grow until spring, by which time a new corm has been formed at its base and provisioned with food reserves for flowering the next season. Species that have adopted this strategy are the only ones in the genus in which the new corm is not produced at the base of the flowering stem, but rather at the base of the shoot that produced the foliage leaf or leaves. This pattern, sometimes termed hysteranthy, is followed by *Gladiolus brevifolius* and its close relatives *G. monticola*, *G. nerineoides* and *G. stokoei* (section *Linearifolius*) and by *G. jonquilliodorus*, *G. martleyi* and *G. subcaeruleus* (section *Homoglossum*).

In the alternative strategy, although no well-developed leaves are associated with the flowering stem, the new corm is produced from its base. The stem and its reduced leaves remain green and alive for months after the plant has finished flowering and fruiting, and the new corm is provisioned by photosynthesis that takes place in these organs. This pattern occurs in *Gladiolus emiliae* (section *Linearifolius*); in several species of section *Homoglossum*, including *G. exilis*, *G. mutabilis* and *G. vaginatus* (series *Mutabilis*) and *G. engysiphon* and *G. nigromontanus* (series *Teretifolius*); and in *G. carmineus* and *G. stefaniae* (section *Blandus*). Plants that produce foliage leaves in one season usually flower the next, so the dry remains of the foliage leaves can sometimes be seen still attached to the corms, as in *G. carmineus*.

In the winter-rainfall region one of the most striking benefits that flowering out of the main growing season confers on plants is their removal from competition for pollinators, which is often a limiting factor during the peak flowering months of August and September. It also permits species to adapt to alternative pollination guilds which are not available during the main growing season. The most obvious examples of this are *Gladiolus nerineoides*, *G. stefaniae* and presumably *G. stokoei*. All have red flowers with fairly long tubes adapted for pollination by the butterfly *Aeropetes tulbaghia*, which is only active after mid-December.

In the summer-rainfall region, flowering early in the spring, before or just after the first rains have fallen, likewise exposes species to less competition for light and perhaps for pollinators as well. At the same time, growth this early in the season exposes plants to limiting water availability and necessitates the rapid development of the flowering stem. As in the winter-rainfall region, the problem is solved either by producing foliage leaves later in the season from separate shoots on the same corm, or by remaining green long after capsules have ripened and seeds dispersed. In section *Ophiolyza* the reduction in the leaf blades is common in tropical African species

(Goldblatt, 1996) such as *Gladiolus dalenii* subsp. *andongensis* and *G. melleri*, but in southern African members of the section it occurs only in *G. aurantiacus* which produces long-bladed foliage leaves on separate shoots after flowering. In *G. brachyphyllus* (section *Densiflorus*) plants simply remain green and photosynthetic until the capsules have ripened and new corms have been provisioned.

Likewise, several species of section *Linearifolius* from the summer-rainfall region, including *Gladiolus pardalinus*, *G. parvulus*, *G. pubigerus* and *G. woodii*, flower in the spring, before or soon after the first rains have fallen and the stems bear bladeless or short-bladed leaves. Like *G. brachyphyllus*, the plants that flowered in that season do not produce foliage on separate shoots. Photosynthesis is carried out by the stems and reduced leaves which persist for several months after the seeds have ripened. The closely related species pair, *G. oatesii* and the tropical African *G. unguiculatus* (section *Heterocolon*), both have the leaf blades of the flowering stem reduced and flower at the end of the dry season, but their growth strategies differ radically. In *G. unguiculatus*, which has large corms, long-bladed foliage leaves are invariably produced from separate shoots after flowering. Small-cormed *G. oatesii*, however, depends on its stem and leaf sheaths to nourish the new corm in much the same way as *G. woodii* and its allies.

In contrast, *Gladiolus vernus* (section *Heterocolon*) and *G. inandensis* (section *Hebea*) produce their flowers after the leaves. They flower at the end of the growing season, postponing the production of a flowering stem until June, July, or August, by which time the long-bladed foliage leaves have completely died. These can often be seen still attached to the base of the flowering stem. By the time flowering is well advanced the new leaves are often emerging from a new shoot. Thus, these two species have achieved the separation of the growth and flowering cycles in a different, purely temporal way from that seen in the other winter- and summer-rainfall zone species, in which the flowering and vegetative phases are separated by different developmental pathways.

FIRE AND THE FLOWERING RESPONSE

Fire is an important aspect of the ecology of southern African species of grasslands, savannas and shrublands, including the fynbos and renosterveld vegetation types of the winter-rainfall zone. For geophytes fire provides an important opportunity to produce flowers and seeds in the absence of competition from larger shrubs and trees. Thus, displays of flowering in *Gladiolus* are frequently striking in the season immediately after a burn and for a year or two thereafter. Mass flowering after fire is particularly evident in the winter-rainfall zone, and more so on the nutrient-poor, sandstone soils that cover much of the region. A secondary aspect of fire is reduced seed mortality from seed-eating beetles which often infest capsules of *Gladiolus*.

Some species are so strongly adapted to fire that they are rarely seen except after a fire. *Gladiolus phoenix* is the most prominent example in the genus. Observations after a fire in the Bain's Kloof Mountains showed that plants flowered extremely well the season after fire and set quantities of seed. The following year not a single plant flowered there and few even produced leaves. Small corms removed from one site two years after mass flowering were induced to flower in cultivation the next season only after the effects of fire had been simulated by burning plant material over planted corms. In the subsequent seasons in cultivation and in the wild no plants flowered. Plants evidently produce assimilatory leaves in the inter-fire period, allowing the corm to enlarge gradually year by year in preparation for mass flowering after the inevitable fire to come. The resources of the corms are exhausted by the flowering and only a small main corm and many tiny cormlets are produced after mass blooming.

SOIL AND EDAPHIC CONSIDERATIONS

Most species of *Gladiolus* show strong edaphic fidelity, that is, they favour a particular soil type over much or all of their range. This is especially evident in areas where soils of sharply different texture and nutrient status occur. In the Cape Flora Region, for example, the two main soil types are nutrient-poor, coarse-grained soils derived from quartzitic sandstone rocks and nutrient-intermediate, fine-grained clays derived from shales. Rarely do species grow locally on both types of soils. Species characteristic of one of these soil types may, however, often be found along the interface between them or on soils derived from granite which is intermediate in its characteristics between the well-drained sandy soils and the poorly drained but richer clay soils.

Thus, *Gladiolus gracilis*, a species restricted to clay soils over much of its range, occurs on granitic sands along the Cape west coast, and even on light sands where there is additional ground water that keeps the soil moist during the growing season. In contrast, *G. carinatus* favours well-drained, sandy soils almost exclusively. It, too, may be found on granitic slopes in well-drained situations. Consequently, these two species, normally found in quite different habitats, grow adjacent to one another on granitic sands. The list of sandstone- versus clay-loving species almost divides the winter-rainfall species in half. In the summer-rainfall region soil preferences between species are also evident. Species favour either coarse-grained soils derived from quartzites or clay soils derived from shales and dolerite.

A number of species of *Gladiolus* are adapted to special soils that have a limited distribution in southern Africa (**see table overleaf**). Dolomite substrates occur along the eastern and northern Drakensberg escarpment in Mpumalanga and Northern Province. There, highly localized species include *G. dolomiticus*, *G. macneilii* and *G. pavonia*. The latter two species are each currently known from a single extended population. *Gladiolus serpenticola* is restricted to serpentine in the Barberton District of Mpumalanga and Swaziland. In the Western Cape, limestone outcrops occur along the southern and west coasts and several species of the genus, most of them endemic, are restricted to these habitats.

Edaphic endemic species of southern African *Gladiolus*

Species	Distribution range
Coastal limestone or calcareous sand	
G. caeruleus	west coast, Western Cape
G. griseus	west coast, Western Cape
G. miniatus	south coast, Western Cape
G. vaginatus	south coast, Western Cape
G. variegatus	south coast, Western Cape
Dolomite	
G. dolomiticus	Northern Province
G. macneilii	Mpumalanga
G. pavonia	Mpumalanga
Serpentine	
G. serpenticola	Mpumalanga and Swaziland

A number of species are restricted to unusual habitats and deserve special mention. Most remarkable is *Gladiolus gueinzii*, a strand plant. It grows in coarse sand at and above the high-tide mark along the southern African coast from Cape Agulhas to near Durban. This habitat is shared in the genus by the Madagascan and Mascarene *G. luteus*, but the two species are not believed to be closely related, despite the fact that both have cormlets with thick, corky, waterproof tunics, an adaptation for water dispersal.

Three species of the winter-rainfall zone, *Gladiolus aquamontanus*, *G. buckerveldii* and *G. cardinalis* (section *Blandus*), are restricted to waterfalls and grow in permanently wet conditions. This permits them to flower out of the normal season, in summer when the surrounding vegetation is mostly dormant. Each of the waterfall species appears to have evolved this habitat preference independently. Rocky cliffs are another specialized habitat and *G. cruentus* (section *Ophiolyza*) and *G. lithicola* and *G. microcarpus* (section *Densiflorus*) are largely restricted to such sites in the mountains of the summer-rainfall zone. Mist and fog as well as relatively high rainfall permit these species to thrive in what would otherwise be sites too dry to support *Gladiolus* species. In the winter-rainfall zone *G. insolens* and occasionally *G. carneus* (both section *Blandus*) are found on damp cliffs. Additional moisture is provided by water percolating down the rock faces from sources on the slopes above.

REPRODUCTIVE BIOLOGY

Vegetative reproduction

Although a few species of *Gladiolus* have much reduced corms, they all have the capacity for annual regeneration from the corm. Many species also have the ability to increase their numbers by vegetative reproduction through the production of cormlets in various ways (see Morphology of the southern African species: Corm, page 15). It may thus be thought that annual regeneration and vegetative reproduction are sufficient for ensuring the continuing survival of the species. This is, however, unlikely. Corms may in theory live to a great age, but

in fact they are highly nutritious sources of food for a variety of animals adapted in special ways to locate and consume them. Major predators are baboons, porcupines, molerats and other rodents, as well as some birds, notably guineafowl, and they account annually for huge reductions in population numbers of mature plants.

The significance of predation on population size and density is illustrated by the situation of introduced *Gladiolus* species in Australia. There, in the absence of predators, species of *Gladiolus* and other corm-bearing Iridaceae may form dense stands and even become noxious weeds. These same species in southern Africa are seldom conspicuous and are often declining in numbers due to impaired reproductive capacity for a variety of reasons, not least an increase in natural predation. Reproduction through the more conventional means of cross pollination and seed production is thus vitally important for geophytic plants such as *Gladiolus*.

In the face of predation pressures, population numbers are maintained in the short term by individuals surviving and even increasing clonally through their cormlets, which are often too small or too inconspicuous to interest a predator. Species may even make a virtue of the inevitable loss of mature individuals by exploiting the behaviour of predators to disperse plants to new habitats. Molerats are particularly important dispersal agents for many species. These animals harvest corms and bulbs in huge numbers and maintain stores of food in their underground burrows. Undamaged mature corms can in this way be carried large distances. Uneaten corms will grow the following season in a new site. Perhaps more important than some forgotten corms in molerat stores is the accidental spreading of cormlets along the underground runs as corms are transported to storage areas. The dispersal of plants of some species may depend to a large extent on molerat activity.

Sexual reproduction

Despite the capacity for vegetative reproduction in *Gladiolus*, conventional sexual reproduction and the recruitment of new plants by seed production and dispersal remain vital for long-term survival and the maintenance of populations, as well as long-distance dispersal and the establishment of new colonies. Sexual reproduction is also essential for the maintenance of genetic variability and the production of novel genotypes through recombination. This may be stating the obvious, but it seems necessary to emphasize the importance of sexual reproduction and seed dispersal in geophytic plants such as *Gladiolus*. The conservation of these plants does not simply require that they be left undisturbed in their habitat. This is certainly a recipe for their ultimate demise and loss. Production of seeds also depends on successful pollination. Conservation thus also implies conservation of pollinators and the maintenance of the environment to the extent that pollinators are available and predator populations are limited. In the absence of large mammal predators over much of the range of *Gladiolus* today, damage by herbivores may be disastrous to plant populations. Likewise, human activities, although they may be distant from plant populations, may affect them through disturbance or loss of habitat for specialist pollinators.

Pollination ecology

As outlined above, sexual reproduction is vital for the long-term regeneration and survival of species of *Gladiolus*. In common with most other plant species, cross pollination must be effected before sexual reproduction can take place. Most species are self-incompatible, and even if pollen is deposited directly on the stigma, either through self- or artificial transfer, it will not accept pollen of the same plant. Cross pollination is essential and the way this is accomplished is one of the most important aspects of plant adaptation and evolution. For most Iridaceae, and particularly *Gladiolus* which has large colourful flowers, pollination is achieved by the transfer of pollen to the stigmas of flowers of different plants by flying animals. Sometimes these are birds, but most often they are insects. They include a variety of bees, night-flying moths, the butterfly *Aeropetes tulbaghia*, short- and long-tongued flies, and in one instance scarab beetles.

The variety of flower colours, shapes and sizes in *Gladiolus* is largely the result of adaptation to attract particular pollinators to visit the flowers. Pollination is accomplished passively by the particular bird or insect species that visits the flowers in search of a food reward. In many Iridaceae the reward is pollen, but it may equally be nectar or even flower parts. In *Gladiolus* species the reward to insect visitors is most often nectar alone. Various insects or birds – and the kind depends on the appearance and phenology of the flower – visit flowers in search of nectar and while probing the flower tube in which the nectar is located, they become dusted with pollen. A few species of *Gladiolus* have prominently displayed pollen and do not offer nectar rewards. Female bees collect pollen to provision their nests and sometimes they actively harvest pollen from the flowers of these species. Despite placing the pollen on specific sites on their bodies for efficient transport to their nests, enough is left dusted about their head and thorax to be transferred to stigmas of other plants, thus effecting fertilization.

POLLINATION AND FLOWER FORM

The genus *Gladiolus* displays a diversity of floral form unmatched elsewhere in the Iridaceae (**Figures 7–11**). This remarkable variability is a direct consequence of specialization for particular pollinators, and the radiation in flower form is paralleled by a radiation in pollination strategy. The greatest diversity in pollination type is undoubtedly evident among the species of the southwestern Cape, many of which have evolved to exploit unusual pollinators. This may be a direct result of the poorer insect diversity in the region relative to other parts of southern Africa, combined with a shorter growing season during which many species are in bloom and competing for the available pollinators.

The six classes of pollinator used by the genus are bees, long-tongued flies, moths, a butterfly, sunbirds and, in a single species, flower-visiting beetles. The floral adaptations which have developed to attract and exploit each of these classes of pollinator are so precise that the shape, colour, markings and scent of the flowers of each species are sufficient to allow one to infer the pollinator without more direct evidence. Because the exploitation of these pollinator classes has occurred independently many times within *Gladiolus*, unrelated species may show a similarity in their flowers which can be quite misleading about their true relationships.

On the basis of our observations of pollination biology in *Gladiolus* we have found that bee pollination is by far the most common in the genus and some 58% of the southern African species have flowers adapted for pollination entirely or partly by bees. Long-tongued flies and birds are next in importance and respectively pollinate some 16% and 17% of the species in the subcontinent. Moths and butterflies each pollinate no more than half as many. Beetle pollination is a recent development in a single species. Bee pollination is probably the ancestral condition in *Gladiolus* and is present in some species of every section of the genus.

BEE POLLINATION

Bees are the most highly developed of the flower-visiting insects and the most skilled and versatile at gathering floral rewards. They are also the most active pollinators. This is because, unlike the other pollinator classes, both adult and larval stages depend entirely on floral products. Adult bees feed on nectar, while the larval stages are sustained on pollen mixed with regurgitated nectar in the form of 'bee bread'. While most other insects have free-living larvae which find their own food, the adult female bee is responsible not only for feeding herself, but also for provisioning her nest with a sufficient store of food for her offspring or the young of the hive. In order to do this she must be able to gather nectar and pollen from a range of flowers of different shapes and colours and with varied fragrances.

Bee flowers are the most diverse of all the floral types. Although bees are able to deal with a variety of flower forms, they tend to visit in turn several individuals of one species before shifting to another. This floral constancy is particularly characteristic of honey bees but also occurs in other bee species. After encountering a particular shape a number of times, foraging animals develop a search image which enables them to recognize this shape more rapidly and deal with it more efficiently in the immediate future. It seems, therefore, that for the flower it is not so much a case of attracting the bee as of being sufficiently distinctive that it will be recognized again among the variety of competing flower shapes and colours. The distinctive and often flamboyant shapes and scents of the flowers of bee-pollinated species of *Gladiolus* (**Figure 7**) can thus be thought of as a form of image enhancement. These range from simple star-like flowers in plain colours, as in *G. stellatus* and *G. quadrangulus* (**Figure 7H, I**), through a large number of species which have more or less two-lipped, gullet-shaped flowers with variously marked tepals, sometimes highlighted by intense scents, to the intricately shaped and unusually coloured flowers of the species in section *Hebea* (**Figure 7E–H**).

Flowers of bee-pollinated species of *Gladiolus* are small or medium-sized with a relatively short, funnel-shaped tube (**Figure 7A, B, E**) that is never more than 20 mm long and has a narrow part never more than 15 mm long. Nectar is usually produced in small quantities, and has a moderate to relatively high sugar concentration. In a few species pollen is offered as a reward, occasionally exclusively as in *G. aureus*, *G. brevitubus* and *G. stellatus*. In these species the flowers have prominent anthers and the pollen is cream or pale yellow. In species which do not offer pollen as a legitimate reward, the anthers are usually concealed below the arched or hooded dorsal tepal and are often the same colour as the tepals.

Bee-pollinated species from the summer-rainfall region have flowers that are rarely scented, usually lack prominent tepal markings, and are usually arranged in many-flowered

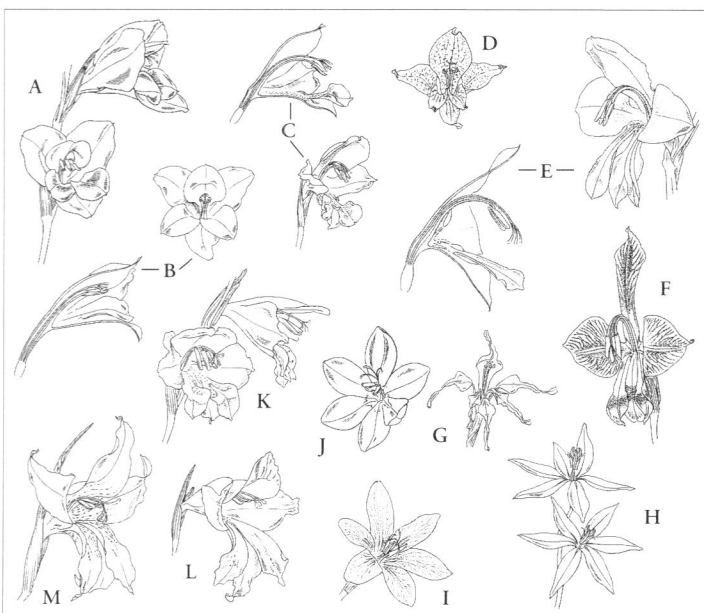

Figure 7 Flowers of southern African *Gladiolus* adapted for pollination by bees.
In section *Densiflorus G. paludosus* (**A**) and *G. exiguus* (**B**), with half flower; in section *Heterocolon G. mostertiae* (**C**) with half flower and *G. rufomarginatus* (**D**); in section *Hebea G. alatus* (**E**) with half flower, *G. uysiae* (**F**), *G. permeabilis* subsp. *edulis* (**G**) and *G. stellatus* (**H**), a completely actinomorphic flower; and in section *Homoglossum G. quadrangulus* (**I**), a completely actinomorphic flower, *G. brevitubus* (**J**), *G. patersoniae* (**K**), *G. exilis* (**L**) and *G. comptonii* (**M**). (approximately x ½)

spikes. Flowers are typically shades of pink to mauve or white, sometimes speckled or mottled as in *Gladiolus ecklonii* and *G. densiflorus*, but the flower form seldom deviates from a simple gullet shape. In the winter-rainfall region, however, due either to nutrient constraints or to enhanced competition for pollinators over the short spring, plants typically produce fewer flowers. The individual ones they do produce are larger and often strongly marked and scented. Flower colour there is extremely varied and includes shades of green and brown, blue, pink, yellow, orange and white, almost always with contrasting nectar guides. The scents are sweet and strong, reminiscent of violets, roses and freesias, among others. Flower shape is also variable, ranging from the gullet shape of many species of series *Gracilis* through the inflated, bell-like perianths of *G. bullatus*, *G. patersoniae* and *G. sufflavus* (**Figure 7K**), to the complex flag flowers in series *Hebea* (**Figure 7E, F**) or the flat, star-like flowers of *G. stellatus* (**Figure 7H**). Despite its apparent simplicity, the flower of *G. stellatus* is highly specialized for bee pollination and the primary reward in the species is pollen, which is prominently displayed in large anthers. In addition, the flowers have a precise daily schedule, opening at about 07h30 and closing again at 12h30. In this way they ensure a fixed period of high visibility during which they are available for visiting, thereby increasing the chances of pollination. This diversity in floral form is evident only in the spring-flowering species. The autumn-flowering species of the winter-rainfall region are invariably gullet-shaped and coloured shades of pink, mauve or cream. Competition for pollinators is apparently much lower at this time of the year.

Bee pollinators are mostly medium-sized species in the family Anthophoridae. In the winter-rainfall region the most important bees are *Anthophora diversipes*, *Amegilla fallax* and *A. spilostoma*, while *A. capensis* and *A. fallax* are important pollinators of *Gladiolus* species in the summer-rainfall parts of southern Africa. Honey bees, *Apis mellifera* (family Apidae), visit less intricately shaped flowers such as those of *G. trichonemifolius* and they are especially important in the pollination of early-flowering species such as *G. griseus* and *G. gracilis*. Living in complex colonies, honey bees are far less seasonal than anthophorid bees which are solitary and have annual life cycles. Other pollinating bees of *Gladiolus* in southern Africa include *Rediviva* species (Melittidae) and small, pollen-collecting bees of the family Halictidae, but these are less important than the anthophorids and honey bees.

LONG-TONGUED FLY POLLINATION

Long-tongued flies are uniquely well represented in southern Africa and have been exploited by a number of late spring- and summer-flowering plant species throughout the subcontinent (Goldblatt et al., 1995; Manning & Goldblatt, 1996, 1997). While various smaller flies in the families Acroceridae, Nemestrinidae, and Tabanidae visit some *Gladiolus* flowers as part of a general suite of visitors that includes bees, five or six of the longer-tongued species of Nemestrinidae and Tabanidae are

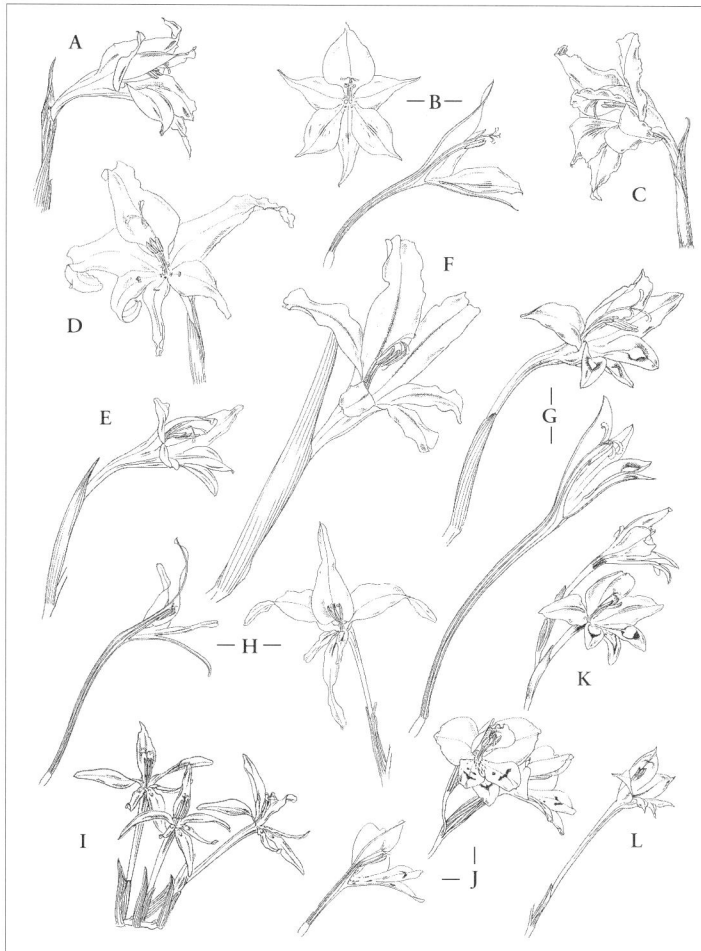

Figure 8 Flowers of southern African *Gladiolus* adapted for pollination by long-tongued flies.
In section *Densiflorus G. varius* (**A**) and *G. microcarpus* (**B**); in section *Ophiolyza G. oppositiflorus* (**C**); in section *Blandus G. undulatus* (**D**), *G. bilineatus* (**E**), *G. floribundus* (**F**) and *G. angustus* (**G**) with half flower showing the extremely long tube of this species; in section *Hebea G. leptosiphon* (**H**) with half flower and *G. lapeirousioides* (**I**); and in section *Homoglossum G. debilis* (**J**) with half flower, *G. vigilans* (**K**) and *G. engysiphon* (**L**). (approximately x ½)

the sole pollinators of several other species. These large-bodied, active insects have proboscis-like mouthparts 20–70 mm long that they use to siphon nectar from flowers. They are particularly well represented in the winter-rainfall zone where *Philoliche gulosa* and *P. rostrata* (Tabanidae), and *Prosoeca longipennis* and *Moegistorhynchus longirostris* (Nemestrinidae) occur. The mouthparts in the first three species are 20–40 mm long, but in *M. longirostris* they are 40–70 mm long. Two long-tongued fly species, *Prosoeca ganglbaueri* and *P. robusta*, occur in the summer-rainfall zone and both are active in the late summer and autumn. In the winter-rainfall zone the adult flies emerge in late spring and early summer, but *P. longipennis* is active in the autumn (Manning & Goldblatt, 1995). Species pollinated by these insects therefore bloom rather late in the season. Nothing is known about the life history of the nemestrinid fly species, but their larval stages are most likely parasitic.

The flowers of *Gladiolus* species pollinated by long-tongued flies (**Figure 8**) are medium-sized to large, odourless, invariably cream or pale pink to deep pink, and the lower tepals are marked with red, usually in the form of longitudinal streaks. The mouth of the tube and the anthers are often red or purple as well. The floral tube is long and fairly large amounts of nectar are produced with moderate sugar concentrations of 25 to 30%. The long tubes, up to 70 mm long in *G. undulatus* and *G. angustus* (**Figure 8D, G**), successfully prevent other insects from removing the nectar, thereby ensuring that suitable rewards are available for the legitimate pollinator. Typical examples of species pollinated by long-tongued flies include most of those in series *Scabridus* of section *Densiflorus* (**Figure 8B**) and several in series *Blandus* of section *Blandus* (**Figure 8D, E, G**) and series *Gracilis* and *Teretifolius* of section *Homoglossum* (**Figure 8J–L**). It is usual to find several fly-pollinated species from different plant families and genera growing nearby, all pollinated by the same species of fly, and the convergence in floral form among the plants is frequently striking. Such functional groups are termed guilds and typical guilds include *G. angustus, Babiana tubulosa, Ixia paniculata, Lapeirousia anceps, L. fabricii* and *Tritonia crispa* on the Cape west coast in late spring; *Gladiolus bilineatus, G. engysiphon, Pelargonium* species and *Tritoniopsis revoluta* along the southern coast in autumn; and *Gladiolus varius, Radinosiphon leptostachya, Watsonia wilmsii, Nerine angulata* and *Disa amoena* along the Mpumalanga escarpment in late summer.

MOTH POLLINATION

Moth-pollinated species differ most significantly from those pollinated by other insects in that they are visited at dusk or during the night. Visual attraction is thus less important than in other flowers and scent becomes the primary attractant. As a result, most moth-pollinated flowers, with the exception of many populations of *Gladiolus hyalinus*, have heavy, usually spicy scents, often redolent of cloves. In series *Tristis*, in three of the four species pollinated by moths, scent is produced only at night. Although colour is a secondary consideration, flowers of moth-pollinated *Gladiolus* species are often white to cream or yellowish, as these shades are more visible than others at low-light intensities. Others are inconspicuously coloured or mottled with green, grey, brown or ochre, possibly to camouflage the flowers during the day from herbivores or illegitimate insect visitors. *Gladiolus liliaceus* is remarkable in having flowers that change colour each evening, turning from dull brown or reddish to pale mauve, the colour change being accompanied by the production of a heavy sweet scent with a strong clove-like component.

The floral tubes in moth-pollinated species (**Figure 9**) are narrow and range in length from quite short (15–25 mm) in species such as *Gladiolus acuminatus* and *G. guthriei* (**Figure 9A**) to extremely long (110 mm) in the northern populations of *G. longicollis* (**Figure 9F**). Fairly large quantities of relatively sweet nectar are produced. Because the loose scales covering the body and wings of moths are an unstable surface, many moth-pollinated species deposit the pollen on the head or proboscis of the insect, and it is common for them to have the anthers partly included in the floral tube.

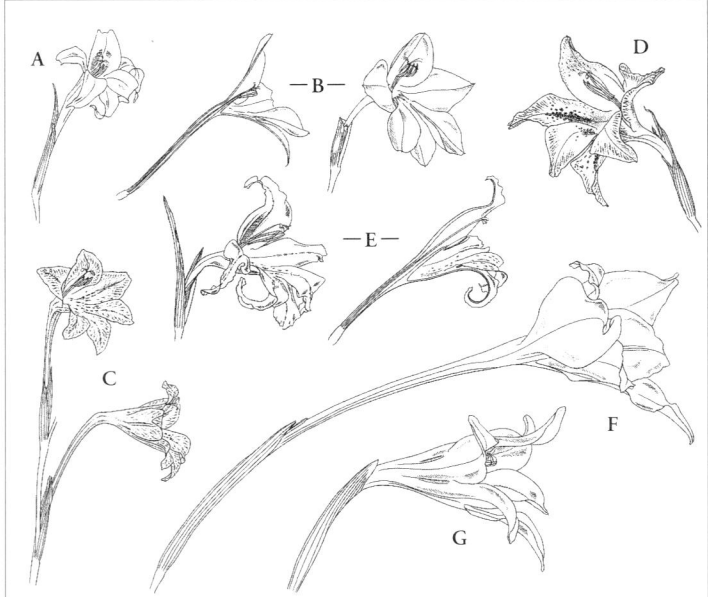

Figure 9 Flowers of southern African *Gladiolus* adapted for moth pollination.
In section *Hebea G. acuminatus* (**A**) and *G. robertsoniae* (**B**) with half flower to show tube length; in section *Linearifolius G. emiliae* (**C**); and in section *Homoglossum* the darkly mottled flower of *G. maculatus* (**D**), *G. recurvus* (**E**) with half flower, the extremely long-tubed flower characteristic of *G. longicollis* (**F**) from Mpumalanga and Swaziland, and *G. tristis* (**G**). (approximately x ½)

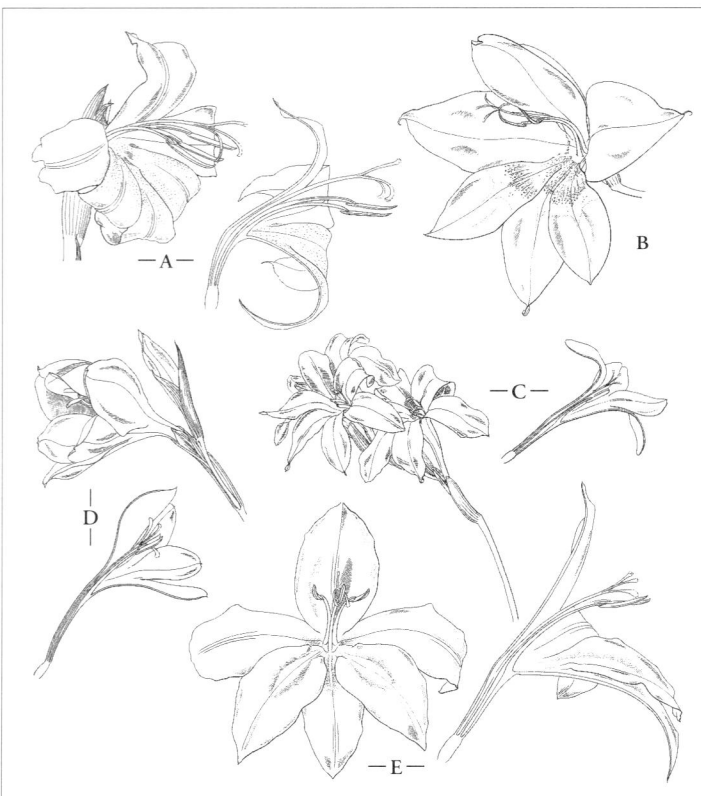

Figure 10 Flowers of southern African *Gladiolus* adapted for pollination by the butterfly *Aeropetes tulbaghia.*
In section *Ophiolyza G. saundersii* (**A**) with half flower, and *G. cruentus* (**B**); in section *Linearifolius G. nerineoides* (**C**) with half flower showing the anthers in the perianth tube; and in section *Blandus G. insolens* (**D**) with half flower, and *G. stefaniae* (**E**) with half flower. (approximately x ½)

The moths involved in the pollination of *Gladiolus* species belong to the families Noctuidae and Sphingidae. The largest is undoubtedly the sphinx moth *Agrius convolvuli*, which has a proboscis some 100 mm long and is believed to pollinate *G. longicollis* in KwaZulu-Natal, Mpumalanga and Swaziland. In the winter-rainfall zone noctuid moths, especially *Cuculia extricata*, and smaller sphinx moths are the usual pollinators.

Possibly present in 12 species, moth pollination is relatively uncommon in southern African *Gladiolus*. It is best developed in the winter-rainfall zone, especially in series *Tristis* of section *Homoglossum*, and is also found in *G. acuminatus* and *G. robertsoniae* of section *Hebea* (**Figure 9A, B**), *G. guthriei* and *G. emiliae* of section *Linearifolius* (**Figure 9C**), and *G. albens*, *G. maculatus* and *G. recurvus* of section *Homoglossum* (**Figure 9D, E**).

BUTTERFLY POLLINATION

Restricted to southern Africa, the satyrid butterfly *Aeropetes tulbaghia* is a large, strong-flying insect of mountain slopes whose range stretches in a more or less continuous arc from the Piketberg in the southwest through the Drakensberg of eastern southern Africa to the Vumba in eastern Zimbabwe. *Aeropetes* is on the wing during the summer months from December to April. Unusually for insects, it is strongly attracted to red and as it is an active insect with a large appetite for nectar, a variety of plant species have evolved flowers adapted for pollination by this butterfly alone (Johnson & Bond, 1994). These include *Disa uniflora*, *Nerine sarniensis* and *Tritoniopsis burchellii*, as well as a number of species of *Gladiolus*.

Species of *Gladiolus* visited by *Aeropetes* all have odourless, bright red flowers, sometimes with white marks on the lower tepals (**Figure 10**). Nectar, contained in a floral tube about 20–30 mm long, is provided as a reward. The nectar in these plants has a fairly low sugar concentration, but is present in large amounts. Because of the importance of visual cues, species of *Gladiolus* pollinated by *Aeropetes* include those with the largest flowers in the genus, for example *G. cardinalis*, *G. sempervirens* and *G. stefaniae* (**Figure 10E**). Alternatively, a number of smaller flowers are massed together to improve the display, as in *G. nerineoides* (**Figure 10C**). In the large-flowered species the anthers are prominently positioned well outside the throat of the flower and pollen is brushed off more or less randomly onto the wings and body of the butterfly as it forages for nectar. In *G. nerineoides*, however, the relatively small flowers have the anthers mostly included in the tube and the pollen is placed precisely on the head of the insect as it feeds.

Aeropetes is completely indiscriminate in its foraging habits, moving randomly between species, sometimes for significant distances, and there is a real danger of hybridization between related species. It is probably this factor which has resulted in an almost perfect geographical sequence in the *Gladiolus* species relying on the butterfly for pollination, with each one replacing another in turn along mountain chains. Because *Aeropetes* is active only during the summer and autumn months, some readjustments in flowering time are necessary in the species of the winter-rainfall region which rely on it. The most simple expedient is for the species to occupy a habitat which supports growth during the dry summers, and *G. cardinalis*, *G. insolens* and *G. sempervirens* grow in seeps, alongside streams or in the streams themselves. Moisture precipitated along the tops of the coastal ranges by strong south-easterly winds ensures a supply of water throughout the dry summer, but such habitats are limited in extent and found only on suitable mountain ranges. Other species, for example *G. nerineoides*, *G. stefaniae* and *G. stokoei*, have minimized water loss by transpiration by producing flowers on leafless stems. In this way they can flower during the summer drought without being restricted to perennially moist habitats.

BIRD POLLINATION

A number of species in both winter- and summer-rainfall regions have flowers pollinated by sunbirds (*Nectarinia* species). The flowers are almost always orange-red or scarlet, but may also be greenish yellow, and the lower tepals are occasionally green or marked with black. The range of colours has less to do with an innate preference by the birds than the fact that these colours are inconspicuous to insects which might otherwise rob or damage the flowers. Bird-pollinated flowers are highly specialized in form and often unusual-looking, especially in series *Involutus* of section *Hebea* and series *Homoglossum* of section *Homoglossum* (**Figure 11**). To accommodate the beak of the pollinator, the upper part of the perianth tube is usually broad and cylindrical. In addition, the dorsal or all three upper tepals are prominent, effectively displaying the flowers, while the lower tepals may be reduced or almost obsolete so as not to obstruct foraging. The flowers are unscented and often borne on stems which are stout, in order to support the weight of the bird. Large amounts of nectar are produced, upward of ten times as much as in bee-pollinated species. The nectar is sometimes also fairly dilute compared with that in most insect-pollinated species. A characteristic of bird-pollinated flowers is for the hexose sugars, glucose and fructose to be dominant in the nectar, in contrast to insect-pollinated species which have sucrose-rich or sucrose-dominant nectar. Bird-pollinated flowers in species of section *Homoglossum*, however, do not follow this pattern and have sucrose-dominant nectar of relatively high concentration.

The bird's sharp claws and bill are a potential hazard to the flower and many species have adaptations to minimize possible damage. These include large, firm bracts which conceal and protect the nectar-containing tube, a thick, firm perianth and particularly tough filaments. In *Gladiolus cunonius* and its allies the floral tube has a curious horny tooth in the throat (**Figure 11 I, J**). Sunbirds are strong fliers capable of moving rapidly over a considerable distance and they visit a variety of suitable flowers indiscriminately. These conditions favour hybridization between related species and it is no accident that the ranges of closely related bird-pollinated species do not

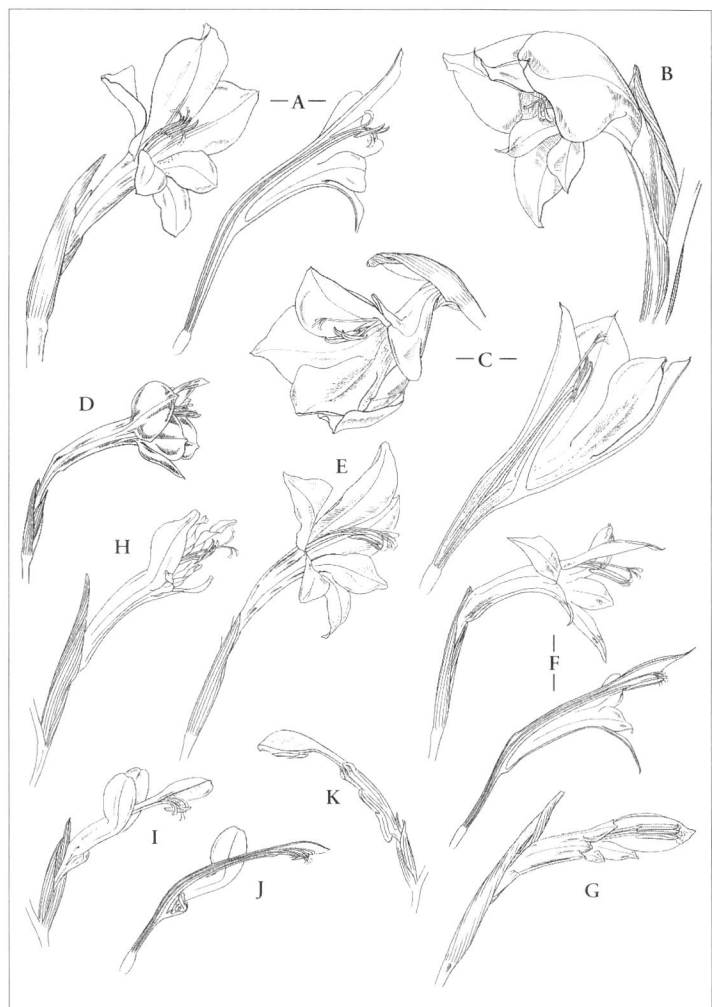

Figure 11 Flowers of southern African *Gladiolus* adapted for pollination by sunbirds.

In section *Ophiolyza G. aurantiacus* (**A**) with half flower showing the tube narrow below and wide and cylindric above, *G. dalenii* (**B**), with its hooded dorsal tepal concealing the stamens, and *G. flanaganii* (**C**) with half flower showing the unusual arrangement of the flower with the lower tepals exceeding the dorsal; in section *Linearifolius G. bonaspei* (**D**); in section *Homoglossum G. huttonii* (**E**), *G. watsonius* (**F**) with half flower showing the long perianth tube, and *G. abbreviatus* (**G**) with the upper lateral tepals reduced in size and the lower tepals vestigial; and in section *Hebea G. vandermerwei* (**H**) with linear lower tepals, *G. cunonius* (**I**) with vestigial lower tepals, *G. splendens* (**J**) with half flower showing the tube, the chamber formed by the reduced lower tepals, and the callus-like tooth separating the tube into two halves, and *G. saccatus* (**K**) showing the upper lateral tepals reduced and the lower tepals vestigial, their bases forming part of the nectar spur. (approximately x ½)

overlap. This is especially evident in series *Involutus* and *Homoglossum* where there is a regular geographic displacement of species.

In the western Cape and eastern Cape the most important bird pollinators are the malachite (*Nectarinia famosa*) and orange-breasted sunbirds (*N. violacea*), and in Namaqualand the dusky sunbird (*N. fusca*) is the most common bird pollinator. In the summer-rainfall zone the greater double-collared (*N. afra*) and malachite sunbirds are the most active pollinators of *Gladiolus* species. Bird pollination has evolved repeatedly in southern African *Gladiolus*: it occurs in species of sections *Blandus*, *Hebea*, *Linearifolius* and *Ophiolyza* and arose twice in section *Homoglossum*.

BEETLE POLLINATION

Although many flower-visiting beetles are destructive, eating parts of the petals and stamens, large numbers of plant species in the southern African winter-rainfall zone have become adapted for pollination by monkey beetles (Scarabaeidae: Rutelinae: Hopliinae). These beetles visit flowers actively and although they eat pollen and forage to a limited extent on floral parts, they are not overly destructive and use the flowers as a platform for assembling and mating as well as for feeding. Their bodies are hairy and while they crawl over the flower

they invariably accumulate a dusting of pollen which is transferred to the stigmas of the next flower they visit. The beetles fly short distances and move actively from flower to flower. Flowers which are adapted for this type of pollination are typically flat, brightly coloured, often red, orange, yellow or sometimes blue, and are usually strongly marked with black or brown blotches, called beetle marks as they presumably imitate beetles. These blotches probably act as lures for the wandering males.

Only one species of *Gladiolus*, *G. meliusculus*, has flowers adapted for pollination by these monkey beetles. The rather rounded, salmon-pink flowers have dark reddish black blotches near the centre of the lower tepals. In colour and markings it resembles *Romulea eximia* and *R. obscura*, amongst which it grows and which are also visited by the beetle *Lepisia rupicola*.

GEOGRAPHY OF SOUTHERN AFRICAN *GLADIOLUS*

Species of *Gladiolus* in southern Africa fall into two broad geographic categories: those from the summer-rainfall and those from the winter-rainfall zones of the subcontinent (**Map 1**). Although there is no clear geographic separation between the two areas, only a few species occur in both. Most of these cross from one zone to the other for only a short distance, and even then respond to the rainfall pattern of their main range. It is thus useful to speak of summer-rainfall or winter-rainfall area species when discussing many aspects of *Gladiolus*.

The climatic distinction is significant in an evolutionary sense as well as ecologically, for whole sections appear to have radiated largely in one or other of the rainfall zones and few species of any section extend outside their main centres of radiation. Sections *Densiflorus* and *Ophiolyza* occur exclusively in the summer-rainfall zone (**Map 2, 3**), with species extending only to the north, into tropical Africa where the same rainfall regime prevails. Likewise, sections *Blandus*, *Hebea* and *Homoglossum* have largely radiated in the winter-rainfall zone (**Map 4, 7, 8**). Just a few species of the latter two sections occur outside the winter-rainfall zone and a scant handful occur exclusively in summer-rainfall southern Africa. Section *Heterocolon* is unusual in having three of the nine species in southern Africa occurring in the winter-rainfall zone and several more in tropical Africa (**Map 6**), and section *Linearifolius* stands out in having a more or less even distribution in summer- and winter-rainfall southern Africa and in tropical Africa (**Map 5**).

SUMMER-RAINFALL ZONE

Section *Densiflorus* (Map 2)

Species of the section occur almost throughout sub-Saharan Africa except in the southern African winter-rainfall zone, and most species exhibit the standard pattern of flowering in

mid- to late summer. *Gladiolus crassifolius* is perhaps the most widespread of the southern African species of the section, with a range extending from Eastern Cape Province, South Africa, to southern Tanzania. Even more widespread is the tropical African *G. gregarius* which ranges across tropical Africa from Malawi and Angola northward to Senegal. The section shows moderate radiation in the northern Drakensberg with several species, including *G. appendiculatus* and *G. ferrugineus*, extending from southern Swaziland and southern Mpumalanga to Northern Province of South Africa. *Gladiolus calcaratus*, *G. lithicola* and *G. macneilii* have very narrow ranges in the central Mpumalanga mountain belt. In the south, *G. mortonius* and *G. ochroleucus* of series *Scabridus* are restricted to the Eastern Cape, but the remaining species of the series extend along the Drakensberg from southern KwaZulu-Natal to Mpumalanga, and *G. brachyphyllus* occurs in the lowveld east of the escarpment.

Section *Ophiolyza* (Map 3)

Entirely restricted to areas of summer rainfall, species of section *Ophiolyza* extend from Eastern Cape Province, South Africa, in the south to Senegal in West Africa, to southern Arabia and to Madagascar. Much of this range is occupied by one species, *Gladiolus dalenii*. Series *Ophiolyza* is shared with tropical Africa and has radiated extensively there as well as in southern Africa, where a few red-flowered species have evolved in the Drakensberg. *Gladiolus flanaganii* is a narrow endemic there, while *G. cruentus* is restricted to a small area of central coastal KwaZulu-Natal. Series *Ecklonii*, with three species, includes *G. rehmannii* which is restricted to the bushveld of Mpumalanga and Northern Province. In the third southern African series, *Oppositiflorus*, *G. sericeovillosus* extends from the Eastern Cape to northern Zimbabwe. Moderate radiation in the series has given rise to two localized, edaphic endemics: *G. dolomiticus* in Northern Province and *G. pole-evansii* in western Mpumalanga. *Gladiolus elliotii* has a relatively narrow range across the interior highveld of Mpumalanga and Gauteng westward into Botswana.

Section *Heterocolon* (Map 6)

The small section *Heterocolon* comprises just nine species in southern Africa and some seven to 11 more in tropical Africa. It shows an odd pattern for species of *Gladiolus* of the summer-rainfall zone in favouring relatively dry habitats. Species of series *Heterocolon* include *G. rubellus* and *G. pretoriensis* from eastern Botswana and adjacent South Africa, *G. filiformis*, a local endemic from North-West Province, and several more from the copperbelt of Congo (Zaïre) and the surrounding plateau, a unique pattern for Iridaceae. Series *Unguiculatus* has only *G. oatesii* in southern Africa, which occurs in Botswana, the northern provinces of South Africa, and Zimbabwe. Its immediate relative, *G. unguiculatus*, is widespread across tropical Africa from northern Zimbabwe to Senegal. The third series of the section, series *Vernus*, comprises two species in Mpumalanga, one in eastern Zimbabwe and three more in western South Africa in the dry interior of the winter-rainfall zone, another disjunct pattern that is unusual for Iridaceae.

WINTER-RAINFALL ZONE

Section *Blandus* (Map 4)

Species of section *Blandus* are concentrated in the southern portion of the southern African winter-rainfall zone and only *Gladiolus floribundus* is widespread, occurring almost throughout the zone. The southern distribution pattern is emphasized by the presence of endemic species in the mountains of the southern Cape. Among other montane endemics, *G. crispulatus* is restricted to the central Langeberg, *G. sempervirens* to the Outeniqua–Tsitsikamma Mountain axis, and *G. stefaniae* to the western Langeberg and Potberg. *Gladiolus undulatus* is unusual in the section in that it extends from the Cape Peninsula northward into Namaqualand, but it is a species of wet habitats and is restricted to permanent streams almost throughout its range. *Gladiolus floribundus* and the closely related *G. grandiflorus* occur as far east as East London and Grahamstown, thus intruding into the western edge of the summer-rainfall zone. The only highly localized endemics are *G. phoenix*, known only from a few populations in Bain's Kloof in the mountains east of Cape Town, and *G. insolens*, from the higher mountains of the Piketberg range.

Section *Linearifolius* (Map 5)

With 22 species extending across central Africa to the Cape, section *Linearifolius* has its centre of diversity in the southern African winter-rainfall zone where 11 species occur. There are only five species in the southern African summer-rainfall zone, and the remainder occur in tropical Africa. The species of the summer-rainfall zone are relatively uniform as regards their flowers and pollination system. *Gladiolus pubigerus* and *G. woodii*

have wide ranges, the former extending from Katberg in the eastern Cape to Pilgrim's Rest in Mpumalanga, and the latter from southern KwaZulu-Natal to Northern Province. *Gladiolus parvulus* occurs in southern KwaZulu-Natal, while *G. malvinus* and *G. pardalinus* are narrow endemics, restricted respectively to the Mpumalanga escarpment near Dullstroom and the bushveld of interior Mpumalanga. The tropical African species extend from eastern Zimbabwe to the Congo (Zaïre). They seem relatively uniform florally, except for *G. curtifolius* from northern Malawi and adjacent Tanzania. The white flower has an elongate tube and is evidently adapted for moth pollination. In contrast, the species of the winter-rainfall zone show a fair degree of evolutionary depth, exhibiting a range of different flowering times and floral specializations. All are centred in the west of the winter-rainfall belt and only *G. caryophyllaceus* extends significantly northward as far as southern Namaqualand. The only common species, *G. hirsutus*, extends from the Cold Bokkeveld in the north to near Mossel Bay in the east. Highly localized endemics include *G. monticola*, which is restricted to the Cape Peninsula, *G. nerineoides* and *G. stokoei*. The last-named is known from a single population on the Riviersonderend Mountains. There is a significant break between the ranges of the species of the two climate zones. No member of the section occurs between Mossel Bay in the south and Katberg in the eastern Cape.

Section *Hebea* (Map 7)

The section is centred in western southern Africa where 26 of the 31 species occur. Few species extend outside the winter-rainfall zone more than locally, for example *Gladiolus involutus* as far east as East London and *G. orchidiflorus* into the Karoo as far east as Kimberley. *Gladiolus permeabilis* subsp. *edulis* is widespread in southern Africa and extends into Zimbabwe and northern Namibia. Only *G. loteniensis*, a local endemic of the KwaZulu-Natal Drakensberg, *G. inandensis*, *G. robertsoniae* and *G. wilsonii* are restricted to the summer-rainfall zone. *Gladiolus saccatus* is unusual in having a range extending from southern Namaqualand northward along the west coast and into northern Namibia, thus bridging winter- and summer-rainfall zones in a north–south direction along the west coast. The remaining 23 species of section *Hebea* are found exclusively within the winter-rainfall zone and, moreover, show a strong concentration in the western half of the Cape Flora Region and adjacent western Karoo and Namaqualand. Species of the section are in general more markedly adapted to semi-arid habitats than those of other sections. Local endemics of section *Hebea* include *G. deserticola* of the Richtersveld, *G. salteri* of interior Namaqualand, and *G. lapeirousioides* of the western Karoo near Loeriesfontein, all belonging to series *Deserticola*.

Section *Homoglossum* (Map 8)

With 51 species, this is the largest section of the genus. Section *Homoglossum* shows a pattern of radiation centred in the southwestern corner of southern Africa, an area of the highest

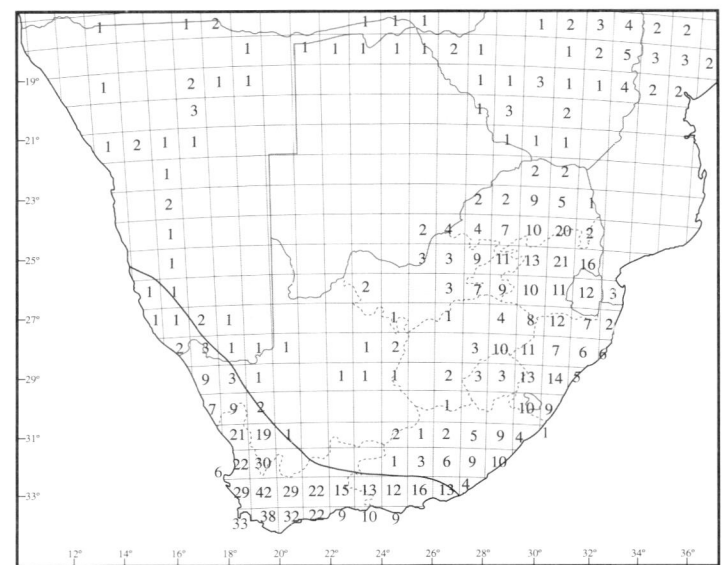

Map 1. Species richness in *Gladiolus* shown by plotting the total number of species occurring in one-degree squares of geographical longitude and latitude. The solid line indicates the division of southern Africa into summer- and winter-rainfall zones, the latter to the south and west of the line. Areas of highest richness are in the southwest, in the centre of the winter-rainfall zone, and in the Mpumalanga highlands and high Drakensberg in the summer-rainfall zone.

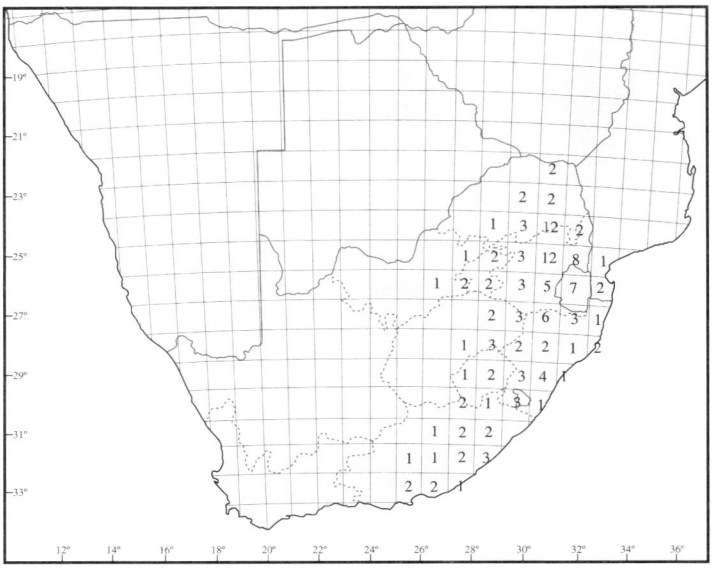

Map 2. Species richness in southern African members of section *Densiflorus.*

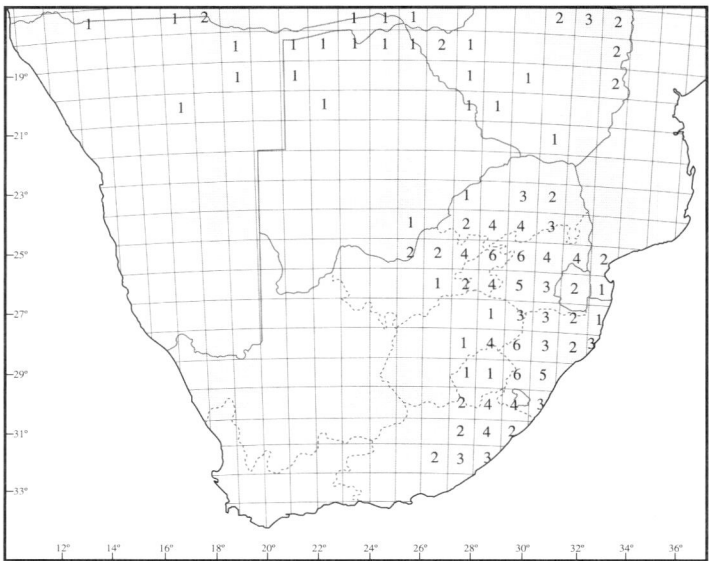

Map 3. Species richness in southern African members of section *Ophiolyza.*

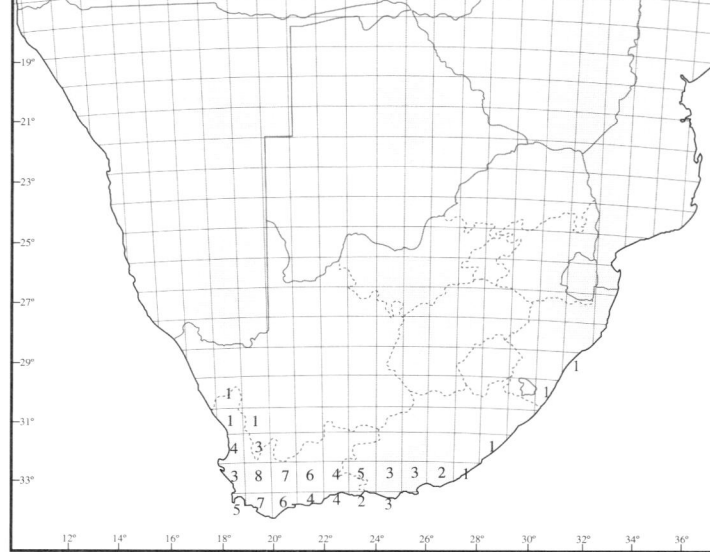

Map 4. Species richness in section *Blandus.*

rainfall and greatest edaphic and topographic diversity in the southern African winter-rainfall zone. Within the section adaptation to particular soils is often marked, and most species can be categorized as being restricted to sandstone-derived soils, clays or limestones. Despite the winter-wet and summer-dry climate, several species have made a shift to flowering in summer or autumn, usually by separating flowering and leafing phases of their life cycle. Flowering spikes produce reduced, usually entirely sheathing leaves during the dry season and foliage leaves in winter and spring. Widespread species of section *Homoglossum* include *Gladiolus liliaceus* and *G. tristis* which occur almost throughout the Cape Flora Region. Only *G. carinatus* and *G. hyalinus* extend northward into Namaqualand, the semi-arid portion of the southern African winter-rainfall zone, and both are widespread within the Cape Flora Region as well. *Gladiolus pritzelii* is the only member of the section that is shared between the Cape Flora Region and the western Karoo. Despite the strong concentration in the winter-rainfall zone, two species occur in the summer-rainfall region: *G. symonsii*, which is endemic to the high Drakensberg of Lesotho and KwaZulu-Natal, and *G. longicollis.* The latter extends from the Swartberg and Kammanassie Mountains in the winter-rainfall zone eastward through the Drakensberg and the plains of the east coast to the Northern Province of South Africa.

Highly local endemics of section *Homoglossum* are numerous, conspicuous among them being *Gladiolus ornatus*, *G. quadrangulus* and *G. sufflavus.* Two of these local endemics, *G. comptonii* and *G. delpierrei*, are known from single populations, geographically isolated from their closest relatives.

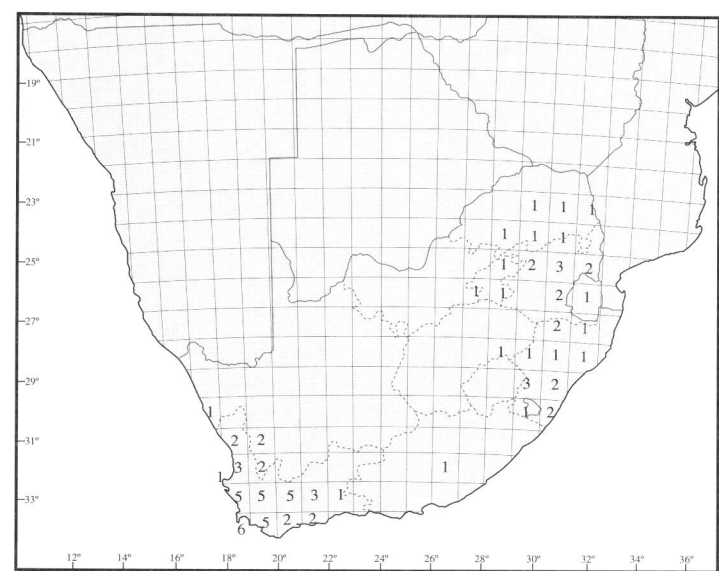

Map 5. Species richness in southern African members of section *Linearifolius.*

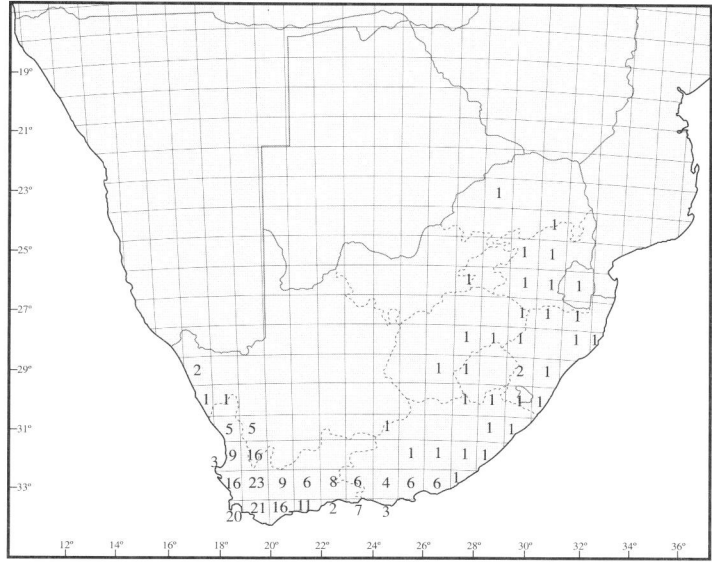

Map 8. Species richness in section *Homoglossum.*

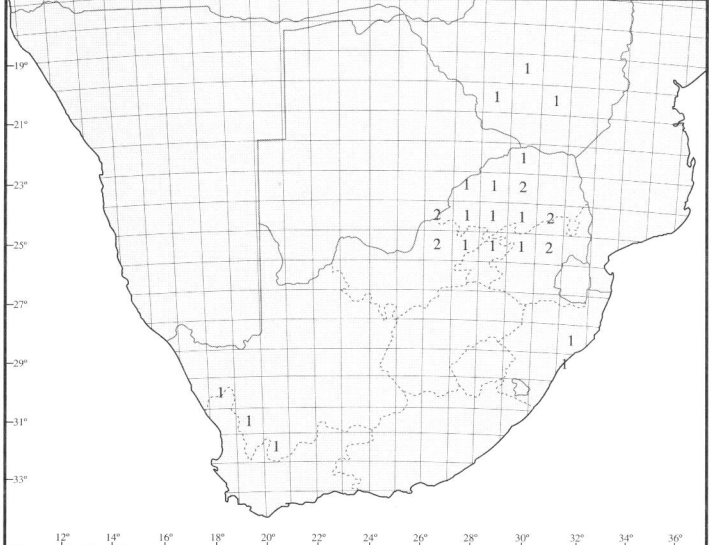

Map 6. Species richness in southern African members of section *Heterocolon.*

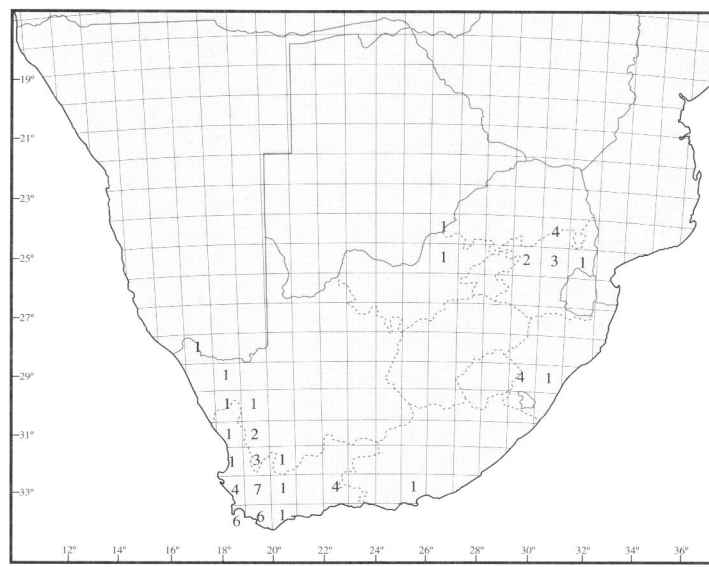

Map 9. Centres of species endemism in southern African *Gladiolus* as shown by plotting numbers of species entirely restricted to one geographical degree square.

CENTRES OF DIVERSITY AND ENDEMISM

Species concentration in southern African *Gladiolus* shows a strongly bimodal pattern (**Map 1**). By far the greatest number of species occur in the southwestern corner of the subcontinent, the geographical centre of the winter-rainfall zone. The Cape Peninsula, a relatively small area, has 33 species and the single degree grids of geographical latitude and longitude lying to the immediate east, the Caledon and Worcester grids, have 38 and 42 species respectively. Species diversity tails off gradually to the north and to the east. The decrease to the north correlates closely with increasing aridity and lower edaphic diversity. The decrease to the east is less pronounced and may be related simply to lower topographic diversity.

A second centre of diversity, lying at the opposite end of

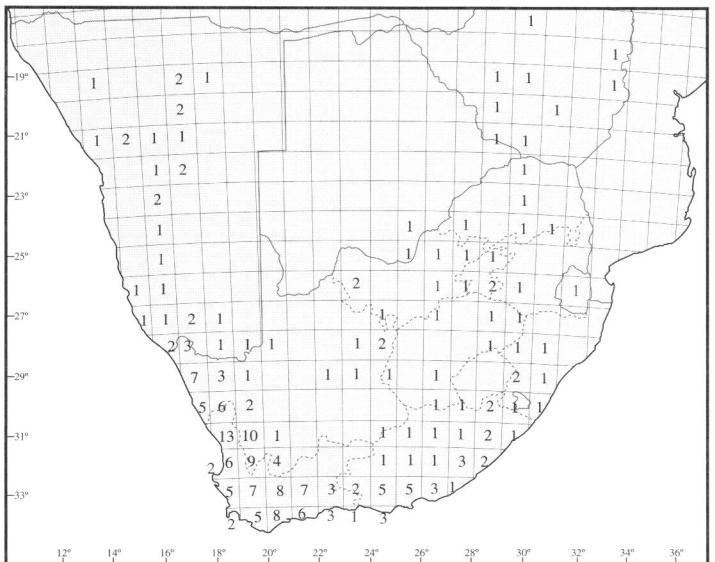

Map 7. Species richness in section *Hebea.*

the subcontinent in the Mpumalanga highlands, is less pronounced. Some 20 and 21 species respectively lie in the Pilgrim's Rest and Lydenburg grids. Species diversity there, associated with the presence of seven local endemics, is primarily the consequence of high edaphic diversity. Substrates include sandstones, shales and dolerite, serpentine and dolomite. The altitudinally higher KwaZulu-Natal Drakensberg has significantly lower species diversity, with 13, 11 and ten species in the grids that include the main Drakensberg range. The relatively lower species diversity in this area is presumably the result of lower edaphic diversity, a factor only partly compensated for by the greater topographic diversity and consequently greater climatic variation.

Overall, species diversity in *Gladiolus* is associated with grassland or shrubland zones of high rainfall, and secondarily with habitat diversity based on both edaphic and topographic variability. The pattern is repeated in tropical Africa (Goldblatt, 1996), where there are several local centres of

diversity and endemism, almost all located in highland areas.

A total of 59 species of *Gladiolus* are restricted to an area smaller than one geographical degree square. Thus by this definition 36% of the genus are local endemics. Centres of endemism (**Map 9**) mirror the centres of diversity based on the geographical degree grids. Seven endemic species occur within the Worcester grid, and six each on the Cape Peninsula and in the Caledon grid. Four endemic species occur in the west coast grid immediately north of Cape Town, and there is an isolated centre in the mountainous central southern Cape in the Oudtshoorn grid that includes the central Swartberg and Outeniqua Mountains. In the summer-rainfall zone local centres of endemism again match the areas of species diversity. There are four endemic species in the Pilgrim's Rest grid and three in the Lydenburg grid, in the heart of the Mpumalanga highlands. The high Drakensberg in KwaZulu-Natal and Lesotho has four endemics, again in a grid that lies in the heart of this centre of diversity.

Within the winter-rainfall zone five local phytogeographic centres are sometimes recognized (Cowling, 1992), and the diversity and endemism of *Gladiolus* within these centres emphasize the impressive radiation that has occurred there. The southwestern centre (including the Cape Peninsula, Caledon grid and near west coast) boasts 66 species of *Gladiolus*, 32 of them endemic. The northwestern centre has 37 species and 11 endemics, and the Langeberg centre 29 species and four endemics. The markedly less diverse Karoo Mountain centre has 14 species, three of them endemic, and the southeastern centre 23 species and five endemics. The local centres of diversity in the winter-rainfall zone thus match or substantially exceed the local diversity in the entire summer-rainfall zone. ♣

Main centres of diversity and endemism in southern African *Gladiolus*

Centre	Total species	Total endemics
Winter-rainfall zone		
Northwestern Cape centre	37	11
Southwestern Cape centre	66	32
Karoo Mountain centre	14	3
Langeberg centre	29	4
Southeastern centre	23	5
Summer-rainfall zone		
Mpumalanga Highland centre	18	8
High Drakensberg centre	16	4

RELATIONSHIPS, SUBGENERIC CLASSIFICATION, PHYLOGENY AND RADIATION

GENERIC RELATIONSHIPS

A member of subfamily Ixioideae, one of four subfamilies currently recognized in the Iridaceae (Goldblatt, 1990a, 1991), *Gladiolus* has all the specialized features that define the subfamily. These include a basally rooting corm and the specialized type of inflorescence found in the Ixioideae which consists of sessile flowers, usually arranged in a spike, each subtended by a pair of opposed floral bracts. The flowers are long-lived (not deliquescing on the same day that they open) and have a well-developed perianth tube. Anatomical specializations of the Ixioideae include pollen exine with a perforate tectum and scabrate sculpturing, and pollen grain apertures with a two-banded operculum (Goldblatt, 1990a, 1991; Goldblatt et al., 1991). *Gladiolus* has leaves with a well-defined midvein and anatomical specializations that include mesophyll cells elongated across the leaf axis and epidermal cells with sinuous margins

and two or more papillae (Rudall & Goldblatt, 1991). These latter features also define two of the three tribes of Ixioideae, Watsonieae and Ixieae (Goldblatt, 1990a, 1991).

Gladiolus is traditionally placed in the Ixieae but its relationships within the tribe are difficult to assess. Following the most recent generic realignments in Iridaceae (Goldblatt & De Vos, 1989; Goldblatt & Manning, 1995, 1996a), Ixieae comprises some 20 genera and is by far the largest of the three tribes of Ixioideae in numbers of genera and species. The patterns of generic relationships within Ixieae are obscure and any subdivision is currently impractical. At least, however, it seems likely that *Gladiolus* is most closely allied to a group of genera that have lightly angular (or more or less rounded) seeds with secondary sculpturing (the cell outlines of the seed coat are well defined) and intact seed vasculature. In addition, this alliance has an unspecialized leaf anatomy in which the leaf margins have epidermal cells of the same shape and size as those of the blade area, and the tissue under the marginal epidermis is heavily thickened into a strand of sclerenchyma that is closely associated with the marginal vein (Rudall & Goldblatt, 1991).

In other genera of Ixieae the seeds are hard and rounded, with a smooth surface in which the cell outlines are obscure or absent, thus lacking secondary sculpturing. These seeds also usually have an excluded vascular trace (Goldblatt & Manning, 1995). Most members of this group also have a derived leaf anatomy in which the epidermal cells at the margins are columnar and have thickened secondary walls, and subepidermal marginal sclerenchyma strands are absent. It seems likely that both the basic seed features of *Gladiolus* and its leaf anatomy are of the ancestral type for Ixioideae and so say nothing about the mutual relationships of the genera with these characters. It does, however, seem unlikely that *Gladiolus* is related to any genera of the *Tritonia–Crocosmia* alliance, or to *Sparaxis*, *Ixia* or *Dierama*, all of which have derived seed features and, except for *Ixia*, specialized leaf margin anatomy. *Freesia* (including *Anomatheca*) can probably also be excluded from the list of genera immediately related to *Gladiolus* because although it has seeds with intact vasculature, the seeds are hard, round and smooth, and the marginal anatomy is of the specialized type.

Just a handful of genera are left as possible close allies of *Gladiolus*. These include the predominantly southern African *Babiana*, *Radinosiphon*, a genus of one or two species of southeastern Africa, and a small group of genera, including *Geissorhiza*, *Hesperantha*, *Melasphaerula* and *Romulea*, which have bell-shaped or asymmetric corms and usually hard, woody corm tunics. The last four genera are all predominantly or exclusively southern African, and probably constitute a clade defined by their asymmetric or sometimes campanulate corms with roots produced from a defined ridge. *Babiana*, an apparently taxonomically isolated genus, has specialized pleated leaves, surface pubescence and smooth seeds with an unusual, apomorphic type of sculpturing. The leaf anatomy accords with that in *Gladiolus*, as does the intact seed vasculature, but these features are both plesiomorphic. Similar apically expanded style branches alone are synapomorphic for these two genera. An additional two genera, the southern African *Syringodea* and the Eurasian *Crocus*, are probably not immediately related to *Gladiolus*, and are not of direct concern here. Species of these two genera are highly specialized, acaulescent plants, most probably allied to *Romulea*. At least *Syringodea* has woody corm tunics and asymmetric corms of the *Romulea* type, and both *Syringodea* and *Crocus* have derived inaperturate pollen grains.

Gladiolus and *Radinosiphon* share no derived characters: the zygomorphic flower, secund flower arrangement and an inclined spike are all most likely plesiomorphic. Moreover, two or all three of these characters may well be linked, and in any event this alone is a weak argument at best for concluding that they are closely related. Just one feature links the two genera: their basic chromosome number. Both are paleotetraploid genera with a

unique basic number for Ixioideae, $x = 15$. More significantly, *Radinosiphon* differs from *Gladiolus* in lacking winged seeds and in having ovoid capsules and filiform style branches.

The best that can be said about the relationships of *Gladiolus* is that the genus probably has no close relatives, and this may be taken as an indication that it diverged from ancestral Ixioid stock at a remote time, perhaps in the early Tertiary. Other genera of Ixieae are probably more recent in origin, and of these *Radinosiphon* seems most nearly related to *Gladiolus*. A more satisfactory answer to these phylogenetic questions will probably only be reached when gene sequencing is used to examine questions about relationships among the genera of the Ixioideae.

SUBGENERIC CLASSIFICATION

Patterns of variation in *Gladiolus* suggest no obvious discontinuities that delineate major species groupings. Hence, there is no widely accepted classification of the genus. Early attempts to produce a useful subgeneric classification of *Gladiolus* are of little more than historical curiosity. Hendrik Persoon (1805), for example, divided *Gladiolus* into five groups of undefined rank, some not given names, and three of the five are now regarded as other genera. Persoon must, however, be credited with the name *Hebea*. The name was later used by J.G. Baker (1877, 1896) and F.W. Klatt (1882) as a subgenus including several southern African species clustered around *G. alatus* that have an enlarged dorsal tepal and both lower and upper tepals abruptly narrowed below, thus more or less clawed. *Hebea* was regarded as a section of *Gladiolus* by G. Bentham & J.D. Hooker (1883), who did not recognize subgenera in *Gladiolus*. Baker recognized one other subgenus, *Schweiggera*, a synonym of the South African genus *Tritoniopsis*, and four groups of no stated rank: *Parviflorae*, *Blandi*, *Cardinales* and *Dracocephalae*. The last is our section *Ophiolyza*, and groups *Blandi* and *Cardinales* are more or less equivalent to series within our section *Blandus*. The species of Baker's *Parviflorae* fall in several of our sections.

Klatt's (1882) classification included four more subgenera in addition to *Hebea* and subgenus *Gladiolus*, which he called *Eugladiolus*, a formulation no longer nomenclaturally admissible. Subgenus *Ophiolyza* included the longer-tubed species with large flowers and smaller lower tepals; subgenus *Hyptissa* included larger-flowered species with nearly equal tepals; subgenus *Ranisia* included large-flowered species with a tube longer than the bracts; and subgenus *Limonia* included both small- and large-flowered species and was poorly defined. The names that Klatt used for his subgenera are taken from the literature, but because he did not cite their sources, even indirectly, they must be treated as new names. Subgenus *Hebea* is obviously Persoon's (1805) group *Hebea*, and subgenera *Ophiolyza*, *Hyptissa* and *Ranisia* are names for genera segregated from *Gladiolus* by the English botanist, R.A. Salisbury (1866), although *Ranisia* included quite different species in Salisbury's treatment. *Limonia* can be traced to *Lomenia*

Pourret, also misspelled *Lemonia* in the literature. Nomenclaturally, *Lomenia* is a synonym of the southern African *Watsonia*, species of which were included in *Gladiolus* by Persoon. Both Baker and Klatt recognized the genus *Acidanthera* and included within it several southern African species now transferred to *Hesperantha* and *Geissorhiza*, as well as several species of *Gladiolus*. Salisbury's (1866) treatment of *Gladiolus*, published long after his death, was highly idiosyncratic and would be no more than a historical curiosity except that some of the six genera he recognized are valid nomenclaturally. Of the five new genera that he described, only *Homoglossum* has ever been recognized.

The provisional division of *Gladiolus* into the four groups *Plurifoliatae*, *Paucifoliatae*, *Unifoliatae* and *Exfoliatae* as proposed by Obermeyer (in Lewis et al., 1972) is admittedly artificial. The system was based on the number of foliage leaves present on the flowering stem. It is useful for identification purposes, but because leaf number has been reduced independently in many lineages, it does not necessarily indicate natural relationships. In a study of the species of *Gladiolus* in tropical Africa, Goldblatt (1996) proposed a classification in which two subgenera, *Gladiolus* and *Ophiolyza*, were recognized. This classification was based largely on flower size and associated features such as bract length, and capsule and seed size. It resulted in the more or less equal division of the species into the two subgenera, and most of these could fairly easily be placed in one or the other of them. Unfortunately this classification does not seem useful when the southern African species are added because too many species fall between the two subgenera. We prefer here to recognize several large, and evidently natural groupings as sections and do not assign the sections to subgenera (**see table, page 14**). In particular, section *Gladiolus* of Goldblatt (1996) has been refined because the majority of the southern African species fall there. Section *Gladiolus* is now restricted to the Eurasian species of *Gladiolus* which include *G. communis*, the type species of the genus. The remaining tropical African species of the section *Gladiolus* now fall into sections *Densiflorus* and *Linearifolius*, both best developed in southern Africa and not recognized at all in the account of the genus for tropical Africa.

Species can usually be referred to the appropriate section on the basis of leaf, corm and sometimes floral features. Adaptive radiation and convergence within several of the sections for particular habitats or pollination strategies may result in some anomalous characters in certain species, but their relationships are usually indicated by a combination of features. Floral characters, in particular, are less reliable than some vegetative ones, and floral features, including flower size, tube length and colour, are not always useful for classification. We have found, however, that the type of nectar guide is often a useful character. Within each section we recognize natural species clusters as informal taxonomic groups which we call series. Each series is, we believe, monophyletic, but the composition and definition are still fluid and likely to change as more information becomes available.

Most sections are easy to recognize and a key to the sections is provided. They are defined as follows, with their specialized character states indicated in ***bold italic***:

Section *Densiflorus*

Corm tunics firm-papery to coriaceous; leaves several in a basal fan, blades plane, sword-shaped with numerous secondary veins and prominent margins and midribs, sometimes scabrid to finely pubescent; stems often branched; spikes usually inclined, ***many-flowered;*** bracts small, often dry above, but large in species with long tubes; flowers small with tubes shorter than the dorsal tepal, but flowers adapted for fly pollination larger, with tubes as long as or longer than the dorsal tepal, never scented; nectar guides a pale blotch in the lower half of each lower tepal, the pale colour sometimes edged distally with dark pigment or, in flowers adapted for long-tongued fly pollination, a reddish median streak; capsules ovoid to oblong, three-lobed above, ***seed wings sometimes weakly developed.***

Section *Ophiolyza*

Corm tunics firm-papery to coriaceous; leaves several in a basal fan, blades plane, sword-shaped with numerous secondary veins and prominent margins and midribs, sometimes scabrid, occasionally pubescent; stems often branched except in species adapted for bird pollination; spikes inclined or erect, often many-flowered; ***flowers medium-sized to large with tubes shorter than to about as long as the dorsal tepal***, never scented; ***perianth colour often streaked or speckled***, nectar guides a pale blotch in the lower half of each of the lower tepals; capsules ovoid to oblong, three-lobed above, seeds broadly and evenly winged.

Section *Blandus*

Corm tunics firm-papery to coriaceous; leaves usually several in a basal fan, few and superposed in some species, blades plane, sword-shaped with numerous secondary veins and prominent margins and midribs, glabrous; spikes inclined or erect, mostly few-flowered; flowers medium-sized to large with tubes shorter than to about as long as the dorsal tepal, but much longer in some species adapted for long-tongued fly pollination, never scented; ***nectar guides lozenge-shaped white longitudinal streaks outlined in dark colour; capsules ovoid to ellipsoid, rarely three-lobed above***, seeds broadly and evenly winged.

Section *Linearifolius*

Corm tunics coriaceous, fragmenting into coarse vertical fibres; leaves several in a basal fan or few and superposed, blades plane, narrowly sword-shaped with several secondary veins and prominent margins and midribs, ***pubescent, rarely more or less glabrous when mature;*** spikes inclined or erect, mostly few-flowered; flowers small to medium-sized with tubes shorter than to about as long as the dorsal tepal, but longer in species adapted for long-tongued fly, moth, butterfly or bird pollination, scented in the moth-pollinated species only; ***nectar guides ill-defined dark longitudinal streaks or veins;*** capsules ovoid and three-lobed above to more or less ellipsoid, seeds broadly and evenly winged, *usually reddish and opaque*.

Section *Heterocolon*

Corm tunics coriaceous, fragmenting into tough wiry fibres, often accumulating with the decayed cataphylls in a neck; leaves in a basal fan or few and superposed, *blades linear, usually very narrow, often oval to round in section with prominent margins and midribs but secondary veins not evident;* spikes inclined, mostly few-flowered; flowers small to medium-sized with tubes shorter than the dorsal tepal but longer in species adapted for long-tongued insect pollination, exceptionally faintly scented; *nectar guides pale transverse bands or blotches edged in dark colour;* capsules ovoid and three-lobed above, rarely ellipsoid, seeds broadly and evenly winged or more often the wings reduced or absent.

Section *Hebea*

Corm tunics often of tough wiry fibres, or thick woody claws, or more or less papery to membranous (rarely thickened and woody); leaves in a basal fan, rarely few and superposed, *leaf blades linear, usually very narrow, blades with prominent midrib area but margins not thickened and secondary veins not evident;* spikes inclined, mostly few-flowered; flowers small to medium-sized with tubes shorter than to about as long as the dorsal tepal, but longer in species adapted for long-tongued fly, moth or bird pollination, *usually sweetly scented; nectar guides solid pale colour in the basal half of each of the lower tepals, occasionally absent; capsules ellipsoid, occasionally inflated, ovoid and three-lobed above*, seeds broadly and evenly winged, *the seed body much darker than the wing*.

Section *Homoglossum*

Corm tunics of papery to coriaceous layers, or *often hard woody layers fragmenting into vertical segments; leaves four or less, superposed, leaf blades linear, very narrow, without secondary veins but usually with prominent midribs and margins, sometimes centric and then usually terete, occasionally both midribs and margins not at all thickened; spikes inclined, usually flexuose and few-flowered;* flowers small to medium-sized with tubes shorter than to about as long as the dorsal tepal, but longer in species adapted for long-tongued fly, moth or bird pollination, *usually sweetly scented; nectar guides mostly irregular dark streaks and spots on a pale background, or longitudinal transverse bands of yellow to cream edged in dark colour, or occasionally absent; capsules ellipsoid, somewhat inflated*, seeds broadly and evenly winged.

In addition to the above sections, we continue to recognize the remaining three sections included in subgenus *Ophiolyza* in tropical Africa (sections *Acidanthera, Decoratus* and *Tenuibracteatus*). As mentioned above, section *Gladiolus*, as circumscribed by Goldblatt (1996), is now subdivided into three sections, two of which are largely southern African and the third, *Gladiolus*, is reserved for the Eurasian species of the genus. We have redefined

section *Blandus* as well. It is now seen as restricted entirely to the southern African winter-rainfall zone. Tropical African species included in the section may best be referred to sections *Densiflorus* or *Ophiolyza*. The position of tropical African species that belong in predominantly southern African sections is indicated in the table on page 14. A full description of each section and a discussion of its characteristics, geography and important adaptations, including pollination biology, are presented at the beginning of the chapter dealing with its species.

Our ideas about species relationships within the section are summarized in **figures 12–18**. These figures are based on the cladistic philosophy of Hennig (1966) in which relationships are assessed using only shared derived characters, i.e. synapomorphies. The apomorphic characters which we use to define clades are indicated on each of the figures.

SPECIATION AND RADIATION IN *GLADIOLUS*

Exploring the reasons for the evolution of species is the key to understanding the diversification or radiation of a group of organisms. Understandably, because of the great time periods required for any speciation event to occur, the causes of the event itself can only be inferred long after its completion. One of the more reliable ways of doing this is to compare the biological attributes of immediately related species pairs or groups in the anticipation that the speciation events which gave rise to them occurred so recently that their ecology and biology still reflects that which prevailed when they diverged. Species of *Gladiolus* that are restricted to lime-enriched habitats fall into this category, limestone and calcareous sands being relatively recent substrates in southern Africa. Although the identification of such related species requires the application of a formal methodology such as cladistic study, for practical purposes they can often be identified as sharing certain unique, derived character states, termed synapomorphies. Applying this method of comparing species, we have identified a number of species pairs or groups in *Gladiolus*. Then, by comparing the biological attributes of these species, we have been able to infer the probable speciation mechanisms which led to their divergence. Although we have not followed the rigorous methodology which we applied to the study of this question in the genus *Lapeirousia* of the Iridaceae (Goldblatt & Manning, 1996b), the picture which emerges is much the same.

Two basic types of speciation are evident. The first follows the classical model of geographic speciation first developed in the 1930s and 1940s and formalized by Ernst Mayr (1942) and G.G. Simpson (1953). In this model a population at one end of the range of a species becomes reproductively isolated from the remaining populations. This may happen through the development of a geographic barrier to gene flow between populations, for example, a mountain range. In an extension of the model, speciation may occur through the chance long-distance dispersal of seed and the subsequent establishment of a founder population, located beyond the range of normal gene flow from the parent population. Should the isolated population(s) then

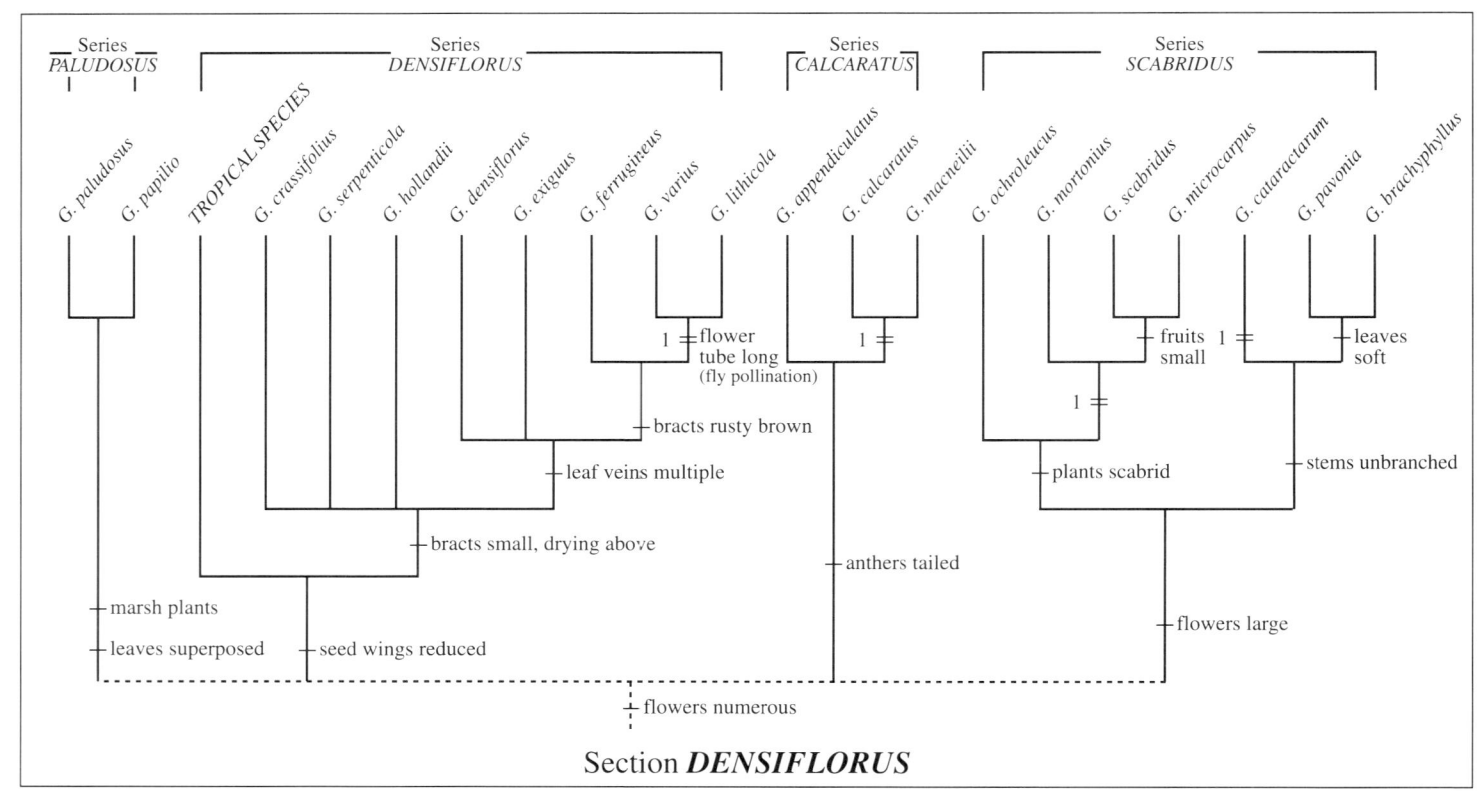

Figure 12

Figures 12–18 Phylogenetic trees showing hypothetical relationships of species in the various sections.
The trees were constructed using the principles of parsimony and the branches or clades are defined only by derived characteristics.
The derived characters defining each clade are marked, but only those recurring on a tree are numbered for clarity. Parallelisms
are indicated by double cross bars and reversals by crosses on the vertical axes.

Figure 13

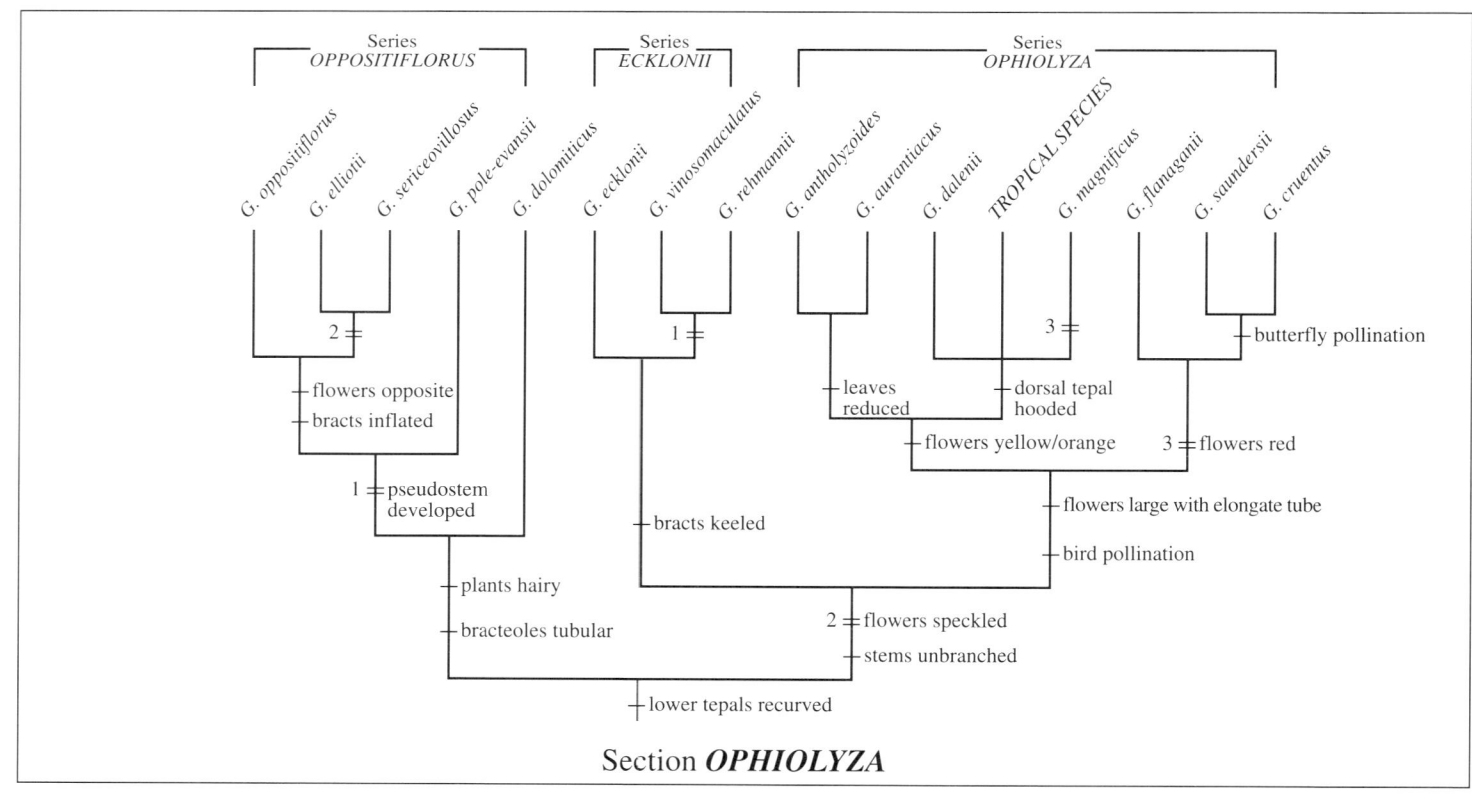

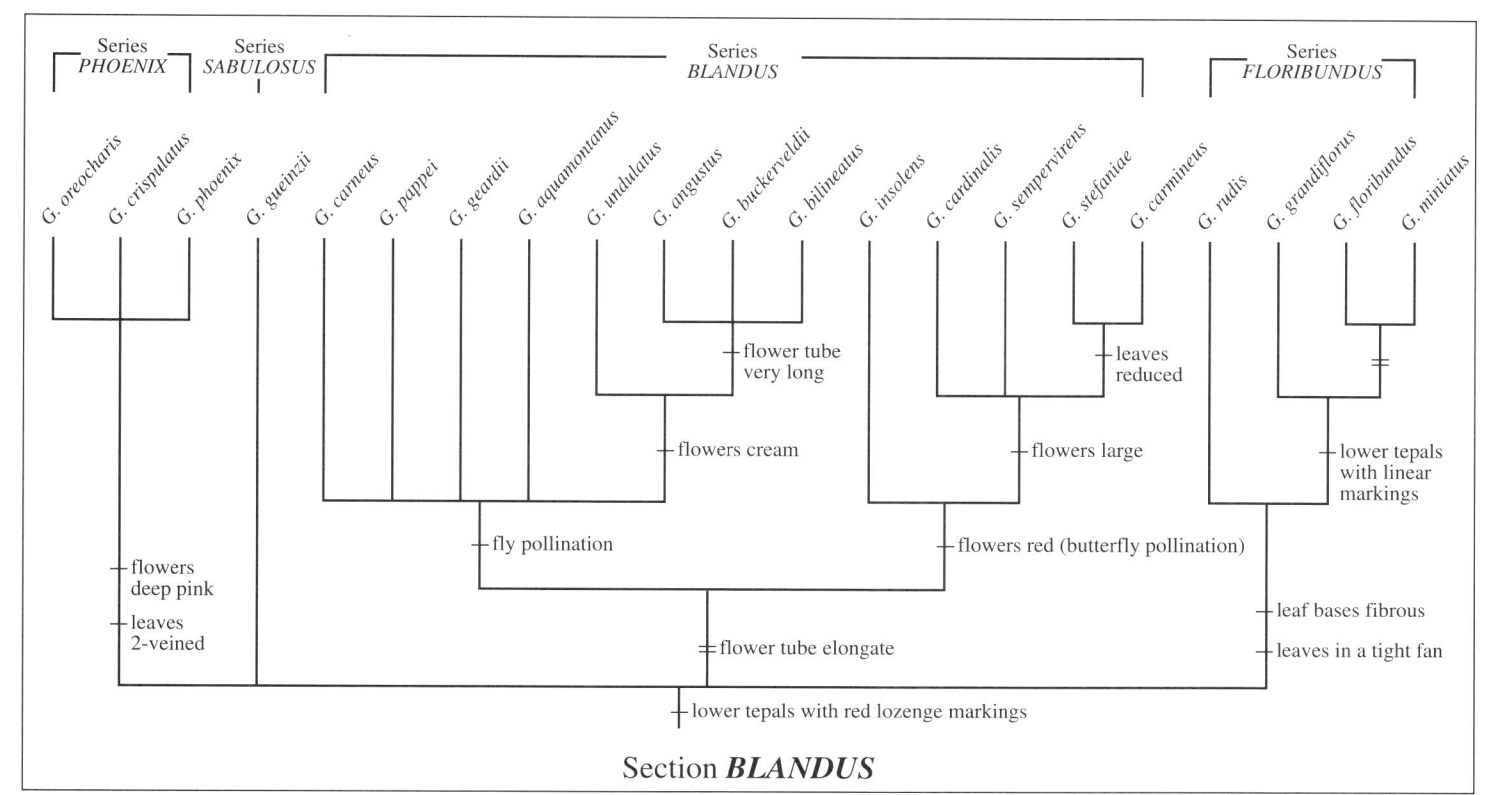

Figure 14

Figure 15

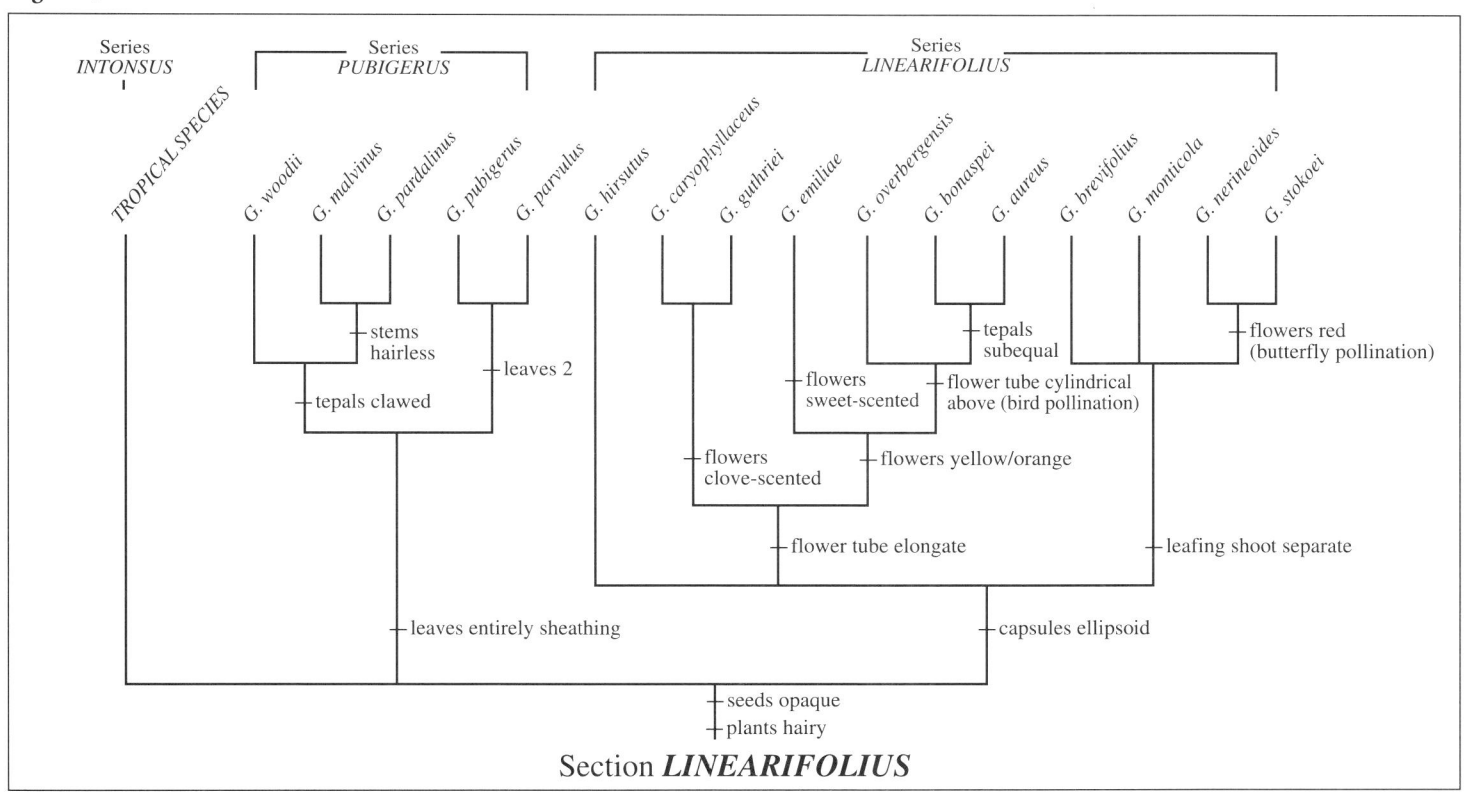

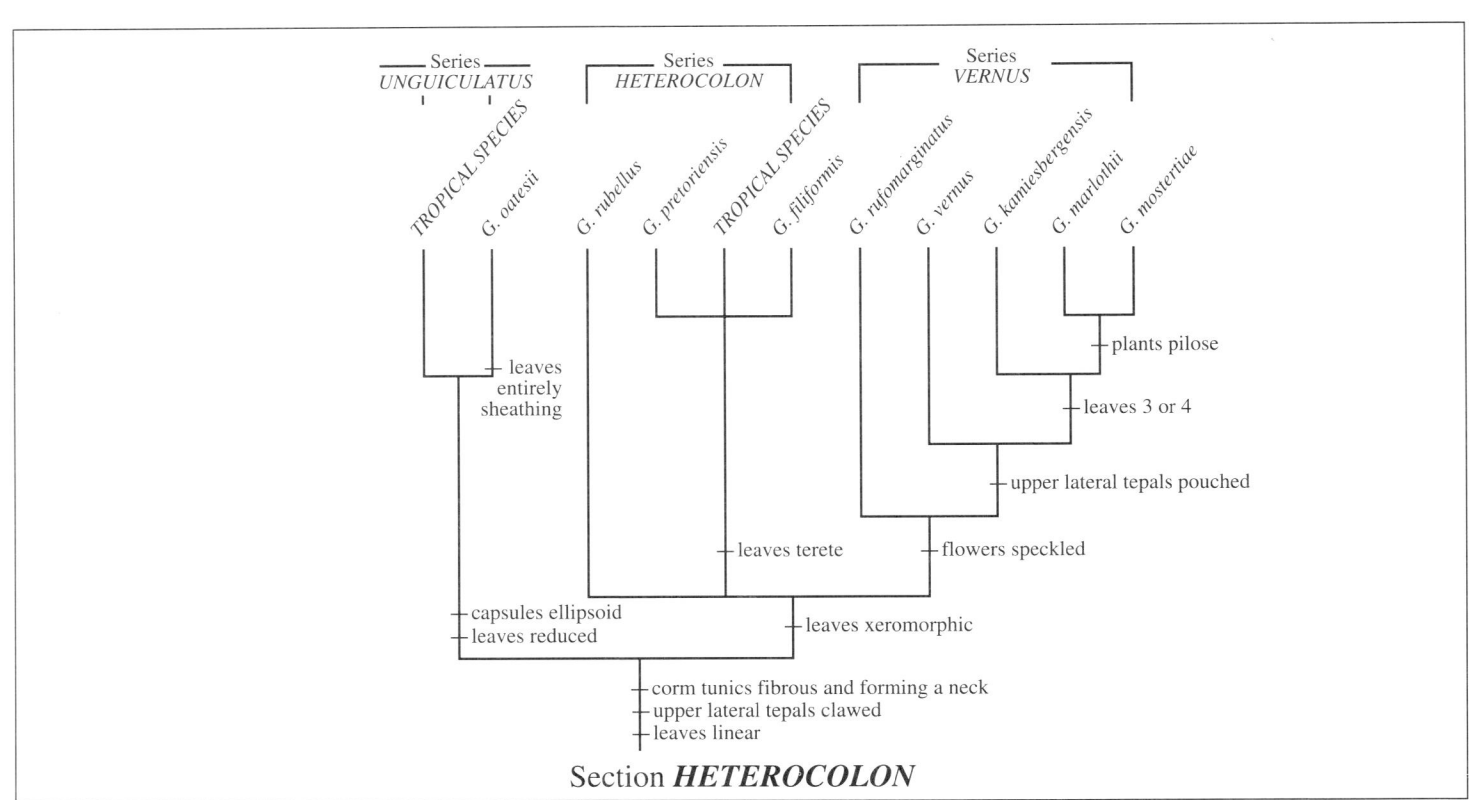

Section *HETEROCOLON*

Figure 16

Figure 17

Series *INVOLUTUS* Series *PERMEABILIS* Series *DESERTICOLA* Series *HEBEA*

G. leptosiphon · G. loteniensis · G. involutus · G. vandermerwei · G. cunonius · G. splendens · G. saccatus · G. permeabilis · G. uitenhagensis · G. acuminatus · G. stellatus · G. wilsonii · G. inandensis · G. robertsoniae · G. arcuatus · G. viridiflorus · G. deserticola · G. scullyi · G. venustus · G. salteri · G. lapeirousioides · G. orchidiflorus · G. watermeyeri · G. virescens · G. ceresianus · G. uysiae · G. equitans · G. speciosus · G. pulcherrimus · G. alatus · G. meliusculus

flowers white

lower tepals deflexed
flowers pink

4 ⊨ dorsal tepal erect, margins revolute

3 ⊨

leaves 2-veined
nectar low in sucrose
upper and lower lateral tepals fused

1 ⊨

capsules squat and 3-lobed above

4 ⊨

flower tube toothed within
anthers tailed
stigmas slender

2 ⊨

corm tunics woody

3 ⊨ leaves ridged
flowers striped

filaments hairy

flowers orange

flowers red (bird pollination)

leaves multiple-veined

2 ⊨ corm tunics papery

leaves soft
stolons well-developed

capsules small

flower spike flexed and scalloped

tepal sutures papillose
outer tepals spade-shaped

1 ⊨ cataphylls mottled

seeds brown

lower median tepal elongate
upper lateral tepals upcurved

upper tepals clawed
flowers scented
flowers dull-coloured

nectar guides yellow-green

seed body dark

corm tunics wiry and fibrous

leaves linear, fleshy, without thickened margins

Section *HEBEA*

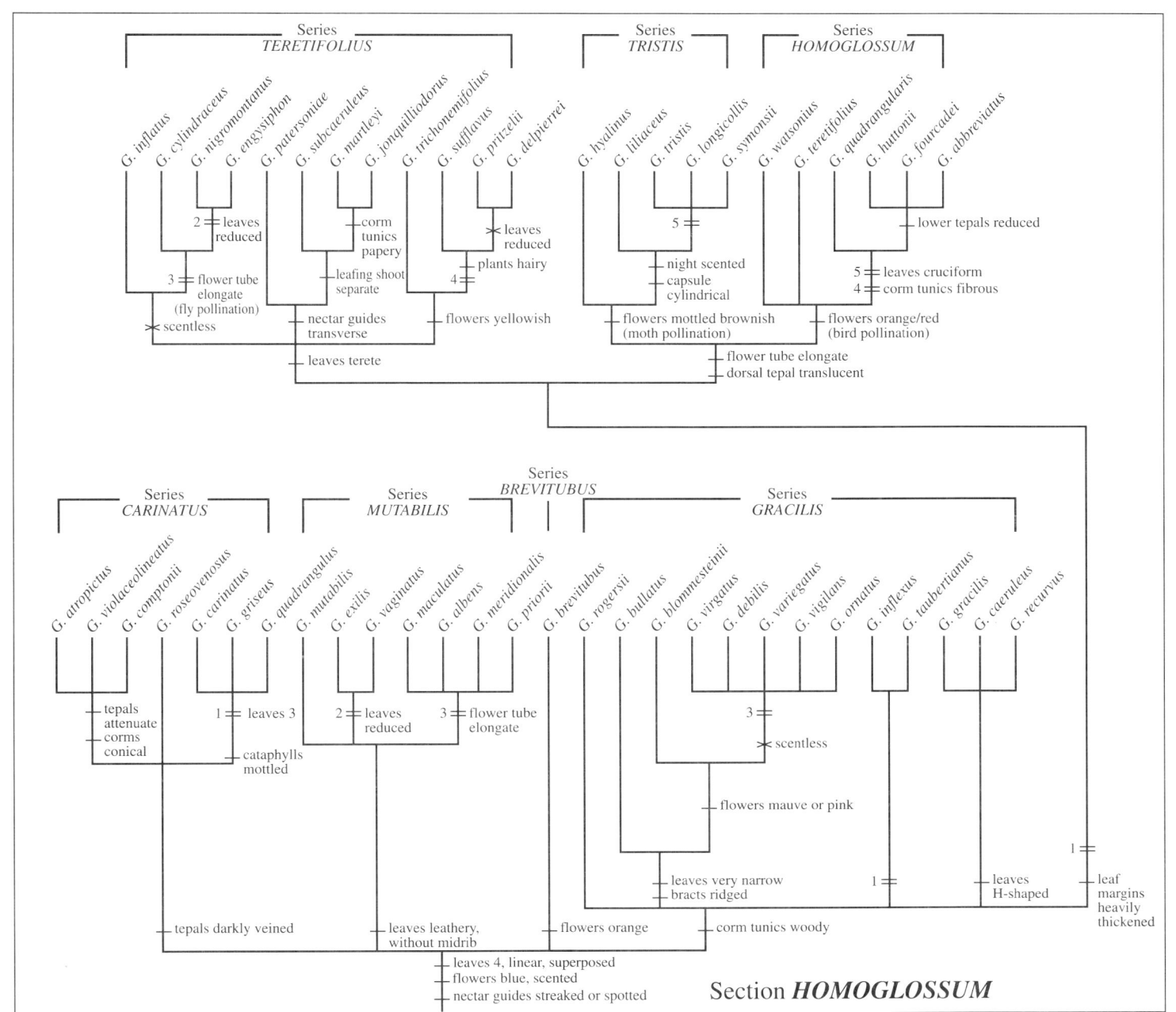

Section **HOMOGLOSSUM**

Figure 18

experience environmental pressures different from those prevailing on the parent populations, they may begin to diverge physiologically and morphologically to the extent that they become separate species. In cases where differentiation is still incomplete, the divergent series of populations may be called subspecies. Thus, for example, the widespread species *Gladiolus permeabilis* and *G. longicollis* are each divided into two subspecies. The plants at opposite ends of the ranges look remarkably dissimilar, but the populations across the range intergrade with weak or obscure morphological discontinuity. These may be examples of incipient speciation.

It is now evident that in plants the distances required to effectively isolate populations of a species from one another may be relatively small, and that divergence can occur across quite short gaps in the presence of sufficiently strong selective

pressures (Linder, 1985). In *Gladiolus* typical examples of geographic speciation include the western Cape species *G. carneus* and its sister from the eastern Cape, *G. geardii*, and the western Cape *G. tristis* and its eastern sister species, *G. longicollis*. The distances involved in these instances are large and the climates experienced at either end rather different, and it is not surprising that speciation has followed. Similarly, but on a smaller scale, the five species of series *Oppositiflorus* (section *Ophiolyza*) have radiated in response to differences in climate and soil on and below the eastern southern African escarpment. On a more localized scale, with the distances measured perhaps in tens of kilometres between suitable mountain tops, are the examples of *G. nerineoides* and *G. stokoei* (section *Linearifolius*) and *G. cardinalis*, *G. stefaniae* and *G. sempervirens* (section *Blandus*). Because there is virtually no possibility of gene flow between these related species, it is not necessary that mechanisms develop to prevent such gene flow from disrupting the newly adapted and integrated genotypes of the founder species. Such related species thus often do not differ from one another in pollinator or flowering time, and in the above examples the species in each group all have the same pollinators.

Where chance long-distance dispersal events lead to the establishment of a daughter population some distance removed from the parent and quite beyond the likelihood of future genetic contact, it can be anticipated that the genetic complement of the daughter population, derived as it is from a single or very few individuals, will differ to a degree from that of the parent population. This is known as the founder effect. This can be expected to result in a daughter population which differs significantly from the parent in appearance. A possible example of this variant of geographic speciation in *Gladiolus* is the species pair *G. violaceolineatus–G. comptonii*. The flowers of the two species are very similar except for colour difference, blue to mauve in one, bright yellow in the other, but the plants can readily be distinguished by their quite different leaves.

The second mode of speciation evident in *Gladiolus* occurs across much shorter distances, and is a combination of microgeographic and edaphic influences. Because short-distance gene flow appears to be the rule for most plants, the existence of a strong selection differential across two nearby populations may be sufficient to initiate divergence in the face of intermittent gene flow. The colonization of different soil types seems to provide just such a strong selection differential and most of the immediately related species pairs or groups which we have identified in *Gladiolus* seem to have differentiated under this influence. Examples are particularly common in the western Cape where a variety of chemically and physically different soil types occur in close juxtaposition to one another. A number of limestone endemics are thus clearly derived from the parent species occurring on nearby clay or sandstone soils. These include the species pairs *G. gracilis–G. caeruleus*, *G. carinatus–G. griseus* and *G. floribundus–G. miniatus*. In fact, whole lineages of species may have diversified from an ancestor which was adapted to a soil type different from its predecessor. The *G. alatus* group of series *Hebea*, for example, appears to represent such a radiation onto sandy or sandstone soils from an ancestor which favoured clay. Because these different soils are often adjacent to one another and the possibility of gene flow between the parent and daughter species exists across such small distances, it is common to find that the daughter species is adapted to a different pollinator or flowers at a different time, thereby minimizing the possibility of gene flow. The following species pairs or species groups favour different soils and may flower concurrently without hybridization: *G. grandiflorus* (bee-pollinated), *G. floribundus* (fly-pollinated) and *G. miniatus* (bird-pollinated); *G. blommesteinii* (bee-pollinated) and *G. virgatus* (fly-pollinated); and *G. maculatus* (moth-pollinated) and *G. meridionalis* (bird-pollinated). Immediately related *G. carinatus* and *G. griseus* are almost completely separated from one another in flowering time and although both species are bee-pollinated, the possibility of hybridization is much reduced.

In *Gladiolus* so-called saltational speciation by some event such as hybridization and rapid stabilization of hybrid populations, or polyploid doubling of chromosomes, does not seem to have occurred. Speciation in *Gladiolus* is thus usually a result of interrupted gene flow, either over longer distances without an associated shift to a different soil type, or over very short distances with the accompaniment of substrate divergence combined with a shift in flowering time or pollinator. With this knowledge it becomes possible to understand the extensive radiation in the eastern southern African escarpment belt and in the western Cape, both because of their geological complexity and their dissected topography. The former encourages the evolution of suitably adapted genotypes while the latter permits both short-distance isolation across valleys or mountain ranges and the maintenance of a range of microclimates which can foster divergence.

RADIATION OF THE SOUTHERN AFRICAN SECTIONS OF *GLADIOLUS*

Section *Densiflorus*

The least specialized section in the genus, *Densiflorus* comprises 20 species in southern Africa, all restricted to areas of summer rainfall. It has radiated extensively only in the eastern southern African highlands of Mpumalanga, a region of complex geology and diverse soils and habitats. Important specializations include the development of scabrid pubescence in some species of series *Scabridus*, the evolution of tailed anthers in series *Calcaratus*, and reduced seed wings in series *Densiflorus*. Several species of series *Densiflorus* also have specialized, finely veined leaves. The consistently unscented flowers are mostly adapted for pollination by bees, the plesiomorphic pollination strategy in the genus. Specialized flowers, adapted for pollination by long-tongued flies, especially *Prosoeca ganglbaueri*, have evolved once in each of three of the four series of the section. Although most species occur on doleritic or shale soils, several species of series *Densiflorus* are found only on sandy soils derived from quartzites. The species of series *Paludosus* are unusual in growing in marshy habitats. Some specialized edaphic endemics include *Gladiolus macneilii* and *G. pavonia* on dolomite, and *G. serpenticola* on serpentine. Flowering in the section typically occurs in the summer or autumn after vegetative growth is complete. In *G. brachyphyllus*, the flowering spikes grow rapidly after the first rains and bear reduced leaf blades, and the plants flower more or less aseasonally in late spring.

Section *Ophiolyza*

Like section *Densiflorus*, section *Ophiolyza* occurs only in areas of summer rainfall, although its range extends across all of sub-Saharan Africa. The section exhibits modest radiation across eastern southern Africa, usually associated with differences in habitat and climate. Pollination by long-tongued bees is ancestral for the section, but series *Ophiolyza* shows a shift to bird pollination and within the series a second shift to butterfly pollination in *Gladiolus cruentus* and *G. saundersii*. It is the only alliance in the genus in the summer-rainfall zone that has exploited either of these pollination strategies. Flowers adapted for long-tongued fly pollination have evolved in *G. oppositi-florus* of series *Oppositiflorus*. Ecological adaptation for early flowering has occurred in *G. antholyzoides* and *G. aurantiacus* in southern Africa. These species have reduced or vestigial leaf blades on the flowering stems and leaves appear later in the growing season on separate shoots.

Section *Blandus*

Centred in the southern African winter-rainfall zone, section *Blandus* comprises 21 species in four series. Species are predominantly adapted to the peculiar sandstone soils of the Cape mountains. The taxonomically isolated *Gladiolus gueinzii* is a coastal strand species that extends into the summer-rainfall zone as far north as Durban in KwaZulu-Natal. Pollination by long-tongued flies is common in the section, although bee pollination is probably ancestral, and various species exploit all the long-tongued flies that occur in the winter-rainfall zone. Adaptations for late flowering, either in the summer or autumn, have occurred in several species which also shift to alternative pollination strategies, mostly to butterfly pollination. Delayed flowering is achieved in two ways, by restriction to perennially moist habitats (*G. cardinalis* and *G. semper-virens*) or by separating flowering and vegetative phases (*G. carmineus* and *G. stefaniae*). Speciation patterns are geographic and lack accompanying shifts in pollination strategy in species of sandstone soils in series *Phoenix* and frequently in series *Blandus*, but the pattern is blurred by habitat and/or pollinator shifts in *G. angustus*, *G. pappei* and *G. undulatus* and in the butterfly-pollinated *G. cardinalis*–*G. sempervirens* pair. Series *Floribundus* exhibits patterns of edaphic shift from sandstone to clay or limestone accompanied by changes in pollinators, from bees to either long-tongued flies or birds. Flowering after fire is a common feature of species of sandstone soils, but this response is particularly pronounced in the small series *Phoenix*, species of which apparently flower very infrequently and only in response to fire.

Section *Linearifolius*

The tropical and southern African section *Linearifolius* comprises some 22 species, 16 restricted to southern Africa. There have been three cycles of speciation in the southern African members of the section, two of them apparently initiated by leaf reduction and aseasonal flowering. In the summer-rainfall series *Pubigerus*, leaves are largely sheathing and have reduced blades, and flowering occurs in early spring. Species of the *Gladiolus brevifolius* alliance of series *Linearifolius* in the winter-rainfall zone also have the leaf blades reduced or absent and flowering takes place in the summer or autumn. In this group a foliage leaf is produced in the winter months. Aseasonal flowering is related to exploitation of pollinators at times when pollinator competition is lowest. Bee pollination is ancestral and is present in all five species of the summer-rainfall zone. There is

considerable floral diversification in the winter-rainfall species, and only two species, *G. hirsutus* and *G. brevifolius*, have flowers adapted for bee pollination. Other species have exploited moths, birds, butterflies or long-tongued flies. We suspect that this reflects their relatively late entry into the winter-rainfall zone. The only species with consistently scented flowers are those pollinated by moths. There has been limited edaphic diversification and all the summer-rainfall species grow on doleritic soils, while those in the winter-rainfall region occur on sandstone-derived substrates.

Section *Heterocolon*

This small section of some 16 to 20 species, including only nine in southern Africa, is primarily adapted to dry habitats, often with nutrient-deficient quartzitic soils. Species show a range of xeromorphic features, including corms with thick fibrous tunics that form a neck around the stem base, and fibrotic leaves with thickened and raised midribs and margins. The leaf blades are sometimes centric or terete and the stomata are restricted to narrow sinuses that run along the length of the blades. Four species, two of them in southern Africa, are adapted to aseasonal flowering. This has been achieved either by delaying flowering until after the end of the rainy season when the leaves are dry (*Gladiolus vernus*) or by flowering before the rainy season and restricting the leaf surface area (*G. oatesii*). Species of the section exhibit limited floral diversification and nearly all have unscented flowers adapted for pollination by bees. Only *G. filiformis* has an elongate perianth tube and is probably pollinated by long-tongued insects. The three species that occur in the winter-rainfall zone (*G. kamiesbergensis*, *G. marlothii* and *G. mostertiae*) are derived in the section and are vegetatively specialized in having either centric or heavily pubescent leaf blades.

Section *Hebea*

Although widespread in southern Africa, section *Hebea* is most diverse in the dry parts of the winter-rainfall zone. Most species favour clay soils or, in Namaqualand, granitic sands. Radiation onto sandstone-derived soils is evident only within series *Hebea*. The diversification in the section is largely floral, but vegetative variation includes the development of hard, woody corm tunics (series *Deserticola*), or very soft, submembranous tunics (series *Hebea*). Leaf specializations include the secondary expansion of the blade, sometimes associated with a reduction in leaf number, and the development of prominent ridges giving the blade a corrugated surface. Leaf specialization is often correlated with sandstone substrates, a trend repeated in section *Homoglossum*. The floral diversity is substantial, involving tepal orientation and colour variation, and is associated with the evolution of scent.

Bee pollination is ancestral and is retained in the majority of species where it is usually accompanied by the development of strong scent, particularly in series *Hebea*. Speciation is apparently seldom accompanied by a shift to alternative pollinators. Notable exceptions are shifts to moth pollination in two unrelated species, *Gladiolus acuminatus* and *G. robertso-niae* (series *Permeabilis*), and to long-tongued fly pollination in *G. leptosiphon* (series *Involutus*) and *G. lapeirousioides* (series *Deserticola*), also unrelated. In the basal series *Involutus* bird pollination appears to have evolved once in *G. vandermerwei* or its immediate ancestor and this pollination strategy charac-terizes a clade of four of the seven species of the series. The four series we recognize in section *Hebea* have diversified in different broad geographic zones: series *Involutus* across the southern Cape (secondarily into the western Karoo and Namaqualand); series *Permeabilis* in the southern and Eastern Cape; series *Deserticola* in interior Namaqualand and the northwest Cape; and series *Hebea* in the west of the winter-rainfall zone, including the western Karoo and Namaqualand.

Section *Homoglossum*

The largest of the southern African sections of *Gladiolus*, sec-tion *Homoglossum* comprises 51 species concentrated in the southern African winter-rainfall zone. Its important specialized features within the section include a blue perianth, a scented flower and superposed, narrow leaves that lack secondary veins. Many of the species have xeromorphic leaf specializations, including the development of thickened margins and midribs, restriction of the stomata to narrow sinuses and a tough, leath-ery texture. Reduction of leaf number or of the blade area has occurred repeatedly, the latter associated with the separation of the flowering and vegetative phases of the life cycle. Many of the species are restricted to sandstone soils, and extreme leaf specialization is correlated with this particular soil type which is acidic and has a low nutrient status and poor water retention. Xeromorphy is a common attribute of taxa which occur on sandstone soils in the Cape Flora Region. The diversification of the section on sandstone soils parallels in many ways the radi-ation of section *Hebea* on clay soils in the same geographical area. Section *Homoglossum*, however, displays a greater degree of adaptation to specialized pollinators and thus a lower pro-portion of species adapted to ancestral bee pollination.

Edaphic shifts from sandstone to clay have accompanied spe-ciation in some lineages, notably the entire series *Homoglossum* and the *Gladiolus gracilis* group of series *Gracilis*. Shifts have also occurred from either sandstone to limestone (*G. griseus, G. vari-egatus*) or granite or clay to limestone (*G. caeruleus*). The section includes a particularly high number of localized geographic endemics, several of which are restricted to single extended populations, for example *G. atropictus* and *G. comptonii*, or very small areas, *G. delpierrei* and *G. sufflavus*. The section appears to be in a state of rapid diversification, with species adapted to some very recent substrates such as coastal sands or limestone. The active radiation is reflected in the high degree of variability in several species and the difficulty in resolving species relationships in many instances. Although it has been possible to define all but one of the series we recognize as monophyletic lineages, we have accepted series *Gracilis* as paraphyletic for practical purposes. Until relationships between species groups are more convincing-ly demonstrated, it seems useful to recognize the distinctive

series *Teretifolius* and *Homoglossum* within the large group of species characterized by having woody corm tunics.

SPECIES CONCEPTS

Species are not immutable entities comprising morphological-ly identical individuals in uniform populations. Once this fact is recognized, the problem arises as to how different does an individual or population have to be from others to qualify for inclusion in a different species. This difficulty led to attempts to define species functionally rather than morphologically. These attempts are numerous and include both practical defi-nitions, often based on the ability of individuals to interbreed, as well as theoretical concepts based on evolutionary pathways or involving genetic history. Although useful in understanding the nature of species, these definitions have very real practical limitations. Some are quite impossible to apply without know-ing the origin or fate of the species, and others at best require experimental data which cannot easily be obtained, if at all. This applies especially when all that the scientist has to work with are dried, pressed herbarium specimens. For practical considerations then, morphological criteria remain the most useful and sometimes the only means of assessing species, although the interpretation of morphological variation is always improved by field knowledge. An added consideration is that whatever criteria are used in defining a species, the users of a taxonomic treatment need some clearly visible way to name the plants they encounter in the field, laboratory or gar-den. For them, morphological characters such as those used in keys remain the only practical solution. Not all morphological features are equally important in defining a species, and those which are affected by environmental conditions, such as plant size or leaf width, are of limited value. Features which have a firm genetic basis and are relatively insensitive to environmen-tal influences are much more valuable. Such morphological features represent functional and genetic differences and often indicate the existence of reproductive isolation between them. As such, they represent profound differences despite their sometimes superficial nature.

Although in practice we base our species circumscriptions primarily on morphological discontinuities between popula-tions, we have used a multidimensional approach to circum-scribe and recognize species. Differences in habitat preference, soil type, pollinator or flowering time, for instance, provide defi-nite grounds for recognizing taxa. In general we have, as far as possible, recognized variant forms as species when they differ in at least two qualitative features and also show geographic, eco-logical or phenological differences from the archetype. Examples will best illustrate how our concepts function in practice. The largely southern Cape *Gladiolus gracilis* is reasonably uniform across its range, flower colour variants excepted, and it typically grows on clay soils, occasionally on granite, and only exception-ally on sandstone in particularly wet sites. What we recognize as *G. caeruleus* (*G. gracilis* subsp. *latifolius* of Lewis et al., 1972) shares with *G. gracilis* a short-tubed blue flower and leaves with margins raised and wing-like. It is restricted to the Western Cape coast where it grows in calcareous sands usually on a limestone substrate. The ecological and geographic distinction is accompanied by visible differences in floral morphology (slightly larger flowers with broader lower tepals), a greater than usual number of flowers per spike, and a fairly short and very broad leaf with the marginal wings held at right angles to the blade surface (versus a narrow leaf with the wings curving over the blade surface). Despite its obvious relationship to *G. gracilis*, the limestone populations seem to us to merit specific recognition.

Similar arguments justify raising to species rank three of the subspecies of *Gladiolus floribundus* sensu Lewis et al. (1972). Typical *G. floribundus* grows on rocky sandstone soils in dry habitats and has flowers with a long tube exceeding the bracts and significantly smaller lower tepals with red median streaks. *Gladiolus grandiflorus* (subsp. *milleri* of Lewis et al., 1972) has flowers with a shorter tube and subequal tepals all with median streaks, and it favours clay soils in renosterveld. A third species, *G. rudis*, has flowers with a tube always shorter than the bracts and subequal tepals, the lower three of which have diamond- to spear-shaped markings, and it grows on sandstone soils in fynbos. *Gladiolus miniatus* (subsp. *miniatus* of Lewis et al., 1972) has orange or reddish flowers with the upper part of the tube elongate and cylindric, and it is restricted to limestone-derived soils or sands in limestone outcrops. The four sets of populations overlap in their overall geographic ranges but maintain their morphological and ecological differences. The floral differences are particularly significant as they represent adaptations to different pollinators, thereby imposing distinct barriers to gene flow between the sibling species. Like the example of *G. gracilis* and *G. caeruleus*, we think that it is appropriate to recognize four species where Lewis et al. recognized one.

These examples illustrate morphological variation between populations associated with more or less obvious differences in ecology or geography or both. The broad species definition adopted by Lewis at al. (1972) which did not take into account the ecological differences between the populations led to their being treated as infraspecific taxa of a single species. Variants that are sympatric and co-blooming, however, provide particularly strong grounds for recognizing separate species. A pertinent example is that of *Gladiolus malvinus* and *G. woodii*. Within the wide range of the eastern southern African *G. woodii*, the typical dark red-flowered form is sympatric with plants that are taller, more robust, and have larger flowers of a mauve colour unknown in *G. woodii*. Yet *G. malvinus* seems to be immediately related to *G. woodii* on the basis of all available criteria. It has leaf blades of the flowering stem vestigial, non-flowering plants produce a single linear, densely pilose leaf, and the flowers have essentially the same structure excepting for the larger size. They also appear to be adapted for pollination by the same agent, anthophorid bees. Although we do not understand the biological foundation for the maintenance of these two forms that are co-blooming and sympatric, we have no hesitation in regarding them as quite separate species.

Because morphology reflects ecology, one must expect more variation in widespread species than in narrow endemics which occupy small areas with low ecological variation. In some cases species are so distinct, that is, they are morphologically so unique or discontinuous from related forms, that there seems no question about their status. Many species, however, especially those with wide distribution ranges, show patterns of variability that raise questions about their taxonomic status, or the usefulness of applying a name to them. Decisions are always made on the basis of available evidence, and all possible information about particular taxa is seldom, if ever, available at one time. Decisions may thus become subject to change in the light of new information.

Minor morphological differences associated with geographic isolation of whatever magnitude are seldom acceptable criteria for taxonomic recognition at any rank when not accompanied by discernible changes in ecology. This question does not arise often in *Gladiolus* because the ranges of species are usually quite coherent. One example, however, is that of *G. caryophyllaceus* which is most common in the Clanwilliam District of the Western Cape, where plants are tall and have clear pink flowers. In the Cedarberg and on the Bokkeveld escarpment plants are relatively short, have few flowers on strongly inclined spikes, and the tepals are dull purple. A few isolated populations occur in the Swartberg and are most like the Cedarberg–Bokkeveld escarpment form, but the flowers are larger and have lilac tepals with purple midlines. This variation pattern is so trivial that we see no need for taxonomic recognition of the variants which, in addition, show no shift in their ecology. They all grow in dry fynbos in rocky soils and are evidently adapted for pollination by the same agents.

We have tried to be consistent in applying our working criteria for recognizing species, a task complicated by the fact that patterns or degrees of variability within species are themselves not consistent. Our approach to this problem has been to apply concepts and parameters derived from species which we understand well and which have posed few difficulties to those which we do not know in the field or which show complex patterns of variation. For example, in the *Gladiolus alatus* group of series *Hebea*, *G. alatus* was treated as one polymorphic species with five varieties by Lewis et al. (1972). We treat the alliance as four separate species based on leaf differences and correlated modes of vegetative reproduction, flowering time, and/or details of flower marking and tepal orientation. Understanding the type and degree of variation within our species and, in particular, of the modes of speciation or diversification within the genus provides a guide to how large morphological discontinuities must be to be evolutionarily and taxonomically significant. One must always keep in mind that species are dynamic entities that have changed in the past and will change in the future and that speciation is a continuous and constantly changing process over geological time. At any moment we are witnessing all stages of speciation, from local differentiation between populations, which may or may not proceed to speciation, through to the final extinction of a species. This revision is our attempt at a snapshot in time of these events. ⚜

THE GENUS *GLADIOLUS* LINNAEUS

Gladiolus Linnaeus *Species Plantarum* 36 (1753). Baker, *Handbook Irideae* 198 (1892); *Fl. Capensis* 6: 135 (1896); *Fl. Trop. Africa* 7: 360 (1898). G. Lewis et al., *J. S. African Bot.*, Suppl. 10 (1972). Marais, *Kew Bull.* 28: 311 (1973). Goldblatt & De Vos, *Bull. Mus. Hist. Nat., Paris*, Sér. 4, Sect. B, *Adansonia* 11: 417 (1989). Goldblatt, *Fl. Zambesiaca* 12(4): 66 (1993); Gladiolus *in Tropical Africa* (1996). Type species: *Gladiolus communis* Linnaeus.

Gladiolus = little sword, referring to the sword-shaped leaves. From Latin *gladius*, a sword.

Synonymy

Antholyza Linnaeus, *Species Plantarum* 37 (1753). Baker, *Handbook Irideae* 229 (1892); *Fl. Capensis* 6: 165 (1896); *Fl. Trop. Africa* 7: 343 (1898). N.E. Brown, *Trans. Roy. Soc. S. Africa* 20: 265 (1932), excluding the type. Type: *A. cunonia* Linnaeus (lectotype designated by Hitchcock & Green, 1929) (= *Gladiolus cunonius* (Linnaeus) Gaertner).

Cunonia Buettner ex Miller, *Figures of Plants* 1: 75 (1756), *Gardeners' Dictionary*, ed. 8 (1768) name rejected in favour of *Cunonia* Linnaeus (1759) (Cunoniaceae), conserved name. Type: *C. antholyza* Miller (= *Gladiolus cunonius* (Linnaeus) Gaertner).

Anisanthus Sweet, *Hortus Britannicus*, ed. 2: 566 (1830). Type: *A. splendens* Sweet (= *Gladiolus splendens* (Sweet) Herbert).

Petamenes Salisbury ex J.W. Loudon, *Ladies' Flower Garden Ornamental Bulbous Plants* 42, pl. 8 (1841). N.E. Brown, *Trans. Roy. Soc. S. Africa* 20: 276 (1932). Phillips, *Bothalia* 4: 44 (1941). Type: *P. abbreviatus* (Andrews) N.E. Brown (= *Gladiolus abbreviatus* Andrews).

Sphaerospora Sweet ex Loudon, *Ladies' Flower Garden Ornamental Bulbous Plants* 66, t. 14 (1841), not *Sphaerospora* Klatt, *Linnaea* 32: 725 (1863) (= *Geissorhiza*). Sweet, *Hortus Britannicus* 398 (1827), name without description. Type: *S. imbricata* (Jacquin) Sweet ex Loudon (= *Gladiolus italicus* Miller).

Acidanthera Hochstetter, *Flora* 27: 25 (1844). Baker, *Handbook Irideae* 185 (1892), in part; *Fl. Capensis* 6: 130 (1896), in part; *Fl. Trop. Africa* 7: 358 (1898). Hutchinson & Dalziel, *Fl. West Trop. Africa* 2: 376 (1936). Hepper, *Fl. West Trop. Africa*, ed. 2, 3(1): 139 (1968). Type: *A. bicolor* Hochstetter (= *Gladiolus murielae* Kelway).

Ballosporum Salisbury, *Genera of Plants* 142 (1866). Type: *Gladiolus segetum* Ker Gawler.

Homoglossum Salisbury, *Genera of Plants* 143 (1866), neither type nor constituent species indicated; Baker, *J. Linn. Soc., Bot.* 16: 161 (1877). De Vos, *J. S. African Bot.* 42: 301

(1976). Type: *H. watsonium* (Thunberg) N.E. Brown, according to De Vos (1976) (= *Gladioius watsonius* Thunberg).

Hyptissa Salisbury, *Genera of Plants* 142 (1866). Type: *H. rosea* Salisbury, name not validly published, identity uncertain.

Ophiolyza Salisbury, *Genera of Plants* 142 (1866). Type: *Gladiolus alatus* Linnaeus (lectotype designated here).

Ranisia Salisbury, *Genera of Plants* 143 (1866). Type: *Gladiolus tristis* Linnaeus.

Symphydolon Salisbury, *Genera of Plants* 142 (1866). Type: *Gladiolus floribundus* Jacquin.

Oenostachys Bullock, *Kew Bull.* 465 (1930). Goldblatt, *J. S. African Bot.* 37: 411, 443 (1971). Type: *O. dichroa* Bullock (= *Gladiolus dichrous* (Bullock) Goldblatt).

Anomalesia N.E. Brown, *Trans. Roy. Soc. S. Africa* 20: 270 (1932). Goldblatt, *J. S. African Bot.* 37: 412 (1971). Type: *A. cunonia* (Linnaeus) N.E. Brown (= *Gladiolus cunonius* (Linnaeus) Gaertner).

Kentrosiphon N.E. Brown, *Trans. Roy. Soc. S. Africa* 20: 271 (1932). Type: *K. saccatus* (Klatt) N.E. Brown (= *Gladiolus saccatus* (Klatt) Goldblatt & De Vos).

Dortania A. Chevalier, *Bull. Mus. Hist. Nat., Paris*, Sér. 2, 9: 402 (1937). Type: *D. amoena* A. Chevalier (= *Gladiolus chevalieranus* W. Marais).

Description

Perennial herbs with deciduous leaves and globose corms with membranous to fibrous, reticulate tunics; the base enclosed by two or three sheathing cataphylls. *Leaves* usually contemporary with the flowers and borne on the same shoot, sometimes borne earlier or later than the flowers and on separate shoots, two to several, basal or some inserted above ground level, with a sheathing base and isobilateral, unifacial blade, sometimes the blade reduced or lacking, thus the entire leaf partly to entirely sheathing, blades linear to lanceolate, either plane and the margins, midrib and sometimes other veins not or only lightly thickened and hyaline, or the margins or midrib strongly raised, sometimes even winged (thus H- or X-shaped in section), or the midribs and margins much thickened and the blade evidently terete but with four narrow longitudinal grooves. *Flowering stem* aerial, terete, simple or branched, erect or flexed downward above, often above the sheath of the uppermost leaf.

Inflorescence a spike (rarely reduced to a single flower), the flowers secund or distichous in a few species; *floral bracts* two, usually green, sometimes dry above or

entirely, usually relatively large, the inner (adaxial) usually slightly smaller than the outer, sometimes much smaller, or rarely slightly longer, usually notched apically for 1–2 mm or entire. *Flowers* bilaterally zygomorphic (actinomorphic in a few tropical African, Madagascan and South African species), tepals united below in a tube, the lower tepals usually with contrasting markings constituting a nectar guide, the stamens arcuate and unilateral, ascending to horizontal, the style arched over them, frequently closing at night; *perianth tube* well developed, usually obliquely funnel-shaped, shorter than or to about as long as the bracts, or sometimes much longer than the bracts; *tepals* usually unequal, the uppermost (dorsal) broader and arched to hooded over the stamens, the lower three narrower, in subgenus *Gladiolus* usually clawed below and united for a short distance, as long as, shorter than, or exceeding the upper tepals. *Filaments* filiform, inserted at the base of the upper part of the perianth tube, usually exserted but occasionally extending only to the mouth of the tube; *anthers* symmetrically disposed around the style in species with actinomorphic flowers, usually unilateral, parallel, ascending to horizontal, lying below the dorsal tepal, dehiscing longitudinally, subbasifixed to centrifixed, rarely with sterile tails, the connective obtusely mucronate above or prolonged into a prominent acute to apiculate appendage (some tropical African species). *Ovary* ovoid to oblong; *style* filiform, dividing opposite to or beyond the anthers, the branches simple, filiform below, expanded gradually to abruptly above and channelled to bilobed. *Capsules* large, usually slightly inflated, ovoid to ellipsoid or globose, rarely elongate and nearly cylindric; *seeds* usually many per locule, discoid with a broad, membranous, circumferential wing or, rarely, few per locule and the seeds wingless and more or less globose to angled by pressure. *Basic chromosome number* $x = 15$.

Distribution

Gladiolus is a genus of approximately 255 species, centred in southern Africa and extending throughout tropical Africa and Madagascar and into the Arabian Peninsula, the Mediterranean basin, Europe and Asia as far east as Iran and Afghanistan; 163 species are recognized in southern Africa.

KEYS TO *GLADIOLUS* IN SOUTHERN AFRICA

The keys deal with *Gladiolus* in the following countries of southern Africa: South Africa, Botswana, Lesotho, Swaziland and Namibia, and Mozambique south of the Save River. Two sets of keys are provided. One consists of natural keys, first to the sections, and then to the species of each section: sections *Densiflorus* and *Ophiolyza*, section *Blandus*, section *Linearifolius*, section *Heterocolon*, section *Hebea* and section *Homoglossum*.

A second series of keys is geographical, for the following areas:
1. The northern provinces of South Africa, Botswana, southern Mozambique, the northern two-thirds of Namibia, and Swaziland.
2. The southeastern provinces of South Africa and Lesotho, excluding the Eastern Cape west of Uitenhage and Port Elizabeth.
3. The southern African winter-rainfall region, including the Western Cape and Northern Cape, the Eastern Cape west of Uitenhage and Port Elizabeth, and southwestern Namibia.

NOTES

1. Measure perianth length from top of ovary to apex of dorsal tepal of mature flower.
2. Measure perianth tube length from top of ovary to first point of separation of any tepal (usually the dorsal).
3. Measure lower tepals from apices to point of fusion with each other.
4. For leaf number *exclude* basal sheathing cataphylls, but *include* both fully developed long-bladed foliage leaves and partly and entirely sheathing (bladeless) leaves present on the flowering stem.
5. Terete leaves usually have four longitudinal grooves, but are described as terete whether 4-grooved or solidly terete.
6. For bract length measure the outer bracts in the middle of the spike.
7. For leaf width choose a well-developed, preferably basal, leaf and measure near midline of blade.

NATURAL KEY TO THE SECTIONS OF *GLADIOLUS*

1. Leaves sword-shaped to lanceolate, usually at least 4 mm wide, often with well-developed secondary veins, sometimes the secondary veins fine and closely set; flowers mostly slightly bilabiate with the lower tepals not narrow and clawed below
  2. Leaves in a distichous fan, glabrous or puberulous
    3. Flowers usually small; capsules oblong, fairly small and less than 15 mm long; seed wings often reduced, usually less than 6 mm long
    . . . . Section 1. *Densiflorus*
    3′. Flowers usually large; capsules oblong, large, usually more than 20 mm long; seeds large with well-developed wings, usually more than 8 mm long
      4. Flowers variously coloured, but rarely white to pink, when red the lower tepals white below or with transverse white banding
      . . . . Section 2. *Ophiolyza*
      4′. Flowers usually shades of cream or pale to deep pink or salmon-orange, the lower tepals with red spade-, diamond- or spear-shaped markings, or a median streak (rarely without markings), sometimes flowers red, then with longitudinal markings . . . Section 3. *Blandus*
  2′. Leaves usually superposed, sheaths and sometimes blades with long hairs, rarely scabrid; leaf blades of the flowering stem often reduced; corm tunics consisting of flattened, parallel vertical fibres, often dark brown . . . . Section 4. *Linearifolius*
1′. Leaves linear, usually less than 5 mm wide, usually without well-developed secondary veins, rarely broader but then without a well-developed midrib; flowers usually strongly bilabiate with the lower tepals narrow and clawed
  5. Leaves usually several, at least three basal
    6. Leaf margins and midribs usually heavily thickened and the blades centric, cross-shaped, oval or round in section, rarely plane and lightly ridged; corm tunics coarsely fibrous, often accumulating in a neck around the base of the stem; flowers rarely lightly fragrant; capsules usually oblong, slightly three-lobed above; seeds often with the wing reduced or seeds unwinged . . . . Section 5. *Heterocolon*
    6′. Leaf margins usually unthickened and the blades plane, slightly raised in the midline, rarely ridged and corrugate; corm tunics membranous, papery, woody or fibrous, not accumulating in a neck; spike drooping in bud; flowers frequently fragrant (except red-flowered species); capsules usually ellipsoid and pointed above, sometimes obovate and deeply three-lobed . . . . Section 6. *Hebea*
  5′. Leaves occasionally two, usually three or four, superposed (inserted serially up the stem); spike usually strongly flexuose; stems never branched; nectar guides when present usually consisting of fine, dark, longitudinal streaks and spots over a pale area or transverse bands of pale colour; capsule almost always ellipsoid to ovoid with acute apices; flowers often fragrant except when red or long-tubed and pink
  . . . . Section 7. *Homoglossum*

KEY TO SECTIONS *DENSIFLORUS* AND *OPHIOLYZA*

1. Anthers with the locules drawn into obvious sterile tails at least 1.5 mm long
  2. Perianth tube 40–45 mm long, about 1.5 times as long as the dorsal tepal; tepals attenuate; flowers cream to salmon, the lower three tepals each with a narrow median red streak . . . . 13. *G. macneilii*
  2′. Perianth tube 15–32(–40) mm long, shorter than to about as long as the dorsal tepal; tepals more or less obtuse
    3. Floral bracts 18–27 mm long; flowers whitish to pale pink with perianth tube 16–22 mm long . . . . 11. *G. appendiculatus*
    3′. Floral bracts 35–55 mm long; flowers white, yellow in the throat, the tube 16–40 mm long . . . . 12. *G. calcaratus*
1′. Anthers without sterile tails

4. Margins of the inner bracts united below and enclosing the ovary, or spikes distichous and erect; bracts usually somewhat inflated below and the apices dry and light brown, attenuate and deflexed; plants usually with a fan of several rigid basal leaves often shortly hairy or at least scabrid
    5. Perianth tube (28–)40–55 mm long; flowers bright pink to mauve or salmon; the Eastern Cape to southern KwaZulu-Natal . . . . 23. **G. oppositiflorus**
    5′. Perianth tube less than 22 mm long; flowers pale pink, cream, greenish or pale blue, often minutely speckled with darker colour; KwaZulu-Natal to Northern Province and Botswana
        6. Leaves forming a tight basal fan usually shorter than the spikes; leaf blades with numerous fine parallel veins and without a prominent midrib; plants never hairy . . . . 25. **G. elliotii**
        6′. Leaves narrow and usually exceeding the spikes, in a lax fan usually arising some distance above the ground; leaf blades always with a prominent midrib and sometimes secondary veins also well developed; plants usually hairy
            7. Perianth tube 18–27 mm long; tepals elliptic, the apices attenuate; bracts 26–30 mm long and with attenuate apices . . . . 21. **G. dolomiticus**
            7′. Perianth tube 10–16 mm long; tepals lanceolate and more or less acute, not attenuate; bracts 15–30 mm long, more or less acute, without slender attenuate apices
                8. Filaments 10–12 mm long; bracts about two internodes long . . . . 24. **G. sericeovillosus**
                8′. Filaments 18–20 mm long, reaching almost to the tepal apices; bracts about one, sometimes one and a half internodes long . . . . 22. **G. pole-evansii**
4′. Margins of inner bracts free and spike never distichous, often inclined; bracts not normally inflated below, green or dry but then the apices never deflexed; plant habit varied; summer- or winter-rainfall area
    9. Flowers carmine to red, sometimes with white median streaks or spots on the lower tepals, or the lower tepals much reduced
        10. Dorsal tepal about twice as long as the upper laterals; lower tepals about two-thirds as long as the upper laterals, c. 10 mm long; Botswana and northern Namibia . . . . 32. **G. magnificus**
        10′. Dorsal tepal about as long as the upper laterals; lower tepals marked with white; the Eastern Cape, KwaZulu-Natal and Lesotho
            11. Flower with a sigmoid perianth tube and the lower three tepals forming a shallow bowl; tepals thick and succulent in texture, 30–42 mm long; lower tepals each with a median white streak, about as long as the upper laterals . . . . . . 33. **G. flanaganii**
            11′. Perianth tube more or less straight and lower three tepals plane and horizontal to down-curved; tepals fairly soft-textured, 45–67 mm long; lower tepals white in the lower half, speckled with red
                12. Stems inclined; leaves soft-textured, as long as or longer than the stems and drooping; coastal KwaZulu-Natal . 35. **G. cruentus**
                12′. Stems erect; leaves firm-textured in an erect basal fan shorter than the stem; interior mountains . . . . . 34. **G. saundersii**
    9′. Flowers various colours except carmine to red, and with various markings on the lower tepals
        13. Flowers uniformly pigmented in shades of pink to mauve, the lower tepals with longitudinal reddish streaks; perianth tube 20–50 mm long
            14. Leaves and/or stems scabrid to minutely pubescent; KwaZulu-Natal, southern Mpumalanga and Swaziland
                15. Stems, spike axis and bracts smooth; dorsal tepal 35–40 mm long; plants of cliffs and rock outcrops, with stems drooping . . . . 16. **G. microcarpus**
                15′. Stems, spike axis and bracts scabrid; dorsal tepal 40–45 mm long; plants of stony slopes, with erect stems . . . 17. **G. scabridus**
            14′. Leaves and stems smooth
                16. Leaves 3–9 mm wide, with prominently thickened margins and midrib and with fine, closely set veins in between; dorsal tepal 26–32 mm long; floral bracts rust-brown above or entirely . . . . 10. **G. varius**
                16′. Leaves broader with prominent secondary veins; dorsal tepal 35–48 mm long; bracts green throughout or dry above
                    17. Leaves leathery and with prominently thickened margins and veins; the Eastern Cape . . . . 15. **G. mortonius**
                    17′. Leaves softer-textured, not leathery, the margins and other veins lightly thickened; eastern Mpumalanga
                        18. Bracts 50–70 mm long, green; perianth tube 40–50 mm long; dorsal tepal 40–48 mm long; flowers pink, the lower tepals each with a prominent red median streak . . . . 18. **G. cataractarum**
                        18′. Bracts 38–48 mm long, rusty above; perianth tube 30–40 mm long; dorsal tepal 35–37 mm long; flowers pale pink with the lower tepals each with an obscure median mark in the midline . . . . 9. **G. lithicola**
    13′. Flowers variously coloured but if uniformly pigmented in shades of pink then perianth tube less than 20 mm long
        19. Perianth tube 25–70 mm long; filaments 20–40 mm long
            20. Leaves superposed, overlapping with blades shorter than to about as long as the sheaths . . . . 30. **G. aurantiacus**
            20′. Leaves clustered at the base forming a distichous fan, the blades well developed
                21. Dorsal tepal 35–50 x 22–30 mm, horizontal and hooded over the stamens; anthers 12–16 mm long . . . 31. **G. dalenii**
                21′. Dorsal tepal 25–30 x 14–16 mm, ascending, not hooded over the stamens; anthers 9–11 mm long . 29. **G. antholyzoides**
        19′. Perianth tube 10–35 mm long; filaments usually 7–18 mm long, rarely to 22 mm but then flowers nodding and bell-shaped and tube 18–20 mm long
            22. Floral bracts imbricate, two to five internodes long, somewhat inflated and folded on the midline, mostly 40–100 mm long; flowers often darkly speckled to spotted over entire surface on a pale background, rarely uniformly coloured but then white
                23. Leaves lanceolate, in a tight basal fan, rarely reaching the base of the spike; filaments 12–15 mm long . . . . 26. **G. ecklonii**
                23′. Leaves narrowly sword-shaped to nearly linear, reaching at least to the base of the spike; filaments 7–12 mm long
                    24. Flowers with large, dark purple to lilac spots over entire surface; anthers 10–13 mm long; style branches 3–4 mm long . . . . 27. **G. vinosomaculatus**
                    24′. Flowers white to pale mauve, uniformly coloured; anthers c. 10 mm long; style branches 6–8 mm long . . . . 28. **G. rehmannii**

22′. Floral bracts slightly less than one to two internodes long, thus sometimes overlapping, hardly or not inflated and not or barely folded on the midline, mostly 12–40 mm long; flowers uniformly pigmented, rarely minutely speckled over entire surface on a pale background

 25. Flowers 40–55 mm long

  26. Leaves of the flowering stem reduced, blades vestigial . . . . 20. *G. brachyphyllus*

  26′. Leaves of the flowering stems with well-developed blades

   27. Perianth tube 18–30 mm long and about as long as or somewhat longer than the dorsal tepal . . . . . .
   . . . . 4. *G. hollandii*

   27′. Perianth tube 10–20 mm long and half to about two-thirds as long as the dorsal tepal

    28. Bracts dry and rust-coloured above; leaves narrow, mostly less than 10 mm wide and with thickened margins and midribs and remaining veins fine and closely set . . . . 8. *G. ferrugineus*

    28′. Bracts green; leaves with thickened secondary veins or these inconspicuous and not closely set

     29. Leaves in a tight, basal distichous fan; flowers secund on a lightly inclined spike; filaments 10–12 mm long . . . . 14. *G. ochroleucus*

     29′. Leaves superposed not forming a tight fan; spike sharply inflexed and strongly inclined; filaments 20–22 mm long . . . . 2. *G. papilio*

 25′. Flowers 23–40 mm long

  30. Leaves with only the midrib and margins thickened, often strongly so, and remaining veins fine and closely set

   31. Leaves several, forming a tight distichous fan shorter than the spike; upper stem and spike strongly inclined, usually at least 45°; filaments c. 15 mm long . . . . 6. *G. exiguus*

   31′. Plants not as above

    32. Bracts green or flushed with dull purple; spike usually with 20–45 flowers; filaments 10–12 mm long . . . . 7. *G. densiflorus*

    32′. Bracts dry and rust-coloured above or entirely; spike usually with 12–15 flowers; filaments 12–15 mm long . . . . 8. *G. ferrugineus*

  30′. Leaves with midrib, margins and either one or more pairs of secondary veins strongly thickened or secondary veins obscure and not closely set

   33. Leaves soft-textured, secondary veins obscure; lower median tepal strongly marked with longitudinal streaks; plants with long stolons . . . . 19. *G. pavonia*

   33′. Leaves firm, leathery; tepal markings obscure or best developed on the lower lateral tepals; plants without long stolons

    34. Floral bracts 12–16 mm long, narrow and attenuate, by anthesis dry and twisted above; flowers pale pink, the lower tepals each with a yellow median streak surrounded by a broad, dark pink blotch . . . . 5. *G. serpenticola*

    34′. Floral bracts 17–40 mm long, green and acute to attenuate, the apices not becoming dry and twisted; flowers various colours and usually the lower tepals with a broad transverse blotch

     35. Flowers larger, the dorsal tepal 20–25 mm long; filaments c. 14 mm long; plants small to medium-sized, mostly 40–50 cm high; leaves without conspicuously thickened secondary veins . . . . 1. *G. paludosus*

     35′. Flowers smaller, the dorsal tepal 18–22 mm long; filaments 12–14 mm long; plants robust, mostly 50–120 cm high; leaves usually fibrotic with margins, midrib and one or more pairs of secondary veins heavily thickened . . . . 3. *G. crassifolius*

KEY TO SECTION *BLANDUS*

1. Flowers carmine to red, sometimes with white median streaks or spots on the lower tepals

 2. Flowers more or less unicoloured; tepals broadly obovate, widest in the upper third, obtuse, 27–30 mm long; filaments c. 20 mm long . . . . 48. *G. insolens*

 2′. Flowers with white markings on the lower tepals; tepals more or less lanceolate, usually acute, 35–67 mm long; filaments 21–50 mm long

  3. Leaves with blades reduced or absent at flowering time

   4. Flowers large, the dorsal tepal 53–58 mm long and tube 35–45 mm long; filaments 30–42 mm long, exserted 20–32 mm from the tube; lower three tepals smaller than the dorsal; anthers dark purplish, or yellow with purple lines . . . . 51. *G. stefaniae*

   4′. Flowers medium-sized, dorsal tepal 35–45 mm long, tube 30–35 mm long; filaments 21–26 mm long and exserted 10–15 mm from the tube; lower tepals hardly smaller than the dorsal; anthers yellow . . . . 52. *G. carmineus*

  3′. Leaf blades well developed at flowering time

   5. Stems erect or inclined; corm vestigial, producing rhizome-like stolons; Outeniqua Mountains, flowering mostly March to May . . . . 50. *G. sempervirens*

   5′. Stems inclined to drooping; corm small but developed, without stolons; southwestern Cape, flowering December and January . . . . 49. *G. cardinalis*

1′. Flowers various colours except carmine to red, and with various markings on the lower tepals

 6. Leaves somewhat succulent, without evident veins and margins unthickened; tepals subequal, blue-mauve, the lower or all the tepals with a white median streak edged with dark red; strand plants growing above the high-tide zone in deep sand . . . . 39. *G. gueinzii*

 6′. Leaves not succulent, the margins usually prominent gnd secondary veins usually evident, often conspicuous; flowers shades of white to cream, greenish cream, or pink to mauve or salmon-orange, the lower tepals with spade-, diamond- or spear-shaped markings, or a single median streak (rarely without markings) in red to purple or pale colour outlined in dark colour

7. Leaves in a tight distichous fan; leaf blades firm-textured with strongly thickened margins and midribs; corm tunics with dense, matted, coarsely fibrous tunics, often interspersed with non-fibrous, stiffly papery fragments, always accumulating with the remains of the leaf bases in a coarse fibrous neck around the underground part of the stem

    8. Perianth tube 18–21 mm long; filaments 12–14 mm long; tepals acute, the lower laterals with spear-shaped yellow markings outlined in dark colour; base of the plant heavily marked with raised white speckles . . . . 53. *G. rudis*

    8′. Perianth tube 22–70 mm long; filaments 14–37 mm long; tepals more or less obtuse to retuse, the lower tepals each with a dark median streak; base of the plant lightly marked but not textured with white or green speckles

        9. Perianth tube 22–35(–55) mm long; lower tepals three-quarters to as long as the dorsal . . . . 54. *G. grandiflorus*

        9′. Perianth tube 40–70 mm long, if less than 50 mm then lower tepals half to three-quarters as long as the dorsal

            10. Filaments 35–37 mm long; flowers salmon to orange-red; perianth tube slender and cylindric in the lower 20–35 mm, wide and cylindric above . . . . 56. *G. miniatus*

            10′. Filaments 20–23 mm long; flowers cream to yellow or greenish; perianth tube slender below for 20–55 mm, flared above . . . . 55. *G. floribundus*

7′. Leaves in a loose distichous fan or superposed; leaf blades soft-textured with margins often lightly thickened; corm tunics of medium-textured fibres not accumulating markedly and not extending upward in a neck

    11. Corms scarcely developed, without accumulated tunics; robust plants with inclined to drooping stems growing on wet rocks and cliffs

        12. Spike usually with 15–20 flowers; lower tepals less than half as long as the dorsal; perianth tube 45–50 mm long, completely enclosed in the bracts . . . . 46. *G. buckerveldii*

        12′. Spike with less than 15 flowers; lower tepals two-thirds to nearly as long as the dorsal; perianth tube (25–)34–40 mm long, exceeding the bracts . . . . 43. *G. aquamontanus*

11′. Corms well developed, with or without accumulated tunics

    13. Perianth tube (45–)50–110 mm long; flowers cream to yellowish or salmon or lilac, with red to purple markings

        14. Tepals attenuate and strongly undulate; lower tepals slightly shorter than the upper . . . . 44. *G. undulatus*

        14′. Tepals obtuse to acute, not attenuate and weakly undulate; lower tepals usually about two-thirds as long as the upper

            15. Leaves four or five; flowers cream to yellowish, with prominent spear-shaped markings outlined in red on the lower tepals; flowering October to November; Western Cape mountains and flats . . . . 45. *G. angustus*

            15′. Leaves three; flowers salmon with red linear markings on the lower tepals; flowering March and April; southern Cape, Swellendam to Albertinia . . . . 47. *G. bilineatus*

13′. Perianth tube (15–)20–50 mm long; flowers shades of pale to deep pink to mauve, purple or rarely white

    16. Flowers large, dorsal tepal 40–55 mm long; east of Uniondale to Plettenberg Bay . . . . 42. *G. geardii*

    16′. Flowers smaller, dorsal tepal 18–40(–50) mm long; west of Knysna

        17. Leaf blades with two prominent veins at least on some of the leaves; corm tunics of fine, more or less reticulate fibres; flowering from mid-November to January

            18. Flowers 48–60 mm long; floral bracts 25–33 mm long; markings on the lower tepals triangular in the middle and large dark spots toward the base and in the throat . . . . 37. *G. crispulatus*

            18′. Flowers 33–46 mm long; floral bracts 13–25 mm long; markings on the lower tepal spear-shaped, white to yellow in the centre surrounded by dark pink to purple . . . . 36. *G. oreocharis*

        17′. Leaf blades either with only the midrib and margins thickened or at least one other pair of veins prominent

            19. Leaves (2–)6–20 mm wide; flowers usually at least five per spike; corm tunics usually accumulating, fibrous or firm-papery

                20. Floral bracts 13–16 mm long; perianth tube 18–20 mm long; lower tepals narrowed below into claws . . . . 38. *G. phoenix*

                20′. Floral bracts 35–45(–65) mm long; perianth tube (15–)25–45 mm long; lower tepals not abruptly narrowed below . . . . 40. *G. carneus*

             19′. Leaves narrow, 1.5–3 mm wide; flowers one or two, occasionally three, per spike; corm tunics weakly developed, membranous, not accumulating . . . . 41. *G. pappei*

KEY TO SECTION *LINEARIFOLIUS*

1. Summer-rainfall zone; perianth tube 4–12 mm long; capsules oblong to globose, lightly three-lobed above and retuse apically; leaf blades usually reduced, shorter than the sheaths or vestigial

    2. Leaves two (sometimes the uppermost scale-like) or three; flowers nearly actinomorphic, the tepals subequal; perianth tube 5–8 mm long

        3. Flowers pale pink with lanceolate obtuse to acute tepals; uppermost leaf a small scale inserted just below the first flower; flowering stem wiry . . . . 61. *G. parvulus*

        3′. Flowers greenish or pale mauve with elliptic attenuate tepals; uppermost leaf not scale-like and inserted some distance below the first flower; flowering stem not wiry . . . . 60. *G. pubigerus*

2′. Leaves three or four, occasionally two; flowers obviously zygomorphic, the stamens unilateral and arcuate, the dorsal tepal often larger than the lower three; the lower lateral tepals distinctly clawed and the claws and often the limbs channelled

    4. Flowers pale yellow, the lower lateral tepals with red to purple markings; outer tepals narrowed below into claws; lower tepals strongly auriculate at the base of the limbs; (non-flowering plants bearing three leaves) . . . . 59. *G. pardalinus*

    4′. Not as above

        5. Flowers maroon, pale yellow, or pale lilac to cream, the upper lateral tepals not discolorous, the lower laterals each with a dark median streak; flowers smaller, the dorsal tepal 17–24(–27) mm long; plants usually densely villous . . . . 57. *G. woodii*

        5′. Flowers mauve-blue, the lower lateral tepals shading to purple below; flowers larger, the dorsal tepal 28–30 mm long; plants sparsely hairy to virtually glabrous . . . . 58. *G. malvinus*

1'. Winter-rainfall zone; perianth tube 11–55 mm long; capsules ovoid to ellipsoid, the apices acute; leaf blades reduced or not
  6. Leaves entirely sheathing, lacking blades
    7. Flowers red to orange-red
      8. Flowers with the tepals cupped, laxly spaced on an erect stem; filaments c. 20 mm long, exserted c. 10 mm from the tube; Riviersonderend Mountains . . . . 72. *G. stokoei*
      8'. Flowers with the tepals spreading at right angles to the tube, crowded apically on an inclined stem; filaments 8–10 mm long, included in the perianth tube; Hottentots Holland Mountains . . . . 71. *G. nerineoides*
    7'. Flowers shades of pink, mauve, white or cream to yellow, greenish or brownish
      9. Perianth tube 11–13 mm long . . . . 69. *G. brevifolius*
      9'. Perianth tube cylindric, expanded only near the apex, 22–45 mm long
        10. Flowers yellow mottled brownish or plain yellow, strongly scented; perianth tube 32–45 mm long; Riviersonderend Mountains and Langeberg and coastal foothills . . . . 65. *G. emiliae*
        10'. Flowers cream to pale pink or apricot, the lower tepals each with a reddish to yellow median streak, unscented; perianth tube 22–30 mm long; Cape Peninsula . . . . 70. *G. monticola*
  6'. Leaf blades well developed, usually at least the lowermost as long as the sheaths or much longer, narrowly lanceolate to linear
    11. Flowers yellow to orange or scarlet; perianth tube longer than the bracts, slender and cylindric below and abruptly expanded into a flared or broad cylindric upper part
      12. Flowers clear yellow; upper part of the perianth tube flared; filaments 9–12 mm long . . . . 68. *G. aureus*
      12'. Flowers scarlet to orange, rarely entirely yellow; upper part of the perianth tube wide and cylindric; filaments 28–32 mm long
        13. Leaves hairy; tepals subequal, broadly ovate . . . . 67. *G. bonaspei*
        13'. Leaves minutely scabrid; tepals unequal, the lower three smaller and elliptic . . . . 66. *G. overbergensis*
    11'. Flowers pink, cream, purple or dull red to brownish; upper part of perianth tube gradually expanded and flared toward the mouth
      14. Flowers shades of brownish purple or dull pink to dull reddish brown, the lower tepals yellowish and mottled with brown; flowering April to June, occasionally to mid-July . . . . 64. *G. guthriei*
      14'. Flowers shades of cream to pink or purple, the lower tepals streaked with dark red to purple; flowering mostly July to October, occasionally in June
        15. Flowers 34–58(–75) mm long, the dorsal tepal 19–27(–35) mm long; odourless or occasionally lightly sweet-scented; leaves with the margins and midribs lightly, or occasionally moderately thickened . . . . 62. *G. hirsutus*
        15'. Flowers large, (67–)75–85 mm long, the dorsal tepal 37–45 mm long; usually strongly scented of cloves; leaves with the margins and midribs strongly thickened . . . . 63. *G. caryophyllaceus*

KEY TO SECTION *HETEROCOLON*

1. Leaves hairy
  2. Leaf blades cross-shaped in section; flowers pale bluish . . . . 80. *G. marlothii*
  2'. Leaf blades plane; flowers pink with the lower tepals marked with pale green . . . . 81. *G. mostertiae*
1'. Leaves glabrous
  3. Leaf blades lanceolate, relatively short, more or less plane, the margins and midribs moderately thickened, but blades not with two narrow grooves on each surface . . . . 73. *G. oatesii*
  3'. Leaves linear, the midribs and margins heavily thickened and each surface narrowly two-grooved
    4. Perianth tube cylindric, expanded only near the apex, 20 mm long . . . . 76. *G. filiformis*
    4'. Perianth tube obliquely funnel-shaped, 11–14 mm long
      5. Floral bracts rust-coloured on the margins; flowers cream densely speckled with red; flowering mainly March and April . . . . 77. *G. rufomarginatus*
      5'. Flowers not as above
        6. Lower tepals spathulate
          7. Leaves dry and dead at flowering time; flowers pink, often very pale; flowering mainly July and August; northern provinces and KwaZulu-Natal . . . . 78. *G. vernus*
          7'. Leaves living at flowering time; flowers pale grey-blue with minute dark spotting; flowering October; Kamiesberg, Namaqualand . . . . 79. *G. kamiesbergensis*
        6'. Lower tepals narrowly oblanceolate
          8. Flowers pink to mauve; seeds not winged . . . . 75. *G. pretoriensis*
          8'. Flowers orange-red with yellow markings; seeds winged, sometimes weakly so . . . . 74. *G. rubellus*

KEY TO SECTION *HEBEA*

1. Flowers zygomorphic with the upper lateral and lower median tepals (the outer whorl) heart-shaped at the base with a short stalk and the filaments long and strongly arched (sometimes shorter but then sparsely pilose); sutures between the tepals forming iridescent papillose ridges (kalkoentjies)
  2. Flowers shades of scarlet to orange or pink; stems angled or lightly winged; floral bracts strongly folded on the midline or carinate
    3. Basal leaf blade plane and broadly lanceolate, 20–40 mm wide, the margins thickened and often reddish; the apices of the lower leaves usually obtuse and apiculate . . . . 108. *G. equitans*
    3'. Basal leaf blade plane or ridged, c. 2–12 mm wide, the margins not, or only lightly, thickened and never reddish; lower leaves gradually tapering to the apices
      4. Leaf blades plane, the midrib and other veins barely evident and not raised; stamens and style lightly pilose or smooth above

5. Dorsal tepals strongly hooded and concave; stamens and style glabrous above; upper lateral tepals weakly spreading above, the limbs virtually sessile, often coloured yellow with orange margins on the reverse . . . . 109. *G. speciosus*

5′. Dorsal tepals more or less plane, inclined or more or less erect; upper lateral tepals spreading, with claws 3–4 mm long, uniformly salmon-orange on the reverse . . . . 110. *G. pulcherrimus*

4′. Leaf blades ribbed, at least the midrib and usually several other veins thickened and raised, the blade thus corrugate; stamens and style lightly pilose

6. Flowers predominantly salmon-pink to brick-red, the lower laterals with small yellow marks broadly edged distally with reddish black; filaments c. 15 mm long . . . . 12. *G. meliusculus*

6′. Flowers predominantly orange to scarlet; filaments 20–26 mm long . . . . 111. *G. alatus*

2′. Flowers shades of purple, green, grey, pale yellow or brownish, occasionally pale pink but then veined with dark colour; stems terete; floral bracts not folded or carinate

7. Blade of the basal leaf plane

8. Dorsal tepal narrow with sides parallel, arched in a semi-circle; corm tunics firm-textured to hard and coarsely fibrous and the corm producing small cormlets around the base . . . . 103. *G. orchidiflorus*

8′. Dorsal tepal fairly broad, lanceolate and erect; corm tunics of soft-textured layers and the corm producing stolons, each with a fairly large terminal cormlet . . . . 107. *G. uysiae*

7′. Blade of the basal leaf ridged (thus corrugate) or with the margins and sometimes other veins thickened and raised, then sometimes terete

9. Dorsal tepal hooded and horizontal, partially translucent; leaf blade with several ridges of equal size (corrugate) . . 104. *G. watermeyeri*

9′. Dorsal tepals erect to slightly inclined, not translucent; leaf blade with a prominent central vein, secondary veins also sometimes thickened, but not uniformly corrugate

10. Leaves imbricate, the stem visible only above the sheathing part of the uppermost leaf; corms with dark, firm-textured, almost woody tunics; Knersvlakte, western Karoo and Cold Bokkeveld . . . . 106. *G. ceresianus*

10′. Leaves not imbricate, the stem visible below the uppermost leaf; corms with soft-textured, almost membranous tunics; Breede River valley and southern Cape east of Bot River to the Langkloof . . . . 105. *G. virescens*

1′. Flowers actinomorphic or zygomorphic with the upper lateral tepals obscurely or not at all heart-shaped, with a short stalk at the base; filaments short to long, horizontal, ascending or strongly arched but then never sparsely pilose; sutures between the tepals never forming iridescent papillose ridges

11. Flowers carmine to red, sometimes the lower tepals much reduced and with yellow and/or green markings

12. Dorsal and upper lateral tepals nearly equal in size, lanceolate; upper part of the perianth tube 20–23 mm long; southern Cape . . . . 85. *G. vandermerwei*

12′. Dorsal tepal about twice as long as the upper laterals or longer, spathulate; upper part of the perianth tube 2–10 mm long

13. Upper part of the perianth tube forming a short spur; upper lateral tepals reduced, linear . . . . 88. *G. saccatus*

13′. Upper part of the perianth tube not forming a spur; upper lateral tepals joined to the dorsal for half their length, broadly ovate and curving upward

14. Lower median tepal shorter than the lower laterals, the limb curving inward; spike inclined sometimes nearly horizontal; Cape coast from Saldanha to Knysna . . . . 86. *G. cunonius*

14′. Lower median tepal more than twice as wide as the lower laterals, the limb directed downward; spike stiffly erect; western Karoo . . . . 87. *G. splendens*

11′. Flowers variously coloured but not carmine to red

15. Corm tunics hard, woody and somewhat gnarled or membranous but then leaf sheaths hairy

16. Uppermost cataphyll purple mottled with white above the ground

17. Perianth strongly zygomorphic; flowers greenish to brownish purple with yellow markings; flowering May to July . . . . 97. *G. viridiflorus*

17′. Perianth actinomorphic or nearly so; flowers blue with dark blue and white markings; flowering (July to) August to September . . . . 98. *G. deserticola*

16′. Uppermost cataphyll not mottled

18. Corm tunics softly membranous, not accumulating with age; sheaths and blades of the lower leaves shortly hairy; capsules ellipsoid, acute at apex . . . . 96. *G. arcuatus*

18′. Corm tunics hard, woody and gnarled, accumulating with age; leaf sheaths not hairy; capsules ovoid to globose, three-lobed and retuse above

19. Perianth tube cylindric, 35–40 mm long; filaments c. 8 mm long; flowers cream . . . . 102. *G. lapeirousioides*

19′. Perianth tube obliquely funnel-shaped, 12–22 mm long; filaments 11–25 mm long

20. Perianth tube 18–22 mm long; filaments c. 25 mm long; lower tepals not conspicuously narrowed below; flowers pink . . . . 101. *G. salteri*

20′. Perianth tube 12–17 mm long; filaments 11–16 mm long; lower tepals abruptly narrowed below and with a sharp bend at the base of the limbs

21. Lower tepals with a conspicuous geniculate bend at the base of the limbs and the upper edges of the claws with broad auriculate lobes; upper lateral tepals twisted to lie partly above the dorsal . . . . 100. *G. venustus*

21′. Lower tepals with a more or less geniculate bend at the base of the limb and the upper edges of the claws not or only weakly auriculate; perianth windowed in profile (gaping between the dorsal and upper lateral tepals) . . . . 99. *G. scullyi*

15′. Corm tunics never hard and woody, but coriaceous, fibrous or membranous

22. Leaves lanceolate to sword-shaped or nearly linear but usually at least 3 mm wide; perianth not windowed in profile; flowers with the upper lateral tepals twisted to overlap the dorsal tepal, the lower lateral tepals strongly involute below and conspicuously shorter than the lower median tepal

23. Perianth tube 16–18 mm long; filaments 12–15 mm long; upper lateral tepals much exceeding the dorsal; flowers white to palest pink, the lower tepals with a green transverse band . . . . 84. *G. involutus*

23′. Perianth tube c. 5 mm long; filaments c. 7 mm long; upper lateral tepals slightly shorter than the dorsal; flowers pale mauve, the lower tepals conspicuously veined with dark purple . . . . 83. *G. loteniensis*

22′. Leaves linear, 1–2 mm wide, or terete to oval in section

24. Flowers actinomorphic, perianth stellate; filaments symmetrically disposed . . . . 92. *G. stellatus*

24′. Flowers zygomorphic, perianth bilabiate; filaments unilateral

25. Perianth tube 28–50 mm long, at least 1.5 times as long as the dorsal tepal; flowers white or cream

26. Flowers white, sometimes flushed with mauve, strongly fragrant; lower lateral tepals without nectar guides, 7.5–9 mm wide; summer-rainfall zone . . . . 95. *G. robertsoniae*

26′. Flowers cream to pale yellow, not scented; lower lateral tepals marked with reddish median streak, c. 3.5 mm wide; winter-rainfall zone . . . . 82. *G. leptosiphon*

25′. Perianth tube 9–28(–35) mm long, less than 1.5 times as long as the dorsal tepal

27. Tepals subequal and lanceolate-acuminate; perianth tube 16–30 mm long; filaments 6–10 mm long and included in the tube (rarely exserted up to 2 mm); flowers yellowish green, often flushed brown to purple on the reverse, the lower tepals unmarked . . . . 91. *G. acuminatus*

27′. Plants not as above

28. In profile perianth windowed (gaping between dorsal and upper lateral tepals); flowers usually sweetly scented

29. Perianth tube 22–28(–35) mm long; dorsal tepal 28–30 mm long . . . . 90. *G. uitenhagensis*

29′. Perianth tube 9–15 mm long; dorsal tepal 16–33 mm long . . . . 89. *G. permeabilis*

28′. In profile perianth not windowed; flowers usually unscented (rarely lightly fragrant)

30. Lower tepals linear, involute, marked with a broad green transverse band edged with pink; upper lateral tepals 25–28 mm long, attenuate, much longer than the dorsal tepal . . . . 84. *G. involutus*

30′. Lower tepals lanceolate, mostly unmarked; upper lateral tepals 15–23 mm long, acute or obtuse, slightly shorter than the dorsal tepal

31. Leaves solidly terete or oval in section, dry and dead at flowering time or those of the new season emergent . . . . 94. *G. inandensis*

31′. Leaves linear, green and prominent at flowering time . . . . 93. *G. wilsonii*

KEY TO SECTION *HOMOGLOSSUM*

1. Plants with only three leaves (including partly to entirely sheathing upper leaves) and the lowermost leaf longest and with a well-developed blade, the blade terete, cruciform in section, or at least with margins and the midrib heavily thickened; corm tunics often cartilaginous to more or less woody layers, with age these becoming regularly notched below into claw-like segments or coarsely fibrous

2. Floral bracts 14–30 mm long or if longer then perianth tube less than 25 mm long; perianth tube short to long, usually shorter than the bracts but if longer then flowers pink with reddish markings on the lower tepals

3. Perianth either almost actinomorphic, the tepals subequal, or zygomorphic but then yellow; tepal markings when present confined to the tepal bases

4. Flowers shades of pink to lilac; odourless; occurring in grassland and rock outcrops in the Drakensberg of KwaZulu-Natal . . . . 157. *G. symonsii*

4′. Flowers shades of yellow to white with dull purple shading and sometimes with dark markings in the centre; usually strongly fragrant (sometimes odourless); seasonally wet habitats in the southwestern Cape . . . . 149. *G. trichonemifolius*

3′. Perianth zygomorphic and white to pink or purple; tepal markings more or less in the middle of the lower tepals

5. Flowers blue to greyish or cream, the lower lateral tepals marked with a transverse band of yellow outlined with dark blue; perianth tube strongly geniculate, short, 10–12(–15) mm long, up to half as long as the dorsal tepal; the stem base always with a fibrous, well-developed neck . . . . 145. *G. patersoniae*

5′. Flowers whitish to pale pink, mauve or purple, the lower lateral tepals with a median streak or diamond-shaped mark in red or yellow, outlined in dark blue to purple; perianth tube usually curved, 11–55 mm long, at least half as long to three times as long as the dorsal tepal; the stem base with or without a fibrous, well-developed neck

6. Flowers pale pink to whitish with red median streaks on the lower tepals; Swartberg . . . . 143. *G. nigromontanus*

6′. Flowers mauve, purple or pink to white with diamond-shaped or transverse markings on the lower tepals; Cedarberg to Langeberg

7. Perianth tube mostly 12–18(–30) mm long; flowers bell- to trumpet-shaped, the dorsal tepals ascending to hooded; filaments 12–18 mm long; the uppermost (third) leaf often reduced to a short scale; flowering mainly September to November, rarely in December . . . . 141. *G. inflatus*

7′. Perianth tube (25–)35–52 mm long; flowers hypocrateriform, the tepals spreading more or less at right angles to the tube; filaments 7–8 mm long; upper (third) leaf short but not scale-like; flowering December to January . . . . 142. *G. cylindraceus*

2′. Bracts (25–)30–80(–115) mm long; perianth tube always fairly long, at least 25 mm long and often about as long as the bracts or up to twice as long

8. Flowers shades of white, cream, brown to dull purple or dull reddish orange; perianth tube widening gradually from base to apex; often sweetly scented in the evenings

9. Flowers mainly shades of brown, dull purplish green or reddish orange; leaf blade oval in cross section with the margins and midrib heavily thickened and with two narrow grooves on each surface

   10. Bracts long and attenuate; tepals long and attenuate, (31–)40–45 mm long; perianth tube 40–53 mm long . . . 154. *G. liliaceus*

   10´. Bracts short and not normally attenuate; tepals obtuse to acute, the upper 22–30 mm long; perianth tube 25–36 mm long

    . . . . 153. *G. hyalinus*

  9´. Flowers mostly white to cream, sometimes streaked or spotted with brown; leaves centric, terete or cross-shaped in section

   11. Perianth tube up to 1.5 times as long as the bracts; filaments (15–)18–25 mm long; winter-rainfall zone, Bokkeveld Mountains to Port Elizabeth . . . . 155. *G. tristis*

   11´. Perianth tube 1.3 to c. 3 times as long as the bracts; filaments 5–13 mm long; winter- and summer-rainfall zones, Oudtshoorn to Northern Province . . . . 156. *G. longicollis*

 8´. Flowers shades of red to deep orange, the tepals sometimes marked with yellow or green; perianth tube narrow and cylindric below, abruptly widened into a long cylindric horizontal upper part

  12. The upper lateral tepals about as long as to slightly longer than the dorsal; leaf blades oblong to oval in transverse section, the margins and midribs thickened; corm tunics hard and more or less woody in texture, decaying into vertical segments from below

   13. Upper and lower lateral tepals lanceolate, more than twice as long as wide; blade of the lowermost leaf linear (oblong in transverse section), (1.5–)3–5 mm at the widest; Breede River valley west of Worcester and west of the Hottentots Holland–Porterville mountain axis . . . . 158. *G. watsonius*

   13´. Upper and lower lateral tepals ovate, up to twice as long as wide; blade of the lowermost leaf subterete (oval in transverse section), 1–2 mm at the widest; east of the Hottentots Holland Mountains and Breede River valley east of Robertson

    . . . . 159. *G. teretifolius*

  12´. The upper lateral tepals somewhat shorter to less than half as long as the dorsal; leaf blades cross-shaped in transverse section, the margins and edges of the midribs thickened; corm tunics cartilaginous to fairly soft in texture, decaying with age into fine vertical fibres

   14. Tepals lanceolate, the upper laterals more than twice as long as wide; flowers with the lower tepals uniformly red or orange toward the base; Cold Bokkeveld and Ceres District . . . . 160. *G. quadrangularis*

   14´. Tepals ovate to elliptic, the upper laterals less than twice as long as wide; flowers with the lower tepals with yellow to greenish markings; Caledon to Grahamstown

    15. Dorsal tepal horizontal, concave, two to three times as long as the upper laterals; lower tepals mostly dark green to reddish black, the median shorter than the laterals, 3–6 mm long . . . . 163. *G. abbreviatus*

    15´. Dorsal tepal erect to lightly inclined, and less than 1.5 times as long as the upper laterals; lower tepals mostly yellow, the lower median about as long as or slightly longer than the laterals, 9–22 mm long

     16. Flowers large, the dorsal tepal 25–40 mm long; filaments 32–45 mm long; flowers mostly orange with the reverse of the upper part of the tube with dark red longitudinal streaks below the upper lateral tepal sinuses . . . . .

     . . . . 161. *G. huttonii*

     16´. Flowers smaller, the dorsal tepal 15–20 mm long; filaments 26–32 mm long; flowers dull red or greenish, the upper tepals with red to grey veining, the tube without red streaks . . . . 162. *G. fourcadei*

1´. Plants with two to many leaves, if only three either the blade of the lowermost vestigial (much shorter than the sheathing part, or lacking) or the blade well developed with the midrib and margins not or only lightly thickened; corm tunics various, woody to more or less membranous, sometimes coarsely to finely fibrous

 17. Leaf sheaths and sometimes the blades hairy

  18. Leaf blades cross-shaped in section; flowers greenish to pale yellow without strongly developed dark markings on the lower tepals . . .

   . . . . 150. *G. sufflavus*

  18´. Leaf blades with one or two raised veins, the margins not thickened; flowers yellow with brown to reddish transverse markings on the lower tepals . . . . 151. *G. pritzelii*

 17´. Leaf sheaths and blades glabrous or minutely scabrid

  19. Leaf blades vestigial, lacking or less than 10 mm long

   20. Perianth tubular, the tube (22–)30–60 mm long; Langeberg foothills . . . . 144. *G. engysiphon*

   20´. Perianth funnel-shaped, the tube 6.5–20 mm long

    21. Leaves two, the lowermost sheathing the lower two-thirds of the stem . . . . 122. *G. vaginatus*

    21´. Leaves two to four, all rather short

     22. Corm tunics fibrous; flowers lacking a strong sweet scent; leaves of non-flowering plants hairy . . . . . . 146. *G. subcaeruleus*

     22´. Corm tunics very soft in texture, membranous and not accumulating; flowers usually with a strong sweet scent; leaves of non-flowering plants glabrous

      23. Flowers shades of mauve or pink; flowering February to May; leaves of non-flowering plants one, rarely two; lower tepals joined to the upper laterals for 2–5 mm . . . . 147. *G. martleyi*

      23´. Flowers cream, yellow or pale lilac; mostly flowering December and January; leaves of non-flowering plants two or three; lower tepals not united basally with the upper laterals . . . . 148. *G. jonquilliodorus*

  19´. Leaf blades usually well developed to fairly short, but always present

   24. Leaves linear, smooth, neither the midrib nor margins thickened

    25. Perianth tube shorter than the bracts, 11–17 mm long; flowers pale blue to brown; filaments 10–15 mm long

     26. Leaves four . . . . 120. *G. mutabilis*

     26´. Leaves three . . . . 121. *G. exilis*

    25´. Perianth tube slightly shorter than the bracts to twice as long, 23–48 mm long; flowers red, pink, brownish or cream to yellow

     27. Perianth tube slender below, abruptly expanded above into a wide cylindrical upper part; flowers cream, pink or red; filaments 25–40 mm long

28. Leaves usually four, rarely three; flowers predominantly red, yellow in the throat . . . . 126. *G. priorii*

28′. Leaves three; flowers yellow, or pale to deep pink but not yellow in the throat, the lower tepals often darkly speckled below . . . . 125. *G. meridionalis*

27′. Perianth tube gradually expanded from base to apex; flowers cream to yellowish or brown to dull lilac, scented; filaments 12–16 mm long

    29. Flowers cream or white; perianth tube 37–65 mm long, well exserted from the bracts . . . . 124. *G. albens*

    29′. Flowers dull yellow to lilac conspicuously mottled with brown to purple; perianth tube 23–37(–48) mm long, not or barely exserted from the bracts . . . . 123. *G. maculatus*

24′. Leaves with either the margins or midrib raised, or both

    30. Lower leaf blade with the margins raised into wings extended at right angles to the blade surface, H-shaped in section

        31. Perianth tube slightly shorter than to about as long as the bracts, 27–36 mm long; tepals attenuate and recurved, mostly cream to pale pink . . . . 140. *G. recurvus*

        31′. Perianth tube shorter than the bracts, 12–18 mm long; tepals not attenuate and recurved, mostly blue (occasionally pink or yellowish to light brown)

            32. Leaves soft and broad with the flanges at right-angles to the blade; lower tepals rounded and marked with discrete spots; flowers eight to 14, occasionally four, per spike . . . . 139. *G. caeruleus*

            32′. Leaves firm and narrow with the flanges arching over and almost concealing the blade; lower tepals more or less acute and marked with streaks; flowers two to seven per spike . . . . 138. *G. gracilis*

    30′. Lower leaf blade with the margins unthickened or thickened but not raised into wings extending at right angles to the surface or the leaves reduced and clasping

        33. Uppermost cataphyll purple mottled with white above the ground

            34. Perianth actinomorphic or nearly so, flower stellate with tepals spreading . . . . 119. *G. quadrangulus*

            34′. Perianth zygomorphic and usually more or less bilabiate, with an arched dorsal tepal and smaller lower tepals

                35. Flowers inflated and bell-like, blue, the lower tepals each with a transverse or median white or yellow marking; leaves usually four, occasionally three; corm tunics firm-textured to almost woody and broken into narrow vertical strips . . . . 128. *G. rogersii*

                35′. Flowers funnel-shaped, blue, pink, yellow or greyish, the lower tepals each pale to deep yellow and lightly streaked with longitudinal lines or dots; leaves three (rarely four); corm tunics soft-textured, irregularly fragmented or somewhat fibrous

                    36. Flowers larger, 37–54 mm long; perianth tube 13–16 mm long; flowering mainly late July to September . . . . 117. *G. carinatus*

                    36′. Flowers smaller, 22–36 mm long; perianth tube 6–10 mm long; flowering May to mid-July . . . 118. *G. griseus*

        33′. Uppermost cataphyll uniformly coloured above the ground or flushed with darker colour but not distinctly mottled

            37. Flowers white to pink or pale blue-mauve with purple to red spots, streaks or diamond-shaped markings on the lower tepals; leaf blades linear and usually with thickened margins

                38. Corm tunics membranous or firm-papery to coriaceous and becoming fibrous with age

                    39. Flowers whitish to pale pink, the lower three tepals feathered with deep red below; corm tunics coriaceous to firmly-papery becoming coarsely fibrous with age; perianth tube 36–44 mm long . . . . 116. *G. roseovenosus*

                    39′. Flowers deep pink, the lower tepals each with an elongate pale median mark edged in darker pink; corm tunics soft-textured, membranous, not accumulating; perianth tube 18–20 mm long . . . . . 135. *G. ornatus*

                38′. Corm tunics woody, splitting below into regular segments

                    40. Perianth tube 35–40 mm long, usually about 1.5 times as long as the bracts; floral bracts with a smooth surface, not ridged . . . . 134. *G. vigilans*

                    40′. Perianth tube 10–24 mm long, slightly shorter than to about 1.5 times as long as the bracts

                        41. Leaves usually three, rarely four; floral bracts smooth, not ridged; dorsal tepals 25–30 mm long; flowers whitish to pale pink with large red spots unevenly spread across the lower half of the lower tepals . . . . 133. *G. variegatus*

                        41′. Leaves four; floral bracts ridged above the veins

                            42. Filaments 4–10 mm long; dorsal tepal 17–27 mm long; flowers white, rarely pale pinkish with red markings on the lower tepals . . . . 132. *G. debilis*

                            42′. Filaments 12–16 mm long; dorsal tepal 28–44 mm long; flowers shades of pale to deep pink or blue-mauve

                              43. Perianth tube narrowly funnel-shaped to cylindric, 22–27 mm long; lower tepals with darker V-shaped transverse markings . . . . 131. *G. virgatus*

                              43′. Perianth tube obliquely funnel-shaped, 13–24 mm long; lower tepals cream to yellow with dark longitudinal lines and spots . . . . 130. *G. blommesteinii*

        37′. Flowers blue, yellow or orange to salmon or scarlet

            44. Flowers yellow or orange to salmon or scarlet

                45. Flowers more or less actinomorphic, orange; perianth tube 2.4–4 mm long . . . . 127. *G. brevitubus*

                45′. Flowers zygomorphic, bilabiate, never orange, salmon or scarlet; perianth tube at least 8 mm long

                    46. Flowers buttercup yellow; upper tepals narrowly elliptic, tapering to long attenuate acute apices; lower tepals marked with fine lines and dots; perianth tube 11–14 mm long . . . . 115. *G. comptonii*

                    46′. Flowers creamy yellow; upper tepals ovate; lower tepals with paired narrow reddish streaks in the lower third; perianth tube c. 8 mm long . . . . 152. *G. delpierrei*

        44′. Flowers shades of blue

47. Leaves three
    48. Flowers irregularly mottled with purple, without transverse yellow to white markings on the lower tepals . . . . 136. *G. inflexus*
    48´. Flowers with transverse yellow markings on the lower tepals, not irregularly mottled with purple . . . . 137. *G. taubertianus*
47´. Leaves four, rarely five; flowers with transverse or lateral yellow to white markings
    49. Upper tepals narrowly elliptic, tapering to long attenuate acute apices; lower tepals marked with fine lines and dots
        50. Leaf blade with the midrib winged only on one side, the blade thus triangular in section; tepals pale blue, streaked, speckled and feathered with violet, the lower tepals white to yellow below and more strongly marked than the upper; northwestern Cape . . . . . . 114. *G. violaceolineatus*
        50´. Leaf blades identical on both surfaces; tepals violet, the lower tepals pale yellow in the lower two-thirds and finely feathered with dark violet; Riviersonderend Mountains . . . . . 113. *G. atropictus*
    49´. Tepals lanceolate to ovate, obtuse to acute, not attenuate
        51. Flowers one or two per spike; floral bracts ridged and attenuate, mostly 30–50 mm long . . . . 129. *G. bullatus*
        51´. Flowers usually three or more per spike; bracts smooth and not attenuate, 15–22 mm long . . . . 128. *G. rogersii*

KEY TO *GLADIOLUS* IN BOTSWANA, NORTHERN NAMIBIA, THE NORTHERN PROVINCES OF SOUTH AFRICA AND SWAZILAND

1. Upper lateral tepals much reduced in relation to the dorsal, less than half as long; flowers red
    2. Upper part of the perianth tube forming a short spur; dorsal tepal spoon-shaped . . . . 88. *G. saccatus*
    2´. Upper part of the perianth tube not forming a spur; dorsal tepal lanceolate . . . . 32. *G. magnificus*
1´. Upper lateral tepals well developed, slightly shorter to slightly longer than the dorsal; flowers various colours but rarely red
    3. Leaf blades with the margins and midrib strongly thickened and raised, terete or oval in section with four narrow longitudinal grooves
        4. Perianth tube cylindric, expanded only near the apex, 20–150 mm long
            5. Flowers mauve; perianth tube c. 20 mm long . . . . 76. *G. filiformis*
            5´. Flowers white to cream, sometimes with brownish markings; perianth tube 100–150 mm long . . . . 156. *G. longicollis*
        4´. Perianth tube obliquely funnel-shaped, 11–14 mm long
            6. Leaves three or four (occasionally five), superposed, thus inserted serially up the stem and the lower three or more not forming a distichous fan
                7. Flowers pink to mauve; seeds not winged . . . . 75. *G. pretoriensis*
                7´. Flowers orange-red with yellow markings; seeds winged, sometimes weakly so . . . . 74. *G. rubellus*
            6´. Leaves four or more, at least the lower three basal
                8. Floral bracts rust-coloured on the margins; flowers cream densely speckled with red; flowering mainly March and April . . . . 77. *G. rufomarginatus*
                8´. Floral bracts green to grey with narrow hyaline margins; flowers pink, often very pale; flowering mainly July and August . . . . 78. *G. vernus*
    3´. Leaves more or less plane in section, the margins and midrib variously thickened
        9. Margins of the inner bracts united below and enclosing the ovary, or spikes distichous and erect; bracts usually somewhat inflated below and the apices dry and light brown, attenuate and deflexed; plants usually with a fan of several rigid basal leaves, often shortly hairy or at least scabrid
            10. Leaves forming a tight basal fan shorter than the spikes; leaf blades with numerous fine parallel veins and without a prominent midrib; plants never hairy . . . . 25. *G. elliotii*
            10´. Leaves narrow and usually exceeding the spikes, in a lax fan usually arising some distance above the ground; leaf blades always with a prominent midrib and sometimes secondary veins also well developed; plants usually hairy
                11. Perianth tube 18–27 mm long; tepals elliptic, the apices attenuate; bracts 26–30 mm long and with attenuate apices . . . . 21. *G. dolomiticus*
                11´. Perianth tube 10–16 mm long; tepals lanceolate and more or less acute, not attenuate; bracts 15–30 mm long, more or less acute, without slender attenuate apices
                    12. Filaments 10–12 mm long; bracts about two internodes long . . . . 24. *G. sericeovillosus*
                    12´. Filaments 18–20 mm long, reaching almost to the tepal apices; bracts about one, sometimes one and a half internodes long . . . . 22. *G. pole-evansii*
        9´. Margins of inner bracts free and spike never distichous, often inclined; bracts not normally inflated below, entirely brown or green to the apex or sometimes the apices dry but never deflexed; plant habit varied
            13. Leaves superposed with the blades reduced or if present shorter than the sheaths
                14. Plants robust, the stems 3–4 mm in diameter below the base of the spike; flowers 70–100 mm long, orange usually with reddish flecks; foliage leaves several, overlapping, the blades variously developed . . . . 30. *G. aurantiacus*
                14´. Plants not as above
                    15. Cataphylls and leaves glabrous
                        16. Flowers small, 28–30 mm long, cream to mauve or purple; perianth tube 10–11 mm long . . . . 73. *G. oatesii*
                        16´. Flowers large, 48–50 mm long, pink with white markings; perianth tube c. 20 mm long . . . . . 20. *G. brachyphyllus*
                    15´. At least the upper cataphyll, usually the leaf sheaths, and sometimes the blades distinctly hairy

17. Leaves two, occasionally three; flowers nearly actinomorphic, the tepals subequal; the lower lateral tepals not clawed, more or less plane throughout; flowers greenish or pale mauve with elliptic attenuate tepals . . . . . . 60. **G. pubigerus**

17´. Leaves three or four, rarely two; flowers obviously zygomorphic, the stamens unilateral and arcuate, the dorsal tepal often larger than the lower three; the lower lateral tepals distinctly clawed and the claws and often the limbs channelled

    18. Flowers pale yellow, the upper lateral tepals with red to purple markings; upper tepals narrowed below into claws; lower tepals strongly auriculate at the base of the limbs; (non-flowering plants bearing three leaves) . . . . 59. **G. pardalinus**

    18´. Not as above

        19. Flowers maroon, pale yellow or pale lilac to cream, the upper lateral tepals not discolorous, the lower laterals each with a dark median streak; flowers smaller, the dorsal tepal 17–27 mm long; plants usually densely villous . . . . 57. **G. woodii**

        19´. Flowers mauve-blue, the lower lateral tepals shading to purple below; flowers fairly large, the dorsal tepal 28–30 mm long; plants sparsely hairy to virtually glabrous . . . . 58. **G. malvinus**

13´. Leaves clustered at the base of the stem and with well-developed blades much exceeding the sheaths

  20. Leaves linear, c. 2 mm wide

    21. Perianth tube narrow and cylindric, 28–44 mm long; flowers white, sweetly clove-scented; tepal apices obtuse to subacute, never drawn into attenuate cusps . . . . 95. **G. robertsoniae**

    21´. Perianth tube funnel-shaped, 9–15 mm long; flowers white to cream with purple markings on the lower tepals; tepal apices drawn into attenuate cusps . . . . 89. **G. permeabilis**

  20´. Leaves lanceolate to sword-shaped, occasionally linear but then more than 3 mm wide

    22. Flowers uniformly pigmented in shades of pink to salmon (never speckled with colour on a paler background), the lower tepals with longitudinal reddish streaks; perianth tube 25–50 mm long

      23. Leaves and/or stems scabrid to minutely pubescent . . . . 17. **G. scabridus**

      23´. Leaves and stems smooth

        24. Leaves 3–9 mm wide, with prominently thickened margins and midribs and with fine, closely set veins in between; dorsal tepal 26–32 mm long . . . . 10. **G. varius**

        24´. Leaves broader with prominent secondary veins; dorsal tepal 35–48 mm long

          25. Bracts 50–70 mm long, green; perianth tube 40–50 mm long; dorsal tepal 40–48 mm long: flowers pink, the lower tepals each with a prominent red median streak . . . . 18. **G. cataractarum**

          25´. Bracts 38–48 mm long, rusty above; perianth tube 30–40 mm long; dorsal tepal 35–37 mm long; flowers pale pink, the lower tepals each with an obscure purple median mark . . . . 9. **G. lithicola**

    22´. Not as above, long-tubed species rarely pink, but then pigment speckled

      26. Anthers with the locules drawn into sterile tails at least 1 mm long

        27. Perianth tube 40–45 mm long, about 1.5 times as long as the dorsal tepal; tepal attenuate; flowers cream to pale salmon, the lower three tepals each with a narrow, red median streak . . . . 13. **G. macneilii**

        27´. Perianth tube 15–32(–40) mm long, shorter than to about as long as the dorsal tepal; tepals more or less obtuse

          28. Floral bracts 18–27 mm long; flowers whitish to pink with perianth tube c. 15 mm long . . . . 11. **G. appendiculatus**

          28´. Floral bracts 35–55 mm long; flowers white, yellow in the throat, the tube (16–)25–40 mm long . . . . 12. **G. calcaratus**

      26´. Anthers without sterile tails

        29. Perianth tube 25–50 mm long; filaments 20–35 mm long

          30. Dorsal tepal 35–50 x 22–30 mm, horizontal and hooded over the stamens; anthers 12–16 mm long . . . . 31. **G. dalenii**

          30´. Dorsal tepal 25–30 x 14–16 mm, ascending, not hooded over the stamens; anthers 9–11 mm long . . . . 29. **G. antholyzoides**

        29´. Perianth tube 10–35 mm long; filaments usually 7–18 mm long, rarely to 22 mm but then flowers nodding and bell-shaped and tube 18–20 mm long

          31. Floral bracts imbricate, two to five internodes long, somewhat inflated and folded on the midline, mostly 40–100 mm long; flowers often darkly speckled to spotted over entire surface on a pale background, rarely uniformly coloured but then white

            32. Leaves lanceolate, in a tight basal fan, rarely reaching the base of the spike; filaments 12–15 mm long . . . . 26. **G. ecklonii**

            32´. Leaves narrowly sword-shaped to nearly linear, reaching at least to the base of the spike; filaments 7–12 mm long

              33. Flowers with large, dark purple to lilac spots over entire surface; anthers 10–13 mm long; style branches 3–4 mm long . . . . 27. **G. vinosomaculatus**

              33´. Flowers white to pale mauve, uniformly coloured; anthers c. 10 mm long; style branches 6–8 mm long . . . . 28. **G. rehmannii**

          31´. Floral bracts slightly less than one or two internodes long, thus sometimes overlapping, hardly or not inflated and not or barely folded on the midline, 12–40 mm long; flowers uniformly pigmented, rarely minutely speckled over entire surface on a pale background

            34. Flowers 40–60 mm long

35. Perianth tube 18–30 mm long and about as long or somewhat longer than the dorsal tepal
. . . . 4. **G. hollandii**

35´. Perianth tube 10–20 mm long and half to about two-thirds as long as the dorsal tepal

  36. Bracts dry and rust-coloured above; leaves narrow, mostly less than 10 mm wide and with thickened margins and midribs, remaining veins fine and closely set . . . . . 8. **G. ferrugineus**

  36´. Bracts green or becoming pale and dry above; leaves with thickened secondary veins or these inconspicuous and not closely set

    37. Flowers not nodding, dorsal tepal erect and lower median tepal strongly marked with longitudinal streaks; plants of dry, stony slopes . . . . 19. **G. pavonia**

    37´. Flowers nodding, dorsal tepals more or less horizontal and lower tepals with discrete dark markings; plants usually of marshy sites . . . . 2. **G. papilio**

34´. Flowers 23–38(–40) mm long

  38. Leaves with only the midrib and margins thickened, often strongly so, and remaining veins fine and closely set

    39. Leaves several, forming a tight distichous fan shorter than spike; upper stem and spike strongly inclined, usually at least to 45°; filaments c. 15 mm long . . . . 6. **G. exiguus**

    39´. Plants not as above

      40. Bracts green or flushed with dull purple; spike usually with 20 to 45 flowers; filaments 10–12 mm long . . . . 7. **G. densiflorus**

      40´. Bracts dry and rust-coloured above or entirely; spike usually with 12 to 15 flowers; filaments 12–15 mm long . . . . 8. **G. ferrugineus**

  38´. Leaves with midrib, margins and either one or more pairs of secondary veins strongly thickened or secondary veins obscure and not closely set

    41. Floral bracts 12–16 mm long, narrow and attenuate, by anthesis dry and twisted above; flowers pale pink, the lower tepals each with a yellow median streak surrounded by a broad, dark pink blotch . . . . 5. **G. serpenticola**

    41´. Floral bracts 17–40 mm long, green and acute to attenuate, the apices not becoming dry and twisted; flowers various colours and usually the lower tepals each with a broad transverse band of dark colour in the upper third

      42. Flowers larger, the dorsal tepal 20–25 mm long; filaments c. 14 mm long; plants small to medium-sized, mostly 30–50 cm high; leaves without conspicuously thickened secondary veins . . . . 1. **G. paludosus**

      42´. Flowers smaller, the dorsal tepal 18–22 mm long; filaments 12–14 mm long; plants robust, mostly 50–120 cm high; leaves usually fibrotic with margins, midrib and one or more pairs of secondary veins heavily thickened . . . . 3. **G. crassifolius**

KEY TO *GLADIOLUS* IN LESOTHO AND THE SOUTHEASTERN PROVINCES OF SOUTH AFRICA
(Free State, KwaZulu-Natal and the Eastern Cape east of Port Elizabeth and Uitenhage)

1. Leaf blades with the margins and midrib strongly thickened and raised, terete or oval in section with four narrow longitudinal grooves

  2. Leaves several, dry and brown at flowering time . . . . 78. **G. vernus**

  2´. Leaves three, the upper two sheathing the stem, green at flowering time

    3. Perianth tube less than 15 mm long; perianth almost actinomorphic, tepals subequal; flowers shades of pink to lilac . . . 157. **G. symonsii**

    3´. Perianth tube more than 40 mm long

      4. Flowers shades of white to cream often with brown markings; perianth tube widening gradually from base to apex; filaments less than 10 mm long; often sweetly scented in the evenings . . . . 156. **G. longicollis**

      4´. Flowers red to orange, the tepals often marked with yellow; perianth tube narrow and cylindric below, widening abruptly into a long cylindric horizontal upper part; filaments more than 20 mm long; unscented . . . . 161. **G. huttonii**

1´. Leaves with the margins and midribs lightly or not at all thickened, more or less plane in section

  5. Leaf sheaths distinctly hairy and the blades reduced, shorter than the sheaths or vestigial; perianth tube up to 1.5 times as long as the bracts; leaves two to four, superposed, not in a basal fan

    6. Leaves three or four, rarely two; flowers obviously zygomorphic, the stamens unilateral and arcuate, the dorsal tepal often larger than the lower three; the lower lateral tepals distinctly clawed and the claws and often the limbs channelled; perianth maroon or pale yellow
. . . . 57. **G. woodii**

    6´. Leaves two, rarely three; flowers nearly actinomorphic, the tepals subequal; the lower lateral tepals not clawed, more or less plane throughout

      7. Flowers pale pink with lanceolate obtuse to acute tepals; uppermost leaf a small scale inserted just below the first flower; flowering stem wiry . . . . 61. **G. parvulus**

      7´. Flowers greenish or pale mauve with elliptic attenuate tepals; uppermost leaf not scale-like and inserted some distance below the first flower; flowering stem not wiry . . . . 60. **G. pubigerus**

  5´. Leaf sheaths glabrous or at most velvety to minutely scabrid, but then perianth tube at least twice as long as the bracts and the blades well developed and longer than the sheaths

    8. Flowers carmine to red with white median streaks or spots on the lower tepals

      9. Flowers with a sigmoid perianth tube and the lower three tepals forming a shallow bowl; tepals thick and succulent in texture, 30–42 mm long; lower tepals each with a narrow white median streak . . . . 33. **G. flanaganii**

9′. Perianth tube more or less straight to sharply curved and lower three tepals plane and horizontal to down-curved; tepals fairly soft-textured, 45–67 mm long; lower tepals white in the lower half, speckled with red

    10. Stems inclined; leaves soft-textured, as long as or longer than the stems and drooping; coastal KwaZulu-Natal
         . . . . 35. *G. cruentus*

    10′. Stems erect; leaves firm-textured in an erect basal fan usually shorter than the stem; interior mountains . . . . 34. *G. saundersii*

8′. Flowers various colours except carmine to red, and with various markings on the lower tepals

    11. Leaves three or four, superposed, thus inserted serially up the stem; leaf blades without thickened margins or midribs, firm in texture, rather short and not twisted, up to 3 mm wide

        12. Perianth tube shorter than the bracts, 11–17 mm long; flowers pale blue to brown; filaments 10–15 mm long
             . . . . 120. *G. mutabilis*

        12′. Perianth tube slightly shorter than the bracts to twice as long, 23–48 mm long; flowers red, pink, brownish or cream to yellow

            13. Perianth tube slender below, abruptly expanded above into a wide cylindrical upper part; flowers yellow or pale to deep pink; filaments 25–40 mm long . . . . 125. *G. meridionalis*

            13′. Perianth tube gradually expanded from base to apex; flowers cream to yellowish or brownish, scented; filaments 12–16 mm long

                14. Flowers cream or white; perianth tube 37–65 mm long, exserted from the bracts . . . . 124. *G. albens*

                14′. Flowers dull yellow to lilac, conspicuously mottled with brown to purple; perianth tube 23–37(–48) mm long, not or barely exserted from the bracts . . . . 123. *G. maculatus*

11′. Leaves various but the blades with at least the midrib thickened, the margins either thickened or not

    15. Flowers uniformly pigmented in shades of pink to mauve or salmon (never speckled with colour on a paler background), the lower tepals with longitudinal reddish streaks; perianth tube 25–50 mm long

        16. Margins of the inner bracts united below around the ovary . . . . 23. *G. oppositiflorus*

        16′. Margins of the inner bracts free

            17. Leaves and stems smooth . . . . 15. *G. mortonius*

            17′. Leaves and or stems scabrid to minutely pubescent

                18. Stems, spike axis and bracts smooth; dorsal tepals 35–40 mm long; plants of cliffs and rock outcrops, the stems drooping . . . . 16. *G. microcarpus*

                18′. Stems, spike axis and bracts scabrid; dorsal tepals 40–45 mm long; plants of stony slopes, the stems erect
                     . . . . 17. *G. scabridus*

15′. Not as above, the long-tubed species white or speckled and never pink

    19. Leaf margins not thickened or raised (rarely lightly thickened but then tepal apices drawn into attenuate cusps); leaves linear, up to 10 mm wide

        20. Perianth tube 28–50 mm long, at least 1.5 times as long as the dorsal tepal; dorsal tepal shorter than the upper laterals; flowers white or cream . . . . 95. *G. robertsoniae*

        20′. Perianth tube 9–28 mm long, less than 1.5 times as long as the dorsal tepal

            21. Tepals equal or subequal, mauve, the lower or all the tepals with a white median streak edged with dark red; leaves somewhat succulent, without evident veins and margins unthickened; strand plants growing above the high-tide zone in deep sand . . . . 39. *G. gueinzii*

            21′. Plants not as above

                22. Filaments strongly arched, 20–25 mm long; dorsal tepal narrowly oblanceolate . . . . . 103. *G. orchidiflorus*

                22′. Filaments ascending to slightly curved, 7–16 mm long; dorsal tepal narrowly obovate

                    23. In profile perianth windowed (gaping between dorsal and upper lateral tepals); flowers usually sweetly scented . . . . 89. *G. permeabilis*

                    23′. In profile perianth not windowed; flowers usually unscented (rarely lightly fragrant)

                        24. Flowers with the upper lateral tepals twisted to overlap the dorsal tepal; lower lateral tepals strongly involute below and conspicuously shorter than the lower median tepal

                            25. Perianth tube 16–18 mm long; filaments 12–15 mm long; upper lateral tepals much exceeding the dorsal; flowers white to palest pink, the lower tepals with a green transverse band . . . . 84. *G. involutus*

                            25.′ Perianth tube c. 5 mm long; filaments c. 7 mm long; upper lateral tepals slightly shorter than the dorsal; flowers pale mauve, the lower tepals conspicuously veined with dark purple . . . . 83. *G. loteniensis*

                      24′. Flowers not as above

                          26. Leaves solidly terete or oval in section, dry and dead at flowering time or those of the new season emergent . . . . 94. *G. inandensis*

                          26′. Leaves linear, green and prominent at flowering time . . . . 93. *G. wilsonii*

19′. Leaf margins thickened or raised; leaves lanceolate, more than 10 mm wide

    27. Perianth tube 25–70 mm long; filaments 20–40 mm long

        28. Leaves superposed, overlapping with blades shorter than to about as long as the sheaths . . . . 30. *G. aurantiacus*

        28′. Leaves clustered at the base forming a distichous fan, the blades well developed

            29. Spike strongly inclined; flowers cream to greenish, the lower tepals each with a dark median streak . . .
                 . . . . 55. *G. floribundus*

            29′. Spike erect; flowers pale yellow, orange or brownish, the lower tepals often yellow below, never with dark median streaks

30. Dorsal tepal 35–50 x 22–30 mm, horizontal and hooded over the stamens; anthers 12–16 mm long
. . . . 31. *G. dalenii*

30′. Dorsal tepal 25–30 x 14–16 mm, ascending, not hooded over the stamens; anthers 9–11 mm long
. . . . 29. *G. antholyzoides*

27′. Perianth tube 10–35 mm long; filaments usually 10–18 mm long, rarely to 22 mm but then flowers nodding and bell-shaped and tube 18–20 mm long

31. Spikes distichous and erect; margins of inner bracts united around the ovary

32. Leaves forming a tight basal fan shorter than the spikes; leaf blades with numerous fine parallel veins and without a prominent midrib; plants never hairy . . . . 25. *G. elliotii*

32′. Leaves narrow and usually exceeding the spikes, in a lax fan usually arising some distance above the ground; leaf blades always with a prominent midrib and sometimes secondary veins also well developed; plants usually hairy . . . . 24. *G. sericeovillosus*

31′. Spike never distichous, often inclined; margins of inner bracts free to the base

33. Floral bracts imbricate, two to three internodes long, somewhat inflated and folded on the midline, 40–80 mm long; flowers darkly speckled to spotted over entire surface on a pale background
. . . . 26. *G. ecklonii*

33′. Floral bracts slightly less than one or two internodes long, thus sometimes overlapping, not or hardly inflated and not or barely folded on the midline, 12–40 mm long; flowers uniformly pigmented, rarely minutely speckled over the entire surface on a pale background

34. Anthers with the locules drawn into sterile tails 3–4 mm long . . . . 11. *G. appendiculatus*

34′. Anthers without sterile tails

35. Flowers 40–55 mm long

36. Leaves in a tight, basal distichous fan; flowers secund on a lightly inclined spike; filaments 10–12 mm long . . . . 14. *G. ochroleucus*

36′. Leaves superposed, not forming a tight fan; spike sharply inflexed and strongly inclined; filaments 20–22 mm long . . . . 2. *G. papilio*

35′. Flowers 23–38(–40) mm long

37. Leaves with only the midrib and margins thickened, often strongly so, and remaining veins fine and closely set . . . . 7. *G. densiflorus*

37′. Leaves fibrotic with midrib, margins and either one or more pairs of secondary veins strongly thickened . . . . 3. *G. crassifolius*

KEY TO *GLADIOLUS* IN THE WINTER-RAINFALL REGION OF SOUTHERN AFRICA
(southwestern Namibia, Namaqualand, and southwestern Cape and western Karoo)

1. Flowers strongly zygomorphic with the upper lateral and lower median tepals (the outer whorl) cordate at the base with a short stalk and the filaments long and strongly arched (sometimes shorter but then sparsely pilose); sutures between the tepals forming iridescent papillose ridges (kalkoentjies)

2. Flowers shades of scarlet to orange or pink; stems angled or lightly winged; floral bracts strongly folded on the midline or carinate

3. Basal leaf blade plane and broadly lanceolate, 20–40 mm wide, the margins thickened and often reddish; the apices of the lower leaves usually obtuse and apiculate . . . . 108. *G. equitans*

3′. Basal leaf blade plane or ridged, c. 2–12 mm wide, the margins not, or only lightly, thickened and never reddish; the lower leaves gradually tapering to the apices

4. Leaf blades plane, the midrib and other veins barely evident and not raised; stamens and style lightly pilose or smooth above

5. Dorsal tepals strongly hooded and concave; stamens and style glabrous above; upper lateral tepals weakly spreading above, the limbs virtually sessile, often coloured yellow with orange margins on the reverse . . . . 109. *G. speciosus*

5′. Dorsal tepals more or less plane, inclined or more or less erect; upper lateral tepals spreading, with claws 3–4 mm long, uniformly salmon-orange on the reverse . . . . 110. *G. pulcherrimus*

4′. Leaf blades ribbed, at least the midrib and usually several other veins thickened and raised, the blade thus corrugate; stamens and style lightly pilose

6. Flowers predominantly salmon-pink to pale brick-red, the lower laterals with small yellow marks broadly edged distally with reddish black; filaments c. 15 mm long . . . . 112. *G. meliusculus*

6′. Flowers predominantly orange to scarlet; filaments 20–26 mm long . . . . 111. *G. alatus*

2′. Flowers shades of purple, green, grey, pale yellow or brownish, occasionally pale pink but then veined with dark colour; stems terete; floral bracts not folded or carinate

7. Blade of the basal leaf plane

8. Dorsal tepal narrow with sides parallel, arched in a semi-circle; corm tunics firm-textured to hard and coarsely fibrous and the corm producing small cormlets around the base . . . . 103. *G. orchidiflorus*

8′. Dorsal tepal fairly broad, lanceolate and erect; corms strongly depressed-globose, tunics of soft-textured layers and the corm producing stolons, each with a fairly large terminal cormlet . . . . 107. *G. uysiae*

7′. Blade of the basal leaf ridged (thus corrugate) or with the margins and sometimes other veins thickened and raised, then sometimes terete

9. Dorsal tepal hooded and horizontal, partially translucent; leaf blade with several ridges of equal size (corrugate) . . . .
. . . . 104. *G. watermeyeri*

9′. Dorsal tepals erect to slightly inclined, not translucent; leaf blade with a prominent central vein, secondary veins also sometimes thickened, but not uniformly corrugate

10. Leaves imbricate, the stem visible only above the sheathing part of the uppermost leaf; corms with dark, firm-textured, almost woody tunics; Knersvlakte, western Karoo and Cold Bokkeveld . . . . 106. *G. ceresianus*

10´. Leaves not imbricate, the stem visible below the uppermost leaf; corms with soft-textured almost membranous tunics; Breede River valley and southern Cape east of Bot River to the Langkloof . . . . 105. *G. virescens*

1´. Flowers actinomorphic or zygomorphic with the upper lateral tepals obscurely or not at all cordate at the base, with a short stalk and the filaments short to long, horizontal, ascending or strongly arched but then never sparsely pilose; sutures between the tepals never forming iridescent papillose ridges

11. Plants with only three leaves (including partly to entirely sheathing upper leaves) and the lowermost leaf longest and with a well-developed blade, the blade terete, cruciform in section, or at least with margins and the midrib heavily thickened; corm tunics often cartilaginous to more or less woody layers, with age these becoming regularly notched below into claw-like segments or coarsely fibrous

12. Floral bracts 14–30 mm or if longer then perianth tube less than 25 mm long; perianth tube short to long, usually shorter than the bracts but if longer then flowers pink with reddish markings on the lower tepals

13. Perianth either almost actinomorphic, tepals subequal, or zygomorphic but then yellow; tepal markings when present confined to the tepal bases; flowers shades of yellow to white with dull purple shading and sometimes with dark markings in the centre; often strongly fragrant (sometimes odourless); seasonally wet habitats in the southwestern Cape . . . . 149. *G. trichonemifolius*

13´. Perianth zygomorphic and white to pink or purple; tepal markings more or less in the middle of the lower tepals

14. Flowers blue to grey or cream, the lower lateral tepals marked with a transverse band of yellow outlined with dark blue; perianth tube strongly geniculate, short, 10–12(–15) mm long, up to half as long as the dorsal tepal; corm always with a well-developed fibrous neck . . . . 145. *G. patersoniae*

14´. Flowers whitish to pale pink, mauve or purple, the lower lateral tepals with a median streak or diamond-shaped mark in red or yellow outlined in dark blue to purple; perianth tube usually curved, 11–55 mm long, at least half as long to three times as long as the dorsal tepal; corm with or without a well-developed fibrous neck

15. Flowers pale pink to whitish with red median streaks on the lower tepals; Swartberg . . . . 143. *G. nigromontanus*

15´. Flowers mauve, purple or pink with diamond-shaped or transverse markings on the lower tepals; Cedarberg to Langeberg

16. Perianth tube mostly 12–18(–30) mm long; flowers bell- to trumpet-shaped, the dorsal tepals ascending to hooded; filaments 12–18 mm long; the uppermost (third) leaf often reduced to a short scale; flowering mainly September to November, rarely in December . . . . 141. *G. inflatus*

16´. Perianth tube (25–)35–52 mm long; flowers hypocrateriform, the tepals spreading more or less at right angles to the tube; filaments 7–8 mm long; upper (third) leaf short but not scale-like; flowering December and January . . . . 142. *G. cylindraceus*

12´. Bracts (25–)30–80(–115) mm long; perianth tube always fairly long, at least 25 mm long and often about as long as the bracts or up to twice as long

17. Flowers shades of white, cream, brown to dull purple or reddish orange; perianth tube widening gradually from base to apex; often sweetly scented in the evenings

18. Flowers mainly shades of brown, dull purplish green or reddish orange; leaf blade oval in cross section, with the margins and midribs heavily thickened and with two narrow grooves on each surface

19. Bracts long and attenuate; tepals long and attenuate, (3–)40–45 mm long; perianth tube 40–53 mm long . . . . 154. *G. liliaceus*

19´. Bracts short and not normally attenuate; tepals obtuse to acute, the upper 22–30 mm long; perianth tube 25–36 mm long . . . . 153. *G. hyalinus*

18´. Flowers mostly white to cream, sometimes streaked or spotted with brown; leaves centric, terete or cross-shaped in section

20. Perianth tube up to 1.5 times as long as the bracts, and flowers not to only lightly mottled with brown; filaments (15–)18–25 mm long . . . . 155. *G. tristis*

20´. Perianth tube 1.3 to c. 3 times as long as the bracts, and flowers usually heavily mottled with brown; filaments 5–13 mm long . . . . 156. *G. longicollis*

17´. Flowers shades of red to deep orange, the tepals sometimes marked with yellow or green; perianth tube narrow and cylindric below, abruptly widened into a long cylindric horizontal upper part

21. Upper lateral tepals about as long as to slightly longer than the dorsal; leaf blades oblong to oval in transverse section, the margins and midribs thickened; corm tunics hard and more or less woody in texture, decaying into vertical segments from below

22. Upper and lower lateral tepals lanceolate, more than twice as long as wide; blade of the lowermost leaf linear (oblong in transverse section), (1.5–)3–5 mm at the widest; Breede River valley west of Worcester and west of the Hottentots Holland–Porterville mountain axis . . . . 158. *G. watsonius*

22´. Upper and lower lateral tepals ovate, up to twice as long as wide; blade of the lowermost leaf subterete (oval in transverse section), 1–2 mm at the widest; east of the Hottentots Holland Mountains and Breede River valley east of Robertson . . . . 159. *G. teretifolius*

21´. Upper lateral tepals somewhat shorter than to less than half as long as the dorsal; leaf blades cross-shaped in transverse section, the margins and edges of the midribs thickened; corm tunics cartilaginous to fairly soft in texture, decaying with age into fine vertical fibres

23. Tepals lanceolate, the upper laterals more than twice as long as wide; flowers uniformly red or orange toward the base; Cold Bokkeveld and Ceres District . . . . 160. *G. quadrangularis*

23´. Tepals ovate to elliptic, the upper laterals less than twice as long as wide; lower tepals with yellow to greenish markings; Caledon to Grahamstown

24. Dorsal tepal horizontal, concave, 2–3 times as long as the upper laterals; lower tepals mostly dark green to reddish black, the median shorter than the laterals, 3–6 mm long . . . . 63. *G. abbreviatus*

24´. Dorsal tepal erect to lightly inclined, less than 1.5 times as long as the upper laterals; lower tepals mostly yellow, the lower median about as long as or slightly longer than the laterals, 9–22 mm long

25. Flowers large, the dorsal tepal 25–40 mm long; filaments 32–45 mm long; flowers mostly orange, the upper part of the tube with dark red longitudinal streaks below the upper lateral tepal sinuses . . . . . 161. **G. huttonii**

25′. Flowers smaller, the dorsal tepal 15–20 mm long; filaments 26–32 mm long; flower dull red or greenish, the upper tepals with red to grey veining, the tube without red streaks . . . . 162. **G. fourcadei**

11′. Plants with two to many leaves, if only three either the blade of the lowermost vestigial much shorter than the sheathing part, or lacking, or the blade well developed, with the midrib and margins not or only lightly thickened; corm tunics various, woody to more or less membranous, sometimes coarsely to finely fibrous

26. At least the upper cataphyll, usually the leaf sheaths and sometimes the blades distinctly hairy; perianth tube up to 1.5 times as long as the bracts; leaves two to four, not in a basal fan

27. Leaf blades glabrous; flowers greenish to yellow; lower tepals oblong

28. Leaf blades cross-shaped in section; flowers greenish to pale yellow without darker markings on the lower tepals . . . . 150. **G. sufflavus**

28′. Leaf blades with one or two raised veins, the margins not thickened; flowers yellow with brown to reddish transverse markings on the lower tepals . . . . 151. **G. pritzelii**

27′. Leaf blades hairy

29. Lower tepals spathulate, narrowly clawed and abruptly expanded into the limbs

30. Leaf blades cross-shaped in section; flowers pale bluish . . . . 80. **G. marlothii**

30′. Leaf blades plane; flowers pink with the lower tepals marked with pale green . . . . 81. **G. mostertiae**

29′. Lower tepals not obviously narrowed below into claws

31. Flowers yellow to orange; perianth tube longer than the bracts, slender and cylindric below and abruptly expanded into a flared or broad cylindric upper part

32. Flowers clear yellow; upper part of the perianth tube flared; filaments 9–12 mm long . . . . 68. **G. aureus**

32′. Flowers red to orange, rarely entirely yellow; upper part of the perianth tube wide and cylindric; filaments 28–32 mm long . . . . 67. **G. bonaspei**

31′. Flowers pink, cream, purple or dull red to brownish; upper part of perianth tube gradually expanded and flaring toward the mouth

33. Flowers shades of brown to dull pink or dull reddish, the lower tepals yellowish and mottled with brown; flowering April to mid-July . . . . 64. **G. guthriei**

33′. Flowers shades of cream to pink or purple, the lower tepals streaked with dark red to purple; flowering mostly July to October, occasionally April to June

34. Flowers 34–58(–75) mm long, the dorsal tepal 19–27(–35) mm long; odourless or lightly sweet-scented; leaves with the margins and midribs lightly, or occasionally moderately, thickened . . . . 62. **G. hirsutus**

34′. Flowers large, (67–)75–85 mm long, the dorsal tepal 37–45 mm long; usually strongly clove-scented; leaves with the margins and midribs strongly thickened . . . . 63. **G. caryophyllaceus**

26′. Cataphylls, leaf sheaths and blades (when present) glabrous or at most velvety to minutely scabrid; leaf number and insertion various

35. Flowers carmine to red, sometimes with white median streaks or spots on the lower tepals or the lower tepals much reduced and with yellow and/or green markings

36. Lower tepals much reduced in relation to the dorsal tepal, differing markedly in shape, sometimes scale-like; lower tepals marked with yellow and/or green

37. Perianth tube slender below, abruptly expanded into a wide horizontal and cylindric upper part; dorsal and upper lateral tepals nearly equal in size, the dorsal lanceolate, not hooded or strongly clawed . . . . 85. **G. vandermerwei**

37′. Perianth tube slender below, abruptly expanded into a short, wide upper part; upper lateral tepals not more than half as long as the dorsal, the dorsal spathulate with a narrow channelled claw and hooded limb

38. Upper part of the perianth tube forming a short spur; upper lateral tepals reduced, linear . . . . 88. **G. saccatus**

38′. Upper part of the perianth tube not forming a spur; upper lateral tepals joined to the dorsal tepal for half its length, broadly ovate and curving upward

39. Lower median tepal shorter than the lower laterals, the limb curving inward; spike inclined, sometimes nearly horizontal; Cape coast from Saldanha to Knysna . . . . 86. **G. cunonius**

39′. Lower median tepal more than twice as wide as the lower laterals, the limb directed downward; spike stiffly erect; western Karoo . . . . 87. **G. splendens**

36′. Lower tepals about as long as to half as long as the dorsal tepal but similar in shape

40. Perianth tube slender and cylindric below, abruptly expanded into a long horizontal cylindric upper part; flowers scarlet, sometimes yellow at the base of the lower tepals and in the throat

41. Leaves narrowly sword-shaped to nearly linear, minutely scabridulous on the veins, the midribs raised; tepals unequal, ovate, the lower three about half as long as the dorsal tepals . . . . 66. **G. overbergensis**

41′. Leaves linear, smooth, neither the midribs nor margins thickened; tepals subequal in size, narrowly lanceolate to ovate

42. Leaves usually four, rarely three; flowers predominantly red, yellow in the throat . . . . 126. **G. priorii**

42′. Leaves three; flowers red without yellow in the throat . . . . 125. **G. meridionalis**

40′. Perianth tube gradually expanded from base to apex; flowers never yellow in the throat

43. Flowers more or less unicoloured, without white markings on the lower tepals; tepals obtuse, 26–30 mm long; filaments 8–20 mm long

44. Leaves with long, well-developed blades at flowering time; Piketberg . . . . 48. **G. insolens**

44′. Leaves entirely sheathing, lacking blades

45. Flowers with the tepals cupped, laxly spaced on an erect stem; filaments c. 20 mm long, exserted c. 10 mm from the tube; Riviersonderend Mountains . . . . 72. **G. stokoei**

45′. Flowers with the tepals spreading at right angles to the tube, crowded apically on an inclined stem; filaments 8–10 mm long, included in the perianth tube; Hottentots Holland Mountains . . . 71. **G. nerineoides**

43′. Flowers with white markings on the lower tepals; tepals 35–67 mm long, acute; filaments 21–50 mm long

46. Leaves with blades reduced or absent at flowering time

47. Flowers large, the dorsal tepal 53–58 mm long and perianth tube 35–45 mm long; filaments 30–42 mm long, exserted 20–32 mm from the tube; lower tepals smaller than the dorsal; anthers dark purplish or yellow with purple lines on the ridges . . . . 51. **G. stefaniae**

47′. Flowers medium-sized, dorsal tepal 35–45 mm long and perianth tube 30–35 mm long; filaments 21–26 mm long, exserted 10–15 mm from the tube; lower tepals hardly smaller than the dorsal; anthers yellow . . . . 52. **G. carmineus**

46′. Leaf blades well developed at flowering time

48. Stems erect or inclined; corm vestigial, producing rhizome-like stolons; Outeniqua Mountains, flowering mostly March to May . . . . 50. **G. sempervirens**

48′. Stems inclined to drooping; corm small but developed, without stolons; southwestern Cape, flowering December and January . . . . 49. **G. cardinalis**

35′. Flowers various colours except carmine to red, and with various markings on the lower tepals

49. Leaves two, three or four (occasionally five), superposed, thus inserted serially up the stem and the lower three or more not forming a distichous fan

50. Lower leaf blade with the margins raised into wings extended at right angles to the blade surface, H-shaped in section

51. Perianth tube slightly shorter than to about as long as the bracts, 27–36 mm long; tepals attenuate and recurved, mostly cream or pale pink . . . . 140. **G. recurvus**

51′. Perianth tube shorter than the bracts, 12–18 mm long; tepals not attenuate and recurved, usually blue (occasionally pink or yellow to light brown)

52. Leaves soft and broad with the flanges at right angles to the blade; lower tepals rounded and marked with discrete spots; flowers eight to 14, occasionally four, per spike . . . . 139. **G. caeruleus**

52′. Leaves firm and narrow with the flanges arching over and almost concealing the blade; lower tepals more or less acute and marked with streaks; flowers two to seven per spike . . . . 138. **G. gracilis**

50′. Lower leaf blade with the margins unthickened or thickened, but not raised into wings extending at right angles to the tube, or the leaves reduced and clasping

53. Leaf blades without thickened margins or midrib, firm in texture, rather short and not twisted

54. Perianth tube shorter than the bracts, 11–17 mm long; flowers pale blue to brown; filaments 10–15 mm long

55. Leaves three . . . . 121. **G. exilis**

55′. Leaves four . . . . 120. **G. mutabilis**

54′. Perianth tube as long as to twice as long as the bracts, 23–48 mm long; flowers red, pink, brownish or cream

56. Perianth tube slender below, abruptly expanded above into a wide cylindrical upper part; flowers cream, pink or red; filaments 25–38 mm long . . . . 125. **G. meridionalis**

56′. Perianth tube gradually expanded from base to apex; flowers cream to brownish, scented; filaments 13–16 mm long

57. Flowers cream or white; perianth tube 37–43 mm long, exserted from the bracts . . . . . . 124. **G. albens**

57′. Flowers mottled brownish; perianth tube 23–37 mm long, not exserted from the bracts . . . . 123. **G. maculatus**

53′. Leaf blades with at least the midrib thickened, the margins either thickened or not, or leaves reduced and blades vestigial

58. Leaves of the flowering stem with vestigial blades, either barely developed or up to 25 mm long; flowering December to May

59. Perianth tube cylindric, expanded only near the apex, 22–45 mm long

60. Flowers yellow mottled brownish or plain yellow, strongly scented; Riviersonderend Mountains and Langeberg, and coastal foothills . . . . 65. **G. emiliae**

60′. Flowers cream to pale pink, lilac or apricot, the lower tepals each with a reddish or yellow median streak, unscented

61. Lowermost leaf not basal; perianth tube 22–30 mm long; Cape Peninsula . . . . 70. **G. monticola**

61′. Lowermost leaf basal; perianth tube (22–)30–60 mm long; Langeberg foothills . 144. **G. engysiphon**

59′. Perianth tube obliquely funnel-shaped, 6.5–17 mm long, rarely to 20 mm but then leaves two

62. Leaves two, the lowermost sheathing the lower two-thirds of the stem . . . . 122. **G. vaginatus**

62′. Leaves two to four, all rather short

63. Corm tunics fibrous; flowers usually lacking a strong sweet scent; leaves of non-flowering plants hairy

64. Flowers pale blue; leaf of non-flowering plants terete with four longitudinal grooves; (the lowest leaf may have vestigial blade); southern Cape . . . . 146. **G. subcaeruleus**

64′. Flowers shades of pink, grey or brownish; leaf of non-flowering plants narrowly lanceolate; southern and Western Cape . . . . 69. **G. brevifolius**

63′. Corm tunics very soft in texture, membranous and not accumulating; flowers with a strong sweet scent; leaves of non-flowering plants glabrous, terete and four-grooved

65. Flowers shades of mauve or pink; flowering February to May; leaves of non-flowering plants one, rarely two; lower tepals joined to the upper laterals for 2–5 mm . . . . 147. ***G. martleyi***

65′. Flowers cream to pale lilac; mostly flowering December and January; leaves of non-flowering plants two or three; lower tepals not united basally with the upper laterals . . .
. . . . 148. ***G. jonquilliodorus***

58′. Leaves of the flowering stem with blades usually well developed to fairly short, but always present

  66. Flowers white to cream or pale to deep pink; lower tepals with red spade-, diamond- or spear-shaped markings, or a median streak (rarely without markings); leaves with well-developed secondary veins
. . . . see couplet 104

  66′. Flowers not cream to pink with red markings on the lower tepals; leaves linear, less than 3 mm wide and without secondary veins

    67. Uppermost cataphyll purple mottled with white above the ground

      68. Flowers greenish to brownish purple with yellow markings; flower spike strongly inclined and flexuose with the upper leaf diverging from the stem; corm tunics tough and woody, clawed below; flowering May to July . . . . 97. ***G. viridiflorus***

      68′. Flowers blue, mauve, pink, yellow, cream or greenish grey; flower spike weakly inclined with the upper leaf clasping the stem; corm tunics fibrous, rarely woody and clawed but then flowers blue; flowering June to September

        69. Perianth actinomorphic or nearly so, more or less rotate; flowers unscented

          70. Flowers mauve; stem erect; corm tunics softly fibrous; southwestern Cape . . . . . .
. . . . 119. ***G. quadrangulus***

          70′. Flowers deep blue with darker blue and white markings; flower spike inclined and flexuose; corm tunics tough and woody, ridged to clawed below; Richtersveld . . . .
. . . . 98. ***G. deserticola***

        69′. Perianth zygomorphic and bilabiate; flowers normally scented

          71. Flowers inflated and bell-like, blue, the lower tepals each with a transverse or median white or yellow marking; leaves usually four, occasionally three; corm tunics firm-textured to almost woody and broken into narrow vertical strips . . . . . . 128. ***G. rogersii***

          71′. Flowers funnel-shaped, blue, pink, yellow or greyish, the lower tepals each marked pale to deep yellow and lightly streaked with longitudinal lines or dots; leaves three (rarely four); corm tunics soft-textured, irregularly fragmented or somewhat fibrous

            72. Flowers larger, 37–54 mm long; perianth tube 13–16 mm long; flowering mainly late July to September . . . . 117. ***G. carinatus***

            72′. Flowers smaller, 22–36 mm long; perianth tube 6–10 mm long; flowering May to mid-July . . . . 118. ***G. griseus***

    67′. Uppermost cataphyll uniformly coloured above the ground or flushed with darker colour but not distinctly mottled

      73. Leaves with the margins and midribs strongly thickened, terete or oval in section; perianth more or less bell-like, pale whitish blue; lower tepals spathulate . . . . 79. ***G. kamiesbergensis***

      73′. Leaves with the margins and midrib lightly thickened or not at all, more or less plane in section

        74. Flowers predominantly blue, the lower tepals usually yellow toward the base and with purple streaks

          75. Leaves three

            76. Flowers without transverse yellow to cream markings on the lower tepals, more or less irregularly spotted with purple . . . . 136. ***G. inflexus***

            76′. Flowers with transverse yellow to cream markings; flowers dark blue with the veins distinctly darker; Cold Bokkeveld and Cedarberg . . . . 137. ***G. taubertianus***

          75′. Leaves four

            77. Upper tepals narrowly elliptic, tapering to long, attenuate, acute apices; lower tepals marked with fine lines and dots

              78. Leaf blade with the midrib winged on only one side, the blade thus triangular in section; tepals pale blue, streaked, speckled and feathered with purple, the lower tepals white to yellow below and more strongly marked than the upper; northwestern Cape . . . . 114. ***G. violaceolineatus***

              78′. Leaf blades identical on both surfaces; tepals violet, the lower tepals pale yellow in the lower two-thirds and finely feathered with dark violet; Riviersonderend Mountains . . . . 113. ***G. atropictus***

            77′. Tepals lanceolate to ovate, obtuse to acute, not attenuate

              79. Flowers one or two per spike; floral bracts ridged and attenuate, mostly 30–50 mm long . . . . 129. ***G. bullatus***

              79′. Flowers usually three or more per spike; bracts smooth and not attenuate, 15–22 mm long . . . . 128. ***G. rogersii***

      74′. Flowers white to pink, yellow or orange

        80. Corm tunics woody, splitting below into regular segments

81. Perianth tube 35–40 mm long, about 1.5 times as long as the bracts; floral bracts with a smooth surface, not ridged . . . . 134. *G. vigilans*

81′. Perianth tube 10–24 mm long, slightly shorter than to about as long as the bracts

82. Leaves usually three, rarely four; floral bracts smooth, not ridged; dorsal tepal 25–30 mm long; flowers whitish to pale pink with large red spots irregularly distributed on the lower half of the lower three tepals . . . 133. *G. variegatus*

82′. Leaves four; floral bracts ridged above the veins

83. Filaments 4–10 mm long; dorsal tepal 17–27 mm long; flowers white with red markings on the lower tepals . . . . 132. *G. debilis*

83′. Filaments 12–16 mm long; dorsal tepal 28–44 mm long; flowers shades of pale to deep pink or blue-mauve

84. Perianth tube narrowly funnel-shaped to cylindric, 22–27 mm long; lower tepals with darker V-shaped transverse markings . . . . 131. *G. virgatus*

84′. Perianth tube obliquely funnel-shaped, 13–24 mm long; lower tepals cream to yellow with dark longitudinal lines and spots . . . . 130. *G. blommesteinii*

80′. Corm tunics coriaceous to fibrous or membranous, sometimes not accumulating annually

85. Flowers more or less actinomorphic, orange, salmon- or brick-red; perianth tube 2.4–4 mm long . . . . 127. *G. brevitubus*

85′. Flowers zygomorphic, bilabiate, never orange to scarlet; perianth tube at least 8 mm long

86. Corm tunics coriaceous to firm-papery, becoming fibrous and accumulating with age

87. Flowers shades of pink mottled with brown and yellow, strongly carnation-scented; perianth tube 20–27 mm long . . . . 64. *G. guthriei*

87′. Flowers pale pink, the lower three tepals feathered with deep red below, flowers unscented; perianth tube 36–44 mm long . . 116. *G. roseovenosus*

86′. Corm tunics soft-textured, membranous, not accumulating; lower tepals each with an elongate pale median mark edged in dark pink

88. Leaf blades with the margins moderately to heavily thickened; perianth tube 18–20 mm long . . . . 135. *G. ornatus*

88′. Leaf blades with the margins not or only lightly thickened; perianth tube c. 35 mm long . . . . 41. *G. pappei*

87′. Flowers bright to pale creamy yellow

89. Upper tepals narrowly elliptic, tapering to long attenuate apices; flowers bright yellow . . . . 115. *G. comptonii*

89′. Upper tepals ovate to lanceolate, not tapering to attenuate acute apices; flowers creamy yellow, the lower tepals with paired narrow reddish streaks in the lower third . . . . 152. *G. delpierrei*

49′. Leaves four or more, at least the lower three basal, either sword-shaped to lanceolate, forming a tight distichous fan or leaves narrow and overlapping only near their bases

90. Leaves linear, rarely more than 2 mm wide (see also *G. venustus*).

91. Flowers actinomorphic; perianth stellate; filaments symmetrically disposed . . . . 92. *G. stellatus*

91′. Flowers zygomorphic; perianth bilabiate; filaments unilateral

92. Perianth tube 28–50 mm long, at least 1.5 times as long as the dorsal tepal; dorsal tepal shorter than the upper laterals; flowers cream to yellow, not scented . . . . 82. *G. leptosiphon*

92′. Perianth tube 9–28(–35) mm long, less than 1.5 times as long as the dorsal tepal

93. Tepals subequal and lanceolate-acuminate; perianth tube 16–30 mm long; filaments 6–10 mm long and included in the tube (rarely exserted up to 2 mm); flowers yellowish green, often flushed brown to purple on the reverse, the lower tepals unmarked . . . . 91. *G. acuminatus*

93′. Plants not as above

94. In profile perianth windowed (gaping between dorsal and upper lateral tepals); flowers usually sweetly scented

95. Perianth tube 22–28(–35) mm long; dorsal tepal 28–30 mm long . . . . 90. *G. uitenhagensis*

95′. Perianth tube 9–15 mm long; dorsal tepal 16–33 mm long . . . . 89. *G. permeabilis*

94′. In profile perianth not windowed; flowers usually unscented (rarely lightly fragrant)

96. Lower tepals linear, involute, marked with a broad green transverse band edged with pink; upper lateral tepals 25–28 mm long, attenuate, much longer than the dorsal tepal . . . . 84. *G. involutus*

96′. Lower tepals lanceolate, mostly unmarked; upper lateral tepals 15–23 mm long, acute or obtuse, slightly shorter than the dorsal tepal . . . . 93. *G. wilsonii*

90′. Leaves sword-shaped to lanceolate, usually at least 3 mm wide

97. Corm tunics hard, woody and somewhat gnarled, rarely membranous but then leaf sheaths minutely hairy; perianth often windowed in profile (gaping between dorsal and upper lateral tepals); leaf margins never heavily thickened

98. Corm tunics softly membranous, not accumulating with age; sheaths and blades of the lower leaves shortly hairy; capsules ellipsoid, acute at apex . . . . 96. **G. arcuatus**

98′. Corm tunics hard, woody and gnarled, accumulating with age; leaf sheaths not hairy; capsules globose to ovoid, three-lobed and retuse above

99. Perianth tube cylindric, 35 – 40 mm long; filaments c. 8 mm long; flowers cream . . . . . 102. **G. lapeirousioides**

99′. Perianth tube obliquely funnel-shaped, 12 – 22 mm long; filaments 11 – 25 mm long

100. Perianth tube 18 – 22 mm long; filaments c. 25 mm long; lower tepals not conspicuously narrowed below; flowers pink . . . . 101. **G. salteri**

100′. Perianth tube 12–17 mm long; filaments 11–16 mm long; lower tepals abruptly narrowed below and with a sharp bend at the base of the limbs

101. Lower tepals with a geniculate bend at the base of the limbs and the upper edges of the claws with broad auriculate lobes; upper lateral tepals twisted to lie partly above the dorsal . . . . 100. **G. venustus**

101′. Lower tepals with a more or less geniculate bend at the base of the limb and the upper edges of the claws not, or only weakly, auriculate; perianth windowed in profile (gaping between the dorsal and upper lateral tepals) . . . . 99. **G. scullyi**

97′. Corm tunics never hard and woody, either coriaceous, fibrous or membranous; perianth not windowed in profile

102. Flowers with the upper lateral tepals twisted to overlap the dorsal tepal, the lower lateral tepals strongly involute below and conspicuously shorter than the lower median tepal; leaf blades with margins unthickened and without visible secondary veins; flowers white to palest pink, the lower tepals with a green transverse band . . . . 84. **G. involutus**

102′. Flowers not as above

103. Leaves somewhat succulent, without evident veins and margins unthickened; tepals subequal, blue-mauve, the lower or all the tepals with a white median streak edged with dark red; strand plants growing above the high tide zone in deep sand . . . . 39. **G. gueinzii**

103′. Leaves not succulent, the margins usually prominent and secondary veins usually evident, often conspicuous; flowers shades of white to cream, greenish cream, or pink or salmon-orange, the lower tepals with red spade-, diamond- or spear-shaped markings, or a single median streak (rarely without markings)

104. Leaves in a tight distichous fan; leaf blades firm-textured with strongly thickened margins and midribs; corm tunics dense, matted and coarsely fibrous, often interspersed with non-fibrous, stiffly papery fragments, always accumulating with the remains of the leaf bases in a coarse fibrous neck around the underground part of the stem

105. Perianth tube 18 – 21 mm long; filaments 12 – 14 mm long; tepals acute, the lower laterals with spear-shaped yellow markings outlined in dark colour; base of the plant marked with raised white speckles . . . . 53. **G. rudis**

105′. Perianth tube 22 – 70 mm long; filaments 14 – 37 mm long; tepals obtuse to retuse, the lower tepals each with a dark median streak; base of the plant lightly marked but not textured with white or green speckles

106. Perianth tube 22 – 35(–55) mm long; lower tepals three-quarters to as long as the dorsal . . . . 54. **G. grandiflorus**

106′. Perianth tube 40 – 70 mm long, if less than 50 mm then lower tepals half to three-quarters as long as the dorsal

107. Filaments 35 – 37 mm long; flowers salmon- to orange-red; perianth tube slender and cylindric in the lower 20 – 35 mm, wide and cylindric above . . . 56. **G. miniatus**

107′. Filaments 20 – 22 mm long; flowers cream to yellow or greenish; perianth tube slender below for 20 – 55 mm, flared above . . . . 55. **G. floribundus**

104′. Leaves in a loose distichous fan or superposed; leaf blades soft-textured with margins often lightly thickened; corm tunics of medium-textured fibres, not accumulating markedly and not extending upward in a neck

108. Corms scarcely developed, without accumulated tunics; robust plants, inclined to drooping on wet rocks and cliffs

109. Flowers usually 15 to 20 per spike; lower tepals less than half as long as the dorsal; perianth tube 45 – 50 mm long, completely enclosed in the bracts . . . . . . 46. **G. buckerveldii**

109′. Flowers fewer than 15 per spike; lower tepals two-thirds to nearly as long as the dorsal; perianth tube (25 –)34 – 40 mm long, exceeding the bracts . . . . . . 43. **G. aquamontanus**

108′. Corms well developed, with or without accumulated tunics

110. Perianth tube (45 –)50 – 110 mm long; flowers cream to salmon with red markings

111. Tepals attenuate and strongly undulate; lower tepals not much shorter than the upper . . . . 44. **G. undulatus**

111′. Tepals obtuse to acute, not attenuate and weakly undulate; lower tepals usually shorter than the upper

112. Leaves four or five; flowers cream to yellowish, the lower tepals with prominent spear-shaped marks outlined in red; flowering October and November; Western Cape mountains and flats . . . . 45. **G. angustus**

112′. Leaves three; flowers salmon, the lower tepals with reddish linear markings; flowering March and April; southern Cape, Swellendam to Albertinia . . . . 47. ***G. bilineatus***

110′. Perianth tube (15–)20–50 mm long; flowers shades of pale to deep pink to mauve or purple, rarely white

    113. Flowers large, dorsal tepal 40–55 mm long; occurring east of Uniondale–Plettenberg Bay . . . . 42. ***G. geardii***

    113′. Flowers smaller, dorsal tepal 18–40(–50) mm long; occurring west of Knysna

        114. Leaf blades with two prominent veins at least on some of the leaves; corm tunics of fine, more or less reticulate fibres; flowering from mid-November to January

            115. Flowers 48–60 mm long; floral bracts 25–33 mm long; markings on the lower tepals triangular in the middle and large dark spots toward the base and in the throat . . . . 37. ***G. crispulatus***

            115′. Flowers 33–46 mm long; floral bracts 13–25 mm long; markings on the lower tepal spear-shaped, white to yellow in the centre surrounded by dark pink to purple . . . . 36. ***G. oreocharis***

        114′. Leaf blades with only the midrib and margins prominent, but at least one other pair of veins evident

            116. Leaves narrow, 1.5–3 mm wide; flowers one to three per spike; corm tunics weakly developed, membranous, not accumulating . . 41. ***G. pappei***

            116′. Leaves (2–)6–20 mm wide; flowers usually at least five per spike; corm tunics accumulating, fibrous or firm-papery

                117. Floral bracts 13–16 mm long; perianth tube 18–20 mm long; lower tepals narrowed below into claws . . . . 38. ***G. phoenix***

                117′. Floral bracts 35–45(–65) mm long; perianth tube (15–)25–45 mm long; lower tepals not abruptly narrowed below . . . . 40. ***G. carneus***

1. Section *Densiflorus*

Goldblatt & Manning

Section *Densiflorus* Goldblatt & Manning, new section. Type species *Gladiolus densiflorus* Baker.

Latin diagnosis

Plantae foliis ensiformibus vel lanceolatis, distichis flabella formantibus, laminis usitate saltem 4 mm latis, spica usitate inclinata multifloraque, bracteis viridibus vel ferrugineis, floribus usitate parvis bilabiatis carneis vel malvinis, inodoris, tubo perianthii campanulato vel elongato, capsulis usitate infra 15 mm longis, oblongis vel obovatis trilobatis apicem versus, seminibus saepe alas reductas gerentibus.

Description

Plants medium-sized to large. *Corm* globose or depressed-globose, the tunics membranous to papery, usually decaying with age into irregularly shaped fragments, sometimes coarsely fibrous. *Cataphylls* glabrous, rarely shortly puberulous. *Leaves* sword-shaped to lanceolate, several forming a basal fan, or sheathing one another for some distance around the stem and the fan above the ground, the blades plane, usually the margins and midrib thickened, sometimes strongly so, usually at least 4 mm wide, either with well-developed secondary veins, or secondary veins numerous, fine and closely set, smooth, rarely shortly hairy. *Stem* branched or unbranched, firm and usually somewhat fleshy.

Spike flexed at the base and inclined or erect, straight or barely flexuose, often with many flowers, internodes usually short and flowers somewhat crowded; *bracts* green throughout or dry above, sometimes rusty above or entirely, often clasping the stem and overlapping. *Flowers* usually small, mostly slightly bilabiate with the lower tepals not narrow and clawed at base, nectar guides sometimes present as transverse bands of dark colour or longitudinal streaks, unscented; *perianth tube* usually obliquely funnel-shaped, occasionally tubular, usually half to two-thirds as long as the dorsal tepal but up to half as long again in a few species; *tepals* unequal, the lower three narrower but usually about as long as the upper, the dorsal arched or hooded over the stamens. *Anthers* in three species with the bases drawn into sterile tails. *Capsules* oblong to subglobose, usually fairly small and less than 15 mm long; *seeds* relatively small, usually less than 6 mm long, the wings often reduced and unevenly developed.

Comprising some 20 species in southern Africa, section *Densiflorus* occurs widely across the summer-rainfall parts of the subcontinent but is absent from the winter-rainfall zone. A handful of species occurs in tropical Africa, including *Gladiolus crassifolius* which is shared with southern Africa, and *G. gregarius*, *G. harmsianus* and *G. microspicatus*. These four species were provisionally included in section *Gladiolus* by Goldblatt (1996). That section is now completely restructured and comprises only the Eurasian species of the genus.

The southern African species are divided among four series. The relatively unspecialized series *Paludosus* includes two species of marshy habitats. They have moderate-sized flowers with fairly short tubes and are unusual in the section in having well-developed stolons and marked vegetative reproduction, traits found elsewhere in the section in *G. pavonia* (series *Scabridus*). Series *Densiflorus* includes species with relatively small flowers and short bracts that are often membranous to dry above by anthesis. The leaves have the margins and midrib moderately thickened. In *G. crassifolius* and its close allies at least one pair of secondary veins is visibly developed and moderately, sometimes heavily thickened. The group of species within series *Densiflorus* which includes *G. densiflorus* and *G. ferrugineus* is recognized on the basis of an unusual leaf venation. The margins and midrib of the blades are usually somewhat thickened, but the remaining veins are fine and closely set. Especially when dry, these leaves are unmistakable. The perianth tube is short, less than two-thirds the length of the dorsal tepal, except in the two species *G. varius* and *G. lithicola* which have an elongate tube. The three species of series *Calcaratus* have flowers with unusual anthers, the bases of which are drawn into long, sterile tails or spurs. The series is restricted to the Drakensberg escarpment of Mpumalanga, South Africa, and Swaziland. Series *Scabridus* includes seven species, all with relatively large and usually pink flowers and, except for *G. ochroleucus*, a long perianth tube. Fine to marked scabrid pubescence on the leaves and sometimes the bracts characterizes most members of the series, but *G. brachyphyllus* and *G. pavonia* have smooth, soft-textured leaves. This lack of xeromorphic features despite their markedly xeric habitat suggests that they are adapted for drought avoidance rather than drought tolerance.

Flowers of most species of section *Densiflorus* are relatively unspecialized and adapted for pollination by long-tongued bees, the ancestral pollination strategy in the genus. A few have long-tubed, pink or white flowers and are adapted for pollination by long-tongued flies of the family Nemestrinidae. These include *Gladiolus microcarpus* and *G. mortonius* (series *Scabridus*), *G. calcaratus* and *G. macneilii* (series *Calcaratus*), *G. varius* and probably *G. lithicola* (series *Densiflorus*).

1. GLADIOLUS PALUDOSUS Baker
PLATE 1

Gladiolus paludosus Baker, *J. Bot.* 29: 70 (1891); *Fl. Capensis* 6: 148 (1896). Type: South Africa, Mpumalanga, Ermelo District, Lake Chrissie, marshy places, Nov. 1888, *Elliot 1588* (K, holotype; BOL, drawing).

paludosus = of marshes, for the habitat which comprises poorly drained grassland, vleis and marshes.

Synonymy

[*Gladiolus crassifolius* sensu Lewis et al., *J. S. African Bot.*, Suppl. 10: 60 (1972), in part, not sensu Baker (1891).]

Description

Plants 36–54 cm high. *Corm* globose, 10–12 mm in diameter, the tunics of soft-papery layers, these rapidly decaying into vertical fibres, the layers not accumulating, bearing one or two cormlets on short stolons at the base. *Cataphylls* pale and membranous, the uppermost reaching 1–3 cm above the ground and then firm-textured and green. *Leaves* four to six, the lower two to four basal and longest, the longest of these reaching to between the middle of the stem and the base of the spike, the blades relatively short, lanceolate, 7–10 mm wide, the margins moderately thickened and hyaline,

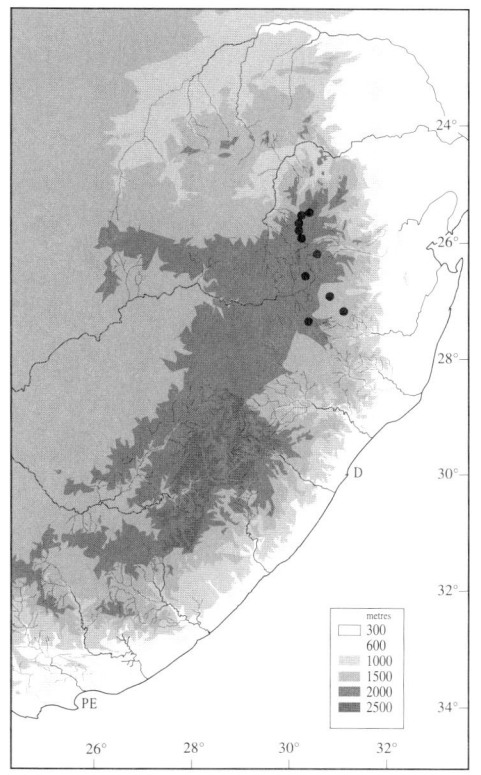

the midrib lightly raised, the remaining two or three leaves inserted on the lower and upper third of the stem respectively, the uppermost smallest, channelled throughout but diverging from the stem, free to the base, the edges sometimes overlapping. *Stem* erect, flexed outward above the sheathing part of the uppermost leaf and becoming erect again, occasionally with one branch, 1.8–2.8 mm in diameter below the spike.

Spike erect, virtually straight, 5- to 9-, occasionally to 20-flowered; *bracts* pale to grey-green, often flushed with grey-purple on the dorsal surfaces, the outer 17–20 mm long, the inner 1–2 mm shorter than the outer and notched apically for c. 2 mm. *Flowers* mauve to pink, or reddish, the lower lateral tepals with a broad diamond-shaped to semicircular band of dark mauve or purple across the midline, unscented; *perianth tube* obliquely funnel-shaped, 10–14 mm long, the lower cylindrical part c. 6 mm long; *tepals* unequal, ovate, the dorsal largest, extending more or less horizontally over the stamens, 20–25 x 14–19 mm, the upper laterals directed forward in the lower half, gradually curving outward in the upper half, 20–25 x 14–17 mm, the lower three tepals sometimes joined to the upper laterals for c. 1 mm, and to one another for 1–2 mm, the lower laterals 18–20 x 7–10 mm, the lower median c. 20 x 14 mm, in profile the lower three tepals extending as far as the upper three. *Filaments* c. 14 mm long, exserted c. 10 mm from the tube; *anthers* 7–8 mm long, yellow or light mauve with dark lines on the ventral side, the pollen cream to yellow. *Ovary* obovoid, c. 4 mm long; *style* dividing near the anther apices, the branches c. 4–5 mm long. *Capsules* and *seeds* unknown. *Chromosome number* unknown.

Flowering time mid-October to mid-November.

Distribution and biology

Restricted to Mpumalanga, South Africa, *Gladiolus paludosus* favours marsh and vlei habitats that either remain wet throughout the year or dry out completely for only a few months. It flowers early in the season, usually after the first soaking rains of October or November have fallen and before the surrounding grass flora has grown to its full height. By the end of December capsules are usually beginning to ripen and shed their

seeds. Populations appear to be centred in the eastern highveld between Middelburg and Belfast, but records show the range extending as far north as Dullstroom and southward to Piet Retief and Wakkerstroom.

The short-tubed pink or mauve flowers are evidently adapted for pollination by long-tongued bees, but the pollination biology of the species has not been studied.

Diagnosis and relationships

Moderate-sized pink to mauve or occasionally red flowers, the dorsal tepal more or less hooded over the stamens, and the lower lateral tepals with prominent dark purple nectar guides are characteristic of the species. It is fairly easy to recognize *Gladiolus paludosus* by its flower, together with its wetland habitat and its relatively soft-textured leaves which have only moderately thickened margins and midrib and secondary veins which are not visible.

The species is most easily confused with the very widespread *Gladiolus crassifolius* of eastern southern and tropical Africa, which has similar, although usually somewhat smaller flowers, and longer leaves that have much more strongly thickened margins and midrib and at least one pair of secondary veins also strongly raised. In general *G. crassifolius* flowers in late summer, from late January until April, and shows a preference for well-drained, often rocky habitats. Although the two species are fairly similar morphologically, we have no doubt that they are separate species, and they can readily be distinguished by their sharply different leaves, as well as their characteristic habitats and flowering times. The relatively soft-textured leaves and prominent, dark nectar guides of *G. paludosus* are reminiscent of *G. papilio*, and *G. paludosus* seems to link that apparently taxonomically isolated species to section *Densiflorus*.

History

The earliest record of *Gladiolus paludosus* is the collection made in 1888 by the English botanist, George Scott Elliot, in marshes near Lake Chrissie in southeastern Mpumalanga. *Gladiolus paludosus* was named by J.G. Baker some three years later and the species was recognized in *Flora Capensis* (Baker, 1896). It remained much misunderstood, however, and herbarium specimens were often confused with the superficially similar *G. crassifolius*. In their revision of *Gladiolus* in South Africa, Lewis et al. (1972) united the two species. Annotations on specimens made by Lewis before her death, however, indicate that she intended to maintain *G. paludosus*. The decision to regard it and *G. crassifolius* as conspecific was presumably made by Amelia Obermeyer during her preparation of the manuscript for publication. Its confusion with the common *G. crassifolius* has probably discouraged collectors from recording *G. paludosus* and although not uncommon, *G. paludosus* is poorly known.

2. GLADIOLUS PAPILIO J.D. Hooker

PLATE 2

Gladiolus papilio J.D. Hooker, *Curtis's Bot. Mag.* 92: pl. 5565 (1866). Baker, *Fl. Capensis* 6: 152 (1896). Lewis et al., *J. S. African Bot.*, Suppl. 10: 53 (1972). Type: South Africa, without precise locality, cultivated at Kew Gardens, 1863, *Cooper s.n.* (K, lectotype here designated, PRE, photograph).

papilio = butterfly(-like), probably referring to the markings on the lower lateral tepals which resemble the wings of a butterfly.

Synonymy

Gladiolus purpureo-auratus J.D. Hooker, *Curtis's Bot. Mag.* 98: pl. 5944 (1872). Baker, *Fl. Capensis* 6: 152 (1896). Type: South Africa, KwaZulu-Natal, without precise locality, cultivated at Bull's Nursery in London, illustration in *Curtis's Bot. Mag.* 98: pl. 5944 (1872).
Gladiolus brachyscyphus Baker, *Handbook Irideae* 210 (1892); *Fl. Capensis* 6: 149 (1896). Type: South Africa, KwaZulu-Natal, Mount Currie near Kokstad, Feb. 1883, *Tyson 1427* (K, holotype; BOL, GRA [not seen], isotypes).
Gladiolus spathulatus Baker, *Handbook Irideae* 223 (1892); *Fl. Capensis* 6: 160 (1896). Type: South Africa, Northern Province, Waterberg Mountains, Zacharias de Beer's farm, near Nylstroom River, Aug. 1880, *Nelson 295* (K, holotype).
Gladiolus schlechteri Baker, *Bull. Herb. Boissier*, Sér. 2, 4: 1006 (1904). Type: South Africa, Gauteng, Donkerhoek east of Pretoria, 4 Jan. 1894, *Schlechter 4132* (Z, presumed holotype; BOL, Z, isotypes).

Description

Plants (40–)50–75(–100) cm high. *Corm* small, hardly developed at flowering time, the rootstock rhizome-like, no more than a slight enlargement at the base of the stem, the tunics membranous, dark brown, fragmenting irregularly with age, producing several slender stolons from the base, each terminating in a single small cormlet, the stolons bearing brown, sheathing scales at the nodes. *Cataphylls* coriaceous, usually dark brown, the uppermost reaching 4–7 cm above the ground and then green or purplish to brown, loosely sheathing or spreading. *Leaves* usually six or seven, sometimes up to nine, the lower four to six more or less basal and longest, reaching to about the middle of the stem, occasionally somewhat longer, the blades narrowly lanceolate, 9–14(–19) mm wide, the margins and midrib lightly thickened, remaining leaves cauline and smaller than the basal, largely to entirely sheathing, overlapping, the margins free to the base and overlapping, the uppermost reaching to the upper third of the stem. *Stem* erect or slightly inclined, flexed outward above the sheath of the uppermost leaf and then inclined at c. 40°, unbranched, 3–4 mm in diameter below the spike.

Spike inclined at 45°–60°, usually strongly flexed at the base, 4- to 8-, occasionally to 13-flowered, lightly flexuose; *bracts* green or flushed grey-purple on the dorsal surfaces, held at 30°–40° from the axis, the outer 20–35 mm long, the inner bracts 3–5 mm shorter than the outer, acute. *Flowers* nodding, cream to greenish, translucent pink or light purple on the inside of the upper tepals, greenish to grey or purple on the reverse, the lower lateral tepals with dark green or light to dark purple blotches in the lower two-thirds, the colour extending into the base of the upper part of the tube, and with obscure to well-defined crescent-shaped yellow to cream bands just below the apices, unscented; *perianth tube* obliquely funnel-shaped, curving outward at the base of the upper part, 18–20 mm long, emerging from just above the middle of the bracts; *tepals* obovoid, unequal, the dorsal largest, 25–35 x 23–30 mm, hooded over the stamens, horizontal or tilted toward the ground, upper laterals directed forward and contiguous with the dorsal and lower tepals to form a wide-mouthed bell, 24–35 x 17–24 mm, lower tepals joined to one another for c. 2 mm, straight and inclined downward, narrowed in the lower third, 22–32 x 14–21 mm, the lower median 20–26 x 13–20 mm, in profile the upper and lower tepals appearing equal. *Filaments* 20–22 mm long, exserted 7–10 mm from the tube; *anthers* 8–10 mm long, purplish on the dorsal surface, cream below, the pollen cream. *Ovary* ovoid, c. 5 mm long; *style* arching over the stamens, dividing opposite the upper third of the anthers, the branches 4–6.5 mm long, very broad in the upper halves. *Capsules* ovate-oblong, three-lobed above, the apices retuse, 18–24 mm long; *seeds* broadly winged, oblong, 8–9 x 4–5 mm, rust-brown. *Chromosome number* unknown.

Flowering time November to February, occasionally later in the year.

Distribution and biology

Gladiolus papilio is widespread in the southern African summer-rainfall area. It extends from near Butterworth in the Eastern Cape in the south, through KwaZulu-Natal, Lesotho, the eastern Free State and across Gauteng to Mpumulanga, Northern Province and Swaziland. Plants are most often found in marshes and seeps, but sometimes also in damp grassland. They occur from close to sea level in southern KwaZulu-Natal to 2200 m at Dullstroom in Mpumalanga. Where it occurs, *G. papilio* is usually fairly common and plants sometimes grow in dense colonies, a habit no doubt enhanced by vegetative reproduction from stolons produced from the base of the corms.

Despite the distinctive nodding flower, which is unusually large for section *Densiflorus*, and the particularly prominent dark markings on the lower tepals, *Gladiolus papilio* appears to be pollinated by the same anthophorid bees that pollinate florally less specialized species of *Gladiolus*, for example, *G. crassifolius* and *G. sericeovillosus*. One of the most common pollinators of the species is the golden-haired *Amegilla capensis* (= *A. natalensis*) which pollinates a range of diverse plants of eastern southern Africa, including species of *Disa* (Orchidaceae) and *Moraea* (Iridaceae) (Goldblatt et al., 1989).

Diagnosis and relationships

Striking in its fairly large flower, usually with dark purple nectar guides on the lower lateral tepals, *Gladiolus papilio* is unique in the genus in having nodding flowers, the whole perianth recurved at the top of the perianth tube so that the tepals are directed downward. The spike itself is inclined too, and normally bears only four to eight flowers, occasionally up to 13 flowers. Although uniform in most of its attributes, *G. papilio* varies to an unusual extent in flower colour. The flowers are most often light mauve or translucent pink, but sometimes cream, yellow or light green. The lower lateral tepals usually have large dark purple, sometimes light purple, blotches in the lower halves, the colour extending into the base of the upper part of the tube, and usually obscure to well-defined crescent-shaped yellow to cream bands just below the tepal apices. The leaves are firm-textured, but except for the midrib the veins are weakly developed, as is usually the case in section *Densiflorus*. The capsules are relatively large for the section, but typical of the entire subgenus in being three-lobed above and with retuse apices. The seeds are broadly winged. Green-flowered populations occur with some frequency in the central Drakensberg and the striking yellow form, also with large purple nectar guides, occurs in the eastern Free State and northern KwaZulu-Natal.

History

The handsome rather than beautiful *Gladiolus papilio* appears to have first been collected in 1832 by J.F. Drège in the eastern part of the then Cape Colony, territory until then never explored botanically. It was not until later in the 1860s, however, when live plants were grown and flowered in Britain, that *G. papilio* began to receive attention. Plants were evidently grown at Kew Gardens after corms were sent there by D. Arnot of Colesberg in 1861. Plants forwarded by the collector, Thomas Cooper, in 1862 and first grown by his patron, W. Wilson Saunders – who maintained a fine garden at Reigate, Surrey – were given to Kew Gardens in 1863. Specimens of the gathering were preserved and now constitute the lectotype of the species. A painting made c. 1863 from one of these two gatherings was published in *Curtis's Botanical Magazine*, accompanied by the protologue by J.D. Hooker.

The golden yellow-flowered form with dark purple markings and highlights was described as a distinct species, *Gladiolus*

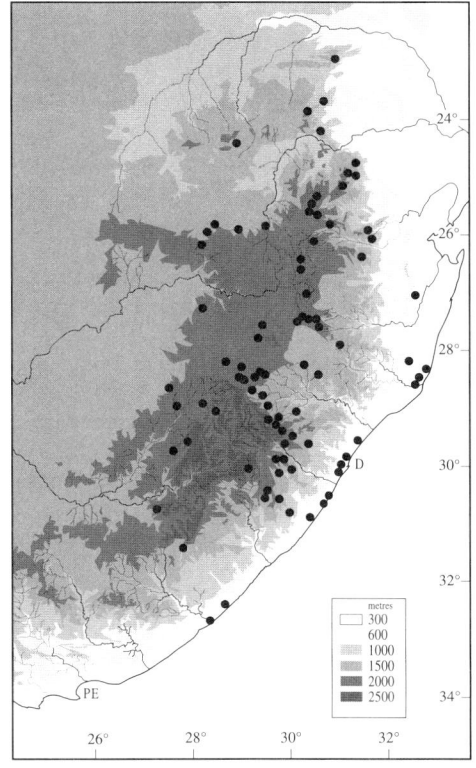

purpureo-auratus, by Hooker in 1872. He described this species from plants imported from KwaZulu-Natal by Mr Bull's Nursery, Chelsea, London. The plants were obtained directly from South Africa, although the collector is unknown. We wonder if the supplier was the same person who sent plants of another KwaZulu-Natal species, *G. cruentus*, to Mr Bull. He flowered plants from which that species was described in 1868 and depicted in *Curtis's Botanical Magazine* in 1869, just three years earlier than *G. purpureo-auratus*. J.G. Baker described three more minor variants of *G. purpureo-auratus*: *G. brachyscyphus* from Mt. Currie in southern KwaZulu-Natal and *G. spathulatus* from Northern Province, both in 1892, and *G. schlechteri* from the Pretoria District, in 1904. Lewis et al. (1972) quite correctly treated these as conspecific with *G. papilio*.

Gladiolus papilio was cultivated fairly widely in Europe in the later nineteenth century and is still sometimes seen in gardens today. One of its attractions is its hardiness outdoors in most parts of Great Britain, France and southern Europe, as well as its ease of cultivation. As the plant is a native of wetland habitats, its corms may be left in the ground when dormant and will not rot in wet conditions. The species is one of several used in the development of the modern *Gladiolus* cultivars. The dark markings on the lower tepals of the so-called 'butterfly' hybrids are reputed to have been derived from this species.

3. GLADIOLUS CRASSIFOLIUS Baker
PLATE 3

Gladiolus crassifolius Baker, *J. Bot.* (London) 14: 334 (1876); *Handbook Irideae* 215 (1892); *Fl. Capensis* 6: 150 (1896). Lewis et al., *J. S. African Bot.,* Suppl. 10: 60 (1972). Goldblatt, *Fl. Zambesiaca* 12: 68 (1993); Gladiolus *in Tropical Africa* 94 (1996). Type: South Africa, Free State, 1862, *Cooper 3185* (K, lectotype designated by Lewis et al., 1972; G, PRE, isolectotypes).

crassifolius = thick-leafed, referring to the thick, leathery, strongly ribbed leaves.

Synonymy

Gladiolus rachidiflorus Klatt, *Abh. Naturforsch. Ges. Halle* 12: 339 (Ergänzungen 5) (1882). Baker, *Fl. Capensis* 6: 143 (1896). Type: South Africa, KwaZulu-Natal, Durban (Port Natal), around the bay, Mar. 1832, *Drège 4537* (B [not seen], holotype presumed lost; P [not seen], isotype according to Lewis et al. (1972) but not located there).

Gladiolus thomsonii Baker, *Handbook Irideae* 223 (1892); *Fl. Trop. Africa* 7: 372 (1898). Type: Tanzania, high plateau north of Lake Malawi (Nyasa), in 1880, *Thomson s.n.* (K, holotype).

Gladiolus masukuensis Baker, *Kew Bull.* 1897: 283 (1897); *Fl. Trop. Africa* 7: 365 (1898). Type: Malawi, 'Mesuku Mts.', [type label], 'Masuku Plateau', [protologue], (cultivated in Zomba), *Whyte s.n.* (K, lectotype designated by Goldblatt, 1993; K, isolectotype).

Gladiolus mosambicensis Baker, *Fl. Trop. Africa* 7: 576 (1898). Type: Mozambique, Beira, without date, *Braga 117* (B, holotype).

Gladiolus tritoniiformis O. Kuntze, *Revis. Gen.* 3, 2: 308 (1898). Type: South Africa, KwaZulu-Natal, Estcourt District, Highlands Station, 1600 m, 15 Mar. 1894, *Kuntze s.n.* (NY, holotype, PRE photo; K, Z, isotypes).

Gladiolus junodii Baker, *Bull. Herb. Boissier,* Sér. 2, 1: 866 (1901). Type: South Africa, KwaZulu-Natal, Lions River District, Howick, 1893, *Junod 320* (Z, holotype, PRE photo; G, Z, isotypes).

Gladiolus conrathii Baker, *Bull. Herb. Boissier,* Sér. 2, 4: 1005 (1904). Type: South Africa, Gauteng, Modderfontein, Feb. 1898, *Conrath 582* (Z, holotype, PRE photo; GZU [not seen], isotype).

Gladiolus gazensis Rendle, *J. Linn. Soc., Bot.* 40: 210 (1911). Plowes & Drummond, *Wild Fl. Rhodesia,* pl. 37 (1976). Type: Zimbabwe, Chimanimani, 1800 m, 23 Sept. 1907, *Swynnerton 779* (BM, lectotype designated by Goldblatt, 1993; K, isolectotype).

Gladiolus dieterlenii Phillips, *Ann. S. Africa Mus.* 16: 282 (1917). Type: Lesotho, Leribe, 5000–6000 ft, Mar. 1913, *Dieterlen 445* (SAM, holotype; GRA [not seen], K, P, PRE, isotypes).

Description

Plants (25–)35–90(–120) cm high, sometimes the leaves emergent or charred by fire, or the leaves reaching to at least the middle of the spikes and even shortly exceeding them, sometimes producing two or three, rarely five stems per corm. *Corm* 18–30 mm in diameter, tunics coriaceous to more or less coarsely fibrous. *Cataphylls* coriaceous, red-brown. *Leaves* four to eight, either few and emergent at flowering time, then the stem bearing only one or two cauline and sheathing leaves and two laminate basal leaves, or with several basal laminate leaves and two or three cauline leaves, these with progressively shorter blades and the upper sometimes entirely sheathing, the blades narrowly lanceolate to linear, exceeding the spike in short-stemmed plants, or reaching to

about the base of the spike in long-stemmed plants, 5–12 mm wide, the midrib and margins and two or more pairs of secondary veins hyaline and moderately to heavily thickened. *Stem* erect below, usually flexed outward above the sheath of the uppermost leaf, simple or with one to three, rarely up to five branches, c. 2 mm in diameter below the spike.

Spike more or less erect or weakly inclined, (12–)16–22-flowered, the branches with fewer flowers than the main axis; *bracts* usually pale green or flushed purple, soft-textured initially, becoming dry and membranous above in fruit (usually translucent light red-brown when capsules ripen in tropical African populations), acute and attenuate (subacute to obtuse and apiculate in tropical African populations), 25–28(–40) mm long, c. two internodes long, the inner about two-thirds as long as the outer, forked apically for c. 1–4 mm. *Flowers* pale to deep pink or light purple, occasionally orange-red or cream, the lower lateral, and sometimes the lowermost, tepals each with a dark band of colour across the lower half of the limbs, the dark band often edged with cream, unscented; *perianth tube* obliquely funnel-shaped, cylindric below, widening and curving outward near the apex, 9–17 mm long; *tepals* broadly ovate, unequal, the dorsal largest and arched to hooded over the stamens, 18–22 x 10–12 mm, the upper laterals directed forward and curving outward above, 18–20 x 16–18 mm, the lower three tepals more or less straight and directed downward, united below for 2–3 mm, narrowed into claws below, the claws channelled, the lower laterals 12–16 x 5–7 mm, the lower median 15–20 x 12 mm, in profile the lower tepals usually slightly exceeding the upper. *Filaments* 12–14 mm long, exserted 6–7 mm from the tube; *anthers* 7–8 mm long, pale lilac to purple, the pollen whitish to pale yellow. *Ovary* obovoid, 3–4 mm long; *style* arching over the filaments, usually dividing opposite the lower half of the anthers, the branches c. 3 mm long, seldom reaching past the middle of the anthers. *Capsules* obovoid, 9–12(–14) mm long, three-lobed above and retuse; *seeds* elliptic, broadly winged, 4–6 x 2.5–4 mm, the wing usually more or less transparent. *Chromosome number* 2*n* = 30.

Flowering time mainly February and March, occasionally later or earlier at higher elevations; in tropical Africa usually April to June, but sometimes August to November, especially after fires.

Distribution and biology

Gladiolus crassifolius is a widespread species of eastern southern Africa, common in South Africa, Swaziland and Lesotho. It is one of the few southern African species shared with tropical Africa, where its distribution extends through southern and eastern Zimbabwe and central Mozambique northward to Malawi and southwestern Tanzania. Isolated populations also occur in western Angola, where it is known from two sites more than 2000 km west of the nearest populations of the species in Zimbabwe (Goldblatt, 1993, 1996). In southern Africa *G. crassifolius* extends from the northeastern edge of the Eastern Cape near Elliot through Swaziland to Northern Province, South Africa. It is one of the few species that extends almost throughout the central highveld, and populations have been recorded as far west as Krugersdorp and Rustenburg. Plants almost always grow in hilly country, where they favour well-drained, rocky grassland habitats, usually in full sun but occasionally in light shade. Typically a late summer-flowering plant, *G. crassifolius* is most commonly seen blooming from the end of February to late April, but plants are occasionally seen in flower in late January, even at higher elevations.

In tropical Africa a series of populations from eastern Zimbabwe and central Mozambique flowers from September to November, before the new season's leaves have developed. These plants usually have stems with two or three branches and more intensely coloured flowers (see Plowes & Drummond, 1976). They evidently represent a distinct genotype and need to be critically investigated to determine if they are taxonomically distinct from *Gladiolus crassifolius*.

Like other small-flowered species of *Gladiolus*, *G. crassifolius* is adapted for pollination by long-tongued bees. Most common visitors to the flowers are species of *Amegilla* (family Anthophoridae). In the Drakensberg the most frequently seen visitor is *A. capensis*, also an important pollinator of *Moraea brevistyla* (Iridaceae) and *Disa versicolor* (Orchidaceae) (Goldblatt et al., 1989; S.D. Johnson, pers. comm.). This and other bee species make rapid visits to flowers to feed on nectar, small quantities of which are located in the lower half of the perianth

tube. Afterward they return to forage for pollen on other flowers. In eastern Zimbabwe a small long-tongued fly, *Prosoeca* sp. (Nemestrinidae), has also been found pollinating *G. crassifolius*. Both the anthophorid bees and the nemestrinid fly have bodies that are snugly accommodated by the flowers of *G. crassifolius*. As the bees and flies search for nectar they climb into the base of the floral cup, almost disappearing from sight, and pollen is then passively deposited on their upper thorax. On a visit to another flower the insect may then inadvertently transfer pollen to the stigma, so accomplishing cross pollination.

Diagnosis and relationships

Although the flower of *Gladiolus crassifolius* is fairly typical of the smaller-flowered members of the genus, the species is readily recognized by the combination of the long, many-flowered, inclined spike and the basal fan of several tough leaves with conspicuously thickened margins, midrib and secondary veins. The flowers vary in colour over its wide range, but are most frequently light pink or salmon with darker pink to purple nectar guides highlighted with yellow. Plants with orange, red and coral-pink flowers occur locally in Mpumalanga. In the KwaZulu-Natal Drakensberg plants have purple rather than pink flowers, while those in the Eastern Cape have unusual pale pink flowers with poorly defined nectar guides.

Gladiolus crassifolius has been confused with *G. paludosus* (Lewis et al., 1972), but that species flowers in the spring and early summer, favours marshy sites and has soft-textured leaves and bracts. *Gladiolus densiflorus* is also mistaken for *G. crassifolius*, largely due to their similarly shaped flowers and their spikes being many-flowered. The leaves of the two species, however, are quite different. Those of *G. crassifolius* have heavily thickened margins and veins, whereas those of *G. densiflorus* have only lightly thickened margins and midrib and the other veins are unusually fine and closely spaced, a feature more pronounced in herbarium specimens. The distribution of these two species is largely complementary. *Gladiolus crassifolius* is typically a highland species and *G. densiflorus* a lowland one, most common in the KwaZulu-Natal coastal plain and extending north into lowland Swaziland and Mozambique. Their ranges do, however, overlap along the edge of the eastern Drakensberg escarpment, but then the two species rarely flower together, *G. crassifolius* blooming a few weeks later than *G. densiflorus*.

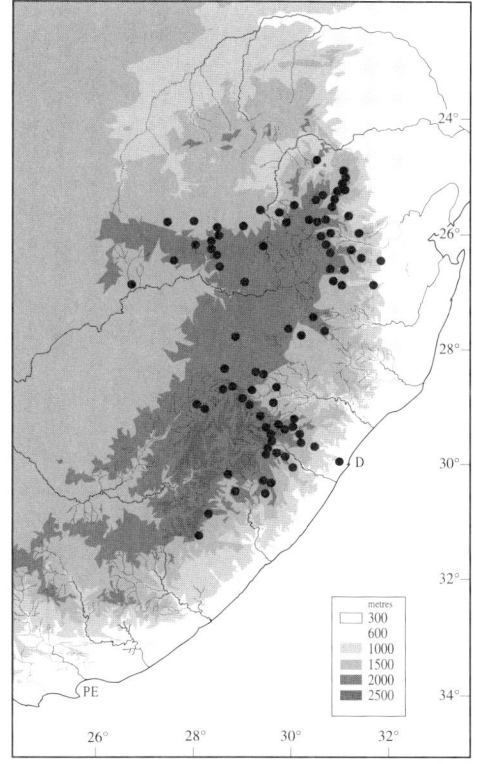

In the field *Gladiolus crassifolius* can usually also be distinguished by the pink to salmon flowers strongly marked with purple on the lower tepals. The flowers of *G. densiflorus* range widely in colour, but typically are finely speckled with dark spots and the lower tepals do not have conspicuous dark markings. *Gladiolus crassifolius* rarely reaches the stature attained by *G. densiflorus* in KwaZulu-Natal and Swaziland, where the latter usually exceeds 1 m. To confuse the issue, plants in populations of *G. densiflorus* in the Sabie–Graskop area of Mpumalanga are always smaller than *G. crassifolius*, seldom reaching 40 cm in height. Such plants can reliably be distinguished from *G. crassifolius* by their somewhat smaller, pale-coloured flowers as well as by their distinctively veined leaves.

History

Although *Gladiolus crassifolius* is not a particularly variable species, and one distinctive in its tough, fibrotic leaves with conspicuously thickened margins, midrib and secondary veins, it was poorly understood for many years. The first collection of the species was made in southern KwaZulu-Natal in 1832 by J.F. Drège in the vicinity of Durban, then called Port Natal. Plants subsequently found by the plant collector, Thomas Cooper, in the eastern Free State in 1862 were described as *G. crassifolius* by J.G. Baker in 1876. Drège's earlier collection formed the basis for F.W. Klatt's *G. rachidiflorus*, described in 1882. We have not been able to locate type material of *G. rachidiflorus*, said by Lewis et al. (1972) to be at the Paris Herbarium, and have to accept their opinion that this is conspecific with *G. crassifolius*, although the locality is more consistent with *G. densiflorus*. Plants from tropical Africa collected as early as 1880 in southwestern Tanzania and from northern Malawi in 1895 were described by Baker as *G. thomsonii* and *G. masukuensis* in 1892 and 1897 respectively, and both match *G. crassifolius*.

from southern Africa fairly closely. The distinctive spring-flowering form of *G. crassifolius* from eastern Zimbabwe and Mozambique was named *G. mosambicensis* by Baker in 1898 and *G. gazensis* by A.B. Rendle in 1911. Other synonyms include *G. tritoniiformis* and *G. conrathii* from South Africa and *G. dieterlenii* from Lesotho. The last, described by E.P. Phillips in 1917, represents a particularly narrow-leafed race of the species. The others do not differ in any

taxonomically significant way from the type of *G. crassifolius* and were described in ignorance of the morphology and variation of that species. All the above-named species, plus *G. paludosus* which is restricted to the highveld of Mpumalanga, were included in *G. crassifolius* by Lewis et al. (1972). There is no doubt that *G. paludosus*, which flowers in the spring and early summer and has soft-textured leaves, is a separate species.

4. GLADIOLUS HOLLANDII L. Bolus
PLATE 4

Gladiolus hollandii L. Bolus, *J. Bot.* 69: 261 (1931). Lewis et al., *J. S. African Bot.*, Suppl. 10: 34 (1972). Goldblatt, *Fl. Zambesiaca* 12(4): 70 (1993). Type: South Africa, Mpumalanga, Tomango near Nelspruit, cultivated at Kirstenbosch Gardens (NBG 493/27), Apr. 1928, *Holland s.n.* (BOL, holotype).

hollandii, named in honour of F.H. Holland, a wild flower enthusiast from Port Elizabeth who collected the plants that were grown at Kirstenbosch and then preserved as the type of the species.

Synonymy

Gladiolus varius var. *elatus* F. Bolus, *Ann. Bolus Herb.* 2: 104 (1917); Verdoorn, *Fl. Pl. S. Africa* 20: pl. 791 (1940). Type: South Africa, Mpumalanga, Barberton, base of the mountains at Barberton, 2800-3000 ft, 17 Mar. 1890, *Galpin 860* (BOL, holotype; K, PRE, SAM, isotypes).

Description

Plants 65–120(–150) cm high. *Corm* depressed-globose, 18–40 mm in diameter, the tunics coriaceous, fragmenting irregularly with age, bearing a few small cormlets at the base. *Cataphylls* membranous and pale, the uppermost reaching 7–10 cm above the ground and then green or flushed with purple. *Leaves* seven to ten, the lower four to six more or less basal and longest, reaching to between the base and apex of the spike, the blades narrowly to broadly lanceolate, 8–18(–28) mm wide, the remaining leaves cauline and shorter, the blades decreasing in size and the uppermost often entirely sheathing, imbricate, enclosing the stem almost to the base of the spike, the margins of the sheaths of the uppermost leaves open to the base, the midvein and margins lightly thickened, sometimes a second pair of veins also lightly thickened, the remaining veins fine and closely set. *Stem* erect, simple or with one or two branches, 2.3–3 mm in diameter below the spike.

Spike straight and erect, 18- to 30(–55)-flowered; *bracts* green, lanceolate, the apices attenuate, becoming dry soon after flowering, the outer 23–30(–35) mm long, the inner about two-thirds as long as the outer, acute or subobtuse. *Flowers* pale pink, the surface usually minutely dotted with dark

pink, all the tepals with a dark pink to reddish median streak running their entire length, this most pronounced on the lower lateral tepals, unscented; *perianth tube* obliquely and narrowly funnel-shaped, 18–25(–30) mm long, the wider upper part c. 9 mm long; *tepals* subequal, lanceolate, the dorsal inclined over the stamens, 20–27 x 12–14 mm, the upper laterals 18–25 x 11–12 mm, the lower three tepals united for 1–2 mm, (12–)18–20 x 10 mm, the lower median slightly longer than the lower laterals. *Filaments* 12–17 mm long, exserted 6–8.5 mm from the tube; *anthers* 6.5–8 mm long, dark purple, the pollen cream. *Ovary* ovoid, c. 4 mm long; *style* arching over the stamens, dividing at or just beyond the anther apices, the branches c. 3 mm long. *Capsules* narrowly obovoid, 10–12 mm long,

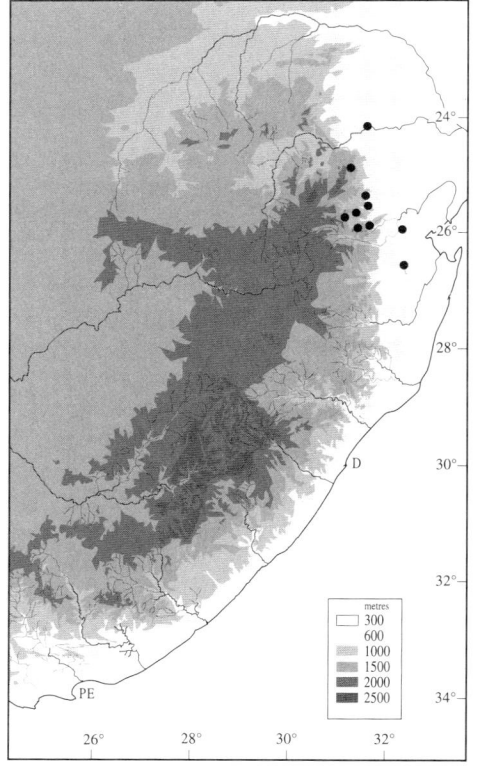

three-lobed above, truncate; *seeds* oblong, poorly and unevenly winged, c. 5 x 3 mm. *Chromosome number* unknown.

Flowering time mid-February to April.

Distribution and biology

A species of the eastern southern African lowveld, *Gladiolus hollandii* extends from the Mthethomusha Nature Reserve in Northern Province southward to Barberton in Mpumalanga. It also extends eastward into southern Mozambique and locally into adjacent Swaziland. Plants grow on the lower slopes of mountains and on granite outcrops, often in well-drained gritty granitic soil in light to moderate shade.

The comparatively long-tubed flower seems likely to have a specialized pollinator. Whereas its relatives have flowers with tubes in the 9–15 mm range, the floral tubes of *Gladiolus hollandii* are 18–30 mm long. We suspect that they are pollinated at least in part by flies of the family Nemestrinidae rather than the long-tongued bees of the family Anthophoridae that are the main

pollinators of shorter-tubed species of section *Densiflorus*.

Diagnosis and relationships

Although a fairly tall species and often very robust vegetatively, *Gladiolus hollandii* has the short floral bracts that soon dry at the tips and the many-flowered spikes typical of series *Densiflorus*. Surprisingly small, the flowers are distinctive mainly in their relatively long perianth tube. They are pale, rather dull pink, the surface usually minutely dotted with dark pink. A reddish median streak along the entire length of the tepals is particularly pronounced on the lower lateral tepals. Although the perianth tube is relatively long, 18–25 mm, occasionally up to 30 mm, the tepals are quite short, the upper three being 18–27 mm long. The relatively small flower with a perianth tube slightly longer than to about twice as long as the upper tepals, and the dark reddish streaks on all the tepals are features diagnostic of the species.

The leaf venation, with margins and

midrib moderately thickened and at least one other pair of veins slightly enlarged, suggests that *Gladiolus hollandii* belongs in series *Densiflorus* and we suspect that it is most closely related to *G. serpenticola*, with which it is almost identical when not flowering.

History

Gladiolus hollandii was first collected by the South African amateur naturalist, Ernest Galpin, in 1890 at Highland Creek, Barberton. All the early records of the species are from this area, where it is common and conspicuous when in flower. The species was first described as var. *elatus* of *G. varius* by Frank Bolus in 1917, based on Galpin's specimens. Plants collected in the 1920s at Tomango by F.H. Holland, the Port Elizabeth wild flower enthusiast, were grown at Kirstenbosch Gardens and formed the basis for *G. hollandii*, described by H.M.L. Bolus in 1931. Bolus seems to have been unaware that the species had been named earlier at varietal rank.

5. GLADIOLUS SERPENTICOLA Goldblatt & Manning

PLATE 5

Gladiolus serpenticola Goldblatt & Manning, *Novon* 6: 173 (1996). Type: South Africa, Mpumalanga, Kaap Valley on Nelspruit–Barberton road near Noordkaap River bridge, serpentine outcrops, 9 Feb. 1994, *Goldblatt & Manning 9844* (holotype, NBG; isotypes, E, K, MO, PRE, WAG).

serpenticola = growing on serpentine.

Description

Plants 75–150 cm high. *Corm* depressed-globose, 25–50 mm in diameter, the tunics initially coriaceous and unbroken, fragmenting irregularly, becoming coarsely fibrous. *Cataphylls* coriaceous and pale, the uppermost reaching 8–12 cm above the ground and then green or flushed with purple. *Leaves* eight to ten, the lower four or five more or less basal and longest, reaching to about the base of the spike, grey-green and somewhat glaucous, the blades narrowly sword-shaped to linear, 7–10 mm wide, the margins and midrib moderately thickened, the remaining leaves cauline, progressively smaller above, imbricate and sheathing the stem to about the base of the spike. *Stem* erect, usually with one or two branches, occasionally more, 2–2.5 mm in diameter below the spike.

Spike slightly inclined, more or less straight, 18- to 30-flowered, the branches with fewer flowers; *bracts* pale green in bud, dry and light brown above by anthesis, narrowly lanceolate, the apices attenuate, dry and twisted, 12–16 mm long, the inner bracts similar, c. two-thirds as long as the outer. *Flowers* pale pink to nearly white, the lower lateral tepals each with a central yellow streak surrounded by a light purple blotch, the yellow fading to purple with age, unscented; *perianth tube* obliquely funnel-shaped, 10–12 mm long; *tepals* broadly ovate, the dorsal inclined to hooded over the stamens, 23 x 14 mm, the upper laterals directed forward, weakly curving upward in the distal third, 23 x 12 mm, lower three tepals united for c. 2 mm, the lower laterals narrowed below, gradually expanding into the limbs, 13–14 x 5.5–6 mm, the lower median 16–18 x c. 8 mm, arching downward, much exceeding the lower laterals. *Filaments* 14 mm long, exserted c. 8 mm from the tube; *anthers* c. 6 mm long, purple, the pollen cream. *Ovary* ovoid, c. 2.5 mm long; *style* arching over the stamens,

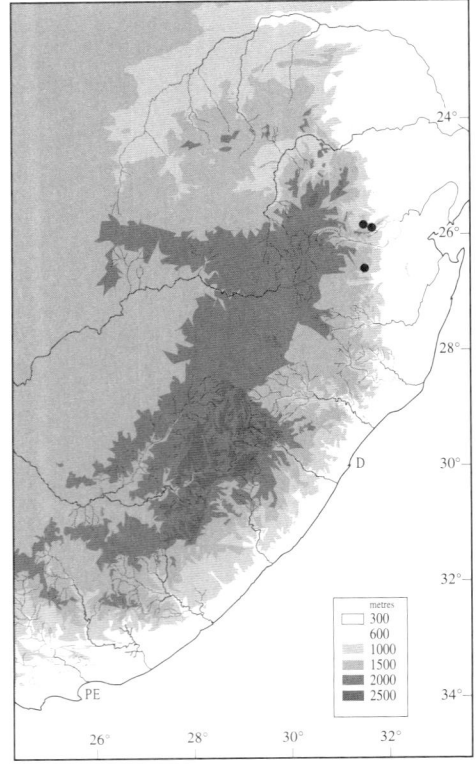

Africa. Plants occur widely across the low-lying Kaap Valley, the floor and lower edges of which have extensive outcrops of verdite, a dark green form of serpentine. Soils derived from this substrate are relatively inhospitable to plants (Morrey et al., 1989) because they have high concentrations of toxic minerals, including unusually high magnesium/calcium ratios. They are often associated with high levels of iron and cobalt, and potentially toxic concentrations of chromium and nickel.

The short-tubed flowers are evidently adapted for pollination by long-tongued bees, but the pollination biology of *Gladiolus serpenticola* has not been studied.

Diagnosis and relationships

The distinguishing features of *Gladiolus serpenticola* are the tall stature, 75–150 cm high, the narrow glaucous leaves which are 7–10 mm wide, and the small, dry, brittle floral bracts which are only 12–16 mm long. The relatively small, pale pink flowers with pale mauve markings on the lower lateral tepals and the short perianth tubes, 10–12 mm long, are unremarkable in the section and closely resemble those of related species such as *G. densiflorus* and *G. crassifolius*. Typical of series *Densiflorus*, *G. serpenticola* has short, ovoid to oblong capsules with broadly trilobed apices and small seeds with narrow, poorly developed wings. The capsules of *G. serpenticola* are particularly small, only 5–8 mm long.

Its relationships appear to lie most closely with a third species of eastern southern Africa, *Gladiolus hollandii*. Also unusually tall and with relatively soft-textured leaves,

G. hollandii has flowers with a perianth tube c. 25 mm long, floral bracts which are longer at 20–30 mm long, and capsules which are larger at 10–20 mm long. It occurs on quartzitic and granitic soils from Pilgrim's Rest to Barberton in Mpumalanga southward to Goba in Mozambique and Hlatikulu in Swaziland.

Until now the few available collections of *Gladiolus serpenticola* have been confused with either *G. crassifolius* or *G. densiflorus*, both short-tubed species with similarly proportioned flowers that broadly resemble those of *G. serpenticola*. *Gladiolus crassifolius* can immediately be distinguished by its coarsely ribbed leaves with thickened and hyaline midrib, secondary veins and margins, a sturdy stem, sharply inclined spike with the flowers usually darkly marked on the lower tepals, and broader floral bracts. *Gladiolus densiflorus* also has distinctive leaves, with the midrib and margins lightly thickened and the remaining veins fine and closely set.

History

Gladiolus serpenticola was first recorded in 1911 by Mabel Stewart who collected specimens near Hlatikulu in Swaziland. Stewart was one of the first people to collect plants in Swaziland and although many of her collections, housed at the Kew Herbarium, are types of eastern southern African species, *G. serpenticola* attracted no attention for many years. The species was described in 1996 based on living plants that we collected near Barberton in 1994 and the illustrations used here are from the same source.

dividing opposite the lower to upper third of the anthers, the branches c. 2 mm long, not reaching the anther apices. *Capsules* ovoid, three-lobed apically, 5–8 mm long; *seeds* ovate, c. 5–6 x 2.5–3 mm, the wing c. 1.5 mm wide, unevenly developed, sometimes lacking on one side or across the middle. *Chromosome number* unknown.

Flowering time early February to late March, rarely in late January.

Distribution and biology

Gladiolus serpenticola is endemic to the Barberton District of Mpumalanga, South

---

6. GLADIOLUS EXIGUUS G. Lewis

PLATE 6

Gladiolus exiguus G. Lewis, *J. S. African Bot.*, Suppl. 10: 58 (1972). Type: South Africa, Transvaal (Mpumalanga), Lydenburg District, Long Tom Pass, 19 km east of Lydenburg, 22 Feb. 1949, *Codd 5173* (PRE, holotype; BM [not seen], K, isotypes).

exiguus = small, slender, describing both the slender plants and the small flowers.

Description

Plants 30–40 cm high. *Corm* depressed-globose, 25–35 mm in diameter, the tunics of coarse reticulate fibres, these accumulating with the decayed remains of the leaf bases to form a thick neck of fibres around the base

of the stem. *Cataphylls* membranous, the uppermost reaching 2–3 cm above the ground and then firm and green. *Leaves* seven to eight, the lower five or six more or less basal and longest, forming a crowded distichous fan, reaching to between the upper third of the stem and the base of the spike, the blades narrowly lanceolate, 9–15 mm wide, the margins moderately thickened, the midrib less so, the upper one or two leaves cauline and largely sheathing, with short blades c. 10 mm long, the margins free to the base, overlapping. *Stem* erect below, flexed outward above the sheath of the penultimate leaf, inclined in the upper third, unbranched,

2.5–3 mm in diameter below the spike.

Spike inclined 30°–50°, 12- to 20-flowered; *bracts* dull green or flushed with grey-purple, the outer 23–33 mm long, one and a half to two internodes long, lightly folded along the midline, acute and somewhat attenuate, the inner bract slightly more than half as long as the outer, pale green, notched apically for up to 0.5 mm. *Flowers* pale creamy pink to deep pink, without noticeable nectar guides, unscented; *perianth tube* obliquely funnel-shaped, 14–16 mm long; *tepals* lanceolate-elliptic, the dorsal arching over the stamens and nearly horizontal, 20–22 x 14 mm, the upper lateral and lower median c. 22 x 12 mm, the lower laterals smallest, 18 x 10 mm. *Filaments* c. 15 mm long, exserted c. 7 mm from the tube; *anthers* c. 6 mm long, brownish purple, the pollen cream. *Ovary* ovoid, c. 4 mm long; *style* arching over the stamens, dividing opposite the middle of the anthers, the branches c. 3 mm long, curving outward between the anther apices. *Capsules* broadly ovoid, (9–)12–14 mm long, rounded apically; *seeds* ovate to triangular, 4–6 mm long, unevenly winged, sometimes winged only at one end. *Chromosome number* unknown.

Flowering time late January to March.

Distribution and biology

Evidently a fairly local endemic, *Gladiolus exiguus* is recorded only from the northern Drakensberg escarpment of Mpumalanga, South Africa. It grows at high elevations

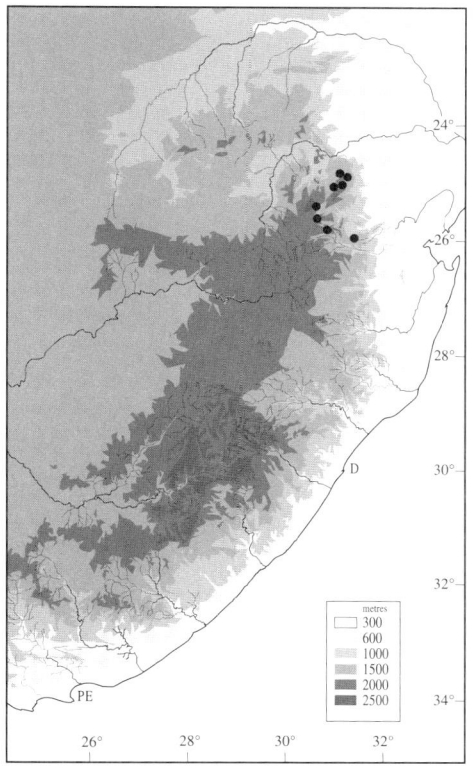

along the upper escarpment, and extends from Mount Sheba near Pilgrim's Rest southward to the heights above Waterval Boven. Plants grow in open grassland in well-drained sites on rocky quartzitic or sandstone slopes. They can be fairly common locally and are conspicuous on the drier, western slopes of Long Tom Pass in January and February.

The moderate-sized, short-tubed flowers are relatively unspecialized and are pollinated by long-tongued bees of the family Anthophoridae. Our observations suggest that *Amegilla fallax* may be the most important pollinator of the species.

Diagnosis and relationships

Difficult to distinguish because of its lack of specialized features, *Gladiolus exiguus* may be recognized by its relatively low stature, seldom more than 40 cm high, spikes inclined 35°–50° to the ground, and pale pinkish salmon flowers with purple-brown anthers. The leaves are relatively short and form a tight fan. The blades are usually slightly twisted and have relatively strongly thickened margins and midrib, but the remaining veins are fine and closely set. This venation pattern is shared with the closely related *G. densiflorus*. The uniformly coloured flowers of *G. exiguus* lack even a suggestion of the minute dark spotting that is characteristic of most populations of *G. densiflorus* which, in addition, has more or less erect spikes. The upper tepals, at 20–22 mm long, and the perianth tube, at 14–16 mm long, are somewhat larger than normally found in *G. densiflorus*, which has upper tepals mostly 15–16 mm, rarely to 20 mm long, and a perianth tube 12–14 mm long.

History

Gladiolus exiguus was first collected in 1886 by the German apothecary and naturalist,

Friedrich Wilms, whose botanical exploration in the Lydenburg District, where he lived in the 1880s, yielded an imposing number of species new to science. This early gathering attracted no scientific interest, as his collections were confused with other species of eastern South Africa such as *G. crassifolius* and *G. densiflorus*. *Gladiolus exiguus* was only recognized as distinct from these species by G.J. Lewis in the course of research for her revision of *Gladiolus* in South Africa, and the species was described in her posthumously published monograph which was completed by A.A. Obermeyer and published in 1972.

7. GLADIOLUS DENSIFLORUS Baker

Gladiolus densiflorus Baker, *Vjschr. Naturforsch. Ges. Zürich*, 49: 178 (1904). Lewis et al., *J. S. African Bot.*, Suppl. 10: 63 (1972). Type: South Africa, Northern Province, Letaba District, Shilouvane, Apr. c. 1900, *Junod 1204* (Z, possible holotype; G, K, PRE, isotypes).

densiflorus = densely flowered, referring to the spikes of numerous and closely spaced flowers characteristic of the species.

Synonymy

Gladiolus invenustus G. Lewis, *J. S. African Bot.*, Suppl. 10: 65 (1972). Type: Swaziland, Little Usutu River at Evelyn Baring Bridge, 13 Dec. 1960, *Compton 30382* (NBG, holotype; PRE, isotypes).

Description

Plants (40–)65–120 cm high. *Corm* depressed-globose, 25–40 mm in diameter, the tunics coriaceous, fragmenting irregularly with age and becoming somewhat fibrous.

Cataphylls pale and more or less coriaceous, the uppermost reaching 3–8 cm above the ground and then green or flushed with red. *Leaves* eight or nine, the lower four or five more or less basal, forming a distichous fan, reaching to about the middle of the stem, sometimes to the base of the spike, the blades lanceolate, sometimes narrowly so, (5–)10–24 mm wide, the margins and midrib thickened and raised, remaining veins poorly developed and closely set, the cauline leaves progressively smaller above, the upper

two usually entirely sheathing, the margins free to the base and overlapping. *Stem* erect below, flexed outward and inclined 20°–30° above the sheaths of the upper one or two leaves, usually unbranched, occasionally with one branch, 2–3.3 mm in diameter below the spike.

Spike inclined 20°–30°, lightly flexuose, 12- to 25(–45)-flowered; *bracts* greenish or purple, the outer 15–20(–28) mm long, acute to attenuate, one to two internodes long, the inner about two-thirds as long as the outer, acute or minutely forked apically. *Flowers* cream, greenish, pink, mauve, slate-grey or occasionally orange, usually minutely speckled with pink to purple spots on the base colour, either without noticeable nectar guides, or the lower three tepals each with a poorly defined pale yellow zone along the midline, unscented; *perianth tube* obliquely funnel-shaped, 12–14 mm long; *tepals* unequal, narrowly to broadly ovate, the dorsal 15–18(–20) x 11–13 mm, ascending and arching over the stamens, the upper

laterals 15–20 x 10–12 mm, directed forward and curving weakly outward in the upper third, the lower three tepals joined to one another for up to 1–1.5 mm, 13–17 x 6–9 mm, weakly clawed in the lower 3 mm, channelled and horizontal below, flexed downward distally. *Filaments* 10–12 mm long, exserted 5–6 mm from the tube; *anthers* 4.5–6 mm long, dark purple to blackish, the pollen white to cream. *Ovary* ovoid, c. 3 mm long; *style* arching over the stamens, dividing opposite the upper halves of the anthers, the branches 2–3 mm long. *Capsules* obovoid, c. 10 mm long, three-lobed above, sometimes retuse; *seeds* ovate to triangular, 4–6 x 3–4 mm, poorly and unevenly winged, sometimes narrowly winged at one end only. *Chromosome number* unknown.

Flowering time December to March over most of its range, but at almost any time of the year on the northern KwaZulu-Natal coastal plain.

Distribution and biology

Gladiolus densiflorus is a widespread species of the lowveld and coast of eastern southern Africa. It extends from Tate Vondo in Northern Province, South Africa, through the Mpumalanga lowveld and Swaziland to coastal southern Mozambique and KwaZulu-Natal as far south as Durban. Although mainly found below 300 m altitude, plants have been recorded near Sabie and Graskop at elevations of up to 1200 m, and at 2000 m in the mountains south of Barberton. In the Sabie and Graskop area plants are usually quite short, possibly a result of the poor sandstone-derived soils there. Plants from similar elevations in Swaziland are as tall as those found along the coast. The species grows in open grassland, usually on deep soils in areas of relatively high rainfall.

The relatively small, short-tubed flowers are adapted for pollination by long-tongued bees, the most common visitor being the anthophorid *Amegilla fallax*, one of the most important pollinators of smaller-flowered *Gladiolus* species in eastern southern Africa.

Diagnosis and relationships

Gladiolus densiflorus is relatively easy to recognize by the combination of a tall stem, a dense spike of 25–45 flowers, occasionally fewer, and a fan of several basal leaves, with only moderately thickened margins and midrib and the remaining veins very fine and closely set. The flowers are fairly small, usually 27–35 mm long, with the perianth

tube 12–14 mm long, and they are usually minutely speckled with dark purple or red on a cream, pink or mauve background. The lower tepals have obscure pale yellow markings or apparently lack markings entirely. Although the specific epithet suggests a very dense arrangement of the flowers on the spike, this is not always so, and although the bracts are usually imbricate and one and a half to two internodes long, they may only be about one internode long. Nevertheless, the many flowers per spike usually give the appearance of being fairly crowded because of their small size.

The unusual and presumably derived venation pattern of only a prominent midrib and numerous fine, closely set secondary veins so characteristic of *Gladiolus densiflorus* is also found in *G. exiguus*, *G. ferrugineus* and a few more species. It is to this alliance within series *Densiflorus* that *G. densiflorus* evidently belongs. It is, however, more often confused with the less closely related *G. crassifolius*. Leaf venation in the latter is quite different, as are the flowers which lack minute dark speckling and usually have well-marked nectar guides of contrasting dark and pale pigment.

Variation

Gladiolus densiflorus exhibits an extraordinary range of variation in plant size and in flower colour. Plants from the Sabie–Graskop area of the eastern Drakensberg escarpment are usually short, rarely more

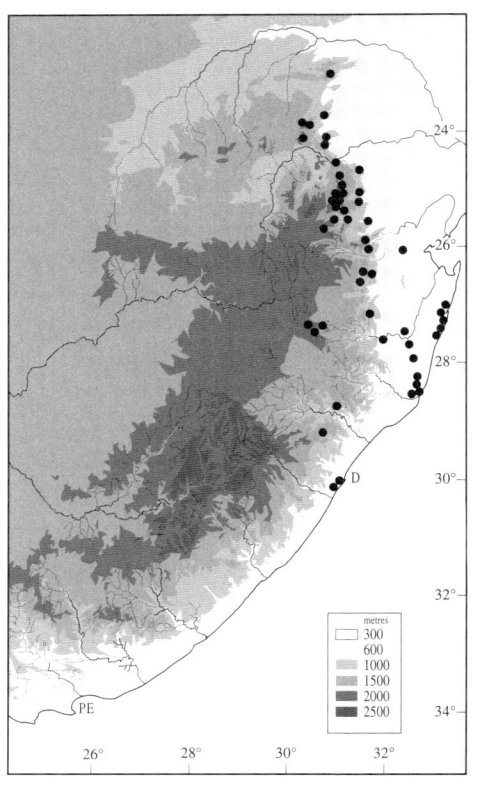

than 45 cm high, and the stems are correspondingly slender. Plants elsewhere are usually taller, sometimes up to 120 cm high, and they may have leaves 20–24 mm wide. The range of variation in populations to the north makes it impossible to assign the robust and the short forms to different taxa. Possibly soil or other ecological conditions are involved in the differentiation of the dwarfed plants. KwaZulu-Natal coastal plants are always robust and are surprisingly variable in flower colour, although not significantly in vegetative features. Plants from the Kosi Bay–Maputa area of KwaZulu-Natal have orange flowers and some from Wakkerstroom in Mpumalanga have reddish flowers. The most common colouring is light pinkish mauve with purple speckles. Sometimes, as at MacMac near Sabie, plants range from reddish to mauve to cream with darker speckles within the same population.

The plants included in *Gladiolus invenustus* by Lewis et al. (1972) form an integral part of the overall variation of *G. densiflorus*. These authors distinguished *G. invenustus* by its smaller, pale pink or whitish flowers without dark speckles and its rather more lax spike. Additional plants collected since the distinction was made render it meaningless. We see literally no difference between plants from Northern Province and Mpumalanga that match the type of *G. densiflorus* and many from Swaziland and KwaZulu-Natal, where *G. invenustus* was said to occur. As noted above, flower colour is particularly variable in KwaZulu-Natal plants, which can have either pale- or dark-coloured flowers.

History

Gladiolus densiflorus was first recorded in 1855 from the coast of KwaZulu-Natal by the journalist, trader and sometime plant collector for Kew Gardens in England, John Sutherland, but this rather fragmentary collection was not perceived to represent a novelty. A later collection made in about 1900 by the missionary Henri-Alexandra Junod near Shilouvane in Northern Province did, however, attract botanical attention and *G. densiflorus* was described by J.G. Baker in 1904 based on Junod's specimen alone. The lowveld of Mpumalanga and northern KwaZulu-Natal was not properly explored botanically until rather late and it was only after the 1930s that the wide distribution of *G. densiflorus* began to be understood. Plants from Swaziland were considered by Lewis et al. (1972) to represent a second species, *G. invenustus*, but as we explain above, there seems to be no significant difference between this and typical *G. densiflorus* from further north.

8. GLADIOLUS FERRUGINEUS Goldblatt & Manning

PLATE 7

Gladiolus ferrugineus Goldblatt & Manning, new name for *Gladiolus micranthus* Baker, *Bull. Herb. Boissier*, Sér. 2, 4: 1005 (1904), an illegitimate homonym, not *G. micranthus* Stapf, 1885 (a European species) nor *G. micranthus* Baker, 1892 (identity uncertain). *Gladiolus varius* var. *micranthus* (Baker) Obermeyer in *J. S. African Bot.*, Suppl. 10: 34 (1972). Type: South Africa, Northern Province, Sanatorium, Apr.–May 1901, *Junod 1716* (G, lectotype; G, Z, isolectotypes).

ferrugineus = rust-brown, referring to the distinctive colour of the upper parts of the floral bracts.

Description

Plants 35–60(–80) cm high. *Corm* globose, 15–20 mm in diameter, the tunics coriaceous, decaying into medium-textured, mostly vertical fibres with age. *Cataphylls* membranous, the uppermost and sometimes the lower rust-coloured when extended above the ground, reaching 5–8 cm above the ground. *Leaves* six to eight, the lower three or four more or less basal and largest, reaching to between the base and apex of the spike, the blades narrowly lanceolate to linear, (3–)5–8(–15) mm wide, twisted several times, fibrotic and tough in texture, the margins and midrib thickened and raised, sometimes heavily so, remaining veins poorly developed, remaining leaves cauline and progressively smaller above, the two upper largely sheathing, the margins free to the base. *Stem* erect below, flexed outward above the sheaths of the cauline leaves and lightly inclined, unbranched, 2–3 mm in diameter below the spike.

Spike inclined, 8- to 15(–27)-flowered; *bracts* greenish in bud, by anthesis becoming dry and rust-coloured in the upper third, rapidly becoming entirely dry, the outer 20–30 mm long, attenuate-apiculate, the inner about two-thirds as long as the outer, yellowish green, forked apically for c. 2 mm. *Flowers* white, cream to pearl-grey or pale pink, the tepals becoming flushed with pink or blue on fading, the lower lateral tepals with faint cream to pale yellow blotches in the midline, the markings sometimes darkening to purple with age, unscented; *perianth tube* obliquely funnel-shaped, 12–16(–25) mm long; *tepals* unequal, ovate, the dorsal arching over the stamens, nearly horizontal, 18–25(–28) x 12–15 mm, broadest in the lower third, the upper laterals directed forward and barely curving outward toward the apices, 18–30 x 15 mm, the lower three tepals united basally for 1.5–6 mm, the lower laterals 12–20 x c. 9 mm, narrowed below for c. 1.5 mm into claws, the limbs abruptly widened and channelled below, the margins curving upward, the limbs flat distally, the lower median tepal 15–25 x c. 11 mm. *Filaments* 12–15 mm long, exserted 7–8 mm from

the tube; *anthers* 6–8 mm long, lilac to purple, the pollen yellow. *Ovary* narrowly obovoid, c. 5 mm long; *style* arching over the stamens, dividing between the middle and apex of the anthers, the branches

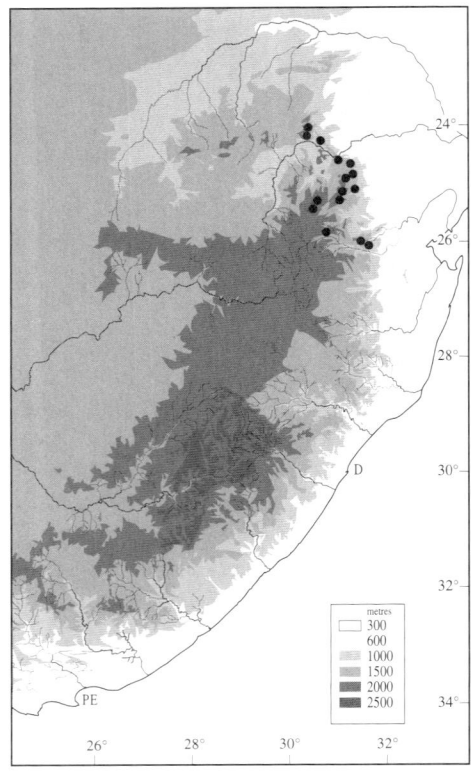

often on peaty sandy ground derived from quartzitic rocks.

The short-tubed and pale-coloured flowers are adapted for pollination by large, long-tongued bees of the family Anthophoridae, including *Amegilla aspergina* and *A. capensis*, which visit the flowers in search of nectar.

Diagnosis and relationships

Long confused with the related, but larger-flowered *Gladiolus varius*, *G. ferrugineus* has relatively small, short-tubed white to pale pink flowers, usually 30–40 mm long. In most populations the flowers have tubes 13–16 mm long, but occasionally, as in plants from the higher parts of the main Mpumalanga escarpment, flowers are some-what larger and have tubes up to 20 mm long. Confusion with the larger-flowered *G. varius* most probably resulted from the rust-coloured floral bracts, a conspicuous feature of both species. Flowers of typical *G. varius* are 55–70 mm long with a tube 30–50 mm long and are always pink with the lower tepals marked with reddish nectar guides. Confusion may arise over hybrids between the two which we have seen on the upper slopes of Long Tom Pass. These plants have flowers 45–50 mm long and pale pink with dark longitudinal streaks on the lower tepals. These hybrids are the result of occa-sional visits by bees to the large-flowered *G. varius* where they fail to find nectar, but nevertheless carry pollen to other species they normally visit, including *G. ferrugineus*.

There are two distinct forms of *Gladiolus ferrugineus*, one from the lower escarpment which flowers in summer, from late November to early February, and another from the upper escarpment which flowers in autumn, in March and April. The early-flowering plants have slightly broader leaves, 6–15 mm wide, and flowers with a short perianth tube, mostly 12–16 mm long. The

later-flowering plants have narrower leaves, 3–6 mm wide, with heavily thickened mar-gins and twisted blades, and the flowers have tubes 18–25 mm long. We suspect that these two series of populations are ecological races, adapted to lower elevations and less exposed sites – hence the softer leaf texture and earlier flowering – compared with the higher elevation plants from sites exposed to greater climatic stress – hence the tougher leaf texture and later flowering. The longer tubes in the latter may be an adaptation to the larger, longer-tongued bees that are active later in the season. The presence of plants intermediate in floral and leaf features prevents us from recognizing more than one taxon.

History

Gladiolus ferrugineus was one of the first of the species of *Gladiolus* from northern South Africa to be discovered. It was first recorded by the German apothecary and naturalist, Friedrich Wilms, who lived in Lydenburg for some 13 years, making valuable collections across this botanically diverse and, at the time, unexplored part of southern Africa. Wilms's collection from the Devil's Knuckles was made in 1889 and was followed by one from Barberton made by Ernest Galpin in 1890, and another by Rudolf Schlechter in 1893. Although all of these collections were available to J.G. Baker at Kew Gardens, only a later collection made by the French missionary Henri-Alexandra Junod in 1901 formed the basis for *G. micranthus* (Baker, 1904). *Gladiolus micranthus* was reduced to the rank of variety in *G. varius* by Lewis et al. (1972), a taxonomic decision which we do not accept. The name *G. micranthus* is a homonym and illegitimate at species rank, and is replaced here by the new name, *G. ferrugineus*.

3–4 mm long. *Capsules* obovoid, three-lobed and retuse above, 12–15 mm long; *seeds* oval, c. 3–4 x 2.5 mm, narrowly and evenly or unevenly winged, pale and nearly transparent, the seed body light brown. *Chromosome number* unknown.

Flowering time mostly January to March, occasionally in late November and December or in April.

Distribution and biology

Occurring along the escarpment of eastern southern Africa, *Gladiolus ferrugineus* has a fairly wide range, from Pigg's Peak in northern Swaziland and the Saddleback Mountains near Barberton, South Africa, in the south to the Wolkberg and Haenerts-burg in the north. Plants occur in marshy grassland or on well-watered stony slopes,

9. GLADIOLUS LITHICOLA Goldblatt & Manning

PLATE 8

Gladiolus lithicola Goldblatt & Manning, new species. Type: South Africa, Mpuma-langa, God's Window, Fanie Botha Hiking Trail, 3 Apr. 1996, *Lötter & Krynauw 86* (NBG, holotype).

lithicola = growing on rocks, referring to the habitat.

Latin diagnosis

Plantae 45–80 cm altae, cormo globoso

20–35 mm in diametro, foliis 6–10 lanceo-latis vel ensiformibus (12–)18–35 mm latis, spica 9–16 florum, floribus pallide malvinis vel carneis, tubo perianthii 30–40 mm longo, tepalis inaequalibus, superioribus 30–37 x 18–20 mm, dorsali inclinato, inferioribus 28–33 x 15–16 mm, filamentis 16–24 mm longis, antheris c. 9 mm longis.

Description

Plants to 45–80 cm high. *Corm* depressed-

globose, 20–35 mm in diameter, the tunics of firm, papery layers, becoming irregularly fragmented with age and partially fibrous. *Cataphylls* membranous and rust-coloured, the uppermost reaching up to 15 cm above the ground and then firm, dry and dark brown. *Leaves* six to ten, the lower four to eight basal and longest, forming a distichous fan, reaching to between the base and middle of the spike, fairly thin but firm-textured, the blades drooping if the stem inclined, lanceolate to narrowly sword-shaped, (12–)18–35 mm wide, the margins and midrib barely thickened, the midrib asymmetrically placed, the uppermost leaf inserted on about the middle of the stem, sheathing the upper part of the stem usually to about the base of the spike, without a blade. *Stem* erect or inclined to horizontal below, lightly flexed in the upper third, apparently always unbranched, 2–3 mm in diameter below the spike.

Spike lightly inclined, straight, 9- to 16-flowered; *bracts* initially greenish grey, soft-textured, becoming dry and rust-coloured at the tips or more or less entirely rust-coloured, especially with age, lightly folded in the midline, the outer 38–48 mm long, about two internodes long, the inner two-thirds to three-quarters as long as the outer, minutely forked apically for c. 1 mm. *Flowers* whitish, flushed with pale mauve or mauve to pink, darkening with age, the lower three tepals each marked with a pale mauve, linear to narrowly diamond-shaped, median longitudinal streak, unscented; *perianth tube* narrowly and obliquely funnel-shaped, 30–40 mm long, flared outward in the upper 10 mm; *tepals* unequal, ovate, the margins slightly undulate, the three upper tepals largest, 30–37 x 18–20 mm, the dorsal inclined, the upper laterals spreading outward in the upper halves, the lower tepals joined to the upper laterals for c. 4 mm and to one another for c. 2 mm, the lower lateral tepals c. 28 x c. 15 mm,

spreading in the upper halves, the lower median c. 33 x 16 mm long. *Filaments* 16–24 mm long, exserted 10–15 mm from the tube; *anthers* c. 9 mm long, yellow, the pollen cream. *Ovary* oblong, c. 6 mm long; *style* arching over the stamens, dividing 1–2 mm below the anther apices, the branches 5–6 mm long. *Capsules* globose, 8–10 mm long, three-lobed above; *seeds* oblong, 4–6 x c. 3 mm, unevenly and weakly winged, pale translucent brown. *Chromosome number* $2n = 30$.

Flowering time mid-March to late April, probably until early May.

Distribution and biology

Gladiolus lithicola is a rare endemic of the southern African summer-rainfall region. It is known only from between Graskop and Mariepskop along the lower Drakensberg escarpment of Mpumalanga. The entire recorded range of the species extends for about 40 km in a north–south line along the escarpment edge. Plants grow in shady places on sandstone rocks and cliffs of black reef quartzite, either on shallow soil usually with leaf litter, in peaty sand on rocky pavement, or in rock cracks on steep cliffs.

The long-tubed, pale pinkish mauve flowers appear to be adapted for pollination by long-tongued flies, most probably of the family Nemestrinidae.

Diagnosis and relationships

Apart from its unusual habitat, *Gladiolus lithicola* is distinguished by its relatively large, pale pink to mauve flowers, the lower tepals with weakly defined pale mauve, linear nectar guides and the floral bracts purplish when young, becoming dry and rust-coloured after anthesis. The flowers, 60–75 mm long, have an obliquely funnel-shaped perianth tube 30–40 mm long, thus about as long as the upper tepals. Plants have relatively broad leaves of unusually soft texture, a consequence of its moist habitat.

Unlike its close allies, *Gladiolus ferrugineus* and *G. varius* which also have rust-coloured floral bracts, *G. lithicola* has leaves that are broad, sometimes up to 35 mm wide, with fairly lightly thickened midrib and margins. The leaf blades lack the pattern of fine, closely set secondary veins that characterize *G. ferrugineus* and *G. varius*. *Gladiolus lithicola* can also be distinguished from *G. ferrugineus* by its longer flowers, those of *G. ferrugineus* usually being whitish and 30–40 mm long with a tube 16–20 mm long. The deep pink flowers of *G. varius* have a slender cylindric perianth tube,

30–50 mm long, and the narrow leaves, rarely more than 4–9 mm wide, have strongly thickened margins and midrib.

History

The first record of *Gladiolus lithicola* appears to have been made in 1964 by Gordon MacNeil, who botanized extensively on the eastern southern African escarpment and also discovered *G. macneilii*. MacNeil's collection, from The Pinnacle near Graskop, was examined by G.J. Lewis when she was completing her study of *Gladiolus*. She realized this might represent a new species and provisionally referred it to the similar, but much misunderstood, *G. varius*. A second collection, made in 1967 by Peter Goldblatt in what is now God's Window Nature Reserve, was likewise tentatively assigned to *G. varius*. Subsequent examination of living plants during field research for this book made it clear that the plant is not a form of *G. varius* but a separate, although allied species. Mervyn Lötter and Sonette Krynauw of the Mpumalanga Department of Nature Conservation have helped in establishing the full range of the species and in obtaining fruiting material so that we now understand it reasonably well.

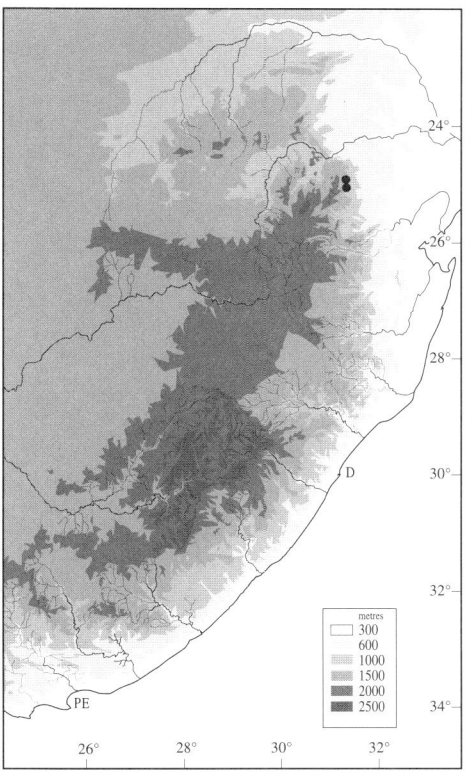

10. GLADIOLUS VARIUS F. Bolus

PLATE 9

Gladiolus varius F. Bolus, *Ann. Bolus Herb.* 2: 104 (1917) in part, excluding var. *elatus.* Lewis et al., *J. S. African Bot.*, Suppl. 10: 33 (1972) in part, excluding var. *micranthus.* Type: South Africa, Mpumalanga, Pilgrim's Rest, Mar. 1915, *Rogers 14601* (BOL, holotype; J, K, isotypes).

varius = variable, for the variability of the species, which when described included three varieties. These are now two different species, *Gladiolus varius* and *G. hollandii.*

Synonymy

Gladiolus varius var. *brevifolius* F. Bolus, *Ann. Bolus Herb.* 2: 104 (1917). Type: South Africa, Mpumalanga, Barberton, Saddleback Mountain, 5000 ft, 22 Feb. 1890, *Galpin 828* (BOL, holotype; K, SAM, Z [not seen], isotypes).

Description

Plants 45–70 cm high. *Corm* depressed-globose, 18–25 mm in diameter, the tunics of cartilaginous layers, fragmenting irregu-larly and becoming coarsely fibrous with age. *Cataphylls* pale and firm-textured, the uppermost reaching 5–7 cm above the ground, brown and more or less dry, often accumulating around the base of the stem in a coarsely fibrous neck. *Leaves* usually seven, the lower four or five more or less basal, forming a lax distichous fan, reaching to about the base of the spike, the blades narrowly lanceolate to nearly lin-ear, (2.5–)4–9 mm wide, the margins and midrib moderately, or sometimes strongly thickened, the remaining leaves cauline and shorter than the basal, the uppermost usually entirely sheathing, the margins free to the base and overlapping. *Stem* erect, flexed outward above the sheath of the penultimate leaf, inclined above, unbranched, c. 3.3 mm in diameter below the spike.

Spike lightly inclined, occasionally 6-, usually 8- to 12-flowered; *bracts* pale greenish brown below, more or less dry and rust-coloured in the upper half by anthesis, the outer (25–)30–40(–70) mm long, overlapping and one and a half or two internodes long, the inner similar to the outer, slightly to c. 5 mm shorter. *Flowers* deep pink, occasionally pale pink to nearly white, the lower three tepals usually each with a dark pink to purple median streak and pale pink in the lower half, unscented; *perianth tube* narrowly and obliquely funnel-shaped, (27–)37–50 mm long, the cylindrical lower part (20–)27–38 mm long, the wider upper part c. 10 mm long; *tepals* lanceolate, unequal, the dorsal largest, horizontal and hooded over the stamens, 26–32 x 17–20 mm (often apparently smaller in preserved specimens), the upper laterals directed forward and curving outward in the upper half, 24–32 x c. 12 mm, the lower three tepals united basally for 4–6 mm, the lower lateral tepals 20–24 x 12 mm, narrowed below into claws c. 2 mm long, the limbs gradually expanded, the lower median tepal 23–26 x 12 mm. *Filaments* 12–16 mm long, exserted 5–8 mm from the tube; *anthers* 8–9 mm long, lilac, the pollen cream. *Ovary* narrowly obovoid, c. 5.5 mm long; *style* arching over the stamens, dividing at or up to 5 mm beyond the anther apices (occasionally dividing opposite the upper third of the anthers), the branches c. 6 mm long. *Capsules* oblong, 12–15 mm long, three-lobed and retuse apically; *seeds* discoid to angular, 2.5–3.5 x c. 2.5 mm, the wing poorly and irregularly developed or vestigial, translucent light brown, the seed body dark brown. *Chromosome number* unknown.

Flowering time February to March, rarely in late January.

Distribution and biology

An endemic of Mpumalanga, South Africa, *Gladiolus varius* is a high mountain species with a fairly restricted range. It extends from the mountains above Barberton in the south to Mount Sheba near Pilgrim's Rest in the north. The distribution is somewhat discon-tinuous as the species occurs only at fairly high elevations along the mountainous eastern escarpment. Plants grow in exposed rocky quartzite on ridges and plateaus, almost always with their corms tightly wedged in rock crevices. This very attractive species is not known in cultivation but we suspect that it will prove a valuable addition to gardens. It is probably fairly hardy and may be left in the ground in frost-prone areas.

The long-tubed flowers are adapted for pollination by long-tongued flies and we have seen and captured the nemestrinid fly *Prosoeca ganglbaueri* foraging for nectar on the flowers. The insects insert their long tongues into the lower part of the tube to reach the nectar, fairly large quantities of which are produced. In its general form,

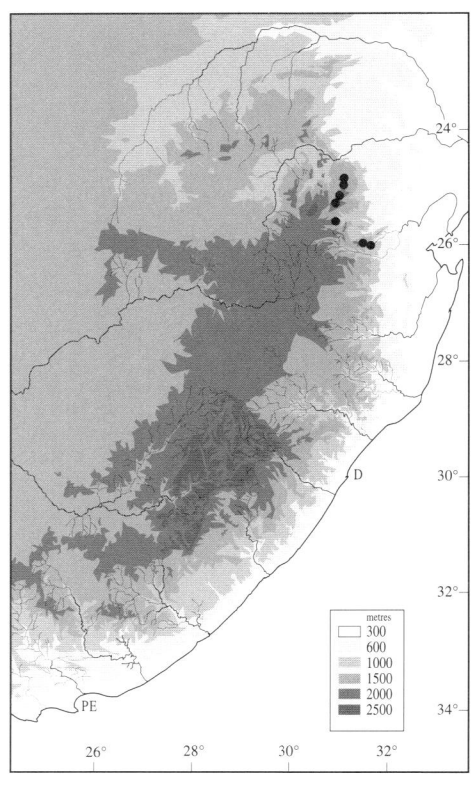

colour and tepal markings, *Gladiolus varius* conforms closely to flowers of other, only distantly related species of *Gladiolus* pollinated by *P. ganglbaueri*, the similarity due no doubt to convergence.

Diagnosis and relationships

Gladiolus varius can be distinguished by the combination of large pink flowers with a perianth tube usually 38–50 mm long, lower tepals each marked with a deep pink streak in the midline and rust-coloured floral bracts. The flowers are typically 65–80 mm long with tubes exceeding 38 mm, but shorter-tubed plants occur in the Barberton mountains with flowers 55–60 mm long and tubes sometimes as short as 27 mm. These plants also have nearly white flowers in contrast to the usual deep pink in plants from elsewhere across its range. They may be hybrids, perhaps with the closely related *G. ferrugineus*.

Bract length correlates closely with tube length, thus shorter-tubed plants also have shorter bracts. The short-tubed plants, sometimes confused with *Gladiolus ferrugineus* (*G. varius* var. *micranthus* of Lewis et al., 1972), are not that species, which has whitish flowers without dark markings on the tepals and a perianth tube 16–20 mm long. *Gladiolus ferrugineus* grows in wet ground, often in somewhat marshy grassland whereas *G. varius* is always found in rocky habitats.

History

Gladiolus varius appears to have first been collected at the southern end of its range near Barberton in 1890 by the amateur but accomplished collector Ernest Galpin. His collection is the type of *G. varius* var. *brevifolius*. The type of var. *varius* was collected by the prolific collector Archdeacon F.A. Rogers in 1915 near Pilgrim's Rest. Both *G. varius* and var. *brevifolius* were described by Frank Bolus in 1917, together with a second variety, var. *elatus*. The latter is *G. hollandii* which Bolus included in *G. varius* because of its long perianth tube. In their revision of *Gladiolus* in South Africa, Lewis et al. (1972) admitted two varieties in *G. varius*, the typical long-tubed form with large flowers and var. *micranthus* with smaller, short-tubed flowers. We treat the latter as a separate species under the new name, *G. ferrugineus*.

Section *Densiflorus*: Series *Calcaratus*

11. GLADIOLUS APPENDICULATUS G. Lewis

PLATE 10

Gladiolus appendiculatus G. Lewis, *J. S. African Bot.*, Suppl. 10: 69 (1972). Type: South Africa, Mpumalanga, Wakkerstroom District, Farm Oshoek, 15 Feb. 1961, *Devenish 582* (PRE, holotype; K, NBG, isotype).

appendiculatus = with small appendages, referring to the long sterile tails or spur-like basal extensions of the anthers.

Synonymy

Gladiolus appendiculatus var. *longifolius* G. Lewis, *J. S. African Bot.*, Suppl. 10: 69 (1972). Type: Swaziland, Hlatikulu, May 1906, *Stewart s.n.* (BOL, 10070, holotype).

Description

Plants 35–60(–85) cm high. *Corm* globose, 15–20 mm in diameter, the tunics of coriaceous to firm papery layers, becoming somewhat fibrous or irregularly broken with age. *Cataphylls* pale and coriaceous, sometimes dry, the uppermost reaching up to 8 cm above the ground, green or dry and brown. *Leaves* six to nine, the lower four or five basal and longest, either narrow and reaching to at least the base of the spike and often shortly exceeding it or broad and reaching only to the middle of the stem, the blades lanceolate or narrowly sword-shaped to linear, (2.5–)4–12(–20) mm wide, sometimes the margins curving toward one another, the surface thus channelled, the midrib usually lightly thickened, margins lightly to fairly heavily thickened, the intercostal surface usually smooth, occasionally shortly hairy, the upper three or four leaves progressively smaller than the basal, the uppermost one or two usually entirely sheathing, the margins free to the base or nearly so. *Stem* evidently erect and more or less straight, unbranched, 2–2.5 mm in diameter below the spike.

Spike more or less erect and straight, 6- to 14-, occasionally to 27-flowered; *bracts* firm-textured, green to greyish purple, by anthesis usually dry and brownish above, the outer 18–30 mm long, overlapping and usually one and a half to about two internodes long, the inner about three-quarters to almost as long as the outer. *Flowers* white or pale to deep pink, the lower lateral or all three lower tepals with a small dark purplish spot or sometimes a dark mauve to purple crescent in the upper third, yellowish in the throat, unscented; *perianth tube* obliquely funnel-shaped, c. 15 mm long; *tepals* unequal, ovate, more or less acute, the upper three largest, 15–18 x 10–12 mm, the dorsal arching over the stamens, strongly channelled, the upper laterals directed forward and curving outward in the upper third, the lower three tepals united for c. 3 mm, 9–12 mm long, channelled and directed forward below, curving downward above. *Filaments* 9–11 mm long, exserted c. 3 mm from the tube; *anthers* 8–9 mm long including the sterile tails

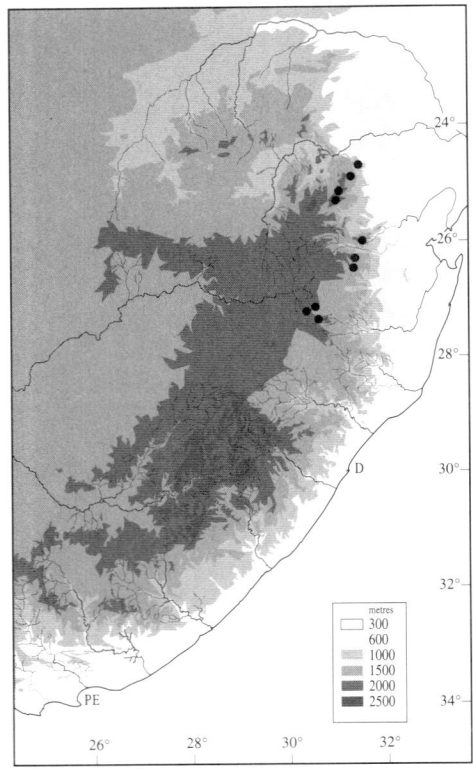

3–4 mm long, appressed to the dorsal tepal, pale yellow fading to mauve on the upper surfaces, the spurs white, held at right angles to the anther thecae, the pollen whitish. *Ovary* ovoid, 4–5 mm long; *style* arching over the stamens, dividing opposite the upper third of the anthers, the branches c. 3 mm long. *Capsules* obovoid, 10–12 mm long, three-lobed above; *seeds* ovate to pear-shaped, 4–5 x 3 mm, narrowly winged, light brown.

Flowering time mainly April and May, occasionally March.

Distribution and biology

Gladiolus appendiculatus has a relatively wide range along the eastern escarpment of southern Africa. It was originally known only from Swaziland and southern Mpumalanga, South Africa, near Barberton and at Wakkerstroom, but recent plant collecting has extended the range well to the north. It is now known from Vryheid in northern KwaZulu-Natal across Mpumalanga and Swaziland to Pilgrim's Rest and Mariepskop in the north. Plants grow in rocky grassland in montane habitats usually above 1800 m, where we have seen it in loamy clay among dolerite rocks.

The short-tubed flowers are adapted for pollination by anthophorid bees. Except for the specialized anthers, the flowers closely resemble in their shape, colour and size those of other eastern southern African species which are also pollinated by these bees. The distinctive, long anther spurs appear to ensure that visiting insects are dusted with pollen when they visit the flowers. As they push themselves into the flower in search of nectar in the perianth tube, they press against the anther spurs, pushing them backward so that the anthers tip downward and brush pollen onto their upper bodies.

Diagnosis and relationships

Relatively small, whitish or pale pink flowers and a straight spike with brown-tipped, overlapping bracts, each up to two internodes long, make *Gladiolus appendiculatus* easy to recognize at a distance. Its most distinctive characteristic is, however, the tailed anthers, a feature shared in the genus only with *G. calcaratus* and *G. macneilii*. All three species have the lower part of the anther locules extended into sterile appendages or spurs.

Both *Gladiolus calcaratus* and *G. macneilii* have larger and longer-tubed flowers, the tubes normally 25–45 mm long. *Gladiolus calcaratus*, which is apparently most closely allied to *G. appendiculatus*, has white flowers, a perianth tube usually 25–40 mm long and upper tepals 28–30 mm long compared with the whitish to pink flowers, a perianth tube c. 15 mm long and upper tepals 15–18 mm long in *G. appendiculatus*. So unusual are anther spurs in the Iridaceae, let alone in *Gladiolus*, that we assume that *G. appendiculatus*, *G. calcaratus* and *G. macneilii* are immediately related. Because of the differences in their flower sizes, perianth tube lengths and corm tunics they are, however, readily distinguished from one another.

There appear to be two distinct forms of *Gladiolus appendiculatus*. One is fairly tall, with stems more than 45 cm high, has narrow leaves, 2.5–8 mm wide, which reach to the middle of the spike, and spikes of up to 25 flowers. The other is shorter and has broader leaves with blades up to 150 mm long and 8–12 mm wide which do not reach the base of the spike, and spikes of 5–12 flowers.

Slender-leafed plants are known from Swaziland and to the north near Lydenburg in Mpumalanga, while those with broader leaves occur both to the southeast in northern KwaZulu-Natal and the Wakkerstroom District of Mpumalanga, and to the north in the mountains around Barberton and in the Pilgrim's Rest section of the eastern escarpment. Lewis et al. (1972) recognized these as two varieties, calling the slender-leafed form var. *longifolius*. There appear to be no floral differences between the two, and the geographical distinction is a weak one. Collections made since that revision was published have, moreover, blurred the distinction between the two varieties. There are now collections from west of Barberton that have broad leaves up to 20 mm wide but tall spikes of up to 25 flowers, and some specimens from near Lydenburg and further north seem intermediate between the two varieties. These have leaves reaching to the middle of the spike, but only about 5 mm wide and spikes of up to nine flowers. The additional information now available makes it undesirable to formally recognize infraspecific taxa in *G. appendiculatus*.

One collection of *Gladiolus appendiculatus* in particular, *Braun 932* from the Malolotja Nature Reserve in Swaziland, seems out of place. The flowers have bracts, 30–40 mm long, and a fan of several broad leaves that have a short, dense pubescence. This plant is so different from other specimens of the species from the same general locality which all have narrow, glabrous leaves, that we wonder if it is not a hybrid.

History

Gladiolus appendiculatus was first recorded by Mabel Stewart near Hlatikulu in Swaziland in 1906. Her specimens were thought to be *G. crassifolius*, which they resemble closely in external appearance, and no attention was paid to them. The broad-leafed form was first recorded in 1924 by H.W. Edwards near Barberton in Mpumalanga. Despite the unusual aspect of the Barberton plant with its broad leaves, it too did not elicit botanical attention until some considerable time later.

Gladiolus appendiculatus was only described in 1972 after plants had been found on the Farm Oshoek in the Wakkerstroom District by N.J. Devenish, a landowner keenly interested in the local flora. It is noteworthy, however, that the English botanist, N.E. Brown, who maintained an interest in southern African, and especially Transvaal plants, in the 1930s, intended to describe the species. The Edwards collection bears the annotation 'G. *succivious* N.E. Br.'

Gladiolus calcaratus G. Lewis, *J. S. African Bot.*, Suppl. 10: 67 (1972). Type: South Africa, Mpumalanga, Lydenburg District, Long Tom Pass, near highest point, 4 Mar. 1968, *Mauve & Leistner 3218* (PRE, lectotype here designated, the specimen with a corm; K, PRE, isolectotypes).

calcaratus = spurred, referring to the long sterile tails or spur-like extensions borne at the base of the anthers.

Description

Plants 30–45 cm high. *Corm* globose, 12–15 mm in diameter, bearing numerous small cormlets c. 2 mm in diameter around the base, the tunics of coriaceous to nearly woody layers, these soon decaying into fairly coarse vertical fibres. *Cataphylls* pale and membranous, the uppermost reaching up to 3 cm above the ground and then purple or becoming dry and brown. *Leaves* six to eight, the lower three to five basal, reaching to between the middle of the stem and the middle of the spike, the blades plane, narrowly sword-shaped to nearly linear, 4–8 mm wide, the midrib and margins moderately thickened, the upper two or three leaves cauline, decreasing in size above and the blades progressively reduced, completely sheathing the stem, the upper one or two more or less entirely sheathing, the margins of the sheathing part usually free to the base. *Stem* erect, simple, usually slightly flexed above the sheaths of the upper two leaves, 2–3 mm in diameter below the spike.

Spike more or less erect and straight, 6- to 10-flowered; *bracts* firm-textured, dark green, dry and brownish above, the outer 35–45(–55) mm long, overlapping, one and a half to two internodes long, lightly folded on the midline, the inner about two-thirds as long as the outer. *Flowers* white, fading to pale pink or lilac, especially in the midlines of the tepals, yellow in the throat, the lower tepals each with a pale yellowish, median longitudinal streak outlined in pale dull pink, unscented; *perianth tube* obliquely funnel-shaped, (16–)25–40 mm long, the lower cylindrical part (12–)20–32 mm long; *tepals* broadly lanceolate, the upper three largest, 28–33 x 12–14 mm, the dorsal extended horizontally, curving upward near the tip, upper laterals sometimes up to 5 mm longer than the dorsal, directed forward, only curving outward in the upper quarter, the lower three tepals basally united for c. 2 mm, the lower laterals 20–22 x c. 12 mm, the median 23–30 x c. 12 mm, in profile the lower median tepal exceeding the upper tepals. *Filaments* 12–14 mm long, exserted c. 7 mm from the tube; *anthers* 9–12 mm long including sterile tails 3–4 mm long, whitish, the pollen cream. *Ovary* oblong, 4–5 mm long; *style* arching over the stamens, dividing at or 1–2 mm beyond the anther apices, the branches 3–4.5 mm long. *Capsules* obovoid, 10–12 mm long, three-lobed above; *seeds* narrowly ovate to pear-shaped, 4–5 x c. 2 mm, unevenly and weakly winged, sometimes one half without a wing, light translucent brown, the seed body darker brown. *Chromosome number* unknown.

Flowering time mid-February to early April.

Distribution and biology

Gladiolus calcaratus is a fairly narrow endemic of the higher mountains of the eastern escarpment of Mpumalanga, South Africa. It extends from Dullstroom and Kemp's Heights in the south to Mount Sheba, near Robber's Pass above Pilgrim's Rest in the north. A single collection from Loskop Dam some 70 km to the west is unexpected and makes it seem likely that additional collecting will extend the range of the species. Not surprisingly, most collections are from Long

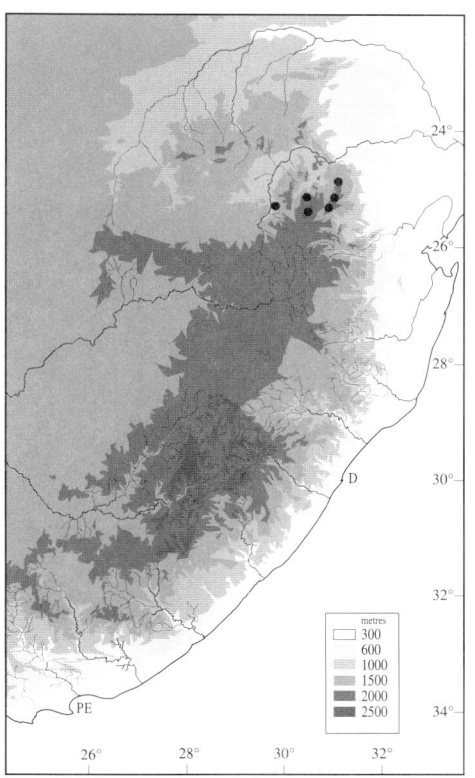

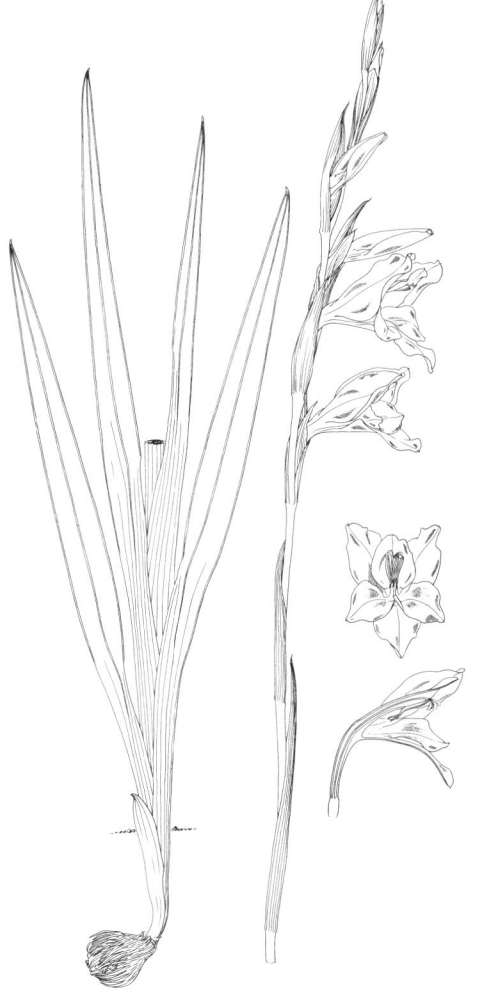

Tom Pass, the most accessible part of the escarpment. Plants appear to favour deeper soils and wetter sites and sometimes grow around the edges of damp depressions. They grow as isolated individuals in grassland and never in dense colonies. Hence, one has the impression that the species is rare but this is probably not the case.

The relatively large flowers with a tube usually 25–40 mm long are evidently most effectively pollinated by the long-tongued fly, *Prosoeca robusta* (Nemestrinidae). Of the insect visitors we have seen visiting the flowers, only this fly has a tongue long enough to reach the nectar in the lower part of the perianth tube and at the same time brush against the anthers and styles. We have also seen the longer-tongued *P. ganglbaueri* visiting the flowers, but the individuals we observed had mouthparts so long that their bodies remained outside the flowers when foraging and thus did not contact either the stigmas or anthers. Other insect visitors include large anthophorid bees, but these insects have tongues too short to reach the nectar. Nevertheless, while searching for nectar they press against the sterile anther tails, causing the anthers to tilt downward and brush against the bee's thorax. Can the role of this remarkable adaptation be a fail-safe mechanism ensuring pollination by insects too small to contact the anthers when they search for nectar? The anther tails are hardly necessary in the case of *P. robusta*, evidently the primary pollinator, as its body is so large that it completely fills the upper part of the flowers when foraging for nectar.

Diagnosis and relationships

One of three species with peculiar anthers that have long sterile tails, *Gladiolus calcaratus* has long-tubed white flowers borne on straight, erect spikes. The flowers are 45–70 mm long with a tube usually at least

25 mm and up to 40 mm long. The lower tepals have pale yellow nectar guides in the lower half, sometimes outlined in pale mauve. The very specialized anthers are 9–12 mm long with prominent tails 3–4 mm long. *Gladiolus calcaratus* and the two other species with tailed anthers, *G. appendiculatus* and *G. macneilii*, are believed to form a clade with *G. calcaratus*, the central species. Both *G. calcaratus* and *G. macneilii* have an elongate perianth tube, and may be immediately related. *Gladiolus macneilii* has flowers with a tube 40–45 mm long and cream to pale salmon tepals, the lower three each with a narrow red median streak and anther tails 1.5–2.5 mm long, somewhat shorter than in *G. calcaratus*. The smaller flowers of *G. appendiculatus* have a

perianth tube c. 15 mm long, making the species appear ancestral to its two long-tubed relatives of series *Calcaratus*.

History

The first record of *Gladiolus calcaratus* was made in 1933 by the energetic amateur botanist and collector of southern African flora, Ernest Galpin. Galpin collected plants on the slopes of Mount Anderson, near Long Tom Pass in Mpumalanga. His well-preserved specimens remained unnamed until 1972, by which time several more collections of the species had accumulated, making it clear that the odd anthers of the Galpin collection were a consistent and highly distinctive feature of the species.

13. GLADIOLUS MACNEILII Obermeyer
PLATE 12

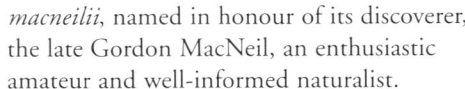

Gladiolus macneilii Obermeyer in *J. S. African Bot.*, Suppl. 10: 84 (1972). Type: South Africa, Mpumalanga, Pilgrim's Rest District, Abel Erasmus Pass, dolomite in red loam, 17 Apr. 1969, *Mauve & MacNeil 4788* (PRE, holotype; BOL, K, NBG, isotypes).

macneilii, named in honour of its discoverer, the late Gordon MacNeil, an enthusiastic amateur and well-informed naturalist.

Description

Plants 70–90 cm high. *Corm* ovoid, 16–22 mm in diameter, the tunics of light brown cartilaginous layers, these soon decaying into pale, medium-textured vertical fibres. *Cataphylls* pale and membranous, the uppermost reaching 3–5 cm above the ground and then usually dry and brown. *Leaves* eight or nine, the lower four basal and longest, reaching to between the base and middle of the spike, the blades plane, mostly 2.5–4 mm wide, the midrib moderately thickened and raised, the margins lightly thickened, the upper leaves cauline and largely sheathing, the blades progressively decreasing in length and width above. *Stem* erect, flexed outward above the sheaths of the upper two or three leaves, simple or with a few short branches, c. 2 mm in diameter below the spike.

Spike usually lightly inclined, lightly flexuose, 7- to 17-flowered; *bracts* narrowly lanceolate, pale green in bud, becoming membranous to dry and light pinkish brown above, dry at the end of flowering, the apices attenuate and dark brown, the outer (25–)30–40(–50) mm long, somewhat less to somewhat more than two internodes long, the inner about half as long to nearly as long as the outer. *Flowers* cream to pale salmon-pink, the lower three tepals each with a narrow, dark red, longitudinal median streak,

unscented; *perianth tube* 40–45 mm long, cylindric below, curved and flared outward in the upper c. 8 mm; *tepals* lanceolate and attenuate, the upper three largest, the dorsal c. 30 x 15 mm, arching over the stamens, the upper laterals directed forward and curving outward lightly in the upper third, c. 27 x 13 mm, the lower three tepals united basally for 4–5 mm, the apices narrow and recurving, the lower laterals c. 18 x 4 mm, the lower median c. 25 x 7 mm, in profile the lower tepals as long as the upper. *Filaments* 12–14 mm long, exserted c. 5 mm from the tube; *anthers* 10–12 mm long, with short apical appendages, the bases of the thecae drawn into sterile tails 1.5–2.5 mm long, purple, the pollen pale yellow. *Ovary* oblong, 4–5 mm long; *style* extending horizontally over the stamens, dividing opposite the anther apices, the branches c. 2 mm long. *Capsules* obovoid, three-lobed and retuse above, 15–20 mm long, brown-speckled or flushed with red; *seeds* oblong, broadly winged, c. 6 mm long. *Chromosome number* unknown.

Flowering time March and April.

Distribution and biology

Gladiolus macneilii is an extremely local endemic of the eastern escarpment of northern Mpumalanga, South Africa. The few collections are all from the upper slopes of the escarpment on Abel Erasmus Pass which links Lydenburg with the eastern lowveld. This is an area of dolomite outcrops and is fairly rich in endemic species apparently adapted to the peculiar soils there, including another species of *Gladiolus*, *G. pavonia*. Plants grow in loamy soil in very rocky ground in grassland and light woodland.

The long perianth tube and cream to pale salmon flower with red lines on the lower tepals suggests that the species is adapted to pollination by long-tongued flies, and we have recorded *Stenobasipteron wiedmannii* (Nemestrinidae) actively visiting the flowers of *Gladiolus macneilii*.

Diagnosis and relationships

The extremely long perianth tube, 40–45 mm long, but relatively short tepals, the upper three 27–30 mm long, and cream to pale salmon flower colour with a longitudinal median streak on each of the three lower tepals make *Gladiolus macneilii* unmistakable. The eight to nine leaves have more or less linear and plane blades, 2–3.5 mm wide, with the margins and midrib moderately thickened and raised.

Long-tubed flowers with colour and markings such as those found in *Gladiolus macneilii* have arisen independently several times in *Gladiolus* and this alone is not very useful in assessing the relationships of the species. The loose fan of several basal leaves, the blades with secondary veins at least moderately developed, is typical of section *Densiflorus* and we assume that the affinities of *G. macneilii* lie here, despite the rarity of long-tubed flowers in the section. Only *G. varius* of section *Densiflorus* has com-

parably long-tubed flowers, but its dry, ferrugineus bracts and large tepals make it seem unlikely that it is immediately allied to *G. macneilii*. The tailed anthers may be most telling in determining the relationships of the species. Such tailed anthers are restricted to only two other species of *Gladiolus*, *G. appendiculatus* and *G. calcaratus*, both of section *Densiflorus*. Both these species have flowers with shorter tubes, but up to 40 mm long in *G. calcaratus*, and are adapted to pollination by long-tongued bees or relatively short-tongued species of nemestrinid flies. We presume that the elongate perianth tube and pale-coloured flower with red nectar guides arose independently in *G. macneilii* in response to selection for long-tongued fly pollination and that its resemblance to other species with similar flowers in different sections of *Gladiolus* is the result of convergence for the same pollination strategy.

History

Gladiolus macneilii was discovered comparatively recently, either in 1967 or the year before, by the keen amateur naturalist, the late Gordon MacNeil. Particularly interested in the flora of the eastern escarpment, he explored the area for many years and took care to draw the attention of botanists to his several plant discoveries. MacNeil sent specimens of the *Gladiolus* he found on the Abel Erasmus Pass to G.J. Lewis at the Compton Herbarium at Kirstenbosch Gardens shortly

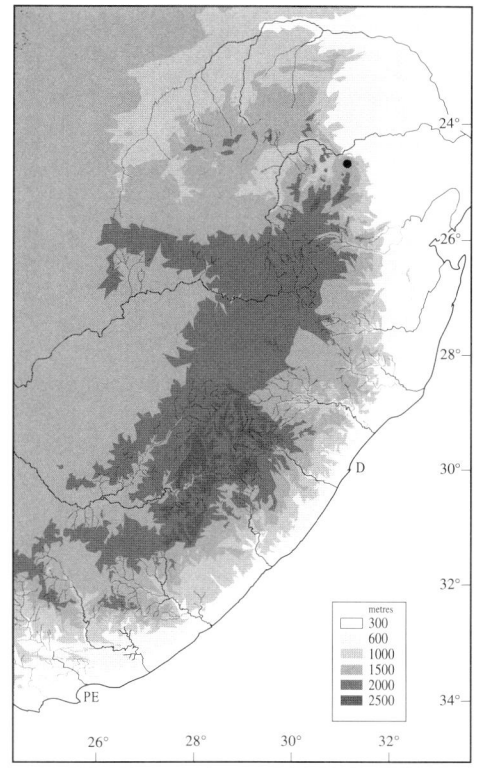

before her death in April 1967, together with a draft description. Clearly an undescribed species, it was left to Amelia Obermeyer to deal with. Obermeyer visited the site where the plants were found in 1969. She then completed the description which was published in 1972, after she had finished Lewis's revision of *Gladiolus* in South Africa.

Section *Densiflorus:* Series *Scabridus*
14. GLADIOLUS OCHROLEUCUS Baker
PLATE 13

Gladiolus ochroleucus Baker, *J. Bot.* 14: 182 (1876); *Fl. Capensis* 6: 151 (1896). Lewis et al., *J. S. African Bot.*, Suppl. 10: 35 (1972). Type: South Africa, Eastern Cape, Umtata District, Baziya Mountain, Dec.–Mar., *Baur 94* (K, holotype; B [not seen], BOL, SAM, isotypes).

ochroleucus = brownish yellow, for the flower colour, with brownish shading, found in plants from the central Transkei where the type collection was made.

Synonymy
Gladiolus kirkii Baker, *Gard. Chron.* 8: 524 (1890); *Fl. Capensis* 6: 152 (1896), not *G. kirkii* Baker (1892) (= *G. decoratus* Baker). Type: South Africa, Eastern Cape, King

William's Town, cultivated in Britain, 30 July 1890, *Kirk s.n.* (K, holotype).
Gladiolus reductus Baker, *Bull. Herb. Boissier*, Sér. 2, 4: 1006 (1904). Type: South Africa, Eastern Cape, East London, Feb. 1898, *Seifert sub Conrath 581* (GZU [not seen], holotype; PRE, photo).
Gladiolus masoniorum C.H. Wright, *Kew Bull.* 1913: 305 (1913). Prain, *Curtis's Bot. Mag.* 140: pl. 8548 (1914). Type: South Africa, Eastern Cape, Tembuland, cultivated in Britain, 30 May 1913, *Mason & Mason s.n.* (K, holotype).
Gladiolus stanfordiae L. Bolus, *S. African Gard.* 17: 293 (1927); Hutchinson, *Curtis's Bot. Mag.* 161: pl. 9522 (1938). Type: South Africa, Eastern Cape, without precise locality, grown in K. Stanford's garden, Stellenbosch, Feb. 1927, *Stanford s.n.* (BOL, holotype).

Gladiolus triangulus G. Lewis, *S. African Gard.* 23: 140, fig. a (1933). Type: South Africa, Eastern Cape, Albany District, Bushman's River, cultivated at Kirstenbosch Gardens, 27 Apr. 1933, *Holland 4006* (BOL, lectotype, with preserved flowers; BOL, isolectotype).

Description
Plants (30–)40–80 cm high, occasionally more or less evergreen. *Corm* globose to depressed-globose, 25–35 mm in diameter, the tunics coriaceous to coarsely papery, decaying into irregularly shaped fragments. *Cataphylls* pale and coriaceous, the upper green and leaf-like, reaching up to 12 cm above the ground. *Leaves* seven to twelve, in a tight distichous fan, mostly basal, about half as long as the stem, the longest reaching

the base of the spike, sometimes minutely ciliate to scabrid on the margins, the blades narrowly lanceolate, (8–)15–25 mm wide, the margins and midrib moderately thickened, the secondary and tertiary veins also lightly thickened, the upper one to three leaves cauline and shorter than the basal. *Stem* erect, simple or with one or two short branches, 4–5 mm in diameter below the spike.

Spike flexed at the base, inclined c. 30°, occasionally 4-, usually 12- to 18-flowered; *bracts* green, the outer 25–40(–50) mm long, acute, sometimes attenuate, the inner slightly shorter than the outer, minutely bifurcate or fairly deeply divided at the apex. *Flowers* either shades of pink, sometimes light purple or reddish, or whitish or yellow flushed brownish on the reverse, the lower tepals each white in the lower half, sometimes with dark red-purple median streaks below, and reddish at the top of the tube, unscented; *perianth tube* obliquely funnel-shaped, 15–20 mm long; *tepals* unequal, ovate, the margins, especially of the upper

laterals, undulate, the three upper largest, 24–28 x 18–20 mm, the dorsal hooded and horizontal, the upper laterals usually slightly longer than the dorsal, usually flaring outward distally or directed forwards, the flower thus appearing closed, the lower lateral tepals smallest, 16–19 x c. 9 mm, narrowly channelled, directed forward, the lowermost tepal curving downward distally. *Filaments* 10–12 mm long, exserted c. 5 mm from the tube; *anthers* c. 9 mm long, pale mauve, the pollen cream. *Ovary* oblong, weakly three-lobed above, 4–5 mm long; *style* arching over the stamens, dividing opposite the upper third of the anthers, the branches c. 4 mm long, expanded distally. *Capsules* narrowly obovoid-oblong, three-lobed and retuse above, (20–)25–30 mm long, shorter than the bracts; *seeds* ovate, 8–9 x c. 5 mm, broadly and evenly winged, light golden brown. *Chromosome number* 2*n* = 30.

Flowering time December to April, occasionally as late as May.

Distribution and biology

The most common, and one of the most attractive, of the Eastern Cape species of the genus, *Gladiolus ochroleucus* extends from the Zuurberg west of Grahamstown and the southern foothills of the Amatola Mountains near King William's Town through the former Transkei to southern KwaZulu-Natal near Byrne. The species grows in grassland and light bush or woodland from close to the coast to elevations of above 1200 m and does not appear to favour any particular soil. Plants seem to be found with equal frequency on coastal sandstone-derived soils and on light clay. Over most of its range the climate is fairly equable and rainfall may occur throughout the year. Under these circumstances plants are sometimes evergreen. They have a long flowering period, extending from December until April and, in some years, even as late as May.

The flowers are typical of those of bee-pollinated species in the section and are adapted for pollination by long-tongued anthophorid bees in the genera *Amegilla* and *Anthophora*.

Diagnosis and relationships

Relatively unspecialized, *Gladiolus ochroleucus* has a basal fan of sword-shaped to lanceolate leaves and simple, or occasionally branched spikes of four to 12, or occasionally up to 18, fairly large, short-tubed flowers. The flowers are moderate in size, the funnel-shaped tube 15–20 mm long, and the upper three tepals 24–28 mm long, thus about half as long again as the tube. The flowers are usually deep pink, the tepal margins are undulate, and the lower tepals have whitish nectar guides, often highlighted with red or purple streaks. The leaves are often lightly ciliate to scabrid, a feature of several species of series *Scabridus* of section *Densiflorus*. The floral bracts are unusually long in *G. ochroleucus* for a species with a relatively short perianth tube, so that in fruit the capsules are completely enclosed and protected by these organs. This fairly widespread species displays some unusual variation both in tepal shape and in colouring. The type collection and other populations from the central Transkei have creamy yellow flowers with darker yellow-brown toward the tips of the tepals. Elsewhere plants are usually bright pink, but locally near East London some populations have quite strikingly coloured flowers with the tepals more rounded and flushed dark purple, as shown in Plate 13.

Except for the floral details, *Gladiolus ochroleucus* very closely resembles a second Eastern Cape species, *G. mortonius*, but that species is readily distinguished by flowers with a much longer perianth tube, usually 30–45 mm long, and often more firmly

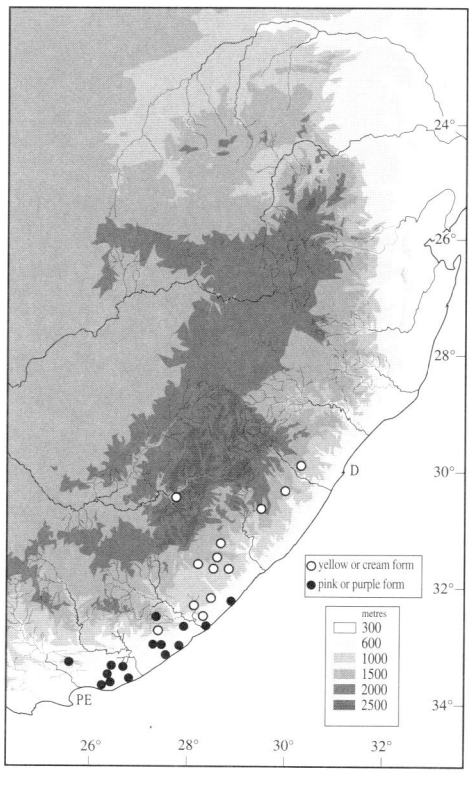

textured leaves with very strongly thickened margins.

History

Gladiolus ochroleucus was apparently first collected in 1830 by C.F. Ecklon on his first expedition to the then eastern Cape Colony. The labels accompanying the specimens are confusing, for both Ecklon and his frequent companion, C.L. Zeyher, are listed as collectors, although Zeyher did not accompany Ecklon on the 1830 expedition (Gunn & Codd, 1981). Later, in 1832, Zeyher together with Ecklon collected

the species again. Some of these early collections are annotated *Neuberia longifolia* (a name without description and probably intended for a western Cape species of *Watsonia*). The early collections were overlooked and specimens sent to Kew Gardens in the mid-1870s by Rev. L.R. Baur were named by J.G. Baker. Baur's plants were gathered on or near Baziya Mountain in the Transkei, where he had established a mission station and was to make many important collections of plants from the area. His plants were the yellowish form of the species, hence the name chosen by Baker,

G. ochroleucus. The more widespread pink-flowered form was grown in Britain later in the nineteenth century and plants originally from King William's Town were described as *G. kirkii* by Baker in 1890. For reasons that are not clear, another collection of the species, this from East London, was also recognized as a new species by Baker, which he called *G. reductus*. Other synonyms for the species are *G. masoniorum* C.H. Wright (1913), *G. stanfordiae* L. Bolus (1927), and *G. triangulus* G. Lewis (1933), none of which represents particularly distinctive plants.

15. GLADIOLUS MORTONIUS Herbert
PLATE 14

Gladiolus mortonius Herbert, *Curtis's Bot. Mag.* 65: pl. 3680 (1838) & sub. pl. 3693 (1838). *Gladiolus blandus* var. *mortonius* (Herbert) Baker, *Handbook Irideae* 217 (1892); *Fl. Capensis* 6: 155 (1896). Type: South Africa, without precise locality, illustration in *Curtis's Bot. Mag.* 65: pl. 3680 (1838).

mortonius, named for a Mr Morton who sent plants from South Africa to Great Britain, where they were flowered and the type illustration was made.

Synonymy

Gladiolus macowanii Baker, *Handbook Irideae* 219 (1892); *Fl. Capensis* 6: 155 (1896), including *G. cardinalis* and *G. saundersii*. Batten & Bokelmann, *Wild Fl. E. Cape Province* pl. 27, fig. 3 (1966). *Gladiolus ochroleucus* var. *macowanii* (Baker) Obermeyer in *J. S. African Bot.*, Suppl. 10: 37 (1972). *Gladiolus massonii* Klatt in Durand & Schinz, *Consp. Fl. Africae* 5: 220 (1895), illegitimate superfluous name for *G. macowanii*. Type: South Africa, Eastern Cape, Somerset East, slopes of Boschberg, Apr. 1869, *MacOwan 236* (K, lectotype designated here; BOL, G, GRA [not seen], S, Z, isolectotypes).
Gladiolus davisonii F. Bolus, *Ann. Bolus Herb.* 2: 98 (1917). Type: South Africa, Eastern Cape, Berlin, Somerset East mountains, cultivated in Cape Town, 17 Mar. 1915, *De Courcy s.n.* (BOL 14762, holotype on two sheets).

Description

Plants 40–72 cm high. *Corm* globose to depressed-globose, (20–)30–50 mm in diameter, the tunics coriaceous to coarsely papery, decaying into irregularly shaped fragments. *Cataphylls* pale and coriaceous, the upper green and leaf-like, reaching up to 15 cm above the ground. *Leaves* seven to twelve, in

a tight distichous fan, mostly basal, about half as long as the stem, the longest reaching to between the middle of the stem and the base of the spike, sometimes minutely ciliate on the margins, the blades narrowly lanceolate, (8–)15–25 mm wide, the margins and midrib moderately thickened, the secondary and tertiary veins also lightly thickened, the upper one to three leaves cauline, shorter than the basal. *Stem* erect, simple, rarely with one branch, 4–5 mm in diameter below the spike.

Spike flexed at the base and inclined c. 30°, (5–)8- to 16-flowered; *bracts* pale green, sometimes glaucous, the outer 40–60 mm long, acute to attenuate, the inner three-quarters to about as long as the outer, usually minutely bifurcate at the apex. *Flowers* pink, the lower tepals each with a red median streak and a few fine reddish longitudinal lines parallel to the midline, the base of the throat dark red, unscented; *perianth tube* cylindric and curving outward above, (20–)30–45 mm long, expanded gradually in the upper 6–8 mm; *tepals* unequal, lanceolate-ovate, the margins undulate, the three upper largest, 32–40 x 20 mm, the dorsal arching over the stamens, the upper laterals usually slightly longer and wider than the dorsal, usually directed forward and flaring outward distally, the lower three tepals joined basally for c. 4 mm, the lower laterals 30–35 x 12–16 mm, the lower median nearly as long as the upper tepals, 32–40 x 12–18 mm, directed forward, the lowermost tepal curving downward distally, in profile the lowermost tepal exceeding the upper by up to 5 mm. *Filaments* c. 14 mm long, exserted 4–7 mm from the tube; *anthers* 12–15 mm long, with short apiculi c. 0.5 mm long, usually purple, the pollen

whitish. *Ovary* oblong, 4–5 mm long; *style* arching over the stamens, dividing opposite the anther apices, the branches 4–6 mm long, expanded distally. *Capsules* narrowly obovoid-oblong, three-lobed and retuse above, 28–32 mm long, shorter than the bracts; *seeds* ovate, c. 8 x 5 mm, broadly and evenly winged, light golden brown.
Chromosome number $2n = 30$.
    Flowering time February to April.

Distribution and biology

A species of the southern African summer-rainfall zone, *Gladiolus mortonius* is restricted to the Eastern Cape, South Africa. It extends from the Boschberg near Somerset East and

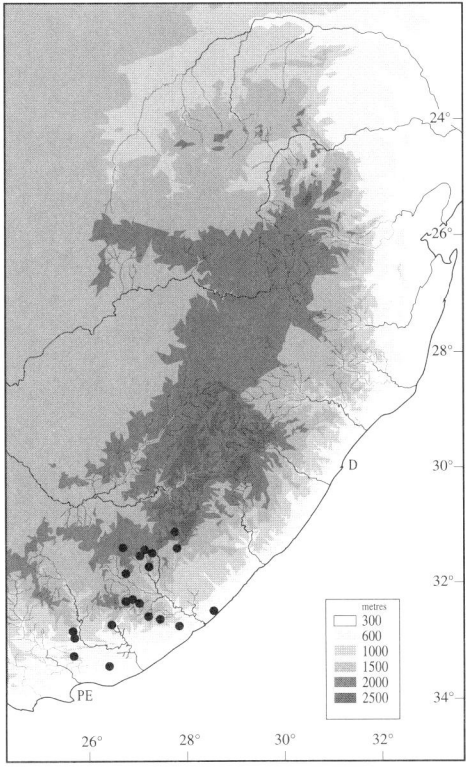

the Zuurberg range in the southwest, through the Amatola Mountains and Winterberg to Elliot, close to the Lesotho frontier. Plants favour open sites such as low, stony grassland where they are sheltered from competition by trees and bush.

The long-tubed pink flowers with widely spreading tepals and reddish nectar guides have the typical appearance of species of the genus adapted for long-tongued fly pollination, and we have seen the nemestrimid fly *Drosoeca ganglbaueri* foraging on the flowers.

Diagnosis and relationships

Gladiolus mortonius is readily recognized by the fairly large flower, 55–95 mm long, with a slender, gently curving perianth tube usually 30–45 mm long, and a basal fan of broad leaves with fairly prominent midrib, margins and secondary veins. It is probably most closely related to *G. ochroleucus* which it closely resembles in vegetative features, even to the frequent light scabrid pubescence on the leaves. That species differs most notably in having smaller flowers with the perianth tube 15–20 mm long and anthers c. 9 mm long, both features considerably smaller than in *G. mortonius*. It is also sometimes difficult to distinguish *G. oppositiflorus* from *G. mortonius* for the general appearance of the two species is very similar, particularly the size and shape of the flowers. As well as having a distichous spike, *G. oppositiflorus* has very distinctive floral bracts, somewhat inflated below. The inner bracts have two long, brown-tipped cusps, and the lower margins unite around the ovary and tightly envelop the flower.

In their revision of *Gladiolus* in South Africa, Lewis et al. (1972) treated *G. mortonius* as a subspecies of *G. ochroleucus* under the name subsp. *macowanii*. Although we concur with their opinion that *G. ochroleucus* and *G. mortonius* are closely related, we

believe that the floral differences between the two warrant specific rank for both.

History

The origin of the plants grown in England in the 1830s on which the name was based are unknown. The species was named in honour of a Mr Morton who sent plants from the Eastern Cape to the botanist, William Herbert, a grower of *Gladiolus* species who produced the first interspecific hybrids, the forerunners of today's hybrid strains. *Gladiolus mortonius* was described in *Curtis's Botanical Magazine* by Herbert in 1838, the protologue being accompanied by a fine illustration of the species. Despite the lack of a type specimen, there should have been no doubt about the identity of the species. Nevertheless, in 1892 J.G. Baker described *G. macowanii*, his description based largely on plants from the Eastern Cape that closely match *G. mortonius*. Baker did not cite any specimens in the protologue but stated that the species was first collected by Francis Masson, the plant collector who sent plants back to Kew Gardens in the late eighteenth century. This left the impression that a Masson collection, or a specimen made later from plants he collected alive, might most appropriately be chosen as the type of the species. Specimens raised from Masson's introduction and preserved in the British Museum Herbarium, however, show this plant to be the Western Cape species, *G. cardinalis*. Then in 1896 Baker cited two fairly recent collections, both from the Eastern Cape, *Cooper 3600* which is *G. saundersii* and *MacOwan 236*, as well as the Masson specimen, there cited as 'G. secundus', Soland. in Herb. Banks!' The citation of the two Eastern Cape specimens explains Baker's name for the species, *macowanii*, as well as his statement that the

species came from the 'Eastern Districts', and we designate the MacOwan specimen the lectotype, thus fixing the application of this name.

In Lewis et al.'s revision of the genus, *Gladiolus mortonius* is treated as subsp. *macowanii* of *G. ochroleucus* and the MacOwan collection at the Kew Herbarium is cited as the holotype. The existence of three syntype collections makes the designation of a holotype incorrect. However, the application of the epithet can be fixed by choosing a lectotype and we here designate that same collection the lectotype. Except in its pink, rather than red flower colour, MacOwan's collection accords fairly well with the description, particularly in the dimensions of the flower, in which the perianth tube is described as exceeding the tepals, and in the rigid texture of the leaves. MacOwan's collection was made in 1869 and may well have been the first record of the species after *G. mortonius* was first described in 1838. An undated collection in the Kew Herbarium made by Henry Hutton from the 'Eastern Frontier' may have been made slightly earlier.

F.W. Klatt (1895) renamed *Gladiolus macowanii*, preferring to call it *G. massonii* to avoid confusion with the similarly named *G. macowanianus* which he had described in 1883. This action is nomenclaturally illegitimate, for the two epithets celebrating MacOwan are different words, hence not homonyms. That they are likely to be confused is a trivial consideration and of no nomenclatural concern. *Gladiolus davisoniae*, described by Frank Bolus in 1917 from plants cultivated in Cape Town, differs in no significant way from the type illustration of *G. mortonius* or the MacOwan collection of *G. macowanii* and is from close to the same locality near Somerset East.

16. GLADIOLUS MICROCARPUS G. Lewis
PLATE 15

Gladiolus microcarpus G. Lewis, *J. S. African Bot.*, Suppl. 10: 85 (1972). Type: South Africa, KwaZulu-Natal, Cathedral Peak Forest Station, cliffs in the Ndumeni River Valley, 7 Feb. 1958, *Killick 1655* (PRE, holotype).

microcarpus = with a small fruit, referring to the relatively small capsule.

Description

Plants up to 1 m high, but usually inclined

or drooping. **Corm** more or less globose, c. 20 mm in diameter, the tunics of membranous layers soon decaying into fine fibres. **Cataphylls** pale and membranous, the uppermost reaching 4–7 cm above the ground and then pale green, with dense short pubescence. **Leaves** five to seven, the lower three more or less basal, about as long as or longer than the stem, but trailing, the blades narrowly lanceolate, 6–12 mm wide, the midrib strongly thickened and raised, a pair of secondary veins, one or two pairs of tertiary

veins and the margins lightly raised, the edges of the raised areas with minute horizontal hairs on the blades and sheaths, upper three to four leaves cauline, decreasing in size above, the blades similar to the basal, the uppermost with a very short blade, the margins of the sheaths united around the stem or open to the base. **Stem** inclined or drooping, unbranched, or robust plants with one to three branches, c. 3 mm in diameter below the spike.

Spike arching upward, ascending to nearly

erect, lightly flexuose, (5–)8- to 12-flow-ered, the branches with two to eight flowers, nearly distichous with the flowers in two ranks 65°–75° apart; *bracts* dull green to dull greenish purple, diverging from the stem, lightly ridged, the outer 25–35 mm long, c. two and a half internodes long, the inner about two-thirds as long as the outer, completely enclosed, pale green, forked api-cally for 1–2 mm. *Flowers* bright pink, the lower three tepals each with a broad white longitudinal zone in the lower two-thirds and a narrow reddish purple streak in the midline, reddish at the base of the throat, unscented; *perianth tube* obliquely funnel-shaped, 35–40 mm long, cylindric in the lower 30–35 mm; *tepals* unequal, lanceo-late, those of the outer whorl smaller than the inner, the dorsal largest, 35–40 x c. 17 mm, ascending and inclined over the stamens, the upper laterals subpatent, curving outward near the apices, 25–30 x c. 10 mm, the lower laterals lightly inclined, the apices curving downward, 32–35 x c. 13 mm, the lower median 24–26 x c. 8 mm. *Filaments* c. 15 mm long, exserted c. 10 mm from the tube; *anthers* c. 8 mm long, whitish to lilac, the apices with small sterile appendages, the pollen white. *Ovary* narrowly obovoid, c. 5.5 mm long; *style* arching over the stamens, divid-ing c. 3 mm beyond the anther apices, the branches c. 4 mm long. *Capsules* oblong, three-lobed and retuse above, 12–15 mm long, enclosed in the bracts; *seeds* oblong,

broadly and evenly winged, c. 5 x 4 mm, light brown. *Chromosome number* unknown.

Flowering time December to late January at lower elevations, to early March at higher elevations.

Distribution and biology

Endemic to the Drakensberg range, *Gladiolus microcarpus* occurs in central KwaZulu-Natal and adjacent Free State, South Africa, and in Lesotho. Its range extends from the summit of the Little Berg near Cathkin Park in the south to Royal Natal National Park in the north, where plants occur in the Mahai Valley and on the high rocky slopes above the Park. The altitudinal range is from 1800 to 2700 m, but plants are more common at higher elevations. They almost always grow on rocks, and often on steep cliffs where the stems hang downward from vertical rock faces, their corms anchored in crevices in basalt or sandstone. During the growing sea-son the habitat is almost constantly moist, the result of almost daily rain or mist, or from water seeping down the rocky faces.

Gladiolus microcarpus belongs to a guild of Drakensberg species which have pink to white or cream flowers with long perianth tubes. These plants are pollinated by the long-tongued nemestrinid fly *Prosoeca gan-glbaueri*. The fly has mouth parts c. 35 mm long which complement perianth tubes 35–40 mm long in the species adapted for this pollination system. Other species in the guild include *Hesperantha rupicola* and *Watsonia wilmsii* (Iridaceae), *Disa* species (Orchidaceae) including *D. amoena* and *D. oreophila*, and *Zaluzianskya microsiphon* (Scrophulariaceae).

Diagnosis and relationships

The distinguishing features of *Gladiolus microcarpus* are the deep pink, long-tubed flowers borne on drooping to pendent spikes and the lightly scabrid leaves with multiple thickened veins.

The flowers have a fairly narrow, slender perianth tube 35–40 mm long, and the tepals are narrowly lanceolate with somewhat attenuate apices. The outer tepals, that is, the dorsal and the lower laterals, are unusual in being more or less equal and substantially larger than the upper lateral and lower median tepals, and this imparts an unusual appearance to the flowers. Both the leaf

sheaths and leaf blades are minutely pubes-cent along the veins.

Within series *Scabridus*, *Gladiolus micro-carpus* is most closely allied to *G. scabridus* which occurs in fairly dry woodland and bush in northern KwaZulu-Natal and south-ern Swaziland. *Gladiolus scabridus*, which was included in *G. microcarpus* by Lewis et al. (1972), has stiffly erect leaves, upright, minutely scabrid stems, scabrid-pubescent bracts, and larger flowers, the upper tepals 40–45 mm long.

History

The first record of *Gladiolus microcarpus* appears to be a collection made in 1932 by Ernest Galpin who found plants growing at the summit of the Little Berg near Cathkin Park. It was, however, not until 1972 that the species was described, the description being based on more recent collections, including those made by Donald Killick and William Trauseld on the main Drakensberg range at Cathedral Peak and Giant's Castle. The plants illustrated here were collected near The Sentinel at the northern end of the range of *G. microcarpus*, and at one of the few sites where this cliff-dwelling plant is easily accessible.

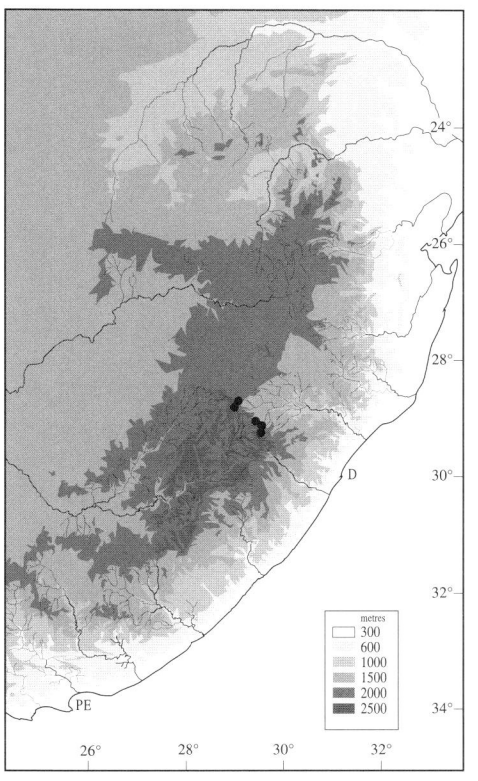

naliter striis albis atrocarneisque, filamentis 15–16 mm longis, antheris 10–13 mm longis.

Description

Plants up to 1 m high. **Corm** more or less globose, 30–35 mm in diameter, the tunics of firm papery layers soon decaying into fibrous fragments. **Cataphylls** pale and membranous, the uppermost reaching 5–8 cm above the ground and then pale green or dull purple, glabrous or with a short, fine pubescence. **Leaves** seven to nine, the lower five or six more or less basal, forming a loose fan, reaching to between the base and apex of the spike, minutely scabrid-papillose, the blades sword-shaped, usually 10–20(–24) mm wide, the midrib and a pair of secondary veins moderately thickened and raised, the margins less so, the upper two or three leaves cauline, decreasing in size above, the blades similar to the basal, the uppermost often entirely sheathing, the margins of the sheaths open to the base. **Stem** erect, scabrid, simple or with one, rarely two branches, c. 3 mm in diameter below the spike.

Spike erect and straight, the axis minutely scabrid, 10- to 16-flowered, the branches 2- to 8-flowered; **bracts** dull green to dull greenish purple, minutely scabrid, often more or less dry above, the apices twisted outward, the outer 25–35 mm long, c. two and a half internodes long, the inner about three-quarters as long as to slightly longer than the outer, forked apically for 1–2 mm. **Flowers** bright pink, the lower three tepals each with a narrow white longitudinal zone in the lower two-thirds and a narrow reddish streak in the midline, unscented; **perianth tube** obliquely funnel-shaped, 35–40 mm long, cylindric in the lower 30–35 mm; **tepals** unequal, lanceolate, the upper three largest and more or less equal, 40–45 mm long, the dorsal 18–20 mm wide, the upper laterals 16–18 mm wide, the lower three tepals joined to the upper laterals for 3–4 mm and to each other for c. 4 mm, more or less straight and tilted toward the ground, 35–40 x 10–12 mm. **Filaments** 15–16 mm long, exserted c. 8 mm from the tube; **anthers** 10–13 mm long, pale yellow, the pollen white. **Ovary** narrowly obovoid, c. 7 mm long; **style** arching over the stamens, dividing at or shortly beyond the anther apices, the branches 5–6 mm long. **Capsules**

oblong, three-lobed above and retuse, 16–22 mm long; **seeds** broadly ovate, 6–8 x 4–5 mm, more or less evenly winged or the wing not developed on one side, rich brown, the seed body relatively large. **Chromosome number** $2n$ = 30, 60 (unpublished counts made at the Jodrell Laboratory, Kew).

Flowering time December to late January at lower elevations, sometimes into late February in cooler, wetter sites.

Distribution and biology

Although the distribution of *Gladiolus scabridus* is not yet well documented, the species is evidently restricted to the mountains of northern KwaZulu-Natal and southern Swaziland. It almost certainly also occurs in parts of southern Mpumalanga immediately adjacent to KwaZulu-Natal, on the opposite bank of the Pongola River. The species grows at elevations of 1000 to 2000 m and is restricted to well-drained, rocky habitats. At Itala it is especially common in quartzite outcrops on the hill slopes above the Pongola River, where the corms become wedged in crevices and are virtually impossible to dig up, thus secure from predation by porcupines and baboons.

We assume that the long-tubed flowers of *Gladiolus scabridus*, with their linear, white nectar guides streaked with dark red, are adapted for pollination by long-tongued flies,

Gladiolus scabridus Goldblatt & Manning, new species. Type: South Africa, Natal (KwaZulu-Natal), Itala Nature Reserve, Craigadam Farm, Mabomvu Ridge, c. 900 m (2950 ft), 5 Jan. 1978, *McDonald 399* (NU, holotype; K, PRE, isotypes).

scabridus = scabrid, the descriptive term for the rough, short pubescence present on the stem and bracts of the species.

Synonymy

Gladiolus microcarpus subsp. *italaensis* Obermeyer, *Bothalia* 13: 451 (1982). Type: South Africa, Natal (KwaZulu-Natal), Itala Nature Reserve, hills above the Pongola River, 10 Jan.1980, *Mauve & Reid 5266* (PRE, holotype).

Latin diagnosis

Plantae ad 1 m altae, cormo globoso 30–35 mm in diametro, tunicis papyraceis, foliis 7–9, inferioribus 5–6 basalibus distichis laminis planis ensiformibus 10–20(–24) mm latis scabridiusculis vel papillosis, spica erecta scabridiuscula 10–16 florum, bracteis 25–35 mm longis, floribus carneis longitudi-

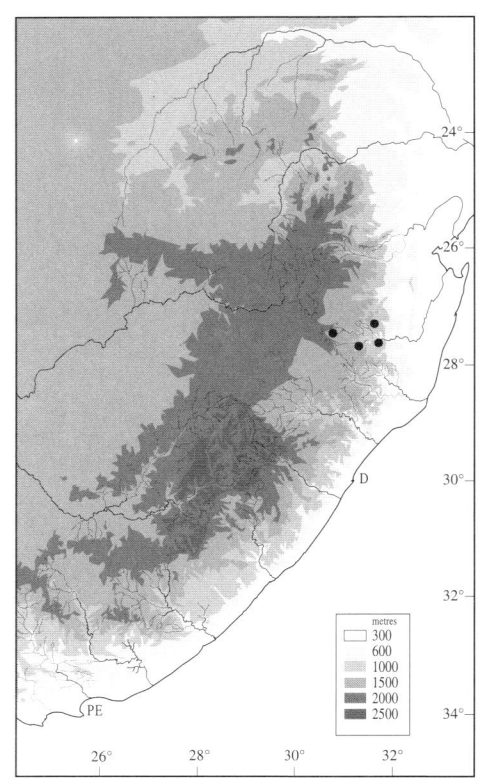

but have not yet had the opportunity to study the pollination biology of the species.

Diagnosis and relationships

Gladiolus scabridus is a member of the eastern southern African complex of series *Scabridus*, a group of species that have fairly large pink flowers with white nectar guides outlined or streaked with red to purple. It can be distinguished from most of the species of the alliance by its particularly long-tubed flowers, the tube about twice as long as the bracts, and the minutely but distinctively scabrid surfaces of the bracts, upper stem and spike axis. It is evidently closely allied to the high Drakensberg species, *G. microcarpus*, which has an equally long perianth tube, but that species differs in having smooth bracts and stems. The perianth of *G. microcarpus* also differs in that the lower lateral tepals are substantially shorter than the lower median and upper lateral tepals, a feature which imparts a quite different appearance to that of *G. scabridus*. The stems and leaves of *G. scabridus* are erect whereas those of *G. microcarpus* are inclined or drooping.

History

The first record of *Gladiolus scabridus* was made by General J.C. Smuts in 1944 in the Wakkerstroom District of Mpumalanga. This and a few other specimens from the area were tentatively assigned to the Drakensberg species, *G. microcarpus*, by Lewis et al. (1972). The species, however, remained unnamed until plants were collected in the Itala Nature Reserve in northern KwaZulu-Natal in 1976. A type collection was made in 1978 and it was formally described as *G. microcarpus* subsp. *italaensis* by A.A. Obermeyer in 1982, after she had examined the plants herself at Itala.

18. GLADIOLUS CATARACTARUM Obermeyer

Gladiolus cataractarum Obermeyer, *Bothalia* 14: 78 (1982). Type: South Africa, Mpumalanga, near Dullstroom, Farm Waterval above Lunsklip Waterfall, Feb. 1981, *Krijt s.n.* (PRE 58472, holotype mounted on two sheets).

cataractarum = of the waterfalls, so named for the Lunsklip Waterfall near where the species was first collected.

Description

Plants to 70 cm high. *Corm* globose, c. 30 mm in diameter, the tunics of coriaceous layers, these becoming irregularly fragmented and partially fibrous with age. *Cataphylls* pale and membranous to firm, the upper reaching up to 12 cm above the ground and then pale green. *Leaves* eight to nine, the lower six basal and longest, reaching to shortly below the base of the spike, the blades narrowly to broadly lanceolate, (20–)35–45 mm wide, the margins and midrib lightly thickened and hyaline, glabrous, the upper two or three leaves progressively shorter above, largely sheathing. *Stem* erect, unbranched, c. 5 mm in diameter below the spike.

Spike straight and erect, 8- to 16-flowered; *bracts* green, firm-textured, the outer 50–60(–70) mm long, attenuate, wrapped around the spike axis below, about two internodes long, the inner bract two-thirds to three-quarters as long as the outer, minutely forked apically for c. 1 mm. *Flowers* pink, the lower three tepals marked with longitudinal reddish lines in the lower midline, unscented; *perianth tube* cylindric below, 40–50 mm long, flared outward in the upper 10 mm; *tepals* unequal, lanceolate, the three upper largest, 44–48 x c. 16 mm, the dorsal arched to nearly horizontal, the upper laterals spreading outward in the upper halves, the lower lateral tepals c. 40 x 14 mm, spreading in the upper halves. *Filaments* c. 12 mm long, exserted c. 4 mm from the tube; *anthers* c. 12 mm long, yellow, the pollen yellow. *Ovary* oblong, c. 6 mm long; *style* arching over the stamens, dividing 1–2 mm beyond the anther apices, the branches c. 5 mm long. *Capsules* oblong, c. 30 mm long, three-lobed above and retuse; *seeds* oblong, 7–8 x 4–5 mm, broadly and evenly winged, light translucent brown, the seed body darker brown. *Chromosome number* unknown.

Flowering time February to mid-March.

Distribution and biology

Gladiolus cataractarum is a rare endemic of the southern African summer-rainfall region, known from a few scattered populations along the eastern escarpment in Mpumalunga. The type locality is on the crest of the upper Drakensberg escarpment close to Lunsklip Waterfall south of the town of Dullstroom. A survey of the distribution of the species made some years later to determine its conservation status showed several more small populations extending along the escarpment edge northward toward Lydenburg for some 10 km. Plants grow mainly on cliffs and steep rocky slopes on quartzite in sheltered, south-facing sites.

The long-tubed pink flowers resemble fairly closely those of other species of *Gladiolus* that are pollinated by long-tongued nemestrinid flies. The pollination biology of *G. cataractarum* awaits investigation.

Diagnosis and relationships

The relatively large pink flower streaked with red on the lower tepals, the long perianth tube, the large green bracts and the fan of broad leaves of *Gladiolus cataractarum* recall the other members of series *Scabridus* very closely, despite the lack of pubescence on the leaves or stem. Within the alliance, a perianth tube as long as 50 mm occurs in *G. microcarpus*, *G. mortonius* and *G. scabridus*, but these species usually have pubescent leaves and sometimes pubescent stems and bracts as well. The general aspect of *G. cataractarum* is of extreme robustness, the leaves being firm and up to 45 mm wide. Thus the species is unlikely to be mistaken for any other. *Gladiolus lithicola*, which grows on the lower Mpumalanga escarpment

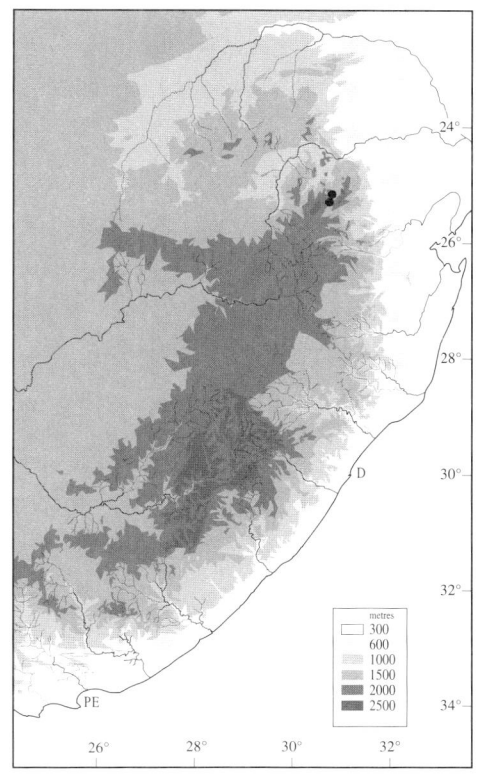

near Graskop, resembles *G. cataractarum*, especially in its broad leaves, but this cliff and rock-outcrop species has mauve-pink flowers without darkly streaked nectar guides and has, in addition, smaller floral bracts that turn rusty brown during anthesis, a feature that seems to place the species in series *Densiflorus* close to *G. ferrugineus*. Additional collections are needed to assess the variation in *G. cataractarum*.

History

Gladiolus cataractarum was first recorded in the late summer of 1981 on the upper escarpment near Lunsklip Waterfall in the Lydenburg District. The species was described from this locality alone the following year by A.A. Obermeyer, who had, some ten years earlier, completed and brought to publication G.J. Lewis's manuscript revision of *Gladiolus* in South Africa. The species remained known from this one site until systematic exploration of suitable habitats along the escarpment was undertaken by the Transvaal Department of Nature Conservation. *Gladiolus cataractarum* was then found to occur at several more places both to the north and to the south along the same range of mountains.

19. GLADIOLUS PAVONIA Goldblatt & Manning
FRONTISPIECE

Gladiolus pavonia Goldblatt & Manning, *Novon* 6: 174 (1996). Type: South Africa, Mpumalanga, dolomite hill slopes between the top of Abel Erasmus Pass and Strydom Tunnel, 5 Dec. 1994, *Goldblatt & Manning 10131* (NBG holotype, K, MO, PRE, isotypes).

pavonia = peacock, alluding to the prominent eye in the centre of the flower, as in a peacock feather.

Description

Plants 45–80 cm high. *Corm* ovoid, 8–13 mm in diameter, with stolons produced from the base, these ultimately producing new plants some distance from the parent, the tunics of more or less papery layers, becoming irregularly broken and somewhat fibrous with age. *Cataphylls* pale and membranous, the uppermost reaching 2–3 cm above the ground and then brownish or purple. *Leaves* six or seven, the lower three or four more or less basal and largest, reaching at least to the base of the spike or sometimes slightly exceeding it, the blades sword-shaped, 8–14 mm wide, usually slightly twisted in the upper half, the remaining two to three leaves cauline and much smaller than the basal, the uppermost largely or entirely sheathing, the margins open to the base. *Stem* erect, sometimes flexed outward above the sheaths of the two upper leaves, but remaining erect, unbranched, 2–2.3 mm in diameter below the spike. *Spike* erect, occasionally 2-, usually 4- to 7-flowered, the flowers in two ranks c. 50° apart; *bracts* pale green, relatively soft-textured, the apices becoming dry and light brown shortly after anthesis, the outer bracts (15–)23–30 mm long, the inner slightly shorter than to about as long as the outer, apiculate or minutely forked apically. *Flowers* pale pink, the upper tepals shading to dark red toward the bases, the dorsal half of the upper tube dark red inside and out, the lower margins of the dorsal tepal with a wide transparent band, the lower lateral tepals lightly streaked with pink longitudinal lines in the lower half, the lower median tepal whitish in the lower half and lined with purple streaks, the lower tepals also with a white zone and central red spot just below the tepal sutures, unscented; *perianth tube* obliquely funnel-shaped, lc. 16 mm long, the lower cylindrical part c. 8 mm long; *tepals* nearly equal and widely spreading, the dorsal slightly larger than the others, 16–25 x c. 16 mm, curving outward below, erect above, with the margins curved back, the upper laterals patent in the upper half, 24–26 x 15 mm, the lower three tepals slightly inclined below, recurved in the upper half, 21–24 x 12 mm. *Filaments* c. 16 mm long, reddish in the lower half, exserted c. 8 mm from the tube; *anthers* c. 8 mm long, tilting below the horizontal, dark purple, the pollen cream. *Ovary* oblong, c. 6 mm long; *style* arching over the stamens, dividing between the base and the middle of the anthers, the branches 5–6 mm long. *Capsules* obovoid, 21–24 mm long, three-lobed above and retuse; *seeds* evidently ovate, c. 7 mm long, broadly and evenly winged, light brown (fully mature seeds not seen).

Flowering time late November and December, occasionally until late January.

Distribution and biology

A rare, narrow endemic, *Gladiolus pavonia* is recorded from the mountains south of the Strydom Tunnel on the slopes of the Abel Erasmus Pass in Mpumalanga, South Africa. Plants grow on semi-arid stony dolomite hills in light woodland, and seem to favour equally shaded or open, exposed sites. They produce stolons from the corm bases and form small clones, several juvenile plants surrounding the larger parent plants. Our observations suggest that *G. pavonia* favours steeper slopes that receive slightly more precipitation than the surrounding hills. Although known from only two sites in the area, we suspect that it is more widespread in the dolomite belt that extends to the north and south along the Drakensberg interior to the escarpment edge (Matthews et al., 1993). The early summer of 1994, when we collected the plants for the illustrations reproduced here, was exceptionally dry and only a small proportion of the many plants we found were in flower. In years of higher rainfall we suspect that the species produces a fine display. *Gladiolus pavonia* appears to be a prime plant for cultivation in a small garden, especially in dry areas of the summer-rainfall area.

The pollination biology of *Gladiolus pavonia* is unknown, but the short-tubed pink flowers with a conspicuous dark eye are probably adapted for pollination by long-tongued bees.

Diagnosis and relationships

Gladiolus pavonia can be distinguished by its pale pink flower with a tube 16 mm long, nearly equal tepals 21–25 mm long, and a circle of dark red colour at the mouth of the perianth tube. The leaves are unusually softly textured, somewhat surprisingly for a plant of a dry habitat. Also unusual are the pale, fairly thick stolons produced from the corm bases, each terminating in a comparatively large cormlet. Within series *Scabridus*, *G. pavonia* is probably most closely related to *G. brachyphyllus* which also has softly textured, glabrous leaves, glabrous stems and bracts and a broadly similar flower with a fairly short perianth tube. That species differs notably in having dark pink flowers with a white median streak on each of the lower tepals and leaves of flowering plants with very short or vestigial blades.

History

Surprisingly for a species that grows within a few metres of a major road, *Gladiolus pavonia* seems to have been completely unknown until 1987, when the first specimens were collected by P. Raal and J. Raal. This first collection and one other made two years later, also on the dolomite ridges near Abel Erasmus Pass, were unknown to us when we coincidentally found plants of an apparently new species of *Gladiolus* in fruit on the northern approach to the Pass in January 1994. Returning there in early December the following summer, we found several plants in flower on the rocky slopes above a small valley, and were then able to

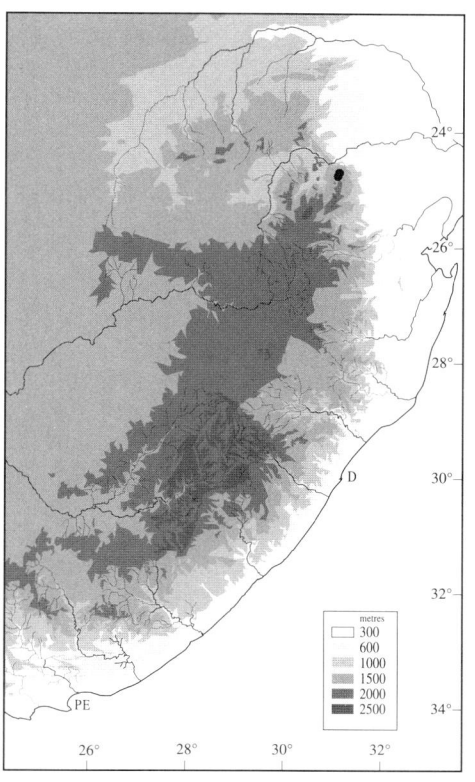

confirm that our discovery was indeed the same species that Raal & Raal had recorded seven years earlier.

20. GLADIOLUS BRACHYPHYLLUS F. Bolus
PLATE 17

Gladiolus brachyphyllus F. Bolus, *Ann. Bolus Herb.* 2: 103 (1917). Lewis et al., *J. S. African Bot.*, Suppl. 10: 291 (1972). Type: South Africa, Mpumalanga, between Komatipoort and the Letaba River on the Selati Railway, c. 31E 24–25S, Nov. 1913, *Rogers 11807* (BOL, holotype; GRA [not seen], K, PRE, isotypes).

brachyphyllus = short-leaved, referring to the short blades on the leaves of the flowering spikes.

Description

Plants 55–80 cm high. *Corm* globose, c. 20–25 mm in diameter, the tunics coriaceous, fragmenting irregularly and becoming somewhat fibrous with age. *Cataphylls* pale and membranous, the uppermost reaching 2.5–5 cm above the ground and then green and coriaceous. *Leaves* three, all largely sheathing, with short or vestigial blades, the lowermost basal, reaching 8–10 cm above the ground, usually overlapping the next leaf, the second leaf inserted on the lower third of the stem, 80–150 mm long, the blades of both the first and second leaves short to vestigial, 5–50 mm long, the third leaf inserted in about the middle of the stem, 50–80 mm long, also entirely sheathing and with a vestigial blade, the margins open to the base; leaves of non-flowering plants present at flowering time but not fully developed, eventually up to four produced, ultimately reaching to about 600 mm long, the blades plane, narrowly lanceolate to nearly linear, 6–12 mm wide, soft-textured, the midrib moderately thickened and raised, the margins lightly thickened. *Stem* erect, scarcely flexed outward above the sheaths of the upper two leaves but becoming straight again, simple or occasionally with one branch, 2.5–3 mm in diameter below the spike.

Spike erect, barely flexuose, 7- to 9-flowered; *bracts* pale green, relatively soft-textured, the outer 25–30 mm long, the inner slightly shorter than to about as long as the outer, notched apically for c. 2 mm. *Flowers* deep pink, the lower three tepals with a broad longitudinal median band of white in the lower two-thirds, and usually a slender dark purple streak in the centre of the white band, the base of the lower tepals becoming dark purple and the throat of the perianth tube dark red-purple, unscented; *perianth tube* obliquely funnel-shaped, c. 20 mm long, the lower cylindrical part c. 10 mm long; *tepals* obovate to lanceolate, bluntly acute or sometimes sharply acute and apiculate, nearly equal, the lower three slightly shorter than the upper, the dorsal extending forward and nearly horizontal, c. 28 x 17 mm, the upper laterals arching gently outward from the base, c. 30 x 14–15 mm, the lower three tepals joined to the upper laterals for c. 2 mm, curving outward from the base and ultimately directed downward, c. 26 x 9–13 mm, the lower median slightly larger than the lower laterals. *Filaments* c. 17 mm long, exserted c. 8 mm from the tube; *anthers* c. 10 mm long, violet, the pollen lilac. *Ovary* oblong, c. 5 mm long; *style* arching over the stamens, dividing opposite the middle and upper third of the anthers, the branches recurving, c. 3.5 mm long, expanded in the upper half. *Capsules* ovoid, 20–25 mm long, three-lobed and retuse above; *seeds* ovate, c. 10 x 6 mm, broadly and evenly winged, translucent light brown, the seed body darker brown.

Chromosome number unknown.
Flowering time October to mid-November.

Distribution and biology

One of the few species of *Gladiolus* to occur in the southern African lowveld, *G. brachyphyllus* is restricted to eastern South Africa and adjacent northern Swaziland and southern Mozambique. Its range extends from Klaserie in the north, through the central and southeastern parts of the Kruger National Park to Tshaneni in Swaziland. This is predominantly a region of dry savanna woodland with well-drained sandy soils, high summer temperatures and low annual rainfall.

The flowering time, October and November, is remarkable in section *Densiflorus*.

This is the end of the dry season, and plants usually come into flower two to three weeks after the first light rains of the wet season have fallen. Nevertheless, they complete their flowering and fruit production under dry, inhospitable conditions. At flowering time the grass cover is dry and the flowering spikes, growing among scattered trees, are striking in the open veld. By the time the seeds are shed the wet season is usually well underway and conditions are suitable for rapid growth of newly germinated seedlings. The habit of flowering at the end of the dry season is one that is fairly common in *Gladiolus* and most species that do this have reduced leaves on the flowering stem. All photosynthetic activity takes place in the stems and leaf sheaths which remain green well into the wet season, and long after the capsules have ripened and the seeds are shed. Non-flowering individuals and immature plants produce leaves with long blades that emerge at about the same time as the flowering spikes of plants that will flower, but the leaves continue to elongate during the wet season and remain green after the aboveground parts of those plants that flowered have dried out.

The pollination biology of *Gladiolus brachyphyllus* is unknown. The perianth tube of intermediate length leads us to believe that it may be a generalist, able to be pollinated by a variety of large bees, tabanid flies with fairly long tongues, and perhaps even butterflies.

Diagnosis and relationships

The relatively large green floral bracts and large pink flower with white longitudinal nectar guides combined with the reduced leaves on the flowering stem are distinctive and make *Gladiolus brachyphyllus* easy to recognize. The size, colour and proportions of the flower, particularly the perianth tube which is about as long as the bracts and the nearly equal lower tepals which curve outward and are ultimately directed downward, are characteristic of the species of series *Scabridus* of section *Densiflorus* to which it is referred here. The soft-textured leaves are unusual in the section, several species of which have firm-textured leaves with a light scabrid pubescence. We suspect that the growth pattern of *G. brachyphyllus* allows it

to dispense with xeromorphic features, instead growing rapidly at the time of year most amenable for a geophyte. It is probably most closely related to the Mpumalanga dolomite endemic, *G. pavonia*, which has light pink flowers of similar shape and size and equally soft-textured, glabrous leaves. Both species also have purple to violet anthers which are not present in other members of the series.

History

Gladiolus brachyphyllus was discovered in 1913 by the prolific plant collector, Archdeacon F.A. Rogers, along the defunct Selati Railway that ran from the Mozambique border at Komatipoort through the Kruger National Park to Letaba. The type locality is a vague one covering virtually the entire length of the railway through the Park. The species was described not long afterwards, in 1917, by Frank Bolus who studied *Gladiolus* in some detail and described several new taxa of the genus at that time. Subsequent collecting has shown that the range of the species extends through the southern half of the Kruger National Park into northern Swaziland and southern Mozambique.

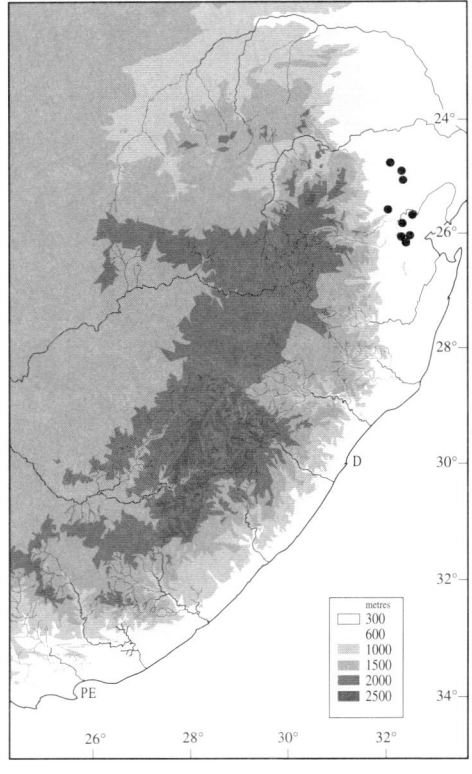

2. SECTION *OPHIOLYZA*

(Klatt) Goldblatt & Manning

Section *Ophiolyza* (Klatt) Goldblatt & Manning, new combination and new rank.

Synonymy

Gladiolus subgenus *Ophiolyza* Klatt, *Abh. Naturforsch. Ges. Halle* 15: 340 (Ergänzungen 6) (1882). Goldblatt, Gladiolus *in Trop. Africa* 174 (1966); not *Ophiolyza* Salisbury (1866) (= *Gladiolus* section *Hebea*). Type: *G. natalensis* Reinwardt ex J.D. Hooker (= *G. dalenii* Van Geel) (lectotype designated by Goldblatt, 1996).

Gladiolus group *Dracocephali* Baker, *J. Linn. Soc. Bot.* 16: 176 (1877). Type: *G. dracocephalus* J.D. Hooker (= *G. dalenii* Van Geel) (lectotype designated by Goldblatt, 1996).

Description

Plants usually medium-sized to large, with stems relatively thick, usually straight and often erect, branched in a few species. *Corm* globose, the tunics coriaceous or firm-papery, decaying into irregularly shaped fragments, or becoming coarsely fibrous. *Leaves* lanceolate to linear, forming a basal distichous fan, the blades usually sword-shaped, plane with margins and midrib moderately thickened, rarely linear or nearly terete (in tropical African *G. pallidus*), the upper leaves smaller and sometimes entirely sheathing, usually contemporary with the flowers but borne on separate shoots after flowering in some species.

Spike erect or inclined, more or less distichous in series *Oppositiflorus*; *bracts* fairly long, (12–)20–80 mm long, slightly shorter than to about as long as the perianth tube or much longer, occasionally pubescent to villous. *Flowers* medium to large, nectar guides usually pale blotches on the lower half of each of the three lower tepals, sometimes narrow median streaks, unscented; *perianth tube* usually slightly shorter than or about as long as the bracts; *tepals* usually unequal, the dorsal largest, strongly arched to hooded, the lower three smallest, not distinctly clawed and hardly united below. *Capsules* usually large, oblong, rounded apically and

often obscurely three-lobed above; *seeds* large, translucent, broadly and evenly winged.

Section *Ophiolyza* is widespread, occurring almost throughout Africa south of the Sahara and extending to Madagascar and southern Arabia, although it is absent from the southern African winter-rainfall zone. There are 15 species in southern Africa, mostly endemic, but *Gladiolus magnificus* and *G. sericeovillosus* extend into Zimbabwe and the former also into southern Angola. One species, *G. dalenii*, occurs throughout the range of the section. An additional 18 species are restricted to tropical Africa. The section is characterized by a fan of basal leaves, usually with narrowly lanceolate to sword-shaped blades, and moderate-sized to large flowers. The dorsal tepal is ascending to strongly hooded and larger than the lower tepals, which have weakly developed nectar guides consisting of a paler blotch on the lower half. The nectar guides are an important character for defining the section. A common trend in the section, and possibly a basal one, is the streaking or speckling of pigment of the tepals on a paler background.

We recognize three series in the section in southern Africa. Series *Oppositiflorus* has the least specialized and the smallest flowers in the section, but is specialized in having the flowers more or less opposed in a distichous spike. The floral bracts are distinctive, the outer somewhat inflated and the inner conspicuously biapiculate and having the lower margins united around the ovary. Several species have well-developed soft pubescence and all are usually at least lightly scabrid on the leaves and sometimes on the outer bracts.

The small series *Ecklonii* is characterized by unusually long, overlapping floral bracts that much exceed the flowers and, except in *Gladiolus rehmannii* which has whitish or pale lilac flowers, the tepals are usually conspicuously speckled with dark colour on a pale background.

The tropical African species fall mainly in series *Ophiolyza*, but we now recommend placing the northeast African *Gladiolus abyssinicus*, *G. watsonioides* and their close allies in a fourth series, *Abyssinicus*. This series would also include the large-bracted species, *G. dichrous*, which was originally placed in a separate genus, *Oenostachys*.

Series *Ophiolyza* comprises particularly large-flowered species, all with straight, erect spikes and flowers with elongate tubes at least as long as and often much longer than the bracts. A strong trend for the reduction in the size of the lower tepals in the series, occasionally to the degree that they are vestigial, is exemplified by *Gladiolus magnificus* from Botswana, Namibia and southern tropical Africa. In tropical Africa this trend has apparently also occurred in the northeast African *G. abyssinicus–G. longispathaceus* alliance (series *Abyssinicus*). Another notable specialization in the series is the temporal separation of the vegetative and reproductive phases of the growth cycle. This occurs in the southern African *G. aurantiacus* and in tropical Africa in *G. dalenii* subsp. *andongensis* and subsp. *welwitschii*, *G. melleri* and *G. roseolus*. The foliage leaves are produced on separate shoots in the wet season, after flowering is completed, and the flowering stems bear leaves with much-reduced or vestigial blades.

The ancestral pollination system in the section is probably that for long-tongued anthophorid bees, for example in *Gladiolus dolomiticus* and *G. sericeovillosus* (series *Oppositiflorus*) and in *G. ecklonii* (series *Ecklonii*). Pollination by sunbirds is evidently basal in series *Ophiolyza* but has been little studied. Observations are available only for *G. dalenii* and *G. flanaganii*. Pollination by the butterfly *Aeropetes tulbaghia* has evolved in *G. saundersii* and is evidently the pollination strategy in *G. cruentus* as well. Both have bright red flowers with white markings on the lower tepals and narrower perianth tubes than in bird-pollinated flowers.

21. GLADIOLUS DOLOMITICUS Obermeyer

PLATE 18

Gladiolus dolomiticus Obermeyer, *Bothalia* 12: 636 (1979). Type: South Africa, Northern Province, Makapansgat Valley, between Research Station and mountain top, 3 Mar. 1976, *Enslin s.n.* (PRE 56899, lectotype here designated, the most complete specimen; PRE, isolectotypes).

dolomiticus = of dolomite, so named for the rock substrate on which the species grows.

Description

Plants 70–100 cm high. *Corm* depressed-globose, c. 25 mm in diameter, the tunics coriaceous, soon decaying to become coarsely fibrous. *Cataphylls* membranous and pale, the uppermost reaching up to 10 cm above the ground and green, pubescent. *Leaves* seven or eight, the lower four or five basal and forming a lax distichous fan, pale green, the lower leaves longest, reaching or slightly exceeding the top of the spike, the blades narrowly lanceolate, 10–15 mm wide, minutely scabrid on the veins, the intercostal surfaces minutely pubescent, the midribs, at least one pair of secondary veins and the margins heavily thickened and raised, the

upper leaves inserted on the stem well above the ground, with fairly short blades, the margins of the uppermost one or two leaves free to the base. *Stem* erect, softly pubescent, often with one to three short branches, lightly compressed and oval in transverse section, 4–5 x 3–4 mm in diameter below the spike.

Spike straight and erect, more or less distichous but flowers secund, 10- to 17-flowered; *bracts* pale green below, becoming dry and light brown above, minutely pubescent, the outer 26–30 mm long, attenuate, the apex often twisted outward, the inner slightly shorter than to about as long as the outer, the margins united below into a tube c. 10 mm long, forked apically for c. 2 mm. *Flowers* pale pink, the lower tepals each with a yellow blotch in the midline, the throat pale yellow, unscented; *perianth tube* narrow, obliquely funnel-shaped, 18–27 mm long; *tepals* unequal, lanceolate-elliptic, the upper three largest, the dorsal arching over the stamens, 29–38 x 10–12 mm, the upper laterals 30–40 x 12–15 mm, the lower three tepals joined below for 1–2 mm, the lower laterals 25–30 x 7 mm, the lower median 27–35 x 7–12 mm. *Filaments* 9–14 mm long, exserted 4–6 mm from the tube; *anthers* (5–)8–10 mm long, with a short apiculus, pale yellow, the pollen pale yellow. *Ovary* ovoid, 5–7 mm long; *style* arching over the stamens, dividing opposite or 1–2 mm beyond the anther apices, the branches 4–6 mm long, barely expanded above. *Capsules* narrowly obovoid-oblong, 25–35 mm long, three-lobed and retuse apically; *seeds* ovate, c. 7 x 4.5 mm, broadly and evenly winged, translucent light brown, the seed body unusually large and dark brown. *Chromosome number* unknown.

Flowering time March and April.

Distribution and biology

A highly localized endemic of the southern African summer-rainfall zone, *Gladiolus dolomiticus* is known for certain only from Makapan's Valley (Makapansgat) in the dolomite hills of Northern Province. The species occurs in fairly dry woodland on rocky dolomite slopes. In this dry environment *G. dolomiticus* flowers relatively late, in March and April, toward the end of the wet season. The finely pubescent leaf surfaces and heavily thickened midrib and secondary veins appear to be adaptations for its dry

habitat. Exploration of the dolomite hills to the north may well extend the range of *G. dolomiticus*, which at present is recorded from a single extended population. Plants collected in fruit in March 1996 some 10 km northwest of Potgietersrus on the low dolomite hill near Ysterberg Siding may well prove to be a second locality for the species. This will not be known for certain until flowering material is available to confirm our preliminary identification.

The flowers of *Gladiolus dolomiticus* are adapted for pollination by long-tongued bees. We have recorded *Amegilla* and *Xylocopa* species, both medium-sized Anthophoridae, foraging for nectar in the large flowers and in so doing accomplishing cross-pollination.

Diagnosis and relationships

The general aspect of *Gladiolus dolomiticus*, with its moderate-sized, short-tubed, pale pink flower, its rigid, pubescent leaves with heavily thickened margins and primary and secondary veins, and its nearly distichous spike, is typical of series *Oppositiflorus* of section *Ophiolyza*. The specialized floral bracts, the inner with the margins united around the base of the flower and the apex deeply forked into dry, filiform cusps, confirm the

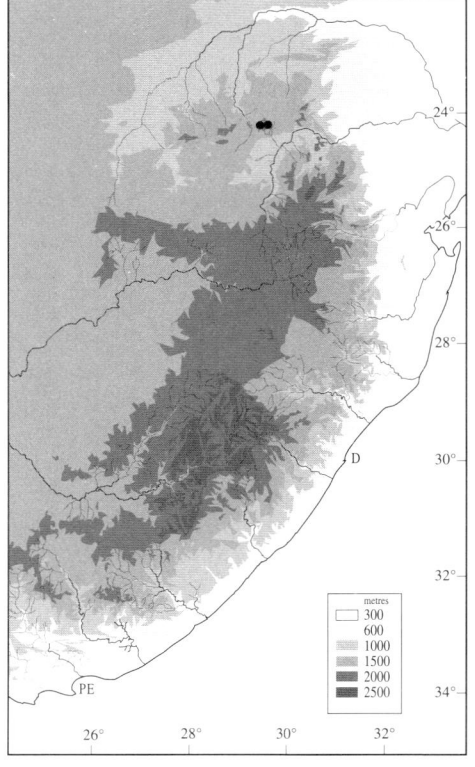

position of *G. dolomiticus* in the series, in which it appears to be a primitive relict.

The relationships of *Gladiolus dolomiticus* probably lie most closely with *G. sericeovillosus* and its near relative, *G. pole-evansii*, an endemic of granite outcrops in the bushveld of central Mpumalanga. In particular, the fine pubescence on the leaf, the large capsules and the seeds with a large, dark body suggest that *G. dolomiticus* and *G. pole-*

evansii are closely allied, and perhaps derived independently from the ancestral stock of the series. Although these three species are broadly similar, they are not likely to be confused. *Gladiolus dolomiticus* is readily recognized by its lanceolate-elliptic tepals and rather short anthers, 4–6 mm long.

History

Gladiolus dolomiticus was discovered by

the anthropologist, Brian Maguire, in 1975 in the course of a plant survey of the Makapansgat (Makapan's Valley) archaeological site. An expedition to the site made by botanists from the Botanical Research Institute the year after Maguire's discovery resulted in the collection of type material, and the species was described in 1979 by A.A. Obermeyer.

22. GLADIOLUS POLE-EVANSII Verdoorn

Gladiolus pole-evansii Verdoorn, *Fl. Pl. Africa* 35: pl. 1373 (1962). Lewis et al., *J. S. African Bot.*, Suppl. 10: 69 (1972). Type: South Africa, Mpumalanga, bushveld near Bronkhorstspruit, Mar. 1936, *Pole Evans 3907* (PRE, lectotype here designated, the specimen with the best preserved flowers; BOL, K, PRE, isolectotypes).

pole-evansii, named for the eminent botanist and first director of the Botanical Research Institute (of South Africa), I.B. Pole Evans, who discovered the species in 1936.

Description

Plants 60–110 cm high. *Corm* globose, 15–22 mm in diameter, the tunics firm-papery, soon decaying to become coarsely fibrous. *Cataphylls* membranous and pale, the uppermost reaching up to 10 cm above the ground and green, pubescent. *Leaves* eight or nine, the lower five or six basal and forming a lax distichous fan, glaucous-green, the lower leaves longest, reaching to about the base, or sometimes to the middle of the spike, the blades narrowly lanceolate, 10–15 mm wide, minutely scabrid on the veins, the intercostal surfaces pubescent, the midrib and at least one other pair of veins heavily thickened and raised, the margins lightly thickened, the upper leaves inserted on the stem well above the ground, with fairly short blades or entirely sheathing, the margins of the sheathing leaves united for half their length. *Stem* erect, minutely scabrid, often with one, rarely up to three branches, lightly compressed, 3 x 2 mm in diameter below the spike.

Spike weakly inclined, often weakly subdistichous, scabridulous, more or less straight, 15- to 20-flowered; *bracts* pale green, minutely papillose-pubescent, the apices usually more or less dry at anthesis, often exuding sugar solution near the apices, 16–25 mm long, shorter than to about as long as an internode, occasionally one and a

half internodes long, the inner about as long as to slightly shorter than the outer, obtuse, sometimes obscurely notched at the apices, the margins united around the flower in the lower half. *Flowers* translucent flesh-pink, becoming almost colourless at the tepal apices, the upper lateral tepals with narrow red lines along the midline, the lower tepals, or sometimes only the lower median, yellow in the lower midline and with a red median streak distal to the yellow mark, the base of the upper part of the tube often dark red; *perianth tube* obliquely funnel-shaped, c. 15 mm long, the wider upper part exserted from the bracts; *tepals* lanceolate, the dorsal largest, extended horizontally over the stamens, c. 28 x 16 mm, the upper laterals extending forward below, curving outward and patent above, 24–26 x c. 12 mm, the lower three tepals straight, inclined toward the ground, 21 x c. 9.5 mm. *Filaments* 18–20 mm long, exserted c. 10 mm from the tube; *anthers* 7.5–9 mm long, light purple, the pollen yellow. *Ovary* narrowly obovoid, c. 5.5 mm long; *style* arching over the stamens, dividing close to the anther apices, the branches c. 4 mm long, extending well beyond the stamens. *Capsules* oblong, three-lobed and retuse above, 20–25 mm long; *seeds* oval to oblong, 8–10 x 4.5–5.5 mm, the wing usually well developed only on one side, grey-brown, the seed body unusually large, dark brown. *Chromosome number* 2n = 30.

Flowering time mid-January to March.

Distribution and biology

A fairly narrowly distributed species, *Gladiolus pole-evansii* occurs in the granite hills between Verena and Groblersdal in

Mpumalanga, South Africa. Plants are quite common in their habitat, but evidently do not occur away from this area of granite basement rock. The plants grow among loose boulders, often with their corms firmly wedged in rock crevices, making it nearly impossible to extract them.

One curious feature of the species is the exudation of sugar solution on the bracts. The sugar evidently attracts large ants, which can normally be seen crawling along the spikes and foraging on this food source. The presence of these ants may play a role in deterring herbivores or robbers of the nectar

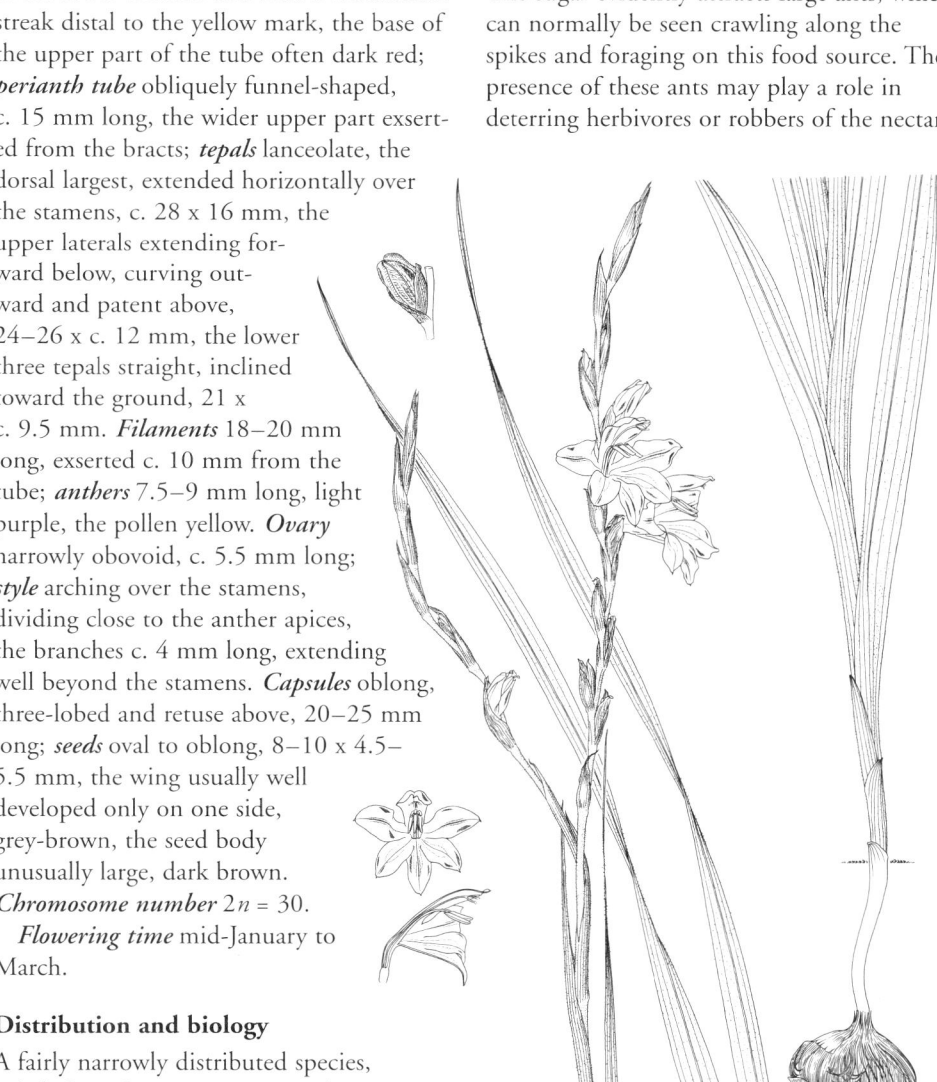

in the perianth tube. The short-tubed flowers are similar in size and general shape to those of related species such as *Gladiolus dolomiticus* and *G. sericeovillosus* and are probably also pollinated by large, long-tongued bees of the family Anthophoridae.

Diagnosis and relationships

Gladiolus pole-evansii is readily distinguished by its tall stature, most plants exceeding 1 m, and the many-flowered spikes of medium-sized translucent pink flowers. The form of the flowers is unremarkable. The tepals are nearly equal, and the dorsal is horizontal and more or less hooded over the stamens. The lower tepals are as long as the upper, and have obscure pale yellow nectar guides toward the bases. The spike is secund or at best weakly two-ranked, but not obviously distichous as suggested by Obermeyer (in Lewis et al., 1972).

Both the weakly distichous spikes and inner floral bracts united basally around the flower confirm that *Gladiolus pole-evansii*

belongs in series *Oppositiflorus*. Its immediate affinities appear to be with large-flowered *G. dolomiticus* and with *G. sericeovillosus*. However, the peculiar extrafloral nectaries on the bracts appear to be a feature unique to this species.

History

Gladiolus pole-evansii was discovered in 1936 by I.B. Pole Evans, then director of the South African Botanical Research Institute. Pole Evans, an avid botanical explorer and collector, found the species in the bushveld of what was then the Transvaal close to the northern edge of the escarpment near Verena in the Bronkhorstspruit District. In 1959, after revisiting sites where the species had been recorded, the Pretoria botanist, Inez Verdoorn, collected living plants for cultivation and illustration. She described *G. pole-evansii* in 1962 in the series, *Flowering Plants of Africa*, where it was accompanied by a painting by the celebrated South African artist, Cythna Letty.

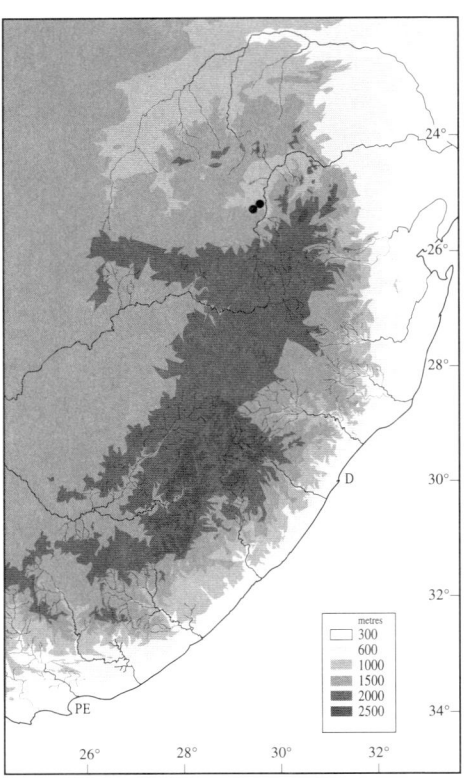

23. GLADIOLUS OPPOSITIFLORUS Herbert
PLATE 19

Gladiolus oppositiflorus Herbert, *Edward's Bot. Reg.* 28: Miscell. 86 (1842). Baker, *Fl. Capensis* 6: 154 (1896). Lewis et al., *J. S. African Bot.*, Suppl. 10: 23 (1972). Type: thought to have come from Madagascar but certainly from South Africa, probably from coastal Eastern Cape, cultivated at Spofforth, Yorkshire, Great Britain, *Herbert s.n.* (K, holotype; PRE, photo).

oppositiflorus = opposite-flowered, for the spike having successive flowers facing in opposite directions, a remarkable feature in *Gladiolus*.

Synonymy

Gladiolus salmoneus Baker, *Handbook Irideae* 217 (1892); *Fl. Capensis* 6: 153 (1896). L. Bolus, *Fl. Pl. S. Africa* 6: pl. 237 (1926). *Gladiolus oppositiflorus* subsp. *salmoneus* (Baker) Obermeyer in Lewis et al., *J. S. African Bot.*, Suppl. 10: 26 (1972). Type: South Africa, KwaZulu-Natal, mountains around Kokstad, 1600 m, Mar. 1883, *Tyson 1180* (K, holotype; BOL, SAM, isotype).
Gladiolus blackwellii L. Bolus, *Ann. Bolus Herb.* 3: 77 (1921). Rendle, *Curtis's Bot. Mag.* 147: pl. 8919–20 (1938). Type: South Africa, Eastern Cape, Tembuland, cultivated in Cape Town, Dec. 1917, *Blackwell s.n.* (BOL 16974, lectotype here designated; K, PRE); Feb. 1919, *Blackwell sub Marloth 7793* (BOL, K, syntypes).

Description

Plants (40–)60–160 cm high. *Corm* depressed-globose, 30–40 mm in diameter, the tunics firm-papery to coriaceous, light brown, fragmenting irregularly but usually extending upward a short distance as a coarse fibrous neck around the stem base. *Cataphylls* pale and membrano-coriaceous, the uppermost reaching about 6–10 cm above the ground and then green and often velutinous. *Leaves* usually seven or eight, mostly basal, reaching at least to the base of the spike or sometimes shortly exceeding it, imbricate and concealing the lower two-thirds of the stem, the blades narrowly sword-shaped to nearly linear, 12–18(–24) mm wide, firm to rigid, the midrib and margins strongly thickened, usually velutinous between the veins, the uppermost leaf inserted in the middle of the stem, like the basal but shorter. *Stem* erect and straight, often sparsely pubescent, unbranched, 5–6 mm in diameter below the spike.

Spike straight and erect, distichous or sub-distichous and then the flowers in two ranks 80°–180° apart, 7- to 15-, occasionally to 26-flowered; *bracts* green, firm-textured, attenuate, sometimes sparsely pubescent to velutinous below, (27–)32–50(–62) mm long, the inner with the margins united below and sheathing the base of the tube,

slightly shorter than the outer, the apices oblique, forked apically into attenuate cusps 2–4 mm long. *Flowers* salmon to pale pink or mauve, the lower three tepals paler in the midline and each with a reddish to purple median streak in the lower half, the tube red or purple in the throat and sometimes on the reverse, unscented; *perianth tube* narrow and curving outward, expanded slightly near the apex, (28–)40–50(–55) mm long; *tepals* unequal, lanceolate-elliptic, the margins undulate, the upper three largest, (37–)45–60 mm long, the dorsal 20–25(–30) mm wide, arching over the stamens, the upper laterals curving outward in the upper half when open, the lower three tepals united for 3–4 mm, the lower laterals shorter than the lower median, c. 13 mm wide, directed forward and arching downward distally, in profile exceeding the upper tepals. *Filaments* 15–17(–20) mm long, exserted 6–10 mm from the tube; *anthers* c. 11 mm long, reaching to about the middle of the dorsal tepal, light mauve, pollen cream. *Ovary* oblong, 5–7 mm long; *style* arching over the stamens, dividing 2–4 mm beyond the anther apices, the branches 6–7 mm long, gradually expanded above. *Capsules* narrowly obovoid-oblong, three-lobed and truncate to retuse apically, 20–25 mm long; *seeds* ovate, c. 6 x 4 mm, broadly and evenly winged, light brown, the seed body darker brown.

Chromosome number 2*n* = 30.

Flowering time mostly February and March, occasionally in December or April to early May.

Distribution and biology

Restricted to the southern part of the southern African summer-rainfall zone, *Gladiolus oppositiflorus* occurs in southern KwaZulu-Natal and the Eastern Cape. Its range extends from the Polela District in KwaZulu-Natal in the north through the mountainous northeastern Transkei southward to the coast near East London. Populations extend inland right to the Lesotho border near Ramatsileso's Gate and Naude's Nek passes, but *G. oppositiflorus* has not yet been recorded in Lesotho itself. Plants grow in open grassland, often in rocky sites which afford the corms some protection from predation.

The large, pink to salmon-coloured flowers with a long, slender perianth tube appear to be adapted for pollination by long-tongued flies, probably *Prosoeca ganglbaueri* of the Nemestrinidae.

Diagnosis and relationships

Gladiolus oppositiflorus can readily be recognized by the two-ranked flower spikes, the flowers either completely opposed or in ranks at least 80° apart, the characteristic inflated floral bracts and the elongate perianth tube. The flowers are the largest encountered in the series, at least 65 mm

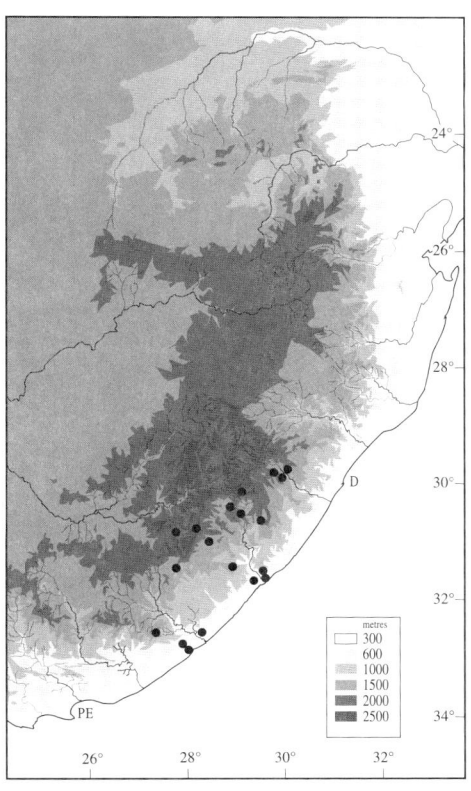

long and mostly 80–110 mm long, and the perianth tube is usually in the 40–55 mm range. The sword-shaped leaves have thickened midribs and secondary veins and, like most other members of series *Oppositiflorus*, are pubescent. In *G. oppositiflorus* the leaves are usually conspicuously velvety between the veins. The inner floral bracts are biapiculate and the lower margins are united in the lower half, thus sheathing the ovary, a feature of all species of the series.

The immediate relationships of *Gladiolus oppositiflorus* most probably lie with *G. elliotii* and *G. sericeovillosus*, the two other species of series *Oppositiflorus* that have strongly two-ranked spikes and the floral bracts inflated. Both *G. elliotii* and *G. sericeovillosus* have smaller flowers that are adapted for pollination by long-tongued anthophorid bees, the ancestral pollination system in section *Ophiolyza*.

There are two main variants of *Gladiolus oppositiflorus*. The typical one is the Eastern Cape coastal series of populations that are usually evergreen and have tall spikes up to 1 m or more with the flowers in opposed ranks, thus completely distichous. The flowers are shades of bright pink to mauve and the tepals tend to open widely. Inland in the hilly country of the interior Eastern Cape and southern KwaZulu-Natal plants are always deciduous and the spikes are shorter, sometimes only 50 cm high or even less. The salmon-coloured flowers are borne in two ranks separated by 100–150°. The two forms were treated by Lewis et al. (1972) as separate subspecies. We have found that they intergrade in the central eastern Cape and the existence of a range of intermediates makes it impossible to treat them as separate taxa.

History

There is no record of who first collected *Gladiolus oppositiflorus*. The species makes its appearance in the literature in 1837 as a plant cultivated in Yorkshire, England. There, at Spofforth in the garden of William Herbert, a pioneer plant breeder and expert on bulbous plants, *G. oppositiflorus* was prized for its many-flowered spikes of large, showy flowers. The species was used by Herbert in his early attempts to produce

hybrid gladiolus for the garden and was formally described by him in 1842. According to Herbert, *G. oppositiflorus* was called, informally perhaps, *G. floribundus*. Thus early references to *G. floribundus* used in plant breeding refer not to the winter-rainfall species of that name, but to *G. oppositiflorus*.

The species was very important in the history of *Gladiolus* breeding. The famous hybrid strain, '*G. gandavensis*' raised in Ghent, Belgium, is reputedly the result of crossing *G. oppositiflorus* and *G. dalenii* (= *G. natalensis*). A second important early garden hybrid, '*G.* x *ramosus*', is either a direct interspecific cross, *G. oppositiflorus* x *G. cardinalis*, or *G. oppositiflorus* x *G. cardinaloblandus*. The latter, of course, is a hybrid between *G. cardinalis* and *G. carneus* (which was long known by the synonym, *G. blandus*).

The shorter-stemmed form of *Gladiolus oppositiflorus* with salmon-coloured flowers was first collected in the later nineteenth century, and plants gathered by William Tyson near Kokstad in southern KwaZulu-Natal were described as *G. salmoneus* by J.G. Baker in 1892. This same form from 'Tembuland', that is, northern Eastern Cape, was sent to the Bolus Herbarium in about 1916 and was described by H.M.L. Bolus in 1921. Both *G. oppositiflorus* and *G. salmoneus* were recognized by Baker in his

account of *Gladiolus* in *Flora Capensis* (Baker, 1896). *Gladiolus salmoneus* was, however, regarded as a subspecies of *G. oppositiflorus* by Lewis et al. (1972). We do

not consider the differences between the two sufficient to warrant even this status and treat them as conspecific.

24. GLADIOLUS SERICEOVILLOSUS J.D. Hooker
PLATE 20

Gladiolus sericeovillosus J. D. Hooker, *Curtis's Bot. Mag.* 90: pl. 5427 (1864). Baker, *Fl. Capensis* 6: 151 (1896). Lewis et al., *J. S. African Bot.*, Suppl. 10: 30 (1972). *Antholyza hirsuta* Klatt, *Linnaea* 35: 379 (1868), an illegitimate name for *G. sericeovillosus.* Type: South Africa, without precise locality 'in the interior of the Cape Colony', c. 1862, cultivated in England, *Cooper s.n.*, illustration in *Curtis's Bot. Mag.* 90: pl. 5427 (1864), no preserved specimens known.

sericeovillosus = silky-haired, referring to the fine, whitish pubescence on the leaves and bracts of typical subspecies *sericeovillosus.*

Synonymy
For full synonymy see under each subspecies.

Description
Plants 35–100 cm high. **Corm** depressed-globose, 25–35 mm in diameter, tunics firm-papery to coriaceous, decaying into vertical fibrous strips, usually drawn into stiff points above. **Cataphylls** pale and membranous below the ground, the upper two green above the ground, the uppermost reaching up to 16 cm above the ground, minutely pubescent. **Leaves** five to seven, the lower four or five basal, exceeding the spike sometimes by twice as much or reaching only to about the base of the spike, villous to microscopically pubescent on the sheaths and blades, sometimes sparsely so on the lower parts of the leaves or entirely glabrous, the blades narrowly sword-shaped to linear, 3–16 mm wide, the margins, midrib and often one or more pairs of secondary veins moderately to strongly thickened and hyaline, the upper two or three leaves inserted on the upper half of the stem, much shorter than the basal, the two uppermost channelled almost to the apices, the margins of the sheathing part open to the base. **Stem** straight and erect, occasionally branched, sparsely pubescent where not sheathed by the leaves, or smooth, 4–6 mm in diameter below the spike.
 Spike distichous, erect and straight, the axis sometimes pubescent, 12- to 20-, some-

times to 40-flowered; **bracts** pale green and soft-textured, becoming membranous toward the apices and usually dry and pale in the upper 5 mm, villous to minutely pubescent or smooth, 15–30 mm long, acute-attenuate, the apices often twisted, the inner with the margins usually united around the base of the flower, usually slightly shorter (occasionally slightly longer) than the outer, forked apically into attenuate cusps 2–4 mm long. **Flowers** greenish to cream, pink, pale lilac or dull red, the tepals unicoloured or minutely dotted with dark red to maroon points, the dots often concentrated in the midline of each tepal, the lower lateral (and sometimes the lowermost) tepals yellow-green in the lower half, or with darker spade-shaped markings distally with a yellow centre, unscented; **perianth tube** obliquely funnel-shaped, widening and curving outward near the apex, 10–16 mm long, extended between the bracts; **tepals** unequal, the dorsal longest and hooded over the stamens, (18–)25–32 x 12–15 mm, the upper laterals directed forward, usually curving outward in the upper third, 22–30 x 13–15 mm, the lower three tepals united for c. 1.5 mm, the lower laterals 16–22 x c. 7 mm, curving outward above, narrowed somewhat below but not abruptly expanded into limbs above, the lower median curving toward the ground, 20–25 x 10 mm. **Filaments** 10–12 mm long, exserted 6–8 mm from the tube; **anthers** 8–10 mm long, reaching the upper third of the dorsal tepal, yellow, the pollen whitish. **Ovary** narrowly oblong, 4–6 mm long; **style** arching over the stamens, dividing between the base and upper third of the anthers, the branches 4–5 mm long. **Capsules** obovoid to oblong, 12–23 mm long, the apices three-lobed and auriculate above; **seeds** ovate, 5–7 x c. 4 mm, broadly and evenly winged, more or less translucent light brown. **Chromosome number** $2n$ = 30.
 Flowering time late summer and autumn, mostly February to April.

Distribution and biology
A widespread species of eastern southern Africa, *Gladiolus sericeovillosus* occurs almost

throughout summer-rainfall South Africa as well as Lesotho and Swaziland, and extends northward into Zimbabwe. There it has been recorded from Belingwe in the south to Inyanga and Mount Darwin in the north, and it is common throughout the eastern half of the country. In South Africa the southernmost station of *G. sericeovillosus* is at Baziya, a short distance west of Umtata. From there it extends across all of KwaZulu-Natal, the eastern Free State and Gauteng into the Soutpansberg in Northern Province. Plants grow in a variety of habitats, but are most often found in light woodland or open bush.

The short-tubed and relatively small flowers are pollinated by anthophorid bees, including *Amegilla fallax.*

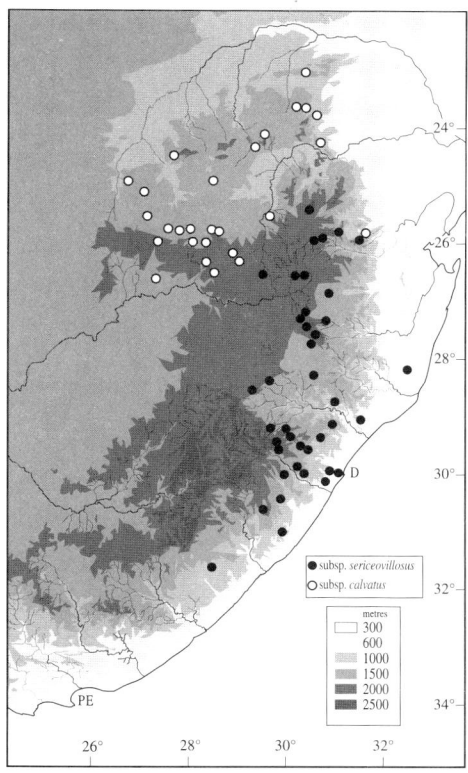

subsp. *sericeovillosus*
subsp. *calvatus*

metres
300
600
1000
1500
2000
2500

Diagnosis and relationships

Gladiolus sericeovillosus is largely recognized by its vegetative features in combination with the inflated floral bracts. The relatively long leaf sheaths form a pseudostem and the leaf blades arise well above ground level and, in subsp. *sericeovillosus*, are always conspicuously hairy, as are the bracts, stem and spike axis. In subsp. *calvatus* plants are always glabrous, but have narrow leaves that usually exceed the stem, and the margins, midrib and other veins are strongly thickened. The flowers of *G. sericeovillosus* are moderate in size and have an arcuate dorsal tepal, as do other members of series *Oppositiflorus*. They are green, yellowish, cream or dull red, often with dark red to purple streaks or flecks, especially on the upper tepals.

The inflated floral bracts ally *Gladiolus sericeovillosus* with the two other species of the series with similar bracts, *G. elliotii* and *G. oppositiflorus*. The latter has large pink flowers with a long perianth tube and is readily distinguished from *G. sericeovillosus*. Shorter in stature, *G. elliotii* lacks the elongate leaf sheaths that form a pseudostem. The blade margins are strongly thickened but the midrib and remaining veins are fine and closely set, giving the leaves a quite different appearance from those of either subspecies of *G. sericeovillosus*.

History

Although widespread in eastern southern Africa, *Gladiolus sericeovillosus* did not become known scientifically until relatively late. The species first appeared in the literature in 1864 when plants that flowered at the Royal Botanic Gardens, Kew, were painted and a description was published by J.D. Hooker in *Curtis's Botanical Magazine*. These plants were originally gathered by Thomas Cooper who collected in South Africa from 1859 to 1862 and sent corms and seeds back to Britain for cultivation. Cooper's collection was not, however, the first one made. J.F. Drège recorded the species during his pioneering botanical expedition from Cape Town to southern KwaZulu-Natal in 1832, and in herbaria there are records of uncertain origin but dated March, 1845. These specimens are attributed to C.W.L. Pappe, the Cape Colonial Botanist, and the sheet at the Kew Herbarium has the annotation *Gladiolus ludwigii* in Pappe's hand. Pappe did not travel so far from Cape Town, and these specimens may have been collected by C.L. Zeyher, whose personal herbarium collection Pappe took charge of after Zeyher's death. The collection is the type of *G. sericeovillosus* var. *ludwigii*, described by J.G. Baker in 1877 and later raised to species rank by Baker in 1892. It is not clear why Baker thought this was distinct from *G. sericeovillosus* itself, nor does there seem any justification for Otto Kuntze describing *G. rubicundus* based on plants with reddish flowers that he collected near Estcourt in 1894. A green-flowered form of *G. sericeovillosus* illustrated in *Curtis's Botanical Magazine* (plate 6291) in 1877 was misidentified as *G. ochroleucus* by Baker who was evidently uncertain of the distinction between the two species at that date. Later, however, he correctly recognized them as separate species.

The narrow-leafed and glabrous subsp. *calvatus* was first collected by Ernest Galpin near Barberton in eastern Mpumalanga in 1890 and was described as *Gladiolus ludwigii* var. *calvatus* by Baker in 1896. A later collection of subsp. *calvatus* from near Johannesburg was described as *G. rigidifolius* by Baker in 1904 and a collection from Marondera in Zimbabwe made by Gretel Dehn was described as *G. dehnianus* by Hermann Merxmüller in 1954. Zimbabwean collections of subsp. *calvatus* were referred to *G. elliotii* by A.A. Obermeyer in the revision of *Gladiolus* in South Africa (Lewis et al., 1972), but these plants are virtually identical to subsp. *calvatus* in South Africa and this decision was clearly a mistake. Subsp. *calvatus* was regard-ed as a forma of *G. sericeovillosus* by Lewis et al. and then as a subspecies by Goldblatt (1993), a rank which we believe fairly reflects its status as a morphologically and geographically well-defined series of populations of *G. sericeovillosus*.

GLADIOLUS SERICEOVILLOSUS subsp. SERICEOVILLOSUS

Synonymy

Gladiolus sericeovillosus var. *ludwigii* Pappe ex Baker, *J. Linn. Soc.* 16: 175 (1877). *Gladiolus ludwigii* (Pappe ex Baker) Baker, *Handbook Irideae* 215 (1892); *Fl. Capensis* 6: 150 (1896). Type: South Africa, KwaZulu-Natal, without precise locality, Mar. 1845, *Pappe s.n.* (K, holotype).
Gladiolus sericeovillosus var. *rubicundus* O. Kuntze, *Revis. Gen.* 3, 2: 308 (1898). Type: South Africa, KwaZulu-Natal, Estcourt District, Highlands Station, 15 Mar. 1894, *Kuntze s.n.* (NY, not seen, holotype; K, isotype).
[*Gladiolus ochroleucus* sensu Baker, *Curtis's Bot. Mag.* 103: t. 6291 (1877), not sensu Baker (1876).]

Description

Plants villous to pubescent on the cataphylls, leaf sheaths, often the blades, and on the stems and bracts. *Leaves* usually reaching to about the base of the spike, lanceolate to sword-shaped, 8–16 mm wide, the margins, midrib and other veins lightly to moderately thickened. *Flowers* usually yellowish green, irregularly flecked or streaked with purple, sometimes creamy yellow or green, rarely reddish, the lower tepals each with cream nectar guides in the lower half.

Flowering time February to April, occasionally in January.

Distribution

Subsp. *sericeovillosus*, the southern subspecies of *Gladiolus sericeovillosus*, is especially common in the KwaZulu-Natal interior but it extends from Baziya in the northern Eastern Cape in the south to the southern Mpumalanga highveld along a line extending from Ermelo to Barberton. Plants favour rich soils in well-watered grassland and are often, at least in KwaZulu-Natal, found in light woodland and bush.

GLADIOLUS SERICEOVILLOSUS subsp. CALVATUS (Baker) Goldblatt

Subsp. *calvatus* Goldblatt, *Fl. Zambesiaca* 12: 88 (1993). Gladiolus *in Tropical Africa* 204 (1996). *Gladiolus ludwigii* var. *calvatus* Baker, *Fl. Capensis* 6: 150 (1896). Phillips,

Fl. Pl. S. Africa 4: t. 125 (1924). *Gladiolus sericeovillosus* forma *calvatus* (Baker) Obermeyer, *J. S. African Bot.*, Suppl. 10: 31 (1972). Type: South Africa, Mpumalanga, Umvoti Creek, Apr. 1890, *Galpin 925* (K, lectotype, designated by Lewis et al., 1972: 31: BOL, SAM, isolectotypes).

calvatus = bald, indicating the absence of pubescence in this subspecies.

Synonymy

Gladiolus rigidifolius Baker, *Bull. Herb. Boissier*, Sér. 2, 4: 1006 (1904). Type: South Africa, Gauteng, Modderfontein, Dec. 1897, *Conrath 577* (Z, holotype; GZU, not seen, K, isotypes, PRE, photo).
Gladiolus sericeovillosus var. *glabrescens* L. Bolus, *S. African Gard.* 18: 213 (1928). Type: South Africa, Mpumalanga, Waterval Boven (grown at Kirstenbosch Botanic Gardens in 1928),

Hutchinson s.n. (BOL, as *Nat. Bot. Gard.* 873/27, holotype on three sheets).
Gladiolus dehnianus Merxmüller, *Proc. & Trans. Rhodesia Sci. Assoc.* 43: 151 (1951) and in Suessenguth & Merxmüller, *Contrib. Fl. Marandellas District* 76 (1951). Type: Zimbabwe, Marondera, 30 Mar. 1941, *Dehn 14* (M, holotype, not seen; K, SRGH, isotypes).
[*Gladiolus elliotii* sensu Lewis et al., *J. S. African Bot.*, Suppl. 10: 27 (1972), in part, not of Baker, *J. Bot.* 29: 70 (1891).]

Description

Plants usually entirely glabrous excepting the cataphylls. *Leaves* usually reaching the top of the spikes or up to twice as long as the stems, the blades linear, stiffly erect, 3–6(–8) mm wide, the margins, midrib and other veins strongly thickened. *Flowers* usually greenish, irregularly flecked or streaked with red, sometimes greenish yellow or dull red, the

lower tepals each with cream nectar guides in the lower half.

Flowering time February to April, occasionally in January.

Distribution

Subsp. *calvatus* extends from Mpumalanga, South Africa, through Northern Province into Zimbabwe where it occurs widely across the eastern half of the country. Plants grow in tall, open grassland or in light woodland in well-drained soils. It seems likely that subspecies *calvatus* also occurs in adjacent parts of Mozambique but we have seen no records from there. Although the ranges of the two subspecies are largely complementary, some overlap occurs in southern Mpumalanga, especially in the Barberton District where both subspecies have been recorded.

25. GLADIOLUS ELLIOTII Baker
PLATE 21

Gladiolus elliotii Baker, *J. Bot.* 29:70 (1891); *Fl. Capensis* 6: 150 (1896). Lewis et al., *J. S. African Bot.*, Suppl. 10: 27 (1972). Goldblatt, *Fl. Zambesiaca* 12(4): 88 (1993). Type: South Africa, Mpumalanga, Bethal District, Steenkoolspruit, marshy places, probably Nov. 1889 but 1891 on the specimen, *Scott Elliot 1557* (K, holotype).

elliotii, named in honour of George Scott Elliot, a British biologist who collected plants in southern Africa in 1888 and 1889 and produced pioneering papers on pollination of the flora.

Description

Plants 40–60 cm high. *Corm* depressed-globose, 20–30 mm in diameter, the tunics firm-papery to coriaceous, decaying into irregularly shaped fragments with age. *Cataphylls* pale and coriaceous, the uppermost reaching 4–6 cm above the ground and then green often flushed with purple. *Leaves* five or six, the lower four basal and forming a fairly tight distichous fan, coriaceous, often minutely puberulous, reaching to about the base of the spike or occasionally somewhat longer, lanceolate, rarely almost linear, 14–20 mm wide, the margins strongly thickened and hyaline, the midrib not or barely distinguishable from the remaining secondary veins, these also prominent but not hyaline, the leaves overlapping one another and the uppermost leaf inserted on the middle of the stem, largely sheathing and

enclosing the stem and sometimes the base of the spike, with a short free apex. *Stem* straight and erect, occasionally branched, slightly compressed, 4–5 mm in diameter below the spike.

Spike straight and erect, 14- to 20-, occasionally to 25-flowered; *bracts* pale green and soft-textured, slightly inflated, the outer (18–)25–37 mm long, slightly more than two internodes long, with two folds or ridges in the lower half, dry toward the apices and folded back, the inner about as long as the outer, the lower margins fused around the ovary. *Flowers* bluish grey to cream, the tepals minutely speckled with blue to red dots, more densely so along the midline of the upper tepals, the lower lateral tepals yellow in the lower half, unscented; *perianth tube* obliquely funnel-shaped, c. 16 mm long, the lower cylindrical part c. 8 mm long; *tepals* ovate-lanceolate, unequal, the dorsal and upper laterals nearly equal, c. 28 x 17 mm, the dorsal arching horizontally over the stamens, the lower laterals directed forward for their entire length or curving outward toward their apices, the lower three tepals united for c. 2 mm, curving outward and downward from the base, the lower laterals c. 23 x 8.5 mm, the lower median c. 25 x 11.5 mm. *Filaments* 11–12 mm long, exserted c. 4–5 mm from the tube; *anthers* c. 12 mm long, mauve, pollen cream. *Ovary* oblong, c. 6 mm long, slightly wider above; *style* dividing between the base and middle of the anthers, the

branches c. 6 mm long. *Capsules* oblong, c. 25 mm long, the apices three-lobed and auriculate above; *seeds* oval to round, 6 x 5 mm, broadly and evenly winged, dark red-brown. *Chromosome number* unknown.

Flowering time late November to early February, occasionally later.

Distribution and biology

A species of the southern African summer-

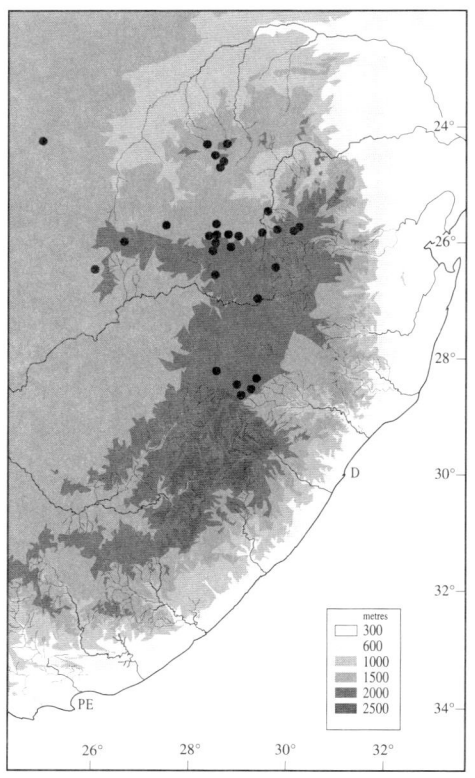

rainfall zone, *Gladiolus elliotii* has a fairly wide distribution across the southern African highveld. Its range extends from eastern Free State in the south to the Waterberg in Northern Province and westward through North-West Province into eastern Botswana. Plants seem to favour moist but well-drained grassland or light woodland and have also been recorded in rocky outcrops.

The short-tubed flowers have the same general shape and size as those of several other eastern southern African species of *Gladiolus* known to be pollinated by long-tongued anthophorid bees and we assume those of *G. elliotii* have the same pollination strategy.

Diagnosis and relationships

The strongly distichous spikes, fan-like cluster of leaves and tubular inner bracts place *Gladiolus elliotii* squarely in series *Oppositiflorus* where is stands out largely on vegetative features. The stems are fairly short and the leaf blades arise shortly above ground level. The short leaf sheaths do not form a pseudostem as they do in *G. sericeovillosus*. The leaf blades are also distinctive in the series in their venation. The margins are strongly thickened and the midrib and other veins are fine and closely set. This is quite unlike other species of the series, which have the midrib and one or more pairs of secondary veins also thickened and the remaining veins not forming a fine parallel pattern. The floral bracts of *G. elliotii* are inflated

below and somewhat dry above and the inner bracts are typical of the series, having the margins united below and the apices drawn into attenuate, dry, brown-tipped cusps. The inflated bracts are particularly characteristic of *G. sericeovillosus* and *G. oppositiflorus*, which must be regarded as most closely allied to *G. elliotii*. The large flowers of *G. oppositiflorus* make confusion with that species unlikely, but it is sometimes difficult to distinguish *G. elliotii* from *G. sericeovillosus* subsp. *calvatus*, which also has glabrous leaves and bracts. The flowers of the two are often very similar, but the leaf venation is so different that they should not be confused if this character is examined carefully. The conspicuous pubescence of the leaf sheaths and blades as well as the bracts and stem make it easy to avoid confusion with subsp. *sericeovillosus*.

History

Gladiolus elliotii was apparently first collected in 1883 by the German apothecary and naturalist, Friedrich Wilms near Bronkhorstspruit. His collection of the species was overlooked by contemporary botanists, but a later gathering, made at Steenkoolspruit near Bethal most likely in 1889 by the English biologist, George Scott Elliot, drew J.G. Baker's attention. Baker described the species in 1891, naming it in Elliot's honour. The species was recorded from Botswana in 1895 by Rudolf Marloth. A fairly conspicuous plant, *G. elliotii* is now

known to be widespread and often common across the interior escarpment of southern Africa.

Section *Ophiolyza:* Series *Ecklonii*

26. GLADIOLUS ECKLONII Lehmann
PLATE 22

Gladiolus ecklonii Lehmann, *Delect. Sem. Hort. Hamburg* 7 (1835). Baker, *Fl. Capensis* 6: 151 (1896). Lewis et al., *J. S. African Bot.*, Suppl. 10: 39 (1972), including *G. rehmannii* and *G. vinosomaculatus.* Type: South Africa, Eastern Cape, Fort Beaufort District, Katberg, Mar. 1830, *Ecklon & Zeyher Irid. 162* (12.3) (S, neotype designated by Lewis et al., 1972: 40; B, C, G, PRE, isoneotypes) (originally cultivated in Hamburg from seed collected by Ecklon, no preserved specimens known).

ecklonii, named in honour of C.F. Ecklon, the prolific early nineteenth-century plant collector in southern Africa. Ecklon sent seeds to the Hamburg Botanic Garden and the description of *Gladiolus ecklonii*

was based on plants that were raised and flowered there.

Synonymy

Gladiolus marmoratus Tausch, *Flora* 19: 142 (1836), an illegitimate homonym, not *G. marmoratus* Lamarck, 1796 (= *Watsonia humilis* Miller). Type: South Africa, without precise locality, cultivated in Vienna, *Ecklon s.n.* (no preserved specimen known).
Gladiolus inclusus F. Bolus, *Ann. Bolus Herb.* 2: 102 (1917). Type: South Africa, Mpumalanga, Pilgrim's Rest, Dec. 1914, *Rogers 14311* (BOL, holotype).

Description

Plants 15–35(–70) cm high, occasionally taller. *Corm* depressed-globose, 25–40 mm in diameter, the tunics brittle-papery to car-

tilaginous, decaying into irregularly shaped fragments. *Cataphylls* membranous, the upper reaching 2–3 cm above the ground and then firm-textured and green. *Leaves* occasionally six, usually seven to nine, rarely up to thirteen, the lower four more or less basal and longest, usually forming a crowded distichous fan, bright green, the margins moderately to strongly thickened, the midrib hardly thickened, reaching to about the middle of the stem, rarely to the base of the spike, the blades lanceolate, 15–30(–53) mm wide, the upper two leaves entirely sheathing or with vestigial blades, resembling the bracts, the margins free to the base, slightly inflated. *Stem* erect below, inclined in the upper two-thirds, unbranched, 3–4 mm in diameter below the spike.

Spike inclined, usually 8- to 12-, occasionally to 26-flowered; *bracts* bright green or flushed with purple below, strongly keeled in the midline, the keels usually reddish, the outer 43–53 mm long, three, sometimes two, internodes long, the inner about half to two-thirds as long (rarely almost as long) as the outer, acute and entire, those of the lower flowers folded or keeled, those of the upper flowers 2-keeled. *Flowers* variously spotted or minutely dotted pink, red or purple on a white field, sometimes so densely spotted as to appear uniformly pink at least on the distal halves of the tepals, the lower three tepals with yellow to cream, broad to narrow, spear-shaped nectar guides on the lower half, closing tightly at night, unscented; *perianth tube* obliquely funnel-shaped, (14–)17–20 mm long; *tepals* unequal, broadly ovate-elliptic, the apices sometimes emarginate-apiculate, the dorsal horizontal and hooded over the stamens, (25–)28–35 x (13–)15–20 mm, the upper laterals usually exceeding the dorsal (24–)28–37 x (11–)16–21 mm, the lower laterals smallest, 18–20 x 9–15 mm, the lower median (22–)28–35 x (9–)13–20 mm, arching outward and downward. *Filaments* 12–15 mm long, exserted 4–5 mm from the tube but included in the lower part of the flower; *anthers* 7–10 mm long, cream, the pollen white. *Ovary* oblong, 6–7 mm long; *style* arching over the stamens, dividing opposite the middle of the anthers, the branches c. 4 mm long,

emerging between the stamens. *Capsules* obovoid, weakly 3-lobed above, the apex retuse, 18–25 mm long; *seeds* discoid, broadly and usually evenly winged, 7–8 x 4.5–6 mm, light beige or reddish brown, the seed body slightly darker. *Chromosome number* unknown.

Flowering time late December to early March, occasionally later.

Distribution and biology

One of the most widespread species of southern African *Gladiolus* in the summer-rainfall zone, *G. ecklonii* extends virtually for the entire length of the mountainous eastern escarpment. It occurs from the Amatola Mountains and Katberg of the Eastern Cape Province in the south through the Drakensberg of Lesotho and KwaZulu-Natal into the northern provinces of South Africa, with the most northern station in the mountains near Haenertsburg. The species favours well-watered low grassland, often in stony places, and occasionally at the edges of marshes and vleis.

Like many other species of *Gladiolus*, the flowers of *G. ecklonii* close at night and only open the following morning after the sun has warmed them. The comparatively small, short-tubed flowers are adapted for pollination by long-tongued bees which visit the flowers for the nectar secreted from septal nectaries into the base of the perianth tube. We have noted two species of anthophorid bees, *Amegilla capensis* and *A. fallax,* foraging for nectar on the flowers, and in so doing pollinating them.

Diagnosis and relationships

Gladiolus ecklonii can usually be recognized by the fairly short stature, relatively short, obtuse leaves forming a basal fan at ground level, and the relatively small flowers, much shorter than the long, imbricate floral bracts. The bracts, like all three species of series *Ecklonii*, are at least two internodes in length, and 45–60 mm long. They thus exceed the flowers which are 40–55 mm long, and have a relatively short tube, 14–20 mm long. The perianth of *G. ecklonii* is typically pale, either whitish or cream, and more or less evenly covered with small, dark pink, reddish or purple spots. There is some variation in the intensity of the spotting and flowers may sometimes appear nearly uniformly pink or dark red, the result of a dense accumulation of small spots. *Gladiolus ecklonii* has in the past often been confused with *G. vinosomaculatus* which resembles it in the basic flower colouring. Although

usually somewhat larger in the flowers, *G. vinosomaculatus* is best separated from *G. ecklonii* on vegetative features. The leaves of *G. vinosomaculatus* are narrowly sword-shaped to almost linear, 4–12 mm wide, and have a whitish waxy bloom. They also have long sheaths that together form a fairly well-developed pseudostem. The midrib and margins are moderately thickened. The bright green leaves of *G. ecklonii* are more or less lanceolate, 15–30(–53) mm wide, and have thickened margins but the midrib is usually hardly thickened at all.

History

Gladiolus ecklonii was first collected at the extreme southern end of its range in the Amatola Mountains near Katberg by C.F. Ecklon in 1830 when he travelled to the eastern part of the Cape Colony. Ecklon collected both herbarium specimens and seeds, and he distributed the latter to several botanical gardens in Europe. Plants were raised at Hamburg and when they flowered there, J.G. Lehmann, Director of the Hamburg Botanical Garden, published a brief description of the species in the seed catalogue of

metres
300
600
1000
1500
2000
2500

the Hamburg Botanical Garden in 1835, naming it after Ecklon. Plants were also grown in Vienna where I.F. Tausch also realized that this was a new species which he named *G. marmoratus* in 1836. Plants named by Lehmann and by Tausch are the common form of the species with pale tepals marked over their entire surface with purple spots, and *G. ecklonii* was known only from this form for many years. When a rather dwarfed specimen collected near Pilgrim's Rest by Archdeacon F.A. Rogers came to Frank Bolus's attention, he believed it to represent a separate species which he called *G. inclusus* (F. Bolus, 1917). The type specimen of *G. ecklonii* was not found by G.J. Lewis when she was studying the nomenclature of *Gladiolus* in preparation for her revision of the genus. In the revision published after her death, specimens collected by Ecklon from Katberg were designated a neotype.

27. GLADIOLUS VINOSOMACULATUS Kies

Gladiolus vinosomaculatus Kies, *Fl. Pl. Africa* 29: pl. 1123 (1952). Verdoorn in Letty, *Wild Flowers Transvaal* pl. 40, fig. 1 (1962). *Gladiolus ecklonii* subsp. *vinosomaculatus* (Kies) Obermeyer in Lewis et al., *J. S. African Bot.*, Suppl. 10: 43 (1972). Type: South Africa, Gauteng, Pretoria, Waverley, 22 Jan. 1948, *Wasserfal s.n.* (PRE 29299, holotype; K, isotype).

vinosomaculatus = wine-spotted, for the tepals which have large, dark red-purple spots on a white background.

Description
Plants 65 cm high. *Corm* depressed-globose, 15–30 mm in diameter, the tunics brittle-papery to cartilaginous, decaying into irregularly shaped fragments. *Cataphylls* firm and pale below, the uppermost reaching well above the ground and then usually dark green, flushed with purple or brownish. *Leaves* five to eight, the lower four or five basal, forming a short pseudostem below, distichous and more or less fan-like above, reaching to between the base and apex of the spike, the blades plane, narrowly sword-shaped to nearly linear, (4–)6–12 mm wide, the margins and midrib moderately thickened, the upper leaves cauline and shorter than the basal. *Stem* erect below, inclined above, occasionally with one or two short branches, 3–4 mm in diameter below the spike.

Spike strongly flexed at the base and inclined, more or less straight, 9- to 14-flowered; *bracts* grey-green, usually conspicuously glaucous, (25–)40–80(–120) mm long, four to five internodes long, the inner about half as long as the outer, acute. *Flowers* variously speckled with red to purple dots on a white ground, unscented; *perianth tube* obliquely funnel-shaped, 15–18 mm long; *tepals* unequal, narrowly lanceolate, usually acute, sometimes obtuse, the dorsal largest, arched to nearly horizontal, 32–35 x 13–15 mm, the upper laterals directed forward, curving outward in the upper third, 30–34 x 8–12 mm, the lower tepals arching downward, the lower laterals 28–30 x 5–8 mm, the lower median 30–33 x 8–11 mm. *Filaments* 7–11 mm long, exserted 2–4 mm from the tube; *anthers* 10–13 mm long, the pollen yellow. *Ovary* oblong, c. 6 mm long; *style* arching over the stamens, dividing between the upper third and apices of the anthers, the branches 3–4 mm long. *Capsules* oblong, 24–30 mm long, the apices rounded and shallowly three-lobed, remaining concealed in the bracts; *seeds* ovate, c. 9.5 x 6 mm, broadly and evenly winged, translucent dark reddish brown. *Chromosome number* unknown.

Flowering time mainly mid-January to March.

Distribution and biology
Fairly widely distributed in southern Africa north of the Vaal River, *Gladiolus vinosomaculatus* extends from the Magaliesberg west of Pretoria in Gauteng eastward to the mountains above Barberton in Mpumalanga. Plants grow in fairly exposed and often dry sites, such as rocky hilltops and slopes in short grassland, and are sometimes encountered in rocky outcrops, tightly wedged in crevices in rock.

We assume that *Gladiolus vinosomaculatus*, like its close relative *G. ecklonii*, is pollinated by large anthophorid bees such as *Amegilla capensis*.

Diagnosis and relationships
One of three closely related species that constitute series *Ecklonii* of section *Ophiolyza*, *Gladiolus vinosomaculatus* has the very long, keeled floral bracts that define the series. The flowers are moderate in size, and always partly concealed by the large bracts. The tepals are covered with irregularly dispersed, large dark purple to reddish spots on a pale background. Unlike those of *G. ecklonii*, which may have similarly coloured flowers, the leaves of *G. vinosomaculatus* have long sheaths that together form a short pseudostem, and the blades are narrowly sword-shaped to almost linear, 4–12 mm wide, with strongly thickened margins and midrib. Unless growing through long grass or bush, the leaf sheaths of *G. ecklonii* do not form a pseudostem, and the leaf blades are usually about half as long as the stems, lanceolate to sword-shaped, and 15–30(–45) mm wide, with only moderately thickened margins and midrib. The other member of series *Ecklonii*, *G. rehmannii*, is unlikely to be confused with *G. vinosomaculatus* because it has larger, uniformly white to lilac flowers, the upper tepals usually 38–52 mm long, slightly longer filaments, 10–12 mm long, and anthers up to 10 mm long. In *G. vinosomaculatus* the filaments are 7–11 mm long and the anthers 12–13 mm long. Another distinction between the two species seems to be the length of the style branches, 3–4 mm long in *G. vinosomaculatus* compared with 6–8 mm in *G. rehmannii*. This feature is known for only a few specimens and needs to be more critically

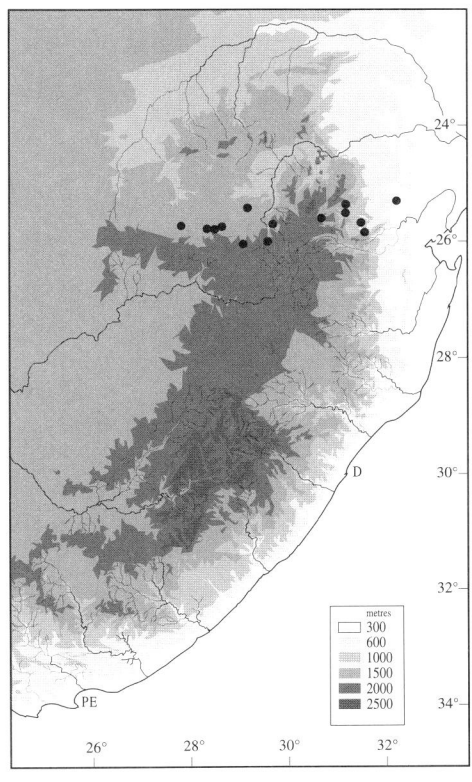

examined to establish whether the difference is consistent.

History

Although known since 1890 when it was first collected by Ernest Galpin near Barberton, *Gladiolus vinosomaculosus* was confused for many years with the closely allied *G. ecklonii*. Only in 1952 was it examined carefully and as a result the Pretoria botanist, Pauline Kies, decided that it was a separate species which she described from plants growing on the hills near Pretoria. The species was reduced to subspecific status in *G. ecklonii* by A.A. Obermeyer in 1972 in G.J. Lewis's posthumously published *Revision of the South African species of* Gladiolus. Lewis, however, originally intended to recognize the species. We concur with her opinion and reverse Obermeyer's treatment here.

28. GLADIOLUS REHMANNII Baker

Gladiolus rehmanni Baker, *Handbook Irideae* 216 (1892); *Fl. Capensis* 6: 153 (1896). Pole Evans, *Fl. Pl. Africa* 1: pl. 20 (1921). *Gladiolus ecklonii* subsp. *rehmannii* (Baker) Obermeyer in *J. S. African Bot.*, Suppl. 10: 44 (1972). Type: South Africa, Mpumalanga, Bronkhorstspruit District, between Elandsrivier and Klippan, 1879-1880, *Rehmann 5096* (K, holotype).

rehmannii, named after the discoverer, Polish botanist Anton Rehmann, who collected plants in the Transvaal in 1879 or 1880.

Synonymy

Gladiolus cymbarius Baker, *Bull. Herb. Boissier,*

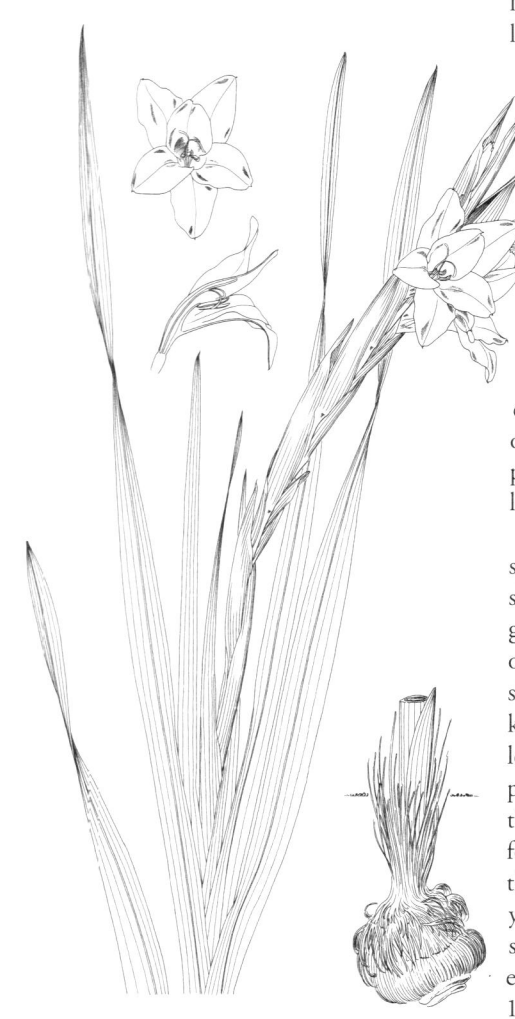

Sér. 2, 1: 866 (1901). Type: South Africa, Mpumalanga, Bronkhorstspruit District, between Elandsrivier and Klippan, probably in 1880, *Rehmann 5100* (Z, holotype, PRE, photograph; Z, isotype).

Description

Plants 30–50(–90) cm high. *Corm* globose, 18–22 mm in diameter, the tunics brittle-papery to cartilaginous, decaying into irregularly shaped fragments. *Cataphylls* coriaceous and pale, the uppermost reaching 12–18 cm above the ground and then green or flushed with dull purple. *Leaves* occasionally as few as six, usually eight or nine, the lower four or five, sometimes three, basal and forming a lax distichous fan, the basal three to five longest, the sheaths long and forming a pseudostem below, reaching at least to the base of the spike, more often to the apex or shortly exceeding the spike, the blades narrowly lanceolate, (4–)8–16 mm wide, the midrib moderately thickened, the margins not noticeably thickened, the cauline leaves overlapping, progressively smaller above, the upper one or two usually entirely sheathing and resembling the bracts, sometimes slightly inflated. *Stem* erect or flexed outward just below the base of the spike, occasionally branched, completely sheathed by the bases of the cauline leaves, 3–4 mm in diameter below the spike. *Spike* inclined, flexuose, 6- to 10-, occasionally to 16-flowered, the internodes fairly short, 12–14(–20) mm long; *bracts* pale green or flushed with purple, often glaucous on the upper surfaces (facing the sun), occasionally lightly folded in the midline (not keeled), the outer (40–)45–60(–115) mm long, three or four internodes long, overlapping, the inner half to two-thirds as long as the outer, acute and entire or forked apically for up to 5 mm. *Flowers* white to pale lilac, the lower lateral or all three lower tepals with yellow nectar guides in the lower midline, sometimes outlined in light purple, unscented; *perianth tube* obliquely funnel-shaped, 15–20 mm long, enclosed in the bracts;

tepals unequal, lanceolate, the dorsal inclined over the stamens, (30–)38–50 x 12–18 mm, the upper laterals usually 32–52 x 10–16 mm, usually longer than the dorsal, directed forward and curving outward in the upper half to one-third, the lower lateral tepals 28–43 x 8–14 mm, lower median usually substantially larger, 30–45 x 8–14 mm, arching downward. *Filaments* 10–12 mm long, exserted 2–5 mm from the tube; *anthers* 9–10 mm long, the pollen pale lilac to blue, or yellow. *Ovary* oblong, c. 6 mm long; *style* arching over the stamens, dividing between the middle and apex of the anthers, the branches 6–8 mm long. *Capsules* oblong, three-lobed above, 25–35 mm long; *seeds* ovate-oblong, 12–14 x 7–8 mm, translucent reddish brown, broadly and evenly winged. *Chromosome number* unknown.

Flowering time mid-January to March.

Distribution and biology

A species of summer-rainfall southern Africa, *Gladiolus rehmannii* is largely restricted to the bushveld of the northern part of South Africa. It extends from Sybrandskraal near Groblersdal in Mpumalanga across Northern Province and the northern part of Northwest Province to eastern Botswana. Collections indicate that it is particularly common in the Waterberg between Nylstroom and Thabazimbi. Plants grow in sandy soils, often in rocky ground in light woodland that is frequently dominated by *Faurea saligna*.

The particularly large flowers of *Gladiolus rehmannii* have a short perianth tube and despite their white colouring and inconspicuous, cream nectar guides, we suspect that they are adapted for pollination by the same anthophorid bees that pollinate most other eastern southern African species of the genus.

Diagnosis and relationships

Largest-flowered of the three species of series *Ecklonii*, *Gladiolus rehmannii* is also often the tallest species of the alliance. Plants may

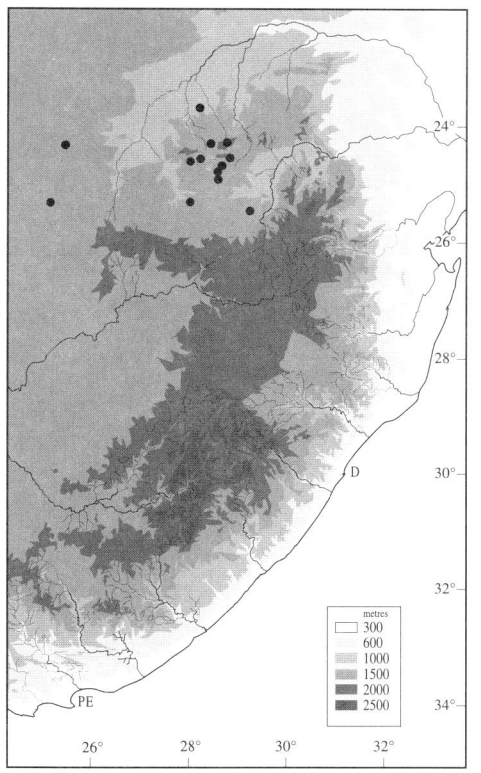

narrowly sword-shaped to linear leaves have long sheaths that form a pseudostem and the blades, carried well above the ground, form a loose fan. The blades are often slightly waxy and are usually 4–12 mm wide, occasionally to 16 mm, with moderately thickened midribs and less strongly thickened margins. The large white to pale mauve flowers are (45–)52–70 mm long and have a relatively short perianth tube, 15–20 mm long. Despite the close relationship with *G. vinoso-maculatus*, the two are unlikely to be confused, mainly because of the conspicuous purple speckling on the perianth of the latter and the nearly uniformly whitish to pale lilac perianth of *G. rehmannii*. They may also be distinguished by small details such as the larger anthers in *G. vinosomaculatus*, 10–13 mm long, and style branches only 3–4 mm long. In *G. rehmannii* the anthers are 9–10 mm long and the style branches 6–8 mm long. The two species also differ in their habitats, and while *G. rehmannii* occurs in Kalahari sands in dry bushveld, *G. vinoso-maculatus* favours stony grassland on the interior southern African escarpment, thus at cooler and somewhat wetter sites.

History

Gladiolus rehmannii was discovered by the Polish botanist Anton Rehmann in 1879 or 1880 when he travelled in the north of what was then the Transvaal Republic (Gunn & Codd, 1981). Rehmann was one of the first plant collectors to venture north of Pretoria and his exploration as far as Woodbush and Magoebaskloof at the northern end of the Drakensberg escarpment yielded numerous species new to science, including in the Iridaceae *Watsonia rehmannii*. Rehmann's collection at the Kew Herbarium in London was described by J.G. Baker in 1892 and named in Rehmann's honour. A second collection from close to the type locality of *G. rehmannii*, and probably made on the same day but bearing a different collection number, was described as *G. cymbarius* by Baker in 1901, based on specimens at the Zurich Herbarium. Predominantly a species of dry bushveld, *G. rehmannii* is now known to occur widely across western Mpumalanga and Northwest Province in South Africa, as well as in eastern Botswana.

reach 90 cm in height, although in some seasons they stand less than 50 cm high. The

Section Ophiolyza: Series Ophiolyza

29. GLADIOLUS ANTHOLYZOIDES Baker
PLATE 23

Gladiolus antholyzoides Baker, *J. Bot.* 29: 70 (1891). *Antholyza laxiflora* Baker, *Fl. Capensis* 6: 170 (1896), as a new name for *G. antholyzoides* in the genus *Antholyza*. Type: South Africa, Gauteng, Apies River near Pretoria, damp ground, Nov. 1890, *Scott Elliot 1447* (K, holotype).

antholyzoides = resembling *Antholyza*, so named for the resemblance of the flower to that of some species of that genus, particularly in the long perianth tube with a wide cylindric upper part.

Synonymy

Antholyza schlechteri Baker, *Bull. Herb. Boissier*, Sér. 2, 4: 1007 (1904). *Gladiolus brachylimbus* Baker, *Viert. Naturforsch. Ges. Zürich* 49: 178 (1905), as a new name for *A. schlechteri* in the genus *Gladiolus*, not *G. schlechteri* Baker (1904) (= *G. papilio* Baker). *Gladiolus magalies-montanum* F. Bolus, *Ann. Bolus Herb.* 2: 108 (1917), an illegitimate superfluous name based on the same type as *G. brachylimbus*. Type: South Africa, Gauteng, Magaliesberg near Apies River, 4 Nov. 1893, *Schlechter 3627* (as *3629*) (Z, holotype, drawing at BOL

and K; BOL, Z, isotypes; NBG, photos).
Gladiolus vogtsii L. Bolus, *J. Bot.* 67: 133 (1929). South Africa, without precise locality, as 'Transvaal', cultivated at Kirstenbosch Botanic Gardens (NBG 617/28), 4 Feb. 1929, *Vogts s.n.* (BOL, holotype).
Gladiolus strictiflorus L. Bolus, *J. Bot.* 69: 15 (1931), an illegitimate homonym, not *G. stric-tiflorus* (Ker Gawler) Delile (1812) (= *Watsonia strictiflora* Ker Gawler). Type: South Africa, Free State, Reitz, cultivated at Kirstenbosch Gardens in 1930, (BOL, holotype).
?*Gladiolus anorthanthus* Ingram, *Gard. Chron.* 94: 273 (1933). Type: South Africa, Northern Province, Potgietersrus, cultivated in England, *Ingram s.n.* (BM, holotype), the identity uncertain because of the absence of mature flowers.

Description

Plants 60–90 cm high. *Corm* depressed-globose, 23–35 mm in diameter, the tunics brittle-papery to membranous, soon decaying into irregularly shaped fragments, always with several cormlets c. 5 mm in diameter around the base, usually bearing 1–3 lateral shoots in addition to the flowering stem. *Cataphylls* pale and membranous, the upper-

most reaching 2–4 cm above the ground and then green and coriaceous. *Leaves* five to eight, the lower three to five basal and usually forming a fairly tight distichous fan, 12–30 cm long and rarely reaching beyond the lower third of the stem, the blades lanceolate, (3–)10–20 mm wide, the margins and midrib lightly raised, remaining leaves smaller than the basal, progressively shorter above and the uppermost sheathing for most of its length. *Stem* erect, unbranched, 3–5 mm in diameter below the spike.

Spike straight and erect, strongly secund, 8- to 12-flowered; *bracts* pale green, smooth and shiny, the outer 30-35(–50) mm long, the inner about two-thirds as long, notched apically for c. 1 mm. *Flowers* pale lemon-yellow to deep yellow, sometimes streaked irregularly with red or orange and with an orange ring around the throat, the reverse of the tube and lower part of the tepals green-ish yellow, the lower tepals slightly darker yellow or yellow-green in the lower half, unscented; *perianth tube* narrowly and obliquely funnel-shaped, 28–40 mm long, the flared upper part 15–25 mm long; *tepals*

lanceolate, the dorsal largest, 25–30 x 14–16 mm, extended more or less horizontally over the stamens, the upper laterals directed forward below, curving outward in the upper third, 25–28 x 16 mm, the lower three tepals arching downward more or less from the base, the lower laterals 17–20 x 9 mm, the lower median 20–23 x 12 mm, in profile the upper and lower tepals about equal or the lower shorter. *Filaments* 20–30 mm long, exserted 2–8 mm from the tube; *anthers* 9–11 mm long, pale yellow, the pollen cream. *Ovary* oblong, c. 4.5 mm long; *style* arching over the stamens, dividing between the middle and apex of the anthers, the branches arching outward, c. 5 mm long, gradually widening in the upper half. *Capsules* ovoid-oblong, three-lobed and retuse above, 20–25 mm long; *seeds* ovate, 7–9 mm long, broadly and evenly winged, light translucent brown. *Chromosome number* unknown.

Flowering time mainly mid-November to early December, occasionally in late October.

Distribution and biology

Gladiolus antholyzoides has a scattered distribution across the central and eastern highveld of South Africa. Populations have been recorded from Reitz in eastern Free State northward through Gauteng to the Waterberg near Nylstroom in Northern Province. Plants flower early in the season, before or soon after the onset of spring rains in November and early December and usually favour sites that retain ground moisture throughout the year. This is, however, not always the case. A population near Sasolburg in northern Free State, from which the plants illustrated here were taken, flowered in fairly dry sandy ground, and before appreciable rain had fallen that particular year.

The pollination of *Gladiolus antholyzoides* is unknown. We suspect that the long and wide perianth tube and ample amounts of nectar are adaptations for bird pollination. The flower colour, usually pale yellow, is unusual for bird-pollinated flowers in southern Africa.

Diagnosis and relationships

Vegetatively, *Gladiolus antholyzoides* closely resembles its near relative, *G. dalenii*, in its fan of sword-shaped leaves with the blades firm in texture and having moderately thickened midrib, secondary veins and margins. The leaves are shorter than is typical for *G. dalenii*, however, and they seldom reach much higher than the upper third of the stem. The flowering spikes thus stand well above the foliage. The flowers of *G. antholyzoides* are relatively large, 55–70 mm long, have an elongate perianth tube 28–40 mm long and lanceolate tepals. The tepals are unequal, the upper three 25–30 mm long and the lower three 17–23 mm long, with the lowermost somewhat larger than the lower laterals. *Gladiolus antholyzoides* is unusual in producing leafy shoots from separate buds on the corms toward the end of flowering and while in fruit. The flowering stems and their associated leaves wither later in the summer but the leafy shoots continue to grow until the plants become dormant at the end of autumn.

Much confused with *Gladiolus dalenii*, and included in that species by Lewis et al. (1972), *G. antholyzoides* is very different in both the growth habit and the structure and size of the flowers. In *G. dalenii*, which has larger flowers 65–100 mm long, the dorsal tepal is broadly ovate, strongly arched and more or less horizontal, thus completely concealing the anthers. The equally broad upper lateral tepals are directed forward below, curving outward only in the upper half. This is quite different from the relatively narrow upper tepals of *G. antholyzoides*, in which the dorsal tepal is horizontal but not noticeably arched and the upper laterals spread outward from just above the base. When seen alive the differences between the two species are so striking that there can be no doubt that they are separate species. The differences are often obscured when dry, hence the confusion between them.

Gladiolus antholyzoides is probably most closely related to a second species of series *Ophiolyza*, *G. aurantiacus*, which flowers early in the season and produces flowering stems that lack foliage leaves, bearing instead sheathing leaves with vestigial blades. Foliage leaves are produced later in the season, usually at the end of flowering, from separate shoots on the same corm that produced the flowering stem. The flowers of *G. aurantiacus* are usually light orange, the perianth irregularly streaked and flecked with scarlet, and 60–100 mm long, with a dorsal tepal usually 35–45 mm long, thus substantially larger than those of *G. antholyzoides*, although similar in basic construction.

History

The first record of *Gladiolus antholyzoides* is evidently the collection made by the English biologist, George Scott Elliot, near Pretoria in 1890. These plants were described by J.G. Baker in 1891 as *Gladiolus antholyzoides*. In *Flora Capensis*, Baker (1896) changed his mind about the correct generic disposition of the species which he there called *Antholyza*

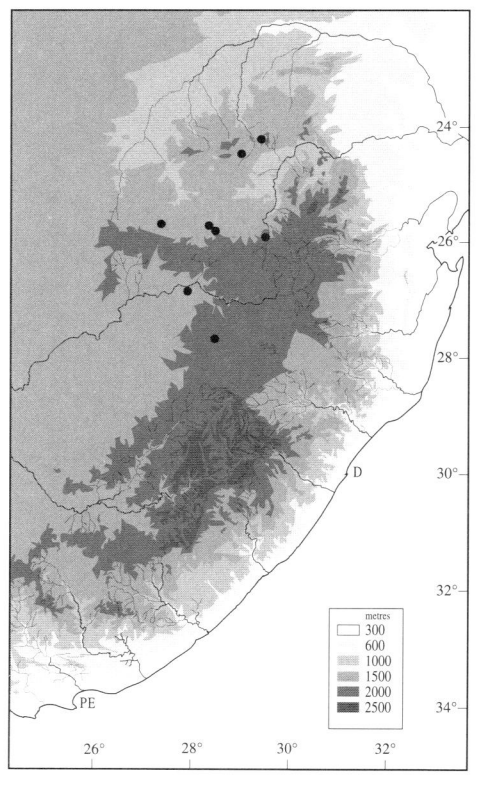

laxiflora, a new specific epithet chosen because the combination *A. antholyzoides* was not thought appropriate at the time. The reason for placing the species in *Antholyza* appears to have been the resemblance of the long perianth tube to that characteristic of the genus, a narrow lower part and a long, cylindric upper part. Baker described a second collection of the species, made by Rudolf Schlechter in 1893, also from the Pretoria area, as *A. schlechteri* in 1904. Then in 1905, he transferred that species to *Gladiolus* where he called it *G. brachylimbus*, the epithet *schlechteri* already being occupied in the genus. Evidently Baker forgot that he had already described the species under a different name.

Both *Antholyza antholyzoides* and *A. schlechteri* / *Gladiolus brachylimbus* were apparently overlooked when Frank Bolus described *G. magaliesmontanum*, which is based on the same Schlechter collection as *G. schlechteri* and hence nomenclaturally superfluous. To add to the confusion surrounding the species, Louisa Bolus described *G. strictiflorus* in 1931, based on cultivated plants. This name is not only a synonym of *G. antholyzoides* but it is invalid, being a homonym of *G. strictiflorus* Delile (1812), itself a synonym of *Watsonia coccinea*.

Gladiolus antholyzoides and two of its synonyms were regarded as conspecific with *G. dalenii* by A.A. Obermeyer (Lewis et al., 1972). *Gladiolus strictiflorus* was, however, excluded because she thought that there was no species in southern Africa that conformed to that plant. Annotations on the type and a few other specimens of this species by Lewis indicate that she planned to recognize it under a new name, *G. rectiflorus*, but her intentions did not reach fruition due to her untimely death. One more synonym of *G. antholyzoides* is tentatively admitted here, *G. anorthanthus*, described by Collingwood Ingram in 1933. Both the illustration published with the protologue and the type specimen lack flowers, so our identification of the plant is based only on the vegetative features which accord better with *G. antholyzoides* than with any other species.

30. GLADIOLUS AURANTIACUS Klatt
PLATE 24

Gladiolus aurantiacus Klatt, *Linnaea* 35: 378 (1867). Baker, *Fl. Capensis* 6: 159 (1896). Lewis et al., *J. S. African Bot.*, Suppl. 10: 289 (1972). Type: South Africa, KwaZulu-Natal, near Pietermaritzburg, Sept. & Oct. 1858, *Sutherland s.n.* (K, S, syntypes).

aurantiacus = orange-coloured, for the bright yellow-orange flowers.

Description

Plants 45–75 cm high. *Corm* depressed-globose, 25–35 mm in diameter, the tunics brittle-papery to membranous, soon decaying into small, irregularly shaped fragments, always with several cormlets c. 5 mm in diameter around the base. *Cataphylls* pale and firm-textured, the uppermost reaching 3–8 cm above the ground and then green and coriaceous.

Leaves of the flowering stem four or five, superposed, imbricate, all with blades usually shorter than or about as long as the sheaths, sometimes the uppermost one or two entirely sheathing, the blades lanceolate to sword-shaped, 6–15 mm wide; foliage leaves five or six, produced from separate shoots on the same corm toward the end of flowering, these forming a distichous fan, ultimately reaching to 60 cm, narrowly lanceolate, 10–15 mm wide, the midrib thickened and raised, the margins barely thickened. *Stem* erect, unbranched, 4–6 mm in diameter below the spike.

Spike straight and erect, 10- to 16-flowered; *bracts* pale green, smooth and shiny, the outer (28–)32–45 mm long, the inner about two-thirds as long, notched apically for c. 2 mm. *Flowers* deep yellow, finely dotted or streaked irregularly with red or orange, sometimes so densely spotted as to appear orange, the lower tepals clear yellow in the lower half, unscented; *perianth tube* narrowly and obliquely funnel-shaped, 44–70 mm long, slender and erect below for 25–33 mm, abruptly expanded into a wide cylindric and horizontal upper part 30–35 mm long; *tepals* lanceolate, unequal, the dorsal largest, (26–)35–45 x c. 15–24 mm, extended more or less horizontally over the stamens, the upper laterals directed forward below, curving outward in the upper half, 25–42 x 14–25 mm, the lower three tepals arching downward more or less from the base, 15–25 x (7–)12–16 mm, the lower median slightly larger than the lower laterals, in profile the upper tepals much exceeding the lower. *Filaments* 28–40 mm long, exserted 12–15 mm from the tube; *anthers* 9–12 mm long, yellow to orange, the pollen yellow. *Ovary* oblong, c. 5 mm long; *style* arching over the stamens, dividing between the base and middle of the anthers, the branches arching outward, 4–6 mm long, gradually widening in the upper halves. *Capsules* ovoid-oblong, three-lobed and retuse above, 18–25 mm long; *seeds* ovate, c. 10 x 7–8 mm, broadly and evenly winged, translucent light brown. *Chromosome number* unknown.

Flowering time mainly in October, occasionally in September or November.

Distribution and biology

A species of the summer-rainfall zone of southern Africa, *Gladiolus aurantiacus* is

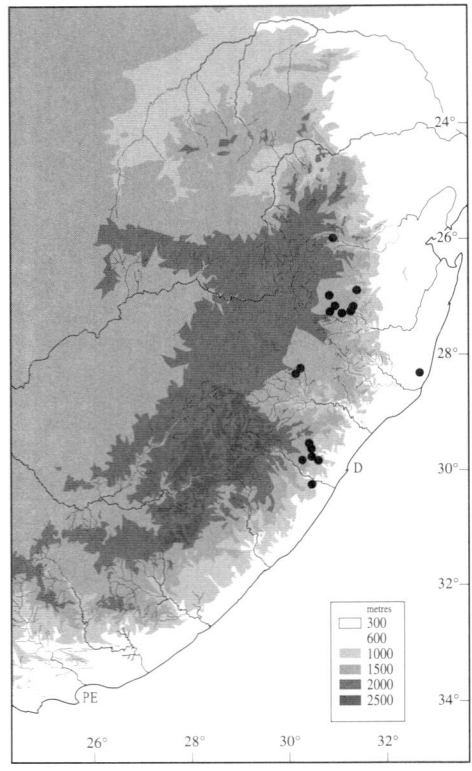

The long-tubed orange flowers with yellow blotches on the lower tepals are clearly adapted for pollination by sunbirds. In shape, size and colour they closely resemble the flowers of *Gladiolus huttonii* of the southern Cape, and to a lesser extent other species of section *Homoglossum*. The nectar characteristics are typical of bird-pollinated flowers, being ample in quantity and relatively low in sugar concentration.

Diagnosis and relationships

The most striking feature of *Gladiolus aurantiacus* is the elongate perianth tube, up to 70 mm long, and the disproportionately small tepals. The perianth tube is slender in the lower 25–33 mm, and abruptly widens into a cylindrical upper part that is up to 30–37 mm long. The dorsal tepal, 26–45 mm long, is slightly ascending to nearly horizontal, while the lower tepals curve sharply downward and are 15–25 mm long, much shorter than the upper tepals. The spikes are straight and stiffly erect, and bear up to 16 flowers, although more often only 10 to 12. The leaves of the flowering stem have relatively short to vestigial blades and foliage leaves with long blades are produced on separate shoots toward the end of flowering or when the capsules begin to ripen.

Gladiolus aurantiacus is closely allied to the common and widespread *G. dalenii* and from a distance is easily mistaken for that species. The shape of the perianth tube in the latter is quite different, as are the much broader upper tepals, the dorsal being broadly ovate, concave and strongly arched, often dipping below the horizontal. At least in southern Africa *G. dalenii* always has long-bladed foliage leaves on the flowering stems.

The appearance of *G. aurantiacus* recalls a second species, *G. antholyzoides*, which it closely resembles in vegetative features and in the early flowering habit. That species also produces long-bladed foliage leaves from separate shoots later in the growing season, but the leaves of the flowering stem have moderately long blades and form a loose basal fan. The flowers of *G. antholyzoides* are 42–65 mm long and have a lanceolate dorsal tepal 25–30 mm long and downward-directed lower tepals 17–23 mm long, the latter rather different from the somewhat recurved lower tepals of *G. aurantiacus*.

History

Gladiolus aurantiacus was first collected by John Sanderson and Peter Cormac Sutherland, both naturalists of the mid-nineteenth century who lived in the Natal Colony and made collections for Kew Gardens in England. The first record was evidently made by Sanderson in 1854. Sutherland's later collection, made in 1858 in the Pietermaritzburg District in the southern part of the colony, is, however, the type of the species which was named in 1867 by the German botanist, F.W. Klatt. Unlike so many eastern southern African species of section *Ophiolyza*, no trace of nomenclatural confusion surrounds *G. aurantiacus*. It was recognized in *Flora Capensis* by J.G. Baker (1896), and interpreted there exactly as Klatt had intended.

unusual in section *Ophiolyza* in flowering early, mostly in September and October, before the main rainy season has begun. It is centred in KwaZulu-Natal, where it occurs from Richmond and Dumisa in the south along the coast and the Midlands through the north of the province and into southern Mpumalanga and Swaziland. It is particularly common in the rolling high grassland around Vryheid and Piet Retief, and makes a striking display in shallow vleis and low or burnt veld in years when there has been ample early rain.

31. GLADIOLUS DALENII Van Geel

Dragon's head lily, parrot lily, Natal lily, sword lily

PLATE 25

Gladiolus dalenii Van Geel, *Sert. Bot.*, Fasc. 28 (1829). Hilliard & Burtt, *Notes Roy. Bot. Gard. Edinburgh* 37: 297 (1979). Goldblatt, *Bull. Mus. Hist. Nat. Paris*, Sér. 4, Sect. B, Adansonia 11: 251 (1989); *Fl. Madagascar famille* 45, Ed. 2, 42 (1991); *Fl. Zambesiaca* 12(4): 89 (1993); Gladiolus in Tropical Africa 207 (1996). Type: South Africa, KwaZulu-Natal, without precise locality, collector unknown, cultivated in Holland (illustration in *Sert. Bot.*, Fasc. 28, lectotype designated by Hilliard & Burtt, 1979).

dalenii, named for Cornelius Dalen, Dutch botanist who was to become director of the

Rotterdam Botanic Garden. Dalen was responsible for the introduction of the species from stock obtained in KwaZulu-Natal and its distribution to gardens in Europe; he also commissioned the painting, published in the protologue, that now serves as the type of *Gladiolus dalenii*.

Synonymy

Three subspecies of *Gladiolus dalenii* are recognized (Goldblatt, 1995); only subsp. *dalenii* occurs in southern Africa. The nomenclature below refers only to southern African synonyms of subsp. *dalenii*. Additional synonyms for *Gladiolus dalenii* based on plants from tropical Africa and Madagascar are cited by Goldblatt, 1996.

Watsonia natalensis Ecklon, *Topographisches Verzeichniss* 34 (1827), not *Gladiolus natalensis* J.D. Hooker (1831). Type (Lewis et al., 1972: 46): South Africa, KwaZulu-Natal coast (cultivated in Cape Town), Nov. 1826, *Ecklon s.n.* (S [not seen], holotype).
Gladiolus psittacinus J.D. Hooker, *Curtis's Bot. Mag.* 57: pl. 3032 (1830). Baker, *Fl. Capensis* 6: 158 (1896). Pole Evans, *Fl. Pl. S. Africa* 3: pl. 116 (1923). Hepper, *Fl. West Trop. Africa*, Ed. 2, 3(1): 141 (1968). *Gladiolus natalensis* Reinwardt ex J.D. Hooker, *Curtis's Bot. Mag.* 58: sub pl. 3084 (1831), an illegitimate name for *G. psittacinus*, although intended as a new name for that species; Loddiges, *Bot. Cab.* 18: pl. 1756 (1831). Lewis et al., *J. S. African Bot.*, Suppl. 10: 44–53 (1972), cited in error as

(Ecklon) Reinwardt ex J.D. Hooker. Geerinck, *Bull. Jard. Bot. Nat. Belgique* 42: 281 (1972), excluding var. *melleri* (Baker) Geerinck. Type: South Africa, KwaZulu-Natal, without precise locality, illustration in *Curtis's Bot. Mag.* 57: pl. 3032 (1830) and specimen cultivated in Hort. Hamilton, without collector (K).

Gladiolus dracocephalus J.D. Hooker, *Curtis's Bot. Mag.* 97: pl. 5884 (1871). Baker, *Fl. Capensis* 6: 157 (1896). Type: South Africa, KwaZulu-Natal, without precise locality or date, 'near the foot of the Drakensberg', cultivated at Kew Gardens, *Cooper 3593* (K, lectotype and isolectotype).

Gladiolus cooperi Baker, *Curtis's Bot. Mag.* 101: pl. 6202 (1875). *G. psittacinus* var. *cooperi* Baker, *Handbook Irideae* 220 (1892); *Fl. Capensis* 6: 158 (1896). Type: South Africa, KwaZulu-Natal, without precise locality, cultivated at Kew Gardens, *Cooper s.n.* (K, holotype).

Gladiolus adlamii Baker, *Gard. Chron.*, Ser. 3, 5: 233 (1880); *Fl. Capensis* 6: 156 (1896). Type: South Africa, without precise locality, as 'Transvaal', cultivated in Cambridge, *Adlam s.n.* (K, holotype, NBG, PRE, photographs).

Gladiolus leichtlinii Baker, *Gard. Chron.*, Ser. 3, 6: 154 (1889). Type: South Africa, without precise locality, as 'Transvaal', cultivated in Hort. Leichtlin, July 1889, *Adlam s.n.* (K, holotype).

Gladiolus tysonii Baker, *Handbook Irideae* 220 (1892); *Fl. Capensis* 6: 158 (1896). Type: South Africa, Eastern Cape, Griqualand East, near Fort Donald, Jan. 1884, *Tyson 1653* (K, holotype; SAM, isotype).

Gladiolus platyphyllus Baker, *Gard. Chron.*, Ser. 2, 14: 456 (1893). Type: South Africa, Eastern Cape, Transkei, lower Kei River, cultivated in Hort. Leichtlin, *Flanagan s.n.* (K, holotype).

Gladiolus fuscoviridis Baker, *Fl. Capensis* 6: 530 (1896). Type: a plant of unknown origin, cultivated at Kew Gardens, without collector (location of type unknown, not at K).

Gladiolus pageae L. Bolus, *S. African Gard.* 18: 213 (1928). Type: Lesotho, near Thaba Tseueu, cultivated in Cape Town, Mar. 1921, *Page s.n.* (BOL 17030, holotype).

Gladiolus leptophyllus L. Bolus, *S. African Gard.* 22: 204 (1932). Type: South Africa, Northern Province, Blouberg, grown at Stellenbosch University Gardens, without date, *Neethling sub Grosskopf s.n.* (BOL, holotype; K, isotype).

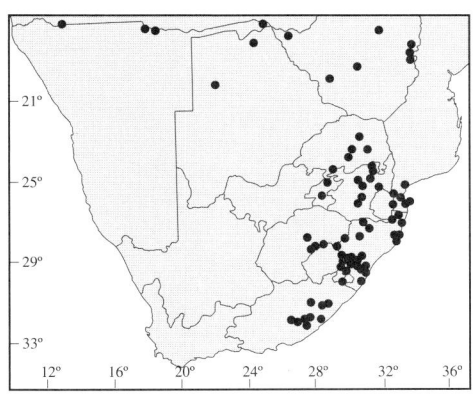

Gladiolus retrocurvus G. Lewis, *S. African Gard.* 22: 204 (1932). Type: South Africa, Northern Province, Letaba District, Selati Estate near Rubbervale, cultivated at Kirstenbosch Gardens (NBG 1361/30), Apr. 1932, without collector (BOL, holotype; K, isotype).

Description

Plants (50–)70–150(–200) cm high. *Corm* (15–)20–30 mm in diameter, tunics of brittle coriaceous layers, the outer becoming irregularly broken, sometimes fibrous, reddish brown, usually bearing numerous tiny cormlets around the base, sometimes with stolons bearing terminal cormlets. *Cataphylls* firm-textured, the upper largest and up to 15 cm long, green or flushed purplish, occasionally densely short-pubescent. *Leaves* contemporary with the flowering stem and four to six, sometimes seven (two to four on the flowering stem in the tropical African subspp. *andongensis* and *welwitschii*, the foliage leaves produced on separate shoots after flowering), when borne on the flowering stem at least the lower two basal or nearly so, usually about half as long as the spike, the blades narrowly lanceolate to more or less linear, (5–)10–30(–70) mm wide, firm-textured with moderately raised and thickened midrib and margins, the upper one or two leaves inserted on the upper half of the stem, often imbricate, sheathing for at least half their length, sometimes entirely sheathing. *Stem* erect and straight, unbranched, 4–6 mm in diameter below the spike.

Spike erect and straight, (2–)3- to 8(–14)-flowered; *bracts* pale green to grey-purple on the upper surfaces, sometimes dry and pale apically, the outer (35–)40–70 mm long, the inner slightly shorter than to two-thirds as long as the outer. *Flowers* either red to orange and then the lower three tepals each with a yellow mark on the lower half, or yellow to greenish and often with red to brown streaks on the upper tepals, unscented; *perianth tube* (25–)35–50 mm long, nearly cylindric below, gradually widening above and curving outward in the upper half; *tepals* unequal, the upper three broadly ovate, the dorsal largest, 35–50 x 22–30 mm, horizontal to downcurved, concave, the stamens concealed within the spoon, the upper laterals about as long as to c. 5 mm shorter than the dorsal, 20–30 mm wide, directed forward, often curving outward distally, the lower three tepals curving downward, 20–25(–30) mm long, 8–12 mm wide, the lowermost somewhat longer and narrower than the lower laterals. *Filaments* 25–35 mm long, exserted 15–20 mm from the tube; *anthers* 12–16 mm long, yellow to brownish, the pollen pale yellow. *Ovary* oblong, 6–9 mm long; *style* arched over the stamens, dividing near the apex of the anthers, the branches (4–)5–6 mm long, broadened above. *Capsules* ovoid-oblong, three-lobed apically, (18–)25–35 mm long; *seeds* 8–12 x 5–9 mm, the wing well developed, lightly undulate, glossy light brown, seed body c. 2 mm in diameter. *Chromosome number* 2n = 30, 60, 90 (45, 75).

Flowering time mainly November to April, but along the KwaZulu-Natal coast there are races that also flower in September and

October; in tropical Africa mid-November to May, rarely in June; May to August in western and central Africa; mostly August and September in Ethiopia; January to March in Eritrea and the Arabian Peninsula.

Distribution and biology

Gladiolus dalenii is by far the most wide-spread and common species of *Gladiolus*. It occurs virtually throughout the grasslands, savannas and woodlands of sub-Saharan Africa, and is absent only in the winter-rainfall zone of southern Africa and the surrounding arid to semi-arid Kalahari and Namib deserts and the Karoo. It also occurs in the highlands of southwestern Arabia and in Madagascar, where it is spread almost throughout the non-forested parts of the island, except in the dry southwest. Thus *G. dalenii* may be found from Senegal in the west, across central Africa to Ethiopia, Eritrea and western Saudi Arabia and Yemen. From there it extends southward through eastern Africa and Mozambique to KwaZulu-Natal and Eastern Cape Province, South Africa, and across southern tropical Africa to the western Angolan highlands. In southern Africa *G. dalenii* occurs in northern Namibia and throughout the eastern half of the sub-continent from Northern Province to the Amatola Mountains of Eastern Province, and inland into eastern Free State and Gauteng. *Gladiolus dalenii* favours moderately moist habitats and is thus most common in hill country and upland grassland, but it also occurs in fairly dry habitats with only a short wet season.

Across its range there are a number of variants, the most important of which are two exclusively tropical African subspecies that flower early, 3–4 weeks after the first soaking rains of the wet season. These plants have reduced, largely to entirely sheathing leaves on the flowering stem. The long-bladed foliage leaves are produced from separate shoots on the same corm later in the growing season. Subsp. *andongensis* occurs widely in tropical Africa and subsp. *welwitschii* is restricted to southwestern Angola.

Although there are few observations recording the pollination biology of *Gladiolus dalenii*, it is almost certain that its primary pollinators are sunbirds. Stefan Vogel (1954) has recorded the greater double-collared sunbird, *Nectarinia afra*, visiting flowers in the Drakensberg and we have seen the malachite sunbird, *N. famosa*, foraging for nectar on plants at Witzieshoek. The volume of nectar produced, up to 20 microlitres per flower, the elongated

perianth tube and the stamens exserted a fair distance from the mouth of the tube are typical adaptations for this pollination strategy. The straight, erect stem and spike and the reddish colour of the flowers of many of the populations are also consistent with bird pollination. It is interesting in this regard that a vernacular name for the species is 'coconut of the sunbird' in coastal East Africa. Local variation in flower colour, including greenish, yellow and even purple, is an unusual feature for a bird-pollinated species. Although populations are nearly always uniform for flower colour, this feature varies across populations from greenish yellow or greenish brown, often with red to brown streaking, to orange or red, and then with a bright yellow to cream or greenish blotch on each of the three lower tepals. Plants with pale to bright yellow and even dull purple flowers are also known, but only from tropical Africa (Goldblatt, 1996). The colour variation is partly responsible for the extraordinary number of synonyms given the species over the course of the first hundred years since its discovery and description in the 1820s.

Diagnosis and relationships

Gladiolus dalenii is readily recognized by the large flowers, 60 to almost 100 mm long, with the upper three tepals 35–50 mm long, much exceeding the recurved lower tepals. The perianth tube is 35–50 mm long, thus slightly exceeding the long, softly textured floral bracts, and is slender and cylindrical below, gradually widening and gently curving outward above. The dorsal tepal is distinctive in being strongly hooded and concave, concealing the well-exserted anthers. The leaves are typically broad and sword-shaped, but in subsp. *andongensis* and subsp. *welwitschii* the flowering stem lacks foliage leaves, and in some populations of subsp. *dalenii* the leaves may be narrow and no more than 5–8 mm wide. Whatever the width of the leaves, the midrib and margins are lightly thickened but seldom especially prominent.

Gladiolus dalenii is unlikely to be confused with any other southern African species of *Gladiolus*, although it may be with some in tropical Africa, especially the western Angolan endemic *G. pallidus* and the southern tropical African *G. melleri* (Goldblatt, 1993, 1996). Perhaps most like *G. dalenii* in southern Africa is *G. antholyzoides*, a species which has large yellow or orange-red flowers with the tube 28–40 mm long and a much narrower dorsal tepal only 25–30 mm long. The similarities suggest a

close relationship between the two species, but differences in bract and flower size, as well as the orientation of the tepals, make confusion unlikely. Also closely related to *G. dalenii* is the eastern southern African *G. aurantiacus*. This species has flowers broadly similar to those of *G. dalenii*, but a noticeably longer perianth tube, 44–70 mm long, abruptly widened in the lower third, with the upper part cylindric and extended horizontally. The flowering stems bear only short-bladed or entirely sheathing leaves, and long-bladed foliage leaves are produced from the same corms later during the flowering season.

The type of *Gladiolus dalenii* is a watercolour illustration based on a plant from KwaZulu-Natal and, like most collections of the species from southern Africa, has rather narrow flowers in which the upper tepals are closely appressed to the hooded dorsal tepal, except toward the recurved apices. In most tropical African *G. dalenii* the upper tepals are more loosely appressed to the dorsal and are sometimes flared outward for half their length, imparting a rather different appearance to the flowers. This does not seem to merit taxonomic recognition for it is not a consistent characteristic. Plants from Ethiopia, for example, may in fact sometimes be so similar to those from South Africa that it is virtually impossible to distinguish them, and taxonomic separation at any rank at all seems unworkable and unwarranted.

Chromosome number is unusually variable in *Gladiolus dalenii*. All African populations of subsp. *dalenii* so far examined are polyploid, thus having somatic numbers of either $2n = 60$ or 90, occasionally $2n = 45$ or 75 (Goldblatt et al., 1993), but Madagascan populations are diploid (Goldblatt, 1989). Variation in the height of the plants and in general vigour may be due at least in part to variation in ploidy level. The Madagascan plants are among the most slender in the species, and the flowers are slightly smaller than those in continental African populations. The only counts for subsp. *andongensis* are also diploid, $2n = 30$.

History

Discovered first near the southern end of its range in KwaZulu-Natal in the late 1820s, this common *Gladiolus* was given four names in the course of as many years after it was first described: *Watsonia natalensis* Ecklon (in 1827), *Gladiolus dalenii* Van Geel (in 1829), *G. psittacinus* J.D. Hooker (in 1830) and *G. natalensis* Reinwardt ex J.D. Hooker (in

1831), the last an illegitimate renaming of *G. psittacinus*. The two first names were ignored for many years and in southern Africa the species came to be known either as *G. psittacinus*, which was based on a striking illustration in the accessible *Curtis's Botanical Magazine*, or by the competing *G. natalensis*, which Hooker substituted for *G. psittacinus*. The name *G. natalensis* had been used for the species in Holland and plants were distributed under this name to growers by Professor C.G.C. Reinwardt at Leyden. There is, however, no record of publication of the name *G. natalensis* prior to Hooker substituting it for *G. psittacinus* in 1831 (Hilliard & Burtt, 1979). Nevertheless, the confusion caused by Hooker's action has rippled through the nomenclature of *Gladiolus* until well into the twentieth century, with authors favouring one or other name.

Just over a decade after its discovery in KwaZulu-Natal, orange-flowered plants were found in Ethiopia by the French collector, Richard Quartin-Dillon. His collections formed the basis for Achille Richard's *Gladiolus quartinianus*, a name that continued to be used for *G. dalenii* in tropical Africa for more than a century. Yellow-flowered plants of *G. dalenii* were first collected in Mozambique by W.C.H. Peters and subsequently described by F.W. Klatt in 1864 as *G. luteolus*. In 1892, J.G. Baker described *G. primulinus*, by which name the yellow-flowered form of subsp. *dalenii* was most often known when it was regarded as separate from typical orange-flowered plants.

No other species of the genus has caused so much taxonomic confusion and misunderstanding. It was given no fewer than 27 synonyms based on plants from tropical Africa and Madagascar, and 14 more based on southern African collections. In his revisions of the genus for southern Africa (1896) and tropical Africa (1898), Baker reduced many of the names until then in use to synonymy under *G. quartinianus* for tropical African plants. Only with the publication of the second edition of *Flora of West Tropical Africa* (Hepper, 1968) was it widely accepted that the tropical African *G. quartinianus* was conspecific with the southern African *G. psittacinus*. Lewis et al. (1972) accepted this conclusion, although they used the name *G. natalensis* for the species under the mistaken impression that it was a combination based on the synonym, *Watsonia natalensis* Ecklon, which predates *G. dalenii*. Hooker's *G. natalensis* is, however, not a combination but a new species (Hilliard & Burtt, 1979) and thus dates only from 1831. The existence of the name *G. natalensis* J.D. Hooker prevents the transfer to *Gladiolus* of Ecklon's *W. natalensis*. The independent choice of the epithet *natalensis* twice for the same species seems to have been coincidental.

The related but smaller-flowered *Gladiolus antholyzoides* and its several synonyms (see the synonymy of that species, page 111) were regarded as conspecific with *G. dalenii* by Lewis et al. (1972). As they did not know *G. antholyzoides* in the field and only had rather poorly preserved herbarium specimens at their disposal, we doubt that they were making an informed decision based on a proper understanding of *G. antholyzoides*.

Ethnobotany

One of the few species of *Gladiolus* important in the human pharmacopeia, *G. dalenii* is recorded (under several of its synonyms) as being used in southern Africa in treating a variety of ailments, including diarrhoea and colds (Watt & Breyer-Brandwijk, 1962; Jacot-Guillarmod, 1971). It is a common component of the African herbalist's medicine horn, the *lenaka*. In parts of West Africa *G. dalenii* is used in preparations to cure both constipation and severe dysentery and is 'highly esteemed in curing snake bites' (Irvine, 1930). At least in West Africa there are records that *G. dalenii* is cultivated on farms in the forest, where it was introduced from the savanna country to the north (Dalziel, 1937). How much of its remarkably wide distribution across Africa is due to deliberate human activity may never be known. Corms of *G. dalenii* are also used as food in southern Congo (Zaïre). The starchy corms are boiled and then leached in water for a week before consumption (Malaisse & Parent, 1985: 62).

A strikingly ornamental plant, *Gladiolus dalenii* is widely cultivated. A southern African form flowering in late summer is perhaps the best known in horticulture. Spikes of the bright orange and yellow flowers of *G. dalenii* itself or a cultivar very like the wild plant are prominent in the exuberant plantings surrounding the terrace at Sainte-Adresse near Le Havre in Monet's painting of 1866, and the species was obviously well established in gardens in Europe by that time. More important, however, than its value as a wild species in gardens is the role of *G. dalenii* in the breeding of the modern *Gladiolus* hybrids. It is one of the species used in making the original crosses that led to the development of the large-flowered *Gladiolus* cultivars which are today among the world's most important cut-flower crops. *Gladiolus dalenii* is the only one of the species used in these crosses that was polyploid, and the resemblance of these cultivars to the species may be the result of its larger genetic contribution to the genome of the hybrid.

32. GLADIOLUS MAGNIFICUS (Harms) Goldblatt

PLATE 26

Gladiolus magnificus (Harms) Goldblatt in Goldblatt & De Vos, *Bull. Mus. Hist. Nat. Paris*, Sér. 4, Sect. B, Adansonia 11: 426 (1989); *Fl. Zambesiaca* 12(4): 94 (1993); Gladiolus *in Tropical Africa* 228 (1996). *Petamenes magnificus* (Harms) R. Foster, *Contrib. Gray. Herb.*, New Ser. 114: 50 (1936). *Antholyza magnifica* Harms in Warburg, *Kunene-Sambesi-Exped.* 201 (1903). Type: Angola, on the Longa near Minnesera, 1200 m, 11 Jan. 1901, *Baum 651* (B, holotype; BM, COI, K, W, isotypes).

magnificus = magnificent, alluding to the handsome appearance of the tall stems and spikes of large, brightly coloured flowers.

Synonymy

Antholyza zambesiaca Baker, *Handbook Irideae* 232 (1892); *Fl. Trop. Africa* 7: 374 (1898), not *Gladiolus zambesiacus* Baker (1892; from Central Africa). *Petamenes zambesiacus* (Baker) N.E. Brown, *Trans. Roy. Soc. S. Africa* 20: 227 (1932); Sölch, *Prodr. Fl. Südwestafrika* 155: 12 (1969). *Oenostachys zambesiacus* (Baker) Goldblatt, *J. S. African Bot.* 37: 443 (1971).

Goldblatt, *Fl. Zambesiaca* 12(4): 87 (1993). Type: Botswana (or northwestern Zimbabwe), south of the Zambezi River, Leshumo valley, in or before 1883, *Holub s.n.* (K, holotype).

Petamenes spectabilis (Schinz) Phillips, *Bothalia* 4: 44 (1941). *Chasmanthe spectabilis* (Schinz) N.E. Brown, *Trans. Roy. Soc. S. Africa* 20: 273 (1932). *Antholyza spectabilis* Schinz, *Mém. Herb. Boissier* 20: 13 (1900), not *Gladiolus spectabilis* Baker, *Bull. Herb. Boissier*, Sér. 2, 4: 1006 (1904) (= *G. saundersii* J.D. Hooker). Type: Namibia, Waterberg Plateau, 19 Apr. 1899, *Dinter s.n.* (possibly destroyed during World War II, evidently not at B, where expected).

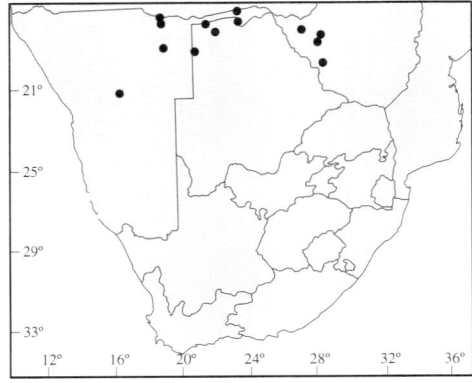

Description

Plants 80–140 cm high. *Corm* depressed-globose, 20–30 mm in diameter, the tunics of fine, wiry, straw-coloured fibres. *Cataphylls* firm-textured, pale below ground, green or flushed purple above ground, the upper reaching up to 10 cm above the ground. *Leaves* five to seven, the lower four or five basal and longest, about half as long as the stem and reaching almost to the base of the spike, the blades narrowly lanceolate, 6–9 mm wide, midrib and margins lightly thickened, the margins hardly raised, the upper leaves progressively smaller and with shorter blades or the blades lacking and then entirely sheathing. *Stem* unbranched, c. 4 mm in diameter below the first flower.

    Spike erect and straight, 6- to 15-flowered; *bracts* green, the outer 25–30(–35) mm long, the inner somewhat shorter than the outer. *Flowers* bright red with yellow markings on the lower tepals and in the throat, the lower three tepals each, at least sometimes, with a dark red-purple blotch in the midline, unscented; *perianth tube* 25–30 mm long, narrow below, expanded near the top of the bracts and curving outward and nearly horizontal, c. 4 mm wide at the mouth; *tepals* very unequal, the dorsal largest, hooded and horizontal, 35–42 mm long, 18–22 mm at the widest, upper laterals broadly lanceolate, 13–15 mm long, directed forward, the apices curving outward, the lower three tepals narrowly lanceolate, to 10 mm long, more or less horizontal, or the lowermost recurving. *Filaments* arched under the dorsal tepal, c. 35 mm long, exserted c. 20 mm from the tube; *anthers* 12–17 mm long, dull yellow, the pollen yellow. *Ovary* oblong to ellipsoid, 5–6 mm long; *style* dividing shortly below the apex of the anthers, the branches about 3 mm long. *Capsules* oblong, 20–30 cm long; *seeds* ovate, c. 13 x 8 mm, broadly and evenly winged, translucent light brown. *Chromosome number* unknown.

    Flowering time January and February, occasionally in March.

Distribution and biology

Largely a southern tropical African species, *Gladiolus magnificus* extends across the subcontinent from central Angola through southern Zambia to western Zimbabwe. Its range also includes the northern edge of the area we include in southern Africa, that is, the northern half of Namibia and northeastern Botswana. In Namibia it occurs in the Waterberg, near Grootfontein and along the northern border, including the Caprivi Strip, while in Botswana it is reported from the Chobe district. *Gladiolus magnificus* is most commonly found in Kalahari sandveld, a light woodland on sandy soil, often with tall grasses.

    The bright scarlet flowers with a long perianth tube and reduced lower tepals with yellow markings are clearly adapted for pollination by sunbirds, but no observations have been recorded on the pollination biology of the species.

Diagnosis and relationships

Superficially resembling a tall, slender variant of the widespread *Gladiolus dalenii*, *G. magnificus* can readily be distinguished by its enlarged, deeply concave dorsal tepal, 35–42 mm long and about twice as long as the upper lateral tepals. The lower three tepals are much reduced in size and are generally somewhat less than a third as long as the dorsal. Apart from these striking floral differences, the vegetative form of *G. magnificus*, including the unbranched stem and large floral bracts, closely resembles that of *G. dalenii*.

History

First collected in Botswana by the Czech naturalist, Emil Holub, in the early 1880s and described by J.G. Baker as *Antholyza zambesiaca* in 1892, this striking species has had a surprisingly complex history. A second collection was made in 1899 by the German botanist, Kurt Dinter, in northern Namibia, and these specimens were described by Hans Schinz in 1900 as *A. spectabilis*. Then a collection made in 1901 in southern Angola by Hermann Baum was described as *A. magnifica*. The confusion over the definition and circumscription of the genus *Antholyza* in the 1930s led to the species being transferred to both *Chasmanthe* and *Petamenes*, genera of Iridaceae based on Cape species not closely related to *Gladiolus magnificus*. In 1971 the species was transferred to the tropical African segregate of *Gladiolus*, *Oenostachys*, and finally in 1989 it came to rest in *Gladiolus* (Goldblatt & De Vos, 1989), where it carries the epithet *magnificus* because the two earlier synonyms were pre-occupied in the genus.

33. GLADIOLUS FLANAGANII Baker

Suicide gladiolus

PLATE 27

Gladiolus flanaganii Baker, *Fl. Capensis* 6: 530 (1896). Hilliard & Burtt, *Notes Roy. Bot. Gard. Edinburgh* 41: 306 (1983). Type: Lesotho, Mont-aux-Sources, near the summit, 8500 ft, Jan. 1894, *Flanagan 1832* (BOL, holotype; PRE, isotype).

flanaganii, named in honour of Henry Flanagan, Eastern Cape naturalist and plant collector, one of the first people to botanize in the high northern Drakensberg, and who collected the type specimen.

Synonymy

[*Gladiolus cruentus* sensu G. Lewis et al., *J. S. African Bot.*, Suppl. 10: 20 (1972), in part, non sensu Moore (1868).]

Description

Plants 35–60 cm high, erect or inclined from cliffs. *Corm* globose, 20–28 mm in diameter, the tunics of soft, lightly cartilaginous layers, fragmenting with age into coarse vertical fibres from the base. *Cataphylls* firm-textured and pale, often suffused with pink, the uppermost reaching up to 4 cm above the ground and then greenish, flushed with pink. *Leaves* usually five, the lower four basal and reaching to between the middle and apex of the spike, the blades lanceolate, 7–14 mm wide, the midrib moderately thickened and hyaline, a pair of secondary veins also somewhat thickened, the margins hardly thickened, the uppermost leaf inserted on the upper third of the stem, much shorter than the basal leaves, sheathing in the lower third, narrowly channelled for almost the entire length, the margins of the sheathing part free to the base. *Stem* usually inclined, occasionally more or less erect, flexed outward above the sheathing part of the uppermost leaf, unbranched, c. 4 mm in diameter below the spike.

Spike lightly flexuose, 4- to 7-flowered; *bracts* firm-textured, green or flushed with dull purple, the outer with a light median fold, 60–70 mm long, the inner lightly two-keeled, about two-thirds as long as the outer, notched apically for c. 2 mm. *Flowers* dark carmine-red, the lower tepals each with a narrow longitudinal white streak in the midline outlined with dark red, the ventral part of the upper part of the tube whitish streaked with red, unscented; *perianth tube* gradually flared from the base, slightly sigmoid in profile, 35–45 mm long; *tepals* more or less cupped above, the flowers somewhat closed and urn-shaped, obovate, those of the outer whorl equal and slightly larger than those of the inner whorl, these also subequal, the dorsal and lower laterals 30–38 x 18–20 mm, the upper laterals and lower median 30–42 x 25 mm. *Filaments* 27–30 mm long, exserted 14–15 mm from the tube; *anthers* c. 10 mm long, dark purple, the pollen pale lilac. *Ovary* oblong, c. 7 mm long; *style* arching over the stamens,

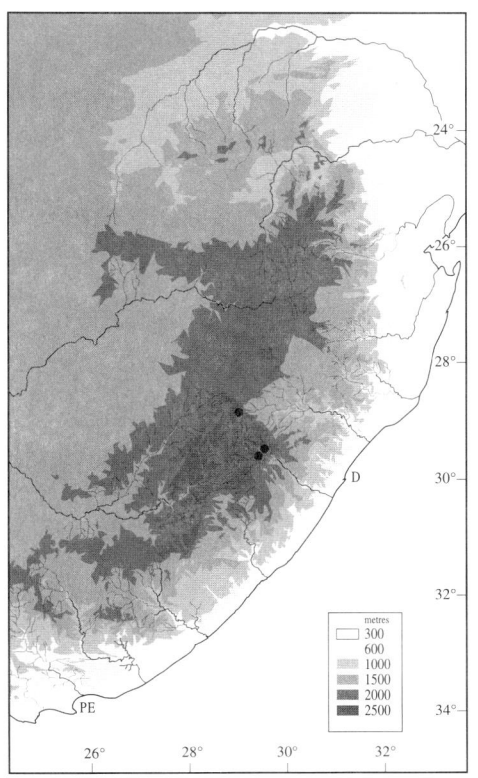

dividing 2–4 mm beyond the anther apices, the branches c. 4.5 mm long. *Capsules* oblong, 16–20 mm long; *seeds* oblong, 6–7 x 2–2.5 mm, the wing developed only at the ends. *Chromosome number* unknown.

Flowering time late November until early January.

Distribution and biology

Restricted to the high Drakensberg escarpment, *Gladiolus flanaganii* occurs along the frontier between Lesotho and KwaZulu-Natal, South Africa. The recorded range extends from near Sani Pass in the south to the slopes of The Sentinel in the north. Plants grow on steep basalt cliffs, their corms wedged in crevices in the rock where water seeping down the cliffs provides almost continuous moisture during the growing season.

The long-tubed, bright red flowers have an unusually firm thick texture and this together with the long perianth tube and copious amounts of nectar accords with pollination by sunbirds. The malachite sunbird, *Nectarinia famosa*, has been recorded foraging on the flowers and it is most probably the sole pollinator of *Gladiolus flanaganii*. The plants illustrated here are from the steep slopes on the ascent to Mont-aux-Sources, close to the type locality.

Diagnosis and relationships

Gladiolus flanaganii has one of the most distinctive flowers of southern African summer-rainfall *Gladiolus*. The foliage is unremarkable, but the long-tubed, carmine-red flowers borne on fairly thick short stems are unmistakable. The perianth tube is 35–45 mm long and slightly sigmoid in profile, and the tepals are somewhat closed above, thus forming an urn-like shape. The unusual shape of the flower is evidently related to the pollination strategy. Unlike most other bird-pollinated species of *Gladiolus*, in which birds perch on the erect stem below the flowers and reach upward into them, in *G. flanaganii* they must grasp the sturdy, inclined inflorescence above the flowers and reach downward.

Gladiolus flanaganii was treated as conspecific with the coastal KwaZulu-Natal species, *G. cruentus*, by Obermeyer in the revision of South African *Gladiolus* by Lewis et al. (1972). Apart from the red flower colour there is, however, very little shared between the two species. *Gladiolus cruentus* has softly

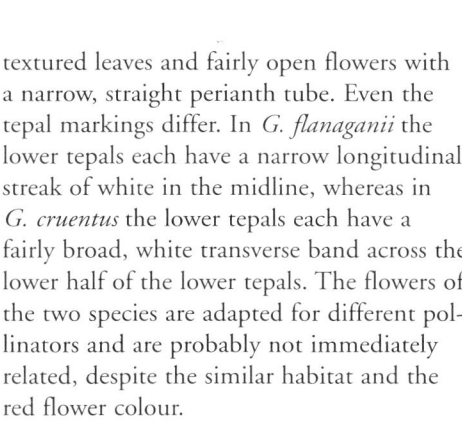

textured leaves and fairly open flowers with a narrow, straight perianth tube. Even the tepal markings differ. In *G. flanaganii* the lower tepals each have a narrow longitudinal streak of white in the midline, whereas in *G. cruentus* the lower tepals each have a fairly broad, white transverse band across the lower half of the lower tepals. The flowers of the two species are adapted for different pollinators and are probably not immediately related, despite the similar habitat and the red flower colour.

History

Gladiolus flanaganii was first collected in 1894 when H.G. Flanagan, accompanied by Harry Bolus, travelled to the eastern Free State. They were the first collectors to climb Mont-aux-Sources, along the slopes of which *G. flanaganii* was found. Flanagan's collection reached the Herbarium at the Royal Botanic Gardens, Kew, in time for the species to be described by J.G. Baker in 1896 and to be included in *Flora Capensis*. Because of its inaccessibility, *G. flanaganii* has not often been collected and it remained much misunderstood for many years. It was included in the coastal KwaZulu-Natal species, *G. cruentus*, which also grows on damp cliffs and has red flowers, by A.A. Obermeyer in the revision of *Gladiolus* in South Africa (Lewis et al., 1972). As pointed out by Hilliard & Burtt (1983), however, these are two quite different species. Their similar habitat, but at vastly different eleva- tions and on different substrates, must be due to parallel evolution.

34. GLADIOLUS SAUNDERSII J.D. Hooker
PLATE 28

Gladiolus saundersii J.D. Hooker, *Curtis's Bot. Mag.* 116: pl. 5873 (1870). Baker, *Fl. Capensis* 6: 158 (1896). Lewis et al., *J. S. African Bot.*, Suppl. 10: 21 (1972). Type: South Africa, Eastern Cape, Herschel District, summit of the Witteberg, 1861, *Cooper 605* (K, holotype; BM, Z, isotypes).

saundersii, named in honour of W. Wilson Saunders, Esq., of Reigate, England, employ- er of the plant collector, Thomas Cooper, who made the type collection of the species.

Synonymy

Gladiolus spectabilis Baker, *Bull. Herb. Boissier*, Sér. 2, 4: 1006 (1904). Type: Lesotho, Maluti Mountains, Mount Machache (c. 32 km east of Maseru), Mar. 1903, *Junod 1927* (Z, possible holotype, PRE, photo; G, SAM 21424, Z, isotypes).

[*Gladiolus cruentus* sensu Pole Evans, *Fl. Pl. S. Africa* 5: pl. 182 (1925), not *G. cruentus* T. Moore.]

Description

Plants (30–)40–60 cm high. *Corm* depressed-globose, (20–)30–50 mm in diameter, the tunics of dry papery layers, these fragmenting irregularly but not becom- ing fibrous, bearing numerous small purplish brown cormlets 4–6 mm long around the base. *Cataphylls* orange to brownish, the uppermost reaching 3–4 cm above the ground and turning green. *Leaves* seven to nine, mostly basal and forming a distichous fan, reaching at least to the base of the spike, sometimes exceeding it, the blades lanceolate, 15–25 mm wide, the midrib strongly thickened, the margins usually less so, one or more pairs of secondary veins also usually prominent, the upper one or two leaves inserted in the middle of the stem, usually sheath- ing the stem to the base of the spike. *Stem* erect, unbranched, c. 5 mm in diameter below the spike.

Spike erect to lightly inclined 10–15° at the stem apex, more or less straight, (2–)4- to 9- flowered; *bracts* green, the apices twisted, the outer 40–60(–70) mm long, lightly folded in the midline, the inner slightly shorter than the outer, forked apically for up to 5 mm into linear cusps. *Flowers* bright red, facing sideways or droop- ing, the lower three tepals lightly to densely speckled in the lower half with red on a white field, unscented; *perianth tube* obliquely funnel-shaped, 33–37 mm long; *tepals* lanceolate, unequal, the three upper largest, 46–67 x 21–30 mm, the dorsal extending horizontally or tilted downward over the stamens, recurving in the upper third, upper laterals curving outward from the base, recurving in the upper half to one-third, the lower three tepals joined to the upper laterals for c. 6 mm, and to one another for c. 2 mm, inclined downward, recurving distally, somewhat smaller than the upper, the lower laterals 34–46 mm long, the lower median 46–53 x 20–28 mm, in profile the dorsal exceeding the lower tepals. *Filaments* 30–50 mm long, exserted 20–30 mm from the tube; *anthers* 15–16 mm long, with a minute apiculus c. 0.5 mm long, tilted 30° below the horizontal, purple, the pollen cream. *Ovary* oblong, c. 6 mm long; *style* arching over the stamens, dividing between the middle and upper third of the anthers, the branches 7–8 mm long, often tangled in the anthers. *Capsules* oblong-obovoid, three- lobed and retuse apically, 25–28 mm long; *seeds* ovate, c. 6.5 x 8.5 mm, broadly and evenly winged, light translucent brown. *Chromosome number* $2n = 30$.

Flowering time mid-January to March.

Distribution and biology

Native to the southern and central Drakensberg, *Gladiolus saundersii* extends from Joubert's Pass and Lady Grey in Eastern Cape Province, through Lesotho to the northeastern Free State near Clarens. Plants seem to favour relatively drier sites that are seasonally wet, and can often be found on rocky outcrops, scree slopes and other fairly exposed habitats. In such places the huge scarlet flowers are a striking sight against the dark rock.

Although species in southern Africa with dark red, long-tubed flowers are often polli- nated by sunbirds, this lovely plant is a member of the *Aeropetes tulbaghia* pollina- tion guild. The small number of southern African species in this guild, mostly montane in distribution and flowering after midsum- mer, are pollinated solely by this large but- terfly. Visits to flowers of *Gladiolus saunder- sii* by *A. tulbaghia* have been reported by J.H.J. Vlok (Johnson & Bond, 1994).

Diagnosis and relationships

Gladiolus saundersii is unmistakable in its large scarlet flower with splashes of white

speckled with red on the lower three tepals, the horizontal to drooping dorsal tepal and the lower tepals strongly arched toward the ground. The vegetative features – a fan of plane, lanceolate leaves reduced in size progressively up the stem, leaf blades with thickened midrib and secondary veins, the large floral bracts and the obovoid, apically three-lobed capsules – place the species in subgenus *Ophiolyza*. Its affinities are clearly with species of section *Ophiolyza* such as the red-flowered coastal cliff species, *G. cruentus*, and possibly *G. dalenii*. Other summer-blooming species of *Gladiolus* with large, red flowers, including *G. cardinalis* and *G. sempervirens* (section *Blandus*), that are also pollinated by *Aeropetes tulbaghia* have flowers of a different shape. The dorsal tepal is upright or recurving, the lower tepals are directed forward, and the nectar guides consist of narrow, longitudinal white streaks. It seems unlikely that *G. saundersii* is closely related to these species. Their superficial floral similarities and shared pollinator must be the result of evolutionary convergence.

History

Although *Gladiolus saundersii* was evidently first collected by J.F. Drège on his expedition

to the Lesotho border in 1833, his collections received no attention. The plant collector, Thomas Cooper, also recorded *G. saundersii* when in 1861 his travels took him to the Witteberg in Eastern Cape Province. Cooper was employed by W. Wilson Saunders, a wealthy Englishman who maintained a large garden at Reigate near London. Plants that were raised there flowered in or before 1870 and were described by the eminent British botanist, J.D. Hooker, the protologue accompanied by a fine painting of the species. A second collection of *G. saundersii* was included in the protologue of *G. macowanii*, a later synonym of *G. mortonius*, when that species was described by J.G. Baker (1892, 1896) and for a time this resulted in some confusion between *G. saundersii* and *G. macowanii*, especially as the latter was described as having a red flower. It was only after 1900 that *G. saundersii* was first recorded in Lesotho, where later collecting has shown it to be widespread and relatively common. Plants collected by the Swiss missionary, Jean-Alexandra Junod, in the Maluti Mountains of Lesotho in 1903 were actually considered by J.G. Baker to be a distinct species which he called *G. spectabilis* in 1904. The type collection

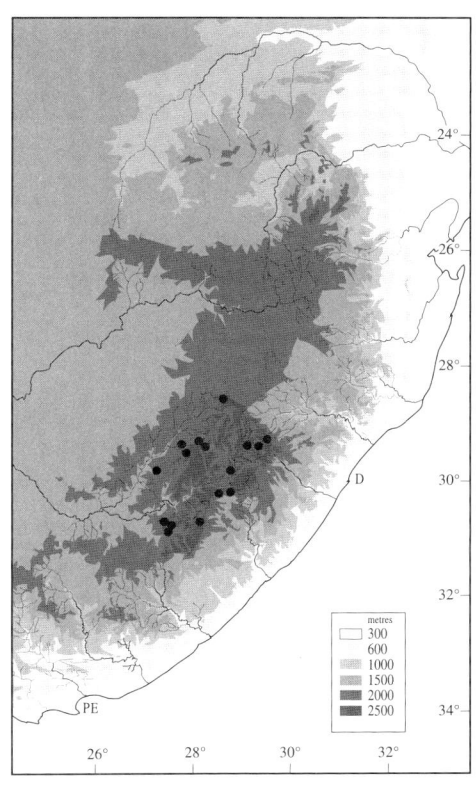

differs hardly at all from plants from Eastern Cape Province that represent typical *G. saundersii*.

---

35. GLADIOLUS CRUENTUS T. Moore
PLATE 29

Gladiolus cruentus T. Moore, *Gard. Chron.* 1868: 1139 (1868). J.D. Hooker, *Curtis's Bot. Mag.* 115: pl. 5810 (1869). Baker, *Fl. Capensis* 6: 157 (1896). Lewis et al., *J. S. African Bot.*, Suppl. 10: 20 (1972), including *G. flanaganii*. Manning & Goldblatt, *Veld & Flora* 81: 4 (1995). Type: South Africa, KwaZulu-Natal, without precise locality, possibly collected by McKen, illustration in *Curtis's Bot. Mag.* 115: pl. 5810 (1869).

cruentus = blood-stained, referring to the bright red flower colour, in particular the white lower tepals irregularly speckled with red.

Synonymy

[*Gladiolus saundersii* sensu Wood, *Natal Plants* 4(2): pl. 342 (1904), not *G. saundersii* J.D. Hooker.]

Description

Plants 50–70 cm long, usually arching horizontally from vertical cliffs. *Corm* globose, 15–20 mm in diameter, the tunics of medium-textured fibres mostly vertically oriented.

Cataphylls coriaceous, pale below ground, the uppermost reaching to about 10 cm above the ground and then green or becoming flushed with purple below. *Leaves* seven or eight, the lower four or five more or less basal, usually slightly longer than the spike but drooping downward, the blades sword-shaped, 10–15 mm wide, the midrib moderately thickened, the margins not at all raised, upper two or three leaves cauline, all except the uppermost with long blades, the uppermost leaf fairly short, the margins of the sheath open to the base and overlapping. *Stem* inclined to drooping below the horizontal, usually arching upward in the upper part, unbranched, c. 5 mm in diameter below the spike.

Spike arching upward and inclined or nearly erect, straight, 4- to 9-flowered; *bracts* green, diverging from the stem, the outer 40–60(–80) mm long, usually about two internodes long, the inner similar, half to two-thirds as long as the outer, often lightly two-keeled, forked apically for c. 1 mm. *Flowers* bright scarlet, the lower tepals each with a transverse white chevron at the base

of the limb, the lower part of the lower tepals and upper part of the tube speckled red on white, unscented; *perianth tube* more or less cylindric, c. 28 mm long, flaring in the upper c. 6 mm; *tepals* lanceolate, the dorsal lightly inclined over the stamens, c. 60 x 30 mm, the remaining tepals patent, the upper laterals c. 55 x 27 mm, the lower three tepals united basally for 2–3 mm, narrowed and ascending below for c. 15 mm, the limbs patent, c. 33 x 19 mm. *Filaments* c. 40 mm long, exserted c. 35 mm from the tube, arcuate; *anthers* c. 12 mm long, diverging, dark purple, the pollen whitish. *Ovary* oblong, c. 6 mm long; *style* more or less erect, adaxial to the stamens, dividing at or shortly beyond the anther apices, the branches c. 5 mm long, only slightly widened above. *Capsules* oblong, 18–25 mm long, rounded to retuse apically; *seeds* narrowly oblong, 7–9 x c. 4 mm, the wing unevenly developed, longest at the micropylar end, light translucent brown. *Chromosome number* unknown.

Flowering time mid-January to early March.

Distribution and biology

A narrow endemic of the southern African summer-rainfall zone, *Gladiolus cruentus* is restricted to central coastal KwaZulu-Natal. Populations have been recorded only along the sandstone belt from Gillitts and Kloof in the south to the Noodsberg and Kranskop in the north. Plants grow on damp cliffs where water drips continuously down the rocks during the growing season, bathing the shallow roots in a constant stream. Because of the specialized habitat, plants are very difficult to collect, but appear to be common wherever suitable wet sandstone cliffs occur.

The flowers of *Gladiolus cruentus* are bright red with splashes of white on the lower tepals and have relatively long, slender tubes. Flowers of this type are usually pollinated by the satyrid butterfly, *Aeropetes tulbaghia*, and those of *G. cruentus* must be assumed to be adapted for pollination by this insect, although no observations have been reported for the species.

Diagnosis and relationships

Consistent with its habitat on wet cliffs, *Gladiolus cruentus* has drooping stems and softly textured, plane leaves without noticeably thickened margins and veins. The flowers are a striking scarlet colour and quite large, with a fairly slender, narrowly funnel-shaped perianth tube, c. 28 mm long, and a dorsal tepal about 60 mm long. Except for the suberect dorsal tepal, the flower is very open, the other tepals outspread from the base, and the lower tepals have conspicuous white transverse banded markings in the centre. Evidently a member of series *Ophiolyza*, *G. cruentus* is probably most closely related to *G. saundersii*. This Eastern Cape and Lesotho montane species of rocky, basalt or dolerite slopes has erect stems and fairly stiff leaves, the margins and veins of which are thickened and prominent. Apart from a more or less horizontal dorsal tepal, the flowers are fairly similar to those of *G. cruentus*.

Gladiolus cruentus and the high Drakensberg species, *G. flanaganii,* have at times been confused and they were considered conspecific by Lewis et al. (1972). However, the resemblance between them is slight and confined mainly to their similarly-coloured red flowers with white markings. The shape of the flower of *G. flanaganii* is different, the perianth tube is longer and the lower tepals arch forward and upward, forming a shallow cup. That species appears to be pollinated by sunbirds and may not be immediately related to *G. cruentus*.

History

The first record of *Gladiolus cruentus* is the collection made by M.J. McKen at Kranskop in 1867. This and a later collection from Inanda near Durban by John Medley Wood in 1881 were confused with *G. saundersii* by J.G. Baker and are cited under that species in *Flora Capensis*. An active collector who supplied herbaria and botanic gardens with Natal plants, McKen may well have supplied Bull's Nursery in Chelsea, London, with the plants that flowered there in September 1868 and were described in the *Gardener's Chronicle* later that year by the British botanist, Thomas Moore. A more complete description of the species, accompanied by a painting, was then published in *Curtis's Botanical Magazine* in 1869, the article having been contributed by J.D. Hooker. Neither Moore nor Hooker mention the collector or precise local origin of the cultivated material, and the plant illustrated was later used for hybridizing and was not preserved.

Specimens from the garden of Mr Veitch in London, apparently also from the same source, are preserved at the Kew Herbarium. Live plants of *G. cruentus* grown in Durban and illustrated in Wood's *Natal Plants* likewise were called *G. saundersii*, although Wood (1904) did mention the discrepancy between the soft-textured leaves of the plants in cultivation compared with the firm, ribbed texture described by Baker. Nevertheless, these two species remained confused and an illustration of *G. saundersii* was published in *Flowering Plants of Africa* (Pole Evans, 1925) under the name *G. cruentus*. In their revision of *Gladiolus* in South Africa, Lewis et al. clearly distinguished *G. saundersii* from *G. cruentus* but included within the latter the high Drakensberg *G. flanaganii*, an error rectified by Hilliard & Burtt (1983).

3. SECTION *BLANDUS*

(Baker) Goldblatt

Section *Blandus* (Baker) Goldblatt, Gladiolus *in Tropical Africa* 176 (1996). Type species: *Gladiolus blandus* Aiton (= *G. carneus* Delaroche).

Synonymy

Gladiolus group *Blandi* Baker, *J. Linn. Soc. Bot.* 16: 176 (1877). Type: *G. blandus* Aiton (= *G. carneus* D. Delaroche), lectotype designated by Goldblatt, 1996.

Subgenus *Hyptissa* Klatt, *Abh. Naturforsch. Ges. Halle* 15: 342 (Ergänzungen 8) (1882), not *Hyptissa* Salisbury (1866), the identity of the latter uncertain. Type: *G. blandus* Aiton (= *G. carneus* Delaroche), lectotype designated by Goldblatt, 1996.

Description

Plants usually medium-sized to large. *Corms* globose, the tunics coriaceous or firm-papery, decaying into irregularly shaped fragments, or becoming coarsely fibrous, rarely the corms not well developed and the tunics not persisting. *Cataphylls* usually glabrous, often mottled purple and green or whitish. *Leaves* usually several and forming a basal fan, occasionally four or only three and then superposed, the blades plane, usually sword-shaped, sometimes linear, the midrib and margins and usually other veins thickened and raised, occasionally distinctly two-veined. *Stem* branched, but consistently unbranched in a few species.

Spike erect or inclined, more or less straight, usually few-flowered; *bracts* small to large, green, firm-textured, usually about as long as the perianth tube, or shorter. *Flowers* medium-sized to large, usually shades of pink to cream, but red in a few species, the lower tepals usually with median, longitudinal, spear- or spade-shaped nectar guides pale in the centre edged with dark pigment, unscented; *perianth tube* narrowly funnel-shaped or more or less cylindric, elongate in a few species, generally about as long as the dorsal tepal or much longer, occasionally somewhat shorter; *tepals* subequal to slightly unequal with the upper three larger than the lower, the dorsal ascending, the lower three forming a horizontal or deflexed lip, not or barely joined to one another or to the upper laterals. *Capsules* oblong-ellipsoid, fairly

large; *seeds* large, broadly and evenly winged, translucent light brown.

Comprising 21 species, section *Blandus* has a geographical range largely restricted to the southern African winter-rainfall zone, and mostly within the Cape Flora Region. One species, *Gladiolus undulatus*, extends north of the Region into the Kamiesberg in Namaqualand. Three more species occur east of the confines of the Cape Flora Region: *G. floribundus* and *G. grandiflorus* extend eastward to Grahamstown and East London, but still respond to a winter–spring growth cycle in this area of both summer and winter rainfall; and only *G. gueinzii* extends significantly into the summer-rainfall zone. Populations of this strand species range eastward from the Agulhas Peninsula along the south-east coast of southern Africa as far north as Durban in KwaZulu-Natal.

The majority of species of section *Blandus* have sword-shaped leaves that usually form a distichous fan at or above ground level. In some species leaf number is reduced to only four or three, and the leaves assume a more or less superposed arrangement. The leaf blades are firm with a well-defined midrib and usually have one or more other prominent pairs of veins. The margins are somewhat thickened, occasionally strongly so. Leaves and flowers are usually produced contemporaneously and well-developed, but in *Gladiolus carmineus* and *G. stefaniae* (series *Blandus*), which flower late in the season – after the middle of summer – the leaves on the flowering stem are reduced and bladeless, and only non-flowering shoots have well-developed leaves.

Most species have pink to cream flowers with a perianth tube at least about as long as the dorsal tepal. The flowers tend to have widely spreading tepals and most often spear- or spade-shaped, white to yellow nectar guides edged with dark red or purple. The predominant pollination system in the species with pink to cream flowers involves long-tongued flies of the families Nemestrinidae and Tabanidae. *Gladiolus angustus*, the flowers of which have a tube 70–110 mm long, is pollinated by the nemestrinid fly *Moegistorhynchus longirostris* which has mouthparts 50–80 mm long.

Other species in which pollination by long-tongued flies has been confirmed are *G. bilineatus*, *G. carneus*, *G. floribundus* and *G. undulatus*. Pollination by long-tongued bees of the family Anthophoridae has been recorded in *G. grandiflorus* and is suspected for other short-tubed species, including *G. oreocharis*, *G. phoenix* and *G. rudis*, all of which have unusually short perianth tubes for the section, 15–25 mm long. *Gladiolus gueinzii*, which has short-tubed and nearly actinomorphic flowers, is autogamous, but whether outcrossing occurs as well has not been established.

Two other pollination strategies are found in the section. *Gladiolus cardinalis* and its allies *G. sempervirens* and *G. stefaniae* have deep red flowers with white markings on the lower tepals and all three flower unusually late in the season, after midsummer until March or even April or May in the case of *G. sempervirens*. These species are adapted for pollination by the butterfly *Aeropetes tulbaghia* (Johnson & Bond, 1984), and pollination by this butterfly has been confirmed for all three species. *Gladiolus insolens*, which has uniformly orange-red flowers, is probably also adapted for pollination by this butterfly. Lastly, *G. miniatus* has flowers adapted for pollination by sunbirds. The flowers are salmon to orange and have a long perianth tube, 50–65 mm long, with the upper half of the tube wide and cylindric. This flower type corresponds closely to that in several other species of southern African Iridaceae that are pollinated by sunbirds (Goldblatt & De Vos, 1989).

We recognize four species groups in section *Blandus*. The majority of the species are included in series *Blandus*, most of which have pink to cream flowers, but five species are red- or purple-flowered. Three fire-adapted species with short-tubed, bee-pollinated flowers are assigned to series *Phoenix*. Four closely related species centred around *G. floribundus* which have unusual, coarsely fibrous corm tunics and short stems are placed in series *Floribundus*. We are uncertain about the immediate relationships of *G. gueinzii* and it is placed alone in series *Sabulosus*.

36. GLADIOLUS OREOCHARIS Schlechter

Gladiolus oreocharis Schlechter, *J. Bot.* 34: 504 (1896). Baker, *Fl. Capensis* 6: 529 (1896). Lewis et al., *J. S. African Bot.*, Suppl. 10: 180 (1972). Type: South Africa, Western Cape, Matroosberg, Dec. 1895, *Marloth 2265* (B, holotype [not seen]; BOL, GRA [not seen], PRE, isotypes).

oreocharis = mountain grace, referring to the habitat, the higher mountains of the southwestern Cape, and the flower's pleasing appearance.

Description
Plants (12–)18–35(–50) cm high. *Corm* globose, 7–10 mm in diameter, the tunics of fine, reticulate fibres. *Cataphylls* pale and membranous, the uppermost reaching shortly above ground and then green, flushed with purple or becoming dry and brown, sometimes hispidulous, often accumulating with the remains of the cataphylls to form a neck around the underground part of the stem. *Leaves* three to five, the lowermost one to three basal, reaching to between the base and apex of the spike, the blades plane, linear, (1–)2–6 mm wide, with two, occasionally one main vein, these and the margins lightly to moderately thickened, the upper two leaves largely to entirely sheathing, the uppermost smallest, inserted in the upper quarter of the stem. *Stem* erect or inclined, flexed outward above the sheathing parts of the two uppermost leaves, c. 1.5 mm in diameter below the spike.

Spike inclined to almost horizontal, 2- to 7-flowered; *bracts* green, firm-textured, the outer 13–20(–25) mm long, purple near the apices, obtuse to truncate, the inner three-quarters as long as to slightly longer than the outer, minutely forked apically. *Flowers* pink to reddish purple or violet, occasionally white, the lower three tepals each with a spear-shaped median dark red to purple mark with a white to yellow centre, unscented; *perianth tube* obliquely funnel-shaped, 15–24 mm long, shortly exserted from the bracts; *tepals* ovate, unequal, the dorsal largest, inclined, 18–22 x 12–14 mm, upper laterals slightly smaller, 15–20 mm long; the lower three tepals joined to the upper laterals for c. 3–4 mm and to one another for 1–2 mm, free parts 14–20 mm long. *Filaments* 15–16 mm long, exserted 7–9 mm from the tube; *anthers* 6–8 mm long, mauve. *Ovary* ovoid, 3–4 mm long; *style* arching over the stamens, dividing at or 1–2 mm beyond the anther apices, the branches c. 4 mm long. *Capsules* obovoid, three-lobed and retuse apically; *seeds* unknown. *Chromosome number* unknown.

Flowering time mostly December, occasionally in early January.

Distribution and biology
Gladiolus oreocharis is a fairly rare species of the higher mountains of Western Cape Province. The type collection is from Matroosberg, one of the highest mountains in the province, but the species extends from the Cedarberg in the north through the Hex River Mountains eastward to the Langeberg at Riversdale and the Klein Swartberg near Ladismith. All records of the species are from south-facing, moist slopes, streamsides or seeps, and above 1000 m, more often from at least 1600 m. *Gladiolus oreocharis* appears to be more common in the Langeberg, but this may simply be the result of more intensive collecting activity there. Although sometimes locally common, the species is most often encountered on slopes that were burned the previous summer. In the second year after a fire plants flower poorly and thereafter are rarely seen until another fire consumes the dense mountain heath vegetation where they grow.

We assume that the pink flowers of *Gladiolus oreocharis*, with a tube of intermediate length, 15–24 mm, are pollinated by long-tongued bees, as are several other species of *Gladiolus* with flowers of a similar shape and colour.

Diagnosis and relationships
Assigned here to series *Phoenix* of section *Blandus*, *Gladiolus oreocharis* is recognized by its relatively small flowers, 33–45 mm long with a tube 15–24 mm long, and characteristically short, obtuse to nearly truncate floral bracts, mostly 13–20 mm long. The leaves are more or less linear and 2–6 mm wide, and usually with a paired midrib. An unusual feature of the species is that the upper cataphyll is often hispidulous. These features combined with its summer-flowering habit, mostly December and January, make it relatively easy to identify the species. It is most often confused with the Langeberg endemic, *G. crispulatus*, which has larger flowers with the tepals crisped along the edges, narrow leaves 1–3 mm wide with a characteristic venation, a double midrib on one surface and single on the other, and smooth cataphylls. The different tepal markings of the two species, discussed under the next species, also aid in distinguishing them.

Gladiolus oreocharis superficially resembles the Western Cape *G. carneus*, but this common species has larger flowers with quite different markings as well as different leaf venation. Almost throughout its range the variable *G. carneus* can be distinguished by broader leaves, usually 6–14 mm wide, but occasionally less, and with the midrib and one or more other vein pairs raised and thickened. The large flowers are usually 50–80 mm long with a tube rarely less than 25 mm and up to 45 mm long.

History
Despite having a relatively wide range across the Cape mountains, *Gladiolus oreocharis* appears to have been first recorded only a century ago when the botanist, Rudolf Marloth, collected the species on the slopes of Matroosberg in 1895. It was named the following year by Rudolf Schlechter who was collecting plants in South Africa at this time. The species remained poorly known until the systematic exploration of the flora of the Cape mountains was undertaken by Cape Town botanist, Elsie Esterhuysen, beginning in the 1940s. She established the northern limits of its range in the Cedarberg and its occurrence throughout the Hex River Mountains as well as in the Langeberg. An unusual population from the Klein Swartberg was only discovered in 1986 by the Cape conservationist, J.H.J. Vlok.

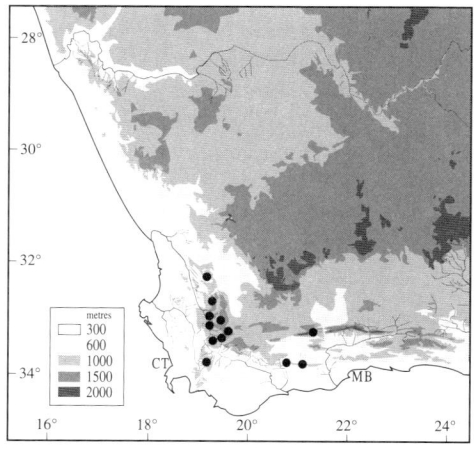

37. GLADIOLUS CRISPULATUS L. Bolus

PLATE 30

Gladiolus crispulatus L. Bolus, *Ann. Bolus Herb.* 3: 182 (1924). Rendle, *Curtis's Bot. Mag.* 147: pl. 8923 (1938). Type: South Africa, Western Cape, Riversdale District, Corente River Farm, Dec. 1908, *Muir sub Galpin 5369* (BOL, lectotype here designated).

crispulatus = with small, tightly curled edges, referring to the distinctive tepal margins.

Synonymy

Gladiolus tenuis Baker, *J. Bot.* 14: 335 (1876), an illegitimate homonym, not *G. tenuis* Bieber (1808); *Fl. Capensis* 6: 145 (1896). Type: South Africa, Western Cape, Langeberg Mountains, Craggy Peak, Swellendam, 14 Jan. 1815, *Burchell 7303* (K, holotype; BOL, tracing).

Description

Plants slender, 30–40 cm high. *Corm* globose, 8–12 mm in diameter, the tunics of fine, netted fibres. *Cataphylls* pale and membranous, the uppermost reaching up to 3 cm above the ground and then dark green. *Leaves* four or five, superposed, the lower two basal and longest, reaching at least the upper third of the stem, sometimes shortly exceeding the spike, the blades linear, 1–3 mm wide, the margins barely raised, the midribs lightly thickened, usually paired on one side, single on the other, the remaining leaves progressively shorter above, sheathing for half their length, the margins of the sheathing part free to the base and overlapping. *Stem* erect or inclined, unbranched, 1–2 mm in diameter below the spike.

Spike flexed at the base and lightly inclined, 2- to 4-flowered; *bracts* pale green, fairly soft-textured, the outer 24–33 mm long, the inner about three-quarters as long as the outer, minutely bifid at the apex.

Flowers deep pink, the lower tepals each with a median, dark red, narrowly triangular streak and speckled with dark red toward the base and in the lower throat, the base of the throat red, unscented; *perianth tube* obliquely funnel-shaped, 20–26 mm long, widening in the upper 8–10 mm; *tepals* unequal, lanceolate, the margins undulate to lightly crisped, the dorsal largest, 28–34 x 15–20 mm, arching over the stamens, curving upward in the upper third, the upper laterals directed forward below, curving outward and spreading above, c. 30 x 16 mm, the lower three tepals joined to the upper laterals for 6–8 mm and to one another for 1.5–3 mm, 20–25 x 8 mm, curving toward the ground in the upper half. *Filaments* 14–19 mm long, exserted 5–9 mm from the upper part of the tube; *anthers* 8–11 mm long, lilac-pink, pollen whitish. *Ovary* ovoid, c. 4.5 mm long; *style* arching over the stamens, pink, dividing between the upper third and slightly beyond the apices of the anthers, the branches 3–5 mm long, spreading. *Capsules* and *seeds* unknown. *Chromosome number* unknown.

Flowering time mainly November and December.

Distribution and biology

Gladiolus crispulatus is a rare endemic of the Langeberg in the southern Western Cape Province, South Africa. The few records from precise localities indicate that it occurs along a short stretch of this range, from Swellendam to Riversdale, a distance of some 80 km. Records indicate that it grows on south-facing slopes from 300 to 900 m. According to Georges Delpierre, who knows the species in the wild, it flowers well the first and second seasons after fires and not or hardly at all after that. This is part of the explanation for its being so poorly known. We also suspect that it is rare and restricted to relatively few sites.

Observations on the pollination biology of this rare and shy-flowering species are not available, but we assume that the deep pink and unusually marked flowers of *Gladiolus crispulatus*, with a perianth tube of intermediate length, 20–26 mm, are adapted for pollination by long-tongued bees or a combination of bees and other long-tongued insects, perhaps including tabanid flies and possibly bombylliid flies.

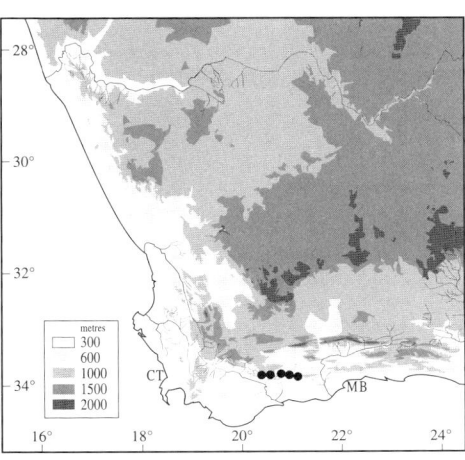

Diagnosis and relationships

Dark pink flowers with a dorsal tepal about as long as the tube and leaves with a double midrib place *Gladiolus crispulatus* in series *Phoenix* of section *Blandus*. Except for its slender habit, especially the narrow, superposed leaves, *G. crispulatus* can easily be mistaken for the allied *G. oreocharis* or for *G. carneus* (series *Blandus*). Careful examination of the leaves shows that the narrow blades, 1–3 mm wide, have the midribs lightly thickened and usually paired on one side, single on the other. *Gladiolus oreocharis*, however, has plane leaf blades mostly 2–6 mm wide with two main veins evident on both surfaces and short, obtuse to truncate floral bracts mostly 13–20 mm long. The flowers are slightly smaller than those of *G. crispulatus*, 33–45 cm long with a tube 15–24 mm long, and the lower tepals have more or less straight margins and spear-shaped, median dark reddish marks with a pale centre. The larger flowers of *G. crispulatus* are 48–60 mm long with a tube 20–26 mm long and the bracts are

20–26 mm long. More telling when seen alive are the lower tepal markings which consist of a bright red triangular mark in the midline and dark red speckles toward the base and in the throat.

Although *Gladiolus crispulatus* is sometimes confused with *G. carneus* and was included in that species by Lewis et al. (1972), the tepal markings in *G. carneus* range from spear-shaped to narrow median lines or are occasionally absent, but are never like those of *G. crispulatus*. The leaves of *G. carneus* are usually 6–14 mm wide, although occasionally less, and never have the unusual double vein on one surface and single on the other. The crisped tepals of *G. crispulatus* are not particularly distinctive, despite its name, and are also characteristic of many populations of *G. carneus*.

History

Gladiolus crispulatus was first collected in 1815 by the early explorer and naturalist, William Burchell. He recorded plants on the Kampscheberg near Riversdale on 9 January and on Craggy Peak near Swellendam on 14 January. The latter collection formed the basis for J.G. Baker's *G. tenuis*, described in 1876, and so named for the long, narrow leaves. Unfortunately, the name is a homonym for *G. tenuis* Bieber, dating from 1805, a Eurasian species of the genus now included in *G. imbricatus*, and cannot be used for the southern African plant. Baker, nevertheless, continued to use the name, and it is included in *Flora Capensis* (Baker, 1896).

The species remained so poorly understood that when it was re-collected nearly a century later by the Riversdale naturalist, John Muir, in 1908 it was thought to be a new discovery. Muir's collections of the species made from the same area for several more years formed the basis for Louisa Bolus's *G. crispulatus* which she described in 1924, unaware that the plant had an earlier, although illegitimate name. Because *G. tenuis* Baker is a later homonyn, Bolus's name must be used for the species. Plants collected some years later were grown in Great Britain and figured in *Curtis's Botanical Magazine* (Rendle, 1938). In their revision of *Gladiolus* in South Africa, Lewis et al. (1972) included the species in a broadly circumscribed *G. carneus*. Plants found in the vegetative state by Georges Delpierre in 1994 were grown and flowered in cultivation in Cape Town the following year, when our illustrations were made.

38. GLADIOLUS PHOENIX Goldblatt & Manning

PLATE 31

Gladiolus phoenix Goldblatt & Manning, new species. Type: South Africa, Western Cape, Bain's Kloof Mountains, Baviaanskloof, Nov. 1944, *Linley s.n.* (SAM, holotype and isotype).

phoenix = like the mythical phoenix, rising

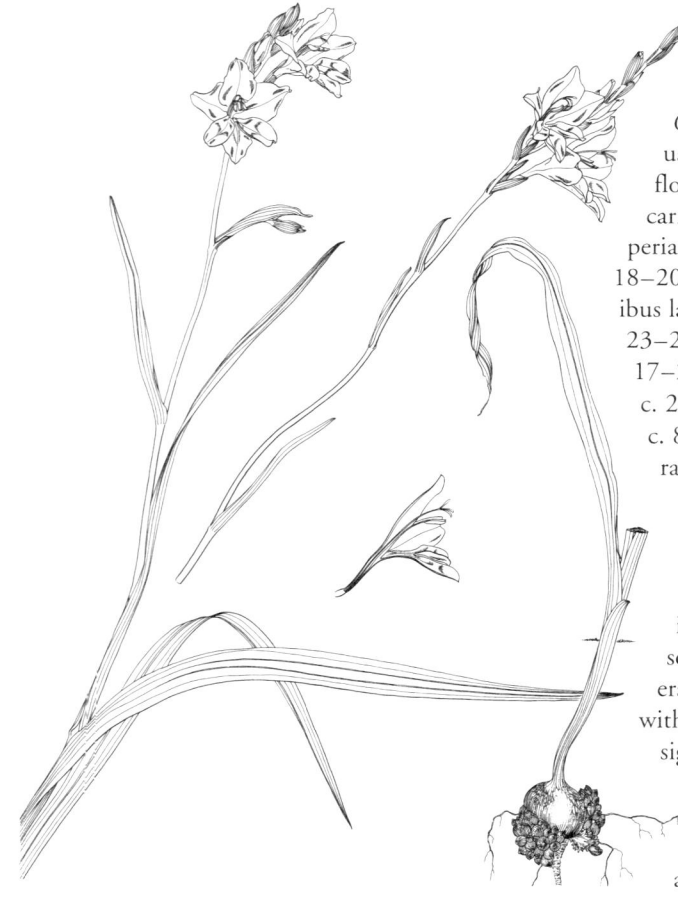

from the ashes, because of its pattern of flowering only after fires.

Latin diagnosis

Plantae 50–75 cm altae, cormo globoso 8–12 mm in diametro tunicis fibrosis cormellos multos stolonibus fasciatis insidentes gerenti, foliis 6–9 inferioribus 3–5 basalibus longioribus laminis planis linearibus 6–12(–20) mm latis, caule usitate ramoso, spica 9–12-florum inclinata, floribus carneis albisque notatis, tubo perianthii infundibuliformi 18–20 mm longo, tepalis inequalibus lanceolatis superioribus 23–28 x 13–15 mm inferioribus 17–20 x 7.5–13 mm, filamentis c. 20 mm longis, antheris c. 8 mm longis atropurpureis, ramis styli c. 6 mm longis.

Description

Plants 50–75 cm high. *Corm* globose, 8–12 mm in diameter, the tunics of soft brown membranous layers, becoming finely fibrous with age but not accumulating significantly, bearing clusters of small cormlets around the base on short fasciated stolons. *Cataphylls* pale and membranous, the

uppermost reaching 5–8 cm above the ground and then green. *Leaves* six to nine (with an additional leaf per branch), the lower three to five basal and longest, reaching to about the base of the spike, the blades narrowly lanceolate to nearly linear, plane, 6–12(–20) mm wide, the midrib lightly raised, the margins scarcely so, the remaining leaves inserted on the stem above the ground, the upper one or two leaves channelled below, not normally sheathing the stem or sometimes sheathing in the lower half, the margins open to the base, sometimes with a short unifacial blade. *Stem* erect, usually with one or two branches, occasionally simple, 2–3 mm in diameter below the spike.

Spike lightly flexed at the base and inclined, the flowers secund, 9- to 12-flowered (flowers fewer on the branches); *bracts* dark green, the outer lightly folded in the upper midline, the apices often slightly hooked, becoming brown toward the end of flowering, 13–16 mm long, the inner two-thirds to nearly as long as the outer, minutely forked apically. *Flowers* deep pink, the lower three tepals each with a white, spear-shaped mark in the midline outlined in deep pink decurrent on the lower half of the throat and with a broad white transverse band on the reverse, unscented; *perianth tube* obliquely funnel-shaped, 18–20 mm long, the cylindrical part c. 10 mm long; *tepals* unequal, the upper lanceolate, the dorsal largest, 26–28 x 15 mm, ascending, the upper laterals c. 23 x 13 mm, directed forward below, curving outward and becoming patent in the upper third, the lower three tepals united basally for c. 4 mm, shortly clawed and channelled below, the free parts of the lower laterals c. 17 x 7.5–13 mm, the lower median c. 20 x 9–13 mm. *Filaments* c. 20 mm long, exserted c. 13 mm from the tube; *anthers* c. 8 mm long, dark purple, the pollen pale yellow. *Ovary* oblong, c. 3.5 mm long; *style* arching over the stamens, dividing just below the anther apices, branches arching outward, when fully expanded c. 6 mm long. *Capsules* narrowly ovoid, 12–15 mm long, lightly three-lobed above and the apices retuse; *seeds* ovate, broadly and evenly winged, c. 8 x 5 mm, light transparent brown. *Chromosome number* $2n = 30$.

Flowering time November to early December.

Distribution and biology

Gladiolus phoenix is evidently a rare and extremely local endemic restricted to moist banks and ravines in the Bain's Kloof Mountains above Wellington in Western Cape Province, South Africa. Although known from only a small area of the Bain's Kloof Mountains, it may occur in other suitable habitats elsewhere along the range.

The biology of *Gladiolus phoenix* is unusual. Plants apparently flower only in the first, or rarely in the second season after a fire. After a fire in Baviaanskloof, a small valley that runs off Bain's Kloof, in February 1991, plants flowered extremely well in 1992 and set quantities of seed. The following year not a single plant flowered there and few even produced leaves. After another fire in the

area early in 1994, a second small colony in Bain's Kloof flowered the following November. Small corms removed from one site, where plants were very common, were induced to flower in cultivation after the effects of fire were simulated by burning plant material over planted corms. After ample artificial feeding, plants flowered later the same year. The illustrations and some details of the description were made from these cultivated plants.

The relatively short-tubed flowers of *Gladiolus phoenix* are most likely adapted for pollination by long-tongued bees and we have captured the anthophorid bee *Amegilla spilostoma* foraging for nectar on the flowers.

Diagnosis and relationships

Relatively unspecialized morphologically, *Gladiolus phoenix* is a fairly large plant, usually with branched stems and several plane leaves forming a basal fan. The flowers are dark pink and the lower tepals each have an oblong white median streak in the midline. Both the floral bracts and the perianth tube are unusually short for species of section *Blandus*, the tube being 16–20 mm long. The short tube and dark flower colour suggest a relationship with *G. oreocharis*, which *G. phoenix* broadly resembles, and we place the two species together with *G. crispulatus* in series *Phoenix*. Unique among the southern African species of the genus, *G. phoenix* produces numerous cormlets on short, fasciated stolons from the corm base. These fasciated stolons are otherwise known in *Gladiolus* only in the distantly related tropical African *G. verdickii* of section *Tenuibracteus* (Goldblatt, 1996).

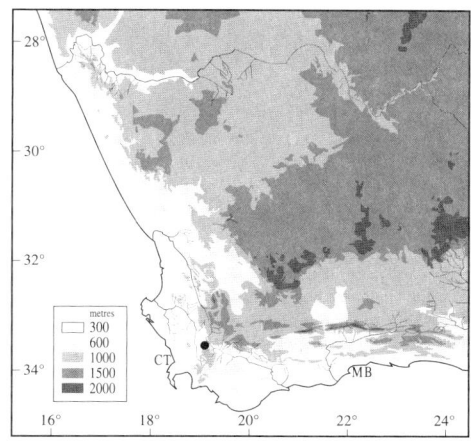

History

The first record of *Gladiolus phoenix* was made in 1944 by the Cape Town pharmacist and keen grower of indigenous plants, J.S. Linley. The collection remained unnamed in the South African Museum Herbarium until we began to critically study the *Gladiolus* species of the southern African winter-rainfall zone. It then became clear that an unidentified species we had collected in fruit in late December, 1992, in Baviaanskloof was the same as the Linley collection, and possibly even from the same population. Small corms collected at the site in 1993 flowered in cultivation in 1994, at about the same time that we discovered a second population flowering in Bain's Kloof. Because of its habit of flowering in the wild only in the season after fire, it has otherwise remained unrecorded, despite growing literally beside the road in Bain's Kloof Pass where generations of botanists have collected plants.

Section *Blandus:* Series *Sabulosus*

39. GLADIOLUS GUEINZII Kuntze

PLATE 32

Gladiolus gueinzii Kuntze, *Linnaea* 18: 510 (1845) & 20: 14 (1847). Lewis et al., *J. S. African Bot.*, Suppl. 10: 74 (1972). Type: South Africa, KwaZulu-Natal, Durban (Port Natal), without date, *Gueinzius 490* (*460* in the protologue) (S [not seen], neotype; NBG, photograph).

gueinzii, named in honour of the German apothecary and naturalist, Wilhelm Gueinzius, who made the earliest collection of the species. He discovered the species at the northern limit of its range, near Durban, in the early 1840s.

Synonymy

Acidanthera brevicollis Baker, *J. Bot.* 14: 339 (1876); *Fl. Capensis* 6: 134 (1896). *Gladiolus sabulosus* G. Lewis, *J. S. African Bot.* 7: 31 (1941), as a new name in *Gladiolus*, not *G. brevicollis* Klatt, 1882 (= *G. brevifolius* Jacquin). Type: South Africa, Eastern Cape, Bathurst District, mouth of the Fish River, Feb. *MacOwan 1890* (K, holotype).

Description

Plants (20–)30–50 cm high. *Corm* globose, 25–35 mm in diameter, the flesh bright yellow, tunics coriaceous, dark brown, breaking

up irregularly into fibrous fragments, forming numerous cormlets around the base; corms near the soil surface producing instead larger cormlets with dark brown, thick coriaceous to corky, fissured tunics, those produced deep underground with thin, lighter brown tunics. *Cataphylls* pale and coriaceous, the upper becoming light brown above the ground, the uppermost reaching up to 5 cm above the soil surface. *Leaves* several, usually four to six, mostly basal, the upper one or two cauline, shorter or longer than the stem, up to 60 cm long, the blades linear, 4–6 mm wide, slightly succulent,

only the midvein evident when live but hardly thickened, when dry the midrib and margins evidently thickened, the upper one or two leaves sometimes entirely sheathing, the sheaths open to the base. *Stem* erect, often branching after the flowers of the main axis have faded and capsules are ripening, c. 3 mm in diameter below the spike, producing clusters of cormlets at the nodes, the cormlets sometimes sprouting on the parent plant.

Spike weakly inclined at the base, not or only weakly flexuose, occasionally 2-, mostly 4- to 8-flowered; *bracts* pale green or greyish pink, the outer 18–27 mm long, acute, the inner slightly shorter, usually bifurcate at the apex for c. 1 mm. *Flowers* zygomorphic or occasionally actinomorphic, mauve to light purple, the upper three tepals with a red-purple median streak, the lower three with white median streaks outlined in red-purple, unscented; *perianth tube* 13–15 mm long, more or less cylindric; *tepals* subequal, obovate, 17–20 x 8–10 mm, or the outer slightly larger, to 13 mm wide, weakly to strongly outcurving distally, the lower three tepals slightly narrowed at the base and weakly clawed. *Filaments* 7–10 mm long,

unilateral or symmetrically disposed, exserted 1.5–3 mm from the tube; *anthers* 5.5–8 mm long, purple, pollen grey-blue to cream. *Ovary* ovoid-oblong, c. 3.5 mm long; *style* dividing between the base and middle of the anthers, the branches c. 5 mm long, weakly expanded above, somewhat tangled in the anthers. *Capsules* obovoid to ellipsoid, 26–32 mm long; *seeds* oval, broadly and evenly winged, 10–14 x 6–10 mm, red-brown. *Chromosome number* $2n$ = 30. *Flowering time* mainly October to December in the winter-rainfall region (the southern half of its range), November to January in the summer-rainfall region (the northern half of its range).

Distribution and biology

Gladiolus gueinzii is distributed along the southeast coast of South Africa where its range extends from Arniston in Western Cape Province to Durban in KwaZulu-Natal. The range is unusual because it straddles both winter- and summer-rainfall regions of the subcontinent. *Gladiolus gueinzii* is also unusual in its habitat, growing on sandy beaches and dunes at and above the tidal high water mark. In this harsh, saline habitat plants thrive and flower well and reproduce freely from seed and cormlets. The distribution pattern is apparently discontinuous, with a marked break between the most westerly station at Arniston and Still Bay to the east and again between Cape St. Francis and Port Alfred. Both breaks in the range are puzzling, given that suitable habitat is available all along the coast.

Vegetative reproduction is important in *Gladiolus gueinzii* and plants produce two kinds of cormlets. Small ones with membranous to papery tunics, such as are found in many other species of the genus, are clustered around the base of buried corms. A second kind, produced when the parent corm is exposed at the soil surface after the sand covering it has been removed, is much larger, sometimes up to 15 mm long and has a thick corky tunic, dark in colour, with a smooth glossy surface. Impermeable to water and buoyant, these cormlets must be assumed to be floating propagules which disperse plants up and down the coast. The only other species of *Gladiolus* that grows on beaches and dunes, the Madagascan *G. luteus*, produces similar cormlets with thick corky

coats, likewise assumed to be dispersed by water. The two species do not appear to be closely related and the convergent development of these unusual cormlets is striking.

Pollination ecology has not been studied, but we assume that the mauve-purple flowers with a fairly short perianth tube are adapted for pollination by long-tongued bees. The strand habitat is poor in insect life, and *Gladiolus gueinzii* is unusual in the genus in having self-compatible and autogamous flowers. Plants set full capsules of well-developed and viable seed without artificial or insect-mediated pollen transfer.

Diagnosis and relationships

The blue-purple, stellate, almost actinomorphic flower with white nectar guides outlined in dark red and the pale grey-green, glaucous, slightly fleshy leaves make it easy to recognize *Gladiolus gueinzii*. In combination with the unusual habitat, they make the species unmistakable. The flowers are relatively small and the tepals subequal, spreading and actinomorphic except for the nectar guides. These are normally confined to the lower three tepals and consist of a narrow median white streak outlined in red or red-purple. The stamens are more or less erect and usually unilateral, but are sometimes symmetrically disposed. The weakly zygomorphic or even completely actinomorphic flower is notable for a species in a genus in which fully zygomorphic flowers are the normal condition. We assume the trend toward actinomorphy is a specialization associated with its facultatively autogamous reproductive biology. The relationships of *G. gueinzii* within section *Blandus* are not clear and we have placed the species in a series by itself, series *Sabulosus*.

History

Gladiolus gueinzii was apparently first collected at its most northern station near Durban, then known as Port Natal, by the German apothecary and naturalist, Wilhelm

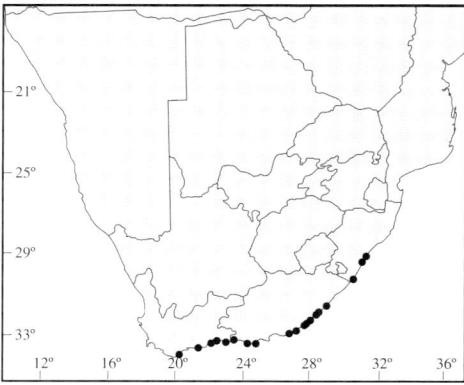

Gueinzius, one of the first people to collect plants in the Natal colony where he lived from 1841 until his death in 1874. The exact date of the collection is uncertain, but must have been between 1841 when Gueinzius arrived in Natal and 1844 when the name first appears in the literature in a seed catalogue from the Leipzig Botanical Garden. The species was described by Otto Kuntze in 1845, based on plants grown at the botanical garden at Leipzig, and was named in Gueinzius's honour. No doubt specimens were made and placed in the Leipzig Herbarium where Kuntze worked, but the specimen was destroyed during World War II and the species is now typified by a neotype at the Stockhom Herbarium, evidently collected by Gueinzius.

Collections of *Gladiolus gueinzii* at the Kew Herbarium, including two from Eastern Cape Province, *Macowan 1890* and *Cooper 3197*, and several from KwaZulu-Natal, one apparently a duplicate of Gueinzius's collection from Port Natal, were described as *Acidanthera brevicollis* by J.G. Baker in 1876. The reason for placing the species in *Acidanthera* is unclear, for that genus was usually reserved for species which had a long perianth tube and pale-coloured flowers. Baker only realized later that his species was the same as *G. gueinzii*, even though one of the specimens he cited in the protologue is apparently an isotype of that species. In his account of *Acidanthera* in *Flora Capensis* (Baker, 1896), however, *G. gueinzii* is cited as a synonym of *A. brevicollis*. G.J. Lewis also seems to have been unaware that *G. gueinzii* and *A. brevicollis* were the same species, and when she transferred the latter to *Gladiolus* in 1941, she renamed it *G. sabulosus*, the name *brevicollis* being preoccupied in the genus.

Section *Blandus:* Series *Blandus*

40. GLADIOLUS CARNEUS Delaroche
Painted lady, white Afrikaner, witbergpypie
PLATE 33

Gladiolus carneus Delaroche, *Plantae Aliquot Novarum* no. 3, fig. 4 (1766). Lewis et al., *J. S. African Bot.*, Suppl. 10: 92 (1972). *Gladiolus blandus* var. *carneus* (Delaroche) Ker Gawler, *Genera Irid.* 140 (1827). Baker, *Fl. Capensis* 6: 155 (1896). Type: South Africa, Western Cape, without precise locality or collector (L, holotype).

carneus = flesh-coloured (literally), thus pink, alluding to the colour of the flowers.

Synonymy
Gladiolus trimaculatus Lamarck, *Encyclop. Méth.* 2: 727 (1786), III 1: 116, pl. 32 (fig. 3) (1791). Type: South Africa, Western Cape, without precise locality or date, *Sonnerat s.n.* (P, holotype).
Gladiolus ventricosus Lamarck, *Encyclop. Méth.* 2: 727 (1786). *Gladiolus cuspidatus* var. *ventricosus* (Lamarck) Baker, *Handbook Irideae* 205 (1892); *Fl. Capensis* 6: 140 (1896). Type: South Africa, Western Cape, without precise locality or date, *Sonnerat s.n.* (P, holotype).
Gladiolus blandus Aiton, *Hort. Kewensis* 1: 6 (1789). Baker, *Fl. Capensis* 6: 154 (1896). Type: South Africa, Western Cape, without precise locality, grown at Kew Gardens, *Masson s.n.* (BM, holotype).
Gladiolus albidus Jacquin, *Icones Pl. Rar.* 2: pl. 256 (1795). *G. blandus* var. *albidus* (Jacquin) Ker Gawler, *Genera Irid.* 140 (1827). Type: South Africa, Western Cape, without precise locality or collector, figure in Jacquin, *Icones Pl. Rar.* 2: pl. 256 (1795).
Gladiolus cordatus Thunberg, *Prodromus Plantae Capensium* 185 (1800). Type: South Africa, Cape, without precise locality, *Thunberg s.n.* (UPS–Herb. Thunberg 1017, holotype).
Gladiolus campanulatus Andrews, *Bot. Repository* 3: pl. 188 (1801). *G. blandus* var. *campanulatus* (Andrews) Ker Gawler, *Genera Iridearum* 140 (1827). Type: South Africa, Western Cape, without precise locality or collector, figure in Andrews *Bot. Repository* 3: pl. 188 (1801).
Gladiolus blandus var. *purpureo-albescens* Ker Gawler, *Curtis's Bot. Mag.* 18: pl. 645 (1803). *G. lunulatus* Klatt, *Abh. Naturforsch. Ges. Halle* 12: 342 (Ergänzungen 8) (1882), new name at species rank for *G. blandus* var. *purpureo-albescens*. Type: South Africa, Western Cape, without precise locality or collector, figure in *Curtis's Bot. Mag.* 18: pl. 645 (1803).
Gladiolus blandus var. *excelsus* Ker Gawler, *Edwards' Bot. Reg.* 7, misc. material (1821).
Gladiolus pictus Sweet, *Hort. Brit.* ed. 1: 397 (1827), new name at species rank for *G. carneus* var. *excelsus*. *G. excelsus* (Ker Gawler) Sweet, *Hort. Brit.* ed. 2: 501 (1830), superfluous name for *G. pictus* Sweet, being based on the same type. Type: South Africa, Western Cape, without precise locality or collector, cultivated in England, illustration in *Curtis's Bot. Mag.* 40: pl. 1665 (1814).
Gladiolus expallescens Schrank, *Denkschr. Bot. Ges. Regensburg* 2: 210 (1822). Type: South Africa, Cape, without precise locality or collector, cultivated in Belgium (BR [not seen], holotype).
Gladiolus vinulus Klatt, *Trans. S. African Philos. Soc.* 3: 199 (1885). Type: South Africa, Western Cape, near Wynberg, Sept. 1884, *MacOwan 2651* (*Herb. Norm. Austr. Afr.* 287) (SAM, said to be the holotype; B [not seen], BM, G, K, P, S [not seen], UPS [not seen]).
Gladiolus macowanianus Klatt, *Trans. S. African Philos. Soc.* 3: 199 (1885). Type: South Africa, Western Cape, Cape Peninsula, grassy slopes behind Hout Bay, Oct. 1884, *MacOwan 2065* (as *2605*) (*Herb. Norm. Austr. Afr.* 284) (SAM, holotype, according to Lewis et al., 1972).
Gladiolus cuspidatus var. *ensifolius* Baker, *Handbook Irideae* 205 (1892). Type: South

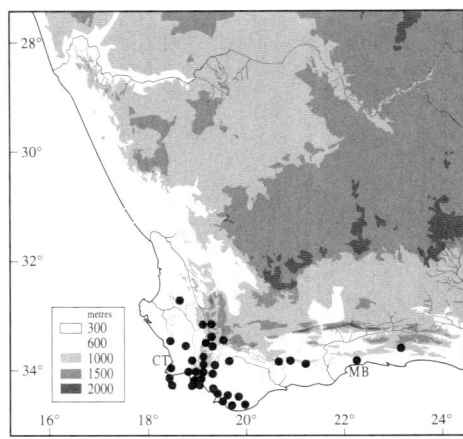

Africa, Western Cape, Houw Hoek, Oct., *Pappe s.n.* (K, holotype).

Gladiolus prismatosiphon Schlechter, *Bot. Jahrb. Syst.* 27: 102 (1900). Type: South Africa, Western Cape, Bredasdorp Mountains near Napier, 9 Dec. 1896, *Schlechter 9656* (B [not seen], holotype; BM, BOL, K, Z [not seen], isotypes).

Gladiolus callistus F. Bolus, *Ann. Bolus Herb.* 2: 105 (1917). Type: South Africa, Western Cape, Worcester District, Orchard Siding, 5 Oct. 1915, *Dicey sub National Botanic Gardens 1217/13* (BOL 14844, lectotype here designated on two sheets; BOL, PRE, isolectotypes).

Gladiolus callistus var. *gracilior* F. Bolus, *Ann. Bolus Herb.* 2: 105 (1917). Type: South Africa, Western Cape, Bredasdorp District, Ratel River, 12 Dec. 1896, *Bolus 8693* (BOL, holotype or lectotype; BOL, K, PRE, isotypes).

Gladiolus eximius Ingram, *Gard. Chron. Ser.* 3, 89: 46 with figure (1931). Type: South Africa, Western Cape, Caledon District, Palmiet River mouth, without date, *Ingram s.n.* (K, lectotype designated here).

[*Gladiolus dubius* Ecklon, *Topographisches Verzeichniss* 41 (1827), name without description, based on plants cultivated in Cape Town, preserved at S (not seen).]

[*Gladiolus lemonia* Pourret ex Steudel, *Nomencl. Bot.* ed. 2, 1: 686 (1840), a name without description, regarded as a synonym of *G. carneus* by Lewis et al. (1972).]

Description

Plants 20–60(–80) cm high. Corm more or less globose, 15–20 mm in diameter, the tunics of firm to soft papery layers, decaying with age and fragmenting irregularly or becoming somewhat fibrous. Cataphylls pale and membranous, the uppermost reaching 5–8 cm above the ground and then dark purple with cream or greenish spots. Leaves four or five, the lower two or three basal, usually forming a distichous fan, sometimes superposed and then not forming a fan, sword-shaped to linear, reaching at least to the base of the spike, sometimes longer, the blades plane, erect or sometimes trailing, narrowly lanceolate to nearly linear, (2–)6–14(–19) mm wide, the midrib and one or more secondary veins lightly thickened, the margins narrowly hyaline, the uppermost one or two leaves inserted on the middle or upper third of the stem, channelled below, either diverging from the stem at the base or sometimes sheathing in the lower half, sometimes with a short unifacial blade, the lower margins free to the base, sometimes overlapping. Stem erect, inclined or drooping, simple, occasionally with one, rarely two branches, 2–4 mm in diameter below the spike.

Spike usually lightly flexed at the base and inclined, or erect, the flowers in two rows separated by 30–60°, thus nearly distichous, occasionally 3-, usually 5- to 8-, rarely to 11-flowered; *bracts* pale green or flushed with grey, the outer 35–45(–65) mm long, the inner two-thirds as long as to only slightly shorter than the outer, acute or forked apically for 1–2 mm. *Flowers* white to pale or deep pink, the lower three tepals usually each with pink to red spade- or diamond-shaped marks, the lower midline often lined in red, the mark decurrent on the tube, sometimes with linear median streaks, or unmarked, or with pale yellow blotches, the tube often spotted or lined with red at the base of the throat, unscented; *perianth tube* narrowly funnel-shaped, (15–)25–38(–45) mm long, the cylindrical part 22–29 mm long, usually just emerging from the bracts; *tepals* lanceolate, the margins usually lightly undulate or sometimes almost crisped, unequal, the upper three largest, 28–40(–50) x 12–20(–30) mm, the dorsal ascending, the upper laterals slightly smaller than the dorsal, gently curving outward from the base, the lower three tepals sometimes joined to the upper laterals for up to 6 mm, 25–30(–40) x 7–15 mm, nearly straight and horizontal or curving slightly toward the ground and upcurved near the apices. *Filaments* 12–17(–25) mm long, exserted 5–10(–15) mm from the tube; *anthers* 6–11 mm long, mauve, the pollen cream or purple. *Ovary* oblong, 5–6 mm long; *style* arching over the stamens, dividing between the bases and apices of the anthers, the branches 4–8(–10) mm long. *Capsules* oblong-ellipsoid, three-sided, 20–30 mm long, obscurely three-lobed above; *seeds* oblong, 7–9 x 5–6 mm, broadly and evenly winged. *Chromosome number* $2n$ = 30.

Flowering time October to mid-November, occasionally later.

Distribution and biology

Restricted to the winter-rainfall region of southern Africa, *Gladiolus carneus* extends from the Cape Peninsula eastward through the southern Cape and Langeberg to the Outeniqua Mountains north of Knysna. It extends to the northwest only as far as Bain's Kloof and the Witsenberg–Skurweberg complex above Ceres. Despite the relatively wide range, it is common only in the west where it is often seen on lower mountain slopes in the passes through the western Cape mountains from Du Toit's Kloof to Tradouw Pass, as well as on the slopes of Table Mountain. Like many species of the Cape mountains, more particularly of nutrient-poor soils, *G. carneus* flowers well after fire or bush clearing and it can make a fine display along the edges of marshes and streams in the southern Cape after a burn. *Gladiolus carneus* is somewhat unusual in growing with apparently equal success on shale or sandstone soils, although it always grows in fynbos or woodland margins.

The long-tubed pink flowers of *Gladiolus carneus* are primarily adapted for pollination by long-tongued flies in the families Nemestrinidae and Tabanidae, which show a strong preference for visiting pale pink to cream flowers with linear reddish markings. The nectar is also typical of flowers pollinated by these flies, having a moderately high sugar concentration and a predominance of sucrose over glucose and fructose. We have recorded both *Philoliche rostrata* (Tabanidae) and *Prosoeca nitidula* (Nemestrinidae) visiting the flowers and evidently accomplishing pollination as they did so.

Diagnosis and relationships

Gladiolus carneus can usually be recognized by its large, pale pink flowers, normally with somewhat undulate tepals, the lower three of which have red linear to spear-shaped markings. The perianth tube is about as long as the dorsal tepal and is usually slightly exserted from the floral bracts. While the leaves are normally 6–14 mm wide and arranged distichously, they are sometimes narrower and superposed, especially in taller specimens. The flowers, too, vary in colour and to a lesser extent in their shape. Plants in the east of the range in the Langeberg and Outeniqua Mountains usually have white flowers with dark red markings and the tepal edges are strongly crisped, whereas those from the west of the range usually have pink or salmon-coloured flowers with less pronounced markings and tepals with undulate to nearly straight edges. A distinctive form, common on the coast between Rooi Els and Kleinmond, has pink flowers with the tepals unmarked except for the tepal bases and the upper part of the tube which is dark red. The lower tepals of these plants are arched downward in the middle, giving them a somewhat spooned appearance. Flowers of just this shape also occur to the north, including the type of *G. callistus* which is from the Breede River valley near Botha. *Gladiolus pictus* (= *G. blandus* var. *excelsus*), based on plants cultivated in Great Britain in the 1820s, is also this form. Another notable variant is the plant named *G. prismatosiphon* from near Gansbaai, which has the typical colouring and markings but a perianth tube considerably longer than elsewhere, and almost twice as long as the bracts. Even among the specimens that comprise the type collection, the tube length varies and we do not believe the longer perianth tube is of taxonomic significance.

Gladiolus carneus can easily be confused with *G. undulatus* but in that species the cream flowers have a perianth tube longer than the dorsal tepal and all the tepals are decidedly attenuate with strongly undulate margins. The local segregate of *G. carneus*

from the Cape Peninsula, *G. pappei*, can be distinguished by its deep pink flowers, slender habit and stems bearing three or four superposed leaves which are never more than 3 mm wide.

History

Known in Europe at least as early as 1688, *Gladiolus carneus* was one of a handful of species of the genus included in the *Prodromi fasciculi rariorum primus et secundus*, compiled from various illustrations owned by the Danzig merchant, Jakob Breyne, by his son Philipp in 1739. There it appears under the polynomial *Gladiolus floribus patentibus, externe carneis, etc.* *Gladiolus carneus* has, in its several forms, become one of the most cherished of the many Cape bulbous plants now in cultivation. No doubt because of its variability and its ease of cultivation, the species has acquired a formidable number of synonyms and has been subjected to a fair amount of confusion. It was first given a binomial, *G. carneus*, by Daniel Delaroche, a student of the Dutch botanist, David van Royen, in 1766. Although the species was clearly described and figured, the name was ignored

for many years. The English botanist, William Aiton, described *G. blandus* in 1789, based on plants collected by Francis Masson and cultivated in England, and it was under this name that *G. carneus* became well known for 180 years.

Gladiolus carneus was not, however, completely forgotten. John Ker Gawler, writing in 1827 under the name John Bellenden Ker, recognized the plant as a variety of *G. blandus*, a treatment that J.G. Baker followed in his various accounts of the genus *Gladiolus*, including that in *Flora Capensis* (1896). When the type specimen of Delaroche's species was examined for G.J. Lewis's revision of the southern African species and its identity confirmed, the earlier name was revived. By 1972 *G. carneus* had acquired at least 15 synonyms at species rank based on different types. Some of the names apply to different forms of *G. carneus* and others correspond closely to variants already named. Amongst the numerous names cited by Lewis et al. (1972) in synonymy, at least *G. crispulatus*, its synonym *G. tenuis* and *G. pappei* are now regarded as distinct species.

41. GLADIOLUS PAPPEI Baker
PLATE 34

Gladiolus pappei Baker, *Handbook Irideae* 208 (1892); *Fl. Capensis* 6: 146 (1896). Type: South Africa, Western Cape, Table Mountain, Dec., *Pappe s.n.* (K, holotype; SAM, isotype).

pappei, named in honour of Ludwig Pappe, Government Botanist at the Cape Colony.

Synonymy

[*Gladiolus carneus* sensu Lewis et al., *J. S. African Bot.*, Suppl. 10: 29 (1972), in part.]

Description

Plants 25–35 cm high. *Corm* globose, c. 5 mm in diameter, the tunics membranous, evidently not accumulating annually. *Cataphylls* pale and membranous, often decayed by flowering time and barely distinguishable, the uppermost sometimes evident above the ground and green. *Leaves* three or four, rarely five, the lower two basal and the lowermost usually fairly small, the second lowest longest and reaching to between the middle of the stem and the base of the spike, the blades linear, 1.6–3 mm wide, neither the midrib nor the margins noticeably thick-

ened, the remaining one to three leaves smaller and sheathing for at least half their length, with short, non-sheathing portions. *Stem* usually erect below, flexed outward above the sheaths of the upper two leaves but usually becoming erect again, unbranched, c. 1.2–2 mm in diameter below the spike.

Spike nearly erect or lightly flexed at the base and weakly inclined, lightly flexuose, 2- to 3- or rarely 4-flowered; *bracts* grey-green, often flushed with purple-grey on the dorsal surfaces, the outer 25–45 mm long, the inner two-thirds to almost as long as the outer, notched apically for c. 2 mm. *Flowers* deep or sometimes pale pink, the lateral tepals usually with median diamond- to heart-shaped markings, white in the centre and reddish on the periphery, the lower median tepals also sometimes marked but less prominently, the base of the upper part of the throat also red, unscented; *perianth tube* obliquely funnel-shaped, 30–35 mm long, the lower cylindrical part c. 12 mm long; *tepals* more or less lanceolate, the dorsal inclined over the stamens, straight or curving upward in the upper quarter, 26–30

x 15–19 mm, the upper laterals gently curving outward from the base, barely fully patent near the apex, 25–28 x 12–15 mm, the lower three tepals joined to the upper laterals for c. 2 mm and to one another for c. 2 mm, horizontal below and curving downward above, 20–28 x 10–11 mm, the lower median slightly longer than the lower laterals, in profile the lower three tepals extending c. 5 mm beyond the upper three. *Filaments* 12–17 mm long, exserted 5–8 mm from the tube; *anthers* 6–8 mm

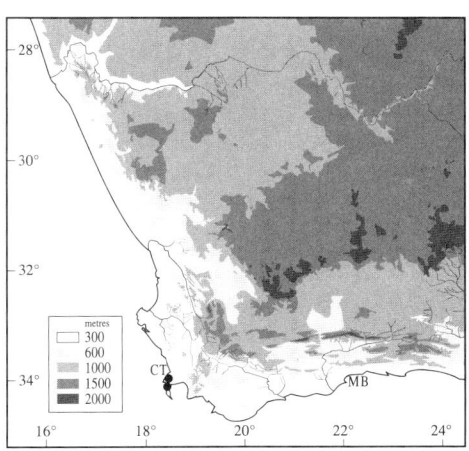

long, pale grey-blue, the pollen pale mauve. *Ovary* cylindric, 6–7 mm long; *style* arching over the stamens, dividing between the middle and shortly beyond the anther apices,

the branches slender, 5–7 mm long, very slightly wider in the upper half. *Capsules* and *seeds* unknown. *Chromosome number* $2n = 30$.

Flowering time mid-October to mid-December.

Distribution and biology

Gladiolus pappei is a narrow endemic of the southern African winter-rainfall zone. The species is apparently restricted to the Cape Peninsula alone, where it extends from the Back Table of Table Mountain to the low flats near Cape Point. Specimens collected on the high plateau above the Jonkershoek Valley near Stellenbosch seem to fairly closely resemble *G. pappei*, but we have been unable to confirm whether they represent this species or simply a narrow-leafed form of *G. carneus* which is common at lower elevations at Jonkershoek. Plants grow in sandy, somewhat peaty soils in marshy habitats which are either perennially moist or at least wet until early summer. Plants seldom flower except in the season immediately following a fire. This pattern is not uncommon in plants of nutrient-poor sandy soils, but is pronounced in *G. pappei*.

The long-tubed flowers of *Gladiolus pappei* conform to the pattern in the genus for species pollinated by long-tongued flies, and the flowers of *G. pappei* must be assumed to be adapted for the same pollination syndrome.

Diagnosis and relationships

Closely related to *Gladiolus carneus* and read-

ily confused with that species, *G. pappei* can be recognized by its deep pink flowers, the lower tepals each with a broad median white mark, and short stems bearing only three or four narrow leaves. The leaves are always superposed and have relatively short blades 1.6–3 mm wide and reaching to between the middle of the stem and the base of the spike. The corms are somewhat unusual for series *Blandus*, being about 5 mm in diameter and having softly membranous tunics that decay rapidly and do not accumulate annually.

Gladiolus pappei was included in *G. carneus* by Lewis et al. (1972) without any reservation, but we feel its unusual ecology and fairly characteristic morphology merit it being treated as a separate species. The more robust *G. carneus* has white, cream or pale pink flowers with red nectar guides usually consisting of linear streaks or small blotches on the lower tepals. It also has much broader leaves and is a much more sturdy plant than *G. pappei*.

History

Although *Gladiolus pappei* was evidently first recorded by C.F. Ecklon and C.L. Zeyher in the 1820s, a later collection made by the Cape Colonial Botanist, C.W.L. Pappe, probably in the 1850s, is the type of the species. Pappe's collection, at the Kew Herbarium, attracted J.G. Baker's attention and he described *Gladiolus pappei* in 1892. The species was not accepted by Lewis et al. (1972), and in their revision of *Gladiolus* in southern Africa *G. pappei* was regarded as a synonym of *G. carneus*.

42. GLADIOLUS GEARDII L. Bolus

PLATE 35

Gladiolus geardii L. Bolus, *Ann. Bolus Herb.* 4: 28 (1925). Type: South Africa, Eastern Cape, Steytlerville District, Farm Hadley near Steytlerville, mountain slopes, cultivated at Kirstenbosch Botanic Gardens, Dec. 1923, *Paterson s.n.* (BOL as Nat. Bot. Gard. 135/23, holotype on three sheets; K, isotype).

geardii, named in honour of Mr C. Geard, owner of the farm Hadley where the species was discovered.

Synonymy

Gladiolus robustus Goldblatt, *J. S. African Bot.* 50: 453 (1984). Type: South Africa, Eastern Cape, Baviaanskloof Mountains, farm Enkeldoorn, 820 m, 16 Nov. 1982, *Vlok 475*

(NBG, holotype on two sheets; MO, PRE, isotypes).
Gladiolus geardii var. *uitenhagensis* L. Bolus, *S. African Gard.* 24: 101 (1934). Type: South Africa, Eastern Cape, Uitenhage District, Kamaehs, Apr. 1932, *Harcourt-Wood s.n.* (BOL 1319, holotype on two sheets; K, SAM, isotypes).

Description

Plants 80–150 cm high. *Corm* depressed-globose, 15–20 mm in diameter, the tunics of membranous layers decaying with age to become softly fibrous. *Cataphylls* firm-textured, pale, the uppermost reaching 10–15 cm above the ground and then purple. *Leaves* seven to nine, mostly basal, about half as long as the stems, the blades plane,

14–28 mm wide, the midrib moderately thickened, the margins lightly thickened. *Stem* erect below, flexed outward below the base of the spike and inclined above, usually with one or two branches, sometimes simple, 4–5 mm in diameter below the spike.

Spike inclined, lightly flexuose, 6- to 10-flowered; *bracts* green, the outer 30–50(–60) mm long, the inner about two-thirds as long as the outer. *Flowers* pinkish purple, the lower three tepals each with a dark purple spear-shaped mark in the lower midline, often pale in the centre, unscented; *perianth tube* obliquely funnel-shaped, 30–40 mm long, the lower cylindrical part 24–32 mm long; *tepals* lanceolate, unequal, the dorsal largest, extending more or less horizontally over the stamens,

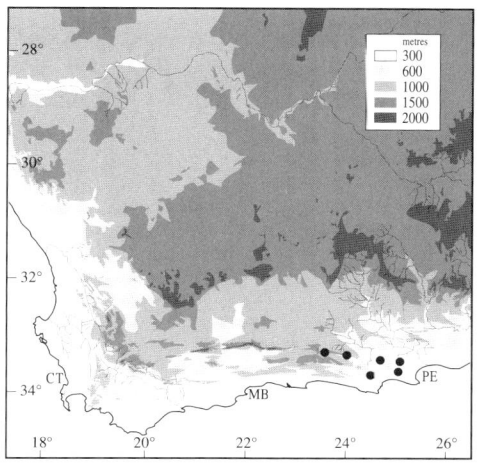

40–55 x 28–30 mm, the upper laterals patent, 45–50 x c. 20 mm, the lower three tepals more or less straight, directed downward c. 45°, 35–40 x 15–18 mm, the lower median slightly longer than the lower laterals. *Filaments* 20–30 mm long, exserted 14–18 mm from the tube; *anthers* 9–14 mm long, purple, the pollen cream. *Ovary* oblong, 6–10 mm long; *style* arching over the stamens, dividing opposite the upper third of the anthers, the branches c. 8 mm long, broadened and bilobed at the apices. *Capsules* ovoid-elliptic, c. 28 mm long; *seeds* ovate, 6.5–7 x 4–5 mm, broadly and evenly winged, opaque dull brown. *Chromosome number* unknown.

Flowering time November to mid-December.

Distribution and biology

Gladiolus geardii is a fairly rare endemic of the southern African winter-rainfall zone. It is known from just a few isolated localities in the Baviaanskloof, Great Winterhoek, Kouga and Van Staden's River Mountains of Eastern Cape Province. Plants grow in fairly moist, montane sites such as seeps, streambanks and shady kloofs in peaty sandstone soil. Although some populations are thought to flower only after fires when the surrounding bush has been removed, in more open habitats *G. geardii* is reported to flower every year. In her efforts to locate living plants for the painting reproduced here, Auriol Batten was able to locate additional populations of the species near Patensie, a fair distance from those previously known. We now suspect that *G. geardii* may be more common than

was supposed. It simply grows in inaccessible sites and in areas that are not well collected.

The long-tubed, pale pink flowers of *Gladiolus geardii* conform to the pattern of other winter-rainfall zone species that are pollinated by long-tongued flies, and we assume that they are likewise adapted for long-tongued fly pollination.

Diagnosis and relationships

The Eastern Cape *Gladiolus geardii* has one of the largest flowers in series *Blandus*, normally 80–100 mm long. Like those of most other members of the section, the tepals are pale pink, the lower three with white longitudinal markings edged in darker pigment. The perianth tube is 30–40 mm long, slightly shorter than the upper tepals, the dorsal of which is 40–55 mm long.

Gladiolus geardii is evidently closely related to *G. carneus* of Western Cape Province and the two are easy to confuse. The latter typically has a flower only 52–78 mm long, the tube 25–38, rarely 45 mm, and the dorsal tepal 28–40(–50) mm long. Apart from details of the flower size, *G. geardii* can be distinguished by its robust habit, normally being 80–150 cm tall and having seven to nine leaves, compared with *G. carneus* which is seldom taller than 50 cm and has only four or five leaves.

History

Although described by H.M.L. Bolus in 1925, *Gladiolus geardii* remained misunderstood until our research was undertaken for the preparation of this book. The type and first collection on record seems to be one

from Hadley Farm near Steytlerville in Eastern Cape Province. A second collection of less robust plants, made at Kamaehs near Uitenhage and named *G. geardii* var. *uitenhagensis* by H.M.L. Bolus, appears to be the same species, and despite the smaller size of both the plant and the flowers it is included here in synonymy. Neither *G. geardii* nor its variety were recognized by Lewis et al. (1972), who included *G. geardii* in *G. floribundus* subsp. *floribundus*.

Unaware that the species had already been described, Goldblatt (1984) named a collection from the Baviaanskloof Mountains, the third record ever, *Gladiolus robustus*. There is no doubt that the types of *G. geardii* and *G. robustus* represent the same species and the later name now falls into synonymy. The less robust population of *G. geardii*, called var. *uitenhagensis* by H.M.L. Bolus, merits further investigation, but for the present we include it in *G. geardii*.

43. GLADIOLUS AQUAMONTANUS Goldblatt & Vlok

PLATE 36

Gladiolus aquamontanus Goldblatt & Vlok, *S. African J. Bot.* 55: 261 (1989). Type: South Africa, Western Cape, Swartberg Mountains, northern slopes of Blesberg, in perennial stream, 1585 m, 15 Dec. 1986, *Vlok 1774* (PRE, holotype; K, MO, NBG, SAAS, isotypes).

aquamontanus = of mountain waters, describing the habitat, perennial mountain streams, waterfalls and wet cliffs.

Description

Plants 40–100 cm high, inclined or drooping. *Corm* vestigial, resembling an erect rhizome, 9–12 mm in diameter, the tunics softly papery, brown, not persisting for more

than a season. *Cataphylls* membranous to firm, rapidly disintegrating, the uppermost up to 12 cm long, often dry and decaying at flowering time. *Leaves* five or six, rarely four, the lower three basal and forming a distichous fan, the blades narrowly lanceolate, 8–15 mm wide, soft-textured, the midrib lightly raised, the remaining leaves inserted on the stem, shorter than the basal, the uppermost bract-like, partly to entirely sheathing, the margins free to the base. *Stem* inclined to drooping, unbranched, 2.5–3 mm in diameter below the spike.

Spike inclined or drooping, usually 4- to 8-flowered, occasionally less; *bracts* green, soft-textured, the outer 25–40 mm long, the inner about two-thirds as long. *Flowers* pale

mauve-pink, the lower three tepals each with a fairly broad, dark purple longitudinal median streak in the lower two-thirds and purple toward the base of the upper part of the tube, unscented; *perianth tube* (25–)34–40 mm long, slender and cylindric below for 20–30 mm, curving outward and flared in the upper 5–10 mm; *tepals* lanceolate, the dorsal largest, inclined to arching over the stamens, 30–35 x 10–12 mm, the upper laterals directed forward below, curving outward in the upper half and subpatent, c. 28 x 7 mm, the lower three tepals extending forward more or less horizontally, 25–30 x c. 10 mm. *Filaments* 15–20 mm long, exserted 10–15 mm from the tube; *anthers* c. 7 mm long, dark purple, the

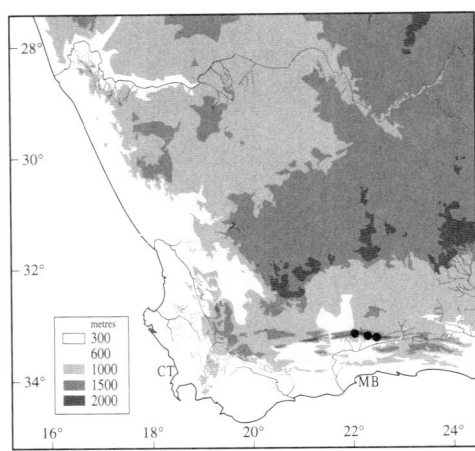

pollen cream. *Ovary* oblong, 6–8 mm long; *style* arching over the stamens, dividing between the base and apex of the anthers, the branches 5–6 mm long. *Capsules* narrowly obovoid, 18–22 mm long; *seeds* narrowly oblong to slightly sigmoid, 7–9 x c. 2.5 mm, unevenly winged, the wing narrow at the sides but well developed at the ends. *Chromosome number* $2n = 30$.

Flowering time mid-November to late December.

Distribution and biology

A very local endemic of the southern African winter-rainfall zone, *Gladiolus aquamontanus* is restricted to the Swartberg of the interior Western Cape Province. The species has now been recorded from four sites, including the Swartberg Pass in the west and Blesberg near Meiringspoort in the east, a distance of some 70 km. Plants grow in perennial mountain streams and on wet cliffs, the roots anchored in minute crevices and clinging precariously to the rock. The plants, which lack properly developed corms, have a rootstock resembling a short, swollen rhizome which is borne on the rock surface. The unusual rootstock is clearly an adaptation to the habitat.

The long-tubed, mauve-pink flowers of *Gladiolus aquamontanus* appear to be adapted for pollination by long-tongued flies.

Diagnosis and relationships

The most distinctive feature of *Gladiolus aquamontanus* is the rootstock which looks like a short, swollen rhizome rather than a corm, the latter being the rootstock of all other species of *Gladiolus* except *G. sempervirens*. *Gladiolus sempervirens*, like *G. aquamontanus*, grows in perennially wet habitats and we assume the corm has been lost as a result of the unusual habitat. In wet situations a corm, essentially an underground perennating organ enabling plants to survive a long dry season, is unnecessary. Like other members of series *Blandus*, *G. aquamontanus* has a loose fan of plane, sword-shaped leaves, the midrib and secondary veins of which are lightly raised and thickened. The flowers are pale mauve-pink, fairly large, 65–75 mm long, and have a tube 34–40 mm long, slightly longer than the dorsal tepal.

When it was first described in 1989, *Gladiolus aquamontanus* was compared with *G. floribundus* in particular, but we now believe that it is most closely allied to the Eastern Cape species, *G. geardii*, which it resembles closely in colour, shape and markings of the flower. The two species differ largely in the habitat and rootstock, the latter having a well-developed corm although it grows in marshes and other moist habitats. The flowers of *G. geardii* are also somewhat larger, 70–95 mm long, with a tube the same length as in *G. aquamontanus* but the dorsal tepal 40–55 mm long, thus as long as or slightly longer than the tube, the reverse of the situation in *G. aquamontanus*. The seeds of *G. aquamontanus* are unusual in shape, being oblong to slightly sigmoid and 7–9 x 2.5 mm, whereas those of *G. geardii* are quite typical of the section, being ovate and c. 7 x 4–5 mm. Another species of series *Blandus*, *G. buckerveldii*, has seeds similar to those of *G. aquamontanus*, and grows in a similar habitat, wet cliffs along waterfalls.

History

The first record of *Gladiolus aquamontanus* that we have traced is a collection of a rather depauperate plant made by the Swedish collector, Erik Wall, in 1934. This specimen had remained completely overlooked, and when plants were collected in 1980 by the South African botanist, J.H.J. Vlok, this was thought to be the first record of an obviously new species. Vlok has since documented *G. aquamontanus* at several more sites in the Swartberg.

44. GLADIOLUS UNDULATUS Linnaeus

Large painted lady, large white Afrikaner

PLATE 37

Gladiolus undulatus Linnaeus, *Mantissa* 1: 27 (1767). Lewis et al., *J. S. African Bot.*, Suppl. 10: 110 (1972). Type: South Africa, Cape, without precise locality, date or collector, specimen sent to Linnaeus by J. and N.L. Burman from Holland and possibly cultivated there (LINN 59/11, holotype).

undulatus = wavy, undulate, referring to the wavy margins of the tepals.

Synonymy

Gladiolus cuspidatus Jacquin, *Icones Pl. Rar.* 2: pl. 257 (1795). Baker, *Fl. Capensis* 6: 140 (1896). Lewis et al., *J. S. African Bot.*, Suppl. 10: 110 (1972). Type: South Africa, Western Cape, without precise locality or collector, illustration in Jacquin, *Icones Pl. Rar.* 2: pl. 257 (1795).

Gladiolus affinis Persoon, *Syn. Pl.* 1: 45 (1805). Type: South Africa, Cape, without precise locality or collector, illustration in Andrews, *Bot. Repository* 4: pl. 219 'G. cuspidatus Jacquin' (1802).

Description

Plants (25–)45–150 cm high. *Corm* depressed-globose, 16–24 mm in diameter, the tunics of soft-textured layers not normally persisting for more than one or two years. *Cataphylls* pale and membranous, reaching 4–6 cm above the ground and then firm-textured and reddish purple speckled with white. *Leaves* four or five, at least the lower two basal, the lowermost longest, usually reaching to about the middle of the spike, the remaining leaves progressively shorter, the blades narrowly lanceolate, 5–12 mm wide, plane, the midrib lightly raised and thickened, the margins hyaline but barely thickened, the uppermost leaf inserted a short distance below the spike, sheathing for half its length, channelled throughout, the margins of the sheathing part free to the base. *Stem* erect below, lightly flexed above the sheaths of the upper three leaves and inclined above, unbranched or robust plants with one or two branches, c. 2.5 mm in diameter below the spike.

Spike inclined, more or less distichous, rarely 3-, usually 6- to 9-, rarely to 12-flowered; *bracts* firm-textured, dull green or more often the dorsal surfaces dull purple, the outer 33–40(–70) mm long, the inner completely enveloped by the outer, about three-quarters as long. *Flowers* pinkish cream to greenish or pale lilac, the lower tepals each with reddish to purple diamond- to spear-shaped or linear markings in the lower mid-line, the base of the upper part of the tube with reddish to purple blotches opposite the bases of the tepals, unscented; *perianth tube* 52–75 mm long, cylindric below for 45–60 mm, flared in the upper 7–9 mm; *tepals* narrowly lanceolate-attenuate, the margins strongly undulate, the dorsal largest, extended almost horizontally over the stamens, arching upward in the upper third, 40–50 x 12–15 mm, the upper laterals gradually curving outward, patent in the upper third, 40–45 x 10–12 mm, the lower three tepals united for c. 2 mm, directed forward below, somewhat twisted and curving downward in the upper half, 33–36 x 6–9 mm, the lower median slightly longer than the lower later-als. *Filaments* (12–)15–28 mm long, exsert-ed 6–13 mm from the tube; *anthers* 8–9 mm long, cream above, purple below, the pollen purple. *Ovary* oblong, c. 6 mm long; *style* arching over the stamens, usually divid-ing shortly beyond the anther apices, some-times dividing opposite the upper third of the anthers, the branches c. 6 mm long, broadly expanded and bilobed in the upper half. *Capsules* oblong-ellipsoid, 20–27 mm long; *seeds* ovate, c. 8 x 5 mm, broadly and evenly winged, light translucent brown. *Chromosome number* 2*n* = 30.

Flowering time mid-November to late December.

Distribution and biology

Gladiolus undulatus extends from Stellen-bosch through the north–south mountain axis of the western Cape to the Bokkeveld Mountains and Kamiesberg in Northern Cape Province. Throughout its range, except in the Kamiesberg, *G. undulatus* grows in stony, coarse, sandstone soils and usually near streams or marshes, often in permanent-ly moist ground. In the Kamiesberg the sub-strate is granite, but the habitat is the same.

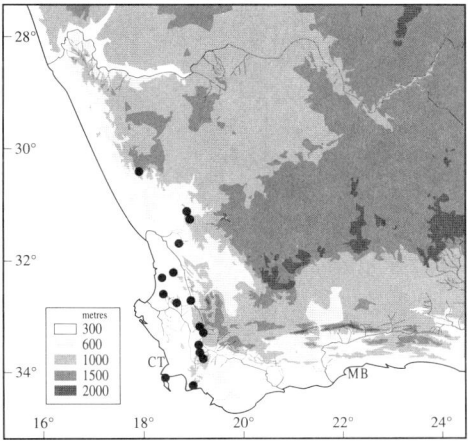

Gladiolus undulatus is one of a hand-ful of species in section *Blandus* that have cream-coloured flowers with extremely long peri-anth tubes and are pollinat-ed by long-tongued flies of the families Nemestrinidae and Tabanidae. In contrast to its close relative, *G. angustus*, which is pollinated by the nemestrinid *Moegistorhynchus longirostris*, *G. undulatus* appears to be pollinated primarily by the tabanid fly *Philoliche rostrata*, but visits by *M. braunsii* have been record-ed at Stellenbosch. The flowers produce moderate quantities of nectar of fairly low sugar concentration, are unscented, and close at night, all features associated with pollination by long-tongued flies.

Diagnosis and relationships

A member of section *Blandus*, *Gladiolus undulatus* bears an obvious resemblance to other members of the *G. carneus* group of species in the nearly distichous spike of large, pale-coloured flowers and plane, narrowly lanceolate leaves. It can be distinguished from the many forms of *G. carneus* by the longer perianth tube, 50–60 mm long, about twice as long as the bracts, and the attenuate and strongly undulate tepals. The flower colour is also different from that of typical *G. carneus*, which has pale pink or white flowers. Some populations of *G. carneus* also have cream or biscuit-coloured flowers, but these have a perianth tube only about as long as or occa-sionally slightly longer than the floral bracts and can thus readily be distinguished from *G. undulatus. Gladiolus angustus* is also broadly similar to *G. undulatus* in having long-tubed, greenish cream to ivory flowers with red markings on the lower tepals. The tube in that species is even longer than in *G. undulatus*, 80–90 mm long, the tepals are lanceolate and not noticeably attenuate, and the lower three tepals are distinctly shorter than the upper three.

History

Gladiolus undulatus was described by Carl Linnaeus in 1767 from plant specimens sent to him from Holland by the botanists J. and N.L. Burman, and perhaps grown there for some years. The original source of the plants was not known to Linnaeus. The identity of Linnaeus' species was evidently puzzling; the

name *G. undulatus* was often applied to *G. floribundus* and in the older literature it most often refers to that species. Thus when Nicholas Jacquin obtained *G. undulatus* in the 1780s he understandably thought this was a new species which he called *G. cuspi-datus.* John Ker Gawler realized that *G. undulatus* and *G. cuspidatus* were the same species, but advocated continued use of the latter name, and the application of the name *G. undulatus* to *G. floribundus* because it was widely known by the epithet. Ker Gawler's advice was followed throughout the nineteenth century and *G. undulatus* is called *G. cuspidatus* in *Flora Capensis* (Baker, 1896). Only when G.J. Lewis had unravelled this nomenclatural knot did *G. undulatus* once again become used in Linnaeus' sense (Lewis et al., 1972). An illustration entitled *G. cuspidatus* published in Henry Andrews' *Botanists Repository* in 1802 is also *G. undu-latus* and it is the type of *G. affinis* described in 1805 by Hendrik Persoon. Persoon evidently thought that this illustration represented a different species from the one figured on Jacquin's plate, the type of *G. cuspidatus*, and the name by which Persoon knew *G. undulatus.*

45. GLADIOLUS ANGUSTUS Linnaeus

Marsh painted lady, katjietee

PLATE 38

Gladiolus angustus Linnaeus, *Species Plantarum* 1: 37 (1753); ed. 2, 1: 53 (1763). Baker, *Fl. Capensis* 6: 140 (1896). Lewis et al., *J. S. African Bot.*, Suppl. 10: 88 (1972). *Gladiolus telifer* Stokes, *Bot. Comment.* 1: 217 (1830), a superfluous name for *G. angustus* Linnaeus. Type: South Africa, Cape, without precise locality or collector, cultivated in Holland, *Herb. Linnaeus* 59.16 (LINN, holotype).

angustus = narrow, evidently alluding to the long, slender perianth tube.

Synonymy

Gladiolus angustifolius Salisbury, *Prod. Stirpium ad Chapel Allerton* 40 (1796), evidently an orthographic variant of *angustus* or a name without description, hence invalid.

Description

Plants 60–120 cm high. *Corm* more or less globose, 14–20 mm in diameter, the tunics of firm to soft papery layers, fragmenting irregularly with age. *Cataphylls* firm-membranous, pale, becoming purple above the ground and obscurely mottled with green or white. *Leaves* four or five, the lower two or three basal and longest, reaching or shortly exceeding the spike, the blades plane, (3–)5–10 mm wide, the midrib lightly thickened, the margins narrowly hyaline, upper two leaves cauline, sheathing for half their length, the margins free to the node and not imbricate. *Stem* erect or inclined, flexed outward above the sheaths of the two cauline leaves, simple or branched, (1.5–)2.5–4 mm in diameter below the spike.

Spike flexed at the base, inclined, weakly flexuose, occasionally 3-, usually 5- to 9-flowered; *bracts* green, the outer (30–)50–65 (–80) mm long, pale green, sometimes flushed reddish near the apices, the inner two-thirds to nearly as long as the outer, minutely forked at the apices for c. 1 mm. *Flowers* cream to pale yellow, the lower three tepals each with a spade-shaped yellow marking in the midline outlined in dark red, unscented; *perianth tube* cylindric, expanded in the upper 10 mm, (45–)60–110 mm long; *tepals* unequal, lanceolate, the dorsal largest, 32–40 x 16–18 mm, extended more or less horizontally over the stamens, the upper laterals 28–38 x 10–13 mm, extending forward, curving outward in the upper third to half, the three lower tepals 25–30 x 9–10 mm, extending forward and nearly horizontal, in profile shorter than the upper tepals. *Filaments* 17–24 mm long, exserted 8–12 mm from the tube; *anthers* 9–11 mm long, whitish, the pollen cream. *Ovary* oblong, 7–10 mm long; *style* extending over the stamens, dividing just below the anther apices, the branches 6–8 mm long, reaching well beyond the anthers. *Capsules* ellipsoid, (25–)35–40 mm long, the apex more or less acute; *seeds* ovate, 8–9 x 5–6 mm, the wing broadly and evenly developed, translucent yellow-brown, the seed body slightly darker. *Chromosome number* 2n = 30.

Flowering time October and November.

Distribution and biology

Native to the southwestern corner of the southern African winter-rainfall region, *Gladiolus angustus* is a fairly narrow endemic of the region. It is best known from the Cape Peninsula and west coast of Western Cape Province, but extends from there northward into the Piketberg and Cedarberg. The species is much rarer to the east and has been recorded at only a few sites along the south coast, with an easterly limit near Elim on the western side of the Agulhas Peninsula.

The flowers of *Gladiolus angustus* are adapted for pollination by long-tongued flies of which *Moegistorhynchus longirostris* is the sole pollinator of the long-tubed populations along the Cape west coast (Manning & Goldblatt, 1997). This is the only insect that has mouthparts long enough to feed on the nectar stored in the lower part of the perianth tube of these populations of *G. angustus*. Whether this fly also pollinates the interior Piketberg and Cedarberg populations which have flowers with slightly shorter perianth tubes remains to be determined.

Diagnosis and relationships

One of the more distinctive species of the genus, *Gladiolus angustus* can immediately be recognized by its very long perianth tube, usually 70–100 mm long, but occasionally only 50 mm. It is clearly allied to *G. carneus* and its allies in series *Blandus* and is virtually identical vegetatively to species such as *G. carneus* and *G. undulatus*. The long perianth tube is not the only unusual feature of the species. The cream to ivory flowers have the lower tepals substantially shorter than the upper and with distinctive dark red, diamond- or spade-shaped markings.

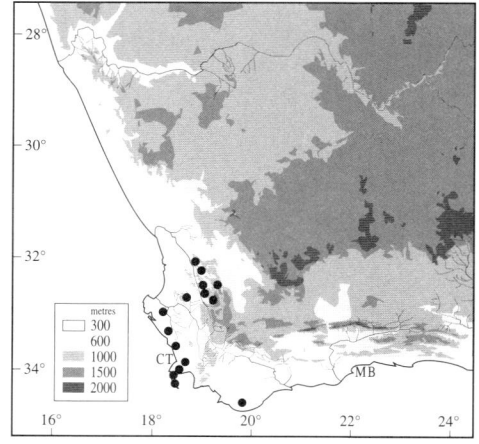

The immediate relationships of *Gladiolus angustus* appear to lie with *G. carneus* and *G. undulatus*, both of which are similar to it in most respects, but have flowers with a shorter perianth tube, 50–70 mm long in *G. undulatus* and usually 25–45 mm long in *G. carneus*, and in the latter normally not much longer than the floral bracts. In both the latter species the lower tepals are more or less the same size or only slightly smaller than the upper and the nectar guides are usually weakly developed – most often consisting of a linear median streak on each tepal – or are absent.

History

One of the first species of *Gladiolus* to be described, *G. angustus* was named by Carl Linnaeus in the *Species Plantarum* in 1753. Like so many Cape plants, it had been grown in Holland for some time, originally having been collected at the Dutch East India settlement at the southwestern tip of Africa, probably not very far from Cape Town. The remarkably long perianth tube, usually in excess of 70 mm, is so striking that *G. angustus* was consistently recognized by contemporary and later botanists. It was thus was not subject to the nomenclatural problems and confusion that have plagued so many other species of the genus first recorded in the eighteenth century. The name *G. angustifolius* of R.A. Salisbury appears to be no more than an unintended, or perhaps intended, variant of the epithet *angustus*.

46. GLADIOLUS BUCKERVELDII (L. Bolus) Goldblatt
PLATE 39

Gladiolus buckerveldii (L. Bolus) Goldblatt, *J. S. African Bot.* 37: 443 (1971). Lewis et al., *J. S. African Bot.*, Suppl. 10: 87 (1972). *Antholyza buckerveldii* L. Bolus, *Ann. Bolus Herb.* 4: 118 (1927). *Petamenes buckerveldii* (L. Bolus) N.E. Brown, *Trans. Roy. Soc. S. Africa* 20: 276 (1932). Type: South Africa, Cape, Clanwilliam District, mountains above Algeria Forest Station, 1000 m, 1926 (cultivated at Kirstenbosch Botanic Garden 75/26), *Buckerveld s.n.* (BOL, holotype on three sheets; K, isotype).

buckerveldii, named to honour M.H. Buckerveld who discovered the plant and sent it to Kirstenbosch Botanic Gardens where the type collection was grown and flowered.

Description

Plants 80–125 cm high. *Corm* globose, 15–24 mm in diameter, the tunics soft-papery, brown, the layers usually decaying rapidly thus not accumulating. *Cataphylls* firm-textured and pale, usually light purple mottled with pale green above the ground. *Leaves* five or six, the three or four lower basal, the upper one or two cauline and smallest, sword-shaped, 25–35 mm wide, more or less trailing distally, the midrib and one or two pairs of secondary veins prominent, the margins not thickened. *Stem* inclined to drooping, unbranched, c. 4 mm in diameter below the spike.

Spike inclined or more or less horizontal, strongly secund, 12- to 20-flowered; *bracts* green, fairly soft-textured, the outer 50–80 mm long, reaching at least the top of the perianth tube, sometimes the top of the flower, the inner about two-thirds as long, entire and acute. *Flowers* ivory to greenish cream, the lower tepals each with a spade- or heart-shaped dark red mark in the centre, unscented; *perianth tube* 45–50 mm long, slender below, widening to 6–7 mm in diameter in the upper half; *tepals* unequal, lanceolate, the dorsal largest, 28–32 x 13 mm, extended horizontally, the upper laterals, c. 21 x 10 mm, curving outward more or less from the base, the lower three much smaller, c. 12 x 6 mm, curving gradually outward. *Filaments* 25–28 mm, exserted c. 10 mm from the tube; *anthers* c. 10 mm long, cream turning to lilac with age, the pollen cream. *Ovary* oblong, c. 6 mm long; *style* arching over the stamens, dividing at or close to the anther apices, the branches c. 5 mm long. *Capsules* oblong, 22–26 mm long, the apices obtuse; *seeds* oblong to slightly sigmoid, 8–9 x 3–4 mm, broadly winged at only one end, sometimes weakly so at the opposite end but hardy at all on the sides, translucent light brown. *Chromosome number* $2n = 30$.

Flowering time January, rarely in late December.

Distribution and biology

Gladiolus buckerveldii is one of the many very narrow endemics of the southern African winter-rainfall zone. It is restricted to just a few sites in the northern Cedarberg of Western Cape Province, best known of which is the waterfall on the steep slopes above Algeria Forest Station. Plants grow out of nearly vertical moss-covered cliffs, the leaves drooping distally and the stems more or less horizontal. Flowering in the summer, usually after the beginning of January, its range is limited to the few permanent streams in this summer-dry and very hot area.

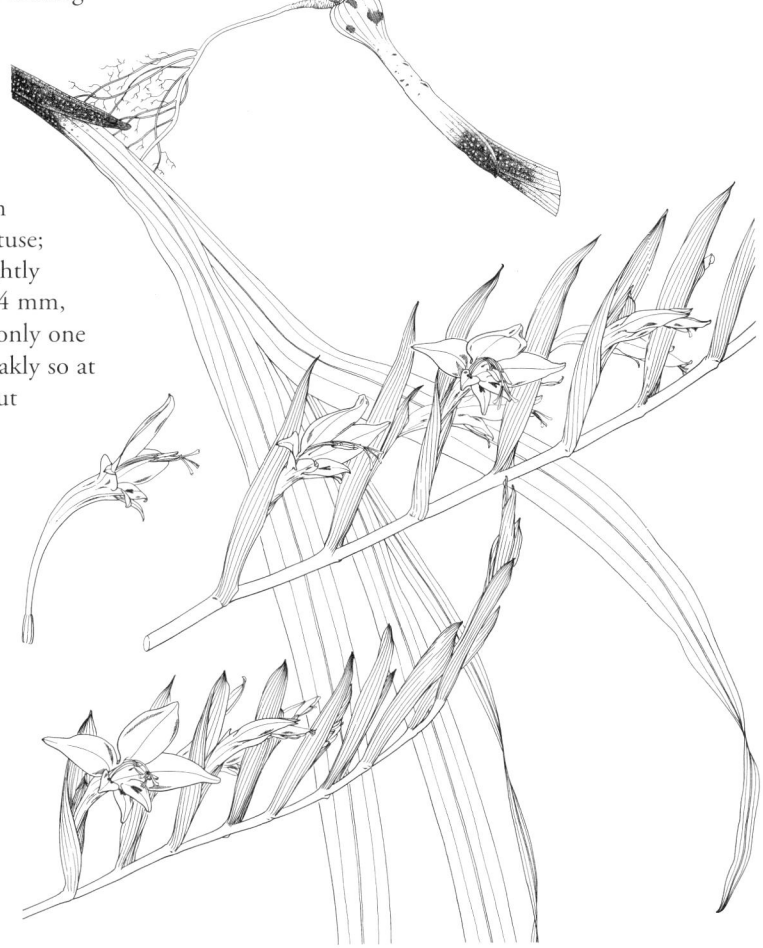

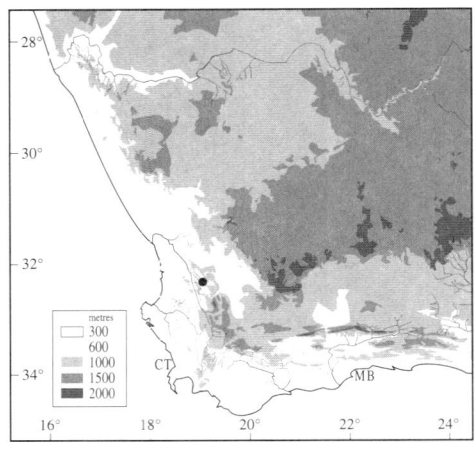

also suggests pollination by birds rather than long-tongued flies. If correct, this is an example of bird pollination associated with flowers that are not red or orange, by far the most common flower colour accompanying this pollination system.

Diagnosis and relationships

The large, ivory flowers with prominent red nectar guides and the elongate perianth tube, 45–50 mm long, combined with the streamside habitat and inclined to decumbent stems, make *Gladiolus buckerveldii* easy to recognize. The flowers are distinctively shaped, the elongate tube narrow below and fairly wide and nearly cylindric in the upper third, and the dorsal and upper lateral tepals much exceeding the lower three tepals.

Apart from the much widened upper part of the perianth tube, the flower colour and shape, including the comparatively short lower tepals with bright red nectar guides, recall particularly *Gladiolus angustus*. It seems likely that the two are immediately related. *Gladiolus angustus* can be distinguished by the shape of the perianth tube, which is fairly slender throughout, and normally 60–100 mm long. Also a species of wet

habitats, it favours marshes, seeps, and streambanks, but does not occur on rock slopes or waterfalls.

History

Discovered in the 1920s by M. Buckerveld, plants of *Gladiolus buckerveldii* were sent to Kirstenbosch Gardens where the species was successfully grown for many years. Plants flowered there in 1926 and the species was described by H.M.L. Bolus a year later. Bolus called it *Antholyza buckerveldii* because plants with the elongate floral tube, slender below and wide and cylindric above, were at that time thought to be appropriately placed in that genus. The British botanist, N.E. Brown, who revised the generic concepts of those Iridaceae with floral tubes of this type, referred *G. buckerveldii* to a third genus, *Petamenes*, largely because it had the lower tepals much shorter than the upper and fairly long floral bracts (Brown, 1932). That taxonomy lasted until 1971, when it became clear that the type species of *Petamenes* was closely related to a group of species now included in *Gladiolus* section *Homoglossum*, and that *P. buckerveldii* was more closely allied to *G. angustus* and *G. undulatus*.

The ivory colour of the flower and the length of the perianth tube suggest that the flowers of *Gladiolus buckerveldii* may be adapted for pollination by the nemestrinid fly *Moegistorhynchus longirostris*, the same insect that pollinates the related species, *G. angustus*. The perianth tube is, however, not as slender as might be expected for this pollination syndrome but is fairly wide in the upper half, a shape more consistent with pollination by long-billed birds such as sunbirds. The large quantity of nectar produced by the flowers, in excess of 40 microlitres,

47. GLADIOLUS BILINEATUS G. Lewis

PLATE 40

Gladiolus bilineatus G. Lewis in *J. S. African Bot.*, Suppl. 10: 268 (1972). Type: South Africa, Western Cape, Riversdale District, foothills of the Langeberg Mountains, 27 Mar. 1934, *Ferguson s.n.* (BOL 21323, holotype; K, SAM, isotypes).

bilineatus = two-lined, referring to the paired dark streaks on each of the three lower tepals.

Description

Plants 20–35 cm high. *Corm* more or less

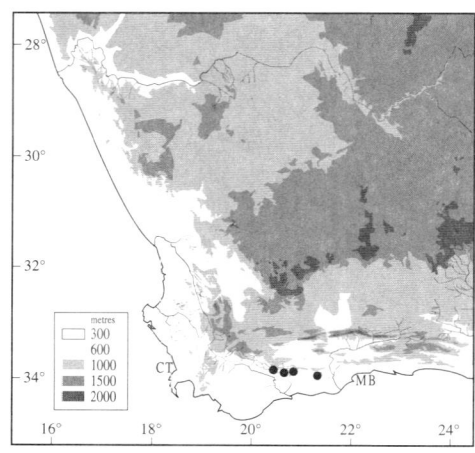

globose, 18–24 mm in diameter, the tunics of fine reticulate fibres. *Cataphylls* membranous and pale, reaching 3–5 cm above the ground and then green or sometimes obscurely speckled with purple. *Leaves* three, the lowermost basal and longest, reaching to about the base of the spike, the blade plane, 6–8 mm wide, the midrib evident but only lightly thickened, the margins narrowly hyaline, upper two leaves cauline, progressively smaller above, sheathing for about half their length, the margins free to the base, usually overlapping. *Stem* erect or inclined, lightly flexed outward above the sheaths of the two upper leaves, unbranched, 2–2.5 mm in diameter below the spike.

Spike lightly inclined, more or less straight, 2- to 5-flowered; *bracts* green or flushed with dull purple, the outer 40–50(–60) mm long, the inner about three-quarters as long as the outer, usually minutely forked at the apices for c. 1 mm. *Flowers* creamy pink to pale salmon, the lower three tepals each with a long pale median streak edged in dark pink to red, unscented; *perianth tube* 50–70 mm long and cylindric in the lower 40–50 mm, flared

in the upper 10–20 mm; *tepals* unequal, lanceolate, the dorsal largest, 23 x 15–16 mm, inclined nearly horizontally over the stamens, the upper laterals 22–23 x 9 mm, extending forward, curving outward in the upper third to half, the three lower tepals joined to the upper laterals for c. 3 mm, 15–17 x 7–8 mm, extending forward and nearly horizontal, in profile usually shorter than the upper tepals. *Filaments* c. 20 mm long, exserted c. 2 mm from the tube; *anthers* c. 9 mm long, purple, the pollen lilac. *Ovary* oblong, c. 7 mm long; *style* arching over the stamens, dividing just beyond the anther apices, the branches c. 4 mm long, very broad and bilobed in the upper halves. *Capsules* ovoid, mature capsules not seen; *seeds* ovate, 5.5–6.5 mm, the wing broadly and evenly developed, translucent reddish brown. *Chromosome number* unknown.

Flowering time March and April.

Distribution and biology

A narrow endemic of the lower southern slopes of the Langeberg, *Gladiolus bilineatus* has been recorded from Tradouw Pass in the west to the fertile flats at the foot of the mountains north of Riversdale and Albertinia in the east. In Tradouw Pass plants grow in true fynbos in a *Leucadendron salignum* community on coarse, sandstone-derived soil, but elsewhere plants occur on the heavier loamy sand along the interface of Cape sandstone strata and Malmesbury shales in a rich renosterveld community.

The cream or pinkish flowers of *Gladiolus bilineatus*, with their long perianth tube, are adapted for pollination by long-tongued nemestrinid flies. The fly *Prosoeca longipennis*, which is on the wing in late summer and autumn, appears to be the sole pollinator of the species (Manning & Goldblatt, 1995).

Diagnosis and relationships

Like so many of the late summer- and autumn-flowering species of *Gladiolus* in the southern African winter-rainfall zone, the flowering stems of *G. bilineatus* bear fairly short, largely sheathing leaves. Non-flowering plants produce a single, long foliage leaf in the winter months, although those that did flower that year do not produce additional foliage leaves later. The flowers are fairly typical of section *Blandus*, being coloured cream to pale pink with linear median streaks on the lower tepals, but the perianth tube is particularly long, 50–70 mm, and in this feature recalls the

spring-flowering *G. angustus*. Among the longer-tubed species of series *Blandus*, *G. bilineatus* is distinguished by the autumn flowering time, March and April, and characteristic vegetative growth habit, as well as the relatively few-flowered spikes of pale pink to salmon flowers, the lower tepals each with a long pale median streak edged in dark pink to red.

History

The first record of this late summer-flowering species is a collection made by the Riversdale physician and naturalist, John Muir, in 1923, from the lower slopes of the Langeberg near Riversdale. A second collection, made in March 1930 by the German naturalist, H.J. Thode, and bearing the vague locality, Swellendam District, extended the range of the species a fair distance. This collection was followed quickly by one made the following year near Riversdale by the naturalist, Emily Ferguson. These collections attracted no immediate notice and it was not until G.J. Lewis began her study of *Gladiolus* in South Africa that *G. bilineatus* was described. The protologue did not appear until 1972, when Lewis's *Revision of the South African species of* Gladiolus was completed and published.

48. GLADIOLUS INSOLENS Goldblatt & Manning
PLATE 41

Gladiolus insolens Goldblatt & Manning, new species. Type: South Africa, Western Cape, Piketberg Mountains, southwestern slopes of Zebrakop, 4 Jan. 1995, *Goldblatt & Manning 10166* (NBG, holotype; K, MO, PRE, isotypes).

insolens = radiant, shining like the sun, for the brilliant orange-red flowers which glow bright red among the rocks and cliffs where they grow.

Latin diagnosis

Plantae 30–50 cm altae, cormo 18–20 mm in diametro tunicis papyraceis aetate fibrescentibus, foliis 5–6, inferioribus 3 basalibus longioribusque laminis linearibus 3.5–5.5 mm latis, spica 1–2(–3)-florum, floribus rubris, tubo perianthii c. 38 mm longo, tepalis subaequalibus late ovatis 27–30 x 18–20 mm leviter cucullatis,

filamentis 20 mm longis, antheris c. 8 mm longis.

Description

Plants 30–50 cm high. *Corm* globose, 18–20 mm in diameter, the tunics of papery layers becoming softly and finely fibrous with age. *Cataphylls* pale and membranous, the uppermost reaching 2.5–2 cm above the ground, then green or becoming dry and brown. *Leaves* five or six, the lower three basal and longest, inclined to trailing, short to fairly long and up to 30 cm long, the blades linear, 3.5–5.5 mm wide, the midrib thickened and lightly raised, the margins narrowly hyaline, not thickened, the remaining leaves inserted on the stem between the lower and upper third, progressively smaller above, the upper one or two often sheathing for almost their entire length, channelled throughout, the margins open to the base,

usually overlapping. *Stem* inclined to trailing, occasionally hanging from rocks, usually unbranched, sometimes with one short branch, 1.2–2 mm in diameter below the spike.

Spike curving upward and inclined, 1- to 2-, rarely 3-flowered, lightly flexuose; *bracts* firm-textured, light green, the outer 30–45 mm long, the apex diverging for c. 5 mm, the inner shorter than the outer by 4–10 mm, minutely notched apically. *Flowers* bright scarlet, the tepals symmetrically disposed and forming a wide bowl with the apices curving inward, the lower tepals each with an obscure darker red median blotch in the upper half; *perianth tube* cylindric below, c. 38 mm long, slightly curved and flared in the upper 10 mm; *tepals* subequal, the upper three slightly larger than the lower three, broadly obovate, arching forward with the apices curving

inward, 27–30 x 18–20 mm, slightly cuccullate, in profile the lower exceeding the upper by c. 8 mm. *Filaments* c. 20 mm long, exserted c. 7 mm from the tube, pale below darkening to light scarlet above; *anthers* c. 8 mm long, scarlet, the pollen pale yellow. *Ovary* oblong, c. 6 mm long; *style* arching over the stamens, dividing opposite the base to lower third of the anthers, the branches c. 12 mm long, arching outward and exceeding the anthers, channelled in the upper half, the lobes unfolding only toward the apices.

Capsules and *seeds* unknown. *Chromosome number* unknown.

Flowering time late December to mid-January.

Distribution and biology

A very narrow endemic of the southern African winter-rainfall zone, *Gladiolus insolens* is restricted to the geographically isolated Piketberg in Western Cape Province, South Africa. In this area of virtually complete summer drought but fairly ample winter precipitation, *G. insolens* grows in a few isolated wet sites along streams and seeps, always in rocky situations on the higher slopes at the northeastern end of the Piketberg massif. These habitats remain wet until at least December and provide enough moisture for plants to grow and flower.

We have, unfortunately, no observations on the pollination biology of *Gladiolus insolens* for no insects were visiting the species in January 1995 when we studied it in the field. We assume, however, that it belongs to a guild of summer- and autumn-flowering species with long-tubed red flowers that are pollinated by *Aeropetes tulbaghia*. Emerging in late December or January, this butterfly is known to visit only scarlet to dark red flowers on which it forages for nectar and coincidentally pollinates (Johnson & Bond, 1994).

Diagnosis and relationships

Gladiolus insolens is easily recognized by its spike of one to three bright scarlet flowers with perianth tube about 38 mm long and slightly cupped, broadly obovate tepals without contrasting nectar guides. The plants have five to seven narrow, grey-green leaves, the lower three of which

are basal and have flaccid blades that trail through the surrounding vegetation.

The only other species of *Gladiolus* that has a flower like that of *G. insolens* is *G. stokoei*, which is restricted to the Riviersonderend Mountains. This species has flowering stems bearing two or three reduced, entirely sheathing leaves and an erect stem with a spike of three to five flowers. Well-developed foliage leaves of *G. stokoei* are produced later in the season in flowering plants. The solitary leaves of these plants are plane, fairly soft-textured and villous, thus quite unlike those of *G. insolens*. The striking similarity of the flowers of these two species is presumably due to convergence for pollination by *Aeropetes tulbaghia*.

History

Discovered by the Cape Town botanist, Peter Linder, in the 1970s, *Gladiolus insolens* was at first thought to be a disjunct population of *G. stokoei* (Goldblatt, 1984) for the unusual flowers of the two species appear to be nearly identical in shape and colour. A search for plants for further study and for illustration in January 1995 showed that *G. insolens* is so different from *G. stokoei* in its vegetative features that they really belong in different sections of the genus.

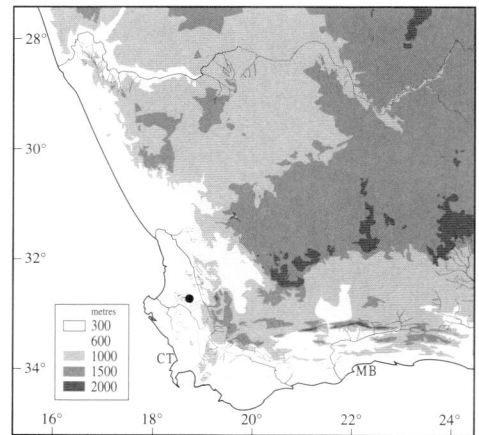

49. GLADIOLUS CARDINALIS Curtis

New Year lily, waterfall gladiolus

PLATE 42

Gladiolus cardinalis Curtis, *Curtis's Bot. Mag.* 4: pl. 135 (1790). Baker, *Fl. Capensis* 6: 156 (1896). Lewis et al., *J. S. African Bot.*, Suppl. 10: 17 (1972). Type: South Africa, without precise locality, cultivated in England, figure in *Curtis's Bot. Mag.* 4: pl. 135 (1790).

cardinalis, named for the bright, cardinal red flower colour.

Synonymy

Gladiolus speciosus Ecklon, *Topographisches Verzeichniss* 41 (1827), an illegitimate homonym, not *G. speciosus* Thunberg (1811). Type: unknown.

Description

Plants 55–90 cm high. *Corm* globose, poorly developed, 12–18 mm in diameter,

attached to rocks, not subterranean, the tunics firm-membranous to papery, decaying into fine fibres but not accumulating. *Cataphylls* membranous or green. *Leaves* five to seven, the lower three or four basal and longest, always drooping but reaching to at least the base of the spike, sometimes slightly exceeding the spike apex, the blade narrowly lanceolate, 11–21 mm wide, the midrib thickened and slightly raised and one other pair of veins prominent, the margins not thickened, the remaining three leaves inserted on the mid- to upper part of the stem, the two uppermost not sheathing, diverging from the stem and channelled throughout. *Stem* inclined to drooping, unbranched, 3–4 mm in diameter below the spike.

Spike inclined or drooping, the flowers borne on the upper side, 10–20 mm apart, occasionally 4-, usually 8- to 11-flowered; *bracts* bright green, diverging sharply from the stem, the outer 45–60(–70) mm long, the inner two-thirds to three-quarters as long as the outer, acute and not normally notched apically. *Flowers* bright red, the lower three tepals each with a broad median white longitudinal spear-shaped mark, unscented; *perianth tube* 32–40 mm long, straight or obliquely funnel-shaped, the lower cylindrical part c. 28 mm long; *tepals* lanceolate, acute or obtuse, unequal, the dorsal largest, held apart from the remaining tepals, generally borne at right angles to the tube or slightly recurved, 45–55 x 24–30 mm, the upper laterals directed forward below, patent above, 48–56 x 20–22 mm, the lower three tepals joined to the upper laterals for 3–7 mm, and to one another for 1–3 mm, arching gently outward, 36–53 x 17–19 mm. *Filaments* (25–)35–42 mm long, exserted 23–29 mm from the tube; *anthers* 11–14 mm, upper surface reddish, the lower surface white, the sutures of the thecae violet, the pollen cream. *Ovary* oblong, c. 7 mm long; *style* arching over the stamens, dividing between the base and upper third of the anthers, the branches 8–14 mm long, channelled and only unfolding in the upper quarter. *Capsules* obovoid, 18–27 mm long; *seeds* ovate to oblong, 6–8 x 3.5–4 mm, the wing evenly developed or reduced at the sides, yellow-brown. *Chromosome number* 2*n* = 30.

Flowering time mid-December to mid-January.

Distribution and biology

Justly famed as the New Year lily (or 'nuwejaarlelie'), *Gladiolus cardinalis* is restricted to the southwestern part of Western Cape Province, although it does not occur on the Cape Peninsula. In this area of predominantly winter rainfall, the species flowers in mid-summer and is confined to sheltered, montane sites. It is usually found on sheer cliffs close to waterfalls where the roots are constantly drenched in water. The corms are tightly wedged in crevices in the sandstone bedrock with the leaves drooping and the

stems arching gracefully downward. The botanist, Rudolf Marloth, tried unsuccessfully to introduce *G. cardinalis* on the Cape Peninsula, planting it on wet cliffs in Disa Gorge on Table Mountain in 1917. These plants persisted until at least 1924 but are no longer there today.

Gladiolus cardinalis is pollinated primarily by the mountain pride butterfly, *Aeropetes tulbaghia*, which can be seen toward the end of December fluttering about the brilliant red flowers and occasionally alighting or brushing against the anthers or style branches as it forages for nectar. This *Gladiolus* belongs to a guild of bright red-flowered species of Cape plants that flower from late December until April and are adapted for pollination exclusively by *Aeropetes tulbaghia*. Other species in the guild include *G. nerineoides*, *G. stefaniae*, the orchids *Disa uniflora* and *D. ferruginea*, *Nerine sarniensis* and *Brunsvigia marginata*, both Amaryllidaceae, and *Crassula coccinea* (Crassulaceae).

Diagnosis and relationships

Gladiolus cardinalis can readily be recognized by its habitat of wet cliffs and waterfalls, combined with its summer-flowering habit and bright red flowers borne on inclined to drooping spikes. The leaves closely resemble those of many other members of section *Blandus*, including *G. carneus* of series *Blandus*, being sword-shaped, grey-green, with plane blades and with the margins, midrib and one or more pairs of secondary veins lightly thickened. The flowers are proportioned much as in the latter species, having a fairly narrow perianth tube 32–40 mm long and shorter than the bracts, and the dorsal tepal about as long as the tube. The lower tepals are only slightly shorter than the dorsal and each has a broad longitudinal splash of white in the upper half. The filaments of *G. cardinalis* are, however, surprisingly long, usually 35–42 mm long and exserted some 23–29 mm from

the tube, a feature associated with the pollination system. One of five species of series *Blandus* sharing a derived predominantly red flower adapted for pollination by the butterfly *Aeropetes*, *G. cardinalis* seems somewhat isolated taxonomically, but it may be most closely related to the southern Cape species, *G. sempervirens*. That species also favours moist habitats, but it is an evergreen plant and has a vestigial, thin, rhizome-like corm, a fan of several dark green, fairly soft leaves and lower tepals with a narrow longitudinal white streak in the midline.

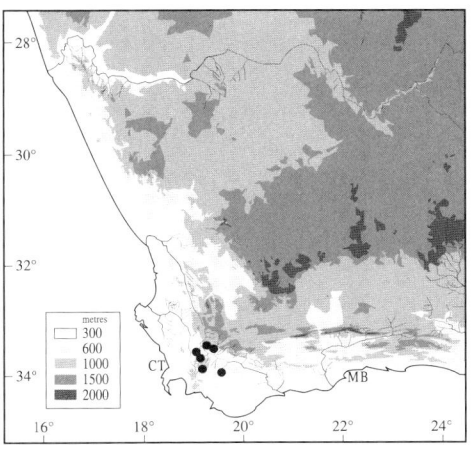

History

Gladiolus cardinalis first appears in the scientific literature in 1790, when plants grown and flowered in England were described by William Curtis and illustrated in *Curtis's Botanical Magazine.* Even before this time, however, there are records of the species being grown at Kew Gardens from material introduced by Francis Masson. In 1892 J.G. Baker cited a specimen grown from Masson's original collection in the protologue of *G. macowanii* under the impression that this plant was the same as two others he had available, a collection of *G. mortonius* from Somerset East made by Peter MacOwan, and a Thomas Cooper collection that is *G. saundersii* from the Witteberg near Aliwal North. Lectotypification has now fixed the application of the name *G. macowanii* to the MacOwan collection.

C.F. Ecklon seems to have been unaware of the existence of the name *Gladiolus cardinalis* for he briefly described the species in 1827, calling it *G. speciosus* and adding the common name 'Caapsch Nuwejahrs Africander.' Ecklon's name is a later homonym, thus illegitimate. It has also been considered a name without description, although the information that he provided about the species amounts to an adequate diagnosis.

50. GLADIOLUS SEMPERVIRENS G. Lewis
Kliplelie (rock lily)
PLATE 43

Gladiolus sempervirens G. Lewis in *J. S. African Bot.*, Suppl. 10: 18 (1972), as a new name for *G. splendens* Baker, *J. Bot.* 14: 333 (1876); *Fl. Capensis* 6: 156 (1896), an illegitimate homonym, not *G. splendens* (Sweet) Herbert (1842). Type: South Africa, Western Cape, George District, mountains near Oakhurst, Apr. 1870, *Dumbleton s.n.* (K, holotype).

sempervirens = always green, because the plants have no truly dormant period, and

have green foliage throughout the spring and summer.

Description

Plants 35–100 cm high. *Corm* 8–12 mm in diameter, shallowly seated, producing rhizome-like stolons from the base, these c. 4 mm in diameter, bearing scales at the nodes, the tunics coriaceous, not much distinguished from the bases of the leaves and not accumulating. *Cataphylls* pale and membranous, the upper green above ground level. *Leaves* seven to twelve, the lower five to ten basal, reaching to about the middle of the stem, the blades lanceolate, 6–18 mm wide, the midrib and one other pair of veins lightly thickened, the margins not thickened. *Stem* erect or inclined, flexed above the sheathing parts of the upper one or two leaves, sometimes with a branch, 2.5–4 mm in diameter below the spike. *Spike* inclined, bearing flowers on the upper side, 4- to 8-flowered; *bracts* green, soft-textured, the outer 35–55(–70) mm long, the inner slightly shorter to about as long as the outer. *Flowers* deep carmine red, the lower three tepals each with a narrow longitudinal white streak in the lower midline, unscented; *perianth tube* funnel-shaped, 25–42 mm long, the lower cylindrical part c. 15 mm long; *tepals* elliptic, the dorsal broadly so and largest, 55–58 x 32–36 mm, the upper laterals c. 54 x 31 mm, the lower three fused to one another for c. 4 mm, c. 50 x 20 mm. *Filaments* 28–42 mm long, exserted 20–24 mm from the tube; *anthers* c. 14 mm long, purple, the pollen yellow; *Ovary* oblong, c. 8 mm long; *style* arching over the stamens, dividing opposite the anther apices, the branches 7–8 mm long. *Capsules* oblong, three-lobed above and retuse, (15–)20–25 mm long; *seeds* oval, c. 8 x 6 mm, the wing well developed, golden brown, the seed body darker. *Chromosome number* 2*n* = 30.
    Flowering time March to May.

Distribution and biology

A rare species of the southern African winter-rainfall zone, *Gladiolus sempervirens* occurs in favoured habitats along the Outeniqua–Tsitsikamma Mountain axis that runs parallel to the southern coast of the Western and Eastern Cape provinces from George in the west to near Kareedouw in the east. It grows in seeps and other perennially damp sites on gravelly sandstone soils at fairly high elevations. Humid southeast winds carry moisture inland from the Indian Ocean in the otherwise largely dry summer months and keep the upper south-facing slopes of the coastal mountains constantly damp. *Gladiolus sempervirens* is so dependent on moisture for its survival that its corm is vestigial. The plants are also evergreen and have fairly soft-textured leaves, unsuited to dry conditions.

The large, bright red, long-tubed flowers of *Gladiolus sempervirens* are adapted for pollination by the butterfly *Aeropetes tulbaghia* which is known to be the sole pollinator of several red-flowered, late summer- and autumn-blooming species of the Cape and eastern southern African mountains. Observations on the pollination system of *G. sempervirens* were first recorded by S.D. Johnson (Johnson & Bond, 1994) in the mountains near George and we have also noted this butterfly visiting the flowers of *G. sempervirens* in the Tsitsikamma Mountains near Witelsbos.

Diagnosis and relationships

The striking *Gladiolus sempervirens* is unmistakable in its large, bright carmine red flowers produced in the autumn on spikes of four to eight flowers carried above a fan of several dark green leaves. As discussed above, the plants grow in constantly moist habitats, are evergreen, and have vestigial corms more like short rhizomes. The plants also produce distinctive thin, rhizome-like stolons from the base which ensure vegetative reproduction locally. The flowers have a perianth tube 25–42 mm long and broadly elliptic tepals, the upper three 54–58 mm long and only slightly larger than the lower. The lower tepals each have a narrow median white streak in the midline, and the long filaments are exserted 20–24 mm from the tube, both features associated with the pollination strategy.

The relationships of *G. sempervirens* lie with two Western Cape species,

G. stefaniae and *G. cardinalis*. The flowers of *G. stefaniae* are virtually identical to those of *G. sempervirens*, but the two have quite different growth strategies and vegetative appearance. Unlike the evergreen *G. sempervirens*, *G. stefaniae* has a marked dormant phase during the hot dry summer. In autumn, usually after the first rains have fallen and the days are cooler, flowering spikes emerge bearing only sheathing leaves in the lower half of the stem. Well-developed leaves, produced by juvenile or non-flowering plants of *G. stefaniae*, also differ from those of *G. sempervirens* in being greyish and with raised and thickened margins, midribs and other veins, quite typical of section *Blandus*.

History

Gladiolus sempervirens was first collected by W.D. Dumbleton in the Outeniqua Mountains above Oakhurst near George in the mid-nineteenth century. Specimens sent to Kew Gardens in London bear an annotation indicating that they were received there in April, 1870. Correspondence in that year with J.D. Hooker, then director at Kew, indicates that the plants were collected some two years earlier, thus in 1868 or 1869. They were recognized by J.G. Baker as representing an undescribed species which he named *G. splendens* in 1876 in one of the first of a long series of

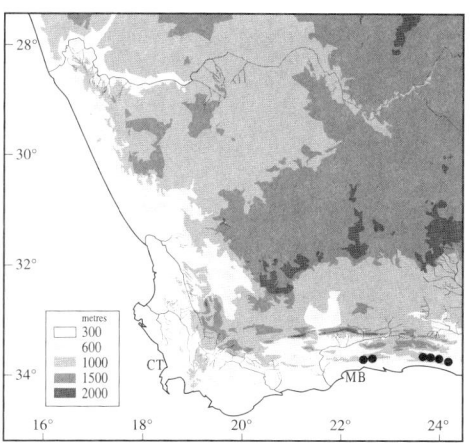

scientific papers and flora accounts he was to publish on African Iridaceae. The name is a homonym for *G. splendens* (Sweet) Herbert, dating from 1842, the western Karoo species recognized by that name today. Baker, however, regarded that species as belonging to another genus, *Antholyza*, and did not see any reason not to use the epithet again in *Gladiolus*, an action now clearly counter to the rules of botanical nomenclature. The name, nevertheless, persisted in use for this rather poorly known southern Cape mountain species for almost 100 years. Only when G.J. Lewis's revision of *Gladiolus* in South Africa was completed and published was the new name *G. sempervirens* proposed to replace *G. splendens* Baker.

51. GLADIOLUS STEFANIAE Obermeyer
PLATE 44

Gladiolus stefaniae Obermeyer in Lewis et al., *J. S. African Bot.*, Suppl. 10: 252 (1972). Type: South Africa, Western Cape, Montagu District, mountains west of Cogmanskloof, Farm Kalkoensnes, 1 Apr. 1970, *Pienaar s.n.* (NBG, holotype).

stefaniae, named in honour of Stefanie Pienaar, daughter of Mr Stefaan Pienaar who assisted Amelia Obermeyer and T.T. Barnard in collecting living plants for the description and preparation of type material.

Description

Plants 40–65 cm high. *Corm* globose, c. 20 mm in diameter, the tunics of papery layers, becoming soft and finely fibrous with age. *Cataphylls* pale and firm-textured, the uppermost reaching up to 8 cm above the ground and then dark purple, usually lightly mottled with cream or green. *Leaves* three to five, rarely six, the lowermost largest,

inserted above the ground, the blade lanceolate, shorter than the sheath, the two upper leaves smaller, usually without a unifacial apex, diverging from the stem; foliage leaves three or four, produced only by plants that did not flower that season, the blades linear, c. 5 mm wide, the midrib fairly strongly thickened, the margins lightly raised, ultimately the blades at least 50–60 cm long. *Stem* more or less erect, slightly flexed outward above the sheaths of the two upper leaves, unbranched, 2.5–3 mm in diameter below the spike.

Spike erect, 2- to 4-flowered; *bracts* pale green or flushed grey on the dorsal side, the outer 55–65 mm long, the margins inrolled in the upper 10–20 mm and somewhat twisted, the inner about three-quarters as long as the outer. *Flowers* brilliant scarlet or carmine, the lower three tepals each with a median whitish streak in the lower two-thirds, the streaks sometimes paler on

the edges and shading to mauve, unscented; *perianth tube* 35–45 mm long, narrowly and obliquely funnel-shaped, the lower cylindrical part 25–30 mm long; *tepals* unequal, those of the outer whorl larger and nearly equal, broadly lanceolate to ovate, the dorsal straight, inclined over the stamens, 53–58 x 25–32 mm, the upper laterals 50–60 x 25–27 mm, directed forward below, spreading and becoming patent above, the lower three tepals joined to the upper laterals for c. 3 mm, arching outward and more or less horizontal, 45–48 x 23–25 mm. *Filaments* 30–42 mm long, exserted 20–32 mm from the tube; *anthers* c. 13 mm long, yellow with purple lines on the sutures of the thecae or uniformly dark purple, the pollen cream. *Ovary* oblong, c. 6 mm long; *style* arching over the filaments, dividing between the base and lower third of the anthers, the branches 10–12 mm long, arching outward

and extending between the anthers. *Capsules* oblong-ellipsoid, 18–25 mm long; *seeds* ovate, broadly and evenly winged, 6.5-8 x c. 5.5 mm, dark translucent brown. *Chromosome number* 2*n* = 30.

Flowering time early March to mid-April.

Distribution and biology

Gladiolus stefaniae is a narrow endemic of the mountains of the southern Cape. Populations occur in the Langeberg south of Montagu and on the Potberg along the south coast. The narrow and disjunct distribution is unexpected, for there is noth-

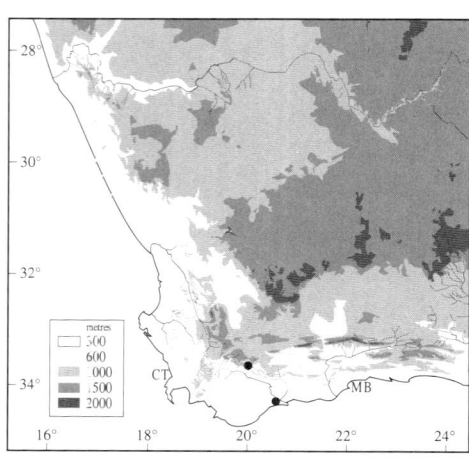

ing remarkable about this portion of the Langeberg, which extends from near Worcester eastward to Mossel Bay. In the mountains near Montagu plants grow on rocky slopes at elevations of between 700 and 1400 m, in gullies that are slightly wetter than the surrounding slopes, and the flowers are bright scarlet. On the Potberg, however, plants are scattered across the open lower slopes of the range on west- or south-facing slopes, and they have carmine flowers. *Gladiolus stefaniae* flowers at the end of the hot and usually dry summer. As soon as the first rains have fallen, or the weather has become cooler, corms sprout and rapidly produce either a flowering spike bearing only sheathing, bract-like leaves, or a leafy shoot that will not flower that year. These plants grow vegetatively, producing long-bladed foliage leaves and later on enlarged corms. Flowering of the leafy plants eventually occurs when sufficient food has been accumulated in the corm to enable the plant to provide enough nutrition for the production of the flowering spike, its flowers and fruit. The sheathing leaves of the flowering stem do not normally have enough photosynthetic capacity to generate sufficient carbohydrate to replace a corm large enough to produce flowers the following year.

Like the related red-flowered species, *Gladiolus cardinalis* and *G. sempervirens*, *G. stefaniae* has flowers adapted for pollination by the mountain pride butterfly, *Aeropetes tulbaghia*. We have seen this butterfly visiting the species at both its known localities, confirming the earlier report of *Aeropetes* visiting *G. stefaniae* by S.D. Johnson. This insect alone pollinates a number of large red-flowered plants in southern Africa (Johnson & Bond, 1994), including at least five species of *Gladiolus*.

Diagnosis and relationships

The bright red flowers of *Gladiolus stefaniae* with splashes of white on the lower tepals

closely resemble those of *G. cardinalis* and *G. sempervirens*. These are both species of moist sites in the Western and Eastern Cape provinces, and despite the difference in habitat and the accompanying difference in growth habit and leaf morphology, it is to these two species that *G. stefaniae* seems most closely related. The leaves of the flowering stem of *G. stefaniae* are bract-like and normally entirely sheathing, and long-bladed foliage leaves are only produced by immature plants or those that will not produce a flowering spike that season. In contrast, both *G. cardinalis* and *G. sempervirens* have a fan of long leaves borne at the base of the flowering stem. *Gladiolus cardinalis* is restricted to steep mountain slopes that are perennially wet, and because of the habitat the stems are inclined to drooping, unlike the erect stems of *G. stefaniae*. Perhaps because of the difference in habitat, the bracts of *G. stefaniae* are fairly soft-textured and partially dry above, whereas those of *G. cardinalis* are green and coriaceous. Other differences between the two species are trivial. We suspect that *G. stefaniae* and *G. cardinalis* are closely related and that their differences represent alternative strategies for survival in a dry summer climate, while maintaining a specialized pollination mode that demands flowering in mid- to late summer, the period when their pollinator is active.

History

The earliest record we have seen of *Gladiolus stefaniae* is a preserved flower collected in 1947 by a Mr Smith in the mountains at Montagu. That collection and a few later ones from Montagu were associated with the southern Cape species, *G. sempervirens*, then known as *G. splendens*. Plants from the Potberg, to the south, were first recorded by the Cape Town botanist, H.C. Taylor, in 1951 and these were identified as *G. carmineus*. When additional collections from Montagu were made by N.J. Myburgh in 1965 and S. Pienaar in 1970, it became clear that this was an undescribed species. Living plants were examined by A.A. Obermeyer in 1971 when she was completing the unfinished manuscript of G.J. Lewis's revision of *Gladiolus*, and *G. stefaniae* was described in 1972 when the revision was published. The Potberg specimens of *G. stefaniae* were still regarded as representing an eastern population of *G. carmineus* in the revision. It was only when we examined these plants ourselves in 1995 that we were able to establish that *G. stefaniae* did, in fact, occur there.

52. GLADIOLUS CARMINEUS C.H. Wright

Cliff gladiolus, Hermanus gladiolus

PLATE 45

Gladiolus carmineus C.H. Wright, *Curtis's Bot. Mag.* 132: pl. 8068 (1906). Lewis et al., *J. S. African Bot.*, Suppl. 10: 256 (1972). Type: South Africa, Western Cape, cliffs at Hermanus, cultivated at Kew Gardens, London, *Abercrombie Smith s.n.,* illustration in *Curtis's Bot. Mag.* 132: pl. 8068 (1906), lectotype here designated; K, flower fragment, isolectotype.

carmineus = carmine, referring to the dark red colour of the flowers.

Description

Plants (16–)30–50 cm high. *Corm* globose, 20–25 mm in diameter, the tunics of soft membranous layers, becoming fibrous with age, extending upwards with the decayed remains of the cataphylls in a neck around the base of the stem. *Cataphylls* pale and firm-textured, the uppermost reaching up to 8 cm above the ground and then dark purple mottled with cream or green. *Leaves* of the flowering stem three to five, the lowermost largest, inserted above the ground, the blade reduced, lanceolate, vestigial, shorter than the sheath, up to 50 mm long, the two upper leaves smaller, usually without a blade, diverging from the stem for a short distance but channelled throughout; foliage leaves usually two, produced only by plants that did not flower that season, dry and decaying by early summer, falcate, long and trailing, 8–10 mm wide, glaucous, the midrib lightly thickened. *Stem* erect, unbranched, flexed outward above the sheaths of the two upper leaves, 2–3 mm in diameter below the spike.

Spike erect, flexuose, 2- to 6-flowered; *bracts* pale green or flushed grey on the upper side, the outer 30–40 mm long, the

inner about two-thirds as long. *Flowers* pale to deep pink, the lower three tepals each with a median whitish streak surrounded by a pale mauve halo, occasionally the upper lateral tepals similarly marked, unscented; *perianth tube* 30–35 mm long, narrowly funnel-shaped; *tepals* unequal, broadly lanceolate, those of the outer whorl larger and nearly equal, the dorsal 35–45 x 18–27 mm, the upper laterals and lower median 33–48 x 16–28 mm, the lower three tepals joined to the upper laterals for 4 mm, spreading more or less horizontally. *Filaments* 21–26 mm long, exserted 10–15 mm from the tube; *anthers* 10–14 mm long, usually yellow, the pollen cream. *Ovary* oblong, 8–10 mm long; *style* arching over the stamens, dividing between the base and middle of the anthers, the branches c. 10 mm long, gradually expanded in the upper half. *Capsules* narrowly obovoid, 20–22 mm long; *seeds* ovate, 8 x 6–7 mm, broadly and evenly winged, light brown and semi-transparent, the seed body darker brown. *Chromosome number* $2n = 30$.

Flowering time mid-February to late March.

Distribution and biology

Although a summer-flowering species, *Gladiolus carmineus* is a narrow endemic of the southern African winter-rainfall zone. It is restricted to the southwestern Cape coast between Pringle Bay in the west and Cape Infanta in the east. Plants grow in rocky sandstone outcrops, often on cliffs along the coast, never out of sight of the sea. The corms usually grow wedged in cracks in the rock, sometimes in virtually no soil at all. Like *G. stefaniae*, which also flowers out of season, the leaves on the flowering stem are reduced and usually bladeless. Fully developed foliage leaves are only produced by plants that did not flower that year. These leaves grow during the wet winter and spring, slowly dying off at the end of October. They are completely dry and often decayed by the time flowers are produced at the end of summer, from February to April.

The large, deep mauve-pink flowers with a relatively long perianth tube so closely resemble those of the red-flowered species of series *Blandus* that we assume they are also adapted for pollination by the butterfly *Aeropetes tulbaghia*. Their distinctive pinkish

mauve colour rather than the brilliant scarlet red typical of this pollination system is approached by *G. sempervirens* and the coastal populations of *G. stefaniae*, both of which have carmine rather than scarlet red flowers.

Diagnosis and relationships

Gladiolus carmineus is fairly closely related to the widespread and relatively common western Cape species, *G. carneus*. It differs from that species most importantly in flowering in February and March, at a time when the foliage leaves are dry and dead, and the flowering stems bear partly to entirely sheathing leaves. Most other species of section *Blandus* have three or more foliage leaves borne contemporaneously with the flowering stem. Apart from the fundamental difference in growth cycle, *G. carmineus* also has flowers of a distinctive mauve-pink colour and characteristic open form with the dorsal tepals somewhat spreading. Most members of the section have the dorsal tepal erect, arched or sometimes extending horizontally over the stamens. *Gladiolus carmineus* has sometimes

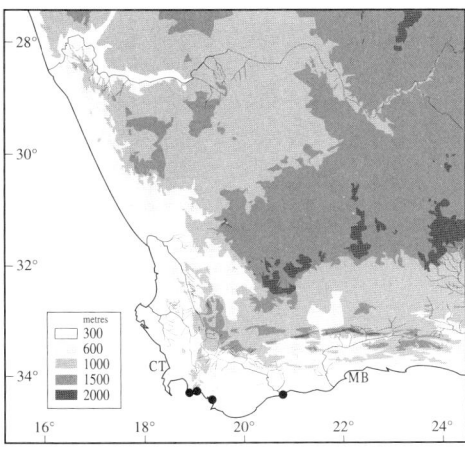

been confused with *G. stefaniae*, and the coastal form of that species from the Potberg, south of Swellendam, was included in *G. carmineus* by Lewis et al. (1972). Although the growth cycle of the two species is virtually identical, *G. stefaniae* has substantially larger flowers with the perianth tube 35–45 mm long and dorsal tepal 53–58 mm long, compared with a tube c. 30 mm long and dorsal tepal c. 35 mm long in *G. carmineus*. This distinction, as well as a different flower shape (compare figures on page 144 and page 145) and colour, help distinguish between the two species.

History

Gladiolus carmineus was apparently first recorded by Sir C. Abercrombie Smith, Controller and Auditor General of the Cape of Good Hope. Smith collected corms from the cliffs at Hermanus and sent them to Kew Gardens in 1903, where they flowered two years later. A watercolour painting, made at the time and published in *Curtis's Botanical Magazine* in 1906, is now the lectotype of the species.

Section *Blandus*: Series *Floribundus*

53. GLADIOLUS RUDIS Lichtenstein ex Roemer & Schultes

PLATE 46

Gladiolus rudis Roemer & Schultes, *Syst. Pl.* 1: 408 (1817). *Gladiolus floribundus* subsp. *rudis* (Lichtenstein ex Roemer & Schultes) Obermeyer in Lewis et al., *J. S. African Bot.*, Suppl. 10: 108 (1972). Type: South Africa, Cape, without precise locality or collector, but probably M. Lichtenstein (B–Herb. Willdenow 917, lectotype designated by Lewis et al., 1972: 108, NBG, photograph).

rudis = rough, raw, uncultivated, also a scoop or ladle, the latter possibly describing the prominent spade- or scoop-shaped nectar guides on the lower tepals; the alternative derivation, rough or uncultivated, makes little sense (the choice of epithet was not explained by the authors of the species).

Synonymy

Gladiolus vomerculus Ker Gawler, *Genera Irid.* 142 (1827). Baker, *Fl. Capensis* 6: 142 (1896). Type: South Africa, Western Cape, without precise locality or collector, cultivated in Great Britain by William Herbert, illustration in *Curtis's Bot. Mag.* 38: pl. 1564 (1813) as *G. hastatus* Thunberg, no preserved specimens known.

Description

Plants 15–55 cm high. *Corm* more or less globose, 20–40 mm in diameter (including the tunics), the tunics of papery layers, decaying into medium- to coarsely textured fibres, often interlayered with brittle papery fragments, accumulating in a thick mass. *Cataphylls* fairly firm-textured and pale, the uppermost reaching up to 10 cm above the ground and then purple, mottled with green or white, roughly textured on the mottling, decaying with age to become fibrous and accumulating upward in a neck around the base of the stem. *Leaves* four to seven, the lower three to five basal and longest, reaching to the base of the spike and often shortly exceeding it, the sheaths often mottled purple and green below, the blades sword-shaped, 7–15(–20) mm wide, the margins and midrib and one or two other veins lightly or sometimes moderately thickened, the margins occasionally minutely crispulate, the upper two leaves cauline and shorter than the basal, the lower of these inserted on the lower third of the stem, the upper inserted on the upper third, both sheathing in the lower half, the sheaths open to the base and the margins overlapping. *Stem* erect below, flexed outward above the sheaths of the two upper leaves, usually unbranched, 2–3 mm in diameter below the spike.

Spike strongly inclined, sometimes horizontal, lightly flexuose, 2- to 5-flowered; *bracts* green or flushed brownish or purple, 35–50(–65) mm long, sometimes lightly folded in the upper midline, often flushed purplish above and on the veins, inner bracts half to two-thirds as long as the outer. *Flowers* cream, whitish or pale pink, the lower half of the tepals darker on the reverse, the lower lateral tepals each with a yellowish, spade-shaped, longitudinal mark outlined in darker colour, the throat yellow, unscented; *perianth tube* obliquely funnel-shaped, 18–21 mm long, the lower cylindrical part 8–10 mm long, emerging from the middle of the bracts, occasionally reaching the bract apices; *tepals* lanceolate to elliptic, acute, subequal, the dorsal slightly larger, ascending, arching upward distally, 32–36 x 18–20 mm, the upper laterals 29–32 x 18–22 mm, curving outward and patent in the distal half, lower tepals united for 2–3 mm, nearly straight below, curving slightly downward distally or inclined c. 30° below the horizontal, 26–32 x 13–15 mm. *Filaments* 12–14 mm long, exserted 5–7 mm from the tube; *anthers* 9–12 mm long, white, the pollen white. *Ovary* oblong, 5–8 mm long; *style* usually arching over the stamens, rarely suberect, dividing opposite the lower half of the anthers, the branches broadly spathulate, 6–8 mm long, held above the stamens. *Capsules* and *seeds* unknown. *Chromosome number* unknown.

Flowering time early September to mid-October.

Distribution and biology

A species of the winter-rainfall zone, *Gladiolus rudis* is a narrow endemic of the Caledon and Bredasdorp Districts of Western Cape Province. It is restricted to rocky sandstone soils on the lower and middle slopes of the Riviersonderend and Bredasdorp mountains and the sandstone ridges south of Elim on the Agulhas Peninsula.

The fairly large, short-tubed flowers of *Gladiolus rudis* have conspicuous spear- to diamond-shaped nectar guides on the lower tepals and conform to the pattern for bee-pollinated flowers in the southern African winter-rainfall zone.

Diagnosis and relationships

Gladiolus rudis has the typical attributes of series *Floribundus* of section *Blandus*. These include a short stem, cataphylls and leaf sheaths speckled purple and white, and coarsely fibrous corm tunics that accumulate together with the leaf bases and cataphylls in a dense mass around the corm and underground part of the stem. The speckling of the cataphylls is so strongly developed in *G. rudis* that the white spots are raised above the surface. The leaf margins are also unusually strongly thickened and sometimes lightly crisped. The flowers are distinctive in the series, having the shortest tube and filaments of any species, the tube 18–21 mm long and the filaments 12–14 mm long. The tepals are widely spread and cream to pink, the lower laterals with well-developed white or yellow, spade- or diamond-shaped markings outlined in red.

Gladiolus rudis is most commonly confused with a second species of the series, *G. grandiflorus*, which has flowers with a longer perianth tube, usually 27–35 mm and occasionally up to 55 mm long, longer filaments, 14–22 mm long, and both the lower and upper tepals usually each with a dark linear stripe in the midline. Not only do the two species differ morphologically, but they favour different habitats. *Gladiolus grandiflorus* is a species of heavy clay soils

where it grows in renosterveld, whereas *G. rudis* favours stony sandstone soils in fynbos. Although the two species were treated as subspecies of *G. floribundus* by Lewis et al. (1972) in their revision of the South African species of the genus, we believe that specific rank better reflects their different morphology and ecology.

History

Although known in Europe since at least the later eighteenth century, *Gladiolus rudis* makes its first appearance in the scientific literature in 1813 when a plant identified as *G. hastatus* was illustrated in *Curtis's Botanical Magazine*. The wild source of these plants was not recorded, but they were raised by the formidable botanist, William Herbert. This collection was evidently associated with C.P. Thunberg's species of that name because of the prominent spear-shaped markings on the tepals, although Thunberg's species is otherwise quite unlike *G. rudis*.

The species was also collected by the explorer-naturalist, Martin Lichtenstein, probably on his journey to Swellendam in September 1804. As were many of the novelties collected by him, the species was described in 1817 by J.J. Roemer and J.A. Schultes, who acknowledged Lichtenstein as having chosen the specific epithet. Ker Gawler was evidently unaware of the identity of *G. rudis*, the type of which was in Berlin, and in 1827 he named the illustration identified as *G. hastatus* in *Curtis's Botanical Magazine G. vomerculus*. It was under the latter name that the species

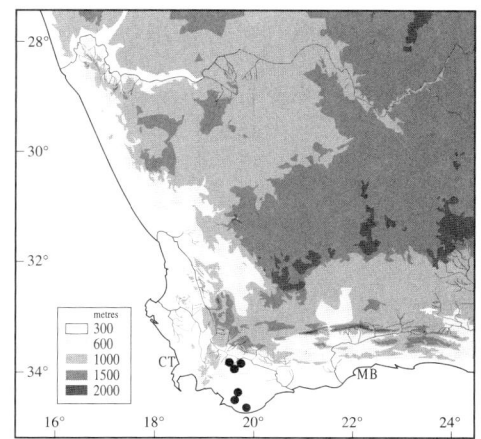

was included in *Flora Capensis*. Collections of *G. rudis* made by C.F. Ecklon and C.L. Zeyher in the 1820s were believed to represent an undescribed species and were distributed to European herbaria under the very apt manuscript name *G. tigrinus* Ecklon. The name was never formally published, although it is often cited as a synonym of *G. rudis*.

In their revision of *Gladiolus* in South Africa, Lewis et al. (1972) recognized that *G. rudis* and *G. vomerculus* were synonyms but they regarded them as a subspecies of the morphologically variable *G. floribundus*. As outlined above, we regard the short perianth tube, short filaments, strong spear-shaped nectar guides, stiff leaves with much thickened margins, and the prominent, raised speckles on the cataphylls and leaf sheaths sufficiently distinctive to warrant recognition of *G. rudis* as a separate species.

54. GLADIOLUS GRANDIFLORUS Andrews

PLATE 47

Gladiolus grandiflorus Andrews, *Bot. Repository* 2: pl. 118 (1800). Type: South Africa, Cape, without precise locality or collector, cultivated in Great Britain, illustration in Andrews, *Bot. Repository* 2: pl. 118 (1800), no preserved specimens known.

grandiflorus = large-flowered, referring to the relatively large flowers, at least compared to many of the species of *Gladiolus* known at the time that *G. grandiflorus* was described.

Synonymy

Gladiolus milleri Ker Gawler, *Curtis's Bot. Mag.* 17: pl. 632 (1803). *Gladiolus floribundus* subsp. *milleri* (Ker Gawler) Obermeyer in Lewis et al., *J. S. African Bot.*, Suppl. 10: 104 (1972). Type: South Africa, Western Cape, without precise

locality or collector, cultivated and flowered in Great Britain at Kensington, figure in *Curtis's Bot. Mag.* 17: pl. 632 (1803), no preserved specimens known.

Gladiolus fasciatus Roemer & Schultes, *Syst. Pl.* 1: 429 (1817), as a new name for *Gladiolus vittatus* sensu Horneman, *Hort. Hafniensis* 2: 950 (1815), not *G. vittatus* Zuccagni (1806), identity uncertain. *G. floribundus* subsp. *fasciatus* (Roemer & Schultes) Obermeyer in Lewis et al., *J. S. African Bot.*, Suppl. 10: 107 (1972). Type: South Africa, Cape, without precise locality, collector or date (C, holotype).

Geissorhiza grandis J.D. Hooker, *Curtis's Bot. Mag.* 96: pl. 5877 (1870). Baker, *Fl. Capensis* 6: 75 (1896). Type: South Africa, Eastern Cape, probably Port Elizabeth District, cultivated in Great Britain, *Wilson s.n.*, illustration in *Curtis's*

Bot. Mag. 96: pl. 5877 (1870), no preserved specimens known.

Gladiolus scaphochlamys Baker, *Handbook Irideae* 217 (1892); *Fl. Capensis* 6: 153 (1896). Type: South Africa, Western Cape, Swellendam District, Sparrbos near Buffeljags River, 30 July 1831, *Drège 8427* (K, lectotype designated here; BM, G, K, L [not seen], MO, P, S [not seen], isolectotypes).

Gladiolus socium L. Bolus, *J. Bot.* 69: 14 (1931). Type: South Africa, Cape, George District, cultivated in Cape Town, Oct. 1929, *De Mole & Kisch s.n.* (BOL 19083, lectotype here designated; BOL, K, isolectotype).

Description

Plants (25–)35–50 cm high. **Corm** more or less globose, 20–40 mm in diameter (including the tunics), the tunics of papery layers,

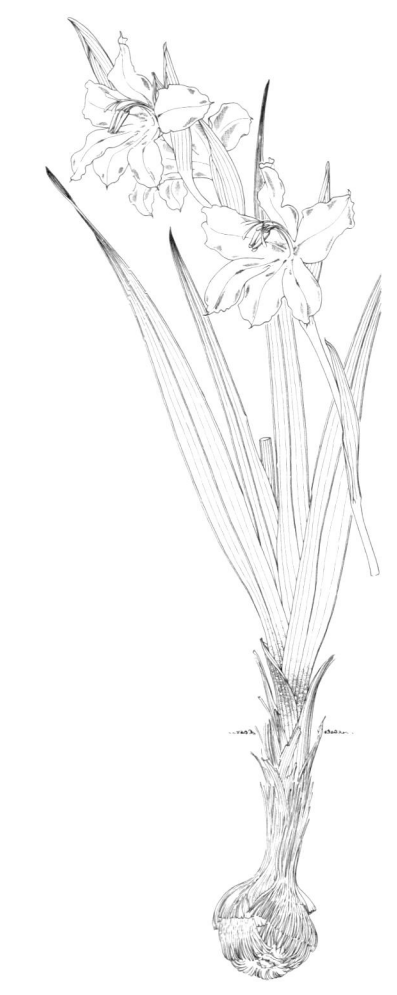

decaying into medium- to coarsely textured fibres, often interlayered with brittle papery fragments, accumulating in a thick mass. *Cataphylls* pale and fairly firm-textured, the uppermost reaching up to 10 cm above the ground and then purple, lightly mottled with green or white, accumulating with age and extending upward in a neck around the base of the stem. *Leaves* four to seven, the lower three to five basal and longest, reaching at least to the base of the spike and often shortly exceeding it, the blades sword-shaped, 6–15(–20) mm wide, the margins and midrib and one or two other veins light-ly or sometimes moderately thickened, the margins occasionally minutely crispulate, the upper two leaves cauline and shorter than the basal, the lower of these inserted on the lower third of the stem, the upper inserted on the upper third, both sheathing in the lower half, the sheaths open to the base and the margins overlapping. *Stem* erect below, flexed outward above the sheaths of the two upper leaves, usually unbranched, c. 3 mm in diameter below the spike.

Spike strongly inclined, sometimes horizontal, lightly flexuose, occasionally 2-, usually 3- to 6-, rarely to 9-flowered; *bracts*

green or flushed brownish or purple, the outer (30–)35–50(–65) mm long, some-times lightly folded in the upper midline, often flushed purplish above and on the veins, the inner half to two-thirds as long as the outer. *Flowers* cream, whitish or pale pink, the lower half of the tepals darker coloured on the reverse, the upper lateral and lower tepals with a pink to reddish median streak, the throat yellow, sometimes with red at the base, occasionally lightly scented; *perianth tube* obliquely funnel-shaped, (22–)27–35(–55) mm long, the lower cylindrical part 13–22(–40) mm long, reaching to between the middle and the apices of the bracts, rarely shortly exceeding them; *tepals* obovate to oblanceolate, the margins sometimes undulate or lightly crisped, obtuse to retuse, subequal or unequal and then the three upper slightly larger or the dorsal largest, ascending, arching upward distally, 35–40(–44) x 13–20 mm, the upper laterals 30–36 x 13–18 mm, curving outward and patent distally, the lower tepals united for 2–3 mm, nearly straight below, curving slightly down-ward distally or inclined c. 30° below the horizontal, 23–31(–37) x 8–13 mm. *Filaments* 14–22 mm long, exserted 4–12 mm from the tube, cream or pink; *anthers* 7–11 mm long, white, cream or purple, the pollen white or mauve. *Ovary* oblong, 5–7 mm long; *style* usually arching over the stamens, rarely suberect, dividing between the middle and apex of the anthers or slightly exceeding them, the branches broadly spathulate, 5–9 mm long, held above the stamens. *Capsules* narrowly obovoid to elliptic, 30–40 mm long; *seeds* oval, 12–14 x 8–10 mm, broadly and evenly winged, light translucent yellow-brown. *Chromosome number* unknown.

Flowering time August to mid-October.

Distribution and biology

Fairly widespread in the southern African winter-rainfall zone, *Gladiolus grandiflorus* extends from Bot River and Villiersdorp in the Caledon District in the west to Port Elizabeth in the east. Plants most often occur in renosterveld vegetation on clay soils that are fairly wet in winter and spring. Along the south coast between Mossel Bay and Cape Infanta a distinct form of *G. grandiflorus* grows in calcareous sand in coastal fynbos.

The flowers are visited by a range of large, long-tongued bees, mostly of the family Anthophoridae, and these insects are the pri-mary pollinators. Although the long anthers contain a large amount of pollen, the attrac-tion offered to visiting insects is the nectar located in the lower half of the perianth tube. The large bee *Anthophora diversipes* appears to be the most common and fre-quent visitor to *Gladiolus grandiflorus*.

Diagnosis and relationships

A member of series *Floribundus*, *Gladiolus grandiflorus* has the typical vegetative attrib-utes of the group, but may be easily recog-nized by its flowers. They are borne on light-ly inclined spikes and are either uniformly white to pink or the lower tepals, and some-times the upper tepals as well, each have a dark pink to red median stripe. The upper and lower tepals are more or less equal in length or the lower may be slightly to about a third shorter. In addition, the perianth tube is relatively short, usually 27–35 mm, rarely to 55 mm long, and typically as long as the bracts (sometimes slightly shorter or longer). A coastal form of the species, regard-ed as a separate subspecies by Lewis et al. (1972), has the tepal margins strongly undu-late to crisped.

Gladiolus grandiflorus is closely related to *G. floribundus* and is sometimes confused with that species, which has deep cream to greenish flowers, usually with a longer peri-anth tube, 40–70 mm long, the anthers 7–9 mm long and the lower tepals half to two-thirds as long as the dorsal. The spike of *G. floribundus* is also typically strongly inclined with the flowers arranged along the upper side, giving it a distinctive appearance.

Although Lewis et al. (1972) treated *Gladiolus grandiflorus* as subsp. *milleri* of *G. floribundus*, we believe it is best treated as a separate species. The differences between them are substantial and include floral morphology, pollination ecology and habitat preference. While *G. floribundus* favours rocky sandstone habitats in dry fynbos, *G. grandiflorus* grows in clay soils

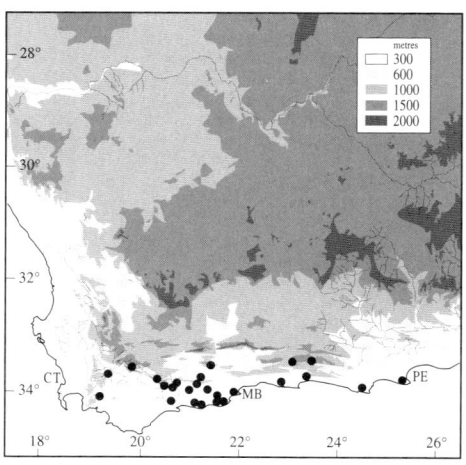

in renosterveld vegetation. *Gladiolus grandiflorus* may be confused with *G. rudis*, a species also regarded by Lewis et al. as a subspecies of *G. floribundus*. *Gladiolus rudis* is readily recognized by the flowers with prominent spade-shaped nectar guides on the lower tepals, the tepals always subequal and the perianth tube only 18–21 mm long. In *G. rudis* the mottling on the cataphylls and sheaths of the lower leaves is always prominent and roughly textured, and the leaf blades are tough and have thickened and sometimes crisped margins.

History

It is puzzling that the fairly common *Gladiolus grandiflorus* was only described as late as 1800. There seems no doubt, however, that the figure published in Andrews' *Botanists Repository*, the type of the species, represents the earliest published record of

this predominantly lowland, southern Cape plant. For reasons that are not clear, the expert on bulbous plants, John Ker Gawler, ignored the name Andrews used and in 1803 in *Curtis's Botanical Magazine* he redescribed the species, calling it *G. milleri*. The latter name commemorates the illustration of what Ker Gawler considered to be the same species that was illustrated in Phillip Miller's *Figures of Plants* published in 1755, but which most likely represents the fairly similar *G. floribundus*. Ker Gawler was the acknowledged authority on the Iridaceae and the name he used was adopted by most botanists at the time. The pink-flowered coastal form of *G. grandiflorus* was named *G. fasciatus* in 1817 by J.J. Roemer and J.A. Schultes who, incidentally, described the closely related *G. rudis* at the same time. *Gladiolus fasciatus* was actually a new name for the plant called *G. vittatus* by the Danish botanist, Jens

Horneman, and was based on a specimen of unknown provenance in the Copenhagen Herbarium.

Plants very similar to the type of *Gladiolus grandiflorus* but with virtually equal tepals and symmetrically disposed stamens were described as *Geissorhiza grandis* by J.D. Hooker in 1870 and the species was recognized by J.G. Baker in *Flora Capensis*. The actinomorphic condition is unstable and not a consistent feature of any known wild population. The typical form of *G. grandiflorus* was treated under two separate names in *Flora Capensis*, *G. milleri* and *G. scaphochlamys* (Baker, 1896), the latter species having been described by Baker in 1892. Baker placed the earlier name, *G. grandiflorus*, in synonymy under *G. floribundus*.

55. GLADIOLUS FLORIBUNDUS Jacquin

PLATE 48

Gladiolus floribundus Jacquin, *Collecteana* 4: 162 (1790) and *Icones Pl. Rar.* 2: pl. 254 (1795). Lewis et al., *J. S. African Bot.*, Suppl. 10: 99 (1972). Type: South Africa, Cape, without precise locality or date, illustration in Jacquin, *Icones Pl. Rar.* 2: pl. 251 (1795).

floribundus = profusely flowering, describing the many flowers produced by well-grown plants which may have three or four branches, some with up to ten or even 13 flowers.

Synonymy

Antholyza spicata Miller, *Gardeners Dict.* ed. 8 (1768), not *Gladiolus spicatus* Linnaeus, 1753 (= *Thereianthus spicatus* (Linneaus) G. Lewis). Type: South Africa, Western Cape, without precise locality or collector, cultivated in Great Britain, illustration in Miller, *Figures of Plants* volume 1: pl. 40 (1755).
Gladiolus striatus Andrews, *Bot. Repository* 2: pl. 111 (1800), an illegitimate homonym, not *G. striatus* Jacquin (1786–1793) (= *G. hyalinus* Jacquin). *G. floribundus* var. *striatus* (Andrews) Persoon, *Syn. Pl.* 1: 45 (1805). Type: South Africa, Western Cape, without precise locality or collector, cultivated at Hammersmith, Great Britain, illustration in Andrews, *Bot. Repository* 2: pl. 111 (1800).
Montbretia pauciflora Baker, *J. Bot.* 14: 336 (1876). *Acidanthera pauciflora* (Baker) Bentham, *Genera Pl.* 3: 706 (1883). Baker, *Fl. Capensis* 6: 132 (1896). *Gladiolus bowkeri* G. Lewis, *J. S. African Bot.* 7: 26 (1941), as a new name for *M. pauciflora* in *Gladiolus*, not *G. pauciflorus* Baker, from tropical Africa. Type:

South Africa, Eastern Cape, Somerset East, without date, *Bowker s.n.* (K, holotype; BOL, PRE, photo).
Acidanthera graminifolia Baker, *J. Bot.* 14: 338 (1876); *Fl. Capensis* 6: 130 (1896). *Gladiolus graminifolius* (Baker) G. Lewis, *J. S. African Bot.* 7: 26 (1941). Type: South Africa, Western Cape, plains between Swellendam and the Gouritz River, in 1818 or 1819, *Bowie s.n.* (BM, holotype).
Acidanthera forsythiana Baker, *Handbook Irideae* 186 (1892); *Fl. Capensis* 6: 131 (1896). Type: South Africa, without precise locality, in 1835, *Forsyth s.n.* (K, holotype).
[*Gladiolus undulatus* sensu Jacquin, *Coll.* 3: 256 (1798) and *Icones Pl. Rar.* 2: pl. 251 (1790). Ker Gawler, *Curtis's Bot. Mag.* 18: pl. 647 (1803). Baker, *Fl. Capensis* 6: 155 (1896), not sensu Linnaeus (1767).]

Description

Plants 25–45 cm high.
Corm more or less globose,

20–40 mm in diameter (including the tunics), the tunics of medium- to coarsely textured fibres, often interlayered with

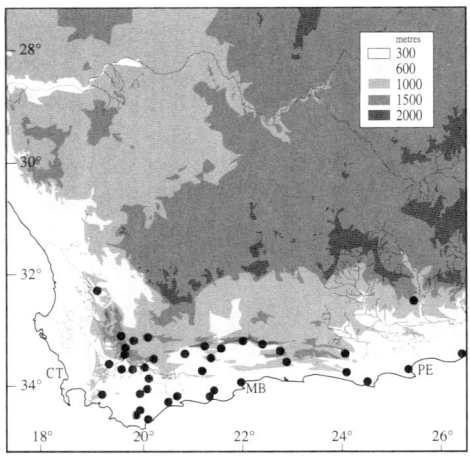

firm unbroken layers. *Cataphylls* pale and firm-textured, the uppermost extending up to 8 cm above the ground and then purplish, usually mottled with green or white, persisting and accumulating with the leaf bases forming a neck around the base of the stem. *Leaves* six to eight, the lower four to six basal, reaching to about the base of the spike but often flaccid and arching toward the ground, the blades sword-shaped, 12–20 mm wide, progressively decreasing in size above, the upper leaves inserted on the stem well above the ground, smaller than the basal, the sheaths open to the base. *Stem* erect below, flexed outward above the sheath of the uppermost leaf, inclined above, usually with one or two branches, occasionally simple, 3–4 mm in diameter below the spike.

Spike strongly inclined to nearly horizontal, flexuose, 3- to 8-, rarely to 13-flowered; *bracts* dull green to grey-purple, the outer (34–)40–50(–65) mm long, the inner two-thirds to nearly as long as the outer, minutely notched apically. *Flowers* cream or pale greenish to ivory, occasionally pale pink, all the tepals or excepting the dorsal with a broad streak of darker colour along the midline, unscented; *perianth tube* obliquely funnel-shaped, 40–60(–70) mm long, the cylindrical lower part 25–35 (–55) mm long; *tepals* unequal, lanceolate to elliptic, obtuse to retuse, the margins usually undulate, the dorsal largest, inclined over the stamens, 30–45 x 16–20 mm, the upper laterals 25–46 x 13–18 mm, the lower tepals 18–29 x 7.5–10 mm, the lower median slightly longer than the lower laterals, to 38 mm long, curving toward the ground in the upper halves. *Filaments* 20–22 mm long, exserted 4–10 mm from the tube; *anthers* 6–9 mm long, purple, the pollen cream to purple or brownish. *Ovary* oblong, c. 5–6 mm long; *style* arching over the stamens, dividing between the middle and

apex of the anthers, the branches c. 6 mm long, arching outward. *Capsules* oblong-ellipsoid, three-sided, 30–40 mm long; *seeds* oval, 9–14 x 7–10 mm, the wing broadly and evenly developed, yellowish brown. *Chromosome number* $2n = 30$.

Flowering time late September to November.

Distribution and biology

A widespread species of the southern African winter-rainfall zone, *Gladiolus floribundus* extends from the Cedarberg in Western Cape Province in the west to the Albany and Alexandria Districts in Eastern Cape Province in the east. It is rare in the west of its range, where it is restricted to the dry interior valleys of the Cedarberg, but relatively common in the south between Worcester and Riversdale. Plants always occur in relatively dry situations, usually in stony places on soils derived from sandstones of the Cape System. Along the southern Cape coast it is also occasionally encountered on limestone outcrops.

Gladiolus floribundus is pollinated by long-tongued flies, the long perianth tube and the narrow, dark median streaks on the tepals being typical of this strategy. The flowers contain moderate quantities of nectar which is only accessible to insects with mouthparts exceeding 15 mm in length. Two tabanid flies, *Philoliche gulosa* and *P. rostrata*, have been recorded on the flowers of some populations, usually carrying heavy loads of the dark pollen.

Diagnosis and relationships

Gladiolus floribundus typically has pale greenish to cream or ivory, or occasionally dull mauve flowers, with lightly to strongly undulate tepals. The lower three tepals are always noticeably shorter than the upper and each has a narrow, dark, longitudinal median streak. The floral bracts are relatively long, usually 40–50 mm, and the perianth tube is as long as or longer than the bracts, usually exceeding them by at least 10–15 mm. The species is closely related to *G. grandiflorus*, *G. miniatus* and *G. rudis*, all of which are similar in general appearance, being fairly short plants with broad leaves shorter than the spikes and distinctive, coarsely fibrous corm tunics that accumulate together with the decayed, fibrous bases of the leaves and cataphylls of past seasons in a dense mass around the corm and underground part of the stem. Despite the vegetative similarity, the flowers of the four species differ substantially in colour, shape and the relative

proportions of the tepals to one another and to the perianth tube. *Gladiolus grandiflorus* and *G. rudis* have a whitish to pink flower, in the former sometimes without markings on the lower tepals. The lower three tepals are slightly shorter to about as long as the upper three, and the perianth tube is shorter than the dorsal tepal and shorter than to nearly as long as the bracts. *Gladiolus miniatus* has salmon to orange flowers with a somewhat darker line on each of the lower tepals. More significantly, the perianth tube is 50–65 mm long and divided into a narrow cylindrical lower part and a wider cylindrical upper part, c. 30 mm long.

We assume that the four species of series *Floribundus* share an immediate common ancestor and that each has become specialized for a different pollinator: long-tongued flies in *Gladiolus floribundus*, bees in *G. grandiflorus* and *G. rudis*, and sunbirds in *G. miniatus*. Not only do the four species have flowers adapted for pollination by different animals, but they are generally found in different habitats. *Gladiolus floribundus* nearly always grows on stony, sandstone-derived soils, and in relatively dry situations. In contrast, *G. grandiflorus* is mostly found on clay soils that are fairly wet in winter and spring; *G. rudis* grows on well-drained rocky, sandstone slopes; and *G. miniatus* is restricted to limestone outcrops along the southern Cape coast. The habitats do overlap to some extent, and we have seen typical *G. floribundus* growing on dry limestone outcrops, and a race of *G. grandiflorus* grows on coastal, lime-enriched sands, but these seem to be exceptions that are not uncommon in plants. Soil type is seldom of such overwhelming importance to plant species that they will not tolerate more than a single soil type.

History

Gladiolus floribundus made its appearance in botanical literature in 1755 in Philip Miller's *Figures of Plants*, that lavishly illustrated record of rare and unusual plants cultivated in the Chelsea Physick Garden, London, of which Miller was curator. In that publication *G. floribundus* was given a polynomial, *Antholyza foliis linearibus sulcatis floribus albis, etc.* In 1768 Miller adopted the binomial system of plant nomenclature and named the plant *Antholyza spicata*. The name cannot, however, be transferred to *Gladiolus* because the epithet had already been used in that genus by Linnaeus for the plant now called *Thereianthus spicatus*. Then in 1790 Nicholas Jacquin published an excellent figure of the species which he identified as

G. undulatus Linnaeus. That Jacquin had misapplied Linnaeus's name was understood by John Ker Gawler but he, nevertheless, continued to use *G. undulatus* in Jacquin's sense because 'it was known to botanists in general by the present title'. This usage was followed throughout the nineteenth century and *G. floribundus* was called *G. undulatus* by J.G. Baker in *Flora Capensis* (Baker, 1896). Throughout this period the true *G. undulatus* was known by the later name *G. cuspidatus*, described by Jacquin in 1795.

Jacquin published a second figure of *Gladiolus floribundus* in 1795 and this is the type of *G. floribundus*. Correct use of the name did not, however, become current until Lewis et al (1972) unravelled the nomenclatural history of the species. *Gladiolus floribundus* acquired several additional synonyms during the nineteenth century, including *G. striatus* Sweet, *Montbretia pauciflora* Baker, and *Acidanthera forsythiana* Baker. The two latter species were transferred to *Gladiolus* by G.J. Lewis in 1941 as *G. bowkeri* and *G. graminifolius* respectively, before she realized that they were conspecific with *G. floribundus*. New names were required because both epithets had already been used in *Gladiolus*.

The decision to include *Gladiolus rudis*, *G. fasciatus*, *G. grandiflorus* and *G. miniatus* as subspecies of *G. floribundus* was taken by Amelia Obermeyer when she completed G.J. Lewis's manuscript after her death. Annotations to specimens in several herbaria show that Lewis intended to recognize *G. grandiflorus*, *G. miniatus* and *G. rudis* as separate species. The inclusion of *G. geardii* and *G. geardii* var. *uitenhagensis*, both described by H.M.L. Bolus, in *G. floribundus* subsp. *floribundus* (Lewis et al., 1972) is a mistake. *Gladiolus geardii*, including var. *uitenhagensis*, is a separate species more closely allied to *G. carneus* than to *G. floribundus* and its allies, and both taxa are now treated as belonging to series *Blandus*.

56. GLADIOLUS MINIATUS Ecklon
PLATE 49

Gladiolus miniatus Ecklon, *Topographisches Verzeichniss* 40. 1827. *Gladiolus floribundus* subsp. *miniatus* (Ecklon) Obermeyer in Lewis et al., *J. S. African Bot.*, Suppl. 10: 103 (1972). Type: South Africa, Western Cape, without precise locality but probably from near Cape Agulhas, 25 Nov., *Ecklon 323* (S [not seen], lectotype designated by Lewis et al., 1972: 103).

miniatus = orange-coloured, for the colour of the flowers.

Description
Plants 15–40 cm high. *Corm* more or less globose, 20–30 mm in diameter including the tunics, the tunics of medium- to coarsely textured fibres interlayered with unbroken, firm-papery tissue. *Cataphylls* pale and firm-textured, the uppermost reaching up to 6 cm above the ground and then dull purple, usually lightly mottled with green or white, persisting and accumulating with the leaf bases in a neck around the base of the stem. *Leaves* usually six, the lower four basal and longest, reaching or slightly exceeding the spike, the blades plane, lanceolate, 7–18 mm wide, the margins and midrib lightly thickened, the upper two leaves cauline and shorter than the basal, the lower of these inserted on the lower third of the stem, the upper inserted on the upper third, both sheathing in the lower half, the margins of the sheath open to the base. *Stem* simple or sometimes with a short branch, flexed outward above the sheathing part of the uppermost leaf or upper two leaves and then strongly inclined to horizontal, c. 4 mm in diameter below the spike.

Spike flexed outward at the base and nearly horizontal, 3- to 7-flowered; *bracts* dull to greyish green, often flushed with purple on the dorsal side, 40–55(–65) mm long, the inner about two-thirds as long as the outer, minutely notched apically. *Flowers* cream to light salmon on opening, soon developing a deep salmon-pink colour, the reverse of the tube reddish, the tepals darkly pigmented along the midlines, the markings decurrent on the tube, unscented; *perianth tube* 50–65 mm long, narrow and cylindric in the lower 20–35 mm, curving outward in the middle and broadly cylindric in the upper 30 mm, in section ovate, 7–8 mm x 6 mm; *tepals* lanceolate, the margins lightly undulate, unequal, the dorsal arching slightly above the horizontal, 32–39 x c. 15 mm, the upper laterals directed forward below, arching outward above, 30–37 x 10–13 mm, the lower three tepals smaller, 28–34 x 8–9 mm, curving downward, the lower median sometimes slightly longer than the lower laterals. *Filaments* 35–37 mm long, exserted 8–11 mm from the tube; *anthers* 5–7 mm long, purple, the pollen purple. *Ovary* oblong, c. 8 mm long; *style* arching over the stamens, dividing opposite the anther apices, the branches c. 5 mm long. *Capsules* ovoid-ellipsoid, 25–30 mm long; *seeds* ovate, c. 9 x 7 mm, broadly winged. *Chromosome number* 2n = 30.

Flowering time October to mid-November, occasionally in December.

Distribution and biology
Gladiolus miniatus is a fairly rare endemic of limestone outcrops along the south coast of Western Cape Province, South Africa. Plants extend from The Downs near Riversdale in the east to the mouth of the Bot River near Hawston in the west, nearly always fairly close to the seashore and within sight of the ocean. It is also found on the limestone hills along the Agulhas coast, notably in the hills east of Gansbaai where it grows together with a second limestone endemic, *G. variegatus*.

The shape of the flower of *Gladiolus miniatus*, with its elongate perianth tube,

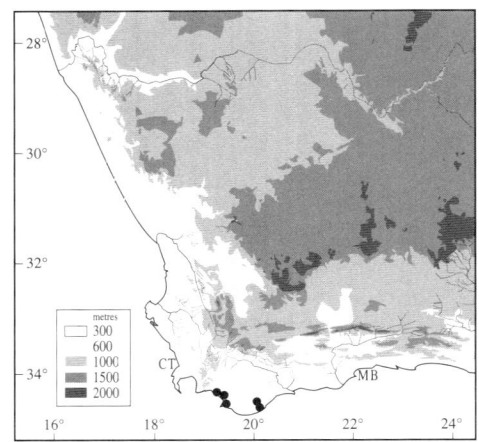

the upper part long and cylindric, and the salmon to orange-red colour, suggest that it is adapted for pollination by sunbirds. The flowers produce copious amounts of nectar, more than 10 microlitres per flower, a volume that in southern Africa is associated only with bird pollination. The nectar also has a relatively low sugar concentration, frequently a feature of bird-pollinated flowers. The bee-pollinated flowers of *G. grandiflorus*, in contrast, produce about half that quantity of considerably sweeter nectar.

Diagnosis and relationships

The general appearance of *Gladiolus miniatus* matches exactly that of *G. floribundus* and *G. grandiflorus*. All three species have a tight fan of short, narrowly lanceolate, firm leaves, distinctive corm tunics of coarse fibres that accumulate with age into a dense mass and, together with the decaying cataphylls and leaf bases, form a thick neck around the base of the stem. The species can only be distinguished by their flowers, and those of *G. miniatus* are unmistakable. The perianth is salmon-orange, usually with darker colouring on the midlines of the tepals and suture lines on the tube. The tube itself is 50–65 mm long and has a narrow lower part c. 25 mm long and a wide cylindrical upper part c. 30 mm long. The perianth tube is 22–40(–55) mm long in *G. grandiflorus* and usually 40–60 mm long in *G. floribundus*. Both these species have cream to ivory, greenish or pink flowers with a more or less funnel-shaped perianth tube, the upper part flared rather than cylindric.

The inclusion of *Gladiolus miniatus* at subspecies rank in a broadly circumscribed *G. floribundus* (Lewis et al., 1972) emphasizes the vegetative similarities of the taxa at the expense of important floral differences that reflect their diverse pollination biology. This, combined with its preference for a coastal limestone habitat which differs from soils favoured by its relatives, has convinced us that species rank is appropriate for *G. miniatus*.

History

Described as early as 1827 by C.F. Ecklon, *Gladiolus miniatus* is assumed to have been collected by Ecklon near Cape Agulhas, although he did not provide this information in the protologue (Lewis et al., 1972). *Gladiolus miniatus* was not recognized by either of the late nineteenth-century specialists, F.W. Klatt or J.G. Baker, in their treatments of African Iridaceae and even the name was completely overlooked in *Flora Capensis* (Baker, 1896).

Amelia Obermeyer (in Lewis et al., 1972) recognized the plant as *G. floribundus* subsp. *miniatus*. However, we are firmly convinced that Lewis herself intended to recognize *G. miniatus* as a separate species.

4. SECTION *LINEARIFOLIUS*

(De Vos) Goldblatt & Manning

Section *Linearifolius* (De Vos) Goldblatt & Manning, new combination. Basionym: *Homoglossum* section *Linearifolius, J. S. African Bot.* 42: 336 (1976). Type species: *H. merianellum* sensu auct. = *Gladiolus bonaspei* Goldblatt & De Vos.

Description

Plants small to medium-sized. *Corm* globose, the tunics coriaceous, soon fragmenting into narrow vertical sections or coarsely fibrous. *Cataphylls* usually shortly puberulous, sometimes evidently glabrous, perhaps only with age. *Leaves* narrowly sword-shaped to linear, either several forming a basal fan or superposed, the blades plane, usually the margins and midrib thickened, and often one or more other veins prominent, often the blades and sheaths villous or puberulous, in several species the leaves of the flowering stem reduced, non-flowering plants then producing long-bladed foliage leaves later in the season in some species. *Stem* unbranched.

Spike either flexed at the base and inclined or erect, straight or barely flexuose, internodes usually short and flowers somewhat crowded; *bracts* green throughout or dry above. *Flowers* usually small, mostly slightly bilabiate, the lower tepals sometimes narrow and clawed at base, scented in some species; *perianth tube* usually obliquely funnel-shaped, occasionally tubular, half to two-thirds as long as the dorsal tepal or much longer; *tepals* subequal or unequal, then the lower three narrower but usually about as long as the upper, the dorsal arching over the stamens. *Capsules* either oblong to subglobose, three-lobed and retuse above or ovoid-elliptic; *seeds* fairly large, broadly and evenly winged, usually opaque and reddish or pinkish brown.

Comprising 16 species in southern Africa, section *Linearifolius* extends widely across southern and south tropical Africa. Eleven species occur in the winter-rainfall zone, mainly in Western Cape Province, South Africa, where all the species seem to favour the acidic, well-drained sandstone soils common in this part of the subcontinent. One of these, *Gladiolus guthriei*, extends into Northern Cape Province where it occurs on the Bokkeveld escarpment, the northernmost occurrence of Table Mountain Sandstone. The winter-rainfall species appear to comprise a single clade sharing ovate-ellipsoid capsules, and are placed in series *Linearifolius*. The most common winter-rainfall species are *G. hirsutus*, which is a widespread spring-flowering species, and *G. brevifolius*, which flowers in the autumn and has the leaf blades of the flowering stem reduced and usually vestigial. In the southern African summer-rainfall zone the most common species is the spring-flowering *G. woodii. Gladiolus pubigerus* has a somewhat wider distribution, but is rare and poorly documented.

Most species have fairly small, short-tubed flowers adapted for bee pollination, but in the winter-rainfall zone pollination strategies are diverse. Two species, both of which have longer floral tubes, *Gladiolus emiliae* and *G. guthriei*, are adapted for pollination by large moths, and *G. monticola* has flowers adapted for long-tongued fly pollination. Two additional species of the area, *G. nerineoides* and *G. stokoei*, have bright red flowers with moderately long, slender tubes adapted for pollination by the butterfly *Aeropetes tulbaghia*. The final two species from the winter-rainfall zone, *G. bonaspei* and *G. overbergensis*, have red (rarely yellow) flowers with elongate tubes widely cylindrical above, which are apparently adapted for pollination by sunbirds. Species of section *Linearifolius* thus show significant floral adaptive radiation.

Several species of the winter-rainfall zone also show major shifts in flowering times. Only *Gladiolus caryophyllaceus* and *G. hirsutus* flower in late winter and spring; *G. monticola, G. nerineoides* and *G. stokoei* flower in late summer; *G. brevifolius* and *G. emiliae* flower in autumn; and *G. guthriei* flowers in winter.

In tropical Africa, several species provisionally assigned to a broadly defined section *Gladiolus* (Goldblatt, 1996) also appear to belong in section *Linearifolius*. These include at least *G. curtifolius, G. fuscoviridis, G. huillensis, G. intonsus, G. pubescens* and *G. zimbabweensis*, all of which have villous leaves, and except for the long-tubed *G. curtifolius* and *G. huillensis* have flowers resembling those of *G. woodii* in general shape and size. *Gladiolus laxiflorus*, which has sparsely hairy leaves and comparatively large flowers, may also belong here.

Section *Linearifolius*: Series *Pubigerus*

57. GLADIOLUS WOODII Baker

PLATE 50

Gladiolus woodii Baker, *Handbook Irideae* 207 (1892); *Fl. Capensis* 6: 144 (1896). Lewis et al., *J. S. African Bot.,* Suppl. 10: 293 (1972), including *G. pardalinus* Goldblatt & Manning and *G. pubigerus* Baker. Type: South Africa, KwaZulu-Natal, Inanda, Oct. 1880, *Wood 618* (K, holotype; BM, SAM, isotypes).

woodii, named in honour of John Medley Wood, first curator of the Natal Botanic Garden, who made a major contribution to the knowledge of the southern African flora.

Synonymy

Gladiolus trichostachys Baker, *Bull. Herb. Boissier,* Sér. 2, 4: 1007 (1904). Pole Evans, *Fl. Pl. S. Africa* 5: pl. 163 (1925). Type: South Africa, Gauteng, Irene, date unknown, *Conrath 579* (GZU [not seen], holotype).
Gladiolus nudus N.E. Brown, *Kew Bull. Misc. Inf.* 1921: 298 (1921). Type: South Africa, Mpumalanga, Barberton (flowers yellow), Oct. 1919, *Thorncroft 1067* (K, holotype).

Description

Plants 20–45(–90) cm high. *Corm* globose, 12–15 mm in diameter, the tunics of fine to moderately coarse-textured, tawny brown vertical fibres, rarely accumulating to any degree. *Cataphylls* pale and membranous, the uppermost usually reaching 3–5 cm above the ground and then coriaceous and

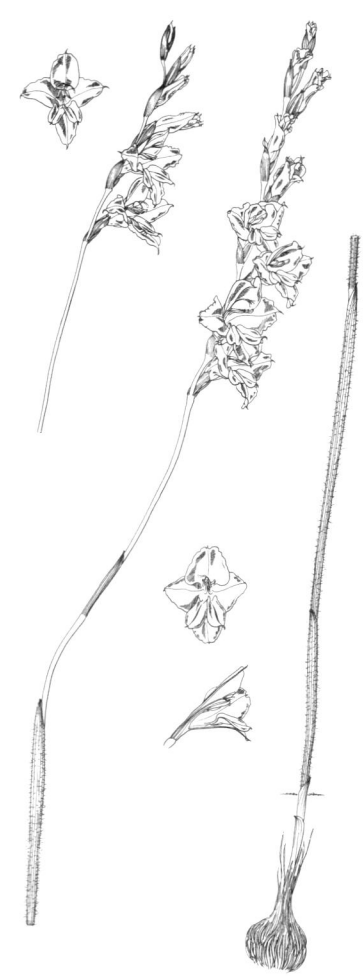

pale to dark green or partly to entirely purple, lightly to densely villous. *Leaves* usually two or three, sometimes four, the lowermost basal and longest, sheathing for their entire length or with a short linear to sword-shaped blade up to 5(–10) cm long, reaching the middle of the stem, sparsely pilose, occasionally overlapping the next leaf, the second leaf entirely sheathing, about as long as the basal when three or four leaves present, smaller if only two present, the remaining leaf or leaves inserted between the middle and upper quarter of the stem, often bract-like, 20–28 mm long, glabrous, the margins free to the base; non-flowering plants with solitary, narrowly lanceolate leaves, at flowering time 7–10 cm long, becoming much longer later in the growing season, the margins and midrib and sometimes a second pair of veins thickened and raised, densely pilose. *Stem* erect below, lightly flexed outward above the sheath of the upper leaf and then inclined, unbranched, 0.8–1.8 mm in diameter below the spike.

Spike drooping in bud, at anthesis straight and lightly flexed below the base, inclined 30°–40°, strongly secund, the flowers placed nearly above one another, sometimes 4-,

usually 6- to 11-flowered, the internodes 7–14(–18) mm long, at least in the upper part of the spike about half as long as the bracts; *bracts* pale grey-green to dark green or the dorsal surface flushed brown to dull purple-brown toward the apices or entirely, by anthesis sometimes becoming dry in the upper half and papery, then flushed pinkish brown, the outer 12–16 mm long, the inner about as long as or slightly longer or shorter than the outer. *Flowers* maroon to reddish brown, pale yellow or pale lilac or oyster-coloured, usually the lower lateral tepals each with a longitudinal dark purple to brown streak along the midline, rarely without markings, or when the upper tepals maroon the lower laterals partly to entirely yellow and with a maroon median line, the top of the tube sometimes marked with purple spots at the tepal sinuses, the lower median tepal without markings, with a faint metallic odour (maroon-flowered plants), a faint daffodil-like odour (oyster- or lilac-flowered plants), or evidently odourless (yellow-flowered plants); *perianth tube* obliquely funnel-shaped, (4–)7–9(–12) mm long; *tepals* unequal, narrowly to broadly ovate, usually conspicuously apiculate, the apiculi sometimes up to 2 mm long, the dorsal tepal largest, inclined over the stamens, 17–27 x 12–19 mm, the margins sometimes lightly crisped, the upper laterals directed forward below, nearly straight or curving outward in the upper half to third, the margins undulate to strongly crisped, the lower three tepals united with the upper laterals for c. 2 mm and to one another for 2–3 mm, narrowed below into claws, the claws 2.5–4 mm long, becoming channelled toward the base, the lower laterals abruptly expanded into limbs 10–12 x 4–7 mm, the limbs auriculate at the base, the lower median limb 12–15 x 6–9 mm, often more or less plane and less strongly clawed, the margins of the lower tepal limbs lightly undulate to lightly crisped, in profile the upper and lower tepals appearing equally long. *Filaments* 9–10(–13) mm long, exserted 5.5–8 mm from the tube; *anthers* 6–9 mm long, blackish, the pollen cream to pale yellow. *Ovary* ovoid, 2–3 mm long; *style* arching over the stamens, dividing between the middle and apex of the anthers, the branches 3.5–5 mm long, arching outward beyond the anther apices, sometimes barely overtopping the anthers. *Capsules* more or less globose, 9–12 mm long, broadly three-lobed above, the apices retuse; *seeds* ovate to oblong, 6–7 x 4 mm, the wing evenly developed,

opaque dark red-brown. *Chromosome number* $2n = 30$.

Flowering time mainly September to early November, occasionally until mid-December.

Distribution and biology

One of the most common of the eastern southern African species of the genus, *Gladiolus woodii* occurs widely across the summer-rainfall zone of the subcontinent. It extends from the southern KwaZulu-Natal Drakensberg and Midlands northward through Swaziland and Mpumalanga to the Soutpansberg in Northern Province. Largely confined to areas of fairly high rainfall and high elevation, it does occur along the KwaZulu-Natal coast, for example at Inanda, near Greytown, and at Hluhluwe. It also extends well into interior southern Africa and is found throughout the Mpumalanga highveld and Gauteng. Plants are most common in stony, low grassland and on a variety of substrates, but most frequently on heavier soils of diabase, dolerite or shale origin. Like the other eastern southern African species of section *Linearifolius*, *G. woodii* flowers very early in the season, usually soon after the first spring rains have fallen and before the surrounding vegetation has grown. The plant's resources are channelled into producing the flowering stem and flowers and the leaves are short and often largely or entirely sheathing. The leaves do, however, remain green for months after the capsules have ripened and

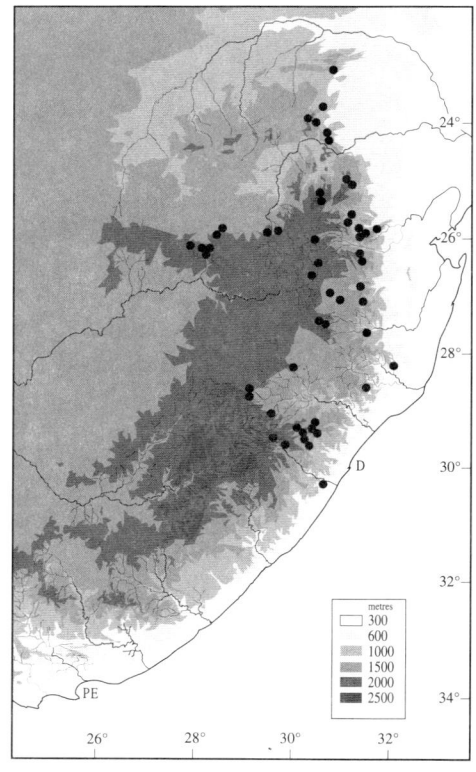

the seeds shed. Only non-flowering plants produce long-bladed foliage leaves, a pattern quite different to the species of section *Linearifolius* from the winter-rainfall zone that flower aseasonally. These species, including the common *G. brevifolius*, have sheathing leaves on the flowering stem and produce a long-bladed leaf during the winter and spring growing period.

The short-tubed flowers of *Gladiolus woodii* conform in size and shape to those of other eastern southern African species of the genus that are pollinated by long-tongued anthophorid bees and the species must be assumed to have the same pollination strategy.

Diagnosis and relationships

The widespread and common *Gladiolus woodii* is readily recognized by the flowering stems that bear two or three, occasionally four, sheathing or short-bladed leaves, villous cataphylls and leaf sheaths, and strongly bilabiate flowers with a more or less hooded dorsal tepal. The upper tepals are 17–27 mm long, ovate and substantially larger than the lower tepals which are conspicuously clawed, channelled below and auriculate at the base of the limbs. In most populations the flowers have the upper tepals conspicuously apiculate, sometimes for up to 2 mm. Flower colour is surprisingly variable. The type form, from Inanda near Durban, flowers with small yellow nectar guides on the lower tepals. Plants with flowers of this colour extend across KwaZulu-Natal into southern Mpumalanga and Gauteng as far

north as the Witwatersrand and also occur locally on the eastern Mpumalanga escarpment in the Dullstroom and Sabie areas. In the eastern Mpumalanga highveld and adjacent KwaZulu-Natal, scattered population have flowers with lilac to oyster-coloured or pale mauve flowers, and along the northern edge of the escarpment between Middelburg and Pretoria plants have yellow flowers. The distribution of the colour forms has no clear pattern and, as far as we can tell, flower colour is not associated with any morphological feature.

The relationships of *Gladiolus woodii* lie on the one hand with the Zimbabwean *G. fuscoviridis* and on the other with three southern African species, *G. malvinus*, *G. pardalinus* and *G. pubigerus*. Specimens of the first two species were actually included in *G. woodii* by Lewis et al. (1972) in their account of that species, while *G. pubigerus* was regarded as a synonym of *G. parvulus*. J.G. Baker, who originally described *G. pubigerus* under the name *G. pubescens*, did, however, recognize the species in *Flora Capensis* (1896). Neither *G. pardalinus* nor *G. malvinus* appears to have been recorded until well after Baker's time. *Gladiolus fuscoviridis*, an endemic of fairly restricted distribution in eastern Zimbabwe (Goldblatt, 1996), has flowers very similar to those of *G. woodii*, but they are produced in the summer. It also has long-bladed, glabrous foliage leaves, the sheaths of which are only lightly pubescent. We suspect that *G. fuscoviridis* is close to the ancestral stock that gave rise to the

eastern southern African members of section *Linearifolius*.

History

Although *Gladiolus woodii* is a fairly inconspicuous plant, it is rather surprising that it remained unknown until the very end of the nineteenth century. The first record of the species we have been able to trace is a collection made in 1880 by the Natal naturalist, later Curator of the Natal Botanic Garden, John Medley Wood, at Inanda near Durban. Collections from the Barberton area of what is now Mpumalanga began to appear in 1889 due to the activities of Ernest Galpin and D.F. Gilfillan. Rudolf Schlechter recorded the species at Six Mile Spruit near Pretoria in 1893. Wood's collection from Inanda was the basis for *G. woodii*, described in 1892 by J.G. Baker. Subsequently, Baker described a collection of yellow-flowered plants from Irene near Pretoria as *G. trichostachys* in 1904. A Thorncroft collection of yellow-flowered plants from near Barberton was described as *G. nudus* by N.E. Brown in 1921. Both *G. trichostachys* and *G. nudus* are seen today as simply colour forms of the widespread *G. woodii* and otherwise differ not at all from the type of the species. As circumscribed by Lewis et al. (1972) in their revision of *Gladiolus* in South Africa, *G. woodii* included *G. pubigerus* as a synonym and specimens of *G. pardalinus* were cited under the species. The treatment was the result of an inadequate knowledge of these species rather than a strong conviction that they belonged in *G. woodii*.

58. GLADIOLUS MALVINUS Goldblatt & Manning
PLATE 51

Gladiolus malvinus Goldblatt & Manning, new species. Type: South Africa, Mpumalanga, dolerite mountain slopes north of Dullstroom, 29 October 1994, *Goldblatt & Manning 10075* (NBG, holotype; K, MO, isotypes).

malvinus = mauve, referring to the distinctive flower colour.

Latin diagnosis

Plantae 55–70 cm altae, cormo globoso 30–40 mm in diametro tunicis fibrosis, cataphyllo summo villoso cataphyllis persistentibus circum basem caulis collum formantibus, foliis 3 raro 2 laminis brevibus 1–5 cm longis vaginis sparsim villosis vel glabrescentibus, spica 9–15 florum, floribus

malvinis inodoratis, tubo perianthii infundibuliformi c. 10 mm longo, tepalis inaequalibus apiculatis superioribus ovatis 28–30 x 14–16 mm, inferioribus unguiculatis, unguibus c. 2.5 mm longis limbis 14–20 mm longis, filamentis c. 10 mm longis, antheris c. 7 mm longis.

Description

Plants 55–70 cm high. *Corm* globose, 30–40 mm in diameter, the tunics of moderately coarse fibres accumulating in a thick mass. *Cataphylls* pale and membranous, the uppermost reaching 4–8 cm above the ground and then bright green, coriaceous, lightly villous, persisting and accumulating with the corm tunics in a thick neck of fibres around the base of the stem. *Leaves* usually

three, rarely two, the lowermost basal, sheathing the lower third of the stem, the sheath usually sparsely villous or sometimes glabrous, the blade 1–5 cm long, plane and narrowly lanceolate, overlapping the second leaf, this entirely sheathing and glabrous, the uppermost leaf shortest, inserted on the upper third of the stem, channelled throughout but sheathing only in the lower half, diverging from the stem above; non-flowering plants with solitary, narrowly lanceolate leaves up to 30 cm long (probably increasing in length later in the season), the blades pubescent, the margins, midrib and one or more other pairs of veins lightly thickened, the midrib not raised. *Stem* erect below, flexed lightly above the sheaths of the two upper leaves and inclined above,

unbranched, 2–2.5 mm in diameter below the spike.

Spike inclined, more or less straight, 9- to 15-flowered, the internodes 15–25 mm long, never shorter than the bracts; *bracts* grey-green, usually flushed purplish on the dorsal side, becoming dry by the time the flowers begin to fade, the outer 20–25 mm long, the inner 1–2 mm shorter than the outer. *Flowers* pale mauve, the lower lateral tepals each with a dark purple streak in the midline and lightly to darkly shaded with purple in the lower half, unscented; *perianth tube* obliquely funnel-shaped, c. 10 mm long; *tepals* broadly lanceolate and apiculate, unequal, the dorsal largest, inclined over the stamens, 28–30 x 14–16 mm, the upper laterals directed forward, curving outward very little or not at all in the upper third, the lower three tepals joined to the upper laterals for c. 2 mm and to one another for c. 3 mm, the lower laterals strongly clawed, the claws channelled, c. 2.5 mm long, the limbs oblong, c. 14 x 6 mm, channelled toward the base, the lower median plane,

obscurely clawed, c. 20 x 12 mm, in profile the lower three tepals about as long as or slightly exceeding the upper. *Filaments* c. 10 mm long, exserted c. 8 mm from the tube; *anthers* c. 7 mm long, blackish, the pollen cream. *Ovary* obovoid, c. 5 mm long; *style* arching over the stamens, dividing opposite the middle to upper third of the anthers, the branches 2.5–3 mm long. *Capsules* obovoid, (10–)12–14 mm long, broadly three-lobed above, the apices retuse; *seeds* ovate, c. 6 x 5–6 mm, the wing evenly developed, opaque reddish brown. *Chromosome number* unknown.

Flowering time October and early November.

Distribution and biology

Restricted to a small part of the southern African summer-rainfall zone, *Gladiolus malvinus* occurs in the hilly upper Mpumalanga escarpment between Dullstroom and Belfast. Plants grow in dolerite outcrops in heavy clay. Like the closely related *G. woodii*, they flower early in the spring in October or November, soon after the first rains have fallen and the surrounding grasses and herbs have barely begun to grow.

The flowers are assumed to be adapted for pollination by the same long-tongued anthophorid bees that pollinate other eastern southern African species of the genus with similar short-tubed and small to moderate-sized flowers.

Diagnosis and relationships

The robust, fairly tall *Gladiolus malvinus* has sturdy flowering stems 55–70 cm high that are produced early in the spring when they stand well above the surrounding vegetation. The two or three leaves borne on the flowering stem have tightly sheathing bases and reduced blades up to 5 cm long, and the blades and sheaths are sparsely pubescent or virtually glabrous. The flowers of *G. malvinus* are more or less bilabiate, relatively large and mauve to lilac with purple markings on the lower tepals.

In general aspect *Gladiolus malvinus* is very similar to *G. woodii* which has the same growth habit, but that species is a much smaller plant with the leaf sheaths usually densely pubescent. The maroon, yellow or pale lilac flowers of *G. woodii* have a tube mostly 7–9 mm long, the upper tepals 17–27 mm long and narrowly clawed lower tepals auriculate at the base of the limbs. Flowers of *G. malvinus* are some-what larger, with a perianth tube c. 10 mm

long and the upper tepals 28–30 mm long. The lower tepals are, like those of *G. woodii*, strongly clawed and auriculate at the limb bases. An additional difference between the two species is that the corms of *G. malvinus* are deep-seated and have coarse fibrous tunics that accumulate with the decayed remains of the cataphylls in a thick neck around the underground part of the stem. This feature is matched among the eastern southern African species of section *Linearifolius* only by *G. pardalinus*.

History

When we first saw *Gladiolus malvinus* flowering in the doleritic hills near Dullstroom in eastern Mpumalanga, we immediately realized that it represented a novelty. Typical *G. woodii*, similar in general aspect and in the basic structure of the flowers, although a much smaller plant, was growing nearby and was just coming into flower and could be compared immediately with the taller and larger-flowered *G. malvinus*. Although we did not at the time know of the existence of the species, later examination of specimens filed under *G. woodii* in the National Herbarium in Pretoria revealed three earlier collections of *G. malvinus*, the first one found near Belfast in 1960 by the horticulturist and well-known plant collector, H.J.E. Schlieben.

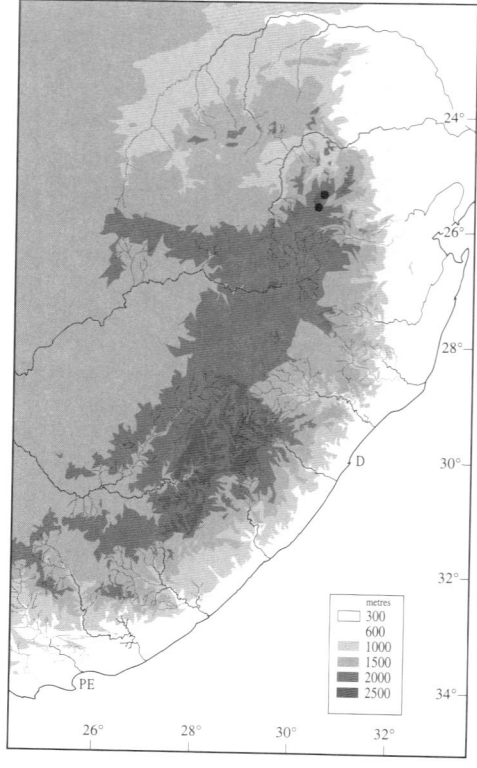

59. GLADIOLUS PARDALINUS Goldblatt & Manning

Gladiolus pardalinus Goldblatt & Manning, new species. Type: South Africa, Mpumalanga, 6 km northeast of Stoffberg, dolerite koppie, 2 Nov. 1994, *Goldblatt & Manning 10094* (NBG, holotype; K, MO, isotypes).

pardalinus = leopard-like, so named for the pale yellow flowers with dark red markings.

Synonymy

[*Gladiolus woodii* sensu Obermeyer, *Fl. Pl. Africa* 35: pl. 1378 (1962). Lewis et al., *J. S. African Bot.*, Suppl. 10: 293 (1972), in part.]

Latin diagnosis

Plantae 35–55 cm altae, cormo conico-ovoideo 18–22 mm in diametro, foliis 3 raro 4 vaginantibus glabris, foliis plantarum efflorentium 2–3 pubescentibus marginibus costisque incrassatis, spica 10–15 florum, floribus pallide flavis rubro notatis, tubo perianthii c. 8 mm longo, tepalis inequalibus unguiculatis apiculatis, superiore majore inclinato 22–24 x c. 10 mm, inferioribus 14–15 x 4 mm canaliculatis auriculatisque basem versus, filamentis c. 15 mm longis, antheris c. 6.5 mm longis rubropurpureis, capsulis obovoideis trilobatis apicem versus 14–20 x c. 10 mm.

Description

Plants 35–55 cm high. *Corm* ovoid to broadly conic, 18–22 mm in diameter, the

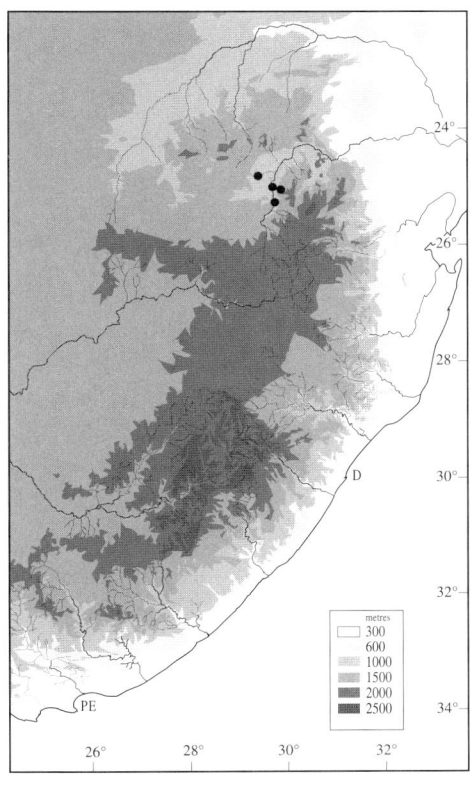

tunics of firm cartilaginous layers, decaying with age into fairly coarse vertical fibres, these often persisting and accumulating in a dense mass around the corm. *Cataphylls* pale and membranous, the uppermost reaching 4–5 cm above the ground and then firm-textured and pale green, lightly villous, the decayed fibrous remains of the cataphylls persisting as a neck around the base of the stem. *Leaves* usually three, occasionally four, all entirely sheathing and enveloping the stem, the lowermost basal and longest, one-quarter the length of the stem, glabrous, normally not overlapping the next leaf, the second leaf inserted on the lower third of the stem, 10–15 cm long, the uppermost leaf inserted on the upper third of the stem, 2–4 cm long, the margins united around the stem or free to the base; non-flowering plants ultimately with three leaves, the lowest longest, sometimes the only one present at the end of flowering, then 20–30 cm, later up to 50 cm long, the remaining leaves shorter and narrower, the sheaths sparsely villous, the blades linear, 5–6 mm wide, lightly to fairly densely villous, the margins and midrib moderately raised and thickened, the secondary veins not or barely thickened. *Stem* erect, more or less straight, unbranched, 2–3 mm in diameter below the spike.

Spike drooping in bud, becoming straight as the flowers open, 10- to 15-flowered, the internodes about as long as the bracts or slightly longer, in fruit inclined up to 30°; *bracts* pale grey-green, often flushed with grey-purple on the dorsal surfaces, the outer 16–20 mm long, the inner 2–3 mm shorter than the outer, notched apically for c. 2 mm. *Flowers* pale yellow, the tepals of the outer whorl densely speckled on the lower half of the limbs and on the claws, the lower lateral tepals irregularly streaked with dark purple lines, with a faint acrid odour; *perianth tube* obliquely funnel-shaped, c. 8 mm long; *tepals* unequal, all distinctly narrowed below into claws, strongly apiculate, the apiculi 2–3 mm long, the dorsal largest, inclined over the stamens, 22–24 x c. 10 mm, the upper laterals directed forward below, lightly curving outward in the upper halves of the limbs, the margins lightly crisped, 19–21 x c. 9 mm, the limbs 11–12 mm long, the lower three tepals united with the upper laterals for c. 2 mm and to one another for c. 3.5 mm, the claws c. 4 mm long, abruptly expanded, the lower laterals with limbs more or less oblong, 10–11 x 4 mm, becoming channelled below and auriculate at the base,

the lower margins raised and almost vertical, the lower median c. 16 mm long, more or less plane, the limb also abruptly expanded and auriculate at the base, the limb 11 x c. 7 mm, in profile the upper and lower tepals appearing the same length. *Filaments* c. 15 mm long, exserted c. 11 mm from the tube; *anthers* c. 6.5 mm long, dark red-purple, the pollen cream. *Ovary* ovoid, c. 3 mm long; *style* arching over the stamens, dividing between the base and lower third of the anthers, the branches c. 3 mm long, arching outward but barely overtopping the anthers. *Capsules* obovoid, three-lobed and retuse above, 14–20 x c. 10 mm; *seeds* ovate-oblong, 7–8 x 4–5 mm, opaque rust-brown, the wing evenly developed around the seed. *Chromosome number* unknown.

Flowering time mid-October and November.

Distribution and biology

Gladiolus pardalinus is restricted to the bushveld of Northern Province and Mpumalanga of South Africa. Not well collected, its range extends from Groblersdal in the east across the low hills and plains of the bushveld to near Nylstroom in the west. Plants grow in dolerite in rocky ground, usually among large boulders, the rocks presumably providing a measure of protection against corm-eating molerats and porcupines. Like other eastern southern African species of section *Linearifolius*, *G. pardalinus* flowers aseasonally, producing flowers early in the spring, in October and November, on flowering stems that bear two or three largely or entirely sheathing leaves.

The strikingly coloured, short-tubed flowers are assumed to be adapted for pollination by long-tongued anthophorid bees.

Diagnosis and relationships

Until now regarded as no more than a minor variant of the widespread eastern southern African *Gladiolus woodii*, *G. pardalinus* differs from that species in several important respects. It has tall, sturdy flowering stems tightly sheathed by leaves, and thus in its general appearance it is not unlike *G. woodii* except for the larger size. The leaves are, however, entirely glabrous and more tightly sheathing than in *G. woodii*, the sheaths of which are densely villous and tend to be somewhat inflated. In addition, non-flowering plants of *G. pardalinus* have two or usually three long leaves, the blades with thickened margins and midrib and the secondary veins not or barely thickened. This is quite different from the single leaf produced in non-flowering plants of *G. woodii*, which has the midrib and one or two other vein pairs raised but the margins unthickened. The flowers of *G. pardalinus* are pale yellow with dark reddish brown markings on all except the dorsal tepal, and both the lower and upper tepals are conspicuously clawed. In *G. woodii* only the lower lateral tepals are marked with a median streak and the upper tepals are not clawed.

History

Discovered in 1957 by the Pretoria botanist and past director of the Botanical Research Institute, L.E. Codd, *Gladiolus pardalinus* immediately provoked controversy. Codd believed that the plant he had discovered was new to science and tried to convince experts of his contention. Amelia Obermeyer would not accept that this was a novelty and when she completed G.J. Lewis's revision of *Gladiolus* in South Africa in the late 1960s she included the few collections of *G. pardalinus* then available in *G. woodii*. We had the opportunity in 1994 to see both *G. woodii* and *G. pardalinus* alive and in full flower, and it is clear that Codd was correct in regarding them as separate species. The collection of fruiting specimens and plants in leaf later in the year provided further evidence that *G. pardalinus* differed significantly from *G. woodii*.

60. GLADIOLUS PUBIGERUS G. Lewis

PLATE 52

Gladiolus pubigerus G. Lewis, *Ann. S. African Mus.* 40: 132 (1954), as a new name for *Gladiolus pubescens* Baker, *J. Bot.* 14: 333 (1876); *Fl. Capensis* 6: 142 (1896), an illegitimate homonym, not *G. pubescens* Lamarck (1791) (= *Babiana pubescens* (Lamarck) G. Lewis). Type: South Africa, Eastern Cape, without precise locality, as 'British Kaffraria,' Oct. 1860, *Cooper 458* (K, holotype).

pubigerus = bearing hairs, referring to the villous character of the cataphylls, lowermost leaf of the flowering stem, and of the leaves of non-flowering plants.

Synonymy

Gladiolus pugioniformis Hilliard & Burtt, *Notes Roy. Bot. Gard. Edinburgh* 37: 299 (1979). Type: South Africa, KwaZulu-Natal, Impendhle District, Farm Umgeni Vlei, 27 Oct. 1976, *Hilliard & Burtt 9077* (NU, holotype; E, K, MO, PRE, isotypes).

Description

Plants (20–)30–50 cm high. *Corm* globose, 12–20 mm in diameter, the tunics of lightly coriaceous layers, soon decaying into softly textured, fairly fine to coarse flat vertical fibres, these usually accumulating in a dense mass. *Cataphylls* pale and membranous, the uppermost reaching 3–6 cm above the ground and then green and coriaceous, the apex often purple, densely villous, usually persisting and forming a light fibrous neck around the stem base. *Leaves* two or three, the lowermost basal and longest, sheathing the lower half of the stem, almost entirely sheathing or with a short blade up to 5(–10) cm long, the sheath pubescent throughout but more densely so below, or pubescent only below, lightly ribbed, slightly inflated and compressed, thus elliptic in transverse section, the blade narrowly lanceolate, 3–4 mm wide, lightly villous, the margins thickened and hyaline, the midrib lightly raised, the second leaf inserted on the upper third of the stem, 15–45 mm long, entirely sheathing, glabrous, the margins free to the base and sometimes overlapping, a third leaf if present like the second but shorter; leaves of non-flowering plants evidently solitary, linear and villous. *Stem* erect below, flexed outward above the upper leaf and then inclined 30°–40°, unbranched, 0.8–1.3 mm in diameter below the spike.

Spike deflexed at the base, lightly to strongly inclined, weakly flexuose to nearly straight, 4- to 9-flowered, the internodes 7–14(–20) mm long, occasionally as long as the bracts, usually almost twice as long; *bracts* pale grey-green or flushed with grey-purple or pink on the dorsal surfaces, the outer (12–)15–18 mm long, the inner two-thirds to three-quarters as long, minutely notched apically. *Flowers* pale lemon-yellow

tinged with green, or bluish mauve, sometimes the tepals lightly flushed with reddish
brown, horizontal to drooping, the tepals
unmarked or the lower median with a dark
streak in the lower midline, often strongly
scented of tuberoses in the mornings;
perianth tube narrowly trumpet-shaped,
widening evenly from the base, 7–8 mm
long, emerging from between the bracts;
tepals nearly equal, narrowly lanceolate-
attenuate, the upper three slightly larger
than the lower and held slightly apart from
them, all directed forward and only lightly
spreading, the upper three 17–23 x 6–
10 mm, the lower three joined to the upper
laterals for 1–2 mm, 17–22 x 5–9 mm.
Filaments 5–10 mm long, exserted 1–5 mm
from the tube; *anthers* 6–7 mm long, dull
yellow or purple, the pollen cream. *Ovary*
obovoid, 2.5–3 mm long; *style* arching over
the stamens, dividing between the upper
third and apex of the anthers, the branches
recurving, 3–4.5 mm long, slender throughout or widening slightly in the upper half.
Capsules globose, lightly three-lobed above,
8–10 mm long; *seeds* oblong, 5–6 x 3–
3.5 mm, the wing evenly developed, opaque
rusty coloured. *Chromosome number*
unknown.

Flowering time September to November,
occasionally as late as December at high
elevations or in late seasons.

Distribution and biology

Fairly widespread in eastern southern Africa,
Gladiolus pubigerus extends from the Katberg
and Hogsback Mountains of Eastern Cape
Province in the south to near Pilgrim's Rest
in Mpumalanga. Although not well collected, it appears to be particularly common in
the KwaZulu-Natal Midlands, but fairly rare
elsewhere. We have seen only two records
from the Eastern Cape and there are none at
all from the Transkei, which lies between
Hogsback and southern KwaZulu-Natal.
There is a second disjunction between the
KwaZulu-Natal populations and Swaziland.
Both breaks in the range may simply reflect
poor collecting. The common form of the
species, with yellow-green flowers, is inconspicuous in fresh green grassland even in full
flower. The Swaziland and some Mpuma-

langa populations have slightly larger, pale
blue-mauve flowers. These plants appear to
represent a distinct race of the species.

The pollination biology of *Gladiolus
pubigerus* has not been studied, but we
assume that the flowers are adapted for pollination by medium-sized long-tongued bees.

Diagnosis and relationships

The vegetative appearance of *Gladiolus
pubigerus* is so distinctive that it can usually
be recognized in the absence of flowers.
Plants have only two, or rarely three, leaves
and the lower one is basal, sheathing the
lower half of the flowering stem. The lower
leaf is pubescent, usually has a short blade
up to 5 cm long, rarely to 10 cm, and the
remaining one or two leaves are much
shorter and entirely sheathing and glabrous.
The flowers have subequal tepals, the upper
and lower of which are similarly oriented
and lanceolate to elliptic, and the short
stamens are held low in the flowers. The
bilabiate flowers of *G. woodii*, with which
G. pubigerus has frequently been confused,
have larger, ovate upper tepals and small,
strongly clawed lower tepals, the limbs
auriculate near the base. Flower colour in
the two species also differs. In *G. woodii*
the flowers range from maroon to yellow
or occasionally lilac, whereas those of
G. pubigerus are often pale green to cream,
or occasionally mauve. *Gladiolus pubigerus*
is closely related to the mauve- to pink-
flowered *G. parvulus* of KwaZulu-Natal
which also has subequal tepals and a virtually
actinomorphic perianth. Like *G. pubigerus*,
plants of *G. parvulus* have two leaves, the
lowermost sheathing the lower half of the
stem much as in *G. pubigerus*, but the
second leaf is scale-like and inconspicuous.
Inserted just below the base of the spike,
it may easily be overlooked or confused
with a floral bract.

History

The first record of the long-misunderstood
Gladiolus pubigerus appears to be a gathering
by C.F. Ecklon and C.L. Zeyher most likely
in October, 1832 (Gunn & Codd, 1981),
when Ecklon was in the Eastern Cape.
When identified at all, that collection was

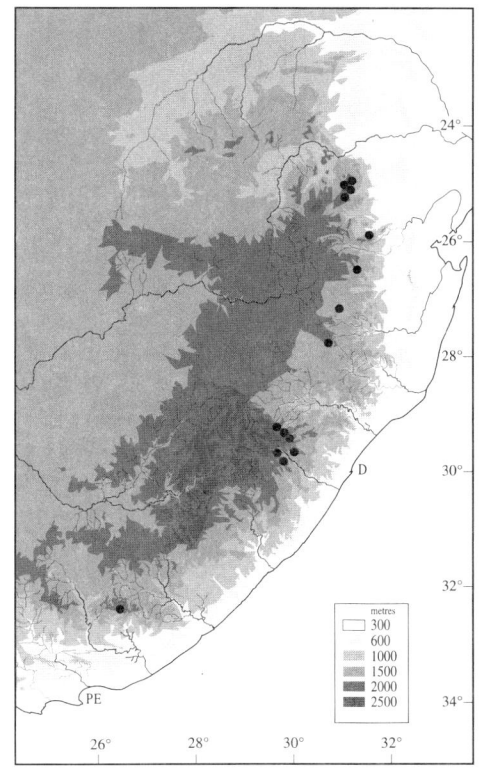

associated with the western Cape species
G. pilosus, now *G. hirsutus*, and did not
attract any botanical attention. A second
collection was made in 1860 in 'British
Kaffraria' by Thomas Cooper, the plant
collector employed by W. Wilson Saunders.
Cooper's well-preserved specimens formed
the basis for J.G. Baker's *G. pubescens*,
described in 1876. The name is unfortunately a homonym for the earlier *G. pubescens*
Lamarck, dating from 1791, now *Babiana
pubescens*. Baker, nevertheless, maintained his
G. pubescens in *Flora Capensis*, although the
species remained poorly known. Realizing
that the name was a homonym, G.J. Lewis
renamed the species *G. pubigerus* in 1954.
Although she apparently intended to recognize *G. pubigerus* in her revision of *Gladiolus*
in South Africa, in A.A. Obermeyer's final
version of the manuscript made after Lewis's
death, the species was included in *G. woodii*.
Still known from just a handful of collections, *G. pubigerus* was found again by
O.M. Hilliard and B.L. Burtt in the 1970s.
Believing it to be an undescribed species,
they named it *G. pugioniformis* in 1979. This
name now falls into synonymy.

61. GLADIOLUS PARVULUS Schlechter

PLATE 53

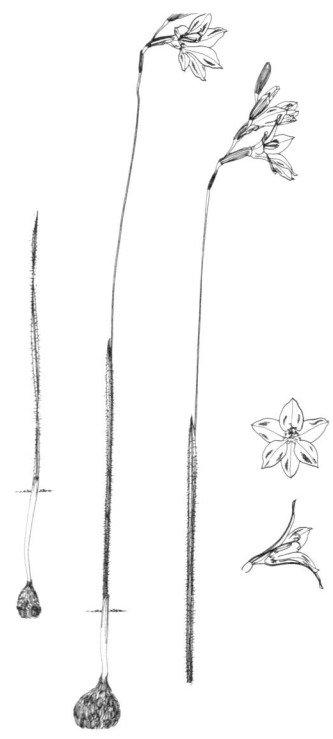

Gladiolus parvulus Schlechter, *Bot. Jahrb. Syst.* 40: 91 (1908). Type: South Africa, KwaZulu-Natal, Umzinto District, short-grass slopes along the Umtwalumi near Fairfield, c. 750 m, Oct. 1905, *Rudatis 122* (not 132 as in the protologue) (B [not seen], holotype).

parvulus = very small, aptly describing the diminutive character of both plant and flowers.

Synonymy

Geissorhiza gracilis Baker, *Handbook Irideae* 155 (1892); *Fl. Capensis* 6: 70 (1896), not *Gladiolus gracilis* Jacquin. Type: South Africa, KwaZulu-Natal, Umzimkulu District, Zuurberg, between Kokstad and Clydesdale, Oct. 1883, *Tyson 1872* (K, holotype; BOL, MO, SAM, isotypes).

Ixia brevifolia Baker, *Handbook Irideae* 165 (1892); *Fl. Capensis* 6: 84 (1896), superfluous name for *Geissorhiza gracilis* Baker (based on the same collection, although on a different sheet when described).

Gladiolus subaphyllus N.E. Brown, *Kew Bull. Misc. Inf.* 1909: 53 (1909). Type: South Africa, KwaZulu-Natal, Pietermaritzburg, Swartkop, 4000-5000 ft, 17 Nov. 1906, *Wylie sub Wood 10153* (K, holotype; BOL, NH [not seen], PRE, Z [not seen], isotypes).

Description

Plants 15–35(–47) cm high. *Corm* globose, 7–14 mm in diameter, the tunics of coria-

ceous to almost woody layers decaying with age into wiry vertical fibres usually somewhat claw-like below. *Cataphylls* pale and membranous, the uppermost not or barely reaching above the ground, sometimes accumulating with age as a finely fibrous neck around the base of the stem. *Leaves* two, the lower one basal and prominent, sheathing the lower half of the stem, almost entirely sheathing, the sheath ribbed and lightly villous, often slightly inflated, the blade vestigial, 0.5-2.5 cm long, plane, with raised margins and midrib, villous, the upper leaf minute, inserted on the stem just below the spike, entirely sheathing, sometimes reaching the base of the bracts of the first flower, (4–)5.5–7 mm long; non-flowering plants with a single linear leaf 8–10 cm long, c. 2 mm wide, sparsely villous, the midrib and margins moderately thickened and hyaline. *Stem* erect, slender and wiry for most of its length, thickened for a short distance from just below the base of the upper leaf to the base of the spike, flexed outward above the sheath of the upper sheathing leaf, unbranched, c. 0.8 mm in diameter below the spike.

Spike flexed outward at the base and inclined c. 45°, lightly flexuose, 2- to 5-flowered; *bracts* grey-purple, sometimes dry and brownish toward the apices, the outer 9–11 mm long, the inner narrower but slightly longer than the outer, minutely notched apically. *Flowers* with the tepals equal in size and disposition but the stamens unilateral and arcuate, pale pink, the throat with narrow paired red lines below the tepal bases, unscented; *perianth tube* funnel-shaped, 5–7 mm long, the flared upper part c. 3.5 mm long; *tepals* lanceolate and equal, 12–17 x 5–6.7 mm, when fully open flared evenly from the base. *Filaments* 5–6 mm long, exserted c. 2 mm from the tube; *anthers* 4–6 mm long, pale pink, the pollen creamy yellow. *Ovary* ovoid, 2.5–3 mm long; *style* arching over the stamens, dividing between the upper third and just beyond the anther apices, the branches 2–3 mm long, recurving, very gradually expanded in the upper half. *Capsules* globose to oblong, three-lobed above, the apices retuse, c. 7 mm long; *seeds* ovate, broadly winged, light brown. *Chromosome number* unknown.

Flowering time October to early December, rarely in September or as late as January.

Distribution and biology

Fairly localized in its geographic range, *Gladiolus parvulus* is restricted to the southern half of KwaZulu-Natal in eastern South Africa and to adjacent eastern Lesotho. Populations extend from the Zuurberg and Ngeli Mountains of KwaZulu-Natal close to the Eastern Cape border in the south to Pietermaritzburg and Howick in the Midlands and along the eastern edge of the Drakensberg from Sehlabathebe to the Cathedral Peak area in the north. Plants grow in thin soil on sandstone pavement or in cracks in sandstone outcrops.

The small, nearly actinomorphic and fairly short-tubed flowers of *Gladiolus parvulus* are most likely adapted for pollination by small bees. They closely resemble the flowers of species of *Dierama* which are known to be pollinated by bees.

Diagnosis and relationships

Most unlike a *Gladiolus*, the flowers of *G. parvulus* are small, horizontal to almost nodding, and the tepals are more or less equal. Despite the actinomorphic perianth, the stamens are unilateral and are held below the dorsal tepal, thus the flower itself is zygomorphic, although this is impossible to tell in dried material. The stems are slender and wiry, and the entire appearance of the plant is like that of one of the dwarf

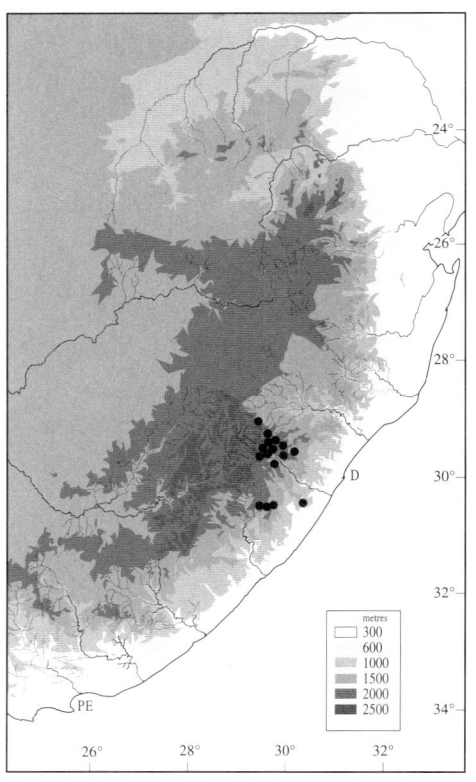

160

species of *Dierama*, for example *D. trichorhizum* or *D. pauciflorum* (Hilliard et al., 1991). Vegetatively, *G. parvulus* is also unusual. Plants have one long, almost entirely sheathing leaf and one tiny, scale-like second leaf inserted on the stem a short distance below the spike. The basal leaf is slightly inflated and the sheath is conspicuously villous. The general appearance is rather like that of *G. pubigerus*, another eastern southern African species and one closely related to *G. parvulus*. The two may be readily distinguished by the flowers which are greenish or mauve in *G. pubigerus* with attenuate tepals and somewhat larger than the pale pink flowers of *G. parvulus*. Vegetatively the two are easy to tell apart for although *G. pubigerus* has a long, basal leaf that sheaths the lower half of the stem, the second leaf and sometimes a third leaf,

are quite large and inserted in the upper third of the stem, not at all scale-like and unlikely to be overlooked.

History

The earliest records of *Gladiolus parvulus* that we have been able to trace are two collections made in October 1883 by William Tyson from the Zuurberg near Kokstad in southern KwaZulu-Natal. Earlier collectors may have missed the species or their collections are in herbaria that we have not examined. One of the Tyson collections was described as *Geissorhiza gracilis* by J.G. Baker in 1892. In the same work, *Handbook of the Irideae*, Baker described a second species, *Ixia brevifolia*, based on the same collection. Neither name can be transferred to *Gladiolus* because both epithets have already been used in the genus. A collection of the species

made in 1905 by the German missionary, A.G.H. Rudatis, near Fairfield in the Alexandria District, formed the basis for Rudolf Schlechter's *G. parvulus*, described in 1908, the name the species bears today. Baker's confusion about the correct genus for *G. parvulus* is easy to understand for without the winged seeds, which were unknown to him, the nearly regular flowers of this very odd species resemble those of both *Geissorhiza* and *Ixia*, especially when pressed. Schlechter showed remarkable insight in placing the species in *Gladiolus*. A collection of the species made by John Medley Wood in 1906 was described as *G. subaphyllus* by the British botanist, N.E. Brown, in 1909. This plant differs in no significant way from *G. parvulus* and we assume Brown simply overlooked the existence of the earlier name for the species.

Section *Linearifolius*: Series *Linearifolius*

62. GLADIOLUS HIRSUTUS Jacquin

Small pink Afrikaner, lapmuis

PLATE 54

Gladiolus hirsutus Jacquin, *Collecteana* 4: 161 (1792); *Icones Pl. Rar.* 2: pl. 250 (1795). Baker, *Fl. Capensis* 6: 153 (1896), the name misapplied to *G. caryophyllaceus*. Pole Evans, *Fl. Pl. S. Africa* 6: pl. 233 (1926). Type: South Africa, Western Cape, without precise locality, grown in Vienna, illustration in Jacquin, *Icones Pl. Rar.* 2: pl. 250 (1795).

hirsutus = hairy, referring to the pilose leaf sheaths and blades.

Synonymy

Gladiolus tristis var. *purpureus* Thunberg, *Dissertatio de Gladiolo* 166 (1784). Type: South Africa, Western Cape, without precise locality, *Thunberg s.n.* (UPS–Herb. Thunberg 1037, holotype).
Gladiolus laccatus Thunberg, *Prodromus Plantarum Capensium* 186 (1800). Type: South Africa, Western Cape, without precise locality, *Thunberg s.n.* (UPS–Herb. Thunberg 1037), illegitimate homonym, not *G. laccatus* Jacquin (1790) (= *Watsonia laccata* (Jacquin) Ker Gawler).
Gladiolus hirsutus var. *villosiusculus* Ker Gawler, *Curtis's Bot. Mag.* 18: sub pl. 727 (1804). Type: South Africa, Cape, without precise locality, grown in Great Britain, *Auge s.n.* (BM, holotype).
Gladiolus biflorus Roemer & Schultes, *Syst. Veg.* 1: 416 (1817), an illegitimate homonym, not *G. biflorus* Thunberg (1784) (= *Olsynium biflorus* (Thunberg) Goldblatt).

Gladiolus punctulatus Schrank, *Bot. Ges. Regensburg* 2: 216 (1822). Lewis et al., *J. S. African Bot.*, Suppl. 10: 207 (1972). Type: South Africa, Western Cape, without precise locality or date, *Brehm s.n.* (BRU [not seen]–Herb Schrank, holotype).
Gladiolus villosus Ker Gawler, *Genera Irid.* 133 (1827). Baker, *Fl. Capensis* 6: 149 (1896), illegitimate homonym not *G. villosus* Burman fil. (1768) (= *Sparaxis villosa* (Burman fil.) Goldblatt). Type: South Africa, Cape without precise locality, grown in Great Britain, *Masson s.n.* (BM, holotype).
Gladiolus pilosus Ecklon, *Topographisches Verzeichniss* 38 (1827). Klatt, *Linnaea* 32: 709 (1863). Type: South Africa, Western Cape, Cape Flats near Wynberg, Aug., *Ecklon s.n.* (S, holotype; SAM, isotype).
Gladiolus punctulatus var. *autumnalis* G. Lewis in *J. S. African Bot.*, Suppl. 10: 211 (1972). Type: South Africa, Western Cape, Riversdale District, south side of Garcia's Pass, 3 Apr. 1959, *Lewis 5380* (NBG, holotype; K, MO, isotypes).

Description

Plants 35–50 cm high. **Corm** more or less globose, 18–30 mm in diameter, the tunics of fairly coarse vertical fibres, often accumulating in a dense mass. **Cataphylls** pale and membranous, the uppermost dark green or purple above the ground, usually pubescent, often dry at flowering time, sometimes accumulating with age as a coarse fibrous neck

around the base of the stem. **Leaves** three, occasionally four, the lower two basal, the second of these sheathing the lower half of the stem, pubescent throughout or only on the sheaths, the blades relatively short, the

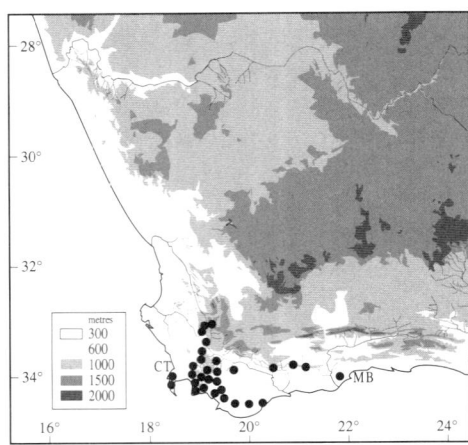

lowermost reaching between the middle and upper third of the stem, occasionally reaching the top of the spike, the second reaching between the upper third of the stem and the base of the spike, the blades plane, 3–8(–13) mm wide, the midrib and margins lightly thickened, the secondary veins also evident, the uppermost leaf inserted on the upper third of the stem, short, sheathing for half its length, pubescent or glabrous, the margins united around the stem. *Stem* erect, flexed outward above the sheath of the second leaf and inclined c. 30°, unbranched, 1–1.5 mm in diameter below the spike.

Spike inclined, lightly flexuose, 3- to 6-occasionally to 8-flowered; *bracts* green or flushed greyish or purple above, the outer 18–26 mm long, the inner two-thirds to about as long as the outer, acute. *Flowers* shades of pale to deep pink or mauve, occasionally almost white, the lower three tepals cream to pale yellow below with nectar guides of irregular parallel spotted or streaked lines in the lower two-thirds, the lines decurrent on the tube, occasionally lightly scented; *perianth tube* obliquely funnel-shaped, 15–26(–40) mm long, the cylindrical lower part enclosed in the tube; *tepals* unequal, obovate to broadly lanceolate, the dorsal largest, arching over the stamens, 19–27(–35) x 13–20 mm, curving upward in the upper quarter, the upper laterals directed forward, curving outward in the upper quarter, 17–23(–34) x 10–15 mm, the lower three tepals joined to the upper laterals for 2–5 mm and to one another for up to 3 mm, narrowly lanceolate, not clearly clawed below, subequal, 15–20(–23) x 7–12 mm, in profile the upper and lower about as long. *Filaments* 11–13 mm long, exserted 4–6 mm from the tube; *anthers* 6–9 mm long, pink, the pollen white. *Ovary* oblong, c. 5 mm long; *style* arching over the stamens, dividing opposite the anther apices, the branches 2–4 mm long. *Capsules* ovoid to more or less ellipsoid,

20–24 mm long, about as long as the bracts; *seeds* broadly ovate, 6–7.5 x c. 5 mm, the wing evenly developed or lacking on one side, dull reddish brown, often opaque, the seed body unusually large. *Chromosome number* 2n = 30.

Flowering time mostly mid-July to September, sometimes as early as June and rarely in March or April.

Distribution and biology

Gladiolus hirsutus is a relatively common and widespread species of Western Cape Province, South Africa. It is frequently seen on stony granite or sandstone mountain slopes in the first year or two following a fire and only occasionally after that. Populations extend from the Cape Peninsula and the Cold Bokkeveld in the west to the Langeberg and its foothills near Robinson's Pass in the east. It is typically an early-flowering species and may be encountered from June to August, or sometimes into September at higher elevations. In the eastern part of its range where summer and autumn rains are frequent, populations of *G. hirsutus* may flower as early as March and April, but this is very unusual. Plants in these populations appear to differ hardly at all from those that flower later in the year and var. *autumnalis*, recognized by Lewis et al. (1972) for early-flowering populations, does not appear to warrant formal taxonomic recognition.

The short-tubed and relatively unspecialized flowers of *Gladiolus hirsutus* are pollinated by long-tongued bees, of which the honey bee, *Apis mellifera*, is the most common visitor early in the season. Plants that flower in August and September are also visited by species of *Anthophora* (Anthophoridae). These bees appear to be foraging for nectar and make only occasional visits to the flowers, interrupting their pollen gathering on other plants. Monkey beetles (Scarabaeidae: Hopliinae) also visit the flowers and may sometimes accomplish pollination. Their visits, however, seem to be opportunistic and the flowers are not in any way adapted for pollination by these insects.

Diagnosis and relationships

A relatively unspecialized species, *Gladiolus hirsutus* has small to moderate-sized flowers of the type found in many species of the genus. Its identification depends to a large extent on vegetative characters. Particularly distinctive are the hairy, superposed leaves with relatively short, narrow blades, rarely reaching beyond the middle of the spike and

3–8 mm, occasionally to 13 mm wide, and the corm tunics of fairly coarse, mostly vertical fibres. The flowers are shades of pale pink to mauve with the lower tepals irregularly streaked and spotted with dark colour over a zone of cream to yellow. While the flowers are typically 32–40 mm long and have a perianth tube 15–20 mm long, barely exceeding the bracts, plants of some populations have larger flowers. The largest flowers in the species, recorded from the Kogelberg, are 75 mm long and have a tube 40 mm long. The form of the flower, the hairy leaves, and the fibrous corm tunics of *G. hirsutus* are typical of section *Linearifolius* and the ovoid-ellipsoid capsules are characteristic of the Cape species of the section, referred here to series *Linearifolius*.

Although there is normally no difficulty in recognizing *Gladiolus hirsutus*, it is sometimes confused with its relative, *G. caryophyllaceus*. The latter has consistently larger flowers, 66–90 mm long with the tube 35–40 mm long, and a strong sweet fragrance, quite unlike the faintly scented to completely unscented *G. hirsutus*. The flowers of the two may sometimes be confused in the absence of scent because of the overlap in flower size, but their leaves differ substantially. Unlike the narrow leaves of *G. hirsutus*, those of *G. caryophyllaceus* are fairly short and broad, 10–27 mm wide, are often arranged in a basal fan, and the blades have the margins, midrib and secondary veins raised, thickened and often flushed pink or purple. Lack of appreciation of the differences in the leaves of the two species led to a history of confusion over the identity of *G. hirsutus* which, because of the large flowers on the type painting, has at times been considered conspecific with *G. caryophyllaceus*.

History

Despite being one of the more common western Cape species of the genus, *Gladiolus hirsutus* was not recognized at species rank until 1792 when it was described by Nicholas Jacquin. The painting associated with the protologue, published only in 1795 and now the type of the species, is of a rather large-flowered plant with superposed leaves that have short, narrow, densely hairy blades. The large flowers resemble somewhat those of *G. caryophyllaceus* and this led to a long history of confusion about the identity of *G. hirsutus* and the redescription of the species by several authors who misunderstood Jacquin's plant. Thus in rapid succession after 1792 the following species were described: *G. laccatus* Thunberg, in 1800 and *G. biflorus* Roemer & Schultes, in 1817, both incidentally homonyms; *G. punctulatus* Schrank, in 1822; and *G. pilosus* Ecklon and *G. villosus* Ker Gawler in 1827. The reason for this proliferation of names is not entirely the result of confusion over the identity of *G. hirsutus*, for the types of all the later synonyms are very similar to one another and none could reasonably be regarded as parti-

cularly distinctive. Confusion between *G. hirsutus* and *G. caryphyllaceus* persisted, and when a fair painting of *G. caryophyllaceus* was published by Andrews in 1896 under the new name, *G. roseus*, some later authorities, including J.G. Baker, regarded the two as conspecific. Baker thus called the larger-flowered *G. caryophyllaceus*, *G. hirsutus*. At the same time Baker used the name *G. villosus* for true *G. hirsutus*,

ignoring the fact that it was a homonym. G.J. Lewis realized that the use of the name *G. villosus* was inadmissable and in her revision of the South African species (Lewis et al., 1972) she used the next available legitimate synonym, *G. punctulatus* for the species. She, too, regarded *G. hirsutus* as a synonym of *G. caryophyllaceus*, but we are confident that we have correctly interpreted Jacquin's painting.

63. GLADIOLUS CARYOPHYLLACEUS (Burman fil.) Poiret

Sandveldlelie, large pink Afrikaner

PLATE 55

Gladiolus caryophyllaceus (Burman fil.) Poiret in Lamarck, *Encycl. Méth.*, Suppl. 2: 795 (1811). Lewis et al., *J. S. African Bot.*, Suppl. 10: 114 (1972). *Antholyza caryophyllacea* Burman fil., *Prodr. Fl. Capensis* 1 (1768). *A. caryophyllea* Panzer, *Pflanzensyst.* 11: 76 (1784), orthographic variant. *Homoglossum caryphyllaceum* (Burman fil.) N.E. Brown, *Trans. Roy. Soc. S. Africa* 20: 227 (1932). Type: South Africa, Cape, without precise locality, date, or collector, grown in Holland by M. Houttuyn and sent to Burman fil., (G – Herb. Burman, lectotype designated by Lewis et al., 1972; K, photo).

caryophyllaceus = referring to the flower colour of the common pink *Dianthus caryophyllaceus* which incidentally has a similar scent.

Synonymy

Gladiolus roseus Andrews, *Bot. Rep.* 1: pl. 11 (1796). *G. hirsutus* var. *roseus* Ker Gawler, *Curtis's Bot. Mag.* 16: pl. 574 (1802). Type: illustration in Andrews, *Bot. Rep.* 1: pl. 11 (1796).

Gladiolus similis Ecklon, *Topographisches Verzeichniss* 40 (1827). Type: South Africa, Western Cape, without precise locality, cultivated in Cape Town, *Ecklon s.n.* (S, lectotype designated by Nordenstam in *J. S. African Bot.* 38: 292, 1972).

[*Gladiolus hirsutus* sensu Baker, *Fl. Capensis* 6: 153 (1896).]

Description

Plants (45–)60–110 cm high. *Corm* globose, 20–40 mm in diameter, the tunics of coarse vertical fibres. *Cataphylls* membranous, the uppermost pale grey-green to purple above the ground and often more or less dry, usually pilose. *Leaves* four or five, the lower one to three basal, usually quite short and sometimes forming a loose basal fan, pilose, particularly on the sheaths, the blades

sword-shaped, 10–27 mm wide, the margins, midrib and secondary veins raised and thickened, often flushed pink or purple, the margins sometimes somewhat crisped, the upper two leaves cauline, similar to the basal but shorter. *Stem* erect, unbranched, 1.5–4 mm in diameter below the spike.

Spike erect, occasionally 2-, usually 4- to 8-flowered; *bracts* pale green or greyish on the upper surface, the outer 30–45(–70) mm long, the inner slightly shorter, twisted to lie within the outer, acute. *Flowers* pale to deep pink, or mauve or cream and then more intensely pigmented along the midlines, the lower tepals and throat streaked with broken longitudinal lines and spots of white to cream and dark pink, usually sweetly carnation-scented; *perianth tube* obliquely funnel-shaped, 30–40 mm long, the lower cylindrical part 16–25 mm long; *tepals* subequal or unequal, lanceolate to elliptic or ovate, the dorsal largest, ascending or weakly hooded over the stamens, broadly oblanceolate, 37–45 x (13–)21–30 mm, the upper laterals 30–35 x 18–25 mm, spreading gradually from the base, the apices flared outward, the margins undulate, the lower three tepals joined to the upper laterals for 7–8 mm and to one another for 2–3 mm, 25–30 x 15–18 mm, more or less horizontal, not distinctly clawed, wider in the distal half, curving gradually downward, the margins undulate. *Filaments* 22–27 mm long, exserted 8–9 mm from the tube; *anthers* 10–14 mm long, cream, the pollen yellow. *Ovary* cylindric, c. 8 mm long; *style* arching over the filaments, dividing slightly below to slightly beyond the anther apices, the branches 4–5 mm long, extending beyond the anthers and not normally in contact with them. *Capsules* unknown; *seeds* ovate, c. 10 x 7–8 mm, the wing broadly and evenly developed, more or less translucent to opaque reddish brown. *Chromosome number* 2*n* = 30.

Flowering time mid-August to the end of September.

Distribution and biology

Gladiolus caryophyllaceus is best known from the hills and valleys of the interior west coast

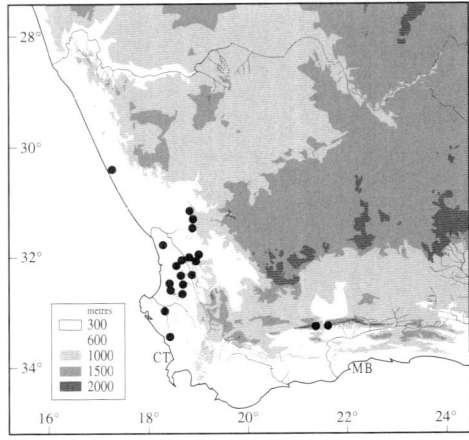

of Western Cape Province and is perhaps most common on the slopes of the Piketberg and Olifants River Mountains. It has a far wider range, however, and extends from the rocky granite ridges inland of the southern Namaqualand coast and the sandstone plateau of the Bokkeveld Escarpment west of Nieuwoudtville southward to Hopefield and Groenekloof, now Mamre, only a short distance north of Cape Town. It also occurs inland in the Cedarberg range and in the Hex River Mountains, and there is an isolated series of populations in the Swartberg well to the east.

The species favours fairly dry habitats, and is found mainly on open stony sandstone slopes and plateaus or in deep, well-drained sandy soils. Plants typically grow in dense clumps of restios, especially of *Willdenowia*, which may afford the corms some protection from predation or simply provide them with slightly moister and cooler growing conditions.

Little is known about the pollination ecology of *Gladiolus caryophyllaceus*. Despite the long perianth tube and sweet, clove-like fragrance, preliminary observations suggest that it is not visited by moths, but may instead be pollinated by long-tongued bees.

This species is thought to be under threat from farming activity in the western part of its range, where large areas of its former habitat have been ploughed, particularly for the cultivation of rooibos tea or potatoes. Nevertheless, the fact that large numbers of plants are routinely collected for local flower shows – especially the Clanwilliam Show – seems to contradict the assumption that they are rare. Although a desirable species for horticulture, both for its large, attractive flowers and for its fragrance, there has been little success in cultivating it, even for garden display. In Western Australia, however, it has become

a weed and grows extremely well in suitable sandy habitats there.

Diagnosis and relationships

Gladiolus caryophyllaceus is readily distinguished by its relatively large flowers, c. 65–80 mm long, with a narrowly funnel-shaped perianth tube, 30–40 mm long, which is slightly shorter than or more or less as long as the dorsal tepal. The leaves are hairy and may form a basal fan, although they are as often superposed. The leaf margins are typically fairly strongly thickened. Stems in well-grown plants from the western Cape lowlands may be 80–110 cm tall. Few other hairy-leaved species of the winter-rainfall area have flowers of this size, and those that do have an elongate perianth tube and substantially shorter tepals.

The flowers of *Gladiolus caryophyllaceus* are usually deep pink and have a strong carnation-like scent. On the rocky Bokkeveld Escarpment, however, the flowers are deep mauve rather than pink, with the tepals darker along the midlines. Plants from this area are usually fairly short, the stems seldom exceeding 60 cm, and the flowers have a weaker scent. Isolated populations in the Swartberg to the east have light creamy mauve flowers, also with the tepal midlines more darkly coloured. *Gladiolus lewisiae*, described by A.A. Obermeyer in 1970 from a single population in Seweweekspoort in the Swartberg, is now understood to be a natural hybrid between this form of *G. caryophyllaceus* and *G. tristis*, and the species is no longer recognized.

The relationships of *Gladiolus caryophyllaceus* lie with the western Cape species of section *Linearifolius* and probably most closely with a second long-tubed species, *G. guthriei*. That species has a smaller flower with a tube 20–27 mm long, the upper tepals 20–30 mm long, and the leaves of the flowering stem have short blades; the blade of the lower leaf is at most only 80 mm long and always vestigial on the upper leaves.

Gladiolus caryophyllaceus is sometimes confused with another species, *G. hirsutus*, which consistently has superposed leaves with linear, hairy blades rarely wider than 3 mm, and without thickened margins. The flowers of *G. hirsutus* are also smaller than those of *G. caryophyllaceus*, usually 33–48 mm long with a tube usually 15–20 mm long. Occasional populations of *G. hirsutus* have flowers which are substantially larger (see that species, page 161). In

such instances plants can best be distinguished from *G. caryophyllaceus* by their leaves which do not have strongly thickened margins and are narrow and superposed.

History

The first documented record of *Gladiolus caryophyllaceus* appears in the report of Simon van der Stel's 1685 journey to Namaqualand (De Wet & Pheiffer, 1979). On folio 79 of this unpublished document there is a fair watercolour painting, under the name *Aquilegia* or Akoleije (the columbine), of a plant which had been collected southeast of present-day Leipoldtville. The accompanying information is interesting: the plant is described as having a purple flower and an edible root ('met een purpere bloem en eetbare wortel'). The illustration is the basis for Leonard Plukenet's *Gladiolus africanus angustissimo folio, dilute purpurascens* (1691).

Gladiolus caryophyllaceus was also grown in Holland in the mid-eighteenth century and a brief description of the species was given by N.L. Burman in 1768 under the name *Antholyza caryophyllacea*, the identity of which puzzled later botanists. Plants were raised in Britain at the end of the eighteenth century, and in 1796 Henry Andrews provided a description of the species, together with a watercolour reproduction under the name *G. roseus*. The expert, John Ker Gawler, saw the close relationship of Andrews' plant to *G. hirsutus* (illustrated in Jacquin's *Icones Plantarum Rariorum* in 1796) and treated *G. roseus* as a variety of that species. Later, in 1827, he simply united them, thus including *G. caryophyllaceus* in *G. hirsutus*. In 1811 J. Poiret, the French botanist who compiled some parts of J.P. Lamarck's *Encyclopédie Méthodique*, concluded that *Antholyza caryophyllacea* was a species of *Gladiolus* and he provided the combination *G. caryophyllaceus*. J.G. Baker included specimens of *G. caryophyllaceus* under the name *G. hirsutus* in *Flora Capensis*, and plants that we believe correspond to true *G. hirsutus* he called *G. villosus*.

The identity of Burman's *Antholyza caryophyllacea* was confirmed in 1928 by N.E. Brown after he had examined the Iridaceae of N.L. Burman's herbarium in Geneva. Unfortunately, he completely misunderstood the relationships of the species, and in 1932 he transferred it to *Homoglossum*, a genus defined at the time by reddish, unscented and tubular flowers, quite unlike the pink and fragrant flowers of *G. caryophyllaceus*.

64. GLADIOLUS GUTHRIEI F. Bolus

PLATE 56

Gladiolus guthriei F. Bolus, *Ann. Bolus Herb.* 2: 101 (1917). Lewis et al., *J. S. African Bot.*, Suppl. 10: 254 (1972). Type: South Africa, Western Cape, Bredasdorp District, Elim, July 1895, *Guthrie 3821* (BOL, holotype).

guthriei, named in honour of Francis Guthrie (1831–1899), British mathematician and botanist who later worked with Harry Bolus in Cape Town and made many important plant collections, especially in the southern Cape.

Synonymy

Gladiolus eulophioides F. Bolus, *Ann. Bolus Herb.* 2: 99 (1917). Type: South Africa, Western Cape, Caledon [District], Oct., *Pappe 461* (GRA, holotype).

Gladiolus odoratus L. Bolus, *S. African Gard.* 17: 293 (1927). Bullock, *Curtis's Bot. Mag.* 170: new series pl. 223 (1954). Lewis et al., *J. S. African Bot.*, Suppl. 10: 270 (1972). Type: South Africa, Northern Cape, Calvinia District, Grasberg near Nieuwoudtville, grown at Kirstenbosch, 27 Apr. 1927, *Buhr s.n.* (National Botanic Gardens 460/26 in NBG, holotype).

Description

Plants 40–70 cm high. *Corm* globose, 20–35 mm in diameter, the tunics of coarse vertical fibres, accumulating in a dense mass. *Cataphylls* of the leafy shoot usually dry and brown by August, of the flowering shoot pale and membranous, the uppermost extending up to 3 cm above the ground and then usually brownish purple, sometimes finely pubescent, accumulating with the corm tunics in a coarse fibrous neck around the underground part of the stem. *Leaves* usually three, sometimes four, glabrous or hairy on the sheaths, the lower two more or less basal, the lowermost with a short, rigid, lanceolate blade 20–80 mm long, 3–6 mm wide, the margins moderately and the midrib and often a second pair of veins lightly thickened, blades of the remaining leaves vestigial or lacking; non-flowering plants with a solitary leaf, sheathed by a dry, purplish cataphyll, sword-shaped to lanceolate, 12–20 mm wide, often trailing, glabrous or pubescent, the margins strongly thickened, the midrib and two or more other vein pairs also prominent, seedling leaves always pubescent. *Stem* erect or inclined, unbranched, 1.5–2.5 mm in diameter below the spike.

Spike usually lightly inclined, 3- to 9-, occasionally to 13-flowered; *bracts* dull green to greyish, soft-textured, the outer 20–40 mm long, the inner about three-quarters as long. *Flowers* shades of dull pink to brownish purple, the upper laterals and lower three tepals dull yellow with a broad, irregular, brownish median streak and lightly speckled with brown, the margins lightly crisped, with a strong, sweet, clove-like scent by day and night; *perianth tube* obliquely funnel-shaped, 20–27 mm long, the lower cylindrical part 12–15 mm long, enclosed or exserted from the bracts, abruptly curved above into the flared part; *tepals* unequal, oblanceolate, the dorsal broadly so, 22–30 x 14–19 mm, inclined over the stamens, the upper laterals 20–30 x 10–13 mm, directed forward, curving outward in the upper quarter, the lower three tepals joined to the upper laterals for 3–5 mm and to one another for 1–2 mm, the free parts 18–25 x 8–11 mm, the lower median as long or slightly longer, directed forward below, curving downwards in the upper half. *Filaments* 12–16 mm long, exserted 5–7 mm from the tube; *anthers* 8–11 mm long, brownish, the pollen yellow. *Ovary* ovoid, c. 3 mm long; *style* arching over the stamens, dividing opposite the upper half of the anthers, the branches 3–4 mm long. *Capsules* ellipsoid, c. 20 mm long; *seeds* broadly ovate, (5–)7–8 x 4–6 mm, the wing broad, evenly developed, more or less opaque dark red-brown. *Chromosome number* $2n = 30$.

Flowering time April to June.

Distribution and biology

Although found widely across the southern African winter-rainfall zone, *Gladiolus guthriei* was described only in 1917, relatively late for a common Cape *Gladiolus*. This is largely because it flowers out of season in late autumn and early winter, a time when little collecting was done before the 1920s. The species extends from the Bokkeveld Escarpment in the north through Western Cape Province to the western end of the Langeberg and Potberg in the east. Surprisingly, it does not occur on the Cape Peninsula. It is actually not uncommon and the characteristic long, broad leaves of sterile plants are often seen growing in rocky crevices in sandstone pavement and rock outcrops in the spring, long after flowering.

The clove-scented flowers with a moderately long perianth tube are adapted for pollination by small, night-flying moths. We have captured the noctuid moth *Cuculia extricata* on the species at Bain's Kloof.

Diagnosis and relationships

Gladiolus guthriei is most easily recognized by its moderate-sized, sweetly scented flowers with a perianth tube 20–27 mm long, leaves of the flowering stem superposed, and the lowermost leaf with a fairly short blade up to 80 mm long. The remaining leaves usually have vestigial blades. The leaves are usually lightly pilose on the sheaths but are sometimes glabrous. Produced in the late autumn and winter, the flowers are usually darkly coloured, either dull pink to brownish or reddish purple, the lower three tepals dull yellow with a broad, irregular, brownish

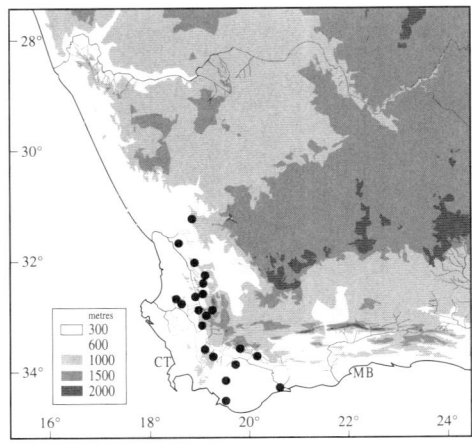

median streak and lightly speckled with brown. Non-flowering plants are readily recognized in the field in the spring and early summer by their long, trailing leaves. These leaves are usually leathery, 12–20 mm wide, and have the margins strongly thickened and hyaline.

Gladiolus odoratus, described by H.M.L. Bolus in 1927 and based on plants from the Bokkeveld Escarpment near Nieuwoudtville in Northern Cape Province, has flowers that closely resemble those of typical *G. guthriei* from the southern Cape in shape, size and colouring. The type of *G. odoratus* and other

plants from the western interior of Western Cape Province and adjacent Northern Cape are either glabrous or have at most the lowermost leaf shortly hairy, whereas the southern populations of *G. guthriei* have the leaves lightly pilose. Seedling plants from both southern and western populations have hairy leaves, and other differences between the adult plants are at best trivial. We have no hesitation in uniting the two species.

History

Until the beginning of the 1930s, *Gladiolus guthriei* was very poorly known, largely because little plant collecting was undertaken in the winter months in the southwestern and western Cape before this time. The first record of the species is an undated one, from the 'Caledon District,' which was ostensibly collected by the Cape Colonial Botanist, Ludwig Pappe. Because Pappe collected few plants east of the Hottentots Holland Mountains, it seems likely that C.L. Zeyher was, in fact, responsible for the gathering. Zeyher's herbarium was acquired by Pappe, who often copied the label information for Zeyher's specimens without indicating the true collector. If this is the case, the collection would have been made before 1850,

when Zeyher ceased to collect actively. The fragmentary specimen is in the Albany Museum Herbarium in Grahamstown, where Frank Bolus worked for a time. In 1917 Bolus described *G. eulophioides* based solely on this specimen, and because of its poor condition its identity remained uncertain (Lewis et al., 1972). At the same time Bolus described *G. guthriei* based on another rather poor specimen collected in 1895 by Francis Guthrie near Elim in the southern Cape. The two specimens represent the same species and we here formally place *G. eulophioides* in synonymy.

The northern and glabrous form of *Gladiolus guthriei* was evidently first collected by Rudolf Marloth in 1916 near Wuppertal. The fragmentary specimen in the Bolus Herbarium was misidentified as *G. hyalinus*. A later collection from the Bokkeveld Escarpment made by Hermann Buhr in 1926 and grown on at Kirstenbosch Gardens caught the attention of H.M.L. Bolus and she described *G. odoratus* in 1927 based on this specimen. *Gladiolus odoratus* was maintained as a separate species by Lewis et al., but is here regarded as conspecific with *G. guthriei*.

65. GLADIOLUS EMILIAE L. Bolus
PLATE 57

Gladiolus emiliae L. Bolus, *J. Bot.* 71: 124 (1933). Lewis et al., *J. S. African Bot.*, Suppl. 10: 269 (1972). Type: South Africa, Western Cape, flats near Riversdale, rocky outcrop, 5 Apr. 1931, *Ferguson s.n.* (BOL 19608, holotype).

emiliae, named in honour of Emily Ferguson, an active plant collector in the Riversdale and Swellendam Districts in the 1920s and 1930s.

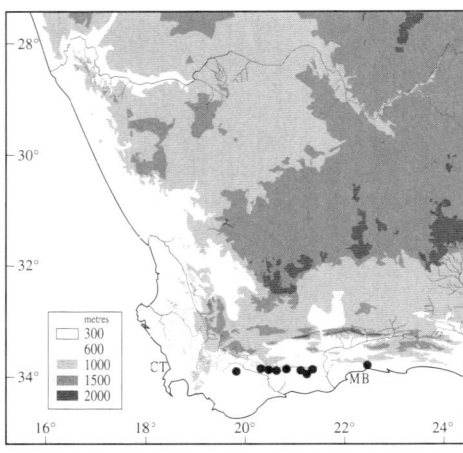

Description

Plants (30–)45–60 cm high. **Corm** globose to depressed-globose, 20–25 mm in diameter, the tunics of unbroken papery layers, or breaking irregularly and becoming fibrous, especially below. **Cataphylls** pale and firm-textured, the uppermost reaching 4–9 cm above the ground and then green, sparsely hairy near the base or glabrous, often the apex becoming dry and brown. **Leaves** two, both entirely sheathing, the lower inserted shortly above the ground, 60–80 mm long, the upper inserted on the upper third of the stem, at least in the lower half the margins united around the stem; leaves of non-flowering plants and seedlings solitary, sheathed by a pubescent, purple cataphyll, leaf blade linear or narrowly lanceolate, ultimately 60–75 cm long, to 7 mm wide, pubescent, the margins and midrib lightly thickened, emerging soon after flowering and growing during the winter and early spring. **Stem** lightly flexed outward above the sheath of the uppermost leaf and gently inclined above, unbranched, 1.8–2.2 mm in diameter below the spike.

Spike inclined, 3- to 8-, occasionally to 10-flowered; *bracts* green, sometimes red to purple on the upper margins, clasping the stem below, lightly diverging above, the outer 20–25 mm long, and about as long as an internode, obtuse to more or less truncate, the inner 1–2 mm longer or shorter than the outer, usually notched apically for 0.5–1 mm. *Flowers* brownish or pale to dull yellow and lightly to heavily dotted or streaked with reddish brown to maroon, or the tepals almost uniformly dark-coloured, usually more intensely pigmented along the tepal midlines, the lower three tepals each with a dark red streak in the lower midline, spicy- to sweet-scented by day and night; *perianth tube* 32–45 mm long, the slender lower part 25–35 mm long, sharply curved at the apex of the cylindrical part into a flared upper part 7–10 mm long; *tepals* ovate-lanceolate, directed forward below, patent above, the dorsal slightly larger than the others, the upper three tepals 14–18 x 8–10 mm, the lower three joined to the upper laterals for 2–4 mm and to one another for 1–2 mm, 14–20 x 5–8 mm, in profile the upper and lower tepals reaching to about the same point. *Filaments* 9–12 mm long, exserted 3–5 mm from the tube; *anthers* 5–7 mm long, brownish dorsally, dark yellow ventrally, the pollen cream. *Ovary* oblong, 5–6 mm long; *style* dividing between the middle and apices of the anthers, the branches c. 3 mm long, shortly to much exceeding the anther apices, the lower stigmatic surfaces occasionally in contact with pollen. *Capsules* obovoid to ellipsoid, triangular in transverse section, 20–27 x c. 10 mm; *seeds* ovate to oblong, 8–9 x 5–6 mm, the wing broadly and evenly developed, light translucent yellow-brown. *Chromosome number* unknown.

Flowering time March and April, rarely as early as mid-February.

Distribution and biology

An uncommon species, *Gladiolus emiliae* is endemic to the southern Cape coastal plain and foothills of the Langeberg and Riviersonderend Mountains of Western Cape Province, South Africa. It has been recorded over a relatively long distance of 250 km, from the foot of the Riviersonderend Mountains in the west to near George in the east. Plants grow in rocky sites, either on clay soils or in loamy sand along the interface between shale basement and sandstone rocks of the Cape System. They are mostly found in low fynbos or in mixed fynbos–renosterveld associations.

The relatively long-tubed flowers have a rich sweet scent, both day and night, and the perianth is lightly to densely speckled with brown on a pale background. The scent and colour patterning are typical of the flowers of *Gladiolus* species pollinated by moths. We have recorded nocturnal visits to the flowers of *G. emiliae* by a variety of moths, including the noctuids *Cuculia extricata* and *C. inaequalis* which have tongues about 30 mm long, a sphinx moth and

several other shorter-tongued species of other families.

Diagnosis and relationships

Unmistakable when seen alive, *Gladiolus emiliae* has dull-coloured flowers, 46–63 mm long, with an elongate perianth tube 32–45 mm long. Plants flower in the autumn, mainly in March and April, and the flowering stem bears two overlapping, entirely sheathing leaves. The markings on the flowers consist of small, brownish speckles which may be lightly dispersed, leaving the pale ground colour obvious, or the speckling may be dense, giving the flower an overall browish colour. The flowers are very strongly scented by both day and night. The species is evidently closely allied to the shorter-tubed but also moth-pollinated *G. guthriei*. Although the leaves of flowering plants of *G. emiliae* are glabrous – unusually for members of section *Linearifolius*, nearly all of which have pubescent leaves, at least on the sheaths – the leaves of non-flowering individuals are pubescent, betraying the affinity of the species to section *Linearifolius*.

History

Gladiolus emiliae was described by Cape Town botanist, H.M.L. Bolus, in 1933 based on plants collected two years earlier by Emily Ferguson of Riversdale. An active collector and correspondent of Louisa Bolus, Ferguson made many important collections from the southern Cape in the 1930s when the flora of the area was still little known. This was not, however, the first record of the species. It had already been collected by the Riversdale physician, John Muir, in 1917, but the poorly preserved specimen available to Louisa Bolus at the Bolus Herbarium had been misidentified as *G. eulophioides* (a synonym of the closely related *G. guthriei*). An even earlier collection of the species, now in the Geneva Herbarium, was made by the

British botanist, John Roxburgh, who collected at the Cape en route to India at the end of the eighteenth and the beginning of the nineteenth centuries. *Gladiolus emiliae* has remained poorly collected and as recently as 1982 was first recorded near George in the east of its range. The several populations in the Langeberg foothills west of Tradouw Pass and near Riviersonderend were only discovered as a result our studies of *Gladiolus* for this monograph.

66. GLADIOLUS OVERBERGENSIS Goldblatt & De Vos

Gladiolus overbergensis Goldblatt & De Vos, *Bull. Mus. Hist. Nat., Paris*, Sér. 4, Sect. B, *Adansonia* 11: 421 (1989), as a new name for *Antholyza guthriei* L. Bolus, *Ann. Bolus Herb.* 3: 12 (1920). *Petamenes guthriei* (L. Bolus) N.E. Brown, *Trans. Roy. Soc. S. Africa* 20: 276 (1932). *Homoglossum guthriei* (L. Bolus) L. Bolus, *S. African Gard.* 23: 47 (1933). De Vos, *J. S. African Bot.* 42: 336 (1976). Type: South Africa, Western Cape, Elim, 250 ft, July 1895, *Guthrie 3827*

(BOL, lectotype designated by De Vos, 1976; K, isolectotype).

overbergensis = from the Overberg, the southern Cape east of the Hottentots Holland Mountains and south of the Breede River.

Description

Plants 35–50 cm high. *Corm* globose, 15–20 mm in diameter, the tunics of coriaceous to woody layers, fragmenting with age

into coarse woody vertical fibres. *Cataphylls* pale and membranous, the uppermost not or barely reaching above the ground and then green and firm-textured, pilose. *Leaves* four, the lower three basal, the lower two longest, reaching to about the middle of the stem, apparently hairless but the sheaths and blades with minute, recurved, prickly-scabrid hairs on the veins, narrowly sword-shaped to nearly linear, 2.5–4 mm wide, the midrib and one other pair of veins lightly raised,

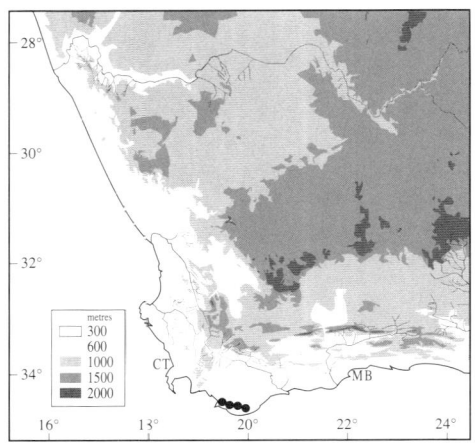

the margins not thickened, the uppermost leaf inserted on about the middle of the stem, usually entirely sheathing, usually channelled throughout, also with recurved prickles. *Stem* erect and straight or slightly flexed above the sheathing parts of the two upper leaves but becoming erect again, occasionally branched, 2–2.5 mm in diameter below the spike.

Spike more or less erect, 2- to 5-flowered; *bracts* pale green or flushed with grey-purple, the outer 28–42 mm long, the inner slightly shorter than to about as long as the outer, minutely notched apically. *Flowers* scarlet, sometimes the lower three tepals orange or yellow, the upper three and sometimes the lower lightly streaked and spotted with darker red toward the base, unscented; *perianth tube* 46–55 mm long, the lower part narrow and cylindric below for 18–20 mm, abruptly expanded into a wide, horizontal, tubular upper part 24–35 mm long, c. 7 x 6 mm in diameter; *tepals* broadly obovate to round, unequal with the upper three larger than the lower, the dorsal largest, directed forward and arched slightly above horizontal, 22–26 x 18–22 mm, the upper laterals slightly smaller, 18–24 x 12–16 mm, much overlapping the dorsal, the lower lateral tepals 10–14 x 8–12 mm, the lower median 13–16 x 8–10 mm, straight and tilted toward the ground, in profile the upper three extending 10–12 mm beyond the

lower tepals. *Filaments* 38–42 mm long, exserted c. 12 mm from the tube; *anthers* 7–8 mm long, yellow, the pollen pale yellow. *Ovary* oblong, 6–7 mm long; *style* extending horizontally over the stamens, dividing shortly below the anther apices, the branches 3–4 mm long, narrow below, broadly expanded and bilobed above. *Capsules* and *seeds* unknown. *Chromosome number* unknown.

Flowering time late July to mid-September.

Distribution and biology

A rare and still poorly known species, *Gladiolus overbergensis* is restricted to the southwestern coast of the southern African winter-rainfall zone. It has been recorded from the western end of the Klein River Mountains near Hermanus in the west to near Elim in the east. It has also been reported growing on the Bredasdorp Mountains and the Soetanysberg on the Agulhas Peninsula, but we have seen no specimens from these last sites. We have been unable to locate populations of the species ourselves, despite carefully searching for them at sites where they have been recorded in the past. This leads us to suspect that the plants flower only in the years following a fire. The vegetation where they have been recorded is either low fynbos or a mixed fynbos–renosterveld association and the soils are sandy loam.

The bright red flowers with a long perianth tube, slender below and abruptly expanded above into a long cylindrical upper part, match closely those of other southern African Iridaceae that are pollinated by sunbirds, and we assume that the flowers of *Gladiolus overbergensis* are likewise adapted for this pollination strategy.

Diagnosis and relationships

Although apparently closely related to *Gladiolus bonaspei* and broadly resembling this Cape Peninsula endemic, *G. overbergensis* is atypical in section *Linearifolius* in lack-

ing the fine, usually long hairs on the leaf blades or at least the leaf sheaths. The leaves do, however, have an unusual abrasive surface, the result of the presence of short, recurved scabrid pubescence on the veins, a feature unique not only in the section but in the genus. The flowers are scarlet and have an elongate perianth tube, 46–55 mm long, with the slender lower part and long, cylindrical upper part associated with sunbird pollination. Like those of *G. bonaspei*, the tepals of *G. overbergensis* are broadly ovate to orbicular, but unlike the Cape Peninsula species, the lower tepals are much shorter than the upper, usually just about half as long and rarely exceeding 15 mm in length.

History

Gladiolus overbergensis was discovered on a plant-collecting expedition to the Agulhas Peninsula by Harry Bolus, Arthur Bodkin and University of Cape Town mathematician and enthusiastic naturalist, Francis Guthrie, who made many important early collections of plants from the southern Cape. Plants were collected under both Bolus's and Guthrie's names near the mission town of Elim in July 1895.

The species was described only in 1920 by H.M.L. (Louisa) Bolus, who had begun to take an active interest in southern African Iridaceae. Bolus initially placed the species in *Antholyza*, naming it *A. guthriei* in Guthrie's honour. It was then transferred to *Petamenes* by N.E. Brown in 1932 after he revised the generic circumscriptions of Iridaceae with long-tubed red flowers that we now regard as adapted for sunbird pollination. Mrs Bolus transferred *P. guthriei* to *Homoglossum* in 1933, where it seemed to her better placed with the closely allied *Gladiolus bonaspei*, which was then *Homoglossum merianellum*. When those species of *Homoglossum* with winged seeds were transferred to *Gladiolus* by Goldblatt & De Vos in 1989, *H. guthriei* was renamed *G. overbergensis* because the epithet *guthriei* was pre-occupied in *Gladiolus*.

67. GLADIOLUS BONASPEI Goldblatt & De Vos

Flames, vlamme

PLATE 58

Gladiolus bonaspei Goldblatt & De Vos, *Bull. Mus. Hist. Nat., Paris* Sér. 4, Sect. B, *Adansonia* 11: 421 (1989) (as *bonaespei*), as a new name for *Watsonia pilosa* Klatt, *Trans. S. African Phil. Soc.* 3: 200 (1885), not *G. pilosus* Ecklon (= *G. hirsutus* Andrews).

Petamenes pilosus (Klatt) Goldblatt, *Ann. Kirstenbosch Bot. Gard.* 19: 144 (1989). Type: South Africa, Western Cape, near Simonstown, May 1882, *MacOwan 2510* (B, holotype; G, isotypes).

bonaspei = of the Cape of Good Hope, referring to the restricted distribution of the species on the southern Cape Peninsula, known to early mariners as the Cape of Good Hope.

Synonymy

Homoglossum merianellum var. *aureum* G. Lewis, *J. S. African Bot.* 14: 34 (1948). De Vos, *J. S. African Bot.* 42: 336 (1976). Type: South Africa, Western Cape, Cape Peninsula, between Scarborough and Klaasjagersberg, burnt marshy ground, Oct. 1945, *Linley s.n.* (holotype, SAM 58293; SAM, isotype).

Gladiolus hirsutus var. *tenuiflorus* Ker Gawler, *Curtis's Bot. Mag.* 16: sub pl. 574 (1802). Type: South Africa, Western Cape, Cape Peninsula, without precise locality, *Thunberg s.n.* (Herb. Thunberg 1037 UPS, holotype).

[*Gladiolus merianellus* sensu Thunberg, *Dissertatio de Gladiolo* no. 11 (1784). *Antholyza merianella* sensu Baker, *Fl. Capensis* 6: 169 (1896). *Homoglossum merianellum* sensu Baker, *J. Linn. Soc., Bot.* 16: 161 (1877) et sensu N.E. Brown, *Trans. Roy. Soc. S. Africa* 20: 279 (1932), not *Antholyza merianella* Linnaeus (= *Watsonia humilis* Miller).]

Description

Plants (25–)35–50 cm high. *Corm* globose, 15–20 mm in diameter, the tunics of coriaceous to woody layers, fragmenting with age into coarse, woody, vertical fibres. *Cataphylls* pale and membranous, the uppermost not or barely reaching above the ground and then green and firm-textured, villous. *Leaves* three, the lower two basal, the lowermost longest, rarely reaching to the middle of the stem, the sheaths of the lower two leaves villous, the blades pubescent, linear, (2–)4–5.5 mm wide, the margins and midrib lightly thickened, the uppermost leaf inserted about the middle of the stem, short and bract-like, sheathing for half its length, usually channelled throughout, sparsely pubescent or glabrous, the margins of the sheathing part united. *Stem* erect, flexed above the sheathing parts of the two upper leaves but becoming erect again, unbranched, 1.3–2 mm in diameter below the spike.

Spike more or less erect, lightly flexuose, 2- to 7-flowered; *bracts* pale green or flushed

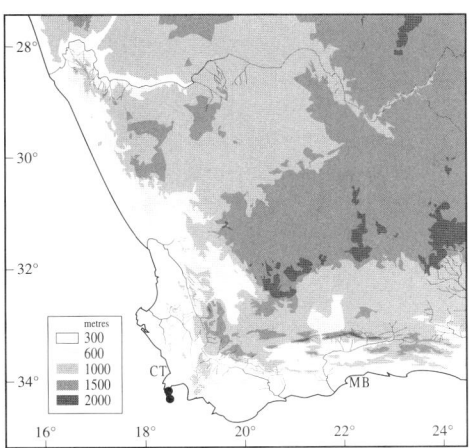

with grey-purple, the outer 15–20(–23) mm long, the inner slightly shorter or slightly longer than the outer, minutely notched apically. *Flowers* bright orange, sometimes the lower three tepals deep yellow and then usually minutely speckled with scarlet towards the bases, rarely entirely yellow, unscented; *perianth tube* 35–43 mm long, with the lower part narrow and cylindric below for 11–18 mm, abruptly expanded into a wide horizontal tubular upper part 22–25 mm long, c. 7–6 mm wide; *tepals* broadly obovate to round, unequal with the upper three larger than the lower, or nearly equal, the dorsal typically largest, directed forward and horizontal, 18–23 x 14–18 mm, the upper laterals slightly smaller, 15–18 x 14–15 mm, much overlapping the dorsal, the lower three tepals 12–20 x 10–18 mm, straight and tilted toward the ground, in profile the dorsal extending 6–8 mm beyond the other tepals. *Filaments* 28–32 mm long, exserted 6–9 mm from the tube; *anthers* 6–8 mm long, dark purple, the pollen pale yellow. *Ovary* oblong, c. 4 mm long; *style* extending horizontally over the stamens, dividing opposite the middle third of the anthers, or sometimes slightly exceeding the apices, the branches 3–4 mm long, narrow below, broadly expanded and bilobed above. *Capsules* ovoid, 15–20 mm long, subacute; *seeds* ovate, 8–10 x 6–7 mm, broadly winged, the seed body asymmetrically placed, translucent golden brown. *Chromosome number* 2*n* = 30.

Flowering time mainly July to September, occasionally as early as April.

Distribution and biology

An extremely local endemic, *Gladiolus bonaspei* is restricted to the Cape Peninsula in Western Cape Province, South Africa. The species is most common on the sandy plains south of Klaasjagersberg, but also occurs to the north on Redhill, Simonstown, and on the Silvermine Plateau. Plants grow in peaty sand both on flats near the coast and at higher elevations.

The orange-red (occasionally yellow) flowers have an elongate perianth tube, the upper part of which is long and cylindric, conforming to the pattern for flowers of *Gladiolus* species adapted for pollination by sunbirds. We assume that *G. bonaspei* also has this pollination strategy.

Diagnosis and relationships

Although included in the genus *Antholyza* by J.G. Baker, amongst others, and referred to *Homoglossum* by N.E. Brown, *Gladiolus*

bonaspei, like its close ally *G. overbergensis*, is not allied to the rest of the species previously included in *Homoglossum* (now series *Homoglossum* of section *Homoglossum*). It has the pubescent leaves, leaf blades with one or more secondary vein pairs thickened and corm tunics of vertical fibres, which are all characteristic of section *Linearifolius*. The ellipsoid capsules are typical of series *Linearifolius* of the section. Within series *Linearifolius*, *G. bonaspei* is readily recognized by the elongate perianth tube, slender below and wide and cylindric above, and the orange-red, rarely yellow, flowers with the tepals ovate and subequal. The closely related *G. overbergensis* has similar flowers, but the tepals are unequal, the lower much smaller than the upper, and the leaves are scabrid rather than pubescent.

History

Known since at least the 1770s when it was collected by C.P. Thunberg, this plant was confused for more than 200 years with *Antholyza merianella* Linnaeus, described in 1774. That species is unambiguously typified by an illustration first published in 1760 in Philip Miller's *Figures of Plants* (Miller,

1756–1759) and the illustration is the type of *Watsonia humilis* Miller, described in 1768 (Goldblatt, 1989b). Thunberg, however, used the name in a different sense, applying it to what is now called *Gladiolus bonaspei*, nevertheless citing Linnaeus's *A. merianella* as the basis for his combination *G. merianellum*. When the species was transferred to *Homoglossum* again, the epithet *merianellum* Linneaus was used as the basionym. Later,

M.P. de Vos treated *G. merianellum* as a species described by Thunberg, thus disassociating the specific epithet from Linneaus's plant. This is, however, nomenclaturally unacceptable. John Ker Gawler realized that Thunberg's plant differed from *A. merianella* Linnaeus and renamed it *G. hirsutus* var. *tenuiflorus* in 1802. That name was, however, never used at species rank and in 1827 Ker Gawler chose to use the name *G. meri-*

anellus in Thunberg's sense for the species.

The only other synonym for *Gladiolus bonaspei*, *Watsonia pilosa* Klatt, cannot be transferred to *Gladiolus* because that epithet has already been used in the genus. *Gladiolus pilosus* Ecklon is a synonym of *G. hirsutus*. It is truly unfortunate that this well-known Cape Peninsula endemic should have had to be given a new name, but it was impossible to avoid this action.

68. GLADIOLUS AUREUS Baker
PLATE 59

Gladiolus aureus Baker, *Fl. Capensis* 6: 530 (1896). Marais, *Curtis's Bot. Mag.* 175: pl. 479 (1965). *Homoglossum aureum* (Baker) Obermeyer, *J. S. African Bot.*, Suppl. 10: 299 (1972). Type: South Africa, Western Cape, Cape Peninsula, near Kommetjie in wet places at the foot of mountains, 300 ft, Aug. 1894, *Fair sub BOL 7951* (K, presumed holotype; B [not seen], BOL, GRA, PRE, Z [not seen], isotypes).

aureus = golden, for the bright yellow flowers.

Description

Plants 35–50 cm high. **Corm** globose, 12–18 mm in diameter, the tunics of coriaceous to woody layers, fragmenting with age into coarse woody fibres. **Cataphylls** pale and membranous, the uppermost reaching 3–6 cm above the ground and then green and firm-textured, villous. **Leaves** three, the lower two basal, the lowermost longest, usually slightly exceeding the spike, the blades and sheaths of the two lower leaves villous, the blades linear, 3–5.5 mm wide, several ribbed, the margins lightly thickened, the uppermost leaf inserted on the upper third of the stem, short and bract-like, sheathing for half its length, usually channelled throughout, sparsely pubescent or glabrous. **Stem** erect, flexed above the sheathing parts of the two upper leaves but becoming erect again, unbranched, 1.3–1.8 mm in diameter below the spike.

Spike more or less erect, 3- to 8-flowered; **bracts** pale green or flushed with dull red or brown, the outer 15–22 mm long, the inner slightly shorter or slightly longer than the outer, minutely notched apically. **Flowers** bright yellow, on fading the upper part of the tube and the base of the tepals flushed with red, without contrasting markings, unscented; **perianth tube** 19–24 mm long, narrow and filiform below for 10–15 mm, ascending, barely exserted from the bracts, abruptly expanded into a slightly flared, more or less horizontal upper part; **tepals** obovate to elliptic, only slightly unequal, the dorsal 20–24 x 12–15 mm, the lower three tepals joined to the upper laterals for 3–4 mm, 20–22 x 9–12 mm, in profile all the tepals appearing the same length. *Filaments* 9–12 mm long, reaching to the mouth of the tube or exserted up to 5 mm; *anthers* 7.5–8.8 mm long, cream, the pollen

white. **Ovary** globose, c. 4 mm long; *style* arching over the stamens, dividing opposite or just beyond the anther apices, the branches c. 4 mm long, arching outward well beyond the anthers. **Capsules** oblong, c. 18 mm long; *seeds* oval, c. 9 x 7 mm, the wing broadly and evenly developed, opaque dark brown. **Chromosome number** unknown.

Flowering time September, sometimes late August.

Distribution and biology

One of the rarest of the southern African species of *Gladiolus*, *G. aureus* is known from just a few populations in the southern Cape Peninsula in Western Cape Province, South Africa. It has been recorded from lower mountain slopes near Kommetjie and Simonstown, but is today known with certainty only from a small reserve at Ocean View near Kommetjie. Plants grow in a specialized habitat, peaty sand in seeps that remain wet well into the spring.

The bright yellow flowers have a short and very narrow perianth tube which is completely blocked by the style. They do not produce nectar and appear to offer pollen as the reward to potential pollinators, of which only honey bees *Apis mellifera* have been recorded on the flowers. The relatively small

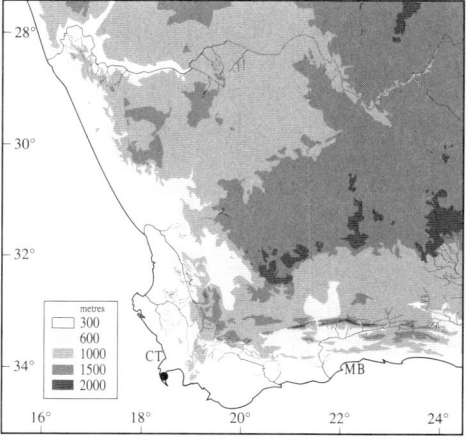

flowers with subequal tepals and short stamens resemble in a general way those of *Oxalis pescaprae*, a favourite pollen source for honey bees.

Diagnosis and relationships

Unusual in *Gladiolus*, particularly in its bright yellow flower, *G. aureus* can readily be recognized by this feature combined with the subequal, more or less elliptic, attenuate tepals, and perianth tube slender below and abruptly expanded into a flared upper part. The leaves are sword-shaped and villous and have the margins, midrib and one or more secondary vein pairs thickened and raised. The general aspect of *G. aureus* is very much like that of *G. bonaspei*, which is also restricted to the Cape Peninsula. The two can really only be distinguished by their flowers. Those of *G. bonaspei* are adapted for sunbird pollination and are orange to reddish, with the upper part of the perianth tube forming a long, wide, horizontal cylinder.

History

Gladiolus aureus appears to have first been recorded by C.B. Fair, an administrator of the Cape Colonial Government who was interested in natural history. His collection, made in August 1894 near Kommetjie, was given to Harry Bolus, founder of the Bolus Herbarium, and distributed under Bolus's name. A duplicate of the collection reached J.G. Baker at Kew Gardens, London, barely in time to be included in his *Flora Capensis* account of the genus. The protologue appears in the addenda to the volume that included the treatment of *Gladiolus* (1896). Because of the narrow perianth tube abruptly expanded in the upper part – a diagnostic character of the genus *Homoglossum* – and the close vegetative similarity to *G. bonaspei* which was then included in *Homoglossum*, A.A. Obermeyer decided to transfer *G. aureus* to *Homoglossum*. Thus the species was not included in the revision of *Gladiolus* in South Africa (Lewis et al., 1972) that she completed after Lewis's death in 1967. The Stellenbosch botanist, M.P. de Vos, who revised *Homoglossum* in 1976, did not accept *H. aureum* as a member of that genus and transferred the species back to *Gladiolus*, where it remains.

69. GLADIOLUS BREVIFOLIUS Jacquin

March pypie, autumn pipes
PLATE 60

Gladiolus brevifolius Jacquin, *Collecteana* 4: 156 (1790); *Icones Pl. Rar.* 2: pl. 249 (1794). Baker, *Fl. Capensis* 6: 143 (1896). Lewis et al., *J. S. African Bot.*, Suppl. 10: 275 (1972). *G. hirsutus* var. *brevifolius* (Jacquin) Ker Gawler, *Curtis's Bot. Mag.* 16: sub pl. 574 (1802) & 19: pl. 727 (1804). Type: South Africa, Cape, without precise locality, illustration in Jacquin, *Icones Pl. Rar.* 2: pl. 249 (1794).

brevifolius = with short leaves, alluding to the reduced leaves on the flowering stems.

Synonymy

Gladiolus tristis var. *aphyllus* Thunberg, *Dissertatio de Gladiolo* no. 8 (1784). *G. hirsutus* var. *aphyllus* (Thunberg) Ker Gawler, *Curtis's Bot. Mag.* 19: sub. pl. 727 (1804). *G. aphyllus* (Thunberg) Ker Gawler, *Genera Irid.* 134 (1827). Type: South Africa, Western Cape, without precise locality, *Thunberg s.n.* (UPS–Herb. Thunberg 1010, holotype).

Gladiolus tristis var. *ruber* Thunberg, *Dissertatio de Gladiolo* 166 (1784). Type: South Africa, Western Cape, without precise locality, *Thunberg s.n.* (UPS–Herb. Thunberg 1011, holotype).

Gladiolus orobanche A. de Candolle in Redouté, *Liliacées* 3: pl. 125 (1805). *G. brevifolius* var. *orobanche* (A. de Candolle) Baker, *J. Linn. Soc., Bot.* 16: 174 (1876). Type: South Africa, Western Cape, without precise locality or collector, grown in France, illustration in Redouté, *Liliacées* 3: pl. 125 (1805).

Gladiolus spilanthus Sprengel ex Klatt, *Linnaea* 32: 711: (1863). *G. andrewsii* Klatt, *Abh. Naturforsch. Ges. Halle* 12: 339 (Ergänzungen 5) (1882), a superfluous name, based on the same type as *G. spilanthus*. Type: South Africa, Western Cape, without precise locality, or collector, grown in Great Britain at Hammersmith, London, illustration in Andrews' *Bot. Repository* 4: pl. 240 (1802), as *G. carneus*.

Gladiolus brevicollis Klatt, *Abh. Naturforsch. Ges. Halle* 12: 339 (Ergänzungen 5) (1882). Type: South Africa, Western Cape, without precise locality or collector, illustration in *Curtis's Bot. Mag.* 19: pl. 727 (1804), as *G. hirsutus* var. *brevifolius*.

Gladiolus brevifolius var. *robustus* G. Lewis, *Ann. S. African Mus.* 40: 126 (1954). Type: South Africa, Western Cape, near Saldanha Bay, 7 Jan. 1935, *Pole Evans s.n.* (BOL 21057, holotype).

Gladiolus brevifolius var. *minor* G. Lewis, *J. S. African Bot.*, Suppl. 10: 279 (1972). Type: South Africa, Western Cape, Bredasdorp, 15 Apr. 1962, *Lewis 5948* (NBG, holotype). *Gladiolus brevifolius* var. *obscurus* G. Lewis, *J. S. African Bot.*, Suppl. 10: 280 (1972). Type: South Africa, Western Cape, Malmesbury Commonage, grown at Kirstenbosch, Apr. 1963, *Lewis 6131* (NBG, holotype).

Description

Plants (12–)25–55(–85) cm high. *Corm* globose, 18–35 mm in diameter, the tunics of coriaceous layers, decaying into fairly coarse vertical fibres, often thicker below. *Cataphylls* pale and membranous, the uppermost reaching shortly above the ground, usually dry and dull purple or becoming dry from the apex, usually minutely pubescent on the veins. *Leaves* of the flowering stem up to three, occasionally two, rarely one, not overlapping, 40–70 mm long, all largely to entirely sheathing, occasionally with short blades up to 40 mm long, these linear, smooth or sparsely and shortly pubescent; foliage leaves produced later in the season,

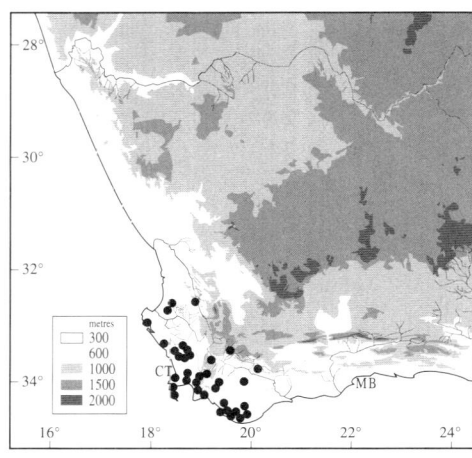

solitary, the blade linear to narrowly lanceolate, pubescent (rarely glabrous), dry and decayed at flowering time or lacking, the cataphyll of the foliage leaf usually dry and brown. *Stem* erect, lightly flexed outward above the sheath of the uppermost leaf, unbranched, 1.2–1.8 mm in diameter below the spike.

Spike slightly inclined, lightly flexuose, 8- to 12(–16)-flowered; *bracts* grey-green, flushed purple above, the outer (15–)18–22 mm long, the inner slightly shorter than the outer, notched apically for c. 0.5 mm. *Flowers* pale to deep pink, occasionally cream, light mauve or greenish grey, usually the tepals darker distally, the lower tepals each with a yellow transverse to obscure median band sometimes edged in darker pink or with a median dark streak when the tepals pale, usually unscented, sometimes with a strong sweet rose scent; *perianth tube* obliquely funnel-shaped, 11–13 mm long, the cylindric lower part 6–7 mm long; *tepals* lanceolate, the dorsal largest, 19–29 x 10–20 mm, the upper laterals 14–25 x 8–15 mm, the lower tepals joined to the upper laterals for 2–5 mm and to one another for 2–3.5 mm, the free parts 12–22 x 5–6 mm, in profile the lower tepals usually slightly exceeding the upper. *Filaments* 10–15 mm long, exserted 7–9 mm from the tube but enclosed by the tepals; *anthers* (4–)6–8 mm long, cream, the pollen pale yellow. *Ovary* oblong, c. 5 mm long; *style* dividing at or just beyond the anther apices, the branches c. 3 mm long. *Capsules* ellipsoid, 16–22 mm long; *seeds* oval, 7–8 x 4 mm, broadly and evenly winged, light yellow-brown, the seed body darker. *Chromosome number* $2n = 30$.

Flowering time March and April, occasionally in February.

Distribution and biology

One of the more common of the autumn-flowering species of *Gladiolus* of the southern African winter-rainfall zone, *G. brevifolius* occurs widely across the west and southwest of Western Cape Province. Its range extends from the western foothills of the Piketberg north of Aurora southward to the Cape Peninsula, and thence eastward to the Agulhas Peninsula and the western end of the Langeberg near Montagu. The plants occur on a variety of soils, including stony sandstone, gritty granitic loam and light to heavy clay, apparently with equal frequency. They are most often seen in flower in March and April, after the heat of summer when nights are cool, but they do not need the first soaking autumn rains to begin their growth. The leaves of the flowering stem are bladeless, and later plants produce a long-bladed foliage leaf from a separate shoot during the wet winter. This leaf carries out most of the photosynthetic activity necessary for growth and the provisioning of a new corm.

The relatively small, pink to cream flowers, usually with yellow nectar guides outlined in red or purple, are adapted for pollination by long-tongued insects, of which the anthophorid bees *Amegilla fallax* and *A. spilostoma* are the most frequent visitors. The fly *Psilodera valida* (Acroceridae) appears to be an important visitor and effective pollinator in some populations on the Cape Peninsula, but has not been observed visiting flowers of *Gladiolus brevifolius* elsewhere. Both this fly and the two *Amegilla* species have mouthparts 8–12 mm long and appear to visit *G. brevifolius* to forage for nectar at a time of the year when few plants of the Western Cape flora are in flower and nectar sources are at a premium.

Diagnosis and relationships

Like those of most other autumn-flowering species of *Gladiolus* of the southern African winter-rainfall zone, the flowering stems of *G. brevifolius* bear short, largely or entirely sheathing leaves superposed on the stem. The flowers are small, 30–40 mm long, with a dorsal tepal 20–29 mm long and a perianth tube 11–13 mm long. The flowers range from deep to pale pink, or cream to mauve, the lower tepals usually with a yellow transverse to obscure median band edged in darker pink, or with a median dark streak when the tepals are pale-coloured. *Gladiolus brevifolius* is apparently unscented over most of the range, but the eastern populations have a strong, sweet scent reminiscent of roses.

Gladiolus brevifolius can be distinguished from other small-flowered species of the genus that have bladeless leaves at flowering time mostly by the corm tunics and by the nature of the foliage leaf or leaves, the dead remains of which may be present next to the current flowering stem. The single foliage leaf of *G. brevifolius* is sword-shaped, or sometimes more or less linear, is usually hairy, and has the margins, midrib and usually one other pair of veins raised and lightly thickened. Almost identical in flower is *G. martleyi*, which usually has scented flowers with spear- or spade-shaped nectar guides. It has a solitary foliage leaf (or occasionally leaves), the blades of which are terete and four-grooved in transverse section. The soft, fleshy corms have membranous to finely fibrous corm tunics. Also with the leaves of the flowering stem reduced, *G. jonquilliodorus* flowers earlier in the year, mostly in December and January, and it, too, has fleshy corms with membranous corm tunics and the two or three foliage leaves are terete. *Gladiolus subcaeruleus*, with its pale blue flowers, is unlikely to be confused with *G. brevifolius*.

A fairly variable species, *Gladiolus brevifolius* was treated by Lewis et al. (1972) as comprising four varieties, none of which we recognize. Var. *robustus* was based on tall, many-flowered plants from near Saldanha Bay; var. *minor* on dwarf plants from Bredasdorp; and var. *obscurus* on dull cream-flowered plants from the Malmesbury and Porterville districts. These are no more than local races of the species and like several more from other parts of its range, do not warrant taxonomic recognition. Another notable variant is the robust form that grows along the foot of the Riviersonderend Mountains. These plants have as many as 16 flowers per spike, and the flowers are deep pink with well-developed nectar guides and a strong, sweet scent. Over most of its range the flowers of *G. brevifolius* have the lower lateral tepals marked with a transverse yellow band more or less outlined with darker colour, but populations from the Cape Peninsula invariably have the markings reduced to a median streak.

History

Gladiolus brevifolius makes its first appearance in the literature as *G. tristis* var. *aphyllus* and var. *ruber*, both of which were described in 1784 by C.P. Thunberg. These were two of 16 species of *Gladiolus* that Thunberg treated as varieties of *G. tristis*, all based on collections that he himself made during the four years he spent studying the Cape flora in the early 1770s. Var. *aphyllus* was raised to species rank as *G. aphyllus* in 1827 by John Ker Gawler, long after Nicholas Jacquin had described *G. brevifolius* in 1790. That species was based on a good illustration that was only published in 1794 and should have left no doubt about its identity. Ker Gawler, however, reduced *G. brevifolius* to varietal rank in *G. hirsutus* in 1802. This decision was ignored by Thunberg in his *Flora Capensis* and much later by J.G. Baker, both of whom recognized *G. brevifolius*. Baker also regarded as synonyms *G. aphyllus* and *G. orobanche*, described by the French botanist, A. de Candolle in 1805. *Gladiolus* brevifolius was circumscribed somewhat narrowly by F.W. Klatt, who also recognized *G. spilanthus* (Klatt, 1863), later renamed *G. andrewsii* by Klatt, and *G. brevicollis* (Klatt, 1882). At best, both of these are only minor variants of *G. brevifolius*.

70. GLADIOLUS MONTICOLA G. Lewis ex Goldblatt & Manning

Autumn painted lady

PLATE 61

Gladiolus monticola Goldblatt & Manning, new species. Type: South Africa, Western Cape, Table Mountain, rocky upper slopes, Mar. 1949, Esterhuysen 15147 (BOL, holotype; BM, K, PRE, S [not seen], isotypes).

monticola = growing on mountains, so named for the habitat.

Synonymy

[*Gladiolus tabularis* sensu Baker, *Handbook Irideae* 207 (1892); *Fl. Capensis* 6: 144 (1896), not *G. tabularis* Ecklon (1827), a nomen nudum (= *G. carneus* Delaroche according to Lewis et al., 1972: 258), and not *G. tabularis* Persoon (= *Tritoniopsis unguicularis* (Lamarck) G. Lewis according to Lewis et al., 1972: 258).]
[*Gladiolus monticola* G. Lewis, *J. S. African Bot.* 14: 85 (1948) (as *monticolus*), an invalid name, without description or type. Dyer, *Fl. Pl. Africa* 34: pl. 1339 (1960). Lewis et al., *J. S. African Bot.*, Suppl. 10: 257 (1972).]

Latin diagnosis

Plantae 30–45 cm altae, cormo globoso 20–25 mm in diametro, tunicis fibrosis verticalibus, foliis reductis elaminatis et vaginantibus, folio plantarum eflorentium ensiformis plano pubescente, spica 3–9 florum, floribus pallide salmoneis vel cremeis, tubo perianthii cylindrico 22–30 mm longo apice curvato, tepalis inaequalibus, dorsali 20–25 x 10–12 mm, inferioribus 15–18 x 6–8 mm, filamentis c. 12 mm longis, antheris 6–7 mm longis, capsulis ovoideo-ellipsoideis 11–13 mm longis.

Description

Plants 30–45 cm high. *Corm* globose, 20–25 mm in diameter, the tunics of fairly coarse to fine vertical fibres. *Cataphylls* pale and membranous, the uppermost reaching (1–)8–15 cm above the ground and then green or dry and reddish brown, glabrous. *Leaves* of the flowering stem two, not overlapping, largely to entirely sheathing, the lowermost inserted shortly above the ground, not overlapped by a cataphyll, occasionally with short blades up to 30 mm long, these linear, smooth or sparsely and shortly pubescent; foliage leaves produced later in the season, solitary, the blade linear to narrowly sword-shaped, pubescent, dry and decayed at flowering time or lacking, the cataphyll of the foliage leaf usually dry and brown. *Stem* erect, lightly flexed outward above the sheath of the uppermost leaf, unbranched, 1.2–1.5 mm in diameter below the spike.

Spike slightly inclined, lightly flexuose, 3- to 9-flowered; *bracts* grey-green, flushed purple on upper surface, the outer 15–24 mm long, the inner three-quarters as long as to only slightly shorter than the outer. *Flowers* pale apricot to cream, the upper tepals flushed pale pink, the lower tepals each with a yellow median stripe often outlined in dark pink, unscented; *perianth tube* 22–30 mm long, cylindric below, curved abruptly near the apex, the upper 6–8 mm horizontal and flaring outward; *tepals* lanceolate, unequal, the dorsal largest, 20–25 x 10–12 mm, the upper laterals 18–20 x 8–10 mm, the three lower tepals united below for c. 5 mm, 15–18 x 6–8 mm, in profile slightly exceeding the upper. *Filaments* c. 12 mm long, exserted c. 4–5 mm from the tube but enclosed by the tepals; *anthers* 6–7 mm long. *Ovary* oblong, 4–5 mm long; *style* dividing at or just beyond the anther apices, the branches c. 4 mm long, extending beyond the anthers. *Capsules* ovoid-ellipsoid, 11–13 mm long; *seeds* oblong, 6–7 x c. 4 mm, the wing evenly developed, translucent golden brown. *Chromosome number* unknown.

Flowering time mid-December to February, occasionally in March.

Distribution and biology

An extremely narrow endemic of the southern African winter-rainfall zone, *Gladiolus monticola* is restricted to the Cape Peninsula, in Western Cape Province, South Africa. It is best known from the Table Mountain range in the northern Peninsula, but it also occurs on Constantiaberg, Silvermine and the mountains above Simonstown in the southern Peninsula. Plants grow in rocky sandstone habitats in fynbos and are often seen along paths and other disturbed sites, or else in veld that was burned two or three years before.

The scentless, pale cream to apricot-coloured flowers with darker pink markings on the lower tepals, together with the long, cylindric perianth tube, indicate that *Gladiolus monticola* is pollinated by long-tongued flies, either Nemestrinidae or Tabanidae. Our observations suggest that the horsefly *Philoliche rostrata* and the nemestrinid *Prosoeca nitudula* are its major pollinators. Their tongues, 18–30 mm long, enable

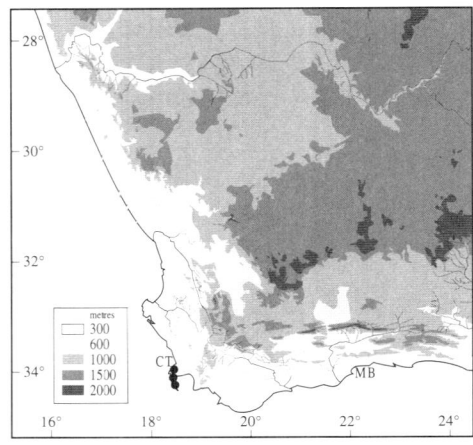

shaped and without the abrupt curve at the base of the wide part characteristic of *G. monticola*. The flowers of the two species are adapted for pollination by different insects, *G. monticola* by long-tongued tabanid flies and *G. brevifolius* by anthophorid bees and other insects with mouth parts of similar length. The flowering time of the two species also differs: whereas *G. brevifolius* flowers in the autumn, usually in March and April, *G. monticola* flowers mainly from mid-December to February and only occasionally in March. We assume that *G. monticola* is a local segregate of the more widespread *G. brevifolius*.

History

Known since at least the mid-eighteenth century, *Gladiolus monticola* was in 1768 treated by the eminent French biologist, J.P. Lamarck, simply as a variety of a plant he called *Gladiolus puniceus*, which is now regarded as a synonym of *G. hirsutus*. Lamarck's plant was sent to him by the French naturalist, Pierre Sonnerat, from the Cape which he visited briefly in 1773 en route to Madagascar and again in 1781 on his return journey to France. The confusion with *G. hirsutus* is difficult to understand for although both species do produce pubescent foliage leaves, those borne on the flowering stem of *G. monticola* are bladeless and quite smooth. Lamarck did, however, correctly note the difference in perianth tube length in the two species. Other early collectors of the species included William Roxburgh at the beginning of the nineteenth century, C.F. Ecklon in the 1820s, and Alexander

Prior in 1846–1847, but despite the numbers of specimens that were known by 1850, the plant remained taxonomically unrecognized.

After 1827, when Ecklon cited, but did not describe, a plant he called *Gladiolus tabularis* (Ecklon, 1827), the species we now know as *G. monticola* became associated with that name and it was so treated by J.G. Baker in *Flora Capensis* (Baker, 1896). Then in 1948 G.J. Lewis pointed out that *G. tabularis* Ecklon was a homonym of *G. tabularis* Persoon (1805), now *Tritoniopsis unguicularis* (Lamarck) G. Lewis. She accordingly renamed the species she thought corresponded to *G. tabularis* Ecklon, calling it *G. monticolus*. Unfortunately, being a name without description, *G. tabularis* Ecklon is nomenclaturally invalid. *Gladiolus monticolus* G. Lewis (correctly *G. monticola*) is thus also invalid. In any event, it later became known that the species that Ecklon called *G. tabularis* is actually *G. carneus* (Lewis et al., 1972). In the treatment of the species by Obermeyer (in Lewis et al., 1972) the name *G. monticola* is still regarded as having been described by Lewis but an early collection by Ludwig Pappe is cited as a 'sensu type!' *Gladiolus monticola* G. Lewis cannot be used for the autumn-flowering Cape Peninsula endemic since there is no formal botanical description of the species. Since *G. monticola* G. Lewis is invalid, the name *G. monticola* may be legitimately used in the genus without becoming a homonym and we have formally described the species here.

these flies to forage on the nectar held in the lower part of the narrow perianth tube, which is 22–30 mm long.

Diagnosis and relationships

Gladiolus monticola is vegetatively identical to *G. brevifolius*, having flowering stems bearing superposed, entirely sheathing leaves and producing solitary, long-bladed, pubescent foliage leaves from a separate shoot during the winter and spring growing season. The flowers also superficially resemble those of *G. brevifolius* but have a perianth tube 22–30 mm long, cylindric below and curved abruptly and widening 6–8 mm below the apex. The flowers are salmon to cream and the lower tepals each have a pale yellow median stripe often outlined in pink. The flowers of *G. brevifolius* range from cream to deep pink or grey, usually with transverse nectar guides on the lower tepals. They always have a perianth tube only 11–13 mm long which is obliquely funnel-

---

71. GLADIOLUS NERINEOIDES G. Lewis

PLATE 62

Gladiolus nerineoides G. Lewis, *Fl. Pl. Africa* 25: pl. 994 (1946). Lewis et al., *J. S. African Bot.*, Suppl. 10: 253 (1972). Type: South Africa, Western Cape, Hottentots Holland Mountains, Somerset Sneeukop, Mar. 1938, *Valpy s.n.* (SAM 54330, holotype).

nerineoides = like *Nerine* [*sarniensis*] (Amaryllidaceae), alluding to the similarity in general appearance and flower colour between the two species.

Description

Plants 35–60 cm high. *Corm* globose, 15–20 mm in diameter, the tunics of pale, finely reticulate fibres, often forming a neck around the base. *Cataphylls* pale and mem-

branous, the uppermost reaching just above the ground and then green and pubescent. *Leaves* two, occasionally one, inserted on the middle and upper part of the stem, widely spaced and entirely sheathing, the lower one 40–70 mm long, pubescent, the upper 2–3.5 mm long; foliage leaf solitary, produced later in the season, the blade linear to narrowly lanceolate, 25–30 cm long, 3–6 mm wide, lightly pubescent, dry at flowering time but often remaining attached to the corm, the midrib and a pair of secondary veins lightly thickened, the margins not thickened. *Stems* generally inclined 30°–40° toward the ground, glabrous, unbranched, 1.5–2 mm in diameter below the spike.

Spike usually flexed outward at the base and inclined to nearly horizontal, occasionally 3-, usually 5- to 10-flowered, flowers crowded on the upper side; *bracts* pale green, the outer 14–18 mm long, obtuse, the inner about as long, minutely forked apically. *Flowers* scarlet red (rarely orange), the lower three tepals each sometimes with a darker median streak towards the base, unscented; *perianth tube* more or less straight, cylindric below, widened somewhat in the upper half, 25–31 mm long; *tepals* subequal, narrowly ovate, held somewhat apart, the lower three directed downward distally, somewhat undulate, the upper sometimes recurving, especially with age, 19–22 x 7–9 mm. *Filaments* 8–10 mm long, entirely included in the tube; *anthers* 6–7 mm long, mostly included in the tube, exserted in the upper third. *Ovary* oblong, 3–4 mm long; *style* dividing at or shortly beyond the anther apices, the branches 3–4 mm long, filiform and notched apically. *Capsules* obovoid, obtuse to subacute, c. 12 mm long (mature capsules not seen); *seeds* unknown. *Chromosome number* unknown.

Flowering time mostly December to February, occasionally in March.

Distribution and biology

A narrow endemic of the southwestern Cape, *Gladiolus nerineoides* is restricted to higher elevations of the mountain ranges that extend north from False Bay, rising abruptly from the low coastal plain. Well known only from the heights surrounding Jonkershoek Valley east of Stellenbosch, *G. nerineoides* extends from Somerset Sneeukop and Helderberg in the south to Simonsberg north of Stellenbosch. An early collection made in 1920 and attributed to Rudolf Marloth is said to have come from Wellington Sneeukop some distance to the north. We suspect that this is erroneous, for there are no other records from the mountains this far north, which are much drier in the summer months when *G. nerineoides* flowers. The species is inconspicuous except in full flower, and it is restricted to rocky outcrops mainly at elevations above 1000 m. Thus, most of its stations are on fairly high slopes that are inaccessible except to the most energetic collectors.

Gladiolus nerineoides flowers at the beginning of summer, in December and into January, a time when even the higher mountain slopes are hot and dry. The growth cycle of the species is adapted to avoid these extreme conditions by flowering after vegetative growth has been completed.

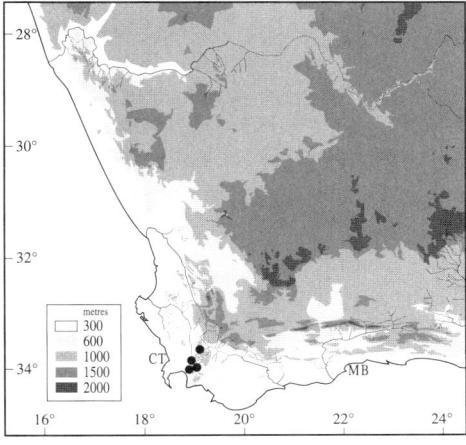

The flowering stems lack foliage leaves and bear only short, entirely sheathing leaves, thus reducing water loss through transpiration. Long-bladed foliage leaves are produced from a separate shoot much later in the year, in April or May, when the weather is cooler and the winter rains have begun. Vegetative growth continues into the late spring when the leaves begin to die back. Only those corms that have accumulated enough reserves begin their flowering cycle some months later.

One of a handful of summer-flowering species of *Gladiolus* with bright red flowers, *G. nerineoides* is, like the others, adapted for pollination by *Aeropetes tulbaghia*, the mountain pride butterfly. This insect is the sole pollinator of a range of red-flowered plant species that flower from mid- to late summer in the mountains extending from the Western Cape through the Drakensberg to the Vumba in eastern Zimbabwe (Johnson & Bond, 1994). The species that use *A. tulbaghia* as their sole pollinator include the orchid, *Disa uniflora*, the amaryllid, *Nerine sarniensis*, and several members of the Iridaceae, among them *Tritoniopsis burchellii*, *T. triticea* and some eight species of *Gladiolus*. Visits to *G. nerineoides* by this butterfly were apparently first reported in 1944 by C.L. Wicht, then a forestry research officer at Jonkershoek Forest Station. We have subsequently confirmed this observation on the Helderberg at Somerset West.

Diagnosis and relationships

Gladiolus nerineoides can readily be recognized by its relatively small, striking scarlet-red flowers, which make a brilliant display crowded together at the end of the inclined to drooping spike. Like the closely related *G. brevifolius* and *G. monticola*, the flowering stems of *G. nerineoides* bear two or three superposed, largely sheathing leaves, and plants produce solitary narrowly sword-shaped pilose leaves later in the season when conditions for vegetative growth are more favourable. The flowers have a relatively long, more or less straight perianth tube, 25–31 mm long, cylindric below and widened somewhat in the upper half. The tepals are subequal, narrowly ovate, and more or less laxly spreading and lightly undulate. The relatively short filaments, 8–10 mm long, are entirely included in the tube, as are the anthers for at least part of their length. The flowers of *G. nerineoides* are somewhat unusual for the species of *Gladiolus* adapted for pollination by

Aeropetes, being relatively small and with partially included anthers. The plants rely on the entire inflorescence instead of a single flower for display and when in bloom *G. nerineoides* cannot be missed against the background of grey, rocky cliffs.

History

The existence of *Gladiolus nerineoides* first became known in 1920 when specimens were collected, apparently by Rudolf Marloth, on Wellington Sneeukop. In the same year, and then again in 1924, plants from Stellenbosch were sent to Cape Town for display at wild-flower shows. The main range of the species was only properly documented in 1936, when plants were collected by the forester, E.J. Borchardt, from the mountains surrounding the Jonkershoek valley near Stellenbosch. Plants were collected on the Helderberg at Somerset West two years later in 1938 by W.P. Valpy and this gathering became the type of the species when it was described by G.J. Lewis in 1946.

Gladiolus stokoei G. Lewis, *Fl. Pl. Africa* 26: pl. 1004 (1947). Lewis et al., *J. S. African Bot.*, Suppl. 10: 252 (1972). Type: South Africa, Cape, Caledon District, Riversonderend Mountains, Mar. 1945, *Stokoe s.n.* (SAM 56457, holotype; BOL, isotype).

stokoei, named in honour of T.P. Stokoe, an amateur plant collector.

Description

Plants 30–45 cm high. *Corm* globose, 12–15 mm in diameter, the tunics of fine brown vertical fibres. *Cataphylls* pale and membranous, the uppermost reaching shortly above the ground and then usually brownish purple, decaying into coarse fibres and these persisting with the corm tunics to form a neck around the underground part of the stem. *Leaves* usually two, occasionally one, not overlapping, largely sheathing, the non-sheathing parts channelled, up to 12 mm long, the uppermost with the margins sometimes united around the stem; foliage leaf solitary, produced after flowering, the blade linear, ultimately to 250 mm long or more, 2–3.5 mm wide, with one or two lightly raised veins, pilose, the margins and midrib lightly raised. *Stem* erect, unbranched, c. 1–1.5 mm in diameter below the spike.

Spike erect, nearly straight, 1- to 3-flowered; *bracts* green and firm-textured, often reddish on the margin, the outer 25–30 mm long, the inner slightly shorter or longer than the outer, minutely forked apically. *Flowers* bright scarlet, the tepals symmetrically disposed, forming a bowl with the apices curving inward, the lower three tepals each with a darker red median streak, unscented; *perianth tube* obliquely funnel-shaped, 30–35 mm long, the lower cylindrical part c. 20 mm long; *tepals* slightly unequal, those of the outer whorl suborbicular, c. 28–30 x 22–28 mm, those of the inner obovate, c. 25–27 x 15 mm. *Filaments* c. 20 mm long, exserted c. 11 mm from the tube; *anthers* 10–12 mm long, yellow, the pollen yellow. *Ovary* oblong, c. 4 mm long; *style* arching over the stamens, dividing close to or slightly beyond the anther apices, the branches 5–7 mm long, slightly expanded toward the apices. *Capsules* and *seeds* unknown. *Chromosome number* unknown.
Flowering time March and April.

Distribution and biology

An extremely local endemic of Western Cape Province, South Africa, *Gladiolus stokoei* is restricted to the eastern end of the Riversonderend Mountains, the higher peaks of which rise to 1200 to 1500 m. Although known from a handful of collections spanning some 60 years since its discovery in 1930 by T.P. Stokoe, all the records of *G. stokoei* are from the slopes of Pilaarkop, highest mountain in the range. Collection information, however, suggests that the species is not restricted to particularly high altitudes, and elevations of 700 to 1400 m are indicated on the few available specimens. The late flowering time of *G. stokoei*, the end of summer in an area of predominantly winter rainfall and summer drought, does not indicate drought tolerance. On the contrary, the southern slopes of the Riversonderend Mountains are fairly cool and well watered in the summer, receiving precipitation from the southeast trade winds that blow inland off the Indian Ocean.

The species has the hallmarks of a specialized pollination syndrome. The bright red flower, a perianth tube of intermediate length, c. 30 mm, exserted stamens, and late summer flowering fit the criteria for pollination by the mountain pride butterfly *Aeropetes tulbaghia*. This butterfly alone is responsible for the pollination of a number of species in the southern and eastern mountains of southern Africa – notably in the Western Cape – that have large bright red to orange flowers (Johnson & Bond, 1994). These species include some eight species of *Gladiolus*. Pollination by *Aeropetes* is best known in the genus in *G. cardinalis* and also occurs in *G. sempervirens*, *G. stefaniae* and *G. nerineoides*. It is suspected for *G. insolens*, another Western Cape species, as well as for *G. cruentus* and *G. saundersii* of eastern southern Africa.

Diagnosis and relationships

Although rather poorly known, *Gladiolus stokoei* is quite unmistakable. The combination of reduced, entirely sheathing leaves on the flowering stem, a straight erect spike of moderate-sized, bright red flowers with orbicular outer tepals, and solitary pilose leaves produced later in the season is unique in the genus. The species that seems closely related to *G. stokoei* is the red-flowered *G. nerineoides*, also of Western Cape mountains, which has a similar habit and a virtually identical, solitary, pilose foliage leaf produced later in the season. Its flowers are, however, quite different from those of *G. stokoei*. Three to ten are borne on an inclined to drooping spike, the tepals are patent, flat and often flaccid, and the stamens are included in the perianth tube.

A collection of *Gladiolus insolens*, a species restricted to high altitudes in the Piketberg, some 200 km to the west of the range of *G. stokoei*, was mistakenly attributed to that species by Goldblatt (1984). The two have remarkably similar flowers, but another collection of the Piketberg plant in January 1995 showed it to be a separate species, and one not even closely related to *G. stokoei*. Described here as *G. insolens*, this species has the long-bladed foliage leaves borne contemporaneously with the flowers, and smooth, plane leaf blades. The foliage differences suggest to us that *G. insolens* and *G. stokoei* belong in different sections of *Gladiolus*, the former in section *Blandus* and the latter in section *Linearifolius*. The similarity of the flowers is presumably the result of selection for the same pollination strategy.

History

Gladiolus stokoei was first collected in 1930 by T.P. Stokoe at Oudebos on the southern slopes of the Riversonderend Mountains. Stokoe knew he had found something novel and sketched the flowers in the field. On this first collection there is an annotation 'G. stokoei L. Bolus', indicating that H.M.L. Bolus recognized the species as distinct and intended to name it after its discoverer. The species was re-collected in March 1945 by Stokoe and this collection, from the same area, was to become the type when *G. stokoei* was formally described by Lewis et al. (1972). Living plants of *G. stokoei* were last seen in 1972 when it was collected by Georges Delpierre and Neil du Plessis for illustration in their handbook on *Gladiolus* in the southern African winter-rainfall zone (Delpierre & Du Plessis, 1973).

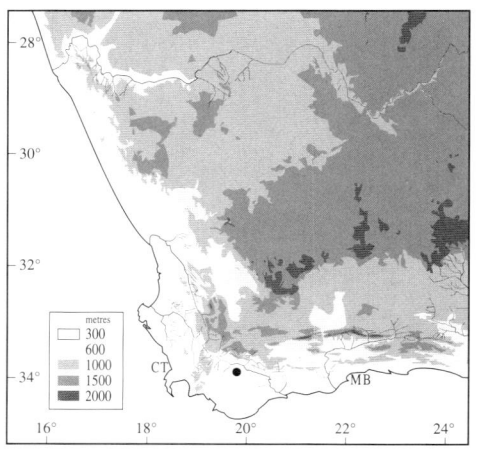

Auriol Batten

Auriol Batten

Fay
Anderson.

Auriol Batten

Muriel Batten

Muriel Batten

Fay Andersson.

Fay Anderson.

Fay
Anderson.

Fay Anderson.

Fay Anderson

Fay
Anderson.

Fay Anderson.

Fay Anderson.

Fay Anderson.

Fay Anderson

Fay Anderson.

Fay Anderson

Fay Anderson.

Fay Anderson.

Fay Anderson.

Fay Anderson

Fay Anderson.

Fay
Anderson

Fay Anderson.

Fay Anderson.

Fay Anderson

Fay Anderson.

Fay Anderson.

Fay Anderson.

Fay Anderson.

Fay Anderson.

Fay Anderson.

Auriol Batten

Fay
Anderson

Fay Anderson.

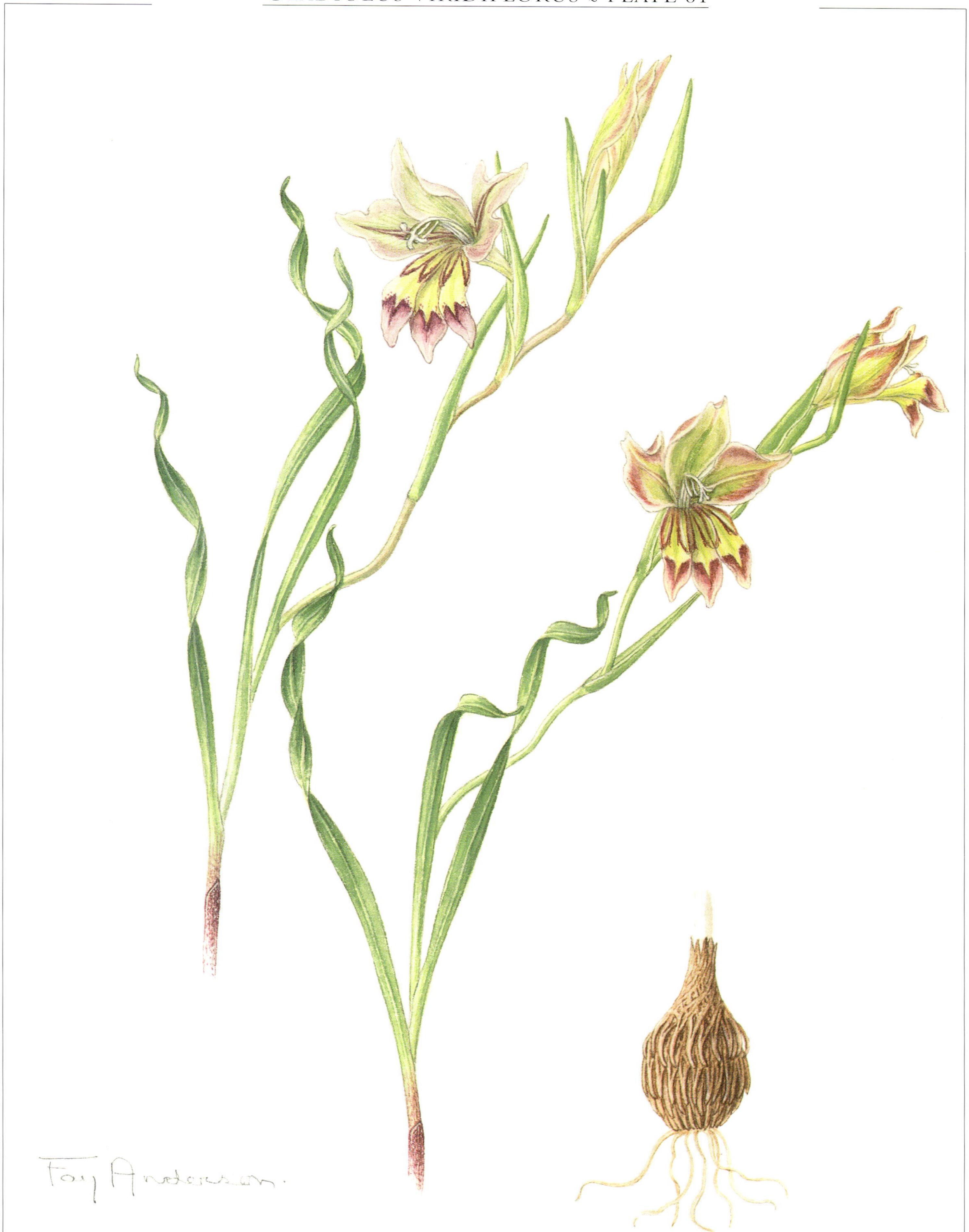
Fay Anderson.

Fay Anderssen.

Fay
Anderson.

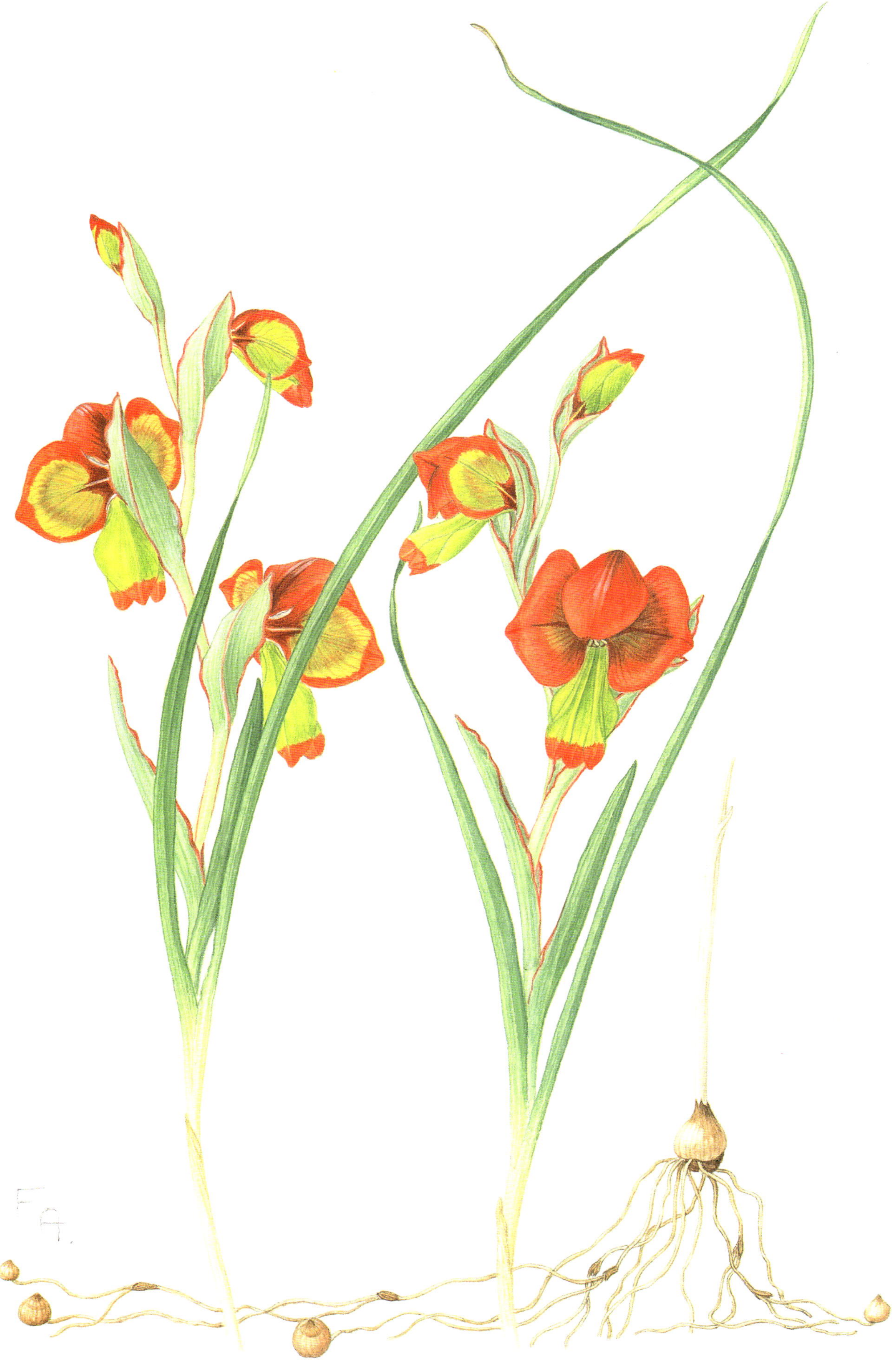

Fay
Anderson

Fay Anderson.

Fay Anderson.

Fay
Anderson.

Fay
Anderson.

Fay Anderson.

Muriol Batten

Fay Anderson.

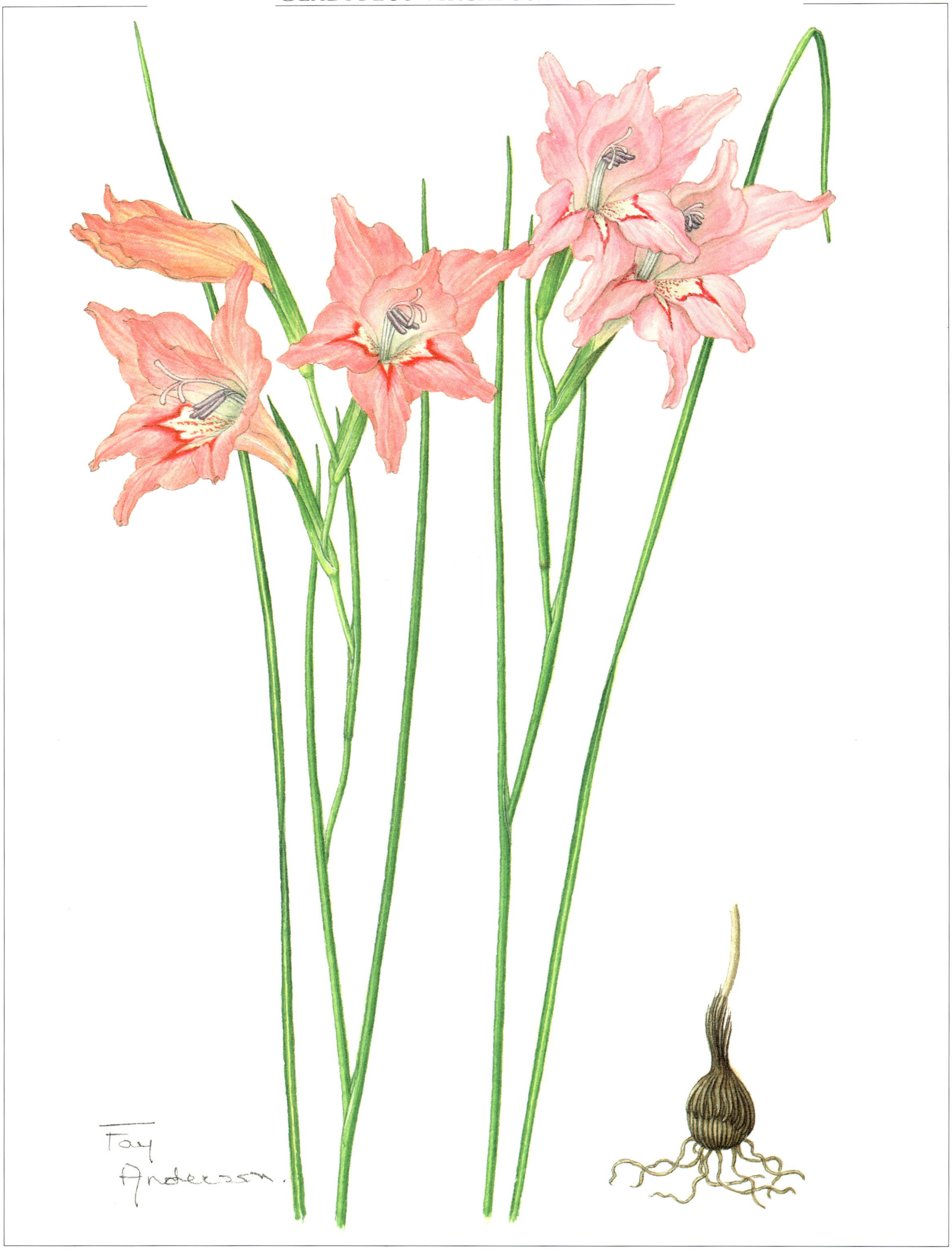

Fay Anderson.

Fay
Anderson.

Fay Anderson.

Fay Anderson.

Fay Anderson

Fay Anderson.

Fay
Anderson.

Fay Anderson.

Fay
Anderson.

Fay Anderson.

Fay Anderson

Fay Andersson.

Fay
Anderson.

Auriol Batten

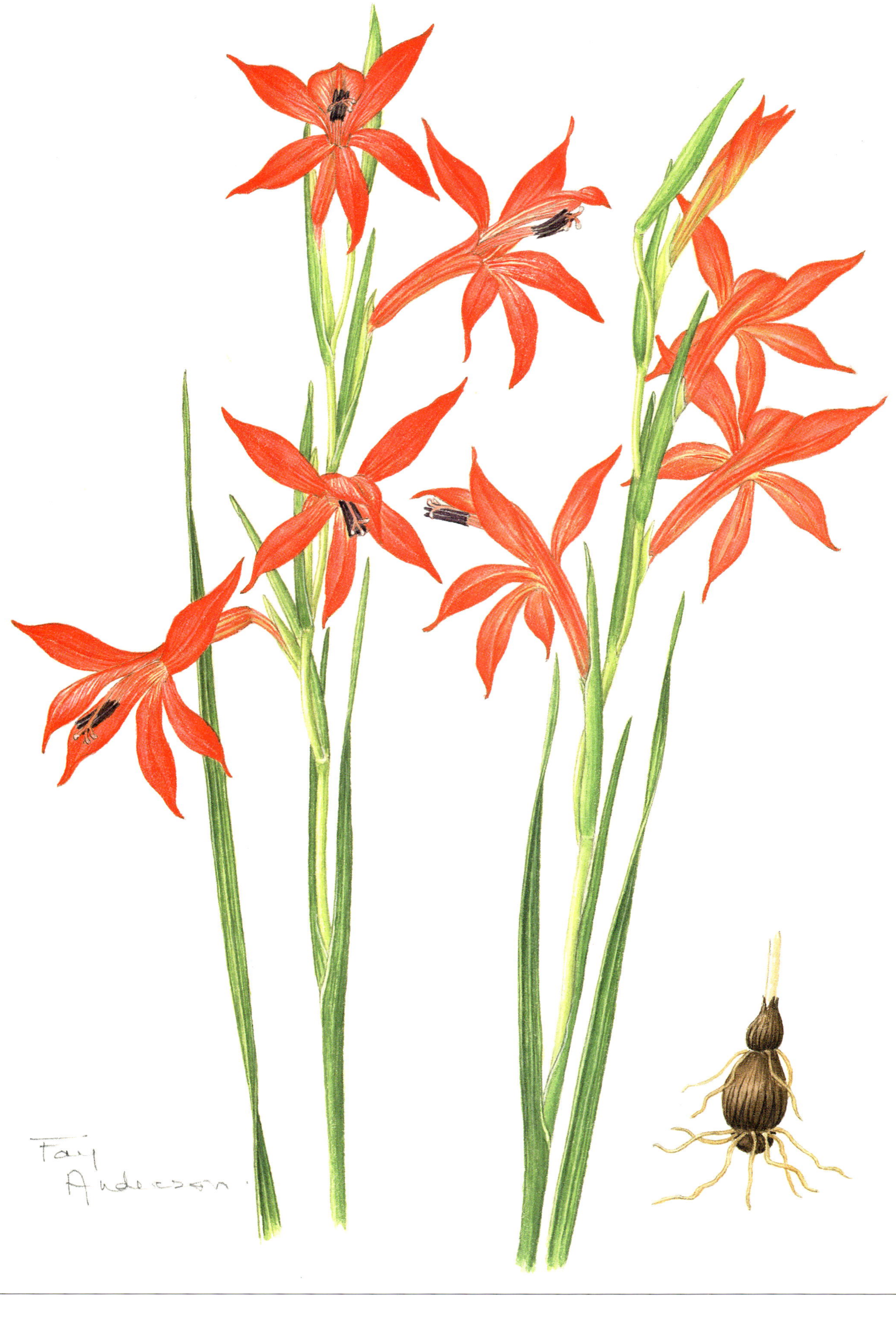
Fay Anderson

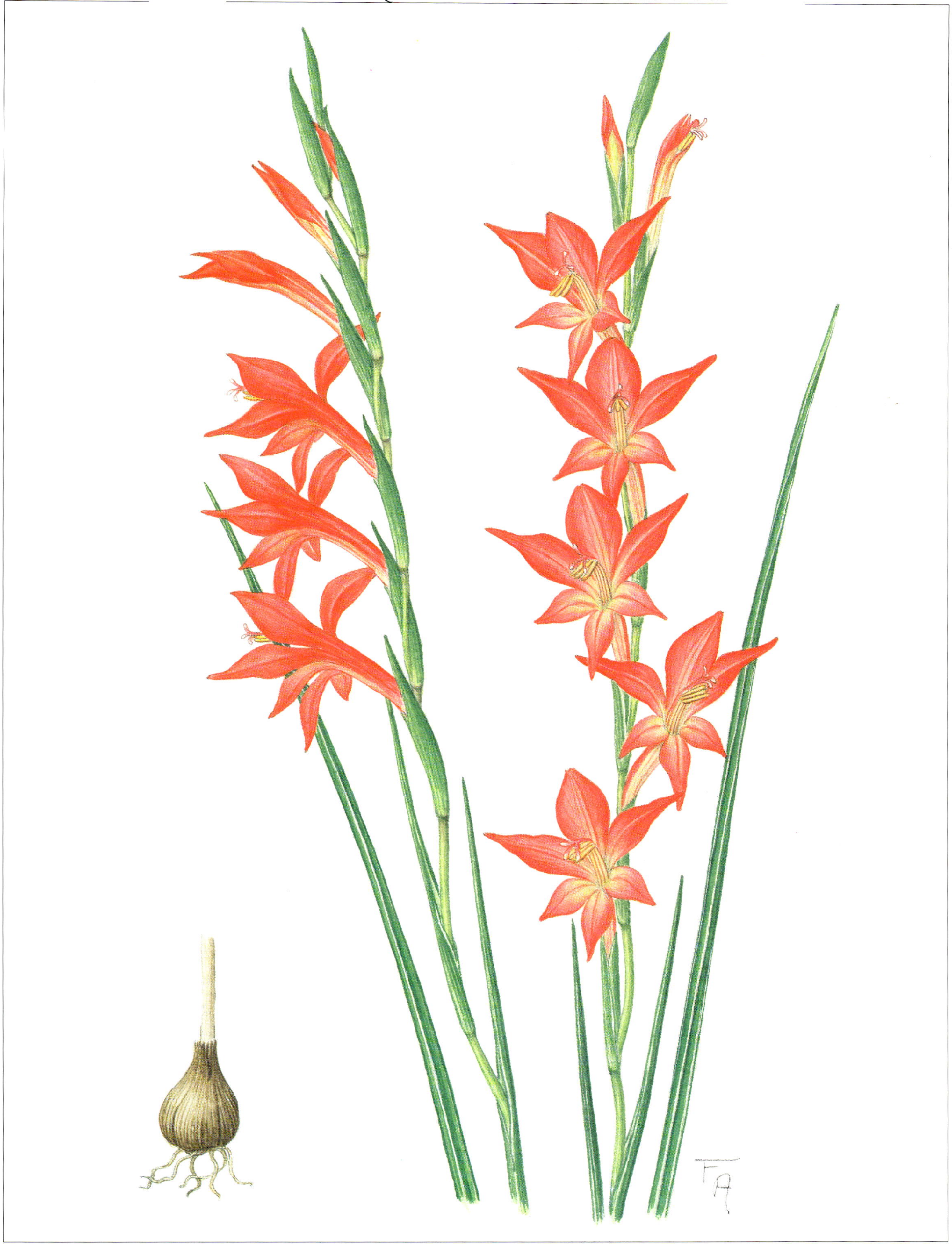

Fay Anderson.

5. Section *HETEROCOLON*

O. Kuntze

Section *Heterocolon* O. Kuntze, *Revisio Generum Plantarum* 3(3): 66 (1898). Goldblatt, *Gladiolus in Tropical Africa* 152 (1996). Type species: *Gladiolus pretoriensis* O. Kuntze.

Description

Plants medium-sized. **Corm** globose, the tunics coarsely fibrous, often accumulating in a neck around the base of the stem. **Cataphylls** glabrous. **Leaves** six or fewer, sometimes only two, usually at least two basal and with long blades present at flowering time, or foliage leaves dry at anthesis and evidently lacking, the blades much reduced or vestigial in a few species (especially of tropical Africa), blades linear to narrowly sword-shaped and more or less plane or with the margins and midribs variously thickened and then oval to terete or cross-shaped in section, less than 4 mm wide, sometimes one or more secondary veins also prominent, occasionally the blades and sheaths hairy. **Stem** occasionally branched.

Spike either flexed at the base and inclined or erect, straight or barely flexuose, internodes usually short and flowers often somewhat crowded; **bracts** green throughout or dry above. **Flowers** usually small, mostly strongly bilabiate, the lower tepals often narrow and clawed at the base, nectar guides consisting of pale colour edged in dark pigment on the distal third of the lower tepals, usually unscented; **perianth tube** obliquely funnel-shaped and about one-third to two-thirds as long as the dorsal tepal, or in one species tubular and about twice as long as the dorsal tepal; **tepals** usually unequal with the dorsal largest and arching over the stamens, the lower three narrower but usually about as long as the upper, usually narrowed below into claws, in one species the flowers nearly actinomorphic and the tepals subequal. **Capsules** either oblong to subglobose and three-lobed above, or in one southern African species ellipsoid to spindle-shaped; **seeds** usually fairly small, the wing either

evenly or unevenly developed or vestigial to lacking.

Section *Heterocolon* was founded by the eccentric German botanist, Otto Kuntze, in 1898 for the southern African *Gladiolus pretoriensis*, but a group of central African species of the subgenus match exactly with the definition of the section. *Heterocolon* comprises some 20 species fairly widely dispersed across southern and south tropical Africa. All the species also have relatively small flowers and small capsules and seeds, the latter in many cases with reduced or vestigial wings. All the species have xeromorphic adaptations, including distinctive brownish corm tunics that form a neck around the base of the plant and leaf blades often with thickened margins and midribs and usually centric or terete. The nine southern African species in the section, divided among three series, show a clear preference for xeric habitats and the section is poorly represented in the Cape Flora Region and the wetter parts of eastern southern Africa.

Species of series *Heterocolon* have slender, usually terete leaves, flowers with transverse bands of colour on the lower tepals, and seeds with reduced or vestigial wings. They occur in rocky habitats in interior southern Africa north of the Vaal River. *Gladiolus pretoriensis*, the type of the section, is fairly common in northern Gauteng and North-West Province, South Africa, while *G. filiformis* and *G. rubellus* are rare local endemics, the former of North-West Province and the latter of eastern Botswana. Several more members of the series – including *G. ledoctei*, *G. actinomorphanthus* and *G. tshombeanus* – occur in the toxic, heavily metal-enriched soils of southern Congo (Goldblatt, 1996). The western Angolan *G. fenestratus*, assigned to section *Hebea* by Goldblatt (1996), also has terete leaves and most likely belongs in this series. Unfortunately, its capsules and seeds are not known.

Series *Unguiculatus* fits uncomfortably in the section, for it has specialized ellipsoid capsules, seeds with well-developed wings, and leaves of the flowering stem with reduced blades. An alternative placement in section *Hebea* seems less likely because the leaf blades have thickened margins and midribs, features that do not occur in section *Hebea*. The series includes *Gladiolus oatesii* of northern South Africa, eastern Botswana and Zimbabwe, and the exclusively tropical *G. unguiculatus*. The latter species is widespread, occurring from northern Zimbabwe to Senegal in West Africa. We suspect that *G. gracillimus* and *G. pusillus*, both centred in Zambia and the Congo may also belong here, and perhaps also *G. atropurpureus* and closely related *G. serapiiflorus* from southern tropical Africa.

Series *Vernus* has an unusual distribution. Species occur on the Mpumalanga escarpment in relatively dry sites and in western South Africa in Namaqualand and the Roggeveld, and along the edge of the Cape Flora Region on the Bokkeveld Escarpment near Nieuwoudtville. We suspect that the eastern Zimbabwean *G. juncifolius* should also be included in this series, but capsules and seeds, needed for a more complete understanding of the species, are not known. Species of series *Vernus* have flowers with the lower tepals joined to the upper laterals for some distance and the lower lateral tepals more or less spathulate and twisted obliquely. Except for *G. vernus*, species of the series also have unusual tepal pigmentation, consisting of minute speckles.

Species of section *Heterocolon* show moderate floral variation, mostly involving colour and tepal markings, and most appear to be adapted for pollination by long-tongued bees. Bee pollination has been confirmed in three species. The elongate, very slender perianth tube of *G. filiformis* is most likely an adaptation for pollination by butterflies or bee-flies (Bombyliidae).

73. GLADIOLUS OATESII Rolfe

Gladiolus oatesii Rolfe in Oates, *Matabeleland,* ed. 2: 410 (1889). Baker, *Handbook of the Irideae* 226 (1892); *Fl. Trop. Africa* 7: 373 (1898). Type: Zimbabwe: Matabeleland, in 1874, *Oates s.n.* (K, holotype).

oatesii, named in honour of the explorer, Frank Oates, who discovered the species near Bulawayo in western Zimbabwe in 1874.

Synonymy

[*Gladiolus unguiculatus* sensu Lewis et al., *J. S. African Bot.,* Suppl. 10: 291 (1972); sensu Goldblatt, *Fl. Zambesiaca* 12(4): 76 (1993); *Gladiolus in Tropical Africa* 119 (1996), in part, not *G. unguiculatus* Baker (1876), a widespread species of tropical Africa.]

[*Gladiolus atropurpureus* sensu Obermeyer, *Fl. Pl. Africa* 44: pl. 1760 (1977), not *G. atropurpureus* Baker (1876), a tropical African species.]

Description

Plants 40–50 cm high. *Corm* globose to obconic, 10–16 mm in diameter, the tunics of matted finely textured fibres usually extending upward for a short distance around the stem base, the inner layers dry and papery, reddish brown. *Cataphylls* pale

and membranous, the uppermost reaching 1.2–3 cm above the ground and with purple veins or entirely purple, drying dark brown and persisting as a dry papery neck around the underground part of the stem. *Leaves* usually three with the uppermost smallest, occasionally two, the uppermost absent, the lowermost longest, sheathing the lower quarter of the stem, the blade short to vestigial, (10–)20–40 mm long, lanceolate, 2.3–4 mm wide, the margins and midrib moderately thickened and hyaline, the second leaf inserted on the lower third of the stem and shortly above the sheathing part of the basal leaf, the blade vestigial, the uppermost leaf entirely sheathing, 20–50 mm long, the margins free to the base but overlapping around the stem; non-flowering plants with solitary, narrowly lanceolate leaves emergent, at flowering time 70–100 mm long, ultimately at least 200 mm long, glabrous, grey-green, 3–4 mm wide, the margins and midrib lightly thickened and raised, borne on corms that did not bear flowering stems. *Stem* erect and more or less straight, unbranched, 1.3–1.6 mm in diameter below the spike.

Spike erect, barely flexuose, 5- to 12-flowered; *bracts* pale green, the apices and upper margins often flushed reddish purple, except for the lowermost flower the bracts about as long as the internodes, the outer 11–18(–21) mm long, the inner about as long as or slightly shorter than the outer. *Flowers* off-white with undertones of mauve or the upper tepals flushed with purple and darker toward the apices, the lower lateral tepal limbs with a transverse band of white (or yellow) outlined with purple across the lower half, unscented; *perianth tube* obliquely funnel-shaped, 10–11 mm long, the cylindrical lower part c. 4 mm long; *tepals* unequal, the three upper lanceolate-elliptic, the dorsal largest, extended horizontally over the stamens, curving upward in the upper quarter, 18–20 x 10–12 mm, the upper laterals directed forward throughout or subpatent in the upper third, c. 15 x 7 mm, sometimes narrowly windowed between the bases of the dorsal and upper lateral tepals, the lower three tepals united with the upper laterals for c. 4 mm and to one another for c. 3 mm, strongly clawed, claws straight and directed forward, the limbs flexed downward and those of the lower laterals diverging from the lower median, the lower laterals

12–14 mm long, the claws c. 6 mm long, the limbs 6–8 x 3.5–4.5 mm, the lower median c. 14 mm long, the claw c. 4 mm long, the limb c. 10 x c. 5 mm, in profile the lower three tepals exceeding the upper by c. 5 mm or more. *Filaments* 12–15 mm long, exserted 5–8 mm from the tube; *anthers* c. 6.5 mm long, greenish or mauve with dark lines on the ventral sides, the pollen yellow. *Ovary* ovoid, 2.5–4.5 mm long; *style* arching over the stamens, dividing opposite the lower third of the anthers, the branches 2–2.5 mm long, expanded in the upper half. *Capsules* ellipsoid, 12–15 mm long; *seeds* ovate, c. 5 x 3.5 mm, the wing broadly and evenly developed, light brown. *Chromosome number* unknown.

Flowering time mid-October to early December.

Distribution and biology

Gladiolus oatesii is native to the bushveld of southern Africa where it extends from eastern Botswana through North-West Province and Gauteng, South Africa, to the Soutpansberg in Northern Province. Outside South Africa it extends across the southern and central parts of Zimbabwe as far north as Harare. Plants typically grow in light woodland, or rarely in open grassland, always in rocky situations where the corms

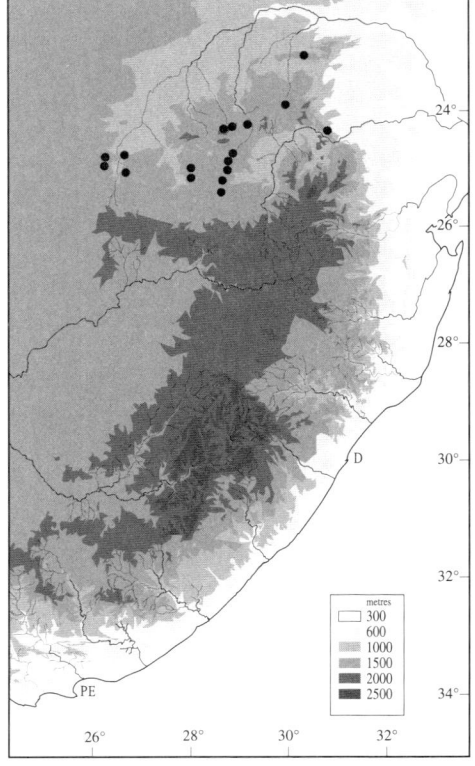

Fay Anderson.

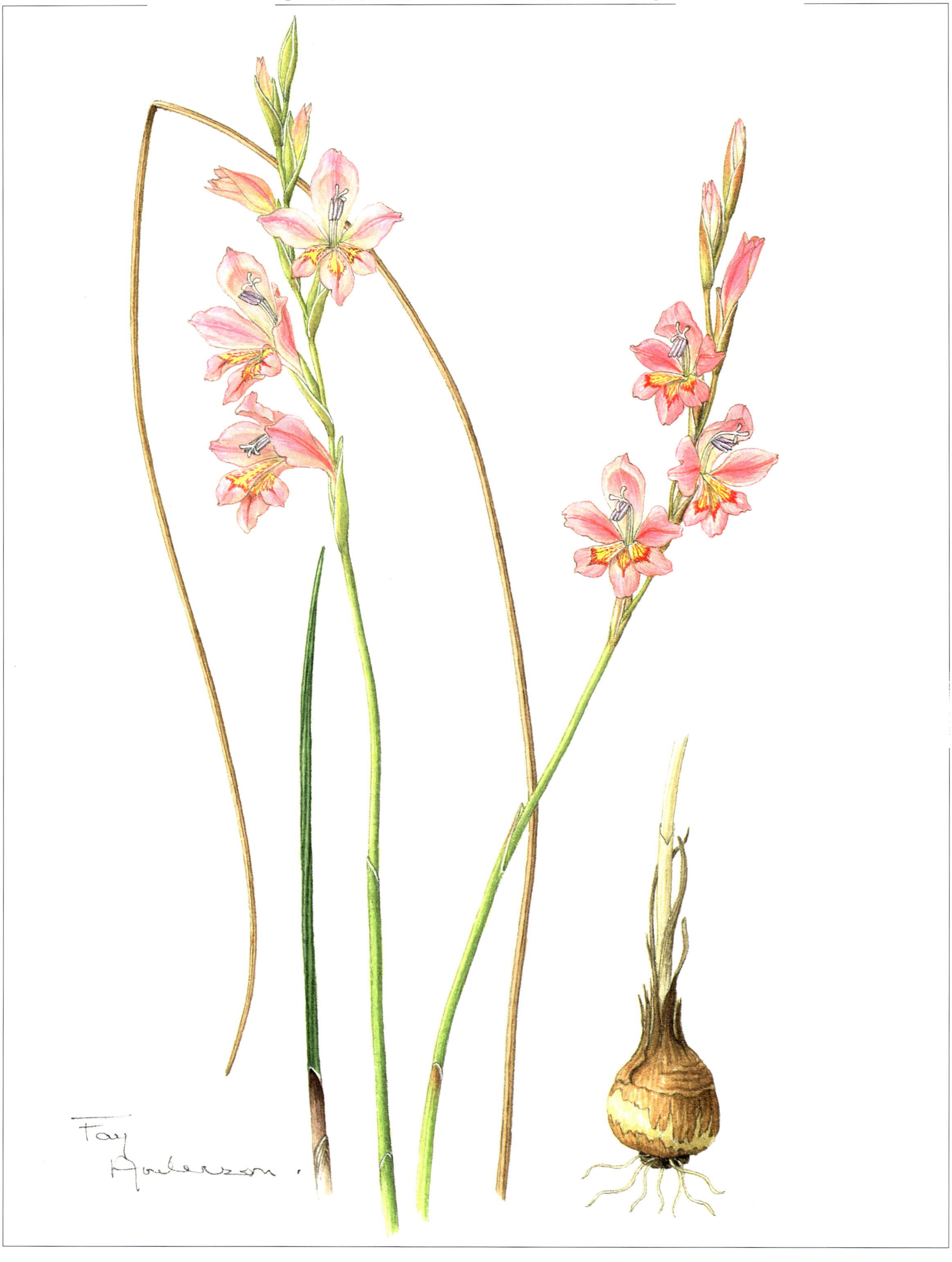
Fay Anderson

GLADIOLUS JONQUILLIODORUS ❧ PLATE 131

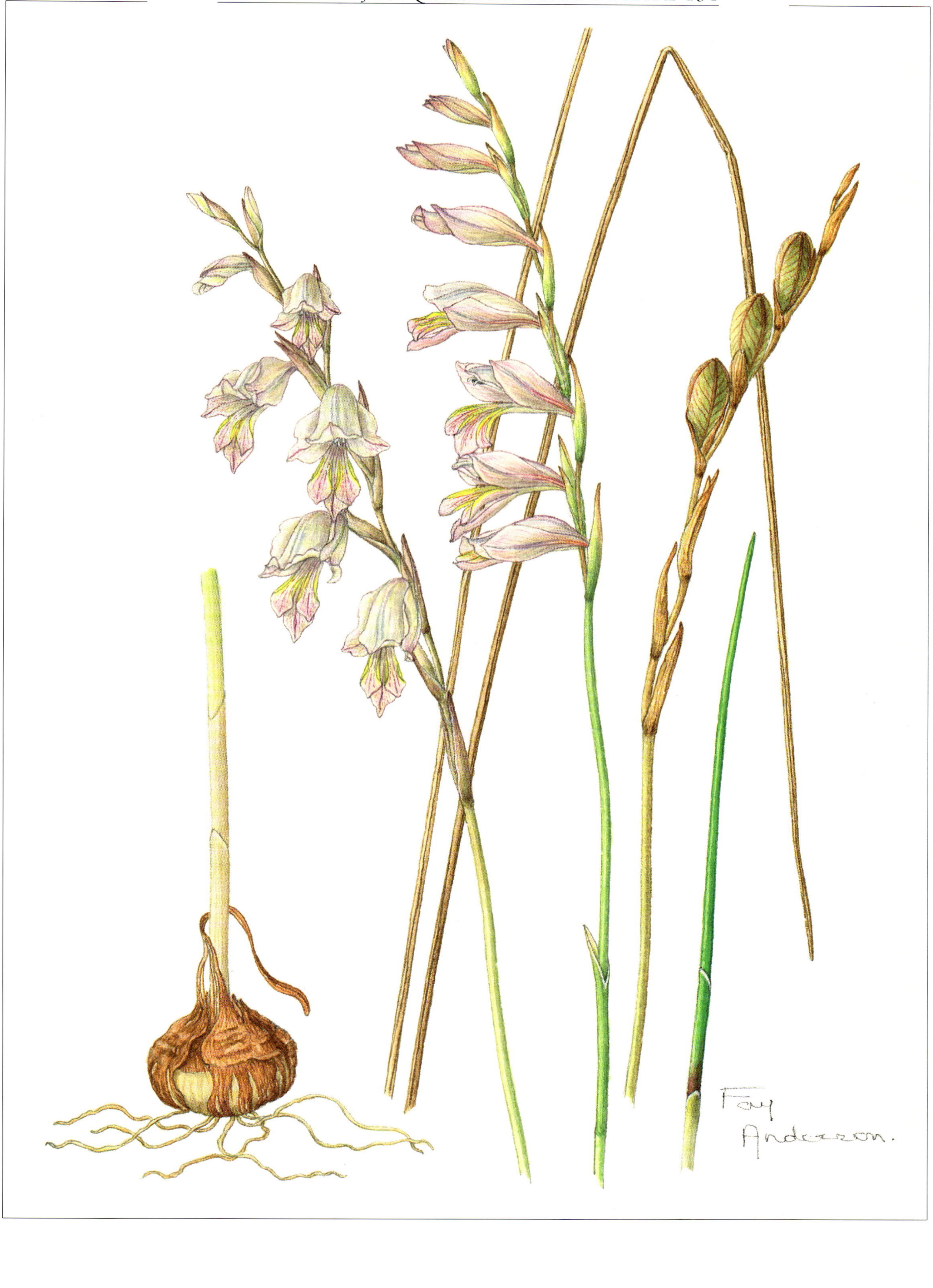

Fay Anderson.

Fay
Anderson.

Fay Anderson

Fay Anderson.

Fay
Anderson.

Fay Anderson.

Fay
Anderson.

are protected from predation by porcupines, baboons and other herbivores.

Like many of the species of *Gladiolus* that have reduced leaf blades on the flowering stem, *G. oatesii* relies on the stem and long-lived leaf sheaths for photosynthetic activity. Plants that flower in the spring do not, as is sometimes believed, produce foliage leaves later in the season. Once seeds have ripened and the new corm has been produced, the flowering stem dies back and plants become dormant through the later summer, autumn and winter. Seedlings and mature plants that do not produce flowering spikes instead bear a single long foliage leaf. This emerges in the late spring or early summer at the same time that flowering takes place or shortly there-after. This growth strategy allows individuals to grow and flower rapidly at the beginning of the wet season before the surrounding trees have leafed out and the understorey of grasses crowd out the slender, fairly short spikes. Production of seeds at the height of the wet season also ensures the next genera-tion the best conditions for germination and continued survival.

The pollination biology of *Gladiolus oatesii* is unknown, but the short-tubed flowers with strongly developed nectar guides are assumed to be adapted for pollination by bees.

Diagnosis and relationships

Among the *Gladiolus* species of the summer-rainfall region of southern Africa that flower early in the season and have the leaves of the flowering stem with short or completely sup-pressed leaf blades, *G. oatesii* is readily recog-nized by its completely smooth cataphylls and leaf sheaths. The presence of pubescence in species such as *G. woodii* and *G. pardali-nus* and their allies (section *Linearifolius*) is so consistent and characteristic that despite

the vegetative similarity, it seems unlikely that *G. oatesii* is related to the species of that section. The flowers of *G. oatesii* are fairly unspecialized, but resemble those of species in section *Heterocolon* most closely in having the lower three tepals extending well beyond the upper three, the upper lateral tepals joined to the lower for some distance, and in the erect rather than inclined spike. The capsules of *G. oatesii* are narrowly ellip-soid and differ from the plesiomorphic ovoid or obovoid and apically rounded and three-lobed capsules of most species of sections *Linearifolius* and *Heterocolon*.

The closest allies of *Gladiolus oatesii* are undoubtedly tropical African and include *G. unguiculatus* and perhaps *G. laxiflorus*. The resemblance between *G. oatesii* and *G. unguiculatus* is so strong that the two have frequently been considered conspecific (Lewis et al., 1972; Goldblatt, 1993, 1996). However, examination of living plants of both species has shown that they differ in their growth cycles and although they are almost certainly closely allied, we have concluded that they must be treated as separate species.

Gladiolus unguiculatus, which is wide-spread across tropical Africa from Senegal to northern Zimbabwe, typically grows in wet-lands or at least seasonal wetlands, and thus generally in flat areas. It produces flowering spikes at the end of the dry season and, as in *G. oatesii*, these bear just three sheathing leaves, but in *G. unguiculatus* almost always without even the shortest blades. Toward the end of flowering or as the capsules are devel-oping, one or more long-bladed foliage leaves are produced from separate shoots on the same corm as the flowering stem. This differs from the pattern in *G. oatesii*, which depends for all photosynthesis on the sheath-

ing leaves of the flowering stem. The corms too, differ in their size, those of *G. unguicu-latus* being large and having reddish internal tissue, unlike the smaller corms of *G. oatesii* which have white internal tissue.

History

Gladiolus oatesii was first collected in the southern spring of 1841 by Joseph Burke, who led a natural history expedition under the patronage of Lord Derby north of the Vaal River accompanied by the well-known collector and traveller in southern Africa, C.L. Zeyher (Gunn & Codd, 1981). Plants were again collected in what was to become the Transvaal in 1847 by Zeyher, this time accompanied by the botanist, Nathaniel Wallich, and in the early 1880s *G. oatesii* was found by the Czeck naturalist, Emil Holub, in eastern Botswana. None of these collections, however, attracted botanical attention, and *G. oatesii* was described in 1889 by the British botanist and orchid specialist, R.A. Rolfe, from plants collected near Bulawayo, Zimbabwe, in 1874 by the naturalist and traveller, Frank Oates.

Although numerous collections of the species have been made since this time, *G. oatesii* has remained poorly understood, largely due to a mistaken interpretation of its life cycle and that of the closely related *G. unguiculatus*, as outlined above. *Gladiolus oatesii* has frequently been regarded as con-specific with *G. unguiculatus,* for example by Goldblatt (1993, 1996). Only as a result of our field research over the past three years has the difference in the life cycles of *G. oatesii* and *G. unguiculatus* been properly understood. With this background knowl-edge, the small, though consistent differ-ences between the two plants are seen as being of taxonomic significance.

Section *Heterocolon:* Series *Heterocolon*

74. GLADIOLUS RUBELLUS Goldblatt
PLATE 63

Gladiolus rubellus Goldblatt, *Fl. Zambesiaca* 12(4): 74 (1993). Type: Botswana, Molepolole, 20 May 1984, *Plowes 7085* (UCBG, holotype; MO, PRE, isotypes).

rubellus = red, referring to the flower colour.

Description

Plants 25–45 cm high. *Corm* obconic, 15–20 mm in diameter, the tunics of medi-um-textured to fairly fine netted fibres.

Cataphylls pale and membranous, the upper-most reaching shortly above the ground and then dark green or flushed with purple, often persisting as a fibrous neck. *Leaves* three or four, at least the lower one or two basal and longest, usually slightly longer than the stem, occasionally shorter, the blades linear, (1.8–)3–4.5 mm wide, the midrib and mar-gins strongly thickened, the margins raised at right angles to the surface, the upper one or two leaves cauline, shorter than the basal

leaves and sometimes the uppermost entirely sheathing. *Stem* often with one or occasion-ally two branches, c. 1.2 mm in diameter below the spike.

Spike inclined, 10- to 16-flowered, the branches with fewer flowers; *bracts* green, the outer 12–14(–16) mm long, the inner usually shorter than the outer. *Flowers* bright orange-red, the lower lateral tepals each with a transverse yellow band outlined in dark red across the distal half, probably unscented;

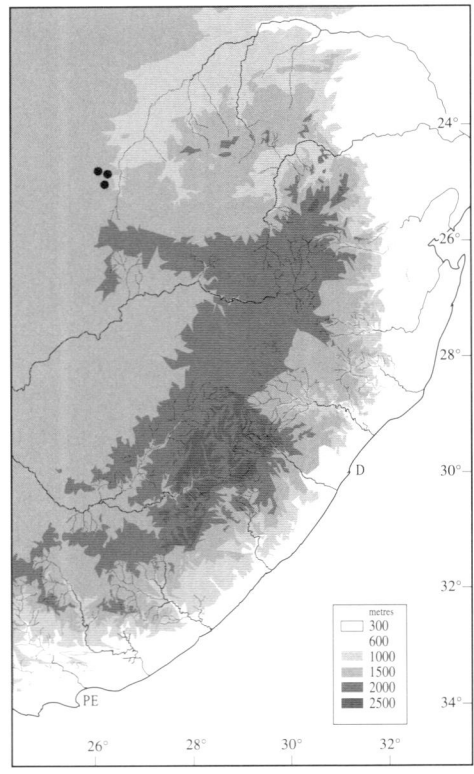

perianth tube obliquely funnel-shaped, c. 14 mm long; *tepals* unequal, lanceolate, the dorsal largest, 16–20 mm long (shorter when dry), the upper laterals slightly smaller than the dorsal, the lower three tepals joined to the upper laterals for 2–4 mm and to one another for c. 3 mm, 12–14 mm long, narrowed below into claws, the limbs abruptly expanded, more or less horizontal, flexed downwards distally, in profile the lower three tepals much exceeding the upper. *Filaments* 12–14 mm long, exserted 8–10 mm from the tube; *anthers* c. 6.5 mm long, yellow. *Ovary* ovoid, c. 2.5 mm long; *style* arching over the stamens, dividing between the middle and apex of the anthers, the branches

2.5–3 mm long. *Capsules* obovoid, three-lobed and retuse apically, 12–14 x 6 mm; *seeds* ovate, 3.5–5 x c. 3 mm, the wing unevenly developed, sometimes somewhat ridged on the angles rather than winged, dark red-brown. *Chromosome number* unknown.

Flowering time January to March.

Distribution and biology

Gladiolus rubellus is restricted to southeastern Botswana and apparently has an extremely narrow distribution. It has been recorded only in a small part of the country between the towns of Lobatse and Molepolole. Plants grow in stony ground, often in *Acacia–Combretum* woodland, and flower in the late summer or autumn.

Like those of most other species of section *Heterocolon*, the fairly small and relatively short-tubed flowers of *G. rubellus* appear to be adapted for pollination by anthophorid bees.

Diagnosis and relationships

Readily recognized by its small, scarlet flowers with bright yellow markings, *Gladiolus rubellus* is unlikely to be mistaken for another species. The flowers are relatively small, 30–35 mm long with a tube about 14 mm long and are, as far as is known, unscented. The three or four leaves are linear, and the lower two are longest and have linear blades, usually 2–4 mm wide. *Gladiolus rubellus* has in the past been confused with *G. pretoriensis* which always has pale pink flowers, but is readily distinguished from that species by flower colour and also by leaf blade structure. Leaf blades of *G. pretoriensis* are terete or oval in transverse section, and have two narrow, almost hairline, longitudinal grooves

on each surface. However, the two species are probably immediately related and both have unusual seeds for the genus. In *G. rubellus* the dark brown seeds have incompletely developed wings which may be present on no more than half the seed body or are unevenly developed at one or opposite ends. The seeds of *G. pretoriensis* lack wings entirely and are almost globose, sometimes with obscurely developed angles or ridges.

History

First recorded in 1927 by a Miss F. ter Horst, *Gladiolus rubellus* was sporadically collected in the following years. It was only in the 1980s, however, that adequate specimens became available and it was not until 1993 that the species was formally described. By this time several well-preserved specimens had accumulated, as well as excellent photographs taken by the naturalist, D.C.H. Plowes. The protologue was published in the account of the Iridaceae for the *Flora Zambesiaca* (Goldblatt, 1993). Early collections were usually mistaken for the Transvaal species *G. pretoriensis*, but in fact it has only a superficial resemblance to that species.

Specimens at the South African National Herbarium were annotated *G. bakeri* Klatt by A.A. Obermeyer, but there is no evidence that she ever saw the type of that species. *Gladiolus bakeri* is a substitute name given by F.W. Klatt in 1895 for J.G. Baker's homonym *G. micranthus* (see Excluded Species, p. 307). The type of the species was collected by the explorer-naturalist, Emil Holub, in the 1870s, but we have not been able to locate it at any of the herbaria we have examined. Until the type is found, the identity of *G. bakeri* will remain a mystery.

75. GLADIOLUS PRETORIENSIS Kuntze

Gladiolus pretoriensis Kuntze, *Revisio Generum Plantarum* 3, 2: 308 (1898). Type: South Africa, Gauteng, Pretoria, Feb. 1894, *Kuntze s.n.* (NY, holotype; K, isotype).

pretoriensis = from Pretoria, the type locality.

Description

Plants 40–70 cm high. *Corm* 14–18 mm in diameter, the tunics light brown, composed of medium-textured fibres, mostly vertically oriented, not extending upward in a neck. *Cataphylls* membranous, the uppermost reaching 4–5 cm above the ground and green or flushed with light purple. *Leaves*

three to five, sometimes up to seven, the lower one or two basal, the remaining cauline and progressively shorter; the lowermost leaf always the longest, reaching at least to the base of the spike, sometimes exceeding it by up to 15 cm, the blade terete or oval in section and c. 1.5–2 x 1.5 mm, the midrib and the margins thickened and raised and the laminar surface concealed within four longitudinal grooves running the length of the leaf, blades of the other leaves similar but shorter and thinner than the basal, becoming almost filiform, the upper one to two cauline leaves often with vestigial blades, the sheaths of the cauline leaves sometimes overlapping

and concealing the stem up to the base of the spike, the margins of the sheathing leaves free to the base. *Stem* simple or with one or two branches, usually flexed above the sheath of the uppermost leaf, c. 1 mm in diameter below the spike.

Spike inclined, lightly flexuose, rarely 5-, usually 9- to 13-flowered; *bracts* green below or purplish, becoming dry and brownish in the upper half at anthesis, completely dry and rust-coloured when the capsules ripen, the outer 9–12(–15) mm long, the inner two-thirds as long as to slightly exceeding the outer, forked at the apex for less than 1 mm. *Flowers* pale lilac to pinkish, the

upper three tepals darker in the midline, the lower lateral tepals each with a spade-shaped yellow mark outlined in purple, broadest in the upper third of the tepals, unscented; *perianth tube* obliquely funnel-shaped, 11–14 mm long, curved outward above; *tepals* unequal, oblanceolate, the dorsal largest and hooded over the stamens, 14–18(–24) x 10–14 mm, the upper laterals directed forward and curving outward distally, 12–16(–23) mm long, the lower three held close together and more or less horizontal or tilted downward, united with the upper laterals for 5–8 mm, the free parts 11–15(–20) mm long, narrowed below into claws, limbs of the lower laterals 2.5–4 mm wide, the lowermost 4.5–6.5 mm wide. *Filaments* 10–13 mm long, exserted 5–7 mm from the tube; *anthers* 4–6 mm long, yellow, the pollen yellow. *Ovary* c. 2 mm long; *style* arching over the filaments, dividing between the lower third and apex of the anthers, the branches 1.5–2 mm long. *Capsules* globose-ovoid, 8–10 mm long, showing the outline of the seeds; *seeds* angular-prismatic, 1.7–2 mm long. *Chromosome number* 2*n* = 30.

Flowering time mid-January and February.

Distribution and biology

Restricted to interior South Africa north of the Vaal River, *Gladiolus pretoriensis* extends from near Krugersdorp and Pretoria in Gauteng westward to Rustenberg and Zeerust in North-West Province. Plants grow on stony hill slopes in open grassland or light open woodland. The soil is often a light clay derived from shale and is well-drained and stony. Plants most often flower in the late summer, from January until late February, but occasional collections of plants in full flower have been made in November and December. We suspect that exploration of the rocky hills of southeastern Botswana will extend the range of the species into that country, but all records of *G. pretoriensis* to date are from the South African side of the frontier, especially in the hills near Zeerust. The collection made by I.B. Pole Evans and Jan Ehrens in 'Bechuanaland' is actually from near Swartruggens in North-West Province.

The small, short-tubed flowers are assumed to be adapted for pollination by bees. The pollination ecology of *Gladiolus pretoriensis* remains to be studied.

Diagnosis and relationships

The type species of section *Heterocolon*, *Gladiolus pretoriensis* stands out among the southern African species of the genus in having globose, completely wingless seeds. The pale pink flowers have nectar guides on the lower tepals consisting of cream transverse bands edged in deep pink and are unusual in their small size, usually 25–32 mm long, with a tube 11–14 mm long. Plants have three to five leaves and coarsely fibrous tunics that usually accumulate with the persistent cataphylls to form a neck around the base. The leaf blades are slender and usually terete, or sometimes oval in section, incorrectly described by Lewis et al. (1972) as having raised margins extended at right angles to the axis of the blade and the midrib not at all or barely raised. Both the margins and the midrib are so heavily thickened that the blade appears to be solid, although it has four narrow grooves, two on each surface. This leaf type is not uncommon in *Gladiolus* and recurs in section after section, mostly in species of xeric habitats or those which are nutrient-poor or have toxic levels of minerals such as cobalt, copper and nickel.

In southern Africa *G. pretoriensis* has been occasionally confused with the eastern Botswana endemic, *G. rubellus* (e.g., by Lewis et al. 1972), which has scarlet flowers with bright yellow, transversely banded nectar guides on the lower tepals. Unlike those of *G. pretoriensis*, the leaves of *G. rubellus* are linear with prominently thickened margins and midribs, and the seeds have poorly developed wings, usually present at least on part of the seed body.

Gladiolus species endemic to the Congo copperbelt, notably *G. ledoctei*, *G. pungens* and *G. robiliartianus*, resemble *G. pretoriensis* closely and are readily confused with it if their origin is not known. These species of soils enriched with heavy metals also have slender, terete leaves, relatively small flowers with dry floral bracts, and rounded capsules with globose, wingless seeds. Plants from the Mafinga Hills in northern Malawi also match *G. pretoriensis* closely, except that they are described as having reddish flowers (Goldblatt, 1996). If these plants prove to have wingless seeds, as seems likely, they should be included in *G. pretoriensis* despite the extraordinary disjunction in distribution.

Lewis et al. (1972) considered *Gladiolus pretoriensis* to be closely related to *G. permeabilis*, but this seems to us unlikely and we regard the two species as belonging in different sections of the genus. *Gladiolus permeabilis* has narrow leaves but without the raised margins that one would expect in an ancestor of a species that has both heavily thickened margins and midribs. It is, moreover, specialized in having a narrow dorsal tepal held well apart from the other tepals and in the peculiar firm, coarsely fibrous corm tunics.

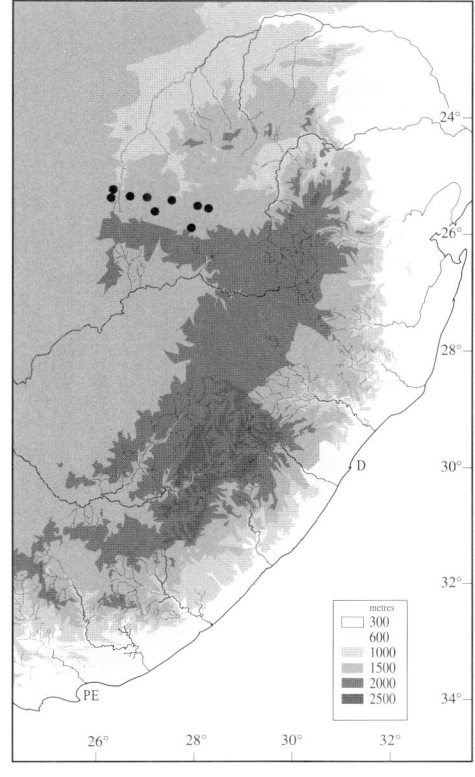

History

The earliest records of *Gladiolus pretoriensis* are collections from Pretoria made at the end of the nineteenth century. Rudolf Schlechter collected plants there in January 1894 and just a month later the German botanist,

Otto Kuntze, also recorded the species. The energetic, but controversial Kuntze wasted little time, naming the species *G. pretoriensis* in 1898, at the same time erecting a new section *Heterocolon* for this single species based on its unusual wingless seeds. Later botanical

exploration has shown the species to be particularly common in the hills around Pretoria, but it ranges westward almost to the Botswana frontier.

76. GLADIOLUS FILIFORMIS Goldblatt & Manning

Gladiolus filiformis Goldblatt & Manning, new species. Type: South Africa, North-West Province, Gopane Mountains, 29 Dec. 1977, *Peeters, Gericke & Burelli 509* (PRE, holotype).

filiformis = extremely slender, thread-like, referring to the long, thin leaf blades.

Latin diagnosis

Plantae 45–50 cm altae, foliis 4 teretibus 4-sulcatis, spica c. 5 florum erecta, bracteis 8–9 mm longis, floribus caeruleo-malvinis, tubo perianthii c. 20 mm longo, tepalis subaequalibus ovatis, c. 10 x 6 mm, filamentis c. 8 mm longis c. 5 mm exsertis, antheris c. 4 mm longis.

Description

Plants 45–50 cm high. *Corm* obconic, with the accumulated tunics c. 20 mm in diameter, the tunics of medium-textured reticulate

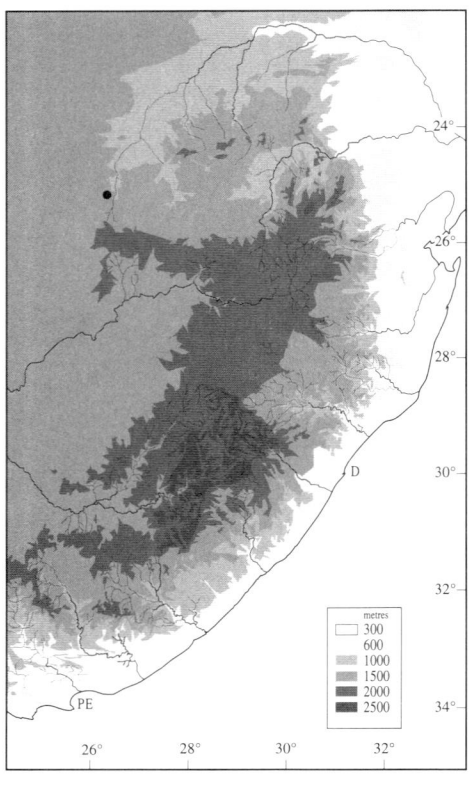

fibres, evidently accumulating in a dense mass. *Cataphylls* membranous and brown, the uppermost reaching c. 2 cm above the ground, dark brown and dry. *Leaves* four, the lower two basal, the lowermost longest, probably reaching to about the base of the spike, the second leaf sheathing the lower third of the stem, the blade shorter than the sheathing part, the blades terete, the margins and midrib heavily thickened and thus with four narrow longitudinal grooves, c. 1 mm in diameter, the upper two leaves entirely sheathing, the uppermost subtending a branch. *Stem* erect, lightly flexed outward above the sheaths of the two upper leaves, apparently usually branched, c. 1.5 mm in diameter below the spike.

Spike more or less straight and erect, c. 5-flowered, the branches with fewer flowers; *bracts* apparently more or less dry and brown at anthesis, the outer 8–9 mm long, the inner about as long as the outer. *Flowers* blue-mauve, markings and presence of scent unknown; *perianth tube* c. 20 mm long, cylindric and slender below for c. 14 mm, flared outward in the upper 6 mm; *tepals* apparently subequal, ovate, c. 10 x 6 mm, their orientation unknown. *Filaments* c. 8 mm long, exserted c. 5 mm from the tube; *anthers* c. 4 mm long, evidently pale yellow. *Ovary* ovoid, c. 2.5 mm long; *style* arching over the stamens, dividing opposite the upper third of the anthers, the branches c. 4 mm long. *Capsules* and *seeds* unknown. *Chromosome number* unknown.

Flowering time early December, but probably highly dependent on the timing of the rains.

Distribution and biology

Known from a single gathering, *Gladiolus filiformis* is a puzzling plant. It has only been found on rocky hills in the Gopane Mountains of North-West Province, South Africa, a series of ridges 200–300 m above the surrounding country. This area, a short distance west of Zeerust and close to the

Botswana–South Africa frontier, is poorly explored botanically. Nevertheless, it is surprising that it harbours such a highly local endemic as *G. filiformis*. The habitat is described as red soil on a rocky koppie, and this appears to be in no way different from the hills that extend north and south for mile upon mile. Additional collections are needed to properly understand the species and to assess its relationships within the genus.

The flowers of *Gladiolus filiformis* are unusual for section *Heterocolon* in having an elongate, filiform perianth tube. We assume that this is an adaptation for pollination by butterflies or bee-flies (Bombyliidae), but further research has still to be done.

Diagnosis and relationships

A terete leaf with four narrow longitudinal grooves located between the heavily thickened margins and midrib, and a long-tubed flower with short, dry floral bracts make *Gladiolus filiformis* unmistakable. The matted, moderately coarsely fibrous corm tunics are much like those of *G. pretoriensis* – as well as several other members of section *Heterocolon* – and it is to this species that *G. filiformis* is perhaps most closely allied. The more or less dry floral bracts and the terete leaf also recall *G. pretoriensis*, supporting our assumption that the two are immediately related. Whether *G. filiformis* has the wingless seeds of *G. pretoriensis* remains to be discovered.

History

Gladiolus filiformis was discovered in 1977 during a survey of the flora of what was then Bophuthatswana and is now part of North-West Province, South Africa, and it has not been re-collected. Despite being known from just a single specimen, its flowers are so distinctive that we feel justified in recognizing it as a separate species. We hope that this decision will encourage further botanical exploration of North-West Province.

77. GLADIOLUS RUFOMARGINATUS G. Lewis

Gladiolus rufomarginatus G. Lewis in *J. S. African Bot.*, Suppl. 10: 145 (1972), as a new name for *Gladiolus marginatus* F. Bolus, *Ann. Bolus Herb.* 1: 108 (1917), an illegitimate homonym, not *G. marginatus* Linnaeus fil., 1772 (= *Watsonia marginata* (Linnaeus fil.) Ker Gawler). Type: South Africa, Mpumalanga, Lydenburg, Jan. 1895, *Wilms s.n.* (PRE 6457, holotype; PRE, isotype, BOL, tracing and flower).

rufomarginatus = with rust-coloured margins, referring to the unusual dry floral bracts with dark rusty brown margins.

Description

Plants 30–50 cm high. *Corm* obconic, 15–18 mm in diameter, the tunics of papery layers, these becoming more or less fibrous with age. *Cataphylls* pale and membranous, the uppermost 5–8 cm above ground and then green or dry and brown above. *Leaves* usually five, the lower two or three basal, shortly exceeding the stem, the blades linear, 2–3 mm wide, the margins and midrib heavily thickened and raised, a secondary pair of veins often evident, sometimes in transverse section more or less oval with narrow longitudinal grooves, the remaining two or three leaves inserted on the upper two-thirds of the stem, decreasing in size above,

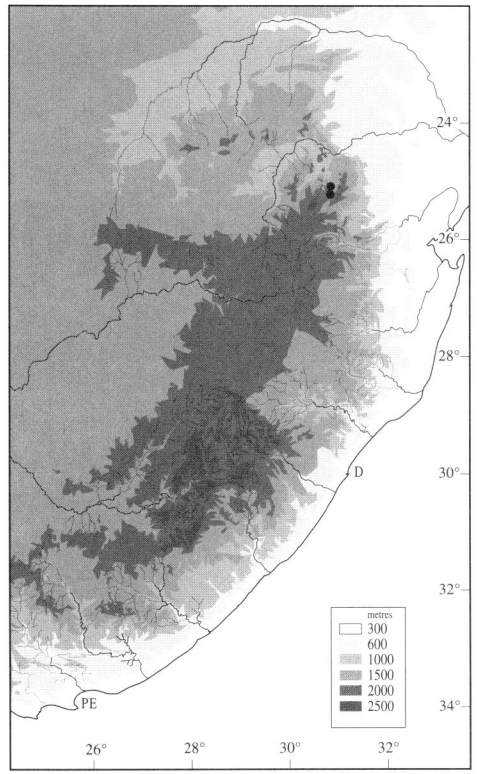

the uppermost bract-like, sheathing for most of its length. *Stems* erect, lightly flexed above the sheaths of the upper leaves, minutely scabridulous, simple or occasionally one-branched, c. 2 mm in diameter below the spike.

Spike straight and erect, 8- to 20(–30)-flowered; *bracts* pale and membranous, dry at anthesis and slightly transparent, flushed pink, the margins rusty brown, the outer rusty and lightly folded along the midline, 14–18 mm long, the inner slightly shorter than to two-thirds as long as the outer, forked apically for c. 2 mm. *Flowers* irregularly mottled dark red on cream, the lower three tepals flushed yellow-green in the lower two-thirds, the veins showing dark red, the tips of the limbs flushed bright pink to red, unscented; *perianth tube* obliquely funnel-shaped, c. 9 mm long, the lower cylindrical part c. 4 mm long; *tepals* unequal, lanceolate, all distinctly narrowed below into claws, the dorsal largest, inclined over the stamens, 18–20 x 12 mm, the upper laterals c. 15 x 5 mm, the claws directed forward, the limbs patent, usually windowed between the bases of the dorsal and upper lateral tepals, the lower tepals joined to the upper laterals for c. 1.5 mm and to one another for c. 2 mm, c. 8 mm long, the claws c. 4 mm long, abruptly expanded at the base of the limbs. *Filaments* c. 13 mm long, exserted c. 9 mm from the tube, pale, spotted with dark red; *anthers* c. 5 mm long, tilting down distally, olive-green, the pollen white. *Ovary* ovoid, c. 2.5 mm long; *style* arching over the stamens, dividing opposite the lower half of the anthers, the branches c. 2 mm long, expanded in the upper half. *Capsules* oblong, 9–12 mm long, three-lobed in the upper half and retuse apically; *seeds* oblong, c. 5 x 2.5 mm, the wing unevenly developed, light translucent brown, the seed body darker brown. *Chromosome number* unknown.

Flowering time mainly March and April, sometimes in May.

Distribution and biology

Gladiolus rufomarginatus is a narrow endemic of the Lydenburg District in Mpumalanga Province, South Africa. Plants occur near Lydenburg itself, and the range extends a short distance northward through the dry hills and valleys on the western side of the Drakensberg escarpment as far as Ohrigstad. *Gladiolus rufomarginatus* grows in grassland communities either in the open or in light shade on stony shale ground, and sometimes in crevices in bare shale outcrops. The species is common locally.

The small, relatively short-tubed flowers are adapted for bee pollination and visits by *Amegilla* bees as well as short-tongued, nectar-eating flies of the genus *Stenobasipteron* (Nemestrinidae) have been recorded.

Diagnosis and relationships

Easily recognized by its remarkable floral bracts that are pale and dry at anthesis and slightly transparent, flushed with pink and with the margins rusty brown, *Gladiolus rufomarginatus* also has distinctive flowers and leaves. The leaf blades are long and slender and have the margins and midrib strongly thickened and raised so that each blade has two narrow grooves running its entire length. The flowers are unusual in their colouring; the tepals are cream to pale straw but densely speckled with small, dark red spots. Within section *Heterocolon* the leaves of *G. rufomarginatus* correspond closely with those of another endemic of eastern southern Africa, *G. vernus*, which has a pale pink perianth and more strongly bilabiate flowers. It flowers in winter and spring when its leaves have withered and died so the flowering stems appear leafless. Because of the difference in the colour and shape of the flowers of the two species and their difference in flowering time, there is no possibility of confusing them.

History

Gladiolus rufomarginatus was first recorded in the 1880s by the German apothecary and naturalist, Friedrich Wilms, who lived in Lydenburg from 1883 to 1896. Wilms collected *G. rufomarginatus* near the town, perhaps to the north along the road to Ohrigstad where the species is quite common in late summer. Based on a later collection made by Wilms in 1895, Frank Bolus described the species in 1917, naming it *G. marginatus*. Unfortunately, this is a homonym for *G. marginatus* Linnaeus fil. (1781), now *Watsonia marginata*, and the new name, *G. rufomarginatus*, was given to the plant by G.J. Lewis (Lewis et al., 1972).

78. GLADIOLUS VERNUS Obermeyer
PLATE 64

Gladiolus vernus Obermeyer in *J. S. African Bot.*, Suppl. 10: 138 (1972). Type: South Africa, Mpumalanga, Blyde River Canyon, lookout point c. 9 km north of Bourke's Luck (DB), 23 Aug. 1963, *Leistner, Thom & Gillham 3314* (PRE, lectotype here designated, the specimen with a corm; BOL, K, PRE, isolectotypes).

vernus = of spring, so named for the flowering period, early spring, an unusual time for flowering in *Gladiolus* species in the summer-rainfall region of southern Africa.

Description

Plants (35–)45–75 cm high. *Corm* more or less globose, 20–40 mm in diameter, the tunics of moderately coarse fibres, persisting and accumulating in a thick mass. *Cataphylls* pale and membranous, the uppermost reaching 2–4 cm above the ground and then green or dry and light brown, persisting and accumulating in a thick fibrous neck around the base of the stem. *Leaves* usually six, the lower four basal, produced the summer before flowering and still present the following spring but partly or completely dry, up to 45 cm long, stiff and gently curving, the blades linear, more or less oval in transverse section, 2–3 mm wide, the margins and midrib raised and thickened, thus 2-grooved on each surface, the uppermost two leaves inserted on the upper half of the stem, largely to entirely sheathing, the margins free to the base, 20–65 cm long, more or less dry, often flushed with pink. *Stem* erect below, inclined above, simple or with up to two branches, c. 2 mm in diameter below the spike.

Spike inclined, lightly flexuose, the main axis usually 12- to 18-flowered; *bracts* partly dry at flowering, often flushed pink above, the outer 10–15(–22) mm long, the inner about three-quarters to as long as the outer, broad, enveloping the flower. *Flowers* pale to deep pink, occasionally cream, the upper tepals darker above, the lower lateral or all three lower tepals pale yellow in the middle third, the yellow zone sometimes edged with dark pink or purple, a cream to yellow blotch in the upper half edged at the base with red or purple, the lower halves of the lower tepals speckled with pink, unscented; *perianth tube* obliquely funnel-shaped, expanded near the apex, 8–13 mm long; *tepals* unequal, the dorsal largest, inclined over the stamens, (17–)20–30 x 10–18 mm, the upper laterals spathulate, asymmetric, narrow at the base, 15–26 x 8–11 mm, usually windowed between the dorsal and upper lateral tepals, the lower three tepals united with the upper laterals for 5–8 mm and to

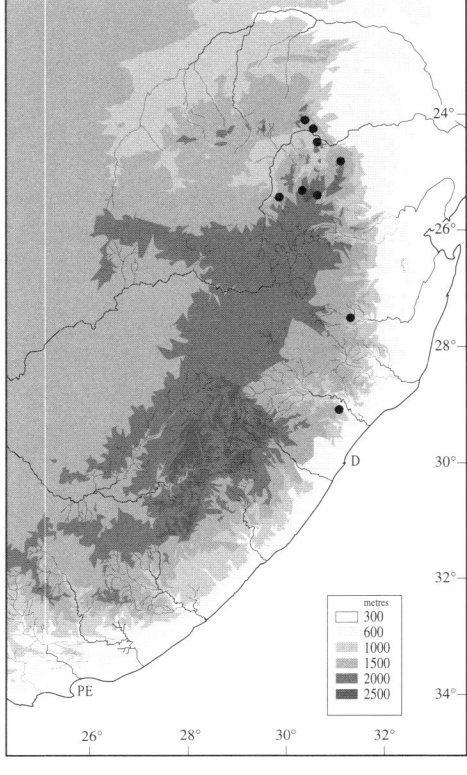

one another for 1.5–3 mm, lower laterals 10–13 mm long, straight, inclined c. 30° below the horizontal, narrowed below into linear channelled claws 2–3 mm long, abruptly expanded into broadly oval limbs c. 7 x 6 mm, the limbs auriculate at the base, the lower median also clawed and the limb auriculate at the base, 11–15 x 6–8 mm, the claw 1–3 mm long, in profile the lower tepals extending c. 5 mm beyond the upper. *Filaments* 10–16 mm long, exserted c. 8–10 mm from the tube; *anthers* 6–8 mm long, purplish, the pollen whitish. *Ovary* ovoid, c. 3 mm long; *style* arching over the stamens, dividing between the base and upper third of the anthers, the branches 2–3 mm long. *Capsules* obovoid, 10–12 mm long, three-lobed and retuse above; *seeds* ovate, c. 6 x 4 mm, the wing well developed, the seed body comparatively large, light brown. *Chromosome number* unknown.

Flowering time July and August.

Distribution and biology

Until recently very poorly collected and still only known from a few isolated sites, *Gladiolus vernus* has a range that extends from Hermansburg near Greytown in KwaZulu-Natal to the Wolkberg in Northern Province, South Africa. The range thus includes almost the entire length of the southern African summer-rainfall zone. Plants favour drier places in the well-watered eastern escarpment and are usually found in rocky grassland where the corms are protected from predation. The growing season, when the foliage leaves are produced, is late spring and summer, the leaves emerging in September or October and remaining green until April or May when they begin to die back. Flowering of the leafy shoot is, however, delayed until late winter or early spring. Flower spikes then emerge through dry grass and the pale pink flowers come into bloom at a time when there are few other plants in flower.

The fairly small flowers are assumed to be pollinated by long-tongued bees.

Diagnosis and relationships

Gladiolus vernus is recognized by its small, pale pink flowers with cream markings on the lower tepals and its flowering in late winter or spring, by which time the leaves are dry. The three to four basal leaves are firm-textured, fibrotic and linear, with the blades having strongly thickened margins and midrib. The blades are thus narrowly two-grooved on each surface. The leaves closely resemble those of the eastern Mpumalanga species, *G. rufomarginatus*, and the two species are probably immediately related. The Zimbabwe endemic, *G. juncifolius*, which also flowers in the late winter and early spring and has the leaves dry at flowering time, may also be closely allied to *G. vernus*. The two species differ in the structure of their leaf blades – linear with narrow grooves on each surface in *G. vernus*, terete and four-grooved in *G. juncifolius* – and the two also have flowers with somewhat differently oriented tepals.

History

Gladiolus vernus appears to have first been recorded in 1917 from the Harmony Block in the former Eastern Transvaal by H.G. Breyer, then Director of the Transvaal Museum in Pretoria. This collection was overlooked for many years, and when *G. vernus* was described in 1972 by Amelia Obermeyer the species was founded on more recent collections from the Drakensberg escarpment of Mpumalanga and Northern Province (Lewis et al., 1972). More recent collecting has filled in many of the gaps in the distribution of the species and extended its known range to KwaZulu-Natal where it was discovered near Hermansburg by the South African botanist, E.G.H. Oliver.

79. GLADIOLUS KAMIESBERGENSIS G. Lewis
PLATE 65

Gladiolus kamiesbergensis G. Lewis in *J. S. African Bot.*, Suppl. 10: 128 (1972). Type: South Africa, Northern Cape, Kamiesberg Mountains, Farm Welkom, 16 Oct. 1954, *Esterhuysen 23716* (BOL, holotype; BOL, K, isotypes).

kamiesbergensis = from the Kamiesberg, a massif in central Namaqualand.

Description

Plants 45–90 cm high. *Corm* globose, 18–22 mm in diameter, the tunics of coriaceous layers, breaking into vertical fibres of moderately thick texture with age. *Cataphylls* membranous below ground, dry and papery above the ground, rich red-brown, accumulating with the leaf bases into a thick neck around the base of the stem. *Leaves* usually four, the lowermost longest, often exceeding the spike, the blade terete with the midrib and margins raised and strongly thickened, with four hairline grooves the length of the blade, c. 1.3 mm in diameter, second leaf much shorter and sheathing the stem below, the blade like the basal but more slender, upper two leaves entirely sheathing or with vestigial blades, the sheaths open to the base and the lower margins overlapping (rarely the sheaths closed near the base). *Stem* erect, unbranched, c. 1.8 mm in diameter below the spike.

Spike more or less erect or lightly flexed at the base and barely inclined, 4- to 9-, occasionally to 12-flowered; *bracts* pale green, the margins hyaline, the outer 13–18 mm long, the inner three-quarters to about as long as the outer. *Flowers* pale lilac, with minute purple dots within and on the reverse of the tepals, especially toward the apices, the lower lateral tepals yellow on the lower half of the limbs, sweetly apple-scented; *perianth tube* obliquely funnel-shaped, 12–13 mm long; *tepals* unequal, the dorsal largest, arching to hooded over the stamens, 16–21 x 10–12 mm, the upper laterals directed forward, the apices curving outward, 16–18 x 8–10 mm, windowed between the bases of the dorsal and upper lateral tepals, the lower tepals united with the upper laterals for 5 mm and to one another for 1–2 mm, lower laterals narrowed below into linear channelled claws 3–4 mm long, abruptly expanded into oval limbs c. 7 x 7 mm, limbs cucullate, inclined 30° below the horizontal, lower median lanceolate, c. 9 x 7 mm, curving toward the ground, in profile the lower tepals slightly exceeding the upper. *Filaments* arched under the dorsal tepal, c. 12 mm long, exserted c. 9 mm from the tube; *anthers* c. 6 mm long, lilac, the pollen whitish. *Ovary* narrowly obovoid, c. 4 mm long; *style* arching over the stamens, dividing opposite the middle of the anthers, the branches c. 2 mm long, not

much expanded above, reaching to about the apices of the anthers. *Capsules* obovoid, the apices obtuse, 15–18 mm long; *seeds* ovate, c. 5.5 x 6 mm, broadly and evenly winged,

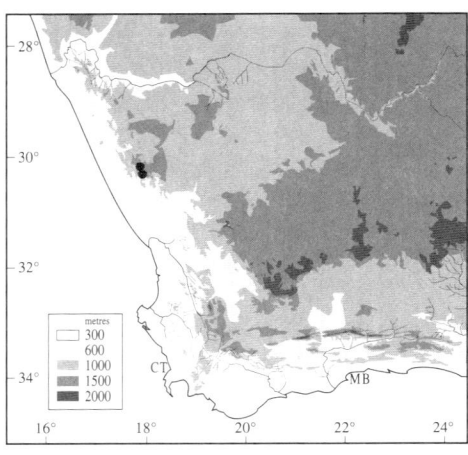

light brown, the seed body reddish brown. *Chromosome number* unknown.

Flowering time occasionally in late September, more often in October.

Distribution and biology

Gladiolus kamiesbergensis is restricted to the Kamiesberg of central Namaqualand in Northern Cape Province, South Africa. It has been recorded once from Sneeukop in the north of the range and several times on Rooiberg and its foothills some distance to the south. Even in full flower the species is inconspicuous, the small, pale flowers hardly visible from more than a few metres away. Plants grow in rocky sites in shrubby fynbos above 1200 m elevation. They may benefit from occasional fires and then flower well, but we have seen them flowering in veld unburned for more than 12 years.

The small flowers of *Gladiolus kamiesbergensis* are believed to be adapted for pollination by long-tongued bees.

Diagnosis and relationships

The small, pale, bluish white flowers of *Gladiolus kamiesbergensis* have a somewhat bell-like appearance, and this together with the awl-shaped, terete leaves with four narrow grooves at first suggest that the species is allied to the bluebells of the southern and western Cape, for example, *G. inflatus*, *G. patersoniae* and *G. rogersii*. We doubt the likelihood of this relationship and consider the similarity the result of convergence. More likely *G. kamiesbergensis* belongs in section *Heterocolon*, where it is perhaps most closely related to the Roggeveld endemic, *G. marlothii*. These two species have terete leaves and multiflowered spikes, and characteristic, minutely speckled flowers. Their position in section *Heterocolon* is suggested by their peculiar oblique upper lateral tepals and narrowly clawed lower tepals with abruptly expanded broad limbs, features shared only with *G. mostertiae*, *G. marlothii* and the Mpumalanga–KwaZulu-Natal species, *G. vernus*.

History

The first record of *Gladiolus kamiesbergensis* appears to be the collection made by the English botanist, John Hutchinson, on Sneeukop in the northern Kamiesberg in 1928. This and later collections by C.L. Leipoldt in 1940 and Elsie Esterhuysen in 1954 – all to the south on the Rooiberg massif – remained undescribed and largely misunderstood. G.J. Lewis realized the species was new and it was described in her posthumously published revision of *Gladiolus* in South Africa (Lewis et al., 1972).

80. GLADIOLUS MARLOTHII G. Lewis

PLATE 66

Gladiolus marlothii G. Lewis in *J. S. African Bot.*, Suppl. 10: 141 (1972). Type: South Africa, Western Cape, Farm Driefontein, north of Sutherland, 1400 m, Oct. 1920, *Marloth 9757* (PRE, holotype; BOL, isotype).

marlothii, named in honour of Rudolf Marloth, botanist and expert on Cape flora at the beginning of the twentieth century.

Description

Plants 45–60 cm high. *Corm* globose, 16–22 mm in diameter, the tunics of coarsely textured, mostly vertical fibres, accumulat-

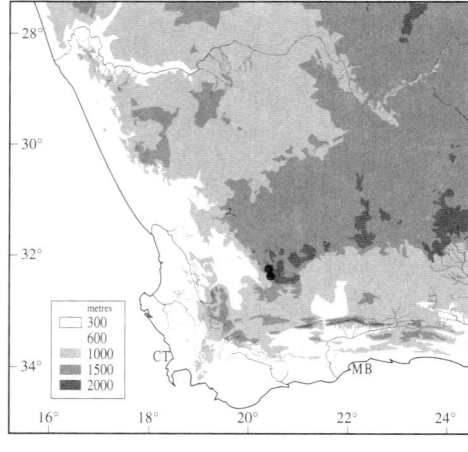

ing with age in a dense mass. *Cataphylls* pale and firm-textured, the upper reaching to 10 cm above the ground and usually dry and brown at anthesis, often purple and velvety when young, accumulating with age as a coarse fibrous neck around the underground part of the stem. *Leaves* four, the lowermost longest, sheathing the lower third of the stem, the sheath shortly pilose, the blade reaching or exceeding the top of the spike, centric, 3–4 mm wide, the midrib raised and forming flanges either side of the blade, thus cross-shaped in section, the margins and midrib flanges thickened and shortly villous on the edges, upper leaves progressively shorter and with villous sheaths, the second leaf with a well-developed blade often reaching at least the base of the spike, uppermost leaf inserted in the upper third of the stem, usually entirely or largely sheathing, the blades of the two upper leaves sometimes reduced, or fairly well developed and similar to the basal but only 1.5–2 mm in diameter. *Stem* erect or slightly inclined, unbranched, c. 2 mm in diameter below the spike.

Spike flexed at the base and inclined, almost straight, 3- to 5-flowered; *bracts* pale green or becoming pale and dry toward the apices, by the end of flowering more or less entirely membranous, the outer 16–24 mm long, the inner nearly as long as the outer. *Flowers* pale blue-lilac, often slightly darker toward the tips of the upper tepals, the lower lateral tepals with a transverse yellow band above the base of the limb and densely speckled with dark purple at the edges of the yellow band, lower median tepal with a yellow blotch in the lower midline, the lower part of the throat also speckled, the tepals forming a slightly inflated bell, unscented; *perianth tube* 9–10 mm long, obliquely funnel-shaped, the lower half erect and cylindric, c. 4 mm long, the upper half bent at right angles and flaring outward; *tepals* unequal, the dorsal largest, arched over the stamens and more or less horizontal, 24–30 x 18–20 mm, upper laterals c. 20 x 17 mm, directed forward and only curving outward near the apices, slightly concave just above the base, the lower tepals united with the upper laterals for 5–8 mm and to one another for c. 2 mm, the lower laterals spathulate, clawed below, the claws c. 6 x 1.3 mm, expanded abruptly into a broadly ovate or oblong limb 12 x 10 mm, straight and tilted c. 20° toward the ground, more or less horizontal, the lower median tepal c. 14 mm long, narrowly clawed in the lower 1–2 mm, in profile the lower tepals exceeding the upper by 5–10 mm. *Filaments* c. 13 mm long, exserted from the tube for c. 9 mm but included in the flower; *anthers* c. 9 mm long, lilac, pollen cream. *Ovary* oblong, c. 4 mm long, style arched over the stamens, dividing opposite the upper third of the anthers, the branches c. 4.5 mm long, recurving. *Capsules* obovoid, three-lobed and retuse above, 16–19 mm long; *seeds* more or less ovate, 6–7 x 4–5 mm, the wing well developed, light brown, the wing paler than the seed body. *Chromosome number* unknown.

Flowering time mid-September to early October.

Distribution and biology

Gladiolus marlothii is endemic to the central part of the Roggeveld Escarpment in the western Karoo of Northern Cape Province, South Africa. First collected by the German-born botanist, Rudolf Marloth in 1920, it has been gathered on only a few occasions since then. It has long been assumed to be rare, but it is relatively common locally. Plants are confined to the escarpment edge and a short distance inland where the rainfall is substantially higher than even a few miles inland, or to the north or south. *Gladiolus marlothii* extends from Sneeukrans in the south to near the farm Rooiwal in the north. It is particularly common in the area around Gannaga Pass. Plants grow among karroid shrubs and grasses, either on open slopes or among rock outcrops, always in heavy clay and usually in rocky situations, and at an elevation of about 1800 m. Night temperatures fall below freezing in winter and substantial snowfalls occur in some years. Because of the relatively high precipitation and cool temperatures during winter and early spring, *G. marlothii* grows in soil that is moist throughout its growing season.

The flowers of *Gladiolus marlothii* are adapted for pollination by long-tongued

bees. Like those of most other species of section *Heterocolon*, the flowers are, however, unscented. During the late spring of 1995 plants flowering near Gannaga Pass were found to be actively visited by the large anthophorid bee, *Anthophora diversipes*.

Diagnosis and relationships

Gladiolus marlothii has distinctive leaves, the sheaths of which are pilose and the blades centric and cross-shaped in transverse section. The thickened margins and midrib are also pilose. The bell-shaped perianth, a pale blue-lilac colour, the short lower median tepal and the yellow transverse bands on the lower lateral tepals are also diagnostic. The upper lateral tepals are notable in being slightly concave just above the base, a feature confined to just a handful of species of section *Heterocolon*. The pale green to membranous floral bracts and three-lobed capsules also conform to that section. The more or less bell-like flower recalls those of the so-called bluebells, a group in section *Homoglossum* which includes *G. bullatus* and *G. inflatus*. Despite the floral resemblance, the soft-textured floral bracts, narrowly clawed lower tepals, concave base of the upper lateral tepals, and three-lobed capsules of *G. marlothii* are inconsistent with an association to these species.

Within section *Heterocolon* the relationship of *Gladiolus marlothii* appears to be closest to the Namaqualand endemic, *G. kamiesbergensis*. The resemblance is so strong that the flowers of the two may be confused until the dimensions are compared. *Gladiolus kamiesbergensis* has smaller flowers, usually paler in colour, and the leaves are glabrous. Like *G. marlothii*, *G. kamiesbergensis* and two other species of section *Heterocolon* – the Nieuwoudtville endemic, *G. mostertiae*, and *G. vernus* of eastern southern Africa – also have unusual, basally concave upper lateral tepals. The disjunct distribution pattern, one species in eastern southern Africa and three restricted to small highland areas along the

interior of the southern African winter-rainfall area, is without parallel in the Iridaceae.

History

Gladiolus marlothii was evidently discovered in 1920 by the German-born botanist, Rudolf Marloth, but this very distinctive species remained very much a mystery for many years. A few later collections from the Roggeveld Escarpment made it clear that this was indeed a distinct species. It remained unnamed until 1972 when it was formally described by Lewis et al.

---

81. GLADIOLUS MOSTERTIAE L. Bolus

PLATE 67

Gladiolus mostertiae L. Bolus, *Ann. Bolus Herb.* 3: 142 (1922). G.J. Lewis et al., *J. S. African Bot.*, Suppl. 10: 144 (1972). Van Wyk, *Veld & Flora* 67: 85 (1981). Type: South Africa, Northern Cape, Calvinia District, Cloudskraal near Nieuwoudtville, cultivated at Kirstenbosch Botanic Gardens, NBG 1609/20, Nov.

1921, *Mostert s.n.* (BOL, holotype).

mostertiae, named in honour of Aletta Johanna Mostert, of the farm Cloudskraal on the Bokkeveld escarpment near Nieuwoudtville, who sent the first recorded specimens of the species to the National Botanical Gardens, Kirstenbosch in 1920.

Description

Plants (20–)30–50 cm high. *Corm* globose, c. 12–14 mm in diameter, the tunics of papyraceous to softly coriaceous layers, soon becoming finely fibrous, not persisting for more than two or three years. *Cataphylls* pale and membranous, the uppermost reaching 0.5–2 cm above the ground and then

purple to brownish and often becoming dry near the apex. *Leaves* usually four, occasionally three, the lower two basal and longest, reaching to between the base and middle of the spike, the blades linear, or occasionally narrowly lanceolate, 2.5–4 mm wide, lightly to densely villous, the veins strongly thickened and the surface ridged, the margins lightly thickened, the upper two leaves inserted in the middle to upper third of the stem, progressively smaller above, usually channelled throughout but sheathing for a short distance or diverging from the stem at the base. *Stem* erect, moderately to strongly flexed above the sheathing parts of the three upper leaves and flexuose, usually with one branch, occasionally with two or three, 1–1.5 mm in diameter below the spike.

Spike erect, nearly straight or lightly flexuose, 4- to 10-flowered; *bracts* bright green or flushed with purple dorsally, firm-textured, broadly truncate, the upper margins usually dull purple, the outer 10–14 mm long, the inner about two-thirds as long, minutely notched apically. *Flowers* pale to deep pink, darker toward the tips of the upper tepals, the lower lateral tepal limbs cream and each with a heart-shaped yellow mark with a greenish centre toward the base, the lower tepal claws and wide part of the tube with minute reddish dots, with a faint acrid

metallic odour; *perianth tube* obliquely funnel-shaped, c. 11 mm long; *tepals* unequal, the dorsal largest, arching to hooded over the stamens, c. 20 x 12 mm, the upper laterals directed forward, curving outward toward the apices, c. 15 x 9 mm, sometimes windowed between the bases of the dorsal and upper lateral tepals or these overlapping, the lower tepals united with the upper laterals for c. 10 mm and to one another for c. 1 mm, the lower laterals c. 11 mm long, straight, narrowed below into linear channelled claws c. 4 mm long, abruptly expanded into broadly oval limbs c. 7 x 6 mm, the limbs cucullate, inclined 30° below the horizontal, the lower median tepal not clawed, broadly lanceolate, c. 9 x 8 mm, straight, in profile the lower tepals extending c. 8 mm beyond the upper. *Filaments* c. 15 mm long, strongly arched, exserted c. 8 mm from the tube; *anthers* c. 4 mm long, lilac, the pollen whitish. *Ovary* narrowly oblong, c. 4.5 mm long; *style* arching over the stamens, dividing between the lower and upper third of the anthers, the branches 3.5 mm long, expanded and channelled in the upper half, the apices reaching or slightly exceeding the anthers when receptive. *Capsules* ellipsoid-trigonous, the locule sutures deeply grooved, (14–)18–20 mm long; *seeds* ovate, 4–5 x c. 3 mm, broadly winged, the wing light brown and semi-transparent, the seed body darker. *Chromosome number* unknown.

Flowering time mid-November to mid-December.

Distribution and biology

Gladiolus mostertiae has a narrow distribution near Nieuwoudtville in Northern Cape Province where it occurs along the edge of the Bokkeveld Escarpment near Vanrhyn's Pass. This is the northernmost extension of the Table Mountain Sandstone Series and the vegetation is typical fynbos. The sandy soils of the Bokkeveld Escarpment harbour an unusually large number of endemic geophytes. Most of these flower in September, and *G. mostertiae* is unusual in its late flowering. The species is rare but fairly common in suitable semi-marshy places that remain moist long into the dry season, from mid-September to April.

The flowers are adapted for pollination by long-tongued bees and we have observed the anthophorid bee *Amegilla obscuriceps* actively visiting them.

Diagnosis and relationships

Immediately recognized by its plane, pilose leaves, small, short-tubed, pale pink flowers and branched stems, *Gladiolus mostertiae* is one of the most distinctive of the *Gladiolus* species of the winter-rainfall region. The flowers have an unusual colouring for the pinkish mauve upper tepals are darker towards the tips, while the limbs of the lower lateral tepals are cream, each with a heart-shaped yellow mark with a greenish centre. The lower tepal claws and abaxial part of the throat are minutely dotted with red. Despite the unusual, plane, pilose leaves which do not match those of other species of section *Heterocolon*, we have no doubt that this is where its affinities lie.

Both the lower tepal markings of *Gladiolus mostertiae* and their spathulate shape with the limb spooned and twisted somewhat obliquely recall the flowers of *G. kamiesbergensis* and *G. marlothii* – also in series *Vernus* – and we believe that the three species are a clade. The similarity in the shape of the flowers of *G. mostertiae* and *G. marlothii* was also noted by Lewis et al. (1972), who placed them next to one another in their treatment of southern African *Gladiolus*. The pale pinkish flowers are unusual in section *Heterocolon* in being lightly scented, a character shared only with *G. kamiesbergensis* in the section.

History

First collected in or about 1920 by Miss A. Mostert at Nieuwoudtville, and only occasionally recorded thereafter, *Gladiolus mostertiae* remained for many years a puzzling species both as regards its exact wild habitat and its relationships. Corms were grown for some years at Kirstenbosch Botanical Gardens in Cape Town, and the type collection was made from cultivated plants. The species was described in 1922 by H.M.L. Bolus. Rediscovered only in 1980 by the botanist, B.-E. van Wyk, the species was relocated for this revision by local farmer Neil MacGregor.

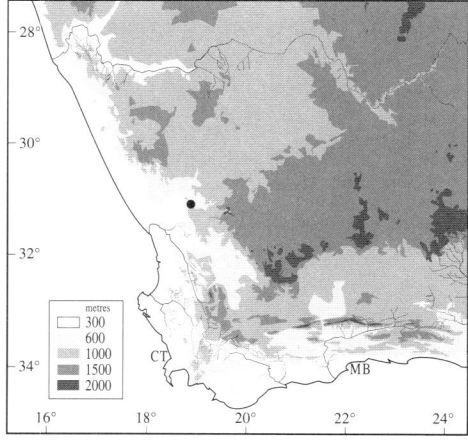

6. SECTION *HEBEA*

(Persoon) Bentham & J.D. Hooker

Section *Hebea* (Persoon) Bentham & J.D. Hooker, *Genera Plantarum* 3(2): 710 (1883). Goldblatt, Gladiolus *in Tropical Africa* 169 (1996). *Gladiolus* group *Hebea* Persoon, *Synopsis Plantarum* 1: 44 (1805). Subgenus *Hebea* (Persoon) Baker, *J. Linn. Soc., Bot.* 16: 177 (1877) – regarded as a combination although the basionym was not directly cited. Type species: *G. alatus* Linnaeus (lectotype designated by Goldblatt, 1996).

Description

Plants usually small, sometimes moderate in size. *Corm* globose to depressed-globose, the tunics variously papery, coriaceous, coarsely fibrous, more or less woody, or membranous and not accumulating with age, often producing cormlets at the base, sometimes these large and produced on long, slender stolons. *Cataphylls* usually glabrous, occasionally mottled purple and white. *Leaves* several to few, sometimes only three or four, usually at least three basal, the blades usually well developed at flowering, linear to narrowly sword-shaped, usually less than 5 mm wide, the margins unthickened and the blades plane and without well-developed secondary veins, raised and thickened in the midline, ridged and corrugate in a few species, rarely leaves dry at flowering and blades terete. *Stem* often branched, especially in robust plants, compressed and angled to winged in a few species.

Spike usually inclined, sometimes nearly horizontal and scalloped; *bracts* short, usually green and soft in texture, membranous or dry above in a few species, generally about as long as or slightly longer than the perianth tube. *Flowers* small to medium-sized, the upper tepals usually conspicuously narrowed below, sometimes abruptly so, nectar guides usually present on the lower tepals, consisting of solid contrasting pale colour edged with darker pigment on the lower halves of the tepals, usually scented, sometimes strongly so; *perianth tube* usually obliquely funnel-shaped, mostly about half as long as the dorsal tepal, or tubular and about twice as long in a few species; *tepals* usually all more or less clawed (at least strongly

narrowed in the lower half), the dorsal inclined, hooded or sometimes erect, lower tepals narrower and usually shorter than the dorsal, usually clawed below and basally united. *Filaments* pubescent in a few species; *anthers* obtuse, in a few species the anthers attached near the midline and the thecae free below this point. *Capsules* mostly oblong-ellipsoid to spindle-shaped, in a few species obovoid-globose and three-lobed above; *seeds* usually broadly and evenly winged, translucent light to golden brown, rarely pinkish, the seed body often blackish.

With 31 species, section *Hebea* is one of the largest infrageneric alliances among the southern African species of *Gladiolus*. It is centred in the drier interior of the winter-rainfall zone and is the only section that has speciated significantly in the semi-arid west coast and near interior. Just a few species extend into the summer-rainfall region of the continent and only *G. permeabilis* occurs outside southern Africa, extending as far north as Zimbabwe. The Angolan *G. fenestratus*, which was placed in section *Hebea* by Goldblatt (1996), is now included in section *Heterocolon*.

The defining features of section *Hebea* are the ellipsoid capsules and the seeds with dark seed bodies. The corm tunics are also specialized in being composed of tough, wiry fibres, but derived lineages have either woody or membranous tunics. The leaves are usually narrow and are remarkable in the genus in having thickened and raised midveins but with the margins completely unthickened. Most species also have unusual nectar guides, the lower halves of the lower tepals being a contrasting solid, usually pale colour, sometimes with a darker edge. Upper tepals that are narrowed below, often abruptly, and thus more or less clawed, are restricted to only three of the four series we recognize, and are not developed in series *Involutus*.

Flowers of section *Hebea* are extremely diverse in colour and tepal orientation. They are usually sweetly scented, especially in species of drier habitats, but members of series *Involutus* lack the freesia-like fragrance typical of so many species of the section.

Apparently scent evolved in the ancestor of the remaining three series. Although the ancestral capsule shape in section *Hebea* is most likely ellipsoid, a few species of series *Deserticola*, including *Gladiolus salteri*, *G. scullyi* and *G. venustus*, have specialized obovoid to rotund capsules which are three-lobed above.

Most notable of the four series of the section is the very specialized series *Hebea* (commonly known as *kalkoentjies*) which is more or less restricted to the winter-rainfall zone. Plants are mostly small but the flowers are often quite large, and either brightly coloured reddish orange to pink with bright yellow nectar guides, or cryptically coloured in shades of dull purple, green or brown. All have strongly scented flowers and with one exception are pollinated by the same large anthophorid bees, *Anthophora diversipes* and its relatives. The exception, *Gladiolus meliusculus*, has unusually dark bands of colour on the lower tepals and appears to be adapted for pollination by monkey beetles (Scarabaeidae: Rutelinae:Hopliinae).

Species of series *Involutus* have unscented flowers and are adapted for pollination by bees or long-tongued flies in three, evidently basal species, or by sunbirds. Flowers of the species that are adapted for bird pollination are bright red, and the dorsal tepal is prominent and often spoon-shaped. Unusually for this pollination syndrome in Iridaceae, the lower tepals – and in three of the species even the upper lateral tepals – show extreme reduction in size. These species also have relatively short, though wide tubes, and may be adapted for pollination specifically by the smaller, shorter-billed sunbirds such as the lesser double-collared sunbird *Nectarinia chalybea* and the dusky sunbird *N. fusca*. Only *Gladiolus robertsoniae* and *G. acuminatus*, both of series *Permeabilis*, have flowers adapted for moth pollination. Long-tongued fly pollination is also poorly represented in the section, evidently occurring only in *G. leptosiphon* (series *Involutus*) and *G. lapeirousioides* (series *Deserticola*).

82. GLADIOLUS LEPTOSIPHON F. Bolus

PLATE 68

Gladiolus leptosiphon F. Bolus, *Ann. Bolus Herb.* 1: 196 (1915). Lewis et al., *J. S. African Bot.*, Suppl. 10: 80 (1972). *Radinosiphon leptosiphon* (F. Bolus) N.E. Brown, *Trans. Roy. Soc. S. Africa* 20: 263 (1932). Type: South Africa, Western Cape, without precise locality, Cape Town Wild Flower Show, among plants probably from the Oudtshoorn or Riversdale Districts, Oct. 1914, (BOL 13754, holotype; K, PRE, isotypes).

leptosiphon = with a slender tube, alluding to the long, thin perianth tube.

Description

Plants 25–45 cm high. *Corm* 14–18 mm in diameter, the tunics composed of fairly coarse fibres, often strongly developed into vertical claw-like ribs below. *Cataphylls* membranous, the upper sometimes purple above the ground. *Leaves* about six, the lower four basal and longest, reaching to about the top of the spike, the blades linear, 2–2.5 mm wide, thickened in the midrib area, the margins not thickened, the two upper leaves cauline and much shorter than the basal, the sheaths open to the base. *Stem* erect, flexed outward above the sheath of the uppermost leaf, simple or branched, c. 1.3–2 mm in diameter below the spike. *Spike* inclined, more or less straight, 6- to 9-flowered; *bracts* fairly soft-textured, light green, usually becoming pale and dry above, the outer (15–)18–22 mm long, the inner slightly shorter (rarely slightly longer), forked apically for c. 1 mm. *Flowers* cream to pale yellow, the lower three tepals each with a dark red to purple median streak in the lower half, the streak extending to the base of the throat, unscented; *perianth tube* 45–50 mm long, the lower part 40–45 mm long, cylindric and curving gently outward, expanding into a wider throat c. 5 mm long; *tepals* unequal, the upper three largest, narrowly lanceolate, attenuate apically, 25–35 x 9–12 mm, directed forward below, arching outward above, the dorsal arched over the stamens, the lower three tepals united with the upper laterals for c. 3 mm, the lower laterals straight, more or less linear, 18–22 x c. 3.5 mm, the lower median narrowly lanceolate, curving below the lower laterals and exceeding them, 25–32 x 4–5 mm. *Filaments* c. 10 mm long, exserted 4–5 mm from the tube, usually light purple; *anthers* 7–8 mm long, dark blue-purple, pollen purplish. *Ovary* oblong, 3–4 mm long; *style* arching over the stamens, dividing close to the anther apices, the branches c. 2 mm long, only slightly exceeding the anthers. *Capsules* obovoid-ellipsoid, c. 15 mm long, the apices obtuse; *seeds* ovate, c. 4.5 x 3 mm, broadly and evenly winged, reddish brown, the seed body blackish. *Chromosome number* $2n = 60$.

Flowering time mid-September to late October.

Distribution and biology

Relatively poorly known, *Gladiolus leptosiphon* appears to be a fairly localized montane endemic of the Western and Eastern Cape provinces of South Africa. Its range extends from the slopes of the Swartberg near Ladismith to the Kouga and Baviaanskloof Mountains of the interior and to the southern Cape coast west of Port Elizabeth. It favours relatively dry habitats and occurs in renosterveld and arid fynbos vegetation, mostly on stony, shale-derived soils, and most often on north-facing slopes at relatively low elevations.

The long perianth tube and pale-coloured, unscented flowers with reddish nectar guides suggest pollination by long-tongued flies of the families Nemestrinidae or Tabanidae, but observations on insect visitors have yet to be made.

Diagnosis and relationships

Gladiolus leptosiphon is readily recognized by its several narrow, whip-like leaves and pale yellowish cream flowers with an elongate tube 45–50 mm long. The flowers are unscented and the lower tepals have unusual nectar guides consisting of dark, reddish median streaks. As in other members of series *Involutus*, the lower median tepal much exceeds the lower laterals.

The corm tunics consist of fairly hard, wiry fibres, most likely the ancestral tunic type in section *Hebea*, and the plants produce a few, fairly large cormlets from the corm bases, sometimes on short stolons or enclosed within the tunics. Except for the long-tubed flowers, the general aspect of *Gladiolus leptosiphon* much resembles that of *G. involutus* and we suspect that the two species are fairly closely related.

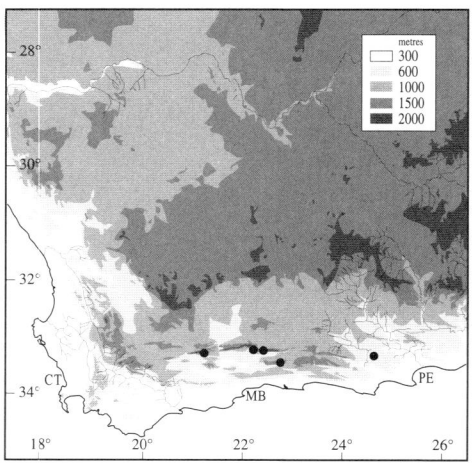

History

Apparently unknown until 1914, *Gladiolus leptosiphon* was found at the Cape Town Wild Flower Show of that year and plants were preserved at the Bolus Herbarium.

Frank Bolus described the species the following year. It was later confirmed to be from the Oudtshoorn District by L.E. Taylor, and was collected from Meiringspoort in the Swartberg in 1941 by F.W. Thorns, who at

the time was the curator of Kirstenbosch Gardens. *Gladiolus leptosiphon* remains a poorly known species, recorded from just a handful of localities in the Eastern Cape and eastern part of Western Cape.

83. GLADIOLUS LOTENIENSIS Hilliard & Burtt

PLATE 69

Gladiolus loteniensis Hilliard & Burtt, *Notes Roy. Bot. Gard. Edinburgh* 43: 206 (1986). Type: South Africa, KwaZulu-Natal, Loteni Nature Reserve, Loteni River Valley near Ash Cave, c. 6000 ft, 13 Jan. 1982, *Hilliard & Burtt 15134* (NU, holotype; E, isotype).

loteniensis = from Loteni, the nature reserve in KwaZulu-Natal where the species was discovered.

Description

Plants 30–60 cm high. *Corm* globose, poorly developed, c. 5–8 mm in diameter, the tunics of papery layers, these decaying to form a fibrous network, producing several fine white stolons from the base, each terminating in a cormlet. *Cataphylls* pale and

membranous, the uppermost reaching 1–2 cm above the ground and then becoming purple. *Leaves* four to seven, the lower three basal and longest, reaching to between the middle of the stem and the base of the spike, the blades narrowly lanceolate to almost linear, 4–8 mm wide, the midrib lightly thickened and raised, the margins narrowly hyaline but not thickened, the remaining leaves inserted between the middle and upper quarter of the stem, progressively smaller above, the uppermost or the two upper leaves sheathing for at least half their length and channelled almost throughout. *Stem* erect below, flexed outward near the base of the spike, occasionally with a short lateral branch, 1–1.5 mm in diameter below the spike.

 Spike inclined, straight to lightly flexuose, usually 6- to 12-, rarely only 3-flowered, the flowers in two ranks c. 60° apart; *bracts* soft-textured, pale green on the ventral side, dull purple dorsally, the outer 14–16 mm long, the inner 1–2 mm shorter than the outer, minutely notched apically, usually twisted out of position to lie almost under the outer bract. *Flowers* pale lilac, on the reverse slightly darker toward the lower centre of the upper three tepals, the lower three tepals each veined violet on a cream to pale yellow background in the lower two-thirds, the upper third pale or dark lilac, the tepal sinuses marked with a dark spot, unscented; *perianth tube* obliquely funnel-shaped, c. 5 mm long; *tepals* unequal, oblanceolate-attenuate, the dorsal largest, horizontal or tilted to the ground, 19–27 x 16–18 mm, the upper laterals twisted upward and appressed to the dorsal in the lower half, curving obliquely upward in the upper third, 17–19 x c. 13 mm, the lower three tepals united for c. 1.5 mm, straight and tilted toward the ground, the lower laterals 12–13.5 x c. 5 mm, narrowed below into claws c. 2 mm long, the limbs abruptly expanded, narrowly channelled in the lower third, becoming flat above, the lower median 15–18 x 5.5–7 mm. *Filaments* c. 7 mm long, exserted c. 5 mm from the tube; *anthers* c. 6 mm long, violet, the pollen pale

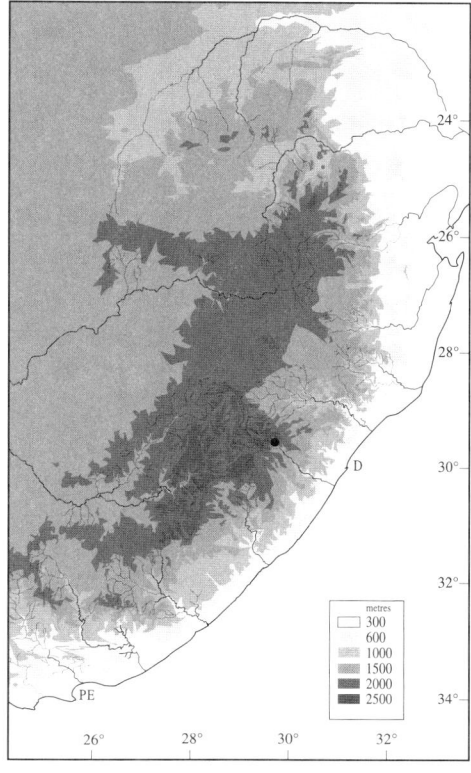

lilac. *Ovary* narrowly obovoid, c. 3 mm long; *style* arching over the stamens, dividing between the middle and upper third of the anthers, the branches c. 2 mm long, arching outward, narrow throughout, channelled in the upper half. *Capsules* and *seeds* unknown. *Chromosome number* unknown.

 Flowering time early December to mid-January.

Distribution and biology

A narrow endemic of the central KwaZulu-Natal Drakensberg, *Gladiolus loteniensis* is known only from the upper Loteni River valley. Plants occur in the sandstone zone at c. 1800 m in flat sites along the riverbanks, growing in black peaty soil and often in the lee of low cliffs or boulders. They are inconspicuous even in full bloom, and it seems likely that the range is wider than the present record suggests. The plants illustrated here were collected at or close to the type locality where, in December 1994, we found a thriving colony just coming into flower. Plants are not restricted to this site for we

191

found individuals scattered downstream for about half a kilometre in similar habitats.

The short-tubed flowers are evidently adapted for pollination by bees.

Diagnosis and relationships

Among *Gladiolus* species of the southern African summer-rainfall region, *G. loteniensis* is readily recognized by its small flower with a pale lilac perianth, its narrow, soft-textured leaves and, in particular, by the unusual dark lilac to violet veining on the cream to pale yellow background of the lower tepals. These nectar guides are confined to the lower two-thirds of the tepals, sometimes almost to the apices of the lower lateral tepals, and the distal portion of the lower tepals is either coloured pale lilac like the rest of the flower,

or dark lilac. Novel as the nectar guides are, the presence of pale colour in the lower parts of the tepals and normal or darker colour distally is a characteristic feature of section *Hebea*. We suggest that the species is most closely allied to *G. involutus* of this section, sharing with it narrowly sword-shaped, soft-textured leaves; flowers with the upper lateral tepals twisted to lie partly over the dorsal tepals; the limbs of the lower lateral tepals involute; and the lower median tepal much longer than the lower laterals. The relatively soft-textured corm tunics and the production of several long stolons are also consistent with *G. involutus* and a handful of other species related to it, all assigned here to series *Involutus*.

Gladiolus loteniensis is geographically

disjunct from *G. involutus*, which occurs from near East London to Swellendam in Western Cape Province. The remaining members of series *Involutus* occur largely in the winter-rainfall region of southern Africa. *Gladiolus loteniensis* and *G. involutus* occur at the eastern end of the section's range.

History

Gladiolus loteniensis was discovered in 1982 by O.M. Hilliard and B.L. Burtt who have made a life-long study of the flora of the KwaZulu-Natal Drakensberg. Realizing that this was a new discovery, they described the species in 1986 from very scanty material. It is essentially known from a single site where we re-collected it in 1994.

84. GLADIOLUS INVOLUTUS Delaroche

PLATE 70

Gladiolus involutus Delaroche, *Plantae Aliquot Novarum* 28, fig. 3 (1766). Baker, *Fl. Capensis* 6: 147 (1896), as to name only, specimens cited are *G. rogersii* Baker. Lewis

et al., *J. S. African Bot.*, Suppl. 10: 118 (1972). Type: South Africa, Cape, without precise locality, collector and date unknown, (L–Herb. Van Royen 904, 137-90, holotype).

involutus = curving inward, referring to the channelled lower tepals with the margins curving upward.

Synonymy

Gladiolus bimaculatus Lamarck, *Encyclopédie Méthodique* 2: 727 (1786). Type: South Africa, Cape, without precise locality, illustration without binomial name in Miller, *Figures of Plants* 158 & pl. 236 fig. 1 (1758).
Gladiolus muirii L. Bolus, *Ann. Bolus Herb.* 1: 132 (1915). Type: South Africa, Western Cape, Riversdale District, Tweekuilen, Oct. 1913, *Muir 588* (BOL, holotype on two sheets; K, SAM, isotypes).

Description

Plants 20–50 cm high. *Corm* depressed-globose, 12–17 mm in diameter, producing large cormlets on slender stolons from the base, the tunics of soft-textured papery layers, rarely somewhat fibrous, these not normally accumulating annually. *Cataphylls* pale and membranous, the uppermost reaching shortly above the ground and then purplish, usually mottled with white. *Leaves* seven to nine, the lower four or five basal and longest, reaching at least to the base of the spike and sometimes exceeding it by up to 10 cm, the blades linear, 1.7–3 mm wide, thickened toward the midline, the margins hyaline, not thickened, the upper three to four leaves

inserted well above the ground, shorter than the basal leaves, progressively smaller above, the uppermost diverging from the stem and not or barely clasping the stem, channelled throughout. *Stem* usually erect below, sometimes inclined, flexed outward above the sheathing part of each cauline leaf, simple or with 2–3 branches, c. 1.2–1.8 mm in diameter below the spike.

Spike lightly inclined and barely flexuose, 4- to 7-flowered; *bracts* pale green flushed greyish purple on the upper surfaces, the outer 25–35(–40) mm long, the inner about three-quarters as long as the outer, obtuse or minutely notched apically. *Flowers* predominantly white, the tepals flushed pink toward the bases or very pale pink, the lower three tepals each pale greenish yellow below, the yellowish colour edged at both ends with a narrow pink or purple line, all three whitish to pale pink distally, the whole flower, especially the nectar guides, turning pink with age, unscented; *perianth tube* obliquely funnel-shaped, curving outward near the apex of the slender part, 16–18 mm long, emerging between the bracts; *tepals* unequal, the upper laterals longest, lanceolate, 25–28 x 12–13.5 mm, directed forward below and twisted to lie over the dorsal, attenuate in the upper third and curving outward, the dorsal extended horizontally and hooded over the stamens, 20–24 x c. 13 mm, the lower three tepals united below for 3–5 mm, the lower laterals linear, 15–19 x 3–5 mm, narrowly clawed below for c. 2 mm, the limb channelled in the lower two-thirds, the margins curving

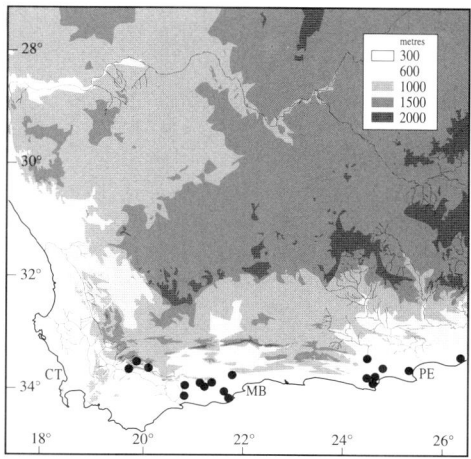

toward one another, thus involute, straight and directed forward, the distal third more or less plane and gently curving outward, the lower median 22–28 x 6–7 mm, obscurely clawed in the lower 1 mm, shallowly channelled in the lower half, arching downward, in profile the lower three tepals exceeding the upper. *Filaments* 12–15 mm long, exserted 6–7 mm from the tube; *anthers* 8–10 mm long, dull purplish, the pollen pale orange. *Ovary* oblong, c. 5 mm long; *style* arching over the stamens, dividing opposite the middle of the anthers, the branches c. 3 mm long. *Capsules* obovoid-ellipsoid, 20–25 mm long, the apex acute or obtuse; *seeds* ovate, 6–7 x c. 4 mm, broadly and evenly winged, light brown, the seed body often dark brown. *Chromosome number* unknown.

Flowering time mid-August to late September, occasionally in October.

Distribution and biology

A fairly common *Gladiolus* of the southern Cape region of South Africa, *G. involutus* extends along the coastal forelands south of the Langeberg–Outeniqua–Vanstadensrivier mountain axis from near Heidelberg in the west to Port Elizabeth and East London in the east. Although its range is relatively poorly documented, *G. involutus* may be found in heavy clay soils in open ground or at the edge of clumps of bush, growing in low grass.

Nothing is known about the biology of the species. The flowers, although unscented, produce large quantities of nectar and we assume that they are pollinated by long-tongued bees, the only insects that can readily access the nectar located in the lower part of the perianth tube and at the same time accomplish cross pollination.

Diagnosis and relationships

Gladiolus involutus is most easily recognized by the moderate-sized pale pink to almost white flowers with pale greenish yellow nectar guides outlined in pink or purple, and by the narrow and channelled lower tepals with the margins curving upward and partly covering the surface. The channelled nature of the lower tepals, particularly the lower laterals, is pronounced. The flower is also notable for the orientation of the prominent upper lateral tepals which are twisted to lie above the shorter, hood-like dorsal tepal. Plants have several leaves, sometimes as many as nine, the blades firm in texture and with lightly thickened midribs, but the margins hyaline and hardly thickened, if at all.

The corm tunics consist of soft, almost membranous layers that do not accumulate with age. An unusual feature of this and several closely related species is the production of slender stolons from the base of the corm. The stolons are 50 mm or more in length and each terminates in a small corm. This distinctive means of vegetative reproduction is shared by a number of species of section *Hebea* here assigned to series *Involutus*, in part because of the stolon character.

The affinities of *Gladiolus involutus* within series *Involutus* lie on the one hand with *G. loteniensis*, a local endemic of the upper Loteni River valley in the southern KwaZulu-Natal Drakensberg, and hence a summer-flowering species, and on the other with the southern Cape endemic, *G. vandermerwei*. The relationship between *G. involutus* and *G. loteniensis* is evident in the very similar shape of the flower, the leaf morphology and the production of long stolons. *Gladiolus loteniensis* differs largely in the markings of the flowers, which are pale lilac with the lower tepals veined with dark purple. The geographic disjunction between the ranges of *G. involutus* and *G. loteniensis* is remarkable. There are few such patterns in the Iridaceae at species level.

A relationship between *Gladiolus involutus* and *G. vandermerwei* is less easy to envision. Although the vegetative morphology is similar in critical taxonomic characters, the flowers of the latter species are quite different. They are bright red and have a long perianth tube, narrow below and abruptly widened into a broad cylindrical upper part. The lower tepals are very much reduced. This type of flower is clearly adapted for bird pollination, hence the major shifts in flower colour and tube length and shape. Once this is understood the very different flowers of *G. involutus* and *G. vandermerwei* can be seen simply as adaptations for different pollination strategies in two closely related species.

History

Gladiolus involutus first appears in the scientific literature in 1758, when it was depicted in Philip Miller's beautifully illustrated series, *Figures of the most beautiful plants, etc.* under the polynomial *Gladiolus foliis linearibus planis, spatha glabra acutiore*. Miller's plants were raised from seed and were grown at the famous Chelsea Physic Garden in London. Evidently this attractive species had recently been introduced into cultivation in Europe, for it was also grown in Holland at this time. Thus it is no coincidence that the species

was illustrated and described again a few years later in Leiden in 1766. In what was in effect his dissertation for a medical degree from the University of Leiden, Daniel Delaroche, a student of the eminent botanist, David van Royen, named the species *G. involutus*. The description and a good ink engraving leave no doubt about the identity of the plant, which is typified by a specimen in the Leiden Herbarium.

Only 20 years later, in 1786, the French biologist, J.F. Lamarck, gave the plant a second name, *Gladiolus bimaculatus*. Lamarck's species was based entirely on the figure in Miller's *Figures of Plants* and we assume that Lamarck simply did not know about Delaroche's *G. involutus,* which is not treated in his comprehensive account of *Gladiolus* in the *Encyclopédie Méthodique*.

The precise origin of the plants grown and illustrated by Miller is unknown, but we assume that his supplier in Holland obtained them directly from what was then a remote part of the Dutch colony at the Cape of Good Hope.

Despite the existence of two available names for the species, both based on published illustrations, J.G. Baker completely misunderstood *Gladiolus involutus*. Specimens that he cited under that name in his *Flora Capensis* account of *Gladiolus* are the slender-leafed and blue-flowered *G. rogersii*.

A third synonym for *Gladiolus involutus* was provided by H.M.L. Bolus in 1915 when she described *G. muirii*. Bolus seems to have been unaware of the identity of *G. involutus* and *G. bimaculatus*, for she gives no reason for distinguishing her species from them. The name *G. involutus* was only used for the species when type specimens in the Leiden Herbarium were examined and Delaroche's dissertation was carefully re-examined in the light of these collections (Goldblatt & Barnard, 1970).

85. GLADIOLUS VANDERMERWEI (L. Bolus) Goldblatt & De Vos

PLATE 71

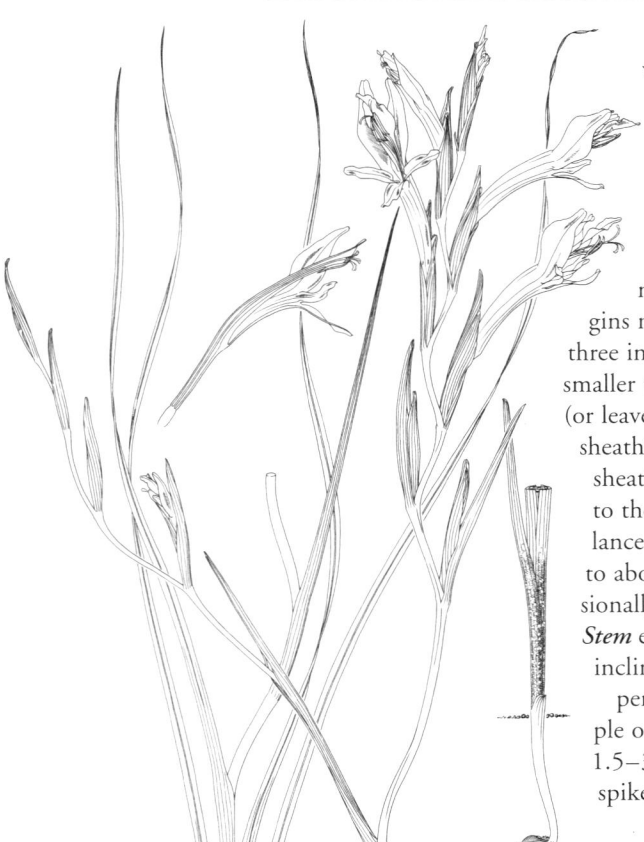

Gladiolus vandermerwei (L. Bolus) Goldblatt & De Vos, *Bull. Mus. Hist. Nat., Paris,* Sér. 4, Sect. B, *Adansonia* 11: 421 (1989). *Antholyza vandermerwei* L. Bolus, *J. Bot. (London)* 69: 14 (1931). *Homoglossum vandermerwei* (L. Bolus) L. Bolus, *S. African Gard.* 23: 47 (1933). De Vos, *J. S. African Bot.* 42: 338 (1976). Type: South Africa, Cape, Swellendam District, cultivated in Cape Town by P. Ross Frames, 19 Sept. 1929, *Van der Merwe s.n.* (BOL 19084, holotype; BOL, K, isotype).

vandermerwei, named in honour of J.S. van der Merwe who collected the plants that were grown in Cape Town and preserved as the type collection.

Description

Plants 30–60 cm high. *Corm* globose, 8–12 mm in diameter, the tunics papery to membranous, decaying into irregularly shaped fragments, light brown, with two to several slender stolons produced from the base, each with a terminal cormlet. *Cataphylls* firm-textured, pale below the ground, the upper extending 10 cm above the ground and then purple mottled with white or cream. *Leaves* five to seven, or more if the stem is branched, the lower four to six basal, the lower sheaths also mottled purple and cream, the blades linear, plane, 2–5 mm wide, the midrib lightly raised, the margins not thickened, the upper two or three inserted above the ground and smaller than the basal, uppermost leaf (or leaves) channelled entirely, often sheathing only near the base, the sheath open and the margins free to the base, the blades narrowly lanceolate to nearly linear, reaching to about the base of the spike (occasionally shortly exceeding the stem). *Stem* erect below, flexed outward and inclined above the sheath of the penultimate or terminal leaf, simple or with one to three branches, 1.5–3 mm in diameter below the spike.

Spike usually erect, lightly flexuose, occasionally 2-, usually 3- to 8-flowered, the lateral branches with fewer flowers; *bracts* dull green to somewhat purplish, the outer 35–50(–60) mm long, the inner two-thirds to nearly as long as the outer, acute or apically notched for c. 0.5 mm. **Flowers** bright scarlet, the lower three tepals yellow in the lower halves, unscented; *perianth tube* 35–45 mm long, narrow and cylindric in the lower 15–22 mm, gradually expanded into a wider cylindric and horizontal upper part, 20–23 mm long, c. 6 mm in diameter; *tepals* unequal, the upper three lanceolate, the lower three linear, the dorsal largest, 23–28 x 8–10 mm, inclined slightly above the horizontal, the upper laterals joined to the dorsal for 2–3 mm, 20–30 x 9–11 mm, the lower lateral tepals spreading at right angles to the tube, straight or twisted through 40°–80°, the lower laterals 12–19 x 2–3 mm, the lower median slightly longer than the lower laterals. *Filaments* 35–40 mm long, extending horizontally, exserted 8–12 mm from the tube; anthers 7–9 mm long, dark purple, the pollen pale yellow. *Ovary* oblong, c. 8 mm long; style extending horizontally over the stamens, dividing opposite the upper third of the anthers, the branches 3–5 mm long. *Capsules* ovoid-ellipsoid, the apices rounded, c. 20 mm long; *seeds* irregularly ovate, c. 6 x 5 mm, evenly or unevenly winged, dark brown, the seed body blackish. *Chromosome number* unknown.

Flowering time September and October, occasionally in August.

Distribution and biology

Gladiolus vandermerwei has a narrow distribution range in the central part of Western Cape Province, South Africa. It occurs in the lower Breede River valley from near Bonnievale in the west to the river mouth near Witsands and Vermaaklikheid. Two records are slightly out of the main range, one from between Bredasdorp and Swellendam, and the other from Cogman's Kloof near Montagu. Plants grow in stony shale ground, mainly on south-facing slopes, in fairly dry habitats.

There are no observations on the pollination biology of *Gladiolus vandermerwei*, but the flowers are clearly adapted for pollination by sunbirds. They have the red colour, elongate perianth tube with the upper portion wide and cylindric, and the ample nectar of relatively low sugar concentration that are the hallmarks of this pollination strategy in Iridaceae.

Diagnosis and relationships

Gladiolus vandermerwei can immediately be recognized by the combination of bright red flowers with an elongate perianth tube – the upper part of which is long and cylindric – and small, linear lower tepals. The flowers resemble fairly closely those in series *Homoglossum* (section *Homoglossum*) and it was in that genus that the species was placed in 1933. As remarked by M.P. de Vos (1976), who revised *Homoglossum*, it is the only species of that genus that has branched stems. This character, as well as the production of

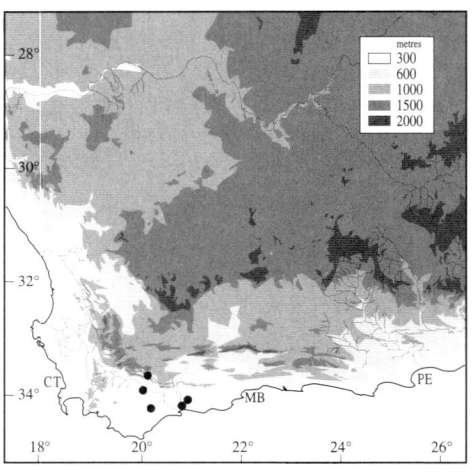

cormlets at the ends of long stolons, the plane leaves and the mottling on the cataphylls, is inconsistent with *Homoglossum* (now series *Homoglossum*). With the exception of the mottled cataphylls, they are also inconsistent with the entire *Homoglossum* section.

We are satisfied that the affinities of *Gladiolus vandermerwei* lie instead with section *Hebea*, and specifically with *G. involutus* and *G. cunonius* of series *Involutus*. Species of the series have relatively broad leaf blades and mostly mottled cataphylls and, with only one exception, they produce cormlets on long stolons. Significantly, the redflowered species of series *Involutus* appear to differ from those of section *Homoglossum* and others in the genus in the nature of the floral pigments, which fail to develop when not exposed to direct sunlight.

The three remaining species of series

Involutus that also have flowers adapted for pollination by birds are, however, rather different from *Gladiolus vandermerwei* in several details of floral morphology, including a rather short upper perianth tube, anthers with long tails, style branches stigmatic only at the tips, and a ridged or tooth-like callus in the base of the upper perianth tube. For these reasons we see *G. vandermerwei* as close to the base of the lineage that gave rise to the more specialized bird-pollinated species of the series. It thus provides a link between them and species like *G. involutus* which have bee-pollinated flowers ancestral for the genus.

History

The first record of *Gladiolus vandermerwei* was made in about 1928 by J.S. van der Merwe of Bonnievale, who collected plants near this small town in the Swellendam

District. Plants were given to P. Ross Frames who grew the species in Cape Town and presented specimens to the Bolus Herbarium, then at Kirstenbosch, in 1929. H.M.L. Bolus immediately recognized the plants as a new species which she described in 1933, first assigning it to the genus *Antholyza* where it would have been placed following J.G. Baker's generic concepts of the Iridaceae in *Flora Capensis*. After N.E. Brown revised the generic concepts of the species included by Baker in *Antholyza*, *G. vandermerwei* was transferred to *Homoglossum*. Only in 1989 was the species placed in *Gladiolus* by Goldblatt & De Vos, who maintained that this was where its affinities lay. The relationship of *G. vandermerwei* to section *Hebea*, especially to *G. involutus*, is a new concept that we propose here.

86. GLADIOLUS CUNONIUS (Linnaeus) Gaertner

Lepelblom

PLATE 72

Gladiolus cunonius (Linnaeus) Gaertner, *Fruct. Semin. Pl.* 1: 37 (1788). Goldblatt & De Vos, *Bull. Mus. Hist. Nat., Paris*, Sér. 4, Sect. B, *Adansonia* 11: 424 (1989). *Antholyza cunonia* Linnaeus, *Sp. Pl.* 1: 37 (1753). Baker, *Fl. Capensis* 6: 168 (1896). *Anisanthus cunonius* Sweet, *Hort. Brit.* ed. 2: 500 (1830). *Anomalesia cunonia* (Linnaeus) N.E. Brown, *Trans. Roy. Soc. S. Africa* 20: 271 (1932). *Gladiolus papilionaceus* Salisbury, *Prod. Stirpium ad Chapel Allerton* (1796), an illegitimate superfluous name for *G. cunonius* Linnaeus. Type: South Africa, Western Cape, without precise locality or collector, illustration in Cuno, *Ode über seinen Garten* (1750).

cunonius, named in honour of J.C. Cuno, an eighteenth-century Dutch botanist.

Synonymy

Cunonia antholyza Miller, *Gardeners' Dictionary*, ed. 8 (1768). Type: South Africa, Western Cape, without precise locality, figure in Miller, *Figures of Plants* 75: pl. 113 (1756).

Description

Plants (12–)30–70 cm high. **Corm** depressed-globose, 6–9 mm in diameter, the tunics papery to membranous, not accumulating to any extent, rapidly decaying and fragmenting into irregularly shaped pieces, with numerous stolons produced from the

base. **Cataphylls** membranous and pale below the ground, the upper usually reaching some distance above the ground and turning purple. **Leaves** occasionally five, usually six to eight, mostly basal, reaching to about the base of the spike or sometimes slightly exceeding it, the blades narrowly lanceolate, plane, 5–12 mm wide, the midrib evident but not or only lightly thickened, at least the two upper leaves cauline and smaller than the basal, the uppermost leaf channelled entirely, usually sheathing only near the base, the sheath open, other cauline leaf sheathing for up to half its length, the sheath open, the lower margins usually overlapping. **Stem** erect, flexed outward above the sheaths of the cauline leaves and inclined above, simple or

with one or two branches, 2–3 mm in diameter below the spike.

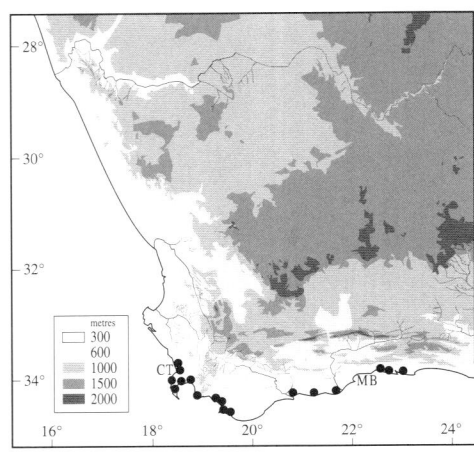

Spike inclined, lightly flexuose, occasionally 3-, usually 5- to 8-flowered; *bracts* soft-textured, green or grey-purple, 20–35(–45) mm long, green to grey-green, the inner about two-thirds as long as the outer, acute or apically notched for c. 0.5 mm. *Flowers* bright red, the lower tepals and tube initially green, fading to yellow and then becoming reddish, unscented; *perianth tube* 12–15 mm long, slender and cylindric below for 10–12 mm, abruptly expanded into a short, more or less tubular upper part; *tepals* unequal, the dorsal 26–29 mm long, extending horizontally, with a linear claw 10–13 mm long, the limb 13–18 x 11 mm, the upper laterals joined to the dorsal for 8–12 mm, the free part directed forward and upward, ovate, 10–13 x 8 mm, the lower tepals united to one another for c. 2 mm, much reduced, the free parts 1–4 mm long, the lower laterals curving upward distally to partially block the mouth of the tube. *Filaments* c. 30 mm long, exserted c. 27 mm from the tube, the lower (abaxial) filament with a small horny tooth at the base; *anthers* 5–7 mm long, attached to the filament near the midline, the thecae free below the filament insertion, red, the pollen yellow. *Ovary* oblong, c. 5 mm long; *style* extending horizontally over the stamens, dividing shortly below the anther bases, the branches 7–8 mm long, filiform, the margins conduplicate, expanded and stigmatic only at the apices. *Capsules* oblong-ellipsoid and three-sided, 27–32 mm long; *seeds* ovate to oblong, 7–8 x c. 5 mm, broadly and evenly winged, pale golden brown. *Chromosome number* 2n = 30.

Flowering time September to mid-November.

Distribution and biology

Gladiolus cunonius occurs along the southwestern and southern Cape coast from Saldanha Bay in the west to Knysna in the

east. Plants typically grow in low coastal scrub in coarse sand, sometimes just above the beach, and on sand dunes a short distance inland. They can often be seen along paths and in clearings in coastal scrub and forest. Presumably corms remain for years under a closed canopy and produce only a few leaves until, as a result of fire or clearance, the sun reaches the ground and they flower and fruit. Occurring in deep sand, the corms are vulnerable to predation by molerats, but survival is ensured by the production of both seed and cormlets, the latter at the ends of the numerous stolons.

The bright red flowers secrete large amounts of nectar and, like other plants with long-tubed, red flowers, are adapted for pollination by birds. We have seen them being visited by the lesser double-collared sunbird *Nectarinia chalybea* at Wolfgat Nature Reserve on the Cape Flats and this bird may be their most important and perhaps sole pollinator. The nectar of *Gladiolus cunonius* and of the closely related *G. spectabilis* and *G. saccatus* is unusual in *Gladiolus* in having low proportions of sucrose, whereas most species in the genus have sucrose-rich nectar. Low-sucrose nectars are often found in bird-pollinated plants and this appears to be the case for these three species. Most other bird-pollinated species of *Gladiolus*, including *G. watsonius* and its allies (series *Homoglossum*) and *G. priorii* and *G. meridionalis* (series *Mutabilis*), have sucrose-rich nectar, as do all insect-pollinated species of the genus.

Diagnosis and relationships

The very peculiar structure of the flower of *Gladiolus cunonius* makes it relatively easy to recognize. The dorsal tepal is elongate, 26–29 mm long, and strongly concave, giving rise to the plant's common name, *lepelblom* (spoon flower). The shorter upper lateral tepals are twisted to lie above the dorsal tepal and the lower three tepals are reduced to short linear scales that are directed forward and form part of the cup that contains the nectar. The flowers are borne on strongly inclined or horizontal spikes and the stems are usually branched. The only other species of *Gladiolus* that has a flower like this is *G. splendens* from the western Karoo, and the two are clearly immediately related. In *G. splendens* the spikes are usually less strongly inclined and the flower differs in having the lower median tepal recurved and directed downward.

Together with *Gladiolus saccatus* from Namaqualand and Namibia, *G. cunonius* and *G. splendens* form a clade within section

Hebea. Apart from the reduced, almost scale-like lower tepals, the three species have unique anthers with long tails and style branches that are stigmatic only at their apices. The small alliance is probably most closely related to another bird-pollinated species, *G. vandermerwei*, of the southern Cape coast, which has red flowers with fairly small lower tepals, but anthers and style branches which conform to the basic type in *Gladiolus.*

History

A relatively common coastal plant, *Gladiolus cunonius* was first recorded in the literature when a volume entitled *Ode über seinen Garten* (Ode to his Garden) by the Dutch botanist, J.C. Cuno, was published in 1750. This work included a section enumerating the plant species, apparently compiled by the German botanist, D.S. Buettner (*Enumeratio methodica plantarum carmine clarissimi J.C. Cuno recensitarum*). There *G. cunonius* is called *Cunonia floribus pedunculatis, spathis minimis* and is accompanied by a woodcut that clearly represents this species. The illustration is entitled *Cunonia floribus sessilibus spathis maximis*, a curious contradiction to the text.

Gladiolus cunonius was formally described in 1753 by Carl Linnaeus, whose knowledge of the species was evidently based on this source, although he also cited published illustrations of a second species, now *Chasmanthe aethiopica*. Linnaeus placed *G. cunonius* in the genus *Antholyza*, curiously mentioning Persia as its native land. In 1759 Linnaeus described *A. aethiopica* and referred to the species some of the plants he had cited under *A. cunonius*. This effectively left the Cuno illustration as the type of the latter species. The appearance of *G. cunonius* also puzzled the celebrated English botanist, Philip Miller, who accepted the genus *Cunonia* in 1756 for this species alone. The generic name obviously honours Cuno whose painting of the species was cited by Miller, although he had specimens of his own, grown from seed that he obtained directly from the Cape. Because Linnaeus preferred to place *G. cunonius* in *Antholyza*, in 1759 he used the generic name *Cunonia* for another plant, the tree *C. capensis* which is the type species of the family Cunoniaceae. Miller ignored Linnaeus' treatment and continued to recognize his genus *Cunonia* with the single species, *C. antholyza*, until his death in 1771. *Cunonia* is now conserved nomenclaturally in Linnaeus' sense.

Later, in 1788, the German botanist Joseph Gaertner, who made pioneering studies of the fruits and seeds of plants, was struck by the similarity of the capsules and seeds of *Antholyza cunonia* to those of *Gladiolus* species. As a result, he transferred *A. cunonia* to that genus. Because of its peculiar flowers, however, few authorities followed Gaertner's decision. A notable exception was the British expert on bulbous plants, William Herbert (1843). *Gladiolus cunonius* was transferred to *Anisanthus* by

Robert Sweet in 1830, who had described this new genus in 1826 for the closely related *G. spectabilis*. Some botanists, including F.W. Klatt, recognized *Anisanthus*, but J.G. Baker regarded the genus as a synonym of the earlier name *Antholyza*. Thus *G. cunonius* is included in *Antholyza* in Baker's account of the Iridaceae for *Flora Capensis* (Baker, 1896).

N.E. Brown, who studied the species of African Iridaceae with flowers we now recognize as adapted for bird pollination, regarded

Sweet's genus *Anisanthus* as a homonym of *Anisanthes* Schultes, described some years earlier, and proposed the substitute name *Anomalesia* for the genus, in effect continuing to regard its two species as belonging to a genus separate from *Gladiolus*. Only in 1989 was *G. cunonius* restored to *Gladiolus* by P. Goldblatt and M.P. de Vos, who did not regard genera allied to *Gladiolus* and having highly specialized flowers adapted for bird pollination as warranting segregation in other genera.

87. GLADIOLUS SPLENDENS (Sweet) Herbert

Roggeveld lepelblom

PLATE 73

Gladiolus splendens (Sweet) Herbert, *Bot. Register*, New Ser., 6: Miscel. 46 (1843). Goldblatt & De Vos, *Bull. Mus. Hist. Nat., Paris*, Sér. 4, Sect. B, *Adansonia* 11: 424 (1989). *Anisanthus splendens* Sweet, *British Fl. Gard.*, Ser. 2, 1: tab. 84 (1831). *Antholyza splendens* (Sweet) Steudel, *Nom. Bot.*, ed. 2, 1: 106 (1840). *Anomalesia splendens* (Sweet) N.E. Brown, *Trans. Roy. Soc. S. Africa* 20: 271 (1932). Type: South Africa, without precise locality, illustration in *British Fl. Gard.*, Ser. 2, 1: tab. 84 (1831).

splendens = brilliant, shining, for the bright red, very attractive flowers.

Description

Plants 50–110 cm high. *Corm* globose, 15–22 mm in diameter, the tunics papery to membranous, not accumulating to any extent, fragmenting into irregularly shaped pieces, light brown, with two to several slender stolons produced from the base, each with a terminal cormlet. *Cataphylls* firm-textured, pale below the ground, the upper usually reaching some distance above the ground and purple mottled with white. *Leaves* five to seven, the lower four to six basal, reaching up to the base of the spike, the blades narrowly lanceolate to nearly linear, 4–9 mm wide, plane, the midrib and sometimes one other vein lightly raised, the margins not thickened, the upper two leaves cauline and smaller than the basal, the uppermost leaf channelled entirely, often sheathing only near the base, the sheath open, the other cauline leaf sheathing for up to half its length, the sheath open, the lower margins usually overlapping. *Stem* erect, usually with one or two branches, sometimes purplish rather than green, 2–5 mm in diameter below the spike.

Spike usually erect, lightly flexuose, the flowers in two ranks separated by more than 90°, occasionally 6-, usually 8- to 14-flowered, flowers fewer on the lateral branches; *bracts* usually purplish, the outer 18–35 mm long, the inner slightly shorter than the outer, acute or apically notched for c. 0.5 mm. *Flowers* bright scarlet, the lower tepals and tube initially green, the lower median tepal sometimes red distally, on fading becoming red throughout, unscented; *perianth tube* 16–18 mm long, narrow and cylindric in the lower 8–9 mm, abruptly expanding into a wider urn-shaped upper part c. 9 mm long; *tepals* unequal, the dorsal largest, the free part spoon-shaped, extended horizontally, 28–34 mm long, the narrow lower part claw-like, 10–14 mm long, the limb 18–20 mm long, ovate and obtuse, the upper laterals joined to the dorsal for 5–8 mm, the free part ovate, 16–17 x 10–12 mm, the lower lateral tepals 4–6 x 1.5 mm, joined to the upper lateral for c. 3 mm, the free parts directed forward, the lower median boat-shaped, the limb directed downward at right angles to the tube, 8–11 x c. 6 mm. *Filaments* 33–36 mm long, extending horizontally, exserted 25–29 mm from the tube, the lower (abaxial) filament with an enlarged, flattened horny tooth at the base; *anthers* 5.5–8 mm long, attached to the filaments near the midline, the thecae free below the

filament insertion, red, the pollen yellow. *Ovary* oblong, c. 6 mm long; *style* extended horizontally over the stamens, dividing c. 5 mm short of the anther bases, the branches diverging, 9–10 mm long, filiform, the margins conduplicate, expanded and stigmatic only at the apices. *Capsules* ellipsoid to obovoid, 18–20 mm long, the apex acute or more or less truncate to retuse; *seeds* ovate, 6–7 x 4–5 mm, broadly and evenly winged. *Chromosome number* unknown.

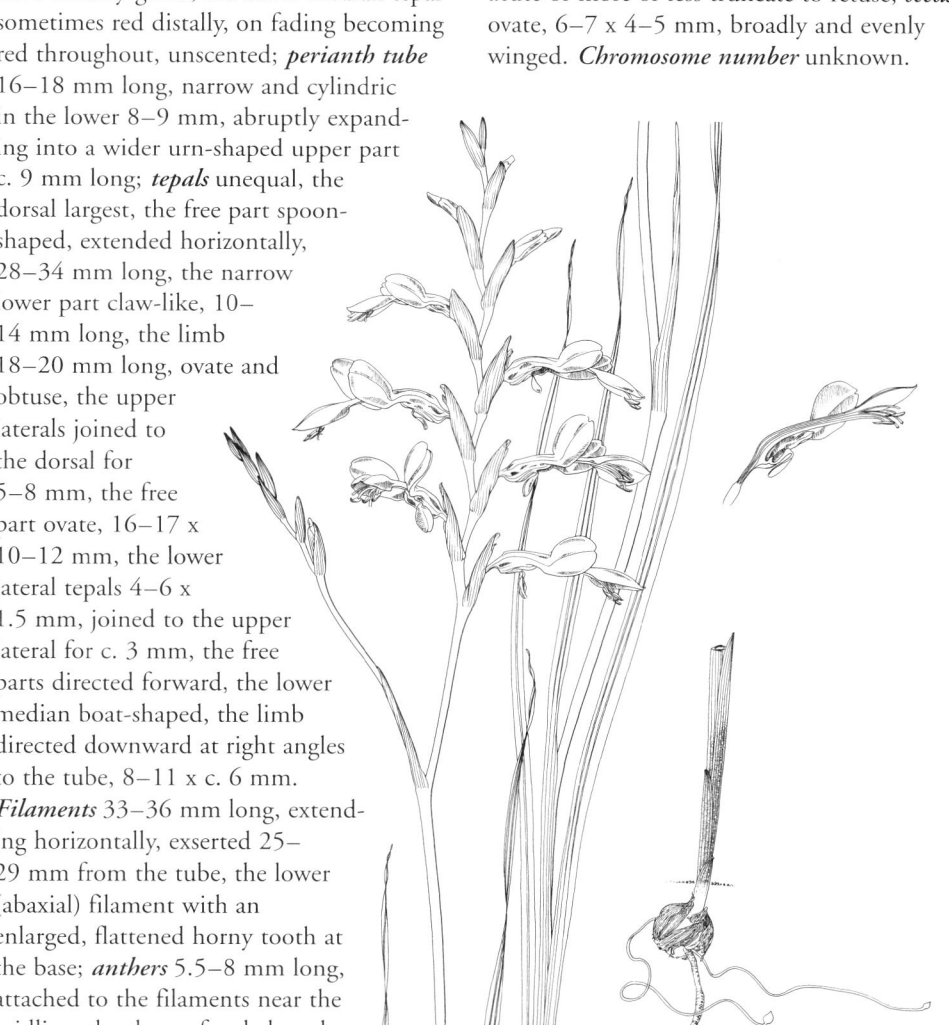

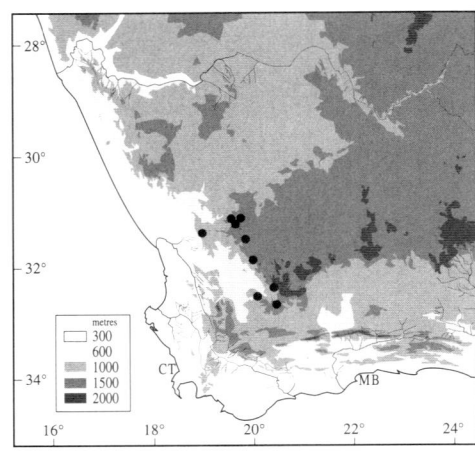

Flowering time mainly October to mid-November, but occasionally in September.

Distribution and biology

A species of the arid interior of the southern African winter-rainfall region, *Gladiolus splendens* occurs in isolated sites in the extreme western Karoo. Its range extends from the Moordenaars Karoo near Laingsburg in the south, northward along the Roggeveld Escarpment to the mountainous country north of Calvinia. In favourable sites where rainfall is higher, plants grow on open slopes among rocks in the shade of small trees, but more often *G. splendens* is found near small seasonal streams in riverine thicket. Although its range is fairly extensive, *G. splendens* is not a common species.

The flowers of *Gladiolus splendens*, like those of *G. cunonius*, are clearly adapted for pollination by sunbirds.

Diagnosis and relationships

Gladiolus splendens is easily distinguished by its remarkable, bright scarlet flowers with grossly disproportionate tepals. The dorsal is elongate and spoon-shaped, the lower laterals are scale-like and directed forward, and the lower median, which is bright green initially and later turns red in the upper half, is directed downward. The upper lateral tepals are joined to the dorsal for 5–8 mm and are twisted upward to lie above the dorsal. This extraordinary flower has specialized anthers with long sterile tails and tubular style branches which are stigmatic only at their tips. Vegetatively *G. splendens* is unexceptional, having a fan of sword-shaped leaves basally enclosed in purple-mottled cataphylls, a branched stem and inclined spikes. The corms have more or less firm-papery tunics, but the production of cormlets on long stolons is notable. The red flower colour, mottled cataphylls and the production of stolons on long cataphylls recall *G. vandermerwei* on the one hand, and *G. cunonius* on the other. The latter species also has the same type of flowers as *G. splendens*, as well as the tailed anthers and tubular stigmas.

History

Gladiolus splendens was apparently discovered by Walter Synnot, deputy landdrost at Clanwilliam in the 1820s, and Synnot was responsible for its introduction into Great Britain in 1825 on his return there from South Africa. The plant was first recorded in Robert Sweet's *Hortus Britannicus* in 1830

and was formally described by Sweet in 1831 in the illustrated magazine *The British Flower Garden*, where the description is accompanied by a watercolor painting. Sweet placed the species in a new genus, *Anisanthus*, which is well-named for the disproportionate tepals. The species was transferred to *Antholyza* in 1840 by the German botanist, E. Steudel, and was thus in the same genus as its close relative, the florally similar *G. cunonius*. The British botanist, William Herbert, preferred to include both species in *Gladiolus* and he transferred them to the genus in 1843.

J.G. Baker's treatment of the red-flowered and long-tubed Iridaceae in *Antholyza*, however, led to the transfer of *Gladiolus splendens* back to *Antholyza*. In his account of the genus in *Flora Capensis* (Baker, 1896), however, *G. splendens* is confused with *A. caffra*, now *Tritoniopsis caffra*. To add to the confusion, N.E. Brown (1932), dealing with the range of disparate species placed by Baker in *Antholyza*, decided that separate generic status was appropriate for *G. splendens* and its close ally, *G. cunonius*. Instead of reviving Sweet's *Anisanthus*, he placed them in a new genus, *Anomalesia*, because he considered *Anisanthus* a later homonym of *Anisanthes* Schultes, and hence nomenclaturally illegitimate. Adaptation for a particular pollination syndrome is a poor criterion for the recognition of genera and the transfer back to *Gladiolus* was recommended by Goldblatt & De Vos (1989) for *A. splendens* and *A. cunonius*.

---

88. GLADIOLUS SACCATUS (Klatt) Goldblatt & De Vos

Suikerkannetjie

PLATE 74

Gladiolus saccatus (Klatt) Goldblatt & De Vos, *Bull. Mus. Hist. Nat., Paris*, Sér. 4, Sect. B, *Adantsonia* 11: 424 (1989). *Anisanthus saccatus* Klatt, *Linnaea* 32: 727 (1863). *Antholyza saccata* (Klatt) Baker, *J. Linn. Soc.*,

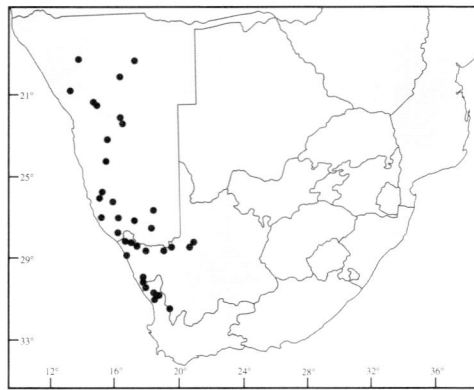

Bot. 16: 180 (1877). *Kentrosiphon saccatus* (Klatt) N.E. Brown, *Trans. Roy. Soc. S. Africa* 20: 271 (1932). *Petamenes saccatus* (Klatt) Phillips, *Bothalia* 4:44 (1941). *Anomalesia saccata* (Klatt) Goldblatt, *J. S. African Bot.* 37: 443 (1971). Type: South Africa, Western Cape, near Hol River, Mierenkasteel, 5 Aug. 1830, *Drège 2646a* (K, P, syntypes).

saccatus = with a sac or pouch, alluding to the prominent spur at the base of the upper part of the perianth tube.

Synonymy

Antholyza duftii Schinz, *Mém. Herb. Boissier* 20: 13 (1900). *Kentrosiphon duftii* (Schinz) N.E. Brown, *Trans. Roy. Soc. S. Africa* 20: 272 (1932). *Petamenes duftii* (Schinz) Phillips,

Bothalia 4: 44 (1941). Type: Namibia, near Rietfontein, Apr. 1899, *Dufi 67* (?Z, holotype [not seen]).

Antholyza steingroeveri Pax, *Bot. Jahrb. Syst.* 15: 156 (1893). Baker, *Fl. Capensis* 6: 530 (1896). *Kentrosiphon steingroeveri* (Pax) N.E. Brown, *Trans. Roy. Soc. S. Africa* 20: 272 (1932). *Petamenes steingroeveri* (Pax) Phillips, *Bothalia* 4: 44 (1941). *Kentrosiphon saccatus* subsp. *steingroeveri* (Pax) Obermeyer, *Fl. Pl. Africa* 34: pl. 1354 (1961). *Gladiolus saccatus* subsp. *steingroeveri* (Klatt) Goldblatt & De Vos, *Bull. Mus. Hist. Nat., Paris*, Sér. 4, Sect. B, 11: 424 (1989). Type: Namibia, Halenberg, 29 Aug. 1929, *Dinter 6618* (K, syntype).

Kentrosiphon gracilis N.E. Brown, *Trans. Roy. Soc. S. Africa* 20: 272 (1932). *Petamenes gracilis* (N.E. Brown) Phillips, *Bothalia* 4: 44 (1941). Type: Namibia, Otjiwarongo, stony koppie,

5 May 1929, *Bradfield 560* (K, holotype; PRE, isotype).

Kentrosiphon propinquus N.E. Brown, *Trans. Roy. S. Africa* 20: 272 (1932). Type: South Africa, Western Cape, Knersvlakte, 600 ft, 16 July 1896, *Schlechter 8163* (K, holotype).

Description

Plants (20–)40–65(–120) cm high. *Corm* globose to conic, 12–15 mm in diameter, the tunics firm-papery, not accumulating to any extent, fragmenting into irregularly shaped pieces or becoming more or less fibrous, light brown, bearing large cormlets on short stolons close to the corm base. *Leaves* five or six, occasionally seven, the lower three basal and longest, reaching to about the upper third of the stem, rarely to the base of the spike, narrowly lanceolate, the blades more or less plane or lightly plicate, 3–15 mm wide, the midrib and usually one other vein well developed, the margins not thickened or raised, the upper two to four leaves inserted in the upper half of the stem, diverging from the stem and not or hardly sheathing, the margins free to the base, at least the uppermost entirely channelled. *Stem* erect below, flexed outward above the first cauline leaf and then inclined, flexed at each of the upper nodes, dark purplish green or dull purple, with two or three, sometimes as many as six branches, 3–4(–9) mm in diameter below the main spike.

Spike inclined at c. 40°, 8- to 12-, occasionally to 15-flowered, flowers in two opposed ranks and alternately facing in opposite directions; *bracts* somewhat succulent, brownish purple, the outer 20–35(–42) mm long, the inner three-quarters to about as long as the outer. *Flowers* bright red, the lower three tepals and the spur green at anthesis, becoming yellow to orange later, unscented; *perianth tube* 11–20 mm long, slender and cylindric in the lower 6–12 mm, abruptly expanded into a wider upper part, 5–8 mm long, with a sac-like spur 2–6 mm long at the lower base; *tepals* unequal, the dorsal largest, spoon-shaped, extending horizontally, 30–45 mm long, the claw 12–16 mm long, the limb 18–26 x 10–15 mm, the upper laterals joined to the dorsal for c. 5 mm, linear, the free part 2–5 x 2–3 mm, the lower lateral tepals joined to the upper laterals on one side, and to one another for c. 4 mm, the free parts 1.5–3 mm long, the lower median c. 1.5 mm long. *Filaments* 31–42 mm long, the lower (abaxial) filament with an enlarged flattened horny tooth at the base; *anthers* 8–10 mm long, attached to the connective shortly below the

centre and versatile, bright red, the pollen yellow. *Ovary* oblong, 5–6 mm long; *style* arching over the stamens, dividing 2–3 mm below the anther bases, the branches c. 10 mm long, the margins conduplicate, flattened toward the apices, stigmatic only apically. *Capsules* ovoid, the apex acute, (20–)25–30 mm long; *seeds* ovate to oblong, 7–11 x 4.5–5.5 mm, broadly and sometimes unevenly winged, semitransparent red-brown, the seed body blackish. *Chromosome number* $2n = 30$.

Flowering time July to mid-August in the winter-rainfall region, rarely later, but occasionally in May when the first rains are early; in the summer-rainfall region mostly March to May.

Distribution and biology

Gladiolus saccatus has a surprisingly wide distribution range, especially for a species of the southern African winter-rainfall region. It extends along the coast and adjacent interior of Western Cape Province from near Vanrhynsdorp northward into Namibia and eastward as far as Pofadder and Kakamas near Upington. Widespread in Namibia, *G. saccatus* occurs almost throughout the country and its most northerly station there is near Grootfontein in the north, not far from Etosha Pan. Although records from Namibia are widely scattered, it is clear that *G. saccatus* shifts from being a winter-rainfall to a summer-rainfall species in the centre of that country, adjusting its growth pattern to respond to rainfall in January and February in the north instead of May and June in the south. In South Africa *G. saccatus* is encountered fairly frequently in the rocky Namaqualand hills in winter and early spring, and in surprising abundance in some seasons, looking rather out of place among vegetation that is in the early stages of growth after the first showers of the winter. In Namibia *G. saccatus* also favours rocky habitats.

One of a number of *Gladiolus* species with flowers adapted for pollination by birds,

G. saccatus is visited by the malachite sunbird *Nectarinia famosa* and the dusky sunbird *N. fusca*. We have seen small flocks of the dusky sunbird, a common resident of the semi-arid west coast and interior of southern Africa, foraging for nectar on colonies of *G. saccatus* near Bitterfontein in Namaqualand. The flowers produce large amounts of nectar that has a fairly low sugar concentration.

Diagnosis and relationships

Gladiolus saccatus has all of the typical adaptations associated with species of Iridaceae pollinated by sunbirds, as well as some unusual ones. The flower is bright red, the dorsal tepal is much enlarged, substantially exceeding the other tepals, and the stamens and style are long and well exserted from the perianth tube. The species is, however, unusual in having a fairly short tube and the nectar accumulates in a spur formed in part by the reduced and partly fused lower tepals. These peculiar adaptations are so different from those typical of bird-pollinated flowers encountered in *Gladiolus* – such as *G. watsonius* and its allies – that *G. saccatus* was assigned to the genus *Antholyza* when first described. Even in that heterogeneous genus it seemed out of place, and N.E. Brown created a separate genus, *Kentrosiphon*, for the species.

That *Gladiolus saccatus* is a true *Gladiolus* is evident from its chromosome number, $2n = 30$, and enlarged ovoid-ellipsoid capsules with broadly winged seeds, synapomorphies for the genus. Its relationships within *Gladiolus* clearly lie with *G. cunonius* and *G. splendens* with which it shares several specialized features, including the almost versatile anthers and filiform style branches expanded only at the apices.

History

The first record of *Gladiolus saccatus* is apparently the collection made by J.F. Drège in August 1830 at Mierenkasteel in southern Namaqualand. These specimens made no

impression on the botanical community until 1863, when F.W. Klatt described the species based on Drège's collection. The strikingly modified flowers do not at all resemble those of other species of *Gladiolus*, and it is not surprising that Klatt did not place the species in that genus. Klatt did, however, see in *G. saccatus* a fair resemblance to *G. splendens* and *G. cunonius*, especially in their flowers. These two species were included in the genus *Anisanthus* for most of the nineteenth century, and he placed *G. saccatus* in that genus as *A. saccatus*. J.G. Baker preferred to place the highly zygomorphic-flowered species in *Antholyza*, and it is in that genus that *G. saccatus* appears in *Flora Capensis* (Baker, 1896). Not content with the situation, N.E. Brown (1932) created a separate genus, *Kentrosiphon*, for *G. saccatus* and two other synonyms, *Antholyza steingroeveri*

and *A. duftii*, both based on collections from central Namibia. Brown also described two more species of *Kentrosiphon*, *K. propinqua* from the Knersvlakte in southern Namaqualand and *K. gracilis* from northern Namibia. These species were all transferred to *Petamenes* by E.P. Phillips in 1941. Then in 1961 A.A. Obermeyer united all four as *K. saccatus*. The genus *Kentrosiphon* was included in *Anomalesia* by P. Goldblatt ten years later, that being the genus in which the related *G. cunonius* and *G. splendens* were accommodated at that period. Finally, in 1989 Goldblatt and M.P. de Vos transferred the three species back to *Gladiolus*, recognizing that they were species of that genus with flowers variously adapted for sunbird pollination.

Section *Hebea*: Series *Permeabilis*

89. GLADIOLUS PERMEABILIS Delaroche

Patrysuintjie, lituin (for subsp. *edulis*)

PLATE 75

Fl. Zambesiaca 12(4): 75 (1993); *Gladiolus in Tropical Africa* 169 (1996). Type: South Africa, Western Cape, without precise locality or collector, cultivated in Holland (L 904,137 – Herb. D. van Royen 45, lectotype designated here, NBG, PRE, photos).

permeabilis = able to pass through, referring to the gap between the dorsal and upper lateral tepals so that in profile there is a window enabling one to see through the flower.

Synonymy

For complete synonymy see nomenclature for the two subspecies.

Description

Plants 15–60 cm high. *Corm* globose to conic, 8–12 mm in diameter, the tunics of relatively hard, medium to coarsely textured fibres. *Cataphylls* pale and membranous, the uppermost reaching a short distance above the ground and then green or flushed with dull purple, sometimes dry and brown. *Leaves* usually four to seven, more if stem more than two-branched, the lower four to six basal and forming a distichous fan, the lower two usually reaching at least to the middle of the spike or shortly exceeding it, the blades linear, 1–2(–3.5) mm wide, rigid, the midrib area moderately thickened, the margins sometimes lightly thickened, the

cauline leaf or leaves shorter than the basal, often without a blade, sheathing only in the lower half and channelled above or entirely sheathing, the lower margins of the two upper leaves free to the base, either sheathing below or diverging from the base. *Stem* usually erect, flexed outward above the sheath of the third leaf and simple or with one or two, occasionally up to four branches, 1–1.8 mm in diameter below the spike.

Spike inclined, lightly flexuose, 4- to 8-flowered; *bracts* pale grey-green, sometimes flushed purplish above, dry and pale near the apices, the outer (7–)12–18(–24) mm long, the inner two-thirds to almost as long as the outer, acute or minutely forked apically. *Flowers* variously whitish to greenish cream, dull blue-grey, dull purple, or yellow-brown, the upper tepals or at least the upper laterals each with a brown to purplish longitudinal median stripe, the lower tepals and throat streaked reddish to purplish, the limbs usually yellow in the lower half, the yellow zone often edged distally in a darker colour, often strongly sweet scented, or unscented; *perianth tube* obliquely funnel-shaped, 9–13(–15) mm long, the slender lower part 7–10 mm long; *tepals* unequal, all narrowed below into claws and more or less spade-shaped, attenuate, sometimes extremely so, the dorsal largest, (16–)20–33 x 10–15 mm, inclined over the stamens, arching upward

Gladiolus permeabilis Delaroche, *Descr. Pl. Nov.* 27, fig. 2 (1766). Baker, *Fl. Cap.* 6: 162 (1896). Lewis in *J. S. African Bot.*, Suppl. 10: 129 (1972). Goldblatt,

near the apex, the upper laterals directed forward, arching outward in the upper half, 16–23 x 7–8 mm, windowed between the bases of the dorsal and upper lateral tepals, the lower three tepals joined to the upper laterals for 2–5 mm and to one another for 3–6 mm, with small thickened knobs at the sinuses between the lower tepals, the free parts 14–16 x 6–8 mm, narrowed below into claws 2–4 mm long, abruptly flexed downward into a broader limb, in profile the lower tepals much exceeding the upper. *Filaments* 12–16 mm long, exserted 8–13 mm from the tube; *anthers* 5–9 mm long, dull cream to grey-purple, the pollen whitish. *Ovary* oblong, 3–5 mm long; *style* arching over the stamens, dividing between the bases and apices of the anthers, the branches 1.5–4 mm long, not or barely exceeding the anthers. *Capsules* ovoid-ellipsoid, lightly 3-lobed and emarginate at the apices, (6–)9–16 x 5–6 mm; *seeds* ovate, 4–6 x 3–4 mm, broadly and sometimes unevenly winged, translucent light brown, the seed body usually dark brown, large in relation to the wing. *Chromosome number* 2n = 30, 28.

Flowering time mainly mid-August to early October in the winter-rainfall zone, October to February in the rest of southern and southern tropical Africa, including Botswana, Namibia and Zimbabwe.

Distribution and biology

Gladiolus permeabilis is widespread across southern and southern tropical Africa, extending from the Caledon District in Western Cape Province, South Africa, to northern Namibia and northeastern Zimbabwe. It is absent only from the western half of the Western Cape and Northern Cape provinces, and from lowland and coastal KwaZulu-Natal, Mpumalanga and Swaziland. Further details about the distribution of *G. permeabilis* are discussed under the two subspecies.

The flowers of *Gladiolus permeabilis* are adapted for pollination by long-tongued bees, and even though they are often dull-coloured and inconspicuous, the strong sweet scent, at least in populations from the southern African winter-rainfall zone, serves to signal their presence. The flowers offer small quantities of sweet nectar, and like those of other species of *Gladiolus* which are similarly proportioned, they are visited at intervals by long-tongued anthophorid bees foraging on other plants for pollen or simply patrolling their territory. Bees collected pollinating *G. permeabilis* include *Anthophora diversipes*

in the winter-rainfall zone and *Amegilla* c.f. *A. fallax* in the summer-rainfall zone.

Diagnosis and relationships

Gladiolus permeabilis is generally easy to recognize by the combination of several basal leaves with linear and fairly firm blades seldom more than 2.5 mm wide with only the midrib strongly raised, and small, usually dull-coloured flowers that viewed in profile have a wide gap or window between the bases of the dorsal and upper lateral tepals. The tepals are all narrowed below and are darkly marked purple to maroon on the keels. The windowed flower is occasionally encountered in other species of *Gladiolus*, but in series *Permeabilis* only in the closely related *G. uitenhagensis*.

The type of leaf found in *Gladiolus permeabilis* is characteristic of several species of section *Hebea* and is probably the basic leaf type for the entire section, as are the distinctive corm tunics of fairly wiry fibres. Other members of the section include *G. wilsonii*, which is restricted to Eastern Cape Province, and the largely KwaZulu-Natal *G. inandensis*. These last two species were included by Lewis et al. in *G. permeabilis* together as subsp. *wilsonii*. We regard them as two distinct species. Both have whitish flowers with purple- or mauve-flushed tepals that lack the window or gap between the dorsal and upper lateral tepals, and the tepals are directed forward, giving the whole flower a rather narrow, tubular shape. In addition, *G. inandensis* has solid, terete leaves, and flowers in the winter and spring after the leaves of the flowering stem have died back.

Across its wide range from the southwestern Cape to Zimbabwe, *Gladiolus permeabilis* seems to fall into two main groups: those populations of the winter-rainfall region with acute to weakly acuminate tepals, often in shades of mauve, greenish brown or purple, and nearly always strongly scented; and those from the summer-rainfall area which have long-acuminate to tailed tepals, mostly in shades of grey, cream or white, and are not often scented. There is little overlap in these features and the two series of populations can almost always be distinguished with ease. We follow Lewis et al. (1972) in recognizing them as separate subspecies.

History

Gladiolus permeabilis was described in 1768 by Daniel Delaroche, a student of the Dutch botanist, David van Royen, from plants

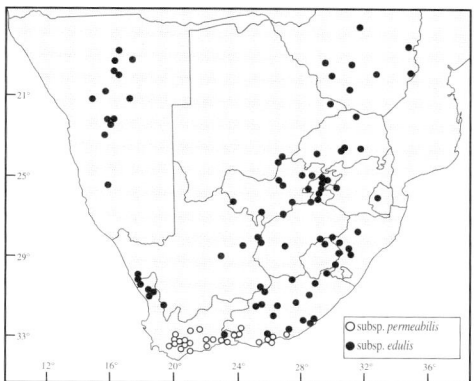

of uncertain origin. Floral morphology, however, makes it clear that they were collected in the winter-rainfall southwestern part of southern Africa. The tepals are acute to slightly attenuate but lack the tail-like extensions characteristic of plants from the summer-rainfall part of the subcontinent. *Gladiolus edulis* was described much later, in 1817, from plants collected by the English botanist and explorer, William Burchell, on his travels into the interior of southern Africa. Plants that he collected in 1812 near Litakun, now in North-West Province, South Africa, were later grown in England, and the species was described in 1817 by John Ker Gawler after a painting had been completed. The similarity between *G. permeabilis* and *G. edulis* was noted by J.G. Baker (1896), for example. Baker distinguished *G. permeabilis* from *G. edulis* largely by the long-cuspidate tepals of the latter. The two were first regarded as conspecific by Lewis et al. (1972), who at the same time referred the related *G. wilsonii* and *G. inandensis* to a third subspecies of a widely circumscribed *G. permeabilis*. Subsp. *edulis* appears to have been recorded in Zimbabwe only in 1909.

Key to the subspecies

1. Tepal apices acute or sometimes barely attenuate; flowers often shades of purple, mauve or greenish brown; sweetly scented .... subsp. *permeabilis*
1'. Tepal apices long-attenuate to tailed; flowers shades of cream to white or grey; scented or unscented .... subsp. *edulis*

GLADIOLUS PERMEABILIS subsp. PERMEABILIS

Description

Plants 15–45 cm high. *Leaves* five to seven, more if branches more than two, usually shorter than the stem, the margins usually unthickened. *Stem* simple or with one or two, occasionally up to four branches.

Spike 5- to 8-flowered; *bracts* pale green, sometimes flushed purplish above, often dry near the apices, the outer 15–24 mm long, the inner bracts about two-thirds as long, acute or minutely forked apically. *Flowers* dull blue-grey, dull purple or yellow-brown, the upper tepals or at least the upper laterals each with a dark longitudinal median stripe, the lower tepals and throat streaked reddish to purplish, the limbs usually yellow in the lower half, the yellow zone often edged distally in a darker colour, often strongly sweet-scented, occasionally unscented; *perianth tube* 10–13(–15) mm long; *tepals* spade-shaped, the dorsal 28–35 x 10–15 mm, the lower three tepals united with the upper laterals for 3–5 mm and to one another for 2–6 mm, the free parts 15–23 x 6–8 mm, the claws 2–3 mm long. *Filaments* 12–16 mm long, exserted 8–13 mm from the tube; *anthers* 6.5–9 mm long. *Ovary* 3–5 mm long; *style* dividing shortly below the anther apices, the branches c. 3–4 mm long. *Capsules* ovoid and rounded apically or ellipsoid and acute, (12–)14–16 mm long.

Flowering time late August to the end of September.

Distribution

Gladiolus permeabilis subsp. *permeabilis* is restricted to the hills and plains of the southern Cape, thus largely within the southern African winter-rainfall zone. It extends from near Caledon in the west to Grahamstown and Alexandria in the east, reaching a short distance into what is generally considered to be a region of predominantly summer rainfall. Plants favour fairly dry habitats and are usually found on stony shale in renosterveld vegetation, not uncommonly on warm, north-facing slopes.

GLADIOLUS PERMEABILIS subsp. EDULIS (Burchell ex Ker Gawler) Obermeyer

Obermeyer in *J. S. African Bot.*, Suppl. 10: 135 (1972); Van Wyk & Malan, *Field Guide Wild Flowers Witwatersrand & Pretoria* 150 t. 352 (1988). Goldblatt, *Fl. Zambesiaca*

12(4): 75 (1993); Gladiolus *in Tropical Africa* 169 (1996). *Gladiolus edulis* Burchell ex Ker Gawler, *Edwards's Bot. Reg.* 2: t. 169 (1817); Baker, *Fl. Capensis* 6: 161 (1896); *Fl. Trop. Africa* 7: 373 (1898). Sölch, *Prodromus Fl. Südwestafrika* 155: 4 (1969). Goldblatt, *Fl. Zambesiaca* 12(4): 75 (1993); Gladiolus *in Tropical Africa* 169 (1996). Type: South Africa, North-West Province, Pellat Plains, Aug. 1812, *Burchell 2240* (K, holotype; B, G, isotypes).

edulis = edible, reflecting the explorer William Burchell's observation that the corms were eaten by some of the tribes he met on his travels in southern Africa.

Synonymy

Gladiolus remotifolius Baker, *Bull. Herb. Boissier*, Sér. 2, 1: 867 (1901). Type: South Africa, 'Transvaal', possibly Mpumalanga, between Porter and Trigardsfontein, probably in 1879, *Rehmann 6618* (Z, holotype).

Description

Plants to 60 cm high, often slender. *Leaves* occasionally three, usually four to six, at least the lower two or three basal, shorter than to about as long as the stem, occasionally longer, 1–2.5(–3) mm wide, the margins sometimes lightly thickened. *Stem* simple or branched, 1–1.8 mm in diameter below the spike.

Spike usually 4- to 8-flowered; *bracts* 7–12 mm long. *Flowers* whitish to cream, sometimes grey or mauve, flushed with grey-blue on the dorsal tepal, the remaining tepals greyish to purple or maroon on the midline sometimes surrounding a yellow streak, the lower lateral tepals yellow in the upper half, unscented or sometimes intensely sweet-scented; *perianth tube* 9–13(–15) mm long; *tepals* long-attenuate or with tail-like appendages up to 11 mm long, twisted and undulate, the dorsal (16–)20–25 mm long with the appendages (often less when dry), the lower three tepals joined for 1–2 mm with the upper laterals, and to one another for 3–5 mm, the free parts 14–16 mm long,

the claws c. 4 x 1 mm. *Filaments* 13–15 mm long, exserted 9–10 mm from the tube; *anthers* 5–7 mm long. *Ovary* c. 3 mm long; *style* dividing between the base and upper third of the anthers, the branches 1.5–2 mm long. *Capsules* nearly globose to obovoid, 6–10 mm long. *Chromosome number* $2n = 30, 28$.

Flowering time mostly February to April, but occasionally at other times.

Distribution

Gladiolus permeabilis subsp. *edulis* is widespread across the summer-rainfall zone of southern Africa and is particularly common in drier areas or in locally dry situations. It extends from Oudtshoorn in Western Cape Province through the Free State to Mpumalanga and Northern Province in South Africa, and into central and northern Namibia, southern and eastern Botswana, and throughout Zimbabwe (Lewis et al., 1972; Goldblatt, 1993, 1996). As might be expected from so wide a distribution, plants occur in a variety of habitats. In Namibia, Botswana and Zimbabwe subsp. *edulis* is most often encountered in deep sandy soils, mostly Kalahari sands. Elsewhere it may be found among low karroid scrub on clay soils or on rock outcrops among low grasses, either in open country or in light woodland. Some populations have flowers that are so strongly fragrant that the presence of the species can often be detected before plants are located, but this appears to be the exception. Curiously, many eastern South African populations, including all those north of the Vaal River that we have examined, have apparently odourless flowers, but plants we have seen in Namibia have wonderfully fragrant flowers.

William Burchell, who first collected subsp. *edulis*, wrote that the corms, which were common on the plains near Litakun, were often eaten by the local people and when roasted had a 'sweet and agreeable taste much like chestnuts'. There are no other reports of the species being eaten by humans.

90. GLADIOLUS UITENHAGENSIS Goldblatt & Vlok

Gladiolus uitenhagensis Goldblatt & Vlok, *J. S. African Bot.* 55: 293 (1989). Type: South Africa, Eastern Cape, Uitenhage District, Groendal State Forest, track along Ten Stop Hill, 12 Sept. 1983, *Vlok 702* (holotype, NBG; isotype, MO).

uitenhagensis, named for the district of Uitenhage in the Eastern Cape Province, South Africa.

Description

Plants 35–60 cm high. *Corm* 15–20 mm in diameter, the tunics of coarse firm, blackish

fibres. *Leaves* seven to eight, the lower four or five basal and longest, reaching to about the base of the spike or sometimes slightly exceeding it, the blades linear, 1.5–2 mm wide, the midrib strongly thickened, the margins barely if at all raised, the upper three to four leaves inserted on the stem, decreasing in size above, the uppermost small and largely sheathing, the margins free to the base. *Stem* erect, usually unbranched, c. 2 mm in diameter below the spike.

Spike 5- to 8-flowered; *bracts* grey-green, soft-textured, the outer 15–20 mm long, the inner 2–3 mm shorter than the outer. *Flowers* mauve, the lower tepals each with a yellow spear-shaped mark outlined in purple in the midline, presence of scent unknown; *perianth tube* obliquely funnel-shaped, (22–)28–35 mm long, well exserted from the bracts; *tepals* unequal, lanceolate, the dorsal largest, inclined over the stamens, 28–30 x 15 mm long, the upper laterals c. 19 x 12 mm, windowed between the bases of the dorsal and upper lateral tepals, the lower three tepals joined to the upper laterals for c. 3 mm and to one another for c. 3 mm, c. 20 x 8 mm. *Filaments* c. 15 mm long, exserted 8–10 mm from the tube; *anthers* c. 8 mm long, yellow, pollen yellow. *Ovary* oblong, c. 4 mm long; *style* arching over the stamens, dividing near the apex of the anthers, the branches c. 2.5 mm long, wider in the upper half. *Capsules* and *seeds* unknown. *Chromosome number* unknown. *Flowering time* September.

Distribution and biology

Gladiolus uitenhagensis, a very local endemic of the southern African winter-rainfall zone, is known only from the eastern end of the Great Winterhoek Mountains in the Groendal State Forest near Uitenhage in Eastern Cape Province. Plants grow on well-drained, stony sandstone slopes in fynbos.

They flower well only one or two seasons following a fire and apparently not at all between fires.

We suspect that, like its close relatives with similarly proportioned flowers and a relatively short perianth tube, *Gladiolus uitenhagensis* is pollinated by long-tongued anthophorid bees.

Diagnosis and relationships

Gladiolus uitenhagensis seems to be most closely related to the widespread *G. permeabilis* of the southern and eastern Cape. It has the general appearance of a robust form of that species, to which it corresponds in having nearly equal floral bracts, three to five slender basal leaves with prominent midribs, and a several-flowered spike. The large flowers of *G. uitenhagensis* are, however, quite different in size, being 50–65 mm long with a tube as long as or longer than the tepals and normally 28–36 mm long. The flowers of *G. permeabilis* are usually smaller, 30–50 mm long, with a maximum tube length of 12 mm. *Gladiolus uitenhagensis* also differs from *G. permeabilis* in its dark brown, firm corm tunics consisting of relatively hard fibres. Of the two subspecies of *G. permeabilis*, *G. uitenhagensis* comes closer to subsp. *permeabilis* in the shape of the flower, but the broad, hooded upper tepal, 20–30 x 15 mm, is more prominent than in any collection of subsp. *permeabilis* that we have examined, although the description provided by Lewis et al. (1972) gives dimensions of 20–30 mm long and 10–14 mm wide for the upper tepal. The general aspect

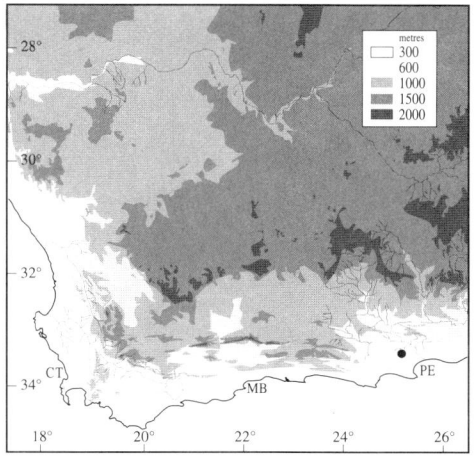

of the flower is very different from that of *G. permeabilis* and although we have considered the possibility that it is no more than a very large-flowered form of that species, more collections are needed to assess the pattern of variation in *G. uitenhagensis* – particularly in flower size – before a final decision can be taken about its status. The habitat preferences of the two species are apparently different, as *G. permeabilis* prefers clay soils and *G. uitenhagensis* grows in rocky sandstone.

History

Gladiolus uitenhagensis was discovered by the South African botanist, J.H.J. Vlok, in September 1983, the season after a fire in the eastern Great Winterhoek Mountains, and it was described a few years later in 1989. It has not been re-collected since, and we have been unable to obtain specimens to examine and illustrate for this monograph.

91. GLADIOLUS ACUMINATUS F. Bolus

PLATE 76

Gladiolus acuminatus F. Bolus, *Ann. Bolus Herb.* 2: 100 (1917). Lewis et al., *J. S. African Bot.*, Suppl. 10: 77 (1972). Type: South Africa, Western Cape, without precise locality or collector, obtained at the Caledon Wild Flower Show, Sept. 1916 (BOL 14793, lectotype here designated, BM, BOL, G, K, PRE, SAM, Z [not seen], isolectotypes).

acuminatus = acuminate, referring to the tepals which have narrowly tapering apices.

Description

Plants 25–50 cm high. *Corm* globose, 12–15 mm in diameter, the tunics of firm-papery layers, breaking into fairly fine vertical fibres with age, sometimes with one to three large cormlets borne at the base on

stolons 10–20 mm long. *Cataphylls* pale and membranous, the upper greenish or turning purple above the ground. *Leaves* five or six, or more if stem branched, the lower three or four basal, at least some of these half as long as the stem or reaching to the base or middle of the spike, linear, 0.8–1.5 mm wide, the midrib somewhat thickened, tapering toward the margins, two upper leaves cauline, usually channelled throughout, the uppermost leaf largely sheathing, sheaths of all the leaves open, the margins of the cauline leaves free to the base. *Stem* erect below, flexed outward above the sheaths of the cauline leaves and inclined above, simple or occasionally branched, 1–2 mm in diameter below the spike.

Spike inclined or erect, more or less

straight, 3- to 10-flowered; *bracts* pale green, the outer 13–25(–35) mm long, rather attenuate, the inner about two-thirds as long

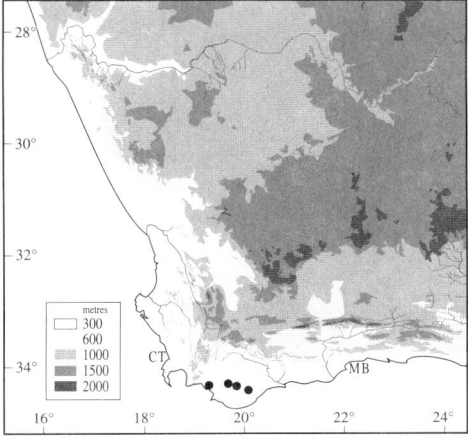

as the outer, the apices entire or notched apically. *Flowers* cream to pale yellow or pale greenish, the tepals often flushed brown to dull purple on the reverse and darkest on the median, especially those of the outer whorl, lightly sweet-scented; *perianth tube* 16–22(–30) mm long, cylindric, the upper half curving outward and slightly wider; *tepals* lanceolate-acuminate, the margins undulate, directed forward below, outward above and ultimately recurving, the apices

lightly twisted, the dorsal slightly larger than the others, 15–21 x 8–10 mm, remaining tepals 12–18 x 6–8 mm. *Filaments* 6–10 mm long, reaching only to the top of the tube, or exserted up to 2 mm; *anthers* 4–6 mm long, exserted from the tube but included in the flower. *Ovary* narrowly ovoid, c. 3 mm long, *style* arching over the stamens, dividing near the anther apices, the branches c. 3.5 mm long, when fully open curving outward beyond the anthers, channelled throughout. *Capsules* ellipsoid, acute at the apices, 10–15 mm long; *seeds* ovate, 3–4 mm long, poorly and unevenly winged, the wing light semi-transparent brown, the seed body dark brown and unusually large. *Chromosome number* unknown.

Flowering time mid-August to late September.

Distribution and biology

Gladiolus acuminatus is a fairly narrow endemic of the southern coast of Western Cape Province, South Africa. Its range extends from southwest of Caledon near Onrus to near Bredasdorp. Plants grow at low elevations in stony shale on flats and north-facing slopes in renosterveld vegetation. The species is poorly recorded, with just a few collections from the wild, and we assume it is rare. We have not seen it in the wild, although we have searched for it at sites where it has been reported in the past.

The pale greenish cream flower with a sweet scent and the relatively long perianth tube suggest that it is pollinated by moths, but nothing is known about its pollination biology.

Diagnosis and relationships

Vegetatively *Gladiolus acuminatus* corresponds closely to other species of series *Permeabilis* in having several basal leaves, the blades linear with a thickened midrib, and corm tunics with coarse, wiry fibres. In the absence of flowers it is virtually identical to species like *G. permeabilis* and *G. stellatus*, to which it is clearly closely allied. The flowers, however, are very different from those of any member of series *Permeabilis* in having a cylindric perianth tube, 22–30 mm, which is fairly long for the series. The flowers are greenish cream to pale yellow, the tepals are subequal, narrow and tapering above, and the perianth tube is narrow and 16–30 mm long. Because of the shape of the tepals and the long tube there should be no difficulty in correctly identifying *G. acuminatus*. In *G. robertsoniae*, the other member of series *Permeabilis* which has flowers adapted for moth pollination, the perianth tube is 28–44 mm long and the tepals are white and substantially larger, the upper being 18–25 x 12–18 mm, thus differing in colour, shape and size from those of *G. acuminatus*.

History

Gladiolus acuminatus first attracted the attention of botanists in 1916 when the renowned Cape botanist, Rudolf Marloth, noticed plants on display at the Caledon Wild Flower Show. Several specimens were preserved by Marloth and the following year Frank Bolus, who maintained a serious interest in *Gladiolus*, described the species. It was not until 1932, however, that *G. acuminatus* was recorded in the wild. Plants were found by the South African botanist, J.L. Sidey, in renosterveld between Napier and Caledon. *Gladiolus acuminatus* remains a rather enigmatic species and has been collected only occasionally since then. It does, however, appear consistently at the Caledon Wild Flower Show, from which we obtained the plants for the illustrations used here.

92. GLADIOLUS STELLATUS G. Lewis

PLATE 77

Gladiolus stellatus G. Lewis, *Bot. Not.* 119: 295 (1966); Lewis et al., *J. S. African Bot.*, Suppl. 10: 70 (1972), as a new name for *G. elongatus* Thunberg, *Prod. Pl. Capensium* 185 (1800), an illegitimate homonym for *G. elongatus* Salisbury (1796) (= *Babiana tubulosa* (Burman fil.) Ker Gawler). Type: South Africa, Western Cape, without precise locality, probably Oct. or Nov. 1772 or 1773, *Thunberg s.n.* (UPS–Herb. Thunberg 1020, holotype, K, photograph).

stellatus = starry, referring to the resemblance of the flowers to small stars, with their equal, spreading tepals.

Synonymy

Geissorhiza patersoniae L. Bolus, *Ann. Bolus Herb.* 1: 32 (1915), not *Gladiolus patersoniae* F. Bolus (1928). Type: South Africa, Eastern Cape, Port Elizabeth District, Redhouse, Sept. 1912, *Paterson 47* (BOL, lectotype designated here; GRA, K, isotypes).

Description

Plants 30–60 cm high. *Corm* globose, 9–15 mm in diameter, the tunics of fine to fairly coarse fibres, often producing stolons from the base. *Cataphylls* pale and membranous, reaching 12–30 mm above the ground and then green or purple. *Leaves* rarely four, usually six to nine, the lower two to seven basal, shorter or slightly longer than the stem, linear, (1–)2–3 mm wide, the midrib area thickened, the margins not raised or prominent, the upper two leaves inserted in the upper half of the stem, much shorter than the basal, usually entirely channelled. *Stem* more or less erect, inclined above the sheath of the uppermost basal leaf, usually with one or two branches, 0.8–1.2 mm in diameter below the spike.

Spike inclined, the flowers weakly secund, 5- to 12-, rarely to 15-flowered; *bracts* pale green, often becoming dry and light brown above after the first flowers open, the outer 10–15 mm long, the inner slightly shorter than the outer. *Flowers* actinomorphic, stellate, white to cream or pale lilac, the tepals each with an obscure narrow brownish or purple median streak, this darker in the lower half, the reverse of the tepals usually lightly feathered with greenish to brown-purple, intensely sweet-scented; *perianth tube* shortly funnel-shaped, 5–7 mm long, included or shortly exserted from the bracts; *tepals* nearly equal, elliptic, 14–20 x 5–8 mm. *Filaments* 5–7 mm long, exserted 3–4 mm from the tube, parallel and contiguous; *anthers* (4–)5–6(–7) mm long, straight and diverging, pale yellow, the pollen yellow. *Ovary* oblong, c. 5 mm long; *style* central, enclosed by the filaments, dividing between the base and upper third of the anthers, the branches 4–5 mm long, usually slender, very little expanded above. *Capsules* elongate-ellipsoid to oblong, 12–17 x 5 mm; *seeds* ovate, c. 6 x 4 mm, broadly and evenly winged, light brown, the seed body dark brown and fairly large. *Chromosome number* unknown.

Flowering time mid-August to late October, occasionally in November. Flowers open at c. 07h00 and close between 11h45 and 12h15.

Distribution and biology

Gladiolus stellatus extends across the southern Cape from a short distance east of Swellendam in the west to Bethelsdorp near Port Elizabeth in the east. Although most

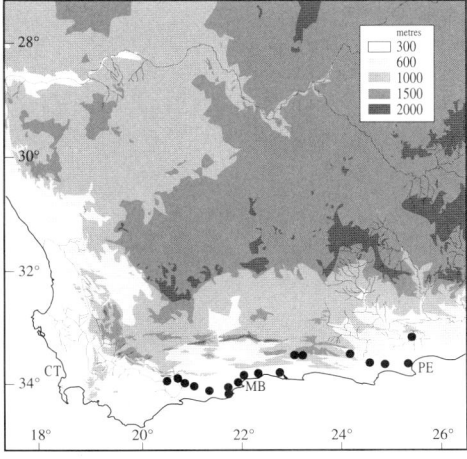

common on the coastal plain and hills south of the Langeberg and Outeniqua Mountain axis, *G. stellatus* occurs locally in the Little Karoo and the Long Kloof, and on the hills south of the Vanstaden's River Mountains. Plants favour exposed, fairly dry sites such as stony grassland and renosterveld on hilltops or north-trending slopes. Although sometimes seen in profusion after fires, burning is not a prerequisite for flowering. Scattered, but often rather dwarfed plants will flower in veld that has not been burned for years, providing that the surrounding bush is not too tall. *Gladiolus stellatus* seems to occur most often on clay soils, but we have also occasionally found it growing on stony sandstone-derived soils.

At a distance the small, stellate flowers can be mistaken for the flowers of either *Chlorophytum* (Anthericaceae) or *Trachyandra* (Asphodelaceae) and, like the species of these genera, *Gladiolus stellatus* appears to be unspecialized in its pollination system. Flowers open early in the morning, at about the time the sun rises, c. 07h00, immediately releasing their characteristic strong, sweet scent. We have observed one species of bee visiting the flowers and assume that *G. stellatus* is pollinated by a range of bees and butterflies seen to visit similar flowers. The main reward to insect visitors is pollen which is prominently displayed and accessible to bees. A small amount of very sweet nectar present in the floral tube is available to nectar-seeking visitors. The flowers of *G. stellatus* close fairly promptly between 11h45 and 12h15, reopening again the next day. At some sites where *G. stellatus* was common we noticed the white-flowered form of *Moraea polyanthos* (Iridaceae) growing nearby and the small flowers of the latter species only began to open after 12h30. These two species appear to be sharing pollinators but dividing pollinator resources between them by not competing for the same insect visitors at the same time.

Diagnosis and relationships

Gladiolus stellatus, as its name implies, has a small, star-like flower with equal spreading tepals. Flower colour ranges from pale cream to light dull purple and the only markings are dark streaks of dull purple colour in the lower half of the tepals along the midline. The tepals, stamens and style are symmetrically arranged and the flower is actinomorphic, thus most unlike the typical *Gladiolus* flower. Despite this, there is no doubt that the species belongs in the genus for it has the circumferentially winged seeds and soft green

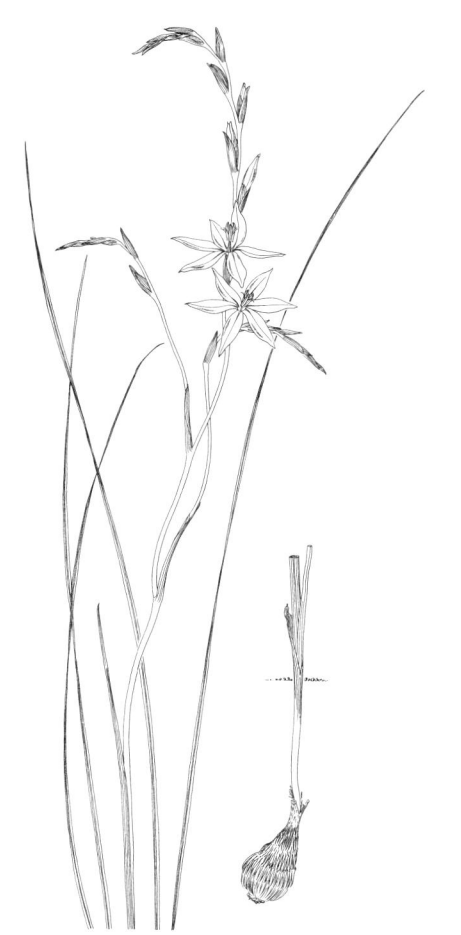

floral bracts that are distinguishing features of the genus. It can also be placed with confidence in section *Hebea*, for it has the distinctive coarse, hard, wiry corm tunics, a cluster of firm narrow leaves with raised midribs and unthickened margins, and the strong, sweet scent so characteristic of many of the species of that section. The fibrous corm tunics and seeds with dark seed bodies are consistent with series *Permeabilis* and it is here that we place *G. stellatus*. Other species in this alliance, all with zygomorphic flowers, include *G. inandensis*, *G. wilsonii* and *G. permeabilis*. The last is perhaps most closely related to *G. stellatus*.

History

Gladiolus stellatus was first recorded in botanical history in 1772 by Carl Peter Thunberg during his travels with J.A. Auge through the southern Cape. Despite the stellate actinomorphic flowers, Thunberg recognized the plants he found as a species of *Gladiolus* which he first named *G. tristis* var. *corollis parvis, etc.* (Thunberg, 1784), and later *G. elongatus* (Thunberg, 1800). This name is unfortunately a homonym for *G. elongatus* Salisbury, dating from 1796, a species of *Babiana*, and cannot be used in *Gladiolus*. Although the species was recognized by some early authorities, for example John Ker Gawler (1827), *G. stellatus* was not recognized in the *Flora Capensis* (Baker, 1896) and the name *G. elongatus* Thunberg was cited under the synonymy of *G. gracilis*. It is no surprise then, that when re-collected after 1900, the species was thought to be new. Louisa Bolus described plants collected by Florence Paterson at Redhouse near Port Elizabeth as *Geissorhiza patersoniae*. The assignment of the species to *Geissorhiza* was no doubt because of the actinomorphic flower. G.J. Lewis (1966) realized that *G. patersoniae* belonged in *Gladiolus* and chose the new name *G. stellatus* for the plant knowing that neither Thunberg's *G. elongatus* nor L. Bolus's *Geissorhiza patersoniae* could be used in *Gladiolus*. The name *Gladiolus patersoniae* was used in 1917 by Frank Bolus for the Karoo Mountain and Eastern Cape bluebell.

93. GLADIOLUS WILSONII (Baker) Goldblatt & Manning
PLATE 78

Gladiolus wilsonii (Baker) Goldblatt & Manning, new combination. *Tritonia wilsonii* Baker, *Gard. Chron.* 26: 38 (1886); *Flora Cap.* 6: 125 (1896). *Gladiolus permeabilis* subsp. *wilsonii* (Baker) Lewis, *J. S. African Bot.*, Suppl. 10: 133 (1972). Type: South Africa, Eastern Cape, vicinity of Port Elizabeth (grown in Britain by Mr John Wilson of St. Andrews, June 1886, *Wilson s.n.* (K, holotype).

wilsonii, named in honour of John Wilson of St. Andrews, Scotland, who grew the plants on which Baker based his description, and/or his brother, Alexander Wilson, who originally collected the species near Port Elizabeth.

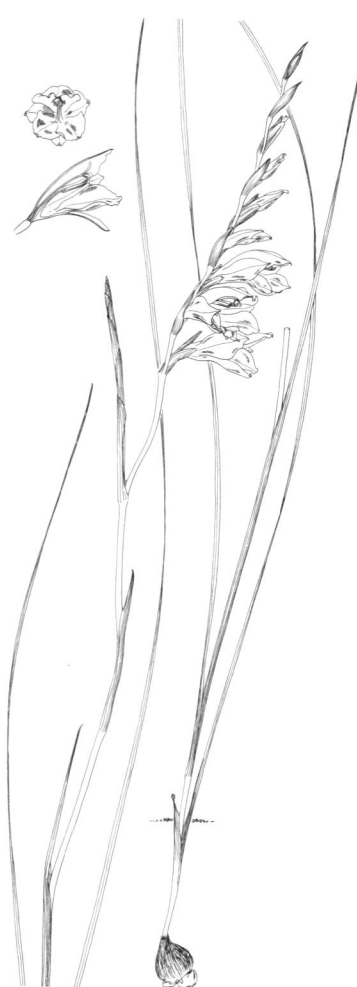

Synonymy

Gladiolus frederickii L. Bolus, *S. African Gard.* 21: 369 and fig. on p. 367 (1931). Type: South Africa, Eastern Cape, Grahamstown Commonage on left of Cradock road (grown in Cape Town), Oct.-Nov. 1929, *Dyer 1682* (BOL, holotype; GRA [not seen], K, PRE, isotypes).

Description

Plants (20–)30–60 cm high. *Corm* globose, 12–22 mm in diameter, the tunics of fine to fairly coarse fibres, often with small cormlets at the base. *Cataphylls* pale and membranous, the uppermost reaching a short distance above the ground and then green or partly to entirely dry. *Leaves* six to eight, at least the lower three to five basal and attached to the flowering stem, reaching to about the base of the spike, occasionally slightly exceeding the stem, the blades linear, 1–3 mm wide, the midrib area thickened, the margins rarely lightly thickened, stem with four to five leaves, these progressively smaller in size, the uppermost two or three entirely sheathing or with vestigial blades, 25–70 mm long, with the margins free to the base, open or overlapping. *Stem* erect, simple or with one or occasionally two branches, usually flexed above the sheath of the uppermost leaf, 1–3 mm in diameter below the spike.

Spike inclined, lightly flexuose, 4- to 12-flowered, the branches normally with fewer flowers; *bracts* green, sometimes flushed purple above, the outer 12–16(–20) mm long, the inner two-thirds to nearly as long as the outer, notched apically for c. 0.5 mm. *Flowers* white to cream, the reverse of the three upper tepals usually flushed pink to dull purple toward the tips, and sometimes also on the midline, sometimes the lower three tepals each with short, pale mauve streaks near the base, often sweetly scented; *perianth tube* obliquely funnel-shaped, 6–13 mm long, the slender cylindrical part 3–6 mm long; *tepals* oblanceolate, unequal, the dorsal largest, 20–26(–30) x

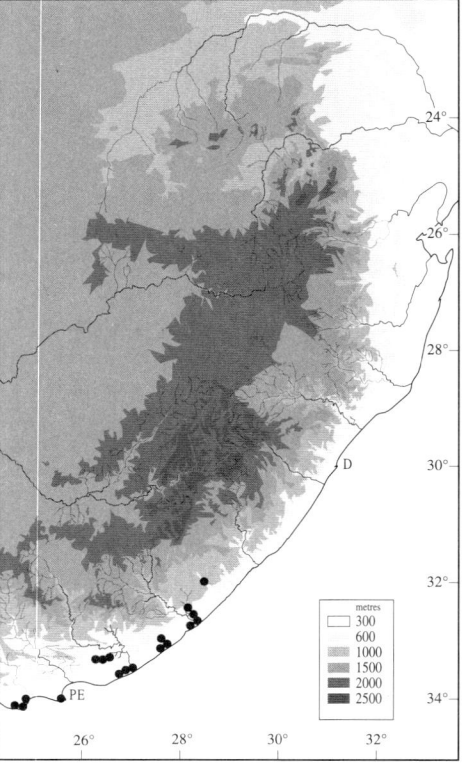

10–13(–18) mm, hooded over the stamens, the upper laterals 16–20(–23) x 8–9 mm long, directed forward, curving outward in the upper half to third, the lower three tepals joined to the upper laterals for 1–2 mm and to one another for c. 2 mm, the lower laterals 16–20 x 6.5–13 mm, the lower median about as long as or slightly shorter and wider or narrower than the laterals. *Filaments* 10–13 mm long, exserted 6–8 mm from the tube; *anthers* 6–9 mm long, purple, the pollen cream. *Ovary* ovoid, c. 4 mm long; *style* arching over the stamens, dividing shortly below the anther apices, the branches 3.5–5 mm long. *Capsules* ellipsoid, 10–15 mm long; *seeds* oblong, 5–6 x c. 3 mm, broadly but somewhat unevenly winged, light brown, the seed body usually darker brown. *Chromosome number* $2n$ = 30.

Flowering time mainly October to mid-November, occasionally in September.

Distribution and biology

An endemic of Eastern Cape Province, South Africa, *Gladiolus wilsonii* occurs in the summer-rainfall zone of southern Africa, but is, nevertheless, a winter-growing species, depending on late autumn and scanty winter rainfall for growth. The species is best known from Grahamstown and East London, but the range extends from Jeffrey's Bay and Cape St. Francis in the west to central Transkei near Engcobo in the east, the latter the most inland record for this predominantly coastal species. Plants are most often found in open grassland in fairly light loamy sand.

Following its pattern of winter growth, *Gladiolus wilsonii* flowers in spring, usually in October and November, but occasionally in September. The short-tubed and usually fragrant flowers are believed to be adapted for bee pollination.

Diagnosis and relationships

Gladiolus wilsonii has the typical vegetative features we associate with series *Permeabilis*, a loose fan of several basal leaves with narrow whip-like blades with thick midribs and unthickened margins, and corms with tunics of tough, wiry fibres. The flowering stems are slender and often branched, and bear several white to cream flowers which are usually, but apparently not always, fragrant. The flowers are relatively small, and unusual in section *Hebea* in having the upper tepals not or hardly narrowed below. As a result, although tepal orientation is exactly the same as in *G. permeabilis*, the flowers lack the window or gap between the bases of the dorsal and upper lateral tepals. This floral difference as well as the flower colour – white to cream in *G. wilsonii*, purple, brownish, grey or cream in *G. permeabilis* – make it easy to distinguish the two species.

We do not accept the treatment of *Gladiolus wilsonii* by Lewis et al. (1972) as a subspecies of *G. permeabilis*, nor their inclusion of a third species, *G. inandensis*, in subsp. *wilsonii*. The differences outlined above warrant the recognition of *G. wilsonii* as a separate species.

We recognize *Gladiolus inandensis* as a separate species for several reasons. Also a species of the southern African summer-rainfall zone, it differs from *G. wilsonii* in having its vegetative growth phase during the summer. Thus the leaves are produced in the late spring and grow throughout the summer, dying back in the autumn. The flowering spikes are produced in the winter or spring when the leaves are more or less dry and dead, or those of the new season are emerging. The flowers of *G. inandensis* closely resemble those of *G. wilsonii* in shape and colour, but they are always unscented, whereas those of *G. wilsonii* usually have a sweet fragrance. The leaf morphology of *G. inandensis* is also distinctive. The leaf blades are more or less oval in transverse section and quite solid in contrast to those of *G. wilsonii* which have the shape typical of section *Hebea*, being linear with a thickened midrib and completely undeveloped margins.

History

According to available records, *Gladiolus wilsonii* was first collected in September 1813 by the explorer-naturalist, William Burchell, near the Kaffir Drift Military Post on the Great Fish River in Eastern Cape Province. Burchell also collected plants a few weeks later in the vicinity of Port Alfred and Rietfontein. The species was also recorded by other early collectors, including C.F. Ecklon and C.L. Zeyher, but despite the availability of well-preserved specimens in several European herbaria, *G. wilsonii* was not described until 1886 when J.G. Baker received sketches and a living plant from John Wilson, grown in Great Britain and originally from Port Elizabeth. Baker named it *Tritonia wilsonii*, honouring either John Wilson or his brother Alexander, the collector of the original stock, and under the mistaken impression that the species belonged in the genus *Tritonia*. In the late 1920s, the South African botanist, R.A. Dyer, collected specimens of a white-flowered species of *Gladiolus* in Grahamstown, and corms grown on in Cape Town flowered in 1929. They seemed to H.M.L. Bolus to represent a new species which she described as *G. frederickii* in 1931. That this last species was conspecific with *Tritonia wilsonii* was only discovered years later when G.J. Lewis began to study the systematics and taxonomy of *Gladiolus*. Lewis intended to recognize *G. wilsonii* but, as outlined above, in her posthumously published account of *Gladiolus* in South Africa, it was reduced to subspecific status in *G. permeabilis* by A.A. Obermeyer.

94. GLADIOLUS INANDENSIS Baker

Gladiolus inandensis Baker, *Handbook Irideae*, 207 (1892); *Fl. Capensis* 6: 144 (1896). Wood, *Natal Plants* 3: pl. 236 (1902). Type: South Africa, KwaZulu-Natal, Inanda, Aug., *Wood 237* (K, lectotype designated by Lewis et al., 1972: 133; BM, K, GRA, isolectotypes).

inandensis = from Inanda, near Durban in KwaZulu-Natal, where the type collection was made.

Synonymy

Tritonia teretifolia Baker, *Handbook Irideae* 194 (1892); *Fl. Capensis* 6: 124 (1896). Type: South Africa, KwaZulu-Natal, foot of Table Mountain near Pietermaritzburg, Aug. 1840, *Krauss 430* (BM, holotype; G, K, OXF [not seen], isotypes).

Gladiolus microphyllus Baker, *Handbook Irideae* 206 (1892); *Fl. Capensis* 6: 143 (1896). Type: South Africa, KwaZulu-Natal, Suurberg Mountains near Kokstad, 5500 ft, Oct. 1883, *Tyson 1852* (K, holotype; GRA [not seen], MO, SAM, isotypes).

Gladiolus stenophyllus Baker, *Kew Bull. Misc. Inform.* 1897: 282 (1897), an illegitimate homonym, not *G. stenophyllus* Schrank (1822) (= *Babiana lineolata* Klatt). Type: South Africa, Eastern Cape, Hangklip Mountain near Queenstown, Jan. 1894, *Galpin 1769* (K, holotype; BOL, GRA [not seen], PRE, isotypes).

Gladiolus microsiphon Baker, *Bull. Herb. Boiss.* Sér. 2, 4: 1006 (1904). Type: South Africa, KwaZulu-Natal, hills near Pinetown, 150 m, 12 Sept. 1893, *Schlechter 3167* (Z, probable holotype; BM, BOL, G, K, SAM, Z, isotypes).

Description

Plants 20–45 cm high, often growing in small clumps. *Corm* globose, 18–25 mm in diameter, the tunics of fine, densely matted fibres, usually with one or two large cormlets at the base. *Cataphylls* pale and membranous below the ground, the uppermost reaching 5–8 cm above the ground and then coriaceous and dark green, accumulating with age as a fibrous neck around the base of the stem. *Leaves* (of the flowering stem) four to six, the lower two to four basal, usually dry or becoming dry at flowering time (or burned and absent), up to 20 cm long, exceeding the spike and trailing, sometimes only reaching to about the top of the spike, the blades somewhat fleshy, oval in transverse section when fresh but with the midrib area thickened and prominent when dry,

1.5–2 mm in diameter, the upper two or three leaves entirely sheathing or with vestigial blades, the margins free to the base and overlapping or not; usually at the end of flowering new leaves emergent, these produced from separate shoots but often intertwined with the old leaves and flowering stem, ultimately three or four basal leaves produced at the end of summer. *Stem* erect or inclined, simple or with one branch, usually flexed above the sheath of the uppermost leaf, 1–1.5 mm in diameter below the spike.

Spike inclined, lightly flexuose, 4- to 12-flowered, the branches normally with fewer flowers; *bracts* green, sometimes flushed purple above, the outer 11–16 mm long, the inner about two-thirds as long as the outer, notched apically for c. 0.5 mm. *Flowers* white to cream, the reverse of the three upper tepals usually flushed pink to dull purple toward the tips, and sometimes also on the midline, evidently without nectar guides, unscented; *perianth tube* 9–12 mm long, narrowly funnel-shaped, the slender cylindrical part 5–8 mm long; *tepals* oblanceolate, unequal, the dorsal largest, 22–23 x 11–15 mm, hooded over the stamens, upper laterals 15–20 x 7–8 mm,

directed forward, curving outward in the upper half to third, the lower three tepals joined to the upper laterals for 2–4 mm and to one another for 2–3 mm, the lower laterals 14–16 x 6–7 mm, the lower median 10–12 x 7 mm. *Filaments* 10–12 mm long, exserted 6–9 mm from the tube; *anthers* 5–6 mm long, mauve, the pollen cream. *Ovary* ovoid, c. 3 mm long; style arching over the stamens, dividing opposite the upper third of the anthers, the branches c. 3.5 mm long. *Capsules* narrowly ovoid-ellipsoid, (10–)12–15 mm long; *seeds* ovate, 6–7 x 3.5–4.5 mm, broadly and evenly winged, the seed body fairly large, golden brown. *Chromosome number* 2*n* = 30.

Flowering time mostly October to mid-November, sometimes in September.

Distribution and biology

Available records indicate that *Gladiolus inandensis* is nearly restricted to quartzitic sandstone soils in KwaZulu-Natal and the adjacent Eastern Cape Province. Most records are from the sandstone system of coastal KwaZulu-Natal, but the species has also been recorded from inland sites in southern Lesotho and the former Transkei, where they presumably grow on the sandy soils derived from the Cave Sandstone system. The species extends from Queenstown in the south to Empangeni in the north. Plants are most often found in rocky grassland on sandstone hillsides where they flower unusually early for plants of the southern African summer-rainfall region, in August and September.

The life history of *Gladiolus inandensis* is unusual. Leaves of the new season emerge in September or early October, just when the spikes produced at the end of the previous season's growth are completing their flowering. Unless burned in veld fires, the leaves belonging to the flowering stalk and basal to it are present and still attached to the flowering stems, but they are quite dry and dead. New green foliage that is present at flowering time is produced from the same corm as the flowering spike, but from a new shoot, lateral to the flowering stem and its associated, but dry leaves. The true relationship of the leaves of the past and coming seasons to the flowering spike is difficult to determine without careful examination and is unusual in *Gladiolus*. The species essentially has no true dormant period when all the above-ground parts are shed.

The flowers of *Gladiolus inandensis* resemble those of several other summer-rainfall species with short floral tubes – for example

in section *Densiflorus* – both in form and in the lack of scent, and are probably also pollinated by long-tongued bees.

Diagnosis and relationships

Gladiolus inandensis is most easily recognized by its small, whitish flowers produced on fairly short flowering stems that have bladeless sheathing leaves and, unless burned off by fire, the dry dead remains of foliage leaves attached to their base. Sometimes, too, plants have short green leaves at flowering time and careful examination shows that these emergent leaves are produced from a new shoot and are not attached to the flowering stem. The leaf blades of the foliage leaves are unusual in being oval to almost terete in transverse section without the longitudinal grooves that characterize the terete leaves of most other species of *Gladiolus*. The relatively small white to cream flowers flushed with purple on the outside of the tepals, the short perianth tube, and the lower tepals without contrasting nectar guides are distinctive in section *Hebea* but differ little, if at all, from those of *G. wilsonii*, a species of the Transkei and East London Districts. Unlike *G. inandensis*, *G. wilsonii* has the flowering stems produced a few months after the new leaves emerge and flowering takes place in November when the leaves are alive and green. The leaves of *G. wilsonii* are more or less plane and linear with the midrib conspicuously raised, thus the leaves of the two species are quite different from one another.

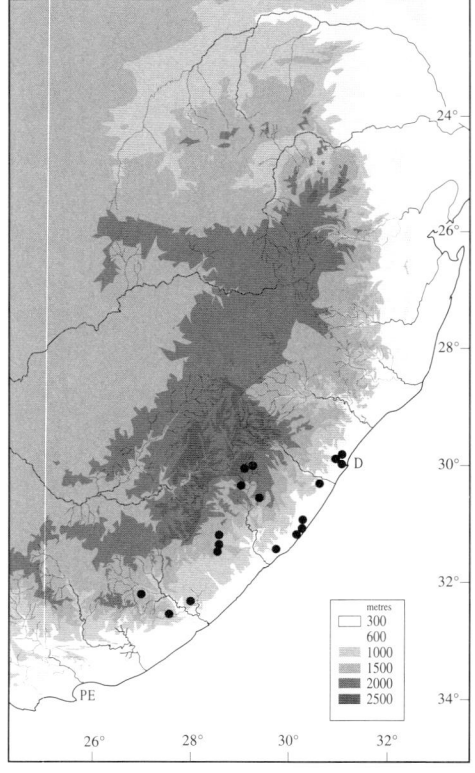

Moreover, unlike *G. inandensis*, *G. wilsonii* does not form clumps, but grows as single plants in the typical fashion of the genus.

History

Gladiolus inandensis appears to have first been recorded in 1840 by the German botanist and collector, C.F.F. Krauss, one of the first people to collect plants in what was then the Natal Colony. Krauss collected the species at the foot of Table Mountain near Pietermaritzburg. The species was also recorded by John Sanderson in 1854 and later in the 1880s and 1890s by John Medley Wood, the Natal botanist and first curator of the Natal Botanic Garden. Wood made ample collections of plants from Botha's Hill, Pinetown and Inanda near Durban. *Gladiolus inandensis* was described by J.G. Baker in 1892, largely based on the plants from Inanda. Baker misunderstood the species so completely that at the same time that he published *G. inandensis* in 1892, he named the Krauss collection of the species *Tritonia teretifolia* and a second from near Kokstad in southern KwaZulu-Natal, *G. microphyllus*. In the following 12 years *G. inandensis* was given two more names by Baker. *Gladiolus stenophyllus* was based on plants from mountains near Queenstown in Eastern Cape Province, and *G. microsiphon* on plants from Pinetown in KwaZulu-Natal.

For a species with so consistent a vegetative and floral morphology, this confusion is difficult to understand.

Subsequently, *Gladiolus inandensis* was confused with the florally similar *G. wilsonii*, and it was regarded as synonymous with that species by Lewis et al. (1972) who treated *G. wilsonii* as a subspecies of *G. permeabilis*. The relationships of *G. inandensis* and *G. wilsonii* no doubt lie with *G. permeabilis*, but as we explain above, vegetative morphology and growth pattern in *G. inandenis* is very different from that in *G. wilsonii* and *G. permeabilis* and we consider the differences sufficient to warrant its recognition as a separate species.

95. GLADIOLUS ROBERTSONIAE F. Bolus
PLATE 79

Gladiolus robertsoniae F. Bolus, *J. Bot. London* 66: 13 (1928). Lewis et al., *J. S. African Bot.*, Suppl. 10: 181 (1972). Type: South Africa, Free State, Frankfort, near Villiers, cultivated in Cape Town, *Robertson s.n.* (BOL 15053, lectotype; GRA, K, isolectotypes).

robertsoniae, named in honour of a certain W. M. Robertson, who made the first recorded collection of the species in 1916.

Description

Plants 20–40 cm high. *Corm* globose, 14–18 mm in diameter, the tunics of medium-textured, somewhat wiry reticulate fibres. *Cataphylls* membranous, the uppermost reaching 1–2 cm above the ground and then green or becoming dry, brown and papery. *Leaves* three, occasionally four, the lower two basal, the lowermost longest, sheathing the base of the stem for a short distance, the blades fairly soft-textured, linear to falcate, 80–160 mm long, 2–4 mm wide, the midrib moderately thickened, the margins lightly raised or not at all, the second leaf sheathing the lower half of the stem and with a short, more or less linear blade, the third leaf shortest, inserted between the middle and upper third of the stem, the lower margins usually sheathing the stem but not united. *Stem* erect, simple or with one branch, usually lightly flexed above the sheaths of the upper two leaves, 1.2–1.7 mm in diameter below the spike.

Spike weakly inclined, more or less straight, 4- to 8-flowered, the branch normally with fewer flowers; *bracts* dark green, sometimes flushed with dark purple, the outer 16–18 mm long, the inner about two-thirds to almost as long as the outer, notched apically for c. 0.5 mm. *Flowers* white, sometimes faintly flushed with violet toward the tepal apices, the reverse of the tepal apices mauve in the bud, on fading the reverse of the tepals and tube becoming pale mauve, the lower side of the throat with fine red lines decurrent onto the base of the lower tepals, remaining open and scented of carnation during day and night; *perianth tube* 28–44 mm long, straight and cylindric below, curving outward c. 6 mm below the mouth and then flaring, two to two-and-a-half times as long as the bracts; *tepals* unequal, broadly lanceolate to ovate, the dorsal smallest, 18–24 x 12–16 mm, lightly inclined, broadly cordate shortly above the narrow base, the upper laterals subpatent, spreading gradually from just above the base, 20–25 x 13–18 mm, the lower three tepals joined to the upper laterals for 4–6 mm, the lower laterals 16–25 x 7.5–9 mm, the lower median 14–22 x c. 11 mm. *Filaments* 8–9 mm long, included in the upper part of the tube or exserted for up to 2 mm; *anthers* 5–9 mm long, white or light mauve, the pollen yellow. *Ovary* ovoid, c. 3 mm long; *style* arching over the stamens, dividing between the upper third and the apices of the anthers, the branches 3.5–4.5 mm long. *Capsules* more or less globose, 10–14 mm long; *seeds* ovate, broadly winged, 5–6 x 3–4 mm. *Chromosome number* unknown.

Flowering time mostly October, occasionally to late November and rarely into early December.

Distribution and biology

A species of the southern African summer-rainfall zone, *Gladiolus robertsoniae* is endemic to a small part of the highveld of southern Mpumalanga and adjacent Free State provinces of South Africa. Plants have been recorded from near Ermelo westward across southern Mpumalanga to Frankfort

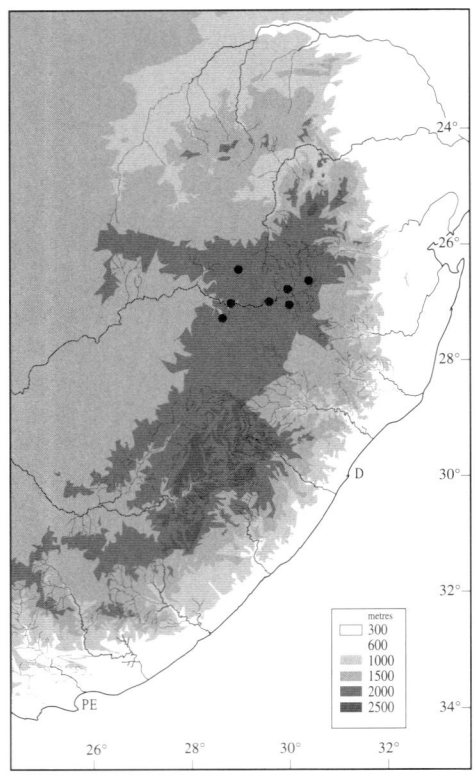

and Villiers in the eastern Free State. The species favours rocky sites, mostly dolerite outcrops, where the corms lie wedged in

crevices in the rock and secure from predation. Plants flower at the end of the dry highveld winter, and are restricted to sites such as seeps or streambanks where moisture is available at the end of the dry season.

The long-tubed and strongly scented, white flowers of *Gladiolus robertsoniae* are almost certainly adapted for pollination by moths.

Diagnosis and relationships

Gladiolus robertsoniae is one of the most distinctive species of section *Hebea*, readily recognized by the large white, strongly fragrant flowers, sometimes flushed with light mauve on fading, a perianth tube 28–44 mm long, and broad, plane tepals. The upper tepals are unusually broad, 18–25 x 12–18 mm, and neither the upper nor the lower tepals are tapered below, a reversal of the usual condition in series *Hebea*. The position of the species in the section is evident from its vegetative morphology, the corm tunics of medium-textured, somewhat wiry fibres, and the narrow leaves with a thickened midrib, but the margins hardly thickened, if at all. We have found it difficult to place *G. robertsoniae* within section *Hebea* because

of its uniquely specialized flowers. Its relatively small capsules and predominantly white flowers, however, recall particularly *G. inandensis* and *G. wilsonii*, both coincidentally species of series *Permeabilis* from summer-rainfall, eastern southern Africa. Both have relatively small tubular flowers with a predominantly white perianth, but are also unusual in the section in having the tepals not much narrowed below.

History

The first collection of *Gladiolus robertsoniae*, made by W.M. Robertson in November 1916, was sent directly to the Bolus Herbarium in Cape Town where the plants arrived in an extremely withered state. Fortunately Robertson's collection included corms, all except one of which were grown on and subsequently flowered the next year. Good herbarium specimens were made the following flowering season and the species was described by Frank Bolus who had access to the living plants when he drew up the description. Rare and inconspicuous, *G. robertsoniae* has not often been collected again, and even today is known from just a handful of sites across the highveld.

Section *Hebea:* Series *Deserticola*

96. GLADIOLUS ARCUATUS Klatt

Kalkoenuintjie, brown kalkoentjie

PLATE 80

Gladiolus arcuatus Klatt, *Abh. Naturforsch. Ges. Halle* 12: 338 (Ergänzungen 4) (1882). Baker, *Fl. Capensis* 6: 161 (1896). Lewis et al., *J. S. African Bot.*, Suppl. 10: 126 (1972).

Type: South Africa, Western Cape, Mierenkasteel, 1830, *Drège 2629* (B [not seen], holotype; BM, G, P, S, isotypes).

arcuatus = arched, perhaps referring to the curved, sickle-shaped leaves or the strongly inclined spike, less likely to the arching dorsal tepal that forms a hood over the stamens.

Description

Plants (8–)12–20(–30) cm high. *Corm* depressed-globose, 12–20 mm in diameter, the tunics of papery to dry membranous layers not accumulating with age. *Cataphylls* pale and membranous, the uppermost green above the ground. *Leaves* five or six, the lower four or five basal, reaching to about the base of the spike, the uppermost of these sheathing for half its length, pubescent to velvety on the sheaths, the blades narrowly lanceolate, 5–10(–12) mm wide, the midrib only lightly thickened, neither the midrib nor the margins raised and prominent, pubescent along the margins and midrib, the

uppermost leaf cauline, inserted in the upper third of the stem, fairly short, diverging sharply from the stem, thus not sheathing at all, the margins open to the base, channelled throughout. *Stem* erect or inclined c. 30°, flexed outward in an arcuate loop above the sheath of the penultimate leaf, flexed again above the next node and at the base of the spike, occasionally with a short branch from the axil of the uppermost leaf, 1.8–2 mm in diameter below the spike.

Spike inclined, 4- to 7-, rarely to 9-flowered; *bracts* pale green, the outer (20–)25–35 mm long, the inner only slightly shorter but much narrower than the outer, minutely forked apically. *Flowers* shades of dull grey-purple, brownish or dark purple, the lower three tepals yellow in the lower two-thirds and papillose below and in the throat, the distal third dark purple fading to light grey-purple toward the apices, the tepals often partly yellowish on the reverse but at least the lower half of the dorsal purple, sweetly scented of apple and the orchid

Pterygodium; **perianth tube** 10–15(–19) mm long, obliquely funnel-shaped; **tepals** unequal, the dorsal largest, arching forward and horizontal in the upper half, narrowed and claw-like in the lower half, 26–32 mm long, the limb c. 18 x 12–15 mm, the claw c. 4.5 mm wide, the upper laterals lanceolate, 25–26 x 10–12 mm, also narrow and almost claw-like below, directed forward, spreading in the upper third, windowed between the bases of the dorsal and upper lateral tepals, the lower three tepals joined to the upper laterals for 3–5 mm and to one another for 2–3 mm, the free parts narrowed below into claws, the claws channelled and more or less horizontal, 8–10 mm long, the limbs lightly channelled and gently curving downward, c. 12–15 x 8–12 mm, in profile the lower tepals exceeding the upper by c. 8 mm. **Filaments** 14–15 mm long, exserted c. 10 mm from the tube; **anthers** 7–10 mm long, pale grey, the pollen cream. **Ovary** ovoid, c. 4 mm long; **style** arching over the stamens, dividing opposite the upper third of the anthers or close to the anther apices, the branches 3.5–5 mm long, ultimately exceeding the anther apices. **Capsules** broadly ellipsoid, 18–25 mm long; **seeds** 8–9 x 7 mm, obliquely saucer-shaped, usually asymmetric with the seed body closer to one side, the wing and seed body dark red-brown. **Chromosome number** unknown.

Flowering time late June to early August at low elevations, late August and September at higher elevations.

Distribution and biology

Gladiolus arcuatus is a fairly widespread species of the dry western portion of the southern African winter-rainfall zone. Its range extends from near Steinkopf in northern Namaqualand through the high country around Springbok and the Kamiesberg to the lower Olifants River valley between Klawer and Vredendal. Plants grow among low shrubs in coarse granite-derived gravel in Namaqualand and in fairly fine-grained silt in the south.

The sweetly scented and short-tubed flowers of *Gladiolus arcuatus* are pollinated by long-tongued bees, including the honey bee *Apis mellifera* (Apidae) and *Anthophora* species (Anthophoridae). The dull and cryptically coloured flowers are difficult to see in the field, but the strong floral scent makes them easy to find and we suspect that scent is a vital part of the floral attraction in this and other species of *Gladiolus* with similar dull-coloured flowers.

Diagnosis and relationships

Relatively unspecialized in its appearance, *Gladiolus arcuatus* is recognized by its soft, papery corm tunics and purple to dull greenish, sweetly scented flowers borne on strongly inclined, flexuose to scalloped spikes. The corms are depressed-globose to almost discoid in shape and have soft-textured, submembranous tunics that do not accumulate with age. The leaves are always lightly pubescent to velvety on the sheaths and blades and this, in combination with the corms and flowers, makes the species easy to identify. The upper three tepals are strongly narrowed below and in profile the flowers are windowed in much the same way as are those of *G. permeabilis* and *G. scullyi*. Differences in the leaves in the case of *G. permeabilis* and in the corm tunics in the case of *G. scullyi* make confusion with either unlikely. *Gladiolus arcuatus* stands close to the base of series *Deserticola* and is unusual in the series in having very soft-textured corm tunics.

History

Gladiolus arcuatus was discovered in 1830 by J.F. Drège on his expedition to Namaqualand and the mouth of the Orange River. He collected the species at two sites on the journey, at Mierenkasteel in southern

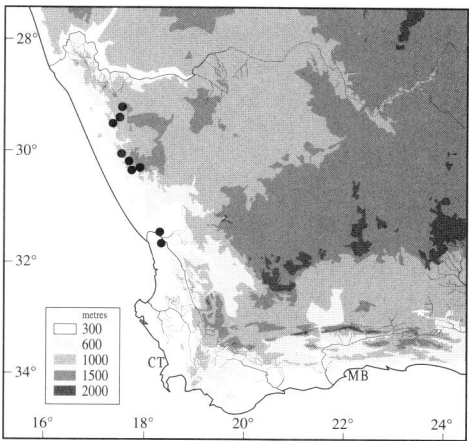

Namaqualand and later near Springbok. Despite the availability of duplicates at several European herbaria, the species did not attract botanical attention, probably being confused with *G. virescens* or *G. orchidiflorus* which were poorly understood at the time. It was only in 1882 that F.W. Klatt realized that the plants that Drège had collected represented an undescribed species. *Gladiolus arcuatus* was included by Baker in *Flora Capensis* but he obviously did not understand the species for specimens of *G. scullyi* are cited along with Drège's collections. The taxonomy of the species we now include in series *Deserticola* was partly clarified by Lewis et al. (1972) and they correctly interpreted *G. arcuatus*, separating it from *G. scullyi* and *G. orchidiflorus* and their respective allies.

Gladiolus arcuatus was apparently not recorded for more than 100 years after its discovery, and when plants flowered at the Botanical Gardens at Stellenbosch in 1931, no one realized their identity. A specimen preserved from this collection at the Bolus Herbarium bears the annotation '*Gladiolus tortuosus* G. Lewis' in Lewis' hand. Evidently she thought that the plant represented a new species, but later realized it was *G. arcuatus*.

97. GLADIOLUS VIRIDIFLORUS G. Lewis

PLATE 81

Gladiolus viridiflorus G. Lewis in *J. S. African Bot.*, Suppl. 10: 217 (1972). Type: South Africa, Northern Cape, Calvinia District, between Nieuwoudtville and Kareebooms Farm, flowered at Kirstenbosch Gardens, 10 May 1963, *Lewis 6132* (NBG, holotype).

viridiflorus = green-flowered, for the flower colour of some plants, although most populations have brownish purple flowers.

Description

Plants 8–15(–30) cm high. **Corm** globose to conic, 12–25 mm in diameter, the tunics of hard woody layers divided into segments below or sometimes coarsely reticulate, the vertical fibres thick, often forming claw-like ridges below. **Cataphylls** membranous, pale below the ground, mottled purple and white above the ground, rarely entirely purple, the upper cataphyll reaching 4–8 cm above the ground, obtuse to more or less truncate, often puberulous. **Leaves** usually four, rarely three, the lower two or three basal, erect, inclined or falcate, the sheaths often puberulous, the blades usually loosely twisted, sometimes lightly coiled, 2.7–6 mm wide, 5–10 cm long and usually shorter than the inclined spike, the midrib evident but neither midrib nor margins thickened, the upper one or two leaves cauline, inserted

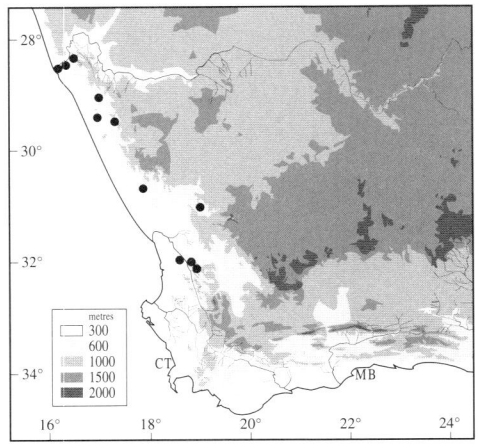

on the upper third of the stem, diverging sharply from the stem, thus not or hardly sheathing, the margins open to the base, channelled throughout. *Stem* erect or slightly inclined below, flexed outward above the sheath of the second leaf and then inclined 40°–60°, flexed at the upper nodes, simple or with one, rarely two branches, 1.3–2 mm in diameter below the spike.

Spike lightly to strongly flexuose and scalloped, inclined 30°–60°, sometimes almost horizontal, occasionally 2-, usually 3- to 8-flowered; *bracts* green and slightly glaucous, the outer (20–)25–40(–75) mm long, the inner 20–22 mm long, about half as long as to just shorter than the outer, the margins broadly transparent. *Flowers* greenish to yellowish brown with purple highlights, the upper tepals more darkly coloured above, the lower tepals each with a broad transverse yellow band in the distal third outlined in purple and often dark near the apices, strongly scented of freesia; *perianth tube* obliquely funnel-shaped, 12–17 mm long, shorter than the bracts; *tepals* unequal, lanceolate, the dorsal longest, 23–31 x 10–15 mm, inclined over the stamens, upper lateral tepals 20–26 x 10–14 mm, narrowly windowed between the dorsal and upper lateral tepals, the lower tepals joined to the upper laterals for 3–5 mm, and to one another for c. 2 mm, narrowly lanceolate, free parts 15–20 x 5–7 mm, arching downward, in profile the lower three exceeding the upper. *Filaments* 12–15 mm long, exserted 6–8 mm from the tube; *anthers* 6–9 mm long, greenish, pollen pale yellow. *Ovary* ovoid, 4–7 mm long; *style* arching over

the stamens, dividing opposite the middle of the anthers, the branches 3–4 mm long. *Capsules* broadly ellipsoid, acute, (12–)15–21 x c. 10 mm; *seeds* broadly ovate, 10–12 x 8–10 mm, broadly and evenly winged, translucent light yellow- to red-brown. *Chromosome number* unknown.

Flowering time May to July.

Distribution and biology

Thought to be rare when first described in 1972, *Gladiolus viridiflorus* has now been recorded from several localities in the north-western part of Western Cape Province and throughout Namaqualand from the mid-Olifants River valley near Clanwilliam in the south to Beauvallon on the banks of the Gariep (Orange) River in the north. Additional collecting in southern Namibia will probably extend its range even further. The early flowering of the species and its low stature and dull-coloured flowers explain why it has remained so poorly understood for so long.

Gladiolus viridiflorus grows among low shrubs on a variety of soils, including clay, granite-derived sand and Kalahari sands. Its habitats include places that are extremely arid, receiving as little as 150 mm per year, and plants are adapted to growing and flowering rapidly so that their life cycle is completed before the soil dries out completely. If the season is a wet one, plants will continue to grow after fruiting is completed and the leaves and stems may remain green and photosynthetically active for months after the seed has been shed.

The very fragrant flowers of *Gladiolus viridiflorus* are assumed to be pollinated by long-tongued bees.

Diagnosis and relationships

Gladiolus viridiflorus has dull-coloured greenish to brown flowers flushed with purple, borne on a strongly inclined and flexuose spike. This, combined with a purple and white mottled cataphyll and only four foliage leaves, the blades of which are relatively short and twisted or sometimes lightly coiled, makes the species easy to recognize. The plant illustrated by Lewis et al. (1972) was drawn from a cultivated specimen and its tall, slender habit is quite atypical of the species. Plants are generally only 8–15 cm tall and have nearly horizontal spikes with strongly scalloped internodes.

The relationships of *Gladiolus viridiflorus* were initially thought to lie with the western and southern Cape species, *G. carinatus*

(section *Homoglossum*), which it resembles mainly in the possession of a mottled cataphyll and strongly scented flowers. These similarities are more likely convergent, and we regard *G. viridiflorus* as a specialized member of section *Hebea*.

The strongly inclined and scalloped spike, dull-coloured flowers and tough-textured, coarsely fibrous corm tunics are features inconsistent with section *Homoglossum*, especially *Gladiolus carinatus* which has softly fibrous corm tunics. These features are, however, characteristic of a group of species of section *Hebea*, notably *G. scullyi*, *G. venustus* and, except for the corm tunics, *G. arcuatus*. In addition, the often puberulous cataphylls recall those of *G. arcuatus* and are unknown elsewhere in the section.

Gladiolus viridiflorus is probably nested among this group of species, and specialized in its possession of only four leaves, short twisted to coiled leaf blades and mottled cataphyll.

History

Although we have established that *Gladiolus viridiflorus* is fairly common along the western southern African interior, it appears not to have been recorded until 1960 when the South African horticulturist and bulb grower, Margaret Thomas, collected plants near Springbok in Namaqualand. After its initial discovery, several more populations were found in Namaqualand and one in the northwestern Karoo.

More recent collecting has shown that the species was most likely overlooked by many botanists because it flowers so early in the season, and has completed flowering and fruiting before collecting expeditions are normally made to Namaqualand and the northern Cape.

98. GLADIOLUS DESERTICOLA Goldblatt

PLATE 86

Gladiolus deserticola Goldblatt, *J. S. African Bot.* 50: 457 (1984). Type: South Africa, Northern Cape, Richtersveld, Stinkfontein Mountains, upper slopes of Cornelsberg on ridge south of beacon, c. 1300 m, 6 Sept. 1977, *Oliver, Tölken, & Venter 706* (PRE, holotype).

deserticola = growing in the wilderness, referring to the remote and wild part of southern Africa where the species grows.

Description

Plants 10–20(–27) cm high. *Corm* more or less conic, the base oblique, 6–12 mm in diameter, the tunics of firm unbroken layers, the outer becoming broken with age, somewhat fibrous, toothed below. *Cataphylls* pale and membranous, the uppermost reaching 3–8 cm above the ground and then purple mottled with white. *Leaves* four, occasionally five if the stem has more than one branch, the lower two basal and longest, reaching at least to the base of the spike, often exceeding it, the blades linear, 1.5–2.5(–3) mm wide, only the midrib raised, usually twisted several times, erect or becoming trailing, the second leaf sheathing the lower half of the stem, remaining leaves inserted on the upper half of the stem, decreasing in size above, the terminal one or two without a sheathing base, sharply diverging from the stem, channelled more or less for their entire length. *Stem* erect or inclined, straight below, sharply flexed above the sheaths of the leaves or above the nodes when leaves without sheaths, simple or with one or two branches, 1.8–2.2 mm in diameter below the main spike.

Spike flexed at the base and usually strongly inclined to horizontal, the axis strongly flexuose, thus scalloped, 4- to 6-flowered,

the branches with fewer flowers; *bracts* green or brownish green, the outer 12–20 mm long, the inner only slightly shorter than the outer, minutely forked apically. *Flowers* virtually actinomorphic except for the asymmetric placement of the nectar guides, dark blue, each tepal with a darker line in the midline, the lower lateral tepals, or sometimes all three lower tepals, cream edged distally with purple in the lower third, weakly rose-scented; *perianth tube* narrowly funnel-shaped, c. 11 mm long, the wider upper part only c. 1.5 mm long; *tepals* nearly equal, lanceolate, the dorsal slightly longer and narrowed below, 14–21 x 4–5 mm, straight and usually upright, remaining tepals spreading from shortly above the bases, 12–18 x 3.5–4.5 mm. *Filaments* 6–9 mm long, exserted c. 4 mm from the tube, symmetrically disposed or unilateral; *anthers* either symmetrically disposed around the style or unilateral, 4–5 mm long, pale yellow, the pollen pale yellow. *Ovary* oblong, 3–4 mm long; *style* more or less central and only slightly arcuate, dividing between the middle and apex of the anthers, the branches c. 4 mm long, divergent and almost straight, not much expanded in the upper third. *Capsules* unknown; *seeds* ovate, c. 6.5 x 5 mm, translucent light brown, the wing well developed. *Chromosome number* unknown.

Flowering time mid-August to mid-September, later at higher elevations.

Distribution and biology

Gladiolus deserticola is a narrow endemic of the dry northern portion of the southern African winter-rainfall zone. It is restricted to the Richtersveld, a mountainous area of northern Namaqualand in Northern Cape Province, South Africa. At present the species is known only from the Stinkfontein Mountains, but additional collecting may well extend its range. The first collection of *G. deserticola* was made on the upper slopes of Cornelsberg at the northern end of the Stinkfontein range and at first the species seemed to be a rare high-mountain endemic. In 1994, a year of good rainfall for the Richtersveld, however, we found it to be common in the foothills of the range from near Eksteenfontein and along the eastern slopes as far as Cornelsberg itself. Plants favour fairly sheltered sites and occur most frequently in clay soils on south-facing slopes in the lee of rocks or under shrubs.

The pollination biology of *Gladiolus deserticola* remains to be fully investigated. We speculate that the flowers are pollinated by short-tongued bees that forage for pollen.

Diagnosis and relationships

An almost regular perianth but with a curved tube, a dark blue flower with white and darker blue nectar guides on one or more of the lower tepals, a purple and white mottled upper cataphyll, and a low stature with an inclined to horizontal, scalloped spike make the species easy to recognize. The flower is unusual, but not unique in *Gladiolus*, in being nearly actinomorphic, although the asymmetric placement of the nectar guides and the frequent asymmetry of the stamens suggest the recent loss of true zygomorphy. The placement of *G. deserticola* in the genus is based on the presence of large, green, soft-textured floral bracts, narrow leaves lacking secondary veins and without raised margins, and corms of the type that are found in most of the species of series *Deserticola* of section *Hebea*. Because of the hard, coarsely fibrous corm tunics we have assigned the species to series *Deserticola* and we suggest that it is most closely related to *G. viridiflorus*, which also has a purple and white mottled cataphyll, a nearly horizontal,

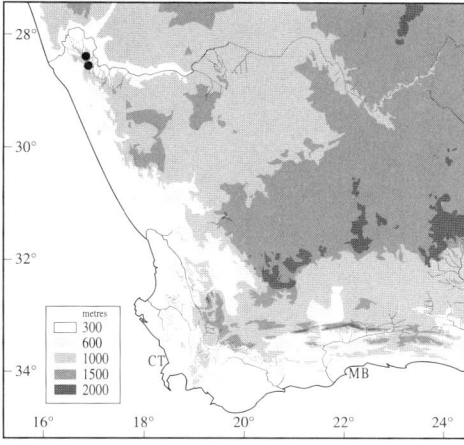

scalloped spike, and twisted to lightly coiled leaf blades. As well as lacking a large inclined to hooded dorsal tepal, *G. deserticola* is unusual in section *Hebea* in having weakly scented flowers. Possibly the brightly coloured flowers provide sufficient attraction for pollinators and scent production has been largely suppressed.

History

Gladiolus deserticola was discovered by the South African botanist, E.G.H. Oliver, in 1977, on the upper slopes of Cornelsberg in the Richtersveld at the northern end of the Stinkfontein Mountains. At first the species seemed to be a high-mountain endemic, but our later investigation has shown that it occurs at several sites along the foot of the Stinkfontein Mountains as well as on the higher slopes of Cornelsberg. As it is still poorly known, we would not be surprised if further exploration shows that *G. deserticola* is somewhat more widespread than presently believed.

99. GLADIOLUS SCULLYI Baker
PLATE 82

Gladiolus scullyi Baker, *Handbook Irideae* 224 (1892); *Fl. Capensis* 6: 162 (1896). Lewis et al., *J. S. African Bot.*, Suppl. 10: 122 (1972), including *G. venustus* G. Lewis. Type: South Africa, Northern Cape, Namaqualand, without precise locality, probably in 1890, *Scully s.n.* (K, holotype).

scullyi, named for William Charles Scully, Magistrate of Namaqualand in the 1890s, whose interest in the natural history of the region led to the discovery of several new plant species.

Description

Plants (12–)20–35 cm high. *Corm* globose-conic, 12–15 mm in diameter, bearing small ellipsoid cormlets at the base, the tunics of hard, more or less woody layers, with age these fragmenting into irregularly shaped segments often drawn into claw-like fibres below, initially orange, maturing to red- to dark brown. *Cataphylls* pale and membranous, the upper reaching shortly above the ground and then pale green or purple. *Leaves* occasionally five, usually six to eight, the lower four to six basal and usually forming a distichous fan, grey-green and often glaucous, reaching to about the base of the spike, occasionally to the middle of the spike, the blades narrowly lanceolate to linear, (2–)3–5 mm wide, fairly thick and succulent, the midrib lightly raised, the margins not raised, the upper one or two leaves inserted on the stem, the uppermost channelled throughout, diverging from the stem or sheathing only near the base, the margins free. *Stem* erect or inclined, flexed outward above the sheaths of the two upper leaves, simple or often with one or two branches, 1.7–2 mm in diameter below the spike.

Spike inclined, lightly flexuose, 5- to 8-flowered, the branches with fewer flowers; *bracts* pale green, translucent on the veins, the outer 20–30(–35) mm long, occasionally dry near the apices, the inner about two-thirds as long as the outer, acute or minutely forked apically. *Flowers* variously greenish cream to yellow-brown or beige, the upper tepals each with a brown to purplish stripe in the midline and sometimes flushed purple above, the lower tepal limbs yellow below, brown to purplish above, the throat irregularly streaked with purple lines, strongly sweet-scented; *perianth tube* obliquely and narrowly funnel-shaped, 12–14 mm long, the slender lower part 8–9 mm long; *tepals* unequal, lanceolate, the dorsal largest, inclined over the stamens, 25–35(–42) x 10–14(–16) mm, the upper laterals curving outward in an arc, 22–28(–35) x 9–15 mm,

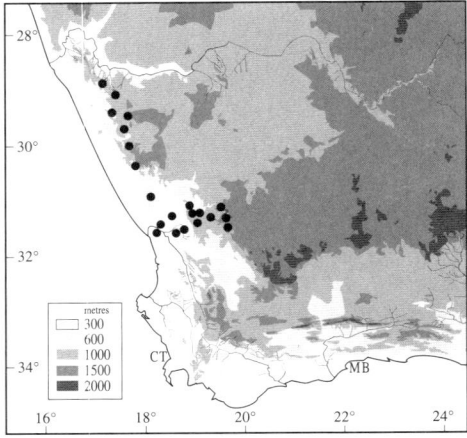

windowed between the dorsal and upper lateral tepals, the lower three tepals united with the upper laterals for 3–4 mm and to one another for 4–7 mm, narrowed into claws below, the claws narrowly channelled and sometimes with weakly developed auricles, 4–6 mm long, the limbs abruptly expanded and deflexed, 10–15(–20) x 8–12 mm, the lowermost slightly larger than the lower laterals. *Filaments* 12–16 mm long, exserted 9–12 mm from the tube; *anthers* 6–9 mm long, light purple to cream, the pollen cream. *Ovary* ovoid, 3–4 mm long; *style* arching over the stamens, dividing opposite or just below the anther apices, the branches 3–4.5 mm long. *Capsules* broadly obovoid, trilobed-retuse above, 9–14 mm long; *seeds* 5–6 x c. 5 mm, obliquely saucer-shaped, usually asymmetric with the seed body closer to one side, the wing and seed body uniformly dark red-brown. *Chromosome number* 2n = probably 30 but only 45 reported.

Flowering time August and September.

Distribution and biology

Gladiolus scullyi is a fairly common species of the more arid, completely summer-dry parts of western southern Africa. It is best known from Namaqualand, but it extends southeastward onto the Bokkeveld Escarpment where it occurs in the Calvinia District of Northern Cape Province. Plants are found on a variety of substrates, including granite-derived sands, shales of the Karoo and the Nama (Malmesbury) Systems, and Dwyka Tillite. In Namaqualand it is not uncommon to see the species growing in crevices in large granite boulders.

The flowers of *Gladiolus scullyi* are strongly scented and despite their rather insignificant colouring they are very attractive to the long-tongued bee, *Anthophora diversipes* (Anthophoridae), which is the primary pollinator of the species.

Diagnosis and relationships

Gladiolus scullyi can usually be recognized by its strongly sweet-scented, fairly dull beige to light brown flowers with yellow nectar guides and darker brown to purplish highlights. The upper lateral tepals are fairly narrow in the lower third, rendering the flowers windowed when viewed in profile. In addition, the lower tepals have narrow claws and the limbs are sharply deflexed at the base.

The clawed upper tepals and the woody, somewhat claw-like corm tunics place the species in series *Deserticola* of section *Hebea*, in which it appears to be closely allied to *G. venustus*, sharing inflated, ovoid and three-lobed capsules. The presence of some plants intermediate between the two

species and an apparently gradual transition between them in the Vanrhynsdorp area and the Bokkeveld Escarpment led Obermeyer (in Lewis et al., 1972) to unite the two species under the oldest legitimate epithet, *G. scullyi*. Except in the narrow zone of geographical overlap, the two species remain quite distinct. Unlike those of *G. scullyi*, the narrower upper tepals of *G. venustus* do not gape below, and instead are twisted upward to lie partly behind the dorsal tepal (compare figures on pages 214 and 216). The lower tepal limbs are deflexed much more strongly than in *G. scullyi*, and the claws are deeply channelled and have prominent auriculate margins, a feature weakly developed or lacking in *G. scullyi*. The tepals of *G. venustus* are flushed or fully coloured purple or pale to dark pink in the upper half, whereas those of *G. scullyi* do not have a strongly contrasting colour and the flowers are much more strongly scented. The combination of different orientation of the upper tepals, tepal pigmentation and the auriculate lobes on the lower tepal claws make it easy to distinguish the species when seen alive. These features are often obscured in pressed specimens, making it difficult – and sometimes impossible – to separate them.

History

This common Namaqualand and western Karoo species appears not to have been recorded botanically until 1883. In that year Harry Bolus made one of the earliest botanical forays into this then virtually unknown area. Reaching Port Nolloth in August of that year, he made his way inland to the copper mines at Okiep and Springbok and thence to Naries in the Spektakel Mountains, where *Gladiolus scullyi* is particularly common on shale outcrops. A later collection made by the Namaqualand magistrate, William Scully, probably in 1890, however, attracted J.G. Baker's attention. The species was described in 1892 and named in Scully's honour. Additional collections, especially in the period between 1930 and 1960, extended the range southward to Vanrhynsdorp, the Bokkeveld Escarpment and Calvinia, where the similar and closely related *G. venustus* also occurs. The presence of some plants intermediate between the two and an apparently gradual transition between them led Obermeyer (in Lewis et al., 1972) to unite the two species under the oldest legitimate epithet, *G. scullyi*. As explained above, we do not accept this decision.

100. GLADIOLUS VENUSTUS G. Lewis

Purple kalkoentjie, perdelelie

PLATE 83

Galdiolus venustus G. Lewis, *J. S. African Bot.* 7: 56 (1941), as a new name for *Gladiolus formosus* Klatt, *Linnaea* 32: 692 (1863), an illegitimate homonym for *G. formosus* Persoon (= *Babiana striata* (Jacquin) G. Lewis). Baker, *Fl. Capensis* 6: 161 (1896). Type: South Africa, Western Cape, Clanwilliam District, Olifants River at Brakfontein, Aug. probably 1829, *Ecklon & Zeyher Irid. 140* (S, lectotype designated by Lewis, 1972; G, SAM, isolectotypes).

venustus = beautiful, alluding to the attractive flowers, often beautifully coloured and marked with deep pink or purple and yellow.

Synonymy

[*Gladiolus scullyi* Baker sensu Lewis et al., *J. S. African Bot.*, Suppl. 10: 122 (1972), in part.]

Description

Plants 12–35 cm high. *Corm* conic, 8–14 mm at the widest diameter, the tunics of hard, more or less woody layers, with age

these fragmenting into irregularly shaped pieces often drawn into claw-like fibres below, initially orange, maturing to reddish brown. *Cataphylls* pale and membranous, the uppermost green above the ground. *Leaves* five or six, the lower three to five basal, reaching at least to the base of the spike and sometimes shortly exceeding it, the blades linear, 1.5–4(–9) mm wide, the midrib lightly raised, the margins not at all, the upper one or two leaves cauline and shorter than the basal, at least the uppermost diverging from the stem or sheathing only near the base, channelled throughout, the margins free to the base. *Stem* erect below, flexed outward above the sheaths of the two upper leaves and then weakly inclined, frequently branched unless depauperate, c. 1.5 mm in diameter below the spike.

Spike inclined, flexuose, often strongly so and then scalloped, 5- to 8-flowered, the branches with fewer flowers; *bracts* pale green, the veins transparent, attenuate, the outer 20–30(–35) mm long, the apices

sometimes becoming dry, the inner two-thirds to nearly as long as the outer, minutely forked apically. *Flowers* shades of blue-purple, deep pink, occasionally entirely dull yellow, the lower three tepals bright yellow in the lower half, usually lightly rose-scented; *perianth tube* obliquely funnel-shaped, 12–17 mm long, the lower cylindrical part 8–11 mm long; *tepals* unequal, lanceolate,

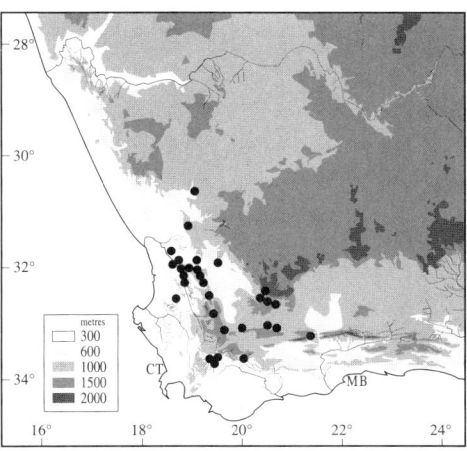

opposite the upper third of the anthers, the branches 2–3 mm long, ultimately exceeding the anther apices. *Capsules* globose, trilobed-retuse above, c. 10 mm long; *seeds* 5–7 x c. 5 mm, obliquely saucer-shaped, usually asymmetric with the seed body closer to one side, the wing and seed body uniformly dark red-brown. *Chromosome number* 2*n* = 30.

Flowering time August to mid-September, sometimes later at higher elevations.

Distribution and biology

Gladiolus venustus has a wide distribution across the interior of the southern African winter-rainfall region. It extends from the Clanwilliam District in Western Cape Province eastward into the dry valleys of the Cedarberg and thence through the Tanqua Karoo and western Roggeveld southward to Worcester, Montagu and the Little Karoo as far east as Bosluiskloof in the northern foothills of the Swartberg. The species is most often encountered in dry habitats on clay and shale substrates, but it is sometimes found on dry, stony sandstone hillsides.

The fairly large and colourful flowers of *Gladiolus venustus* are visited by a variety of long-tongued bees, mostly Anthophoridae, and these insects are the primary pollinators of the species. The sweet scent is an additional attractant. The bees appear to visit *G. venustus* to forage for nectar contained within the fairly long perianth tube.

Diagnosis and relationships

Gladiolus venustus belongs to a small group of species of section *Hebea* that have conic corms with hard, woody tunics and strongly inclined, flexuose to scalloped spikes with the flowers borne on the upper side. The capsules, too, are distinctive, being broad and squat, and three-lobed above with retuse apices. The species stands out in the group in having the upper lateral tepals twisted to lie over the dorsal and the lower lateral tepals with the limbs sharply deflexed and conspicuously auriculate at the base. It is likely to be confused only with *G. scullyi* which is broadly similar, and virtually identical vegetatively. In that species the dorsal tepal is almost horizontal and the upper laterals

extend laterally, and both are almost always narrowed below so that in profile there is a gap or window between them. The lower tepal limbs are less sharply deflexed and not or only weakly auriculate at the base of the limbs. Flower colouring and scent also differ between the two. The upper three tepals and the distal halves of the lower tepals of *G. venustus* are purple or pink, the lower three are clear yellow in the lower half, and the flowers are lightly scented. In *G. scullyi* the tepals are usually beige to light brown, or sometimes flushed or speckled with dull purple above, but never so strongly discolorous as in *G. venustus*, and the flowers have a strong, sweet, heady fragrance. The distinctions between the species are often difficult to see in herbarium specimens, and identification is complicated by a degree of introgression. The two species do grow near one another along the interface between their respective ranges in southern Namaqualand and on the Bokkeveld Escarpment.

History

That this extremely common dry-country species was not described earlier than 1863 when F.W. Klatt named it *Gladiolus formosus* is surprising. The early collectors who encountered the species no doubt confused it with others, in particular, *G. orchidiflorus* or *G. virescens*, but C.F. Ecklon certainly regarded it as a distinct species, which he associated with *G. formosus* Persoon. The identity of Persoon's species is uncertain, but as he associated it with *G. striatus* Jacquin, a species of *Babiana*, Persoon's *G. formosus* is illegitimate and superfluous. Klatt's name, which he evidently intended as new, is a homonym and likewise is illegitimate, although it was accepted by several later authors, including J.G. Baker. G.J. Lewis supplied the new name, *G. venustus*, for the species in 1941. Later, Amelia Obermeyer (in Lewis et al., 1972) concluded that *G. venustus* was conspecific with the Namaqualand *G. scullyi* and she united them under the latter name. As we have explained above, they are separate, although closely related species, more likely to be confused when dry than when examined alive.

the dorsal largest, lightly inclined, 30–37 x 10–16 mm, the upper laterals slightly shorter than the dorsal, directed upward and lying at least partly behind the dorsal, curving outward in the upper third, the lower three tepals united for 3–5 mm, the free parts narrowed below into claws, the claws 2–4 mm long, deeply channelled, the margins raised into auriculate lobes, these ascending, the limbs sharply flexed at right angles to the claws and pinched together at the point of flexure, 15–22 x 9–11 mm, more or less straight. *Filaments* 11–15 mm long, exserted 5–7 mm from the tube; *anthers* 6–8 mm long, greyish or yellow, the pollen yellow. *Ovary* oblong, c. 4 mm long; *style* arching over the stamens, dividing

101. GLADIOLUS SALTERI G. Lewis
PLATE 84

Gladiolus salteri G. Lewis in *J. S. African Bot.*, Suppl. 10: 127 (1972). Type: South Africa, Cape, Namaqualand, 28 km east of Springbok, 19 Sept. 1933, *Salter 3798* (BOL, holotype; BM, K, isotypes).

salteri, named in honour of the amateur but accomplished botanist, Captain T.M. Salter, who collected the plants on which the description was based.

Description

Plants 10–20 cm high. *Corm* globose-conic, 18–22 mm in diameter or more if the tunics accumulate thickly, the tunics coarsely fibrous, bearing a few cormlets around the base. *Cataphylls* membranous and pale, reaching just to ground level. *Leaves* usually four, the lower three basal, falcate, all arching to one side, generally as long as the spike, the blade 4–7 mm wide, grey-green, the uppermost leaf inserted in the middle of the stem. *Stem* inclined, flexed above the sheath of the uppermost leaf, simple or with one or two branches, c. 2 mm in diameter below the spike.

Spike 5- to 8-flowered, the branches with fewer flowers, inclined to nearly horizontal, flexuose; *bracts* pale grey-green, the outer 15–20 mm long, dry and brown toward the apices, the inner slightly shorter than the outer. *Flowers* pale pink, the lower tepals cream with deep pink markings in the midline and toward the apices, papillose on the claws and the lower throat, lightly rose-scented; *perianth tube* 18–22 mm long, narrowly and obliquely funnel-shaped, normally exserted from the bracts, the wider portion 3–4 mm long; *tepals* unequal, the dorsal longest, initially arcuate, the distal half becoming erect with age, narrowed below into a slender claw, 28–32 mm long,

c. 8 mm at the widest, upper lateral tepals 20–22 mm long, narrowed below into claws c. 7 mm long, directed forward, the limbs abruptly expanded, triangular, curving outward and spreading at right angles to the claws, the lower tepals strongly clawed, united basally for 4–5 mm, channelled, the free parts c. 15 mm long, directed forward, the limbs curving downward, in profile exceeding the upper tepals. *Filaments* c. 25 mm long, papillose toward the bases, exserted 18–20 mm from the upper part of the tube; *anthers* c. 6 mm long, cream, the pollen pale yellow. *Ovary* obovoid, c. 3 mm long; *style* arching over the stamens, dividing opposite the lower half of the anthers, the branches c. 4 mm long, extending beyond the anther apices. *Capsules* broadly ovoid, trilobed-retuse above, 9–11 x 10 mm; *seeds* 4–5 x 3.5–4 mm, saucer-shaped, the seed body asymmetrically placed, broadly winged, translucent yellow-brown, the seed body dark yellow-brown. *Chromosome number* unknown.

Flowering time mid-August to mid-September.

Distribution and biology

Restricted to the interior mountains of Namaqualand in Northern Cape Province, South Africa, *Gladiolus salteri* in known from just a handful of collections. Since its discovery it has been recorded only occasionally, always in the granitic hills east of Springbok on the western edge of Bushmanland. Plants grow in coarse, gritty, decomposed granite and usually along the edges of drainage lines.

The flowers of *Gladiolus salteri* are most likely adapted for pollination by long-tongued bees.

Diagnosis and relationships

The general aspect of *Gladiolus salteri* is exactly like that of *G. scullyi* and the two species can only be distinguished when in flower. Flowers of *G. salteri* are an attractive pale pink and the lower tepals have reddish markings at the edge of the cream colour in the lower half, an unusual patterning for section *Hebea*.

Apart from flower colour, *Gladiolus salteri* may be distinguished from other species of series *Deserticola* that have similar hard, more or less woody corm tunics by the relatively long perianth tube, 18–22 mm long, and fairly narrow dorsal tepal 28–32 x 8 mm. The dull-coloured, but intensely sweet-

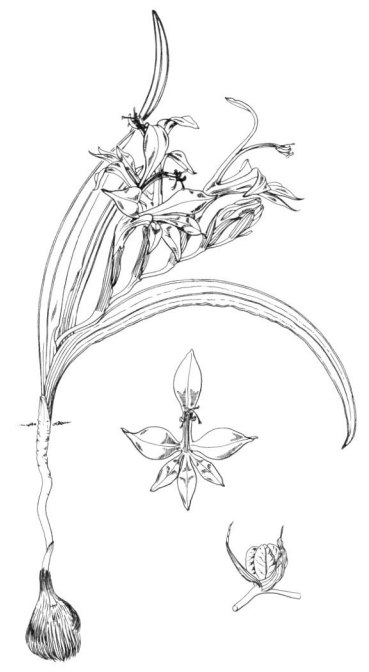

scented flowers of *G. scullyi* have a perianth tube 12–14 mm long and the strongly arched dorsal tepal is 25–42 x 10–16 mm.

The capsules of *Gladiolus salteri* are more or less obovoid and conspicuously three-lobed above, and match exactly those of three other species of series *Deserticola*, including *G. scullyi* and *G. venustus*. These apparently specialized capsules indicate that these species form a clade within the series and are closely related.

History

Gladiolus salteri was first collected by the mountaineer and sometime collector, T.P. Stokoe, in 1929. His collection at the Bolus Herbarium in Cape Town has no precise locality. Four years later, in 1933, the botanist, T.M. Salter, recorded the species from an exact site in the rocky granite dome country east of Springbok, probably following directions given him by Stokoe.

Obviously an undescribed species, *Gladiolus salteri* nevertheless remained unnamed for nearly 40 years at a time when new species of *Gladiolus* were being described by both H.M.L. Bolus and G.J. Lewis. It was only in 1972, when Lewis's revision of the South African species of the genus was published, that *G. salteri* was formally named. During the more than 60 years since its discovery and location, the range of *G. salteri* has been very little extended and it seems clear that it is indeed a highly localized endemic.

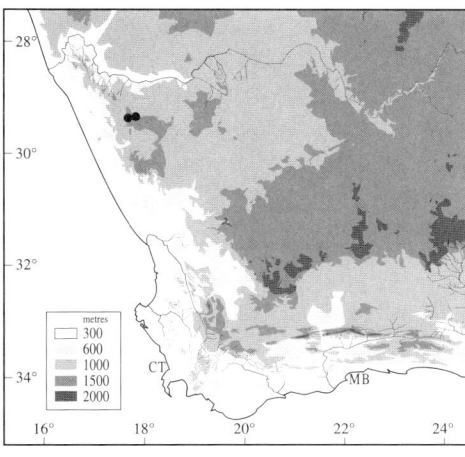

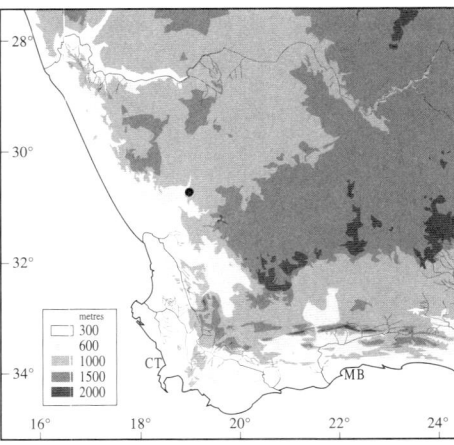

Gladiolus lapeirousioides Goldblatt, *J. S. African Bot.* 37: 229 (1971). Lewis et al., *J. S. African Bot.*, Suppl. 10: 79 (1972). Type: South Africa, Northern Cape, Loeriesfontein District, along the road to Kliprand, 23 Sept. 1970, *Goldblatt 540* (BOL, holotype; NBG, PRE, isotypes).

lapeirousioides = resembling a *Lapeirousia*, referring to the similarity of the shape and colour of the flowers to those of some species of the genus *Lapeirousia*, especially *L. fabricii*.

Description

Plants 8–14 cm high. *Corm* more or less conic, 9–12(–15) mm in diameter, the tunics of bright yellow cartilaginous layers, becoming brown and breaking up into coarsely textured vertical fibres with age, sometimes accumulating in a dense mass, usually with a few cormlets around the base, the cormlets borne on short stolons 5–8 mm long. *Cataphylls* pale and membranous, the upper reaching shortly above the ground and then green or purple. *Leaves* four to six, forming a basal distichous fan, often all arching to one side, only the uppermost inserted on the stem above ground, usually shortly exceeding the stem, falcate, the blades plane, 3–5 mm wide, the midrib and margins lightly thickened, uppermost leaf smallest, sheathing below, the lower margins open to the base. *Stem* erect below, strongly flexed outward above the sheath of the uppermost leaf, sometimes almost horizontal above, simple or with one or two

branches, c. 1.5 mm in diameter below the spike.

Spike nearly horizontal, lightly flexuose, sometimes 5-, usually 7- to 12-flowered, the branches with fewer flowers than the main axis; *bracts* green to greenish grey, 15–20 mm long, the inner about three-quarters as long as the outer. *Flowers* creamy white, the lower three tepals each with a yellow spot near the base bordered with red, unscented; *perianth tube* cylindric, 35–40 mm long; *tepals* unequal, when fully open spreading at right angles to the tube and all held in the same plane, the three upper narrowly elliptic, 20–22 mm long, the lower tepals nearly linear, the lower laterals 8–10 x 1.5–2 mm, the lower median c. 15 x 4 mm. *Filaments* c. 8 mm long, exserted c. 4 mm from the tube; *anthers* c. 3 mm long, purplish, the pollen yellow. *Ovary* ovoid, c. 3 mm long; *style* more or less erect, dividing opposite the anther apices, the branches c. 2 mm long, arching above the anthers. *Capsules* broadly obovoid, trilobed-retuse above, 8–12 mm long, the surface lightly verrucose; *seeds* shallowly saucer-shaped, c. 4 x 3 mm, broadly but unevenly winged, the seed body eccentric, dark reddish brown. *Chromosome number* $2n$ = 30.

Flowering time late August to early October.

Distribution and biology

Gladiolus lapeirousioides is a narrow endemic of the South African winter-rainfall zone. The only known population occurs in the western Karoo, near Loeriesfontein, in the low hills west of the town toward Kliprand.

Plants grow in dry stony shale ground and the corms are usually deeply wedged in bedrock which retains moisture for a long time after penetrating rains have fallen. Rainfall is extremely erratic in this area and *G. lapeirousioides* evidently does not flower every year. Moreover, flowering time is dependent on the timing of the rains and plants have been collected in flower as early as mid-August and as late as the end of September, when the type collection was made.

The flowers have the typical features associated with long-tongued tabanid fly pollination: an elongate, slender and straight perianth tube and a cream to pale pink colour with red nectar guides.

Thus we assume that flowers of *Gladiolus lapeirousioides* are adapted for pollination by these flies.

Diagnosis and relationships

Hard, woody corm tunics, a fan of several basal leaves and firmly textured, somewhat fleshy leaves with unthickened margins clearly place *Gladiolus lapeirousioides* in section *Hebea*. Within the section its cream flowers with an elongate perianth tube are anomalous and it is clear that these flowers, which are apparently adapted for long-tongued fly pollination, are an isolated specialization.

The nearly horizontal and somewhat scalloped spike and the broadly ovoid and slightly inflated capsules that are conspicuously lobed above indicate that the affinities of *Gladiolus lapeirousioides* lie with *G. salteri*, *G. scullyi* and *G. venustus* of series *Deserticola*. These four species apparently comprise a clade defined by the unusual capsules. Other members of section *Hebea* have narrowly to broadly ellipsoid capsules that are not, or hardly at all, lobed above.

History

Gladiolus lapeirousioides was discovered in the spring of 1971 by Peter Goldblatt, who was then studying the genus *Lapeirousia* in the field. From a short distance the flowers of *G. lapeirousioides* were so similar to those of *L. fabricii* that it was only when the plants were examined closely and corms extracted from the stony ground that it was found to be a species of *Gladiolus* and not, in fact, a *Lapeirousia*. No additional sites for the plant have been discovered.

103. GLADIOLUS ORCHIDIFLORUS Andrews

Groenkalkoentjie, green kalkoentjie, vaalkalkoentjie, grey kalkoentjie

PLATE 87

Gladiolus orchidiflorus Andrews, *Bot. Repository* 4: pl. 241 (1802). Baker, *Fl. Capensis* 6: 160 (1896). Phillips, *Fl. Pl. Africa* 5: pl. 165 (1925). Lewis et al., *J. S. African Bot.,* Suppl. 10: 146 (1972). Type: South Africa, without precise locality, most likely from Namaqualand, cultivated in Great Britain in 1802, *Niven s.n.,* illustration in Andrews, *Bot. Repository* 4: pl. 241 (1802).

orchidiflorus = orchid-flowered, referring to the rather elaborate and complex flower that has a somewhat orchid-like appearance.

Synonymy

Gladiolus viperatus Ker Gawler, *Curtis's Bot. Mag.* 18: pl. 688 (1803). Type: South Africa, Western Cape, without precise locality or collector, cultivated in Great Britain from plants sent from the Cape, illustration in *Curtis's Bot. Mag.* 18: pl. 688 (1803).

Gladiolus dregei Klatt, *Linnaea* 32: 693 (1863). Baker, *Fl. Capensis* 6: 162 (1896). Type: South Africa, Northern Cape, Namaqualand, between Pedroskloof and Leliefontein, 1830, *Drège 2631* (B [not seen], presumed holotype; G, K, P, isotypes).

Description

Plants (8–)18–45(–80) cm high. *Corm* globose to conic, 12–20 mm in diameter, the tunics with moderate to very coarse fibres or lightly coriaceous to firmly papery, entire layers, producing numerous small cormlets around the base. *Cataphylls* pale and membranous, the uppermost 2–3 cm above the ground and then green or purplish. *Leaves* occasionally four, usually five to eight, rarely

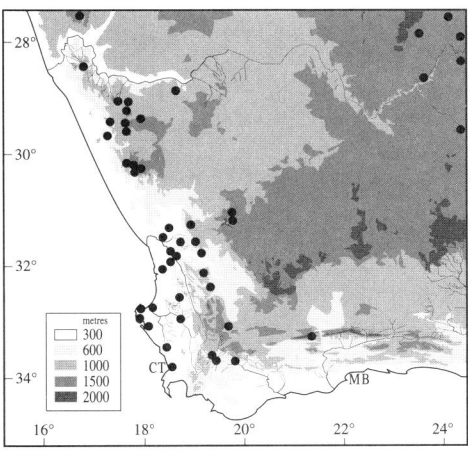

up to ten, the lower three to seven basal and longest, often in a distichous fan, reaching to about the middle of the stem, the blades sword-shaped to linear, sometimes falcate, 2–7(–15) mm wide, soft-textured or sometimes somewhat fleshy, the midrib and sometimes one or more secondary veins lightly thickened, sometimes the midrib not more prominent than the secondary veins, the margins occasionally lightly thickened, rarely lightly crisped. *Stem* erect below, flexed outward above the sheaths of the upper one or two leaves, simple or with one or two branches, (1–)2–3 mm in diameter below the spike.

Spike inclined, lightly flexuose, flowers in two ranks separated by c. 60°, the main spike 5- to 12-flowered, the lateral branches with fewer flowers; *bracts* greenish, often purplish grey on the upper surfaces, the outer (13–)18–23(–30) mm long, the inner slightly shorter than to about as long as the outer, notched apically for up to 1.5 mm. *Flowers* grey-green to dull purple, the dorsal darker apically, the upper laterals darker along the midline, the lower three tepals each with a bright yellow to greenish transverse band in the upper third edged distally in darker purple, the sutures between the tepals thickened into iridescent papillose ridges, with an intensely sweet, violet-like scent; *perianth tube* 9–14 mm long, narrowly and obliquely funnel-shaped; *tepals* unequal, the dorsal longest, narrowly oblong, strongly arched in a semicircle, 20–35 mm long, 2–3 mm wide in the midline, 3–6 mm wide near the apex, the upper laterals narrowly spade-shaped, with claws 2–4 mm wide, the limbs triangular to lanceolate, 12–16 x 6–17 mm, curving outward in the upper half, the apices somewhat attenuate, the lower three tepals joined to the upper laterals for 2–4 mm and to one another for 1–2 mm, 11–20 mm long, arching outward and downward, the lower median slightly longer or shorter than the lower laterals but about twice as wide. *Filaments* 20–25 mm long, exserted 6–20 mm from the tube; *anthers* 5–11 mm long, lilac to purple, pollen cream. *Ovary* oblong, 3–6 mm long; *style* arching over the stamens, dividing opposite the middle to upper half of the anthers, the branches 3.5–5 mm long,

extending well beyond the anthers. *Capsules* ellipsoid, occasionally globose and more or less trilobed and retuse above, 10–23 mm long; *seeds* ovate to rotund, 6–8 x 5–6 mm, broadly and evenly winged, translucent light brown, the seed body often darker brown. *Chromosome number* 2n = 45 (a triploid probably not representative of the species).

Flowering time August and September, occasionally in October or November in Kamiesberg.

Distribution and biology

One of the most widespread species of *Gladiolus* in the southern African winter-rainfall region, *G. orchidiflorus* extends from the Cape Peninsula in the south to northern Namaqualand and extreme southern Namibia. Eastward its range extends across Bushmanland and the Karoo to Kimberley and Fauresmith, although it appears to be decidedly rare outside Namaqualand and the southwestern Cape. It does not occur in the southern and eastern Cape, which is rather surprising in view of the wide range elsewhere in the winter-rainfall area. Possibly the rainfall pattern of summer and winter precipitation there is unsuitable for its growth. *Gladiolus orchidiflorus* is most commonly found in sandy soils, either those of the Cape System or Kalahari sands, but it may also be encountered on light clays or on granite-derived sand, as in Namaqualand. In the west of its range *G. orchidiflorus* is fairly common and it was especially so in the past. It is not often seen because of the dull colour of the flowers, which are difficult to see from even a short distance away.

The inconspicuously coloured flowers are, however, wonderfully scented of a combination of violet and freesia. They are extremely attractive to bees in the families Anthophoridae and Apidae which at times visit them avidly in search of nectar and in so doing accomplish pollination.

Diagnosis and relationships

An apparently basal member of series *Hebea* of section *Hebea*, *Gladiolus orchidiflorus* has the several apomorphic features that define the series: the triangular upper lateral tepals with narrow bases, the strongly arched filaments and the curious papillose ridges in the throat of the perianth tube along the sutures of the tepals. It also often has the soft-textured corm tunics of many of the other members of the series, but that feature is variable, the tunics sometimes being composed of firm-textured fibres. The outstanding diagnostic feature of *G. orchidiflorus* is the narrow dorsal tepal, 3–6 mm wide, that describes a semicircular arc over the stamens. The grey-green or purplish grey flower colour with the upper lateral tepals darkly marked and veined along the midline are also distinctive, but these features are shared with several other species of series *Hebea*.

Gladiolus orchidiflorus is somewhat isolated taxonomically within series *Hebea*, but is perhaps most closely allied to *G. watermeyeri* and *G. virescens*.

History

That the relatively common and very distinctive *Gladiolus orchidiflorus* was not described any earlier than 1802 is probably explained by confusion about the identity of some of its named relatives. C.P. Thunberg knew *G. orchidiflorus* in the field but at first regarded it as a colour form of *G. alatus* and then, in 1811 when he described the related *G. virescens*, he included it in that species. Likewise, in Nicholas Jacquin's *Icones Plantarum Rariorum* in 1795, *G. orchidiflorus* was regarded as no more than a variant of *G. alatus*. The date of discovery of this once common species of the southern African winter-rainfall region appears to be lost. Certainly, it was recorded by all the important eighteenth- and early nineteenth-century collectors of Cape plants, including Francis Masson, James Niven, Franz Oldenburg and William Paterson, as well as Thunberg. It is to Niven that the honour goes of having made the collection that resulted in the naming of *Gladiolus orchidiflorus*. Seeds or possibly corms that he collected in 1799 or 1800 were grown in the garden of his employer, George Hibbert, at Clapham, London. There they first flowered in 1802 and the species was described by Henry Andrews later the same year.

It is not entirely surprising that despite the publication in 1802 of the excellent painting that accompanied the description of the species in the *Botanist's Repository*, the plant should have become known for many years by the later name *G. viperatus*. This species was described by John Ker Gawler in 1803, based on plants raised from corms sent from the Cape earlier that year. *Gladiolus viperatus*, which Ker Gawler called the 'perfumed cornflag', represents the southwestern Cape form of what we consider to be

G. orchidiflorus and its large, bluish green and grey flowers are rather different from the central Namaqualand form of the species. As is accurately shown in Andrews's painting, these plants are fairly slender and have smallish flowers of a dull purple and greyish colour. Plants corresponding to this form were recorded by William Burchell in 1812 on the stony Litakun plains near Kuruman, now in North-West Province, South Africa. The specimens that Burchell collected have flowers of about the same size as *G. permeabilis* subsp. *edulis* which Burchell collected in the same area and with which they have often been associated. J.G. Baker (1896) for example, considered the plant a variant of subsp. *edulis* (which he called *G. edulis*). Close examination shows that they lack the attenuate tails of the tepals and despite the rather small flowers and narrow tepals, they do have papillose ridges in the throat of the tube.

The Namaqualand form of *Gladiolus orchidiflorus* was also collected by J.F. Drège in 1830, incidentally a year of notably low rainfall. The plants are rather depauperate and have particularly small flowers, the dorsal tepal only 20 mm long and 2.5 mm at the widest. The Drège collection is the type of *G. dregei*, described by F.W. Klatt in 1863, and in fact differs in no significant way from the type figure of *G. orchidiflorus*, illustrated by Andrews in 1802. Klatt, nevertheless, regarded *G. orchidiflorus* as a synonym for *G. viperatus* and distinct from his *G. dregei*, although his reasons were not clearly stated. *Gladiolus dregei* continued to be recognized throughout the nineteenth century and was only relegated to synonymy in *G. orchidiflorus* by Lewis et al. in 1972. Baker was the first botanist to consider that *G. orchidiflorus* and *G. viperatus* were conspecific and he united the two in his account of *Gladiolus* in *Flora Capensis*, a decision that has been followed ever since.

104. GLADIOLUS WATERMEYERI L. Bolus

Soetkalkoentjie, sweet kalkoentjie

PLATE 88

Gladiolus watermeyeri L. Bolus, *S. African Gard.* 17: 294 (1927). Lewis et al., *J. S. African Bot.*, Suppl. 10: 148 (1972). Type: South Africa, Northern Cape, Calvinia District, near Nieuwoudtville, Aug. 1913, *Marloth 5569* (BOL, holotype; PRE, isotype).

watermeyeri, named for E.B. Watermeyer, who sent corms of the species to Kirstenbosch Gardens. The species flowered there and was painted in 1917.

Description

Plants (10–)16–35 cm high. *Corm* globose to depressed-globose, 15–22 mm in

diameter, the tunics soft-papery to membranous, not persisting, decaying into irregular fragments. *Cataphylls* pale and membranous, the uppermost reaching shortly above the ground and then usually purple. *Leaves* four to five, the lower three basal, the lowermost longest and usually exceeding the spike, the others seldom reaching beyond the middle of the spike, sword-shaped to linear, 3–11 mm wide, the blades strongly 2- to 5-ribbed, the margins thickened and raised, the sheaths and blades scabridulous on the veins, the upper one or two leaves diverging from the stem, channelled throughout. *Stem* erect below, flexed outward above the sheathing part of the third leaf and inclined above, occasionally robust plants with one branch, c. 2–3 mm in diameter below the spike.

Spike inclined, lightly flexuose, rarely 1-, usually 2- to 6-flowered; *bracts* bright green, transparent along the veins, the outer 40–45 mm long, the inner slightly shorter than the outer, not keeled, forked apically for c. 1 mm. *Flowers* shades of green to cream, the dorsal tepal purplish below, in the mid-line, and apically, elsewhere grey-translucent, the upper lateral tepals conspicuously veined with maroon to purple or pink, the lower three tepals dark olive green, paler at the tips and then lightly veined, the sutures between the tepals thickened into iridescent papillose ridges, strongly sweet-scented (rather like freesia); *perianth tube* 14–16 mm long, narrowly and obliquely funnel-shaped, the lower cylindrical part c. 10 mm long, bent almost at 90° at the base of the flared part; *tepals* unequal, upper laterals largest, the dorsal hooded over the filaments and concave, 27–35 x 10–15 mm, lanceolate, the upper laterals spade-shaped, narrowed near the base into short claws 1–2 mm long, curving outward from just above the base of the limbs, 20–30 x 15–25 mm, the lower three tepals united below for 2–4 mm, the free parts 20–25 mm long, the lower laterals channelled and narrowed below into claws, the limbs boat-shaped, c. 9 mm long, the lowermost triangular, narrowed below into a short claw, c. 22 mm wide. *Filaments* strongly arched, 20–22 mm long, exserted for 15–16 mm from the tube; *anthers* 9–13 mm long, dark green, the pollen yellowish. *Ovary* oblong, 7–9 mm long, style arching over the

stamens, dividing between the middle and apex of the anthers, the branches 4–8 mm long, curving outward between the anthers. *Capsules* broadly ellipsoid and three sided, (20–)27–35 mm long; *seeds* broadly ovate, (7–)9–11 x (5–)7–9 mm, light to dark pinkish brown, broadly and evenly winged, the seed body small and dark brown. *Chromosome number* $2n = 30$.

Flowering time August to mid-September.

Distribution and biology

A narrow endemic of the southwestern part of Northern Cape Province and the adjacent Western Cape Province, *Gladiolus watermeyeri* is best known from the narrow strip of the Cape Sandstone System that runs along the edge of the Bokkeveld Escarpment west of the town of Nieuwoudtville. It is most common there and can usually be found in quantity in early August in seasonally wet sandy soil, especially after fire or land clearing. Its range is considerably wider, however, and extends southward along the escarpment to Lokenberg and thence to the Gifberg and the mountains and valleys at the northern end of the Cedarberg, at least as far south as Wuppertal. It may well occur even more widely in this area of high and not easily accessible mountains. The short plants with dull-coloured flowers are fairly inconspicuous, but their intense sweet fragrance makes their presence known long before they themselves are located.

Despite its rather strange-looking flowers, *Gladiolus watermeyeri* is pollinated by large anthophorid bees, mainly *Anthophora diversipes*, the same pollinator for so many of the short-tubed species of *Gladiolus* in the winter-rainfall region of South Africa.

Diagnosis and relationships

Gladiolus watermeyeri is one of a group of five species of series *Hebea* that have dull mauve, brownish or green flowers and it stands out in the group in having a translucent, strongly hooded dorsal tepal and broad, ridged leaf blades. Of these five species, only three have strongly ridged leaf blades but the other two, *G. ceresianus* and *G. virescens*, are distinctive in having an erect dorsal tepal. The combination of flower colouring, tepal orientation and leaf blade morphology thus makes *G. watermeyeri* unmistakable. It is one of just two species in this group that have a firmly arcuate dorsal tepal, an ancestral feature for series *Hebea*. The other species is *G. orchidiflorus*.

The plants themselves are fairly low-growing, seldom reaching more than 10 cm

high, and the stems are flexed forward so the spikes are strongly inclined, adding to the inconspicuous appearance of the species. *Gladiolus watermeyeri* is also distinctive in having particularly large and inflated capsules, sometimes slightly more than 20 mm long, and unusually large seeds with a pinkish brown colour, apparently unique in *Gladiolus*. The flowers of *G. watermeyeri* have one of the most intense and pleasant fragrances in the entire genus and the species is thus a most desirable garden or pot subject if fragrance is wanted. We wonder if the species might not be useful in the development of scented *Gladiolus* hybrids.

History

Although collected in the early nineteenth century by J.F. Drège in 1831 and by the Scottish collector, James Forsythe, in 1835, and then at the end of the nineteenth century by Rudolf Schlechter, *Gladiolus watermeyeri* did not attract botanical attention until H.M.L. Bolus began to systematically explore the botanically rich area around Nieuwoudtville in the 1920s. Collections made by the botanist, Rudolf Marloth, in 1913 and farmer and land surveyor, E.B. Watermeyer, in 1917 made it amply

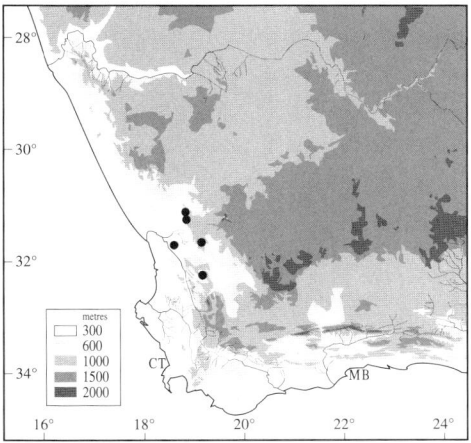

clear to Mrs Bolus that this was not merely a variant of one of the other species of series *Hebea* and finally, in 1927, she described the species, honouring the collector of the plants in cultivation at Kirstenbosch Gardens. Although *G. watermeyeri* is poorly collected, our research in preparation for this book has shown the species to be common along the Bokkeveld Escarpment, but one not very often seen because of its short stature and inconspicuous, well-camouflaged flowers.

105. GLADIOLUS VIRESCENS Thunberg

Geelkalkoentjie, yellow kalkoentjie, bokspoortjie

PLATE 89

Gladiolus virescens Thunberg, *Fl. Capensis* 196 (1811). N.E. Brown, *J. Linn. Soc.* 48: 33 (1928). Lewis et al., *J. S. African Bot.*, Suppl. 10: 150 (1972). Type: South Africa, Western Cape, without precise locality or date, *Thunberg s.n.* (UPS–Herb. Thunberg 1090, holotype).

virescens = greenish, so named for the colour of the flowers of plants that were collected by C.P. Thunberg, who named the species.

Synonymy

Gladiolus luridus Horneman, *Hort. Hafniensis*, Suppl. 113 (1819). Type: South Africa, without precise locality or collector (possibly *C.P. Thunberg*), (C, holotype).
Gladiolus luteus Klatt, *Linnaea* 32: 694 (1863), an illegitimate homonym, not *G. luteus* Lamarck (1786). Type: South Africa, Western Cape,

Caledon District, perhaps in 1831, *Ecklon & Zeyher Irid. 135* (C, holotype; BR [not seen], LD, P, PRE, S, Z, isotypes).
Gladiolus pulchellus Ecklon ex Klatt, *Linnaea* 32: 693 (1863). Baker, *Fl. Capensis* 6: 160 (1896). *Gladiolus virescens* var. *lepidus* G. Lewis in *J. S. African Bot.*, Suppl. 10: 153 (1972), as a new name for *G. pulchellus* at varietal rank. Type: South Africa, Western Cape, Swellendam District, without precise locality, Oct. 1831, *Ecklon & Zeyher Irid 137* (B [not seen], holotype; G, isotype).
Gladiolus bicolor Ecklon ex Baker, *J. Linn. Soc. Bot.* 16: 178 (1877); *Fl. Capensis* 6: 161 (1896), illegitimate homonym, not *G. bicolor* Thunberg (1784) (= *Sparaxis villosa* (Burman fil.) Goldblatt). Type: South Africa, Western Cape, near Caledon, *Ecklon s.n.* (S [not seen], holotype).
Gladiolus templemannii Klatt, *Trans. S. African Phil. Soc.* 3: 196 (1885). Type: South Africa, Western Cape, Caledon District, Zwartberg near Hot Springs, Sept. 1884, *Templemann sub MacOwan 2608* (S, possible holotype; BM, BOL, K, SAM, isotypes).
Gladiolus virescens var. *roseovenosus* G. Lewis, *J. S. African Bot.*, Suppl. 10: 153 (1972). Type: South Africa, Western Cape, Worcester District, Kweekkraal near Brandvlei, 4 Sept. 1962, *Bayliss 642* (NBG, holotype).

Description

Plants relatively short, 12–20 cm high excluding the basal leaf. *Corm* globose to depressed-globose, 12–15 mm in diameter, the tunics soft-papery becoming fibrous, seldom persisting for more than a season. *Cataphylls* membranous and pale, the uppermost not normally reaching above the ground. *Leaves* usually three, rarely four or five, the lower one much longer than the others, linear to terete, 1–3 mm wide, about as long as to twice as long as the stem, oval to round in section with the margins raised and the edges arching toward the midrib, the midrib and often one other vein also raised, the leaf thus two- or three-grooved on each surface, remaining leaves short, the lower one inserted close to the ground, sheathing the stem in the lower half, the sheath with the lower margins free and overlapping, the upper inserted in the upper half of the stem,

the uppermost one usually without a blade, thus channelled and resembling the bracts, leaf sheaths and sometimes the blades sparsely to densely scabrid to pubescent on the veins. *Stem* more or less erect below, flexed outward and inclined in the middle but sometimes becoming erect below the spike, seldom branched.

Spike flexed at the base, inclined, 3- to 7-, occasionally to 12-flowered, fairly straight or lightly flexuose; *bracts* green, the outer 20–30 mm long, the veins more or less transparent, the inner about three-quarters as long as the outer, and more or less enveloped by it. *Flowers* yellow to yellow-green or pink, with the upper three tepals flushed or feathered green, brownish purple or dark pink, the lower three tepals each with a deep yellow to white band across the base of the limb and pale distally, sometimes mauve or pink at the apices, the sutures between the tepals thickened and forming densely papillose iridescent ridges, strongly violet-scented or unscented (some pink-flowered plants); *perianth tube* 9–12(–15) mm long, narrowly and obliquely funnel-shaped; *tepals* unequal, the dorsal longest, more or less spathulate, straight and inclined or erect, the margins sometimes recurved, 20–40(–48) x 10–16(–20) mm, narrowed below and claw-like for c. 18 mm, the upper third abruptly expanded and triangular, the upper laterals spade-shaped, the base narrow and claw-like, 1–3 mm long, the limbs abruptly expanded, triangular, curving outward in the upper half, 14–25(–30) mm long, the lower three tepals joined to the upper laterals for c. 5 mm and to one another for 1–3 mm, the free parts 16–20(–30) mm long, clawed and narrow and grooved below, abruptly expanded into limbs 10–12(–20) mm long. *Filaments* 18–27 mm long, exserted for 15–22 mm, strongly arched; *anthers* 6–8 mm long, pale or dark, the pollen yellow. *Ovary* oblong, c. 4 mm long; *style* arching over the stamens, dividing opposite the upper third of the anthers, the branches 4–6 mm long, the apices reaching beyond the anthers. *Capsules* oblong-ellipsoid, 18–25(–30) mm long; *seeds* ovate to round,

5–7 x 5–6 mm, broadly and evenly winged, light translucent yellow-brown, the seed body dark brown. **Chromosome number** unknown.

Flowering time mid-August through September.

Distribution and biology
Restricted to the southern half of the southern African winter-rainfall zone, *Gladiolus virescens* extends from Bot River in the Caledon District in Western Cape Province to near Humansdorp in Eastern Cape Province. Although most often seen in the southern Cape, populations occur inland in the Little Karoo near Montagu, in the Warm Bokkeveld and on the Witteberg. Plants typically favour stony shale slopes and a heavy clay soil, but are occasionally found in alluvial fans in loamy soil covered with rounded sandstone rocks. In both situations the corms are protected from predation by mole-rats and porcupines that feed on underground plant storage organs.

The very fragrant, short-tubed flowers of *Gladiolus virescens* are adapted for pollination by long-tongued bees. They are pollinated by the same bees, mostly species of *Anthophora*, that pollinate most other species of section *Hebea* with short-tubed and fragrant flowers.

Diagnosis and relationships
The flowers of *Gladiolus virescens* are typical of series *Hebea* in having the upper tepals narrowed below into claws, a short perianth tube and glistening papillose ridges in the throat along the line of union of the tepals. Within the series it is distinguished by the erect dorsal tepal, the margins often recurved, the yellow to pinkish tepal colour and the narrow leaves with the midrib and usually one or two other pairs of veins strongly raised. Flower colour is variable, ranging from predominantly yellow to pale pink or sometimes yellowish brown or even creamy white, the upper tepals variously veined with green, brown, purple or pink.

Gladiolus virescens is most closely related to *G. ceresianus*, which typically has shorter

stems and dark purple flowers with the upper tepal margins strongly recurved and the tepals veined with purple. Although the two species are sometimes difficult to distinguish, they have quite different corms, those of *G. virescens* being depressed-globose with softly membranous to lightly fibrous tunics, while the rounded corms of *G. ceresianus* have hard, blackish tunics. In addition, the leaves of *G. ceresianus* are imbricate below and completely conceal the stem, whereas in *G. virescens* the internode between the upper two leaves is always exposed. Plants recognized as var. *roseovenosus* and var. *lepidus* seem to us no more than minor variants of typical *G. virescens*, with the flowers either smaller and more numerous or with more pronounced pink veins or overall pink colouring, and thus do not merit taxonomic recognition. Var. *roseovenosus*, restricted to Brandvlei near Worcester, has glabrous leaf sheaths and lacks floral scent, but is not the only population of *G. virescens* with unscented flowers. Absence of fragrance is not uncommon in species of *Gladiolus* that normally have scented flowers.

History
One of the first of the species of series *Hebea* with dull-coloured flowers known, *Gladiolus virescens* was evidently first recorded by C.P. Thunberg in the early 1770s and by his compatriot, Swedish naturalist Anders Sparrman, probably in 1775. According to Brown (1928) and Lewis et al. (1972) Thunberg collected two forms of the species, the typical one, most likely from the Breede River valley near Worcester (Herb. Thunberg 1090), and the southern Cape form, called by Lewis et al. var. *lepidus*, which has spikes of more numerous and smaller flowers. Thunberg did not at first distinguish *G. virescens* from *G. orchidiflorus* and specimens he collected were initially identified as *G. alatus*, which is the name he used for that species. In 1811, after he had begun to refine his taxonomy of southern African *Gladiolus* species, he described *G. virescens* based on one of his own collections. A sheet

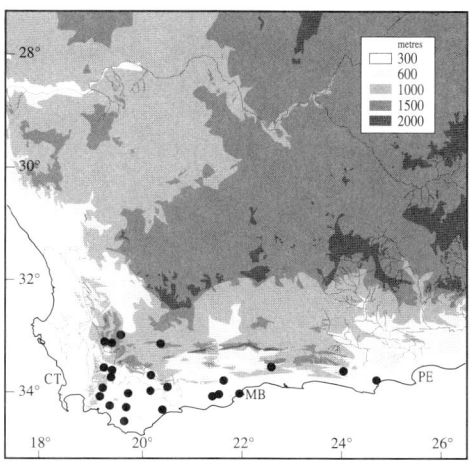

of *G. virescens* in the Copenhagen Herbarium without any collection information, but possibly also collected by Thunberg, is the type of *G. luridus*, described by the Danish botanist, Jens Horneman, in 1819. This is probably the southern Cape form of the species with spikes of fairly small flowers, later called *G. pulchellus* by C.F. Ecklon on the labels accompanying his collections of the plant. This last species was only formally described by F.W. Klatt in 1863 and was also recognized by J.G. Baker in his accounts of *Gladiolus* in 1882 and 1896. This variant of *G. virescens* was reduced to varietal rank by Lewis et al. in 1972, who used the name var. *lepidus* for the plant.

Baker completely misunderstood the identity of *Gladiolus virescens* and treated the name as conspecific with *G. orchidiflorus*, for example, in *Flora Capensis*. In 1877, however, he described *G. bicolor*, adopting another of Ecklon's names, *Hebea bicolor*, published by Ecklon in 1827 without an accompanying description. The type of this species corresponds closely to typical *G. virescens*. Baker continued to recognize *G. bicolor* in his account of *Gladiolus* in *Flora Capensis*, and the name is also, incidentally, a later homonym in *Gladiolus*. Klatt's *G. templemannii* is based on plants from near Caledon that closely match the types of both *G. virescens* and *G. bicolor*.

106. GLADIOLUS CERESIANUS L. Bolus
PLATE 90

Gladiolus ceresianus L. Bolus, *S. African Gard.* 21: 369 (1931). Lewis et al., *J. S. African Bot.*, Suppl. 10: 155 (1972). Type: South Africa, Western Cape, Ceres District (Ceres Wild Flower Show), 1931, *Compton s.n.* (BOL 19883, holotype; K, isotype).

ceresianus = from Ceres, a town in the western part of Western Cape Province, South Africa, near to which the type collection was made.

Description
Plants relatively short, 10–18(–30) cm high. *Corm* conic, 9–12 mm in diameter, the tunics of medium to coarsely textured blackish fibres, with a few cormlets produced at the base. *Cataphylls* pale and membranous,

the uppermost reaching shortly above the ground and then dark green to reddish. *Leaves* three, always overlapping, the lowermost longest and basal, 14–20(–30) mm long, usually longer than the stem, the blade terete and four-grooved, straight or falcate, 1–2 mm in diameter, second leaf sheathing the lower half of the stem, reaching to about the middle of the spike, third leaf shortest, channelled throughout but sheathing only near the base. *Stem* inclined throughout or erect below, flexed outward above the sheath of the uppermost leaf, unbranched, c. 1.3 mm in diameter below the spike.

Spike inclined, rarely 1-, usually 2- to 4-flowered, lightly flexuose; *bracts* green or flushed reddish, the outer 25–35(–40) mm long, the inner slightly shorter than the outer, minutely notched apically. *Flowers* predominantly purplish brown or greenish grey, the three upper tepals with dark purple veins radiating from the midline, the lower tepals each clear yellow in the lower two-thirds, the upper part of the tube with the sutures between the tepals forming glistening papillose ridges, intensely sweet-scented; *perianth tube* 12–16 mm long, narrowly and obliquely funnel-shaped, the slender basal part 8–10 mm long; *tepals* unequal, the dorsal longest, 28–36(–45) x c. 11 mm, erect or inclined below and erect above, oblanceolate, the margins recurved, the

upper lateral tepals spade-shaped, the bases abruptly narrowed into claws c. 4 mm long, the limbs triangular, the apices acuminate, curving outward in the upper half, 20–25 x 14–16 mm, the lower three tepals united with the upper laterals for 3–4 mm, the free parts c. 16–25 mm long, clawed and narrow and channelled below, the lower laterals with limbs 4–6 mm wide, the lower median 9–10 mm wide, the limbs directed downward. *Filaments* 20–22 mm long, exserted for c. 18 mm, arched in a half circle; *anthers* 8–9 mm long, light purple, the pollen yellow. *Ovary* oblong, c. 6 mm long; *style* arching over the stamens, dividing between the middle and upper third of the anthers, the branches c. 6 mm long, extending between the anther apices. *Capsules* and *seeds* unknown. *Chromosome number* 2n = 30.

Flowering time mid-August to the end of September.

Distribution and biology

Gladiolus ceresianus is a fairly widespread endemic of the western Cape interior. Although the type collection was made near Ceres, some 120 km from Cape Town and in an area of relatively high rainfall, the species is more common farther east in the arid western Karoo. Plants grow on rocky slopes and flats in clay ground either in the open or among low bushes, typical of the mountain vegetation of the western Karoo. The range of *G. ceresianus* extends from the Cold Bokkeveld north of Ceres eastward to Matjiesfontein and thence north through the Klein Roggeveld to the high Roggeveld Escarpment at Ganagga Pass near Middelpos. An outlying population occurs at Hol River, near Vredendal, at the southern edge of Namaqualand. This surprising disjunction appears to be correct and not the result of inadequate collecting.

The short-tubed and sweetly scented flowers of *Gladiolus ceresianus* are adapted for pollination by solitary bees. The strong floral fragrance plays an important role in attracting pollinators and the flowers themselves are rather dull- and cryptically coloured, a possible adaptation to avoid predation by herbivorous beetles that are attracted to brightly coloured flowers.

Diagnosis and relationships

One of the most diminutive species in the genus, *Gladiolus ceresianus* is seldom taller than 15 cm high. Its small size and the dark-coloured flowers make it very inconspicuous and it is often found only after its strong fragrance is detected in the veld. The flowers

are typical of series *Hebea* in having a relatively short perianth tube, upper tepals narrowed below into claws and iridescent papillose ridges in the throat. The dorsal tepal is held stiffly erect with the margins recurved, and the dull purplish grey to brownish tepals are veined with dull purple. This fairly distinctive combination is matched in section *Hebea* only by *G. uysiae*. The latter is readily distinguished by its flattened corm with softly membranous tunics, cormlets produced on long stolons and plane, leathery, somewhat fleshy leaves. All these features contrast strongly with the rounded corms with blackish, hard tunics, cormlets borne around the corm base and more or less terete leaves with a strongly raised midrib that are characteristic of *G. ceresianus*. The latter may also be confused with *G. virescens*, but that species also has flattened corms and flowers that are usually pink or yellowish with dark veins on the upper tepals. *Gladiolus ceresianus* can also be recognized in the absence of corms by the overlapping leaf bases, whereas the internode between the upper two leaves is always exposed in *G. virescens*.

History

The first collection of *Gladiolus ceresianus* appears to be that made by J.F. Drège, perhaps in 1830 when he journeyed to Namaqualand. His collection, numbered 1840, is unfortunately without date or locality. This, the only early record of the species, was confused with *G. orchidiflorus* by the botanists of the time. A hundred years later *G. ceresianus* was rediscovered in wild flower displays at the Ceres Agricultural Show in 1930. Specimens were brought back to Cape Town and the species was described by H.M.L. Bolus in 1931. A year later plants were found in the wild by G.J. Lewis near Tweedside at the southern edge of the western Karoo, some distance from Ceres.

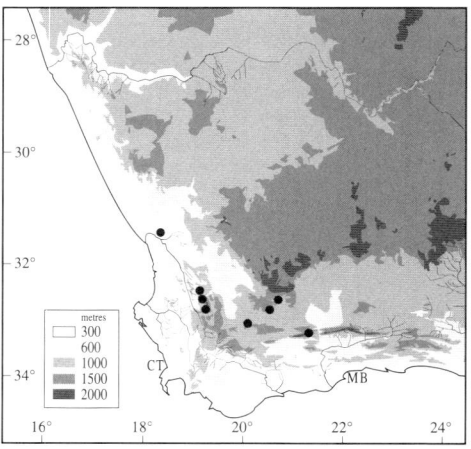

107. GLADIOLUS UYSIAE L. Bolus ex G. Lewis

Perskalkoentjie, purple kalkoentjie

PLATE 91

Gladiolus uysiae G. Lewis in *J. S. African Bot.*, Suppl. 10: 156 (1972). L. Bolus, *S. African Gard.* 21: 367 (1931), provisional name. Type: South Africa, Northern Cape, Calvinia District, without precise locality, cultivated at Nieuwoudtville, Sept. 1930, *Buhr s.n.* (BOL 19894, holotype).

uysiae, named in honour of Mrs G.J. Uys of Nieuwoudtville, who was responsible for bringing this plant to the attention of Mr H. Buhr; he in turn gave a specimen to the Cape Town botanist, H.M.L. Bolus.

Description

Plants relatively short, 7–18 cm high. *Corm* depressed-globose, 10–14 mm in diameter, the tunics of membranous to papery layers not accumulating annually, producing long stolons from the base, each terminating in a fairly large cormlet. *Cataphylls* pale and membranous, reaching shortly above the ground. *Leaves* three, imbricate, the lower one much longer than the others, falcate, about as long as or slightly longer than the stem, 3–6(–9) mm wide, the central and one, sometimes two other pairs of veins evident but not thickened, remaining leaves reaching to between the middle of the stem and the base of the spike, the uppermost one usually mostly channelled, the unifacial part of the blade not more than 1–2 cm long. *Stem* more or less erect below, flexed outward and inclined in the middle, rarely with a short branch, c. 1–1.5 mm in diameter below the spike.

  Spike flexed at the base, inclined, 1- to 3-flowered, lightly flexuose; *bracts* green, the outer 20–25 mm long, the margins transparent, the inner about three-quarters as long

as, and more or less enveloped by, the outer. *Flowers* reddish purple, distally becoming veined purple on a pale background, lower tepals yellow, each with a dark yellow-green band shortly below the apex, sometimes purple-veined at the apices, the sutures between the tepals thickened into iridescent papillose ridges, strongly and sweetly scented; *perianth tube* 10–12 mm long, narrowly and obliquely funnel-shaped; *tepals* unequal, the dorsal longest, straight and erect, 28–35 mm long, widest in the upper third, 10–12 mm wide, the margins recurved when fully open, upper laterals spade-shaped, the bases abruptly narrowed and claw-like for 2–3 mm, the limbs triangular, acute, curving outward in the upper half, 15–18 mm long, the lower three tepals joined to the upper laterals for c. 1 mm and to one another for c. 1 mm, c. 15 mm long, clawed and narrow and channelled below, abruptly expanded into limbs 8–9 mm long, these directed downward. *Filaments* strongly arched, 20–25 mm long, exserted 18–20 mm from the tube; *anthers* 6–8 mm long, purple, pollen yellow. *Ovary* oblong, c. 5 mm long; *style* arching over the filaments, dividing opposite the upper third of the anthers, the branches 5–6 mm long, extending well beyond the anther apices. *Capsules* obovoid, the apices rounded, obscurely three-lobed, 12–18 mm long; *seeds* round to broadly ovate, c. 6 x 4–5 mm, broadly and evenly winged, translucent brown, the seed body slightly darker brown. *Chromosome number* 2*n* = 30.

  Flowering time mid-August to the end of September.

Distribution and biology

Gladiolus uysiae is a relatively poorly collected species restricted to the winter-rainfall part of southwestern South Africa. It extends from the shale plains at the foot of the Hantamsberg in the Calvinia District in the western Karoo westward to the Bokkeveld Escarpment near Nieuwoudtville and southward across the Roggeveld Escarpment to the Klein Roggeveld, Touws River and Hottentot's Kloof in the Ceres District in the south. Plants grow in mountain renosterveld under shrubs or in relatively open ground where they often form fairly dense colonies, the result of vegetative reproduction from cormlets produced on long stolons from the

base of parent corms. Despite a rather small number of collections, *G. uysiae* is probably not at all rare. It is extremely inconspicuous and even in full flower may be overlooked. The flowers are dull coloured and well camouflaged when viewed against bare stony ground. Often their strong, sweet, freesia-like scent reveals their presence before they are visually located. Plants flower well only in open ground, but grow vegetatively in the shade of low shrubs, flowering only after fires, heavy grazing, or clearing of the bush cover.

  The well-camouflaged flowers of *Gladiolus uysiae* are pollinated by large long-tongued bees, including *Anthophora diversipes* which is the main pollinator of the related species, *G. orchidiflorus* and *G. virescens*.

Diagnosis and relationships

The flowers of *Gladiolus uysiae* are virtually identical to those of *G. ceresianus* and some forms of *G. virescens*, having a relatively narrow, erect dorsal tepal with the margins recurved, spade-shaped upper lateral tepals and narrow, deflexed lower tepals. Like *G. ceresianus*, *G. uysiae* has dull-coloured, brownish to dull purple flowers, the upper tepals with dark veining and the lower tepals dull yellow-green tipped with brown or purple. The major differences between the species lie in their leaves and corms. Leaves of *G. uysiae* have plane, sword-shaped blades without raised veins and the corms are flattened and have membranous tunics that rapidly decay and thus do not accumulate. The production of several long, slender stolons from the corm base, each ending

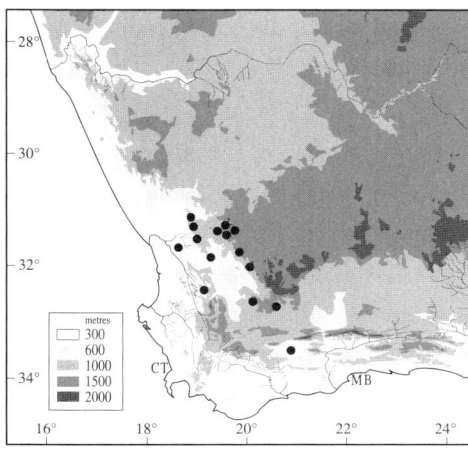

225

in a small corm, is also typical of this species and does not occur in either *G. ceresianus* or *G. virescens*. These latter two species have leaves with strongly raised midribs and, in the case of *G. virescens*, often one or two other vein pairs are also raised. Both can easily be distinguished from *G. uysiae* once the differences between them are understood. They are no doubt a closely related monophyletic lineage and share a common ancestor. The similar, but less closely allied *G. watermeyeri* has strongly ribbed leaves but an arched dorsal tepal, and is readily distinguished from the *G. uysiae* group of species with erect dorsal tepals.

History

Gladiolus uysiae was apparently first collected near Calvinia by Heinrich Meyer, the German physician who practised at the town from 1860 to 1886. It was also recorded by Rudolf Schlechter in 1897 in the hills near the Doring River. Neither collection attracted botanical attention, but then in the 1920s plants were sent to Cape Town by H. Buhr of Nieuwoudtville who had been shown specimens of the species by a neighbour, Mrs G.J. Uys. Louisa Bolus realized that this was an undescribed species for which she proposed the name *G. uysiae* when she published the protologue of a second species,

G. ceresianus, in 1931. Bolus decided to wait for more material before formally naming *G. uysiae*, but none ever came to hand. Plants grown in Britain by T.T. Barnard were illustrated in *Curtis's Botanical Magazine* in 1950 under the name *G. ceresianus*. J.R. Sealy, who wrote the text accompanying the painting, expressed reservations about its identity and suggested that it might be a new species or a hybrid between the related *G. watermeyeri* and *G. ceresianus*. The protologue of *G. uysiae* was published only in 1972 by G.J. Lewis in the revision of *Gladiolus* in South Africa that appeared after her death (Lewis et al., 1972).

108. GLADIOLUS EQUITANS Thunberg

Groot rooikalkoentjie, large red kalkoentjie, Namaqua kalkoentjie

PLATE 92

Gladiolus equitans Thunberg, *Prod. Pl. Capensium* 186 (1800). N.E. Brown, *J. Linn. Soc.* 48: 21 (1928). Lewis et al., *J. S. African Bot.*, Suppl. 10: 169 (1972). Type: South Africa, Cape, without precise locality or collector (UPS–Herb. Thunberg 1021, holotype).

equitans = riding, referring to the broad leaves, the sheaths of which clasp (or ride) the base of the leaf above them.

Synonymy

Gladiolus namaquensis Ker Gawler, *Curtis's Bot. Mag.* 16: pl. 592 (1802). *Gladiolus alatus* var. *namaquensis* (Ker Gawler) Baker, *J. Linn. Soc. Bot.* 16: 177 (1876); *Fl. Capensis* 6: 159 (1896). Pole Evans, *Fl. Pl. S. Africa* 2: pl. 63 (1922). Type: South Africa, without precise locality or date, raised from plants collected by James Niven and grown in Great Britain, illustration in *Curtis's Bot. Mag.* 16: pl. 592 (1802): specimens cultivated in Britain and preserved at BM, annotated *G. namaquensis* Ker Gawler may be type material.
[*Gladiolus galeatus* sensu Andrews, *Bot. Rep.* 2: pl. 122 (1800), not *G. galeatus* Burman fil. (1768) (= *G. alatus* Linnaeus) nor *G. galeatus* Jacquin, 1790 (= *Sparaxis galeata* (Jacquin) Ker Gawler).]

Description

Plants 15–30 cm high. *Corm* globose, 20–40 mm in diameter, the tunics of firm-membranous to coriaceous layers, these decaying into irregularly shaped fragments, occasionally more or less coarsely fibrous, often accumulating in a dense mass. *Leaves* usually four, occasionally five, the lower two or three more or less basal, forming a distichous fan, the lowermost largest, reaching to just below the base of the spike, the remaining leaves progressively shorter, the

lower three falcate, the blades 20–40 mm wide, obtuse-apiculate, the margins heavily thickened, hyaline or reddish. *Stem* erect, compressed, ridged and usually three-winged, with two wide and one narrow face, unbranched, 3–5 mm in diameter below the spike.

Spike flexed at the base and inclined, flexuose, nearly distichous, 3- to 7-flowered, the flowers facing in opposite directions; *bracts* green, the outer prominently keeled, the keels decurrent on the stem angles, keels and margins reddish, 25–40(–50) mm long, the inner about two-thirds as long as the outer, two-keeled, minutely forked apically. *Flowers* orange to scarlet, the lower lateral tepals yellow with the apices also red, the lower median tepal yellow in the lower half, the throat with glistening papillose ridges on the sutures between the tepals, lightly sweet-scented in the mornings (rather like the orchid *Pterygodium*); *perianth tube* obliquely funnel-shaped, 10–15 mm long, entirely included in the bracts; *tepals* unequal, the dorsal largest, 27–38 x 18–20 mm, obovate, concave and hooded over the stamens, the upper laterals broadly heart-shaped, narrowed near the bases, curving gradually outward in the upper half and spreading, 23–26 x 23–30 mm, broadest just above the base, the lower three tepals united below for 3–5 mm, the free parts c. 25 x 9–10 mm, clawed below, the lower laterals channelled up to 3 mm short of the apices, the lower median c. 18 x 12–15 mm, the lower laterals inclined toward the ground and more or less straight. *Filaments* 17–25 mm long, strongly arched, exserted 13–20 mm from the tube; *anthers* 8–14 mm long, cream with red on the lines of dehiscence, the pollen cream. *Ovary* oblong, 8–10 mm long, style arching over the stamens, dividing between the middle and upper third of the anthers, the branches 5–7 mm long, extending beyond the anthers, expanded and bilobed only near the apices. *Capsules* oblong-ellipsoid, 25–30 mm long; *seeds* broadly ovate, 8–9 x c. 7 mm, broadly and evenly winged, translucent dark reddish brown. *Chromosome number* 2*n* = 30.

Flowering time mainly late August to mid-September, occasionally earlier or later, especially at high elevations.

Distribution and biology

Native to Northern Cape Province, South Africa, *Gladiolus equitans* occurs exclusively in Namaqualand in the western part of the province. Records of *G. equitans* show it to be most common in the Kamiesberg and the lower hill country to the west, but it actually extends from the Knersvlakte in the south to the mountains around Springbok and Okiep in the north. The species occurs only on the granitic soils that are common in Namaqualand, and is most often seen in granite outcrops where the corms are wedged in rock crevices. Plants in such sites are sheltered from predation by porcupines and molerats, which feed avidly on corms of Iridaceae and the underground organs of other geophytes.

The extremely large and brilliantly coloured flowers of *Gladiolus equitans* are, according to our observations, pollinated by the large bees of the family Anthophoridae that also visit many of the less conspicuous species of the Namaqualand flora.

Diagnosis and relationships

One of five species of series *Hebea* that have bright orange to scarlet or pinkish flowers, angular stems and keeled floral bracts,

Gladiolus equitans can immediately be recognized by its broad, more or less falcate leaves, usually 20–40 mm wide. In general form and colour the flowers match closely those of the southwestern Cape species, *G. speciosus*, and are also similar to those of *G. alatus* and *G. pulcherrimus*. Both *G. equitans* and *G. speciosus* have flowers with an arcuate dorsal tepal, but the latter has a somewhat smaller flower and narrow leaves without the thickened margins so characteristic of *G. equitans*. *Gladiolus alatus* and *G. pulcherrimus* have much narrower leaves, rarely exceeding 15 mm wide — and in the former they are also ridged — and the dorsal tepals are usually erect, although sometimes arched forward in young flowers. Both *G. alatus* and *G. pulcherrimus* also have flowers with hairy filaments easily distinguished from the glabrous filaments of *G. equitans*. A hooded dorsal tepal and glabrous filaments are ancestral features and place *G. equitans* at the base of the *G. alatus* group of series *Hebea* that all have brightly coloured orange, scarlet or pink flowers.

History

Although *Gladiolus equitans* was described as early as 1800 by C.P. Thunberg based on specimens in his own herbarium, the species remained poorly understood throughout the next century. Plants collected by the Scottish plant collector, James Niven, in 1799 and flowered three years later in England were described independently as *G. namaquensis* by John Ker Gawler in 1802. In the later literature *G. equitans* was ignored or forgotten and *G. namaquensis* remained poorly understood, probably owing to the scant material of the species then available in herbaria. Thus, it is easy to understand why J.G. Baker reduced it to varietal rank in *G. alatus*, which it resembles closely in

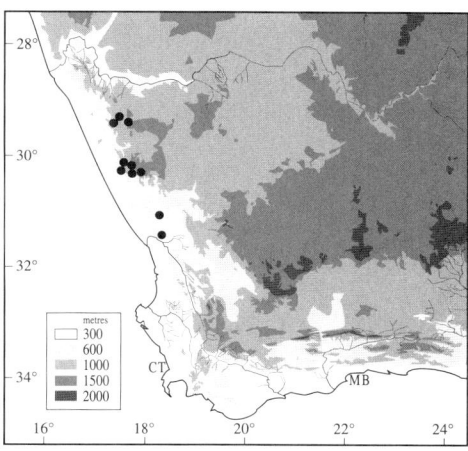

form and flower colour. After examining Thunberg's herbarium, maintained as a separate collection, at Uppsala, Sweden, N.E. Brown (1928) realized that *G. equitans* was indeed the plant then known as *G. alatus* var. *namaquensis*. Brown decided that it was distinct from *G. alatus* and recommended its recognition as a separate species, a decision that has subsequently been followed.

The source of the specimen in the Thunberg Herbarium is puzzling. Few of the early explorers and plant collectors at the Cape ventured into Namaqualand, and Thunberg did not travel farther north than the southern edge of the Knersvlakte on his way to the Bokkeveld Plateau and the Roggeveld in 1774 (Gunn & Codd, 1981). Did he receive specimens from contemporary collectors who did reach Namaqualand? The explorer, William Paterson, certainly encountered *Gladiolus equitans* in the Kamiesberg in 1789 and it is painted in his diary (Forbes & Rourke, 1980). Masson also collected the species when he went to Namaqualand, apparently in 1786 and in 1793 (Gunn & Codd, 1981), and perhaps provided Thunberg with his specimen of the species. History is mute on this matter.

109. GLADIOLUS SPECIOSUS Thunberg

Bontkalkoentjie

PLATE 93

Gladiolus speciosus Thunberg, *Fl. Capensis* 196 (1811); ed. 2, 48 (1823). *Gladiolus alatus* var. *speciosus* (Thunberg) G. Lewis in *J. S. African Bot.*, Suppl. 10: 167 (1972). Type: South Africa, Western Cape, without precise locality, probably Sept. 1773, *Thunberg s.n.* (UPS–Herb. Thunberg 1071, holotype).

speciosus = beautiful, for the appearance of the flowers.

Description

Plants (8–)12–20 cm high. *Corm* depressed-globose, 7–12 mm in diameter, the tunics membranous to dry and softly papery, generally not accumulating, decaying into irregular fragments, bearing several long, slender, often branched stolons from the base, each terminating in a small corm. *Cataphylls* pale and membranous, the uppermost reaching a short distance above the ground and often dry and grey-brown. *Leaves* four to six, occa-

sionally to eight, the lower three or four, sometimes six, basal, the lowermost longest, about twice as long as the stems, the remaining leaves progressively shorter, the blades plane, linear to falcate, (2–)3–7 mm wide, the midrib and sometimes one other pair of veins evident but not raised, the margins sometimes slightly thickened, the upper two or three leaves inserted on the middle to upper part of the stem, at least the upper two diverging from the stem and sheathing

only at their bases, channelled for most or all of their length. *Stem* erect below or inclined above the ground, flexed outward above the sheathing part of the uppermost basal leaf and then inclined c. 45°, compressed and two-angled below, three-sided and winged above, occasionally with one short branch, 4–6 mm wide below the spike (wide side).

Spike inclined, lightly flexuose, 2- to 5-, occasionally to 8-flowered on the main axis, lateral branches 1- to 2-flowered; *bracts* pale glaucous green, the outer 35–40(–50) mm long, keeled and usually red on the keel and margins, the inner about three-quarters to nearly as long as the outer, dark red on the ridges. *Flowers* predominantly orange, the dorsal tepal whitish on the inside, the upper laterals usually mostly yellow outside with orange margins, occasionally completely orange, orange or dull red to purple or yellowish inside broadly edged with orange, lower tepals mostly yellow, the apices orange, the sutures between the tepals forming iridescent papillose ridges, strongly scented of the orchid *Pterygodium*; *perianth tube* obliquely funnel-shaped, 11–13 mm long; *tepals* unequal, the dorsal largest, 25–40 x 18–26 mm, obovate to oblanceolate, nearly horizontal and concave, hooded over the stamens, the upper laterals broadly spade-shaped, strongly narrowed near the bases into short claws 1–3 mm long, the limbs

triangular, 20–25 x 20–28 mm, directed forward below, gently curving outward in the upper two-thirds, lower three tepals united with the upper laterals for 3–5 mm, the free parts 20–25 mm long, the lower laterals spathulate, narrowed below into channelled claws 6–8 mm long, gradually expanding into plane limbs, these 8–10 mm wide, the lowermost with a claw c. 5 mm long, the limb broadly triangular, 17–24 mm wide. *Filaments* ascending to arched, c. 22 mm long, exserted for c. 15 mm from the tube, minutely pubescent toward the bases, glabrous above; *anthers* 8.5–10 mm long, dull greenish yellow, sometimes reddish to purple on the lines of dehiscence, the pollen yellowish. *Ovary* narrowly oblong and three-sided, c. 7 mm long; *style* arching over the stamens, dividing opposite the middle third of the anthers, the branches c. 7 mm long, not reaching the anther apices, reddish, the apices white. *Capsules* ellipsoid, (20–)30–35 x c. 11 mm, three-sided, sometimes the septa inflated and pithy, often drawn into a short beak at the apex; *seeds* ovoid, many or sometimes only one or two per locule, the remaining ovules aborted, 9–12 x 8–11 mm, broadly and evenly winged, light translucent brown. *Chromosome number* unknown.

Flowering time mid-September to mid-October.

Distribution and biology

Although not well collected, *Gladiolus speciosus* appears to have a relatively wide range across the western half of Western Cape Province, South Africa. It extends from Mamre, just north of Cape Town, through the coastal sandveld at least as far north as Leipoldtville, near Lambert's Bay. It also extends inland into the drier valleys and hills of the Pakhuis and Nardouw Mountains, and on the Bokkeveld escarpment between Botterkloof and Nieuwoudtville. Its low stature makes it hard to see except from close by. The intense sweet scent, however, a combination of freesia and the orchid *Pterygodium*, usually makes the plants' presence known long before they are located. *Gladiolus speciosus* grows in deep sandy soils among Restionaceae and fynbos shrubs. Despite its sandy, well-drained habitat, it is one of the last of the Iridaceae along the west coast to flower. Plants may still be found in bloom at the end of the spring season, in the middle of October.

The capsules and seeds of *Gladiolus speciosus* populations from the west coast are unusual in the genus and suggest an

unexpected dispersal mechanism. As the capsules mature, the walls between the seed chambers become filled with light, pithy material that makes the dry capsules appear inflated. The seeds are also unusual. Although broadly winged as in nearly all species of *Gladiolus*, they are extremely large, 9–12 mm long, and there are only one to three seeds per chamber. This extraordinary adaptation is puzzling. Could the light, inflated capsules act as sails, buoying the dry, dead stems and allowing the whole aerial part of the plant to be dispersed as a unit by wind, rather than the seeds alone? On the Bokkeveld escarpment capsules of *G. speciosus* are quite normal and contain numerous seeds slightly smaller than those of the west coast plants.

The sweetly scented flowers of *Gladiolus speciosus* are most likely adapted for pollination by long-tongued anthophorid bees, as are most related species.

Diagnosis and relationships

The compressed and winged stems, large, keeled outer bracts and two-keeled inner bracts and the predominantly orange flowers with yellow-green and orange lower tepals make it clear that *Gladiolus speciosus* is a member of the *G. alatus* complex, a group of species that extend from the Bredasdorp District in the southern Cape to northern Namaqualand. It has seldom been accorded taxonomic recognition, although it is consistently different from other members of the complex, and is itself fairly uniform across its range. This seems to be an example of strong similarities in a very specialized species outweighing particular differences. *Gladiolus speciosus* can easily be distinguished from *G. alatus* in the narrow sense by its plane leaf blades, strongly hooded concave dorsal tepal, filaments and style glabrous except for a light basal pubescence, the rather closed flower with the upper lateral tepals curving outward

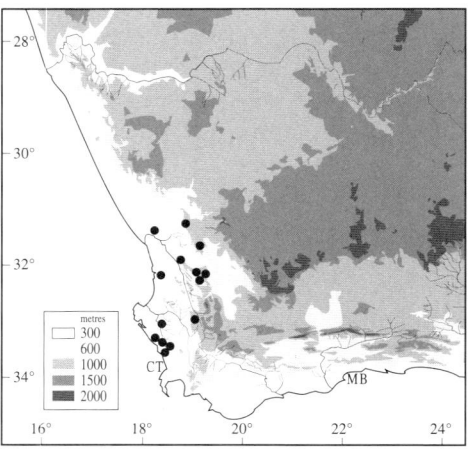

very weakly, and often by the tepal markings. The upper lateral and median lower tepal are usually yellow on the reverse with broad orange margins. In addition, the corms of *G. speciosus* invariably produce long, slender stolons from the base while *G. alatus* very rarely does so, instead bearing cormlets in the leaf axils and at the base of the corm. The morphological differences combined with different habitat preferences and a flowering time somewhat later than *G. alatus* make a convincing argument for according species status to the plant.

History

The first record of *Gladiolus speciosus* is an illustration in Simon van der Stel's diary that recounts his expedition to the copper mountains of Namaqualand in 1685 (De Wet & Pheiffer, 1979). In this unpublished document, several versions of which exist, there is a fair watercolour painting on folio 79 of a plant collected southeast of present-day Leipoldtville under the name *Gladiolus esculentus*. Copies of the diary were available to early botanists who in turn illustrated the plant. Thus in pre-Linnaean literature *G. speciosus* was figured in Leonard Plukenet's *Phytogeographia* in 1691 as *Siyrinchium viperatum*. This, and a manuscript name for the plant, *S. capense, monanthos, flore cucullato* Petiver, were cited by Linneaus in the protologue of *G. alatus* which he described in 1762. That species is now regarded as based primarily on a specimen in the Burman Herbarium, housed in Geneva, which is true *G. alatus*. *Gladiolus speciosus* was re-collected by C.P. Thunberg on the coast north of Cape Town, probably in September 1773, and described by him in 1811 in the first edition of his *Flora Capensis*. It has seldom been regarded as distinct from the related *G. alatus*, although it certainly was by Thunberg who contrasted the striate leaves of that species with those of *G. speciosus*. *Gladiolus speciosus* was included in *G. alatus* by Baker (1896) in *Flora Capensis* and was treated as var. *speciosus* of *G. alatus* by Lewis et al. (1972).

110. GLADIOLUS PULCHERRIMUS (G. Lewis) Goldblatt & Manning

Pronkkalkoentjie

PLATE 94

Gladiolus pulcherrimus (G. Lewis) Goldblatt & Manning, new combination. *Gladiolus alatus* var. *pulcherrimus* G. Lewis in *J. S. African Bot.*, Suppl. 10: 168 (1972). Type: South Africa, Western Cape, Olifants River Valley 18 km (12 miles) south of Klawer, 28 Aug. 1957, *Lewis 5220* (NBG, holotype).

pulcherrimus = most beautiful, so named for the strikingly lovely, bright salmon-coloured flowers.

Description

Plants 20–40(–65) cm high. *Corm* depressed-globose, 10–15 mm in diameter, the tunics membranous, generally not accumulating, decaying into irregular fragments. *Cataphylls* pale and membranous, the uppermost reaching a short distance above the ground and often dry and grey-brown. *Leaves* five to seven, sometimes eight, the lower four or five more or less basal, the lowermost sometimes slightly longer than the others, reaching to about the base of the spike, uppermost often more or less bract-like and sometimes without a blade, grey-green, with a waxy bloom, blades lanceolate to sword-shaped, (3–)6–12 mm wide, plane, the midrib scarcely evident in fresh plants, the margins lightly thickened, sometimes lightly crisped. *Stem* erect below, flexed outward above the sheathing parts of the cauline leaves, inclined 20° to 40° above, usually with one or two branches, these diverging from the main axis, compressed and three-angled, the angles winged, 4–8 mm wide below the base of the spike (wide side).

Spike inclined up to 40°, flexuose, usually 5- to 8-flowered, the lateral branches with fewer flowers; *bracts* pale grey-green, glaucous, the outer 35–50 mm long, with prominent keels, the inner about three-quarters as long, usually two-keeled, minutely notched apically. *Flowers* predominantly salmon, the upper three tepals feathered reddish below, the dorsal often whitish on the inner surface, the lower three tepals pale greenish, the apices salmon, the sutures between the tepals thickened into iridescent papillose ridges, lightly rose-scented; *perianth tube* obliquely funnel-shaped, c. 11 mm long, the lower cylindric part c. 5 mm long; *tepals* unequal, the dorsal largest, c. 40 x 19 mm, lanceolate, initially arching over the stamens, when fully open ascending, curving slightly upward in the distal third, narrowed below into a claw-like base c. 8 mm long, the upper laterals more or less spade-shaped, c. 27 x 23 mm, with a narrow base 3–4 mm long, the limb broadly triangular, directed forward below, patent in the upper two-thirds, the lower three tepals united for c. 2–3 mm, the free parts c. 24 mm long, the lower laterals narrowly spathulate, narrowed below and channelled, not clearly clawed, gradually expanding above and plane, c. 5 mm at the widest, the lower median with a claw c. 8 mm long, the limb narrowly triangular, c. 12 mm wide. *Filaments* strongly arched, c. 22 mm long, exserted for c. 18 mm from the tube, shortly pilose; *anthers* c. 8.5 mm long, dull greenish yellow, reddish on the lines of dehiscence, the pollen yellowish. *Ovary* narrowly oblong and three-sided, c. 7 mm long; *style* arching over the stamens, dividing opposite the middle of the anthers, the branches c. 6 mm long, reddish, the apices white, spreading over the anthers. *Capsules* ellipsoid, three-sided, (20–)25–40 mm long; *seeds* ovate to fiddle-shaped, c. 8 x 6.5 mm,

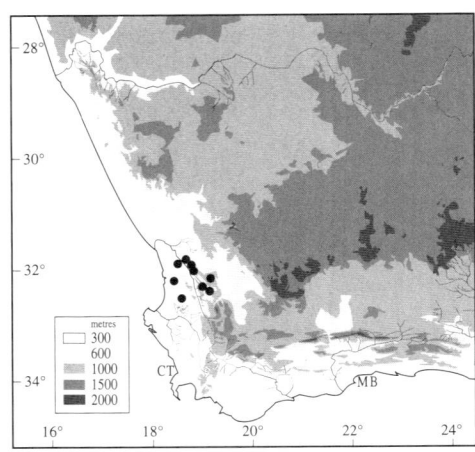

among tufted clumps of Restionaceae. The closely related *G. speciosus* often grows in the same habitats and sometimes at the same sites as *G. pulcherrimus*. When they co-occur their flowering times are separated, *G. speciosus* usually having finished flowering as *G. pulcherrimus* comes into bloom.

Nothing is known about the pollination biology of either of these two species, but we assume that they are both bee-pollinated like their close relative, *Gladiolus alatus*, which has similarly coloured and marked flowers and is equally strongly scented.

broadly and evenly winged, light translucent brown. ***Chromosome number*** unknown.

Flowering time late August to mid-October.

Distribution and biology

Gladiolus pulcherrimus occurs across the northwestern portion of Western Cape Province, South Africa, an area of low, exclusively winter rainfall. Relatively poorly collected, the species has been recorded along the lower Olifants River valley downstream from Clanwilliam, in the mountains south of Leipoldtville near Lambert's Bay, and in the dry valleys and slopes of the Cedarberg, Gifberg and Nardouw Mountains. Plants always grow in deep, well-drained, coarse sandy soil in arid fynbos vegetation, and are often found growing

Diagnosis and relationships

The bright salmon to orange flowers with the lower tepals lime-green in the lower half and minutely pubescent filaments place *Gladiolus pulcherrimus* in the *G. alatus* group of series *Hebea*. Within the alliance it stands out in having relatively broad, sword-shaped leaves with plane blades that lack visible venation, and flowers with an erect to suberect dorsal tepal that makes them look particularly large and showy. In a good season healthy plants are the largest in series *Hebea* and we have seen robust individuals standing more than 60 cm high and with two or three lateral branches. This species forms a link between the less specialized species of the *G. alatus* group with plane leaves and the remaining orange- to pink-flowered species with ridged leaves and suberect to erect dorsal tepals and pubescent filaments.

Gladiolus pulcherrimus is most likely to be confused with *G. alatus*, but can always be distinguished by its plane, glabrous leaf blades that lack prominent venation. The leaves of *G. alatus* always have strongly raised veins with scabridulous edges, and the plants themselves are rarely as robust, seldom reaching more than 20 cm high. Superficially, *G. pulcherrimus* can be confused with *G. equitans* because both have fairly broad leaves with plane blades. In *G. equitans* the leaves are falcate and have thickened margins and the flowers have an arcuate dorsal tepal and perfectly glabrous filaments.

History

Gladiolus pulcherrimus appears to have been first collected in the 1870s by P.A. Mader, a land surveyor in the Clanwilliam District who made several important early collections in the area. The first modern collections were made much later, in 1934 when plants were recorded at two sites in the Cedarberg by the South African botanist, Frances Leighton. These and a few other collections made subsequently were understandably confused with *G. alatus*, and it was only in 1972 that the species was accorded taxonomic recognition, as *G. alatus* var. *pulcherrimus* by G.J. Lewis.

111. GLADIOLUS ALATUS Linnaeus

Kalkoentjie, red kalkoentjie

PLATE 95

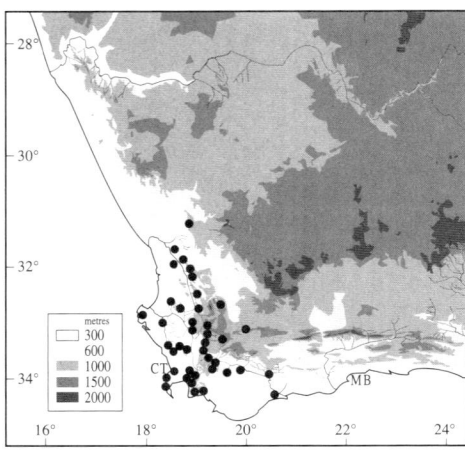

Gladiolus alatus Linnaeus, *Pl. Rar. Africae* 2 (1760). Baker, *Fl. Capensis* 6: 159 (1896). Lewis et al., *J. S. African Bot.*, Suppl. 10: 160 (1972). Type: South Africa, Cape, without precise locality, date or collector

(G—Herb Burman, lectotype designated by Lewis et al., 1972: 163).

alatus = winged, alluding to the flanged ridges or wings on the stem.

Synonymy

Gladiolus galeatus Burman fil., *Prodr. Pl. Cap.* 2 (1768). *Hebea galeata* (Burman fil.) Ecklon, *Topographisches Verzeichniss* 41 (1827), including *G. meliusculus*. Type: South Africa, Western Cape, without precise locality or collector (Herb. Burman, G [not seen], holotype – so indicated by Lewis et al., 1972).
Gladiolus papilionaceus Lichtenstein ex Roemer & Schultes, *Syst. Veg.* 1: 408 (1817), an illegitimate homonym, not *G. papilionaceus* Salisbury (1796) (= *G. cunonius* (Linnaeus) Gaertner). *Gladiolus alatus* var. *uniflorus* Klatt, *Abh. Naturforsch. Ges. Halle* 15: 337 (Ergänzungen 3) (1882), as a new name for *G. papilionaceus*

at varietal rank. Type: South Africa, Western Cape, without precise locality or collector (possibly *M.H. Lichtenstein* (Herb. Willdenow 916, B [not seen], holotype).
Gladiolus alatus var. *algoensis* Herbert, *Curtis's Bot. Mag.* 53: pl. 2608 (1825). Lewis et al., *J. S. African Bot.*, Suppl. 10: 166 (1972). *Gladiolus algoensis* (Herbert) Sweet, *Hort. Brit.* 397 (1827). Type: South Africa, Eastern Cape, vicinity of Algoa Bay (Port Elizabeth), without collector, grown in England, illustration in *Curtis's Bot. Mag.* 53: pl. 2608 (1825).

Description

Plants (8–)15–20 cm high. ***Corm*** depressed-globose, 12–16 mm in diameter, the tunics membranous, generally not accumulating, decaying into irregular fragments, occasionally more or less fibrous. ***Leaves*** three or four, rarely five, the lower two or three more or less basal, the lowermost longest and reaching to about the top of the spike, the others progressively smaller, the uppermost often more or less bract-like and sometimes without a blade, blades narrowly lanceolate to linear or falcate, 2–6(–9) mm wide, lightly two- to several-ribbed to markedly corrugate, rarely one-ribbed, often lightly scabrid-papillose on the veins or blade surface. *Stem* more or less erect below, flexed outward above the sheathing part of the third leaf, slightly compressed, ridged to winged at least on one side, or 2- or 3-ridged, simple or rarely with a short branch, often with a small cormlet in each of the leaf axils below ground, 3–5 mm in diameter below the spike.

Spike 2- to 5-, rarely to 8-flowered, flexed at the base and inclined, lightly flexuose; *bracts* 25–35 mm long, strongly keeled, the margins and keels purple, the inner slightly shorter than the outer, two-keeled, the keels often purple, apices entire. *Flowers* orange to scarlet, the lower tepals light yellow-green with the apices also orange, upper lateral tepals feathered dark red in the lower half, the sutures between the tepals thickened into iridescent papillose ridges, sweetly scented (rather like the orchid *Pterygodium*); *perianth tube* 10–14 mm long, obliquely funnel-shaped; *tepals* unequal, the dorsal largest, 35–40(–50) x 16–25 mm, broadly fiddle-shaped, ascending to nearly erect to weakly arching toward the stamens, occasionally almost horizontal, the margins undulate below, the upper laterals spade-shaped, 22–30 mm long, narrowed near the bases into short claws 1–3 mm long, abruptly expanded into triangular limbs 18–28 mm wide, curving outward in the upper half, lower three tepals united below for 3–5 mm, the free parts 20–26 mm long, the lower laterals narrowly oblanceolate, channelled below or throughout, widest just below the apices, 8–12 mm wide, the lowermost narrowed below into a claw-like base c. 8 mm long, the limb broadly triangular, 12–18 x 16-19 mm, as wide as or wider than both the lower laterals together. *Filaments* 20–26 mm long, exserted for c. 16–18 mm from the tube, strongly arched, lightly to densely pubescent in the upper half, minutely so below; *anthers* 6–10 mm long, dull green to yellowish with purple lines, pollen yellowish. *Ovary* oblong, 6–7 mm long, style arching over the stamens, sparsely pilose, dividing near the anther apices, the branches c. 5 mm long, extending well beyond the anthers. *Capsules* oblong-ellipsoid, (16–)20–25 mm long; *seeds* ovate to round, mostly 5–8 x 4–7 mm,

broadly and evenly winged, light glossy brown, the seed body darker brown.
Chromosome number 2n = 30.

Flowering time August and September, at higher elevations as late as early October.

Distribution and biology

Gladiolus alatus is a commonly encountered species of the southern African winter-rainfall zone. Its range extends from the Bokkeveld Escarpment near Nieuwoudtville in the north to the Cape Peninsula in the south and thence through the Caledon District to Bredasdorp. Plants favour stony sandstone or granitic soils and are often found growing in thin soils or rock crevices that remain wet during the growing season, but dry out quickly once the spring rains have ended. Its drought tolerance makes it a ready pioneer and plants can quickly colonize disturbed sites such as road verges and the edges of ploughed fields.

The brightly coloured and sweetly scented flowers of *Gladiolus alatus* are apparently pollinated primarily by the solitary bee, *Anthophora diversipes*. We have observed this bee at several sites foraging for nectar on flowers of *G. alatus*, and in so doing transferring pollen from one plant to another.

Diagnosis and relationships

A much-loved wild flower, *Gladiolus alatus* is readily recognized by its large, orange-red flower with yellow-green lower tepals tipped with orange, the short inclined stem, keeled floral bracts and long, ridged, corrugate leaf blades. It is one of the few really well-known wild flowers of the southern African winter-rainfall zone, almost everywhere called the *kalkoentjie* (little turkey), because of the fancied resemblance of the reddish tepals to the wattles of a turkey. The distinctive colour of the flowers is consistent across the range of the species, and is shared in *Gladiolus* by four other species, all closely related, that also have winged stems and keeled bracts. Only two members of this group, *G. alatus* and *G. meliusculus*, have leaves with ribbed blades. The remaining species, *G. equitans*, *G. pulcherrimus* and *G. speciosus*, have more or less plane leaf blades with the midribs and other veins scarcely evident in the live state and not at all thickened. Both *G. alatus* and *G. meliusculus* have specialized pilose filaments and styles, a character also shared by *G. pulcherrimus*, and all three also have an inclined rather than hooded and concave dorsal tepal. *Gladiolus pulcherrimus* is easily distinguished by its plane leaves and often robust habit, but *G. meliusculus* is more

difficult to recognize. It has shorter filaments, c. 15 mm long, compared with those of *G. alatus* which are about 22 mm long, and the lower tepals are typically broader and have dark reddish markings at the edge of the yellow zone. Except for the Namaqualand *G. equitans*, treatment of the members of this alliance has been controversial. We believe that their complex patterns of characters and overlapping ranges but subtly different ecological preferences – including different habitats, vegetative propagation strategies, flowering times and/or pollinators – make it important to distinguish them at species rank.

History

The history of the *Gladiolus alatus* complex is an extremely confused one. This is largely because early botanists either misapplied the name *G. alatus* to what we now know as *G. orchidiflorus*, or they failed to distinguish the several allied species in ignorance of their ecology, geography and subtle character differences. A common plant of the southwestern Cape, *G. alatus* was recorded by all the early plant collectors and was one of the first Cape plants recorded in the older botanical literature, appearing first in Jacob Breyne's

Prodromus fasciculi rariorum plantarum published in 1739. *Gladiolus alatus* was formally described by Carl Linnaeus in 1760 and was most probably based primarily on a specimen of unknown provenance in the Burman Herbarium (Lewis et al., 1972: 163). Linnaeus did, however, cite an early black and white illustration in the protologue, Plukenet's *G. viperatus*, which is actually a plant with plane leaves and a hooded dorsal tepal. Plukenet's figure is a copy of a painting in the *Codex Witsenii*, the report of Simon van der Stel's 1685 expedition to the copper mountains of Namaqualand. Entitled *G. esculentus* in that volume, the species is *G. speciosus*.

N.L. Burman evidently believed that the plant he described as *Gladiolus galeatus* in 1768 was distinct from *G. alatus*, the type of which was in his own collection. Burman cited Breyne's illustration in the *Prodromus, etc.* in the protologue of *G. galeatus* but oddly, Linnaeus did not mention this illustration when he described *G. alatus*. The name *G. alatus*, however, became associated with another species, *G. viperatus*, probably because Plukenet's illustration was available for comparison whereas the type specimen of

G. alatus, for many years in the Leiden Herbarium, was not. To add to the confusion, the name *G. alatus* was applied to a third species, now *G. orchidiflorus*, which also has plane leaves. Thus the fine painting of *G. alatus* in Jacquin's *Icones Plantarum Rariorum* is identified as *G. orchidiflorus*.

C.P. Thunberg, however, clearly distinguished *Gladiolus alatus* with orange flowers and ribbed leaves from both *G. orchidiflorus* and the plane-leafed *G. speciosus* which he described in 1811. Any remaining confusion was resolved when *G. orchidiflorus* and then its synonym, Ker Gawler's *G. viperatus*, were formally described in 1802 and 1803 respectively, and the name *G. alatus* was applied to orange-flowered species of series *Hebea* by most careful workers after this time. It is not at all clear, however, why the botanists J.J. Roemer and J.A. Schultes should have described *G. papilionaceus* in 1817 for a plant most likely collected by the explorer M.H. Lichtenstein. The type appears to differ in no significant way from *G. alatus*, although it is unusually low in stature. The name is, moreover, a homonym for *G. papilionaceus* Persoon, a later synonym of *G. cunonius*. *Gladiolus papilionaceus* was

recognized as a variety of *G. alatus* by F.W. Klatt, who chose the new epithet, *uniflorus*, at varietal rank.

J.G. Baker (1892, 1896) admitted only one species in the orange-flowered *Gladiolus alatus* complex, but he did recognize *G. equitans* as var. *namaquensis*. The latter was treated as a separate species by G.J. Lewis et al. (1972) who included five varieties in *G. alatus*. One of these, var. *algoensis*, originally described by William Herbert in 1825, was accorded species rank by Robert Sweet in 1927. Except for var. *algoensis*, the varieties recognized by Lewis et al. are here treated as separate species. The identity of var. *algoensis* is puzzling. Although said to have been found near Port Elizabeth, there are no reliable records of any member of the *G. alatus* group from this far east of Cape Town. The leaves are described as hard, rigid and densely ribbed, and the dorsal tepal is arched and rather weakly hooded. It appears to be nothing more than a variant of *G. alatus* with a slightly pinker flower. We suspect that if, indeed, this plant did come from the Port Elizabeth area, it was either cultivated or accidentally introduced there.

112. GLADIOLUS MELIUSCULUS (G. Lewis) Goldblatt & Manning

Pink kalkoentjie, Darling kalkoentjie

PLATE 96

Gladiolus meliusculus (G. Lewis) Goldblatt & Manning, new combination and rank. *Gladiolus alatus* var. *meliusculus* G. Lewis in *J. S. African Bot.*, Suppl. 10: 166 (1972). Type: South Africa, Western Cape, Malmesbury District, Ganzekraal road, 15 Sept. 1940, *Barker 2000* (NBG, holotype).

meliusculus = a little better, presumably because it was thought somewhat prettier

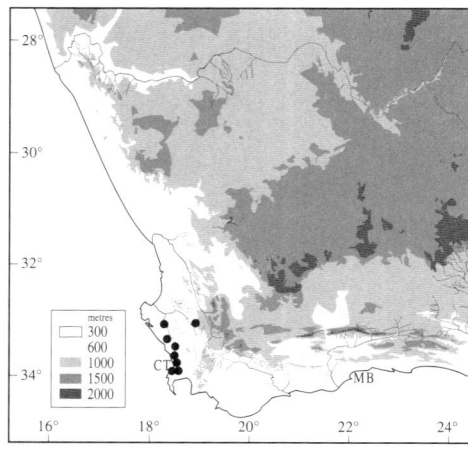

than *Gladiolus alatus*, in which species it was originally described as a variety.

Description

Plants 12–20(–27) cm high. *Corm* depressed-globose, 6–10 mm in diameter, the tunics membranous, light brown, usually bearing 1–2 cormlets on short stolons up to 10 mm long. *Cataphylls* membranous, the uppermost often purple above the ground. *Leaves* four, the lower two or three basal, the lowermost longest, as long as or longer than the spike, blades of the basal leaves linear, (1.8–)3–5 mm wide, strongly ridged, the uppermost leaf inserted on the stem usually above the middle, usually channelled throughout, the surface smooth, diverging from the stem and sheathing the stem only near the base. *Stem* usually inclined from the base or erect below, flexed outward above the sheath of the third leaf, unbranched, slightly compressed, two-ridged to winged below the uppermost leaf, usually three-ridged or winged below the first flower, usually with a small cormlet in each of the axils below ground, 2–3 mm in diameter below the spike.

Spike inclined, occasionally 2-, usually 3- to 6-flowered; *bracts* pale green, the outer 17–30 mm long, keeled on the midline, the keel and margins red, the inner about three-quarters to nearly as long as the outer, usually two-keeled, the keels also reddish, minutely forked apically. *Flowers* pink to brick-red or orange, the lower tepals each yellow to greenish below, the yellow usually broadly edged with dark purple, sometimes the purple only a narrow line, salmon to orange in the distal third to half, the throat with glistening papillose ridges on the sutures of the tepals, sweetly honey-scented; *perianth tube* obliquely funnel-shaped, c. 11 mm long, mostly included in the bracts; *tepals* unequal, the dorsal largest, oblanceolate, 30–38 x 15–18 mm, arching forward or more or less erect when fully open, upper lateral tepals spade-shaped, shortly clawed for c. 2 mm, the limbs abruptly expanded, 20–30 x c. 20 mm, directed forward below, gently curving outward, spreading in the upper third to half, the lower tepals united for 3–4 mm, clawed below, the lower laterals c. 23 mm long, the claws channelled, the limbs 12–16 x c. 12 mm, deflexed, lower median with shorter claws, the limbs 15–18 x 14–16 mm. *Filaments* c. 15 mm long, exserted c. 10 mm from the tube, strongly arched, minutely pilose; *anthers* 5–7 mm long, purple or buff, pollen cream to yellowish. *Ovary* oblong, 6–7 mm long; *style* arching over the stamens, minutely pubescent, dividing between the base and middle of the anthers, the branches c. 6 mm long. *Capsules* ovoid-ellipsoid, 20–26 mm long; *seeds* ovate, broadly and evenly winged, 6–8 x 5–6 mm, light pinkish brown. *Chromosome number* unknown.

Flowering time mid-August to early October.

Distribution and biology

Gladiolus meliusculus is restricted to low hills and flats on the western Cape coastal plain. In past times it extended from the northern end of the Cape Peninsula along the west coast to Hopefield and inland to Malmesbury and Porterville. It no longer grows on the Peninsula and what little of its native habitat remains is much disturbed, threatened by urban expansion, expanding agriculture and weedy alien plants. Only in the Darling District on farms such as Waylands and Oudepost is it likely to persist into the twenty-first century. There *G. meliusculus* is protected on privately owned nature reserves that are large enough and appropriately managed through judicious grazing practices to allow this and several other rare species to maintain viable populations and even to increase their numbers.

Plants favour sandy sites with soils derived from decomposed granite or sandstone where they remain waterlogged for much of their growing period.

Preliminary observations on the pollination biology of the species suggest that unlike its close relative, *Gladiolus alatus*, which is an exclusively bee-pollinated species, *G. meliusculus* may be pollinated largely by monkey beetles (Scarabaeidae: Hopliini). The dark red-purple markings distal to the yellow band of colour at the bases of the lower tepals are characteristic of flowers pollinated by monkey beetles. The markings on the lower tepals of

G. meliusculus closely resemble in colour those of beetle-pollinated species of *Romulea* (*R. obscura, R. eximia*) that often grow nearby. Two beetles, in particular, have been encountered on flowers of *G. meliusculus*, both sufficiently hairy to have pollen accumulate on their bodies and be transported to other flowers as the beetles fly from plant to plant foraging for pollen or in search of mates. The flowers are also occasionally visited by short-tongued bees of the family Andrenidae, apparently searching for accessible nectar.

Diagnosis and relationships

Easily confused with the closely related *Gladiolus alatus, G. meliusculus* has similarly shaped flowers and the same general appearance. The leaves of the two species are identical, being narrow and linear to falcate and strongly ribbed, and the flowers have the same colouring and overall appearance. They are salmon, brick-red or orange with yellow markings on the lower tepals. Close examination of the markings of the two species, however, reveals subtle differences between them. While *G. alatus* has the lower tepals yellowish green for the lower two-thirds of their length and are salmon to orange only toward their apices, *G. meliusculus* has the yellow colour less pronounced and edged by a narrow to broad band of red-purple, and the upper halves are pink to orange. The colour differences are enhanced by the broader tepals of *G. meliusculus*, the lower median being 12–16 mm wide and the lower laterals c. 12 mm wide.

More important than the colour differences is the floral structure. The stamens of *Gladiolus meliusculus* are comparatively short, with the filaments about 15 mm long and the anthers 5–7 mm long. This is markedly different from *G. alatus* which has filaments 20–26 mm long and anthers usually more than 7 mm and up to 10 mm long. Attention to these differences in flower size make it easy to distinguish the two species, even when colour variation in both renders identification doubtful.

The geographical range of *Gladiolus meliusculus* lies entirely within that of *G. alatus* and the two species may grow only

short distances apart and in what appear to be identical habitats. When this is the case, *G. alatus* has usually completed its flowering and is in fruit when *G. meliusculus* begins to flower. The shift in flowering time, the apparently different pollination system and the differences in flower structure were important considerations in our decision to recognize *G. meliusculus* as separate from *G. alatus*.

History

Gladiolus meliusculus has been known since at least the later eighteenth century, when it was found in the 1770s by C.P. Thunberg and in the 1780s by Franz Boos and Georg Schol, who collected plants for Nicholas Jacquin. It was also recorded by J.F. Drège in the 1820s. Drège's gathering was, according to the collection information, made on Lion's Head on the Cape Peninsula where it no longer grows. Despite its obvious differences in flower colour and tepal markings and the less obvious disparity in stamen length, the plant did not attract the attention of botanists. It was only in 1972 that *G. meliusculus* was recognized taxonomically by G.J. Lewis, and then only at varietal rank in *G. alatus*.

7. Section *HOMOGLOSSUM*

(Salisbury) Goldblatt & Manning

Section *Homoglossum* (Salisbury) Goldblatt & Manning, new combination. *Genera Plants* 143 (1866). *Homoglossum* section *Homoglossum* De Vos, *J. S. African Bot.* 42: 339 (1976). Type species: *H. watsonium* (Thunberg) N.E. Brown = *G. watsonius* Thunberg.

Synonymy

Homoglossum section *Quadrangulifolium* De Vos, *J. S. African Bot.* 42: 346 (1976). Type: *Homoglossum quadrangulare* (Burman fil.) N.E. Brown.

Description

Plants usually medium-sized, sometimes small. *Corm* globose to conic, often hard and woody, consisting of concentric layers fragmenting above and below, or coriaceous to fibrous, occasionally membranous. *Cataphylls* usually green above the ground, smooth or rarely minutely puberulous to scabrid, sometimes mottled with purple. *Leaves* two to four, superposed, the lowermost always longest and often sheathing the stem for some distance above the ground, in a few species the leaves of the flowering stem reduced but non-flowering plants producing long-bladed foliage leaves, the blades either linear, sometimes narrowly sword-shaped, or terete with four narrow or wide longitudinal grooves, the margins and midrib usually somewhat thickened and other veins not evident, sometimes the margins and midrib so thickened that the blade is cross-shaped in section, in a few species the margins raised into flat wings, or both midrib and margins not evident and the blade somewhat fleshy or leathery. *Stem* unbranched, usually flexed near the base of the spike.

Spike either flexed at the base and inclined or less often erect, usually flexuose, but straight in some species; *bracts* green throughout or flushed with grey to purple above. *Flowers* medium-sized to large, usually bilabiate, the lower tepals often narrow and clawed at the base, with nectar guides various but primitively consisting of dark longitudinal lines and spots over a paler background, sometimes consisting of pale transverse bands edged in dark colour or a dark median streak, scented in most species; *perianth tube* usually obliquely funnel-shaped, sometimes tubular, half to two-thirds as long as the dorsal tepal or much

longer; *tepals* subequal or unequal, then the upper three broader but usually about as long as the lower, the dorsal largest, arching over the stamens, sometimes more or less horizontal, the lower three usually joined to the upper laterals and to one another for a short distance. *Capsules* oblong-ellipsoid; *seeds* fairly large, broadly and evenly winged, translucent light or dark brown.

A large section comprising 51 species, *Homoglossum* is almost entirely restricted to the winter-rainfall region of southern Africa. Only a handful of species occur outside the Cape Flora Region, and most of these extend only locally eastward to Grahamstown and East London. The exceptions are *Gladiolus longicollis* which extends from Oudtshoorn in the southern Cape to Northern Province, South Africa, and Swaziland, and the high Drakensberg endemic, *G. symonsii*. We divide the section into seven informal series, all but one of which we believe are natural, that is, monophyletic alliances. The exception, series *Carinatus*, may be a residual group comprising two diverse species groups.

The basic features of section *Homoglossum* are an unbranched stem and a reduced number of foliage leaves, no more than four and sometimes only two. The leaves are always superposed (i.e., inserted distantly from one another), instead of forming a distichous fan, and the blades themselves are narrow, and linear, terete or narrowly sword-shaped, with only the midveins visible externally. Corm tunics are frequently hard and woody or coarsely fibrous, but finely fibrous or membranous tunics are thought to be the primitive state. The flowers, diverse in shape and colour, are adapted for different pollination strategies, often independently evolved in different series. Scented flowers are common and probably the basic state for the section.

Species of series *Carinatus* and series *Mutabilis* have plesiomorphic papery or fibrous corm tunics. In series *Carinatus* there appear to be two assemblages, the *Gladiolus atropictus* and *G. carinatus* groups. The latter has derived mottled cataphylls and only three foliage leaves. The *G. atropictus* group, including *G. violaceolineatus* and *G. comptonii*, has elongate-conic corms and attenuate tepals, the lower of which are strongly channelled. The leaf blades in series *Mutabilis* are unique in the genus, being thick and leathery and lacking thickened margins or

midribs. There are often only three leaves per plant in species of the series and just two in *G. vaginatus*.

The taxonomically isolated *Gladiolus brevitubus*, which probably belongs in section *Homoglossum*, is placed alone in series *Brevitubus*. It has no close relatives and the very short-tubed orange flowers and fine fibrous corm tunics do not help to place the species in any of the other series of the section.

Hard woody corms tunics characterize most of the remaining species of section *Homoglossum*. Species of series *Gracilis* comprise two assemblages: the *Gladiolus gracilis* group, in which the margins of the leaf blades are raised into wings extending at right angles to the blade surface; and the *G. debilis* group, which all have flowers in shades of pink with dark, usually red, nectar guides. Most of the 11 species of series *Teretifolius* have leaf blades cross-shaped in transverse section with both the midrib and margins thickened. The longitudinal grooves between the margins and midrib are usually narrow, thus rendering the leaves terete and quite thin. Three autumn-flowering species, *G. jonquilliodorus*, *G. martleyi* and *G. subcaeruleus*, have the leaf blades of the flowering stems reduced or absent, but foliage leaves are produced later in the season. Species of series *Homoglossum* and series *Tristis* consistently have only three foliage leaves, the blades of which have heavily thickened midrib and margins, the former sometimes conspicuously raised so that the leaves are cross-shaped in transverse section.

The most common pollination system in section *Homoglossum* is that for long-tongued bees. Nectar is the usual reward and both colour and scent fulfil the role of attracting insect visitors. Bees most often involved are species of *Anthophora* and *Amegilla* (Anthophoridae) or *Apis mellifera* (Apidae). Flowers are often shades of blue, or occasionally pink or even yellow, and the lower tepals have nectar guides consisting of dark spots or streaks on a pale background. A few species, including *G. debilis*, *G. vigilans* and *G. virgatus* (series *Gracilis*) and *G. cylindraceus* and *G. engysiphon* (series *Teretifolius*), have white or pale pink flowers with red markings and elongate perianth tubes. Pollination by long-tongued flies (Nemestrinidae or Tabanidae) has been confirmed for *G. engysiphon*, *G. vigilans* and *G. virgatus*. Again nectar, fairly large amounts of which are produced, is the reward offered to these flies. Moth pollination has arisen in some series, including *G. recurvus* (series *Gracilis*), *G. albens* and *G. maculatus* (series *Mutabilis*), and all except *G. symonsii* of series *Tristis*, which

has specialized flowers producing scent only in the evening. Flowers adapted for moth pollination are often quite large, have long tubes, produce quantities of comparatively sweet nectar, and have heavy scents with a strong clove-like component. Flowers adapted for sunbird pollination occur in *G. meridionalis* and *G. priorii* (series

Mutabilis) and series *Homoglossum*, but the syndrome has been poorly studied. Flowers of these species are red to orange, have elongate tubes which are wide and cylindric in the upper half, and they normally lack scent. In series *Homoglossum* the stems are thick and erect and the spikes straight, features associated with sunbird pollination.

Section *Homoglossum*: Series *Carinatus*

113. GLADIOLUS ATROPICTUS Goldblatt & Manning

Gladiolus atropictus Goldblatt & Manning, new species. Type: South Africa, Western Cape, MacGregor, pass to Die Galg, 25 July 1993, *Manning 2010* (NBG, holotype; PRE, isotype).

atropictus = darkly painted, for the dark purple network of lines on the lower tepals that constitute the nectar guide.

Latin diagnosis

Plantae 34–60 cm altae, cormo conico c. 15 mm in diametro, tunicis papyraceis, foliis 4, duobus infimis basalibus laminis linearibus, 1.5–2 mm latis marginibus costisque leviter incrassatis, caule eramoso, spica flexuosa 1–2 florum, bracteis 26–35 mm longis, floribus violaceis tepalis inferioribus atrolineatis, tubo perianthii infundibuliformi c. 13 mm longo, tepalis inaequalibus lanceolato-attentuatis dorsali 30–33 x c. 16 mm, inferioribus 18–21 x 8–9 mm, filamentis c. 12 mm longis, antheris 10–11 mm longis.

Description

Plants 34–60 cm high. *Corm* conic, c. 15 mm in diameter, the tunics of concentric, firm-papery layers, fragmenting with age into firm longitudinal strips or fairly coarse fibres, bearing one or two small cormlets at the base. *Cataphylls* pale and membranous, the uppermost reaching up to 15 cm above the ground and then green or purple. *Leaves* four, the lower two basal, the lowermost longest, reaching to between the middle and upper third of the stem, sheathing the lower half of the stem, the blade linear, 1.5–2 mm wide, the margins and midrib moderately thickened and raised, the second leaf sheathing the stem to about the upper third, the blade shorter than the basal, the remaining two leaves much shorter, sheathing for half their length, the blades short but like the basal, reaching to about the base of the spike, the lower margins united around the stem. *Stem* erect, flexed

outward above the sheaths of the three upper leaves and inclined c. 30° in the upper third, unbranched, c. 1 mm in diameter below the spike.

Spike flexuose, 1- or 2-flowered, inclined c. 30°; *bracts* dull greenish grey with broad membranous margins, the outer 26–35 mm long, the inner slightly shorter than the outer, acute or minutely notched apically. *Flowers* violet, the lower three tepals pale yellow in the lower two-thirds and feathered with dark violet, strongly scented of freesia and violet; *perianth tube* obliquely funnel-shaped, c. 13 mm long, the lower cylindrical part enclosed in the bracts; *tepals* unequal, the upper three subequal or the dorsal largest, larger than the lower, lanceolate-attenuate, the dorsal 30–33 x c. 16 mm, horizontal and the margins strongly curving downward, the upper laterals 28–30 x 12–14 mm, directed forward below, curving outward in the distal third, the lower three tepals joined to the upper laterals for c. 10 mm, not notably narrowed into claws below but channelled in the lower halves, the lower laterals 18 x 8 mm, the lower median longer than the lower laterals, c. 21 x 9 mm, all curving outward below, and downward distally, in profile the lower tepals exceeding the upper by c. 8 mm. *Filaments* c. 12 mm long, exserted 5–8 mm from the tube but concealed in the flower; *anthers* 10–11 mm long, extended horizontally, appressed to the dorsal tepal and concealed, pale cream, the pollen whitish. *Ovary* oblong, c. 5 mm long; *style* arching over the stamens, just below the apex of the anthers, the branches c. 3.5 mm long, curving outward beyond the anthers. *Capsules* and *seeds* unknown. *Chromosome number* unknown.

Flowering time mid-July to mid-August.

Distribution and biology

Native to the winter-rainfall region of South Africa, *Gladiolus atropictus* is known only from two sites in the southwestern Cape.

Populations occur on the fairly dry northern slopes of the Riviersonderend Mountains below Jonaskop and near Die Galg, a short distance from the town of McGregor. The entire known range thus encompasses some 20 km. Plants appear to be rare and scattered, growing in pockets of deeper soil on rocky sandstone slopes.

The flowers of *Gladiolus atropictus* have all the marks of being adapted for pollination by long-tongued bees and have a particularly strong, sweet scent.

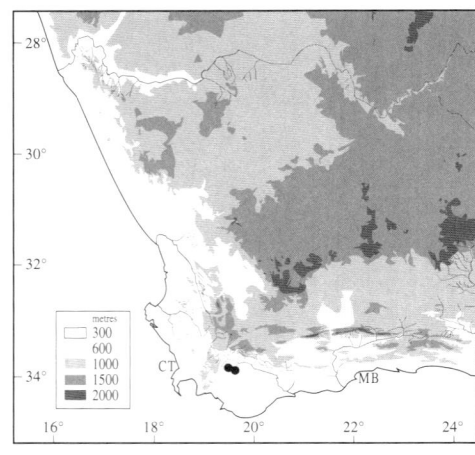

Diagnosis and relationships

Gladiolus atropictus is readily recognized by its unusually dark blue-violet flower, the lower lateral tepals pale yellow below and distinctively marked with dark purple lines in a feathery pattern. The rather weak, inclined stems bear only one or two flowers, the tepals of which are lanceolate and attenuate, with the dorsal tepal more or less horizontal, deeply channelled and concealing the stamens. The flowers are very strongly scented of a combination of freesia and violet, perhaps the finest scented species of section *Homoglossum*.

The stems bear four foliage leaves, the lower two with fairly short blades with thickened margins and midrib. The corms of *G. atropictus* are unusual in that they are elongate-conical with firm-papery tunics. The relatively unspecialized leaves and the corm tunics which are firm but not woody place the species in section *Carinatus* of section *Homoglossum*. Both the lanceolate-attenuate tepals and conical corms suggest a close affinity with the two western Cape species, *G. violaceolineatus* and *G. comptonii*.

History

The first record of this rare *Gladiolus* was apparently made by one of the authors, Peter Goldblatt, on the upper slopes of Jonaskop near the western end of the Riviersonderend Mountains in August 1978. The collection was then identified as *G. gracilis*. A second population was discovered in 1993 when the potter, Jane Banks, of MacGregor in the Worcester District, told us about the existence of an unusual species of *Gladiolus* growing on the lower slopes of the Riviersonderend Mountains nearby, some 20 km east of Jonaskop. After visiting the site then and again in 1996, we came to the conclusion that this plant did, indeed, represent a new species of *Gladiolus*. The existence of this species, apparently first collected in 1978 and only once since then, is surprising coming from what is considered to be a fairly well-botanized area.

114. GLADIOLUS VIOLACEOLINEATUS G. Lewis

Streep pypie

PLATE 97

Gladiolus violaceolineatus G. Lewis in *J. S. African Bot.*, Suppl. 10: 233 (1972). Type: South Africa, Western Cape, Clanwilliam District, Alpha farm, west slope, 28 July 1963, *Lewis 6142* (NBG, holotype; K, isotype).

violaceolineatus = lined or streaked with violet, referring to the dark violet veins on all the tepals.

Description

Plants 30–50 cm high. *Corm* conic, 8–10 mm in diameter, the tunics of concentric, cartilaginous to papery layers, fragmenting with age into firm longitudinal strips or fairly coarse fibres. *Cataphylls* pale and membranous, the uppermost reaching up to 10 cm above the ground and then green or purple, sometimes minutely puberulous. *Leaves* four, the lower one or two basal, the lowermost longest, reaching to between the upper third of the stem and the base of the spike, sheathing the lower half of the stem, the blade linear, 1–2(–4) mm wide, straight, three-angled, the abaxial margin with raised edges extending outward at right angles to the laminar surface, the midrib and adaxial margin not thickened or raised, the remaining leaves progressively shorter, sheathing for half their length, the blades short, more or less plane, the uppermost leaf usually channelled throughout, reaching to the base of the spike, the lower margins united around the stem. *Stem* erect, flexed outward above the sheaths of the three upper leaves and inclined c. 30° in the upper third, unbranched, 0.8–1 mm in diameter below the spike.

Spike flexuose, occasionally 1-, usually 2- to 4-flowered, inclined c. 30°; *bracts* green to dark grey-green with broad membranous margins, the outer (20–)28–33(–50) mm long, the inner about two-thirds as long as the outer and lightly notched apically. *Flowers* pale blue, all the tepals streaked, speckled and feathered with dark violet, the lower three tepals white to yellow below and more strongly marked than the upper, the base of the tepals and the upper part of the tube speckled with red spots, lightly sweet-scented; *perianth tube* obliquely funnel-shaped, 12–15 mm long, the lower cylindrical part enclosed in the bracts; *tepals* unequal, the upper three subequal or the dorsal largest, larger than the lower, narrowly lanceolate-attenuate, the dorsal 26–39 x 9–18 mm, the upper laterals c. 26–37 x 8–11 mm, directed forward below, curving outward in the distal third, the lower three tepals joined to the upper laterals for c. 5 mm and to one another for c. 3 mm, not noticeably narrowed into claws below but deeply channelled in the lower halves, the lower laterals 18–30 mm long, the lower median longer than the lower laterals, 24–32 mm long, all curving outward and downward in the distal third, in profile the lower tepals exceeding the upper by c. 10 mm. *Filaments* 13–15 mm long, exserted 5–10 mm from the tube but included in the flower; *anthers* 9–11 mm long, extended horizontally, light grey, the pollen whitish. *Ovary* oblong, c. 6 mm long; *style* arching over the stamens, dividing between the upper third and apices of the anthers, the branches c. 3.5 mm long, curving outward beyond the anthers. *Capsules* unknown; *seeds* ovate, 7–8 x 5–6 mm, broadly and evenly winged, opaque light brown. *Chromosome number* unknown.

Flowering time July to mid-August, occasionally later at higher elevations.

Distribution and biology

Endemic to the southern African winter-rainfall zone, *Gladiolus violaceolineatus* has a relatively narrow range in the interior mountain ranges of the northwestern part of Western Cape Province, South Africa. Nowhere common, the species has been recorded from Elandskloof in the south through the Cedarberg and Pakhuis Mountains to the Gifberg in the north. Plants grow on rocky sandstone slopes, most often occurring on south-facing slopes where soils retain moisture throughout the growing season, from May to August. Plants typically flower relatively early in the season, usually in August, but flowering specimens have been seen at higher elevations in the Cedarberg, as late as early October.

We assume that the flowers of *Gladiolus violaceolineatus* are pollinated by the same suite of long-tongued bees that pollinate other Cape species of *Gladiolus* with similarly shaped and coloured flowers. Like many of these species, flowers of *G. violaceolineatus* produce a sweet, pleasant scent.

Diagnosis and relationships

Gladiolus violaceolineatus is readily recognized by its unusual flowers and is distinctive as well in its leaves and corms. The flowers are pale blue and have elongate, tapering and attentuate tepals, both the upper and lower of which are characteristically marked with dark violet lines, the feature for which the species is named. The flowers are also sweetly scented, as are most members of section *Homoglossum*. The corms have unspecialized coriaceous tunics, but are unusual for *Gladiolus* in being rather elongate and conical, a feature shared with only a few other species, all in section *Homoglossum*. These include *G. atropictus* and *G. comptonii* to which we assume it is immediately related, not only because of the corm shape, but also because all three have unusually elongate, tapering tepals.

The leaf blades of *Gladiolus violaceolineatus* are perhaps its most remarkable feature. They are linear, and apparently plane on casual examination. The abaxial margin, however, has raised edges extending outward at right angles to the laminar surface, but the midrib and adaxial margin are not thickened or raised, the result being a more or less three-sided blade. Relatively unspecialized in section *Homoglossum*, except for its peculiar

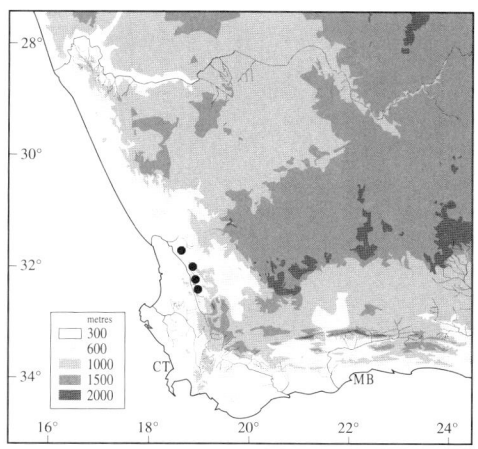

leaves, *G. violaceolineatus* is assigned to the basal series *Carinatus* of the section.

History

Comparatively rare, at least at sites readily accessible from passable roads, *Gladiolus violaceolineatus* was apparently first recorded on Nieuwoudt Pass in the Cedarberg in July 1937 by B.E. Martin, then a horticulturist at Kirstenbosch Botanic Gardens in Cape Town. C.L. Leipoldt, who collected extensively in the northwest Cape in the 1940s, found plants on the Gifberg a few years later, extending its known range well to the north. Despite its unusual appearance and obvious differences to any known species of the genus, *G. violaceolineatus* did not attract botanical attention for many years. It was only in 1972 that the protologue was published in the revision of the South African species of *Gladiolus* by Lewis et al. (1972).

115. GLADIOLUS COMPTONII G. Lewis
PLATE 98

Gladiolus comptonii G. Lewis in *J. S. African Bot.*, Suppl. 10: 235 (1972). Type: South Africa, Western Cape, Heerenlogement Mountain, 26 July 1963, *Lewis 6144* (NBG, holotype; K, MO, isotypes).

comptonii, named in honour of Professor R.H. Compton, founder of the Compton Herbarium at Kirstenbosch Gardens, and a major collector of southern African flora.

Description

Plants 30–45 cm high. *Corm* globose-ovoid, 15–20 mm in diameter, the tunics of coriaceous layers, fragmenting with age to become fairly coarsely fibrous. *Cataphylls* pale and membranous, the uppermost reaching up to 10 cm above the ground and then green or purple and minutely puberulous.

Leaves four, the lower one or two basal, the lowermost longest, reaching up to the base of the spike, sheathing the lower half of the stem, the sheath minutely puberulous, blade linear, 1.5–3 mm wide, straight, the midrib and margins lightly raised, the remaining leaves progressively shorter, sheathing for at least half their length, the blades like the basal but short, the uppermost leaf usually channelled throughout, reaching to the base of the spike, the lower margins united around the stem. *Stem* erect, flexed outward above the sheaths of the two or three upper leaves and inclined in the upper third, unbranched, 1–1.5 mm in diameter below the spike.

Spike flexuose, occasionally 1-, usually 2- to 3-flowered, inclined c. 20°; *bracts* green to dark grey-green with broad membranous margins, the apices attenuate, the outer 35–60 mm long, the inner half to two-thirds as long as the outer and lightly notched apically. *Flowers* bright yellow, the upper tepals streaked and spotted reddish in the lower third, the lower lateral tepals streaked and spotted with brown in the lower halves, lightly sweet-scented; *perianth tube* obliquely funnel-shaped, 11–14 mm long, the lower cylindrical part enclosed in the bracts; *tepals* unequal, narrowly lanceolate-attenuate, the upper three subequal or the dorsal slightly larger, the dorsal 40–50 x 13–15 mm, the upper laterals 36–44 mm long, directed forward below, curving outward in the distal third, the lower three tepals joined to the upper laterals for c. 4 mm and to one another for c. 4 mm, slightly narrowed and deeply channelled below, the lower laterals 33–40 x

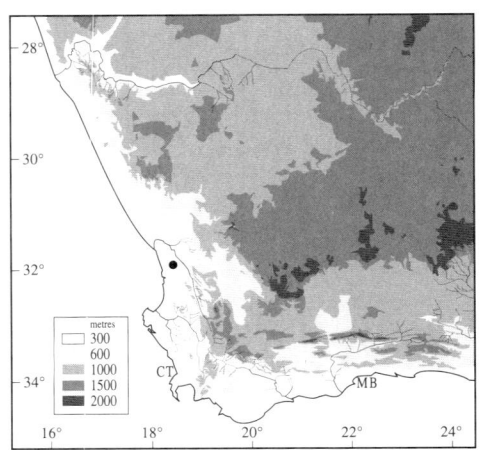

Flowering time mid- to late July, possibly also in late June.

Distribution and biology

Gladiolus comptonii is a highly localized endemic of the mountains of the northwestern Cape, South Africa. The few records of the species are all from the Heerenlogementberg at the northwestern end of the Olifants River Mountains. Plants grow on the upper slopes of the mountain at 600–700 m elevation in rocky sandstone soil in typical mountain fynbos vegetation. They flower relatively well in mature vegetation, and there is no information available on the flowering pattern in response to fire. Because *G. comptonii* is known from a single population, probably of less than one hundred individuals, it must be considered an endangered species. Exploration of nearby mountains may reveal additional populations, but there can be no doubt that the range of *G. comptonii* is not much wider than the present record indicates. The adjacent higher peaks of the Olifants River Mountains as well as the Cedarberg and Nardouw Mountains where suitable habitats for *G. comptonii* occur are fairly well explored botanically.

We assume that the short-tubed flowers of *Gladiolus comptonii* are pollinated by long-tongued bees, as are most species of *Gladiolus* with flowers of a similar size and shape.

Diagnosis and relationships

Gladiolus comptonii is readily recognized by its slender habit, plane, linear leaves with thickened margins and midrib, and striking bright yellow, short-tubed flowers. The lower tepals have what we consider the primitive nectar guides for section *Homoglossum*, consisting of dark longitudinal streaks and lines in the lower halves running onto the lower throat. The long, narrow, attenuate tepals and strongly defined longitudinal nectar guides recall, in particular, *G. violaceolineatus*, which grows nearby in the Olifants River Mountains and Cedarberg. In addition, the two share unusual elongate-conical

corms with leathery tunics. We believe they are immediately related to one another, and that they are allied to *G. atropictus* which has the same flower type and corm. *Gladiolus comptonii* can readily be distinguished from blue- to mauve-flowered *G. violaceolineatus* not only by the difference in flower colour, but by the shape of the leaf blades. Unlike the narrow blades of *G. comptonii*, which have thickened margins and midrib, the leaves of *G. violaceolineatus* are more or less triangular in transverse section, the margin on one side being raised.

History

Gladiolus comptonii was first collected by Professor R.H. Compton in July 1941 at its only known locality, the Heerenlogementberg, an isolated mountain at the northern end of the Olifants River Mountains. A large cave there provided shelter for early travellers on the route north to Namaqualand and hence the mountain was probably explored by many of the early plant collectors that visited the northern Cape. *Gladiolus comptonii* was immediately recognized as a novelty, but it took some 30 years before it was described in the *Revision of the South African Species of Gladiolus* (Lewis et al., 1972). Authorship there is attributed to Lewis alone. Lewis re-collected the species in 1963 when she made the type gathering.

8–9 mm, the lower median slightly longer than the lower laterals, 34–42 x 11–13 mm, all curving outward and downward distally, in profile the lower tepals exceeding the upper by c. 10 mm. *Filaments* c. 13 mm long, exserted 8–10 mm from the tube but concealed in the flower; *anthers* 9–11 mm long, yellow, the pollen yellow. *Ovary* oblong, c. 5 mm long; *style* arching over the stamens, dividing between the upper third and apices of the anthers, the branches 4–5 mm long, curving outward beyond the anthers. *Capsules* ellipsoid, 12–15 mm long, subacute; *seeds* ovoid, 7–8 x 4–5 mm, broadly and evenly winged, translucent light brown. *Chromosome number* unknown.

---

116. GLADIOLUS ROSEOVENOSUS Goldblatt & Manning

PLATE 99

Gladiolus roseovenosus Goldblatt & Manning, new species. Type: South Africa, Western Cape, near Ruitersbos Forest Station, 1 Mar. 1982, *Vlok 347* (MO, holotype).

roseovenosus = red-veined, referring to the nectar guide on the lower tepals of the flower.

Latin diagnosis

Plantae 30–50 cm altae, cormo globoso c. 15 mm in diametro, tunicis papyraceis, foliis 4, infimo basali lamina brevi vel vestigiali lineari, 1.5–2 mm lata marginibus costaque leviter incrassatis, caule eramoso, spica flexuosa 2–4 florum, bracteis

25– 35(– 40) mm longis, floribus carnescentibus tepalis inferioribus rubro-lineatis, tubo perianthii anguste infundibuliformi 36–44 mm longo, tepalis inaequalibus lanceolatis dorsali c. 37 x 17–19 mm, inferioribus c. 27 x 11 mm, filamentis c. 17 mm longis, antheris 8–11 mm longis.

Description

Plants 35–50 cm high. *Corm* more or less globose, c. 15 mm in diameter, the tunics of firm-papery layers, fragmenting with age into narrow vertical fibres. *Cataphylls* pale and membranous, the uppermost dark green or purple above the ground, sometimes lightly mottled with white. *Leaves* four, the lowermost basal and longest, reaching to between the lower third and middle of the stem, the blade short to vestigial, up to 7.5 cm long, linear, plane, 1.5–2 mm wide, the margins and midrib lightly thickened, the remaining leaves progressively shorter above, largely sheathing, the uppermost with the lower margins united around the stem. *Stem* erect, flexed outward above the sheath of uppermost or upper two leaves, becoming erect above, unbranched, 1–1.5 mm in diameter below the spike.

Spike erect, lightly flexuose, 2- to 4-flowered; *bracts* pale green, the outer 25–45 mm long, the inner slightly shorter than the outer, forked apically for up to 5 mm. *Flowers* creamy pink, the lower three tepals feathered with red on the veins, sometimes becoming almost solid red below, mottled and streaked with dark pink on the reverse, unscented; *perianth tube* narrowly funnel-shaped, 36–44 mm long, curving forward and widened above, the cylindrical lower part enclosed in the bracts, 26–34 mm long; *tepals* unequal, lanceolate, the margins undulate, the dorsal largest, arching over the stamens, 34–37 x 18–26 mm, curving upward in the upper quarter, the upper laterals directed forward, curving outward in the upper quarter, 33–38 x 12–14 mm, the lower three tepals joined to the upper laterals for 9 mm and to one another for c. 2 mm, subequal, c. 27 x 11 mm, the lower median slightly larger than the lower laterals, in profile the upper and lower about the same length. *Filaments* 15–17 mm long, exserted 6–8 mm from the tube; *anthers* 8–11 mm long, purple, reaching only to the middle of the dorsal tepal, the pollen cream. *Ovary* oblong, c. 6 mm long; *style* arching over the stamens, dividing at or up to 5 mm beyond the anther apices, the branches 6–8 mm long. *Capsules* and *seeds* unknown. *Chromosome number* unknown.

Flowering time mid-February to April.

Distribution and biology

Gladiolus roseovenosus is a poorly known and rare species of the southern African winter-rainfall zone. It has evidently been recorded at two localities along the coastal foothills of the Outeniqua Mountains in the southern Cape. The range presumably extends from near George, one known site, to the foot of Robinson's Pass near Ruitersbos, the type locality. Plants grow in peaty sandstone soil on well-drained slopes in what is now grassland, but was originally grassy fynbos.

The long-tubed, creamy pink flowers with red nectar guides are unscented and are most likely adapted for pollination by long-tongued flies.

Diagnosis and relationships

Seemingly not particularly distinctive, *Gladiolus roseovenosus* has superposed, linear leaves with lightly thickened margins and midrib which place it in section *Homoglossum*. Within the section its four plane leaves and relatively soft-textured corm tunics suggest a relationship with species of series *Carinatus*, none of which it resembles very closely. The short or vestigial leaf blades are unusual in the series, as are the large pinkish flowers which have a long perianth tube 36–44 mm long and lower tepals lined and feathered with dark red in the lower half. The flowers conform to the general pattern of Iridaceae adapted for pollination by long-tongued flies and in this respect *G. roseovenosus* stands out as being the only species in the series with such flowers, the remainder having unspecialized flowers adapted for bee pollination.

History

The first record of *Gladiolus roseovenosus* was made by the South African botanist, J.H.J. Vlok in March 1982 at Ruitersbos Forest Reserve on the southern slopes of the Outeniqua Mountains. The collection was confused in herbaria with the autumn-flowering race of *G. hirsutus* (as *G. punctulatus* var. *autumnalis*) and only attracted our attention when we critically examined herbarium holdings for this monograph. It then became clear that the collections represented an undescribed species.

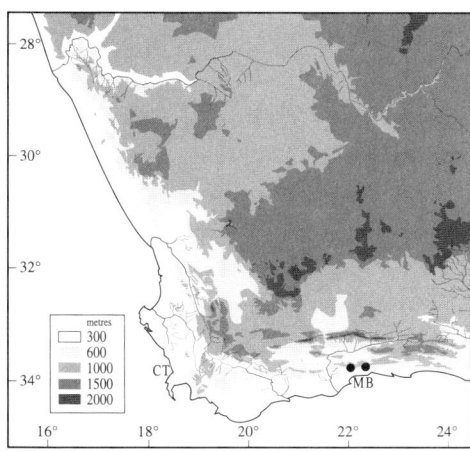

117. GLADIOLUS CARINATUS Aiton

Sandpypie, blue Afrikaner

PLATE 100

Gladiolus carinatus Aiton, *Hortus Kewensis* 1: 64 (1789). N.E. Brown, *Gard. Chron.* Ser. 3, 93: 290 (1933). Lewis et al., *J. S. African Bot.*, Suppl. 10: 212 (1972). *G. odorus* Salisbury, *Prodr.* 40 (1796), a superfluous name for *G. carinatus* Aiton. Type: South Africa, Cape, without precise locality, ex *Hort. Kew*, 1781, *Masson s.n.* (BM, holotype).

carinatus = with a keel, referring the prominent midvein on the leaf blades.

Synonymy

Gladiolus tenellus Jacquin, *Collecteana* 3: 255 (1791, as 1789) & *Icones Pl. Rar.* pl. 248 (1795). Lewis et al., *J. S. African Bot.*, Suppl. 10: 177 (1972), misapplied to *G. trichonemiflorus*. Type: South Africa, Western Cape, without precise locality or collector, illustration in Jacquin, *Icones Pl. Rar.* pl. 248 (1795).
Gladiolus tristis var. *inodorus* Thunberg, *Dissertatio de Gladiolo* no. 8d (1784). *Gladiolus laevis* Thunberg, *Prodromus Pl. Capensium* 184 (1800). Type: South Africa, Western Cape, without precise locality, *Thunberg s.n.* (UPS–Herb. Thunberg 1038, holotype).
Gladiolus tristis var. *humilis* Thunberg, *Dissertatio*

de Gladiolo no. 8n (1784). Type: South Africa, Western Cape, without precise locality, *Thunberg s.n.* (UPS–Herb. Thunberg 1075, holotype).
Gladiolus tristis var. *luteus* Thunberg, *Dissertatio de Gladiolo* no. 8h (1784). Type: South Africa, Western Cape, without precise locality, *Thunberg s.n.* (UPS–Herb. Thunberg 1076, holotype).
Gladiolus ringens Andrews, *Bot. Repository* 1: pl. 27 (1790). Type: South Africa, Western Cape, without precise locality or collector, illustration in Andrews, *Bot. Repository* 1: 27, 1790).
Gladiolus suaveolens Zeyher ex Klatt, *Linnaea* 32: 711 (1863), an illegitimate homonym, not *G. suaveolens* Ker Gawler (= *G. maculatus* Sweet). Type: South Africa, Western Cape, near Constantia, *Ecklon & Zeyher Irid. 150* (G, MO, S, isotypes).
Gladiolus nivenii Baker, *Handbook Irideae* 210 (1892); *Fl. Capensis* 6: 149 (1896), a new name for *G. ringens* var. *undulatus* Andrews, *Bot. Repository* 4: pl. 275 (1803). Type: South Africa, Western Cape, without precise locality, originally collected by James Niven and introduced in 1800, cultivated in England, illustration in Andrews, *Bot. Repository* 4: pl. 275, 1803).
Gladiolus superans N.E. Brown, *Gard. Chron.* Ser. 3, 92: 221 (1932). Type: South Africa, Western Cape, Riversdale District, 1932, *Muir 4850* (K, holotype).
[*Gladiolus recurvus* sensu Ker Gawler, *Curtis's Bot. Mag.* 16; pl. 578 (1802); sensu Baker, *Fl. Capensis* 6: 139 (1896); sensu Marloth, *Fl. South Africa* 4: pl. 47 (1915), not *G. recurvus* Linnaeus.]

Description

Plants 20–50(–100) cm high. *Corm* globose, 12–18 mm in diameter, the tunics of coriaceous to more or less woody layers, breaking with age into regular segments below or into coarse vertical fibres. *Cataphylls* pale and membranous, the upper reaching 3–9(–18) cm above the ground and then purple mottled with white. *Leaves* three, the lowermost longest, usually shortly exceeding the spike, the blade linear, (1–)3–9 mm wide, the midrib lightly to moderately thickened, the margins not or very little thickened, the second leaf sheathing the lower half of the stem, the blade channelled or with short unifacial blades like the basal leaf, uppermost leaf shortest and inserted on the upper third of the stem, the margins united in the lower half of the sheathing part. *Stem* erect below, flexed outward above the sheaths of the two upper

leaves, unbranched, 1–1.5 mm in diameter below the spike.

Spike inclined, lightly flexuose, occasionally 1-, usually 4- to 6-, rarely to 10-flowered; *bracts* pale green or greyish green or flushed with purple, translucent on the veins, the outer 25–35(–65) mm long, the margins transparent, the inner bracts half to three-quarters as long as the outer, lightly notched apically. *Flowers* shades of blue to grey or pale yellow, occasionally purple or pink, the lower tepals each with a yellow band across the upper third or yellow in the lower two-thirds, sometimes also streaked below and above with dark purple to blue lines, yellow-flowered plants with the lower tepals greenish in the middle third and becoming yellow then brownish with age, usually intensely scented of freesias or violets; *perianth tube* obliquely funnel-shaped, 13–16 mm long, the lower cylindrical part 5–7 mm long; *tepals* unequal, ovate to lanceolate, the dorsal largest, arching forward, recurved near the apex, 24–30(–38) x (11–)17–20(–26) mm, the upper laterals about equal or slightly smaller, 22–34 x 14–18 mm, directed forward and curving outward in the upper third, the lower three tepals joined to the upper laterals for up to 5 mm and to one another for 4–6 mm, 18–25 x 7–13 mm, widest in the upper third, the lower laterals deeply channelled below with the margins more or less vertical, the lower median curving downward in the upper half, in profile the lower tepals exceeding the dorsal. *Filaments* 11–15 mm long, exserted 5–6 mm from the tube; *anthers* 8–10 mm long, cream, the pollen cream. *Ovary* ovoid, c. 4 mm long; *style* arching over the stamens, dividing near the anther apices, the branches

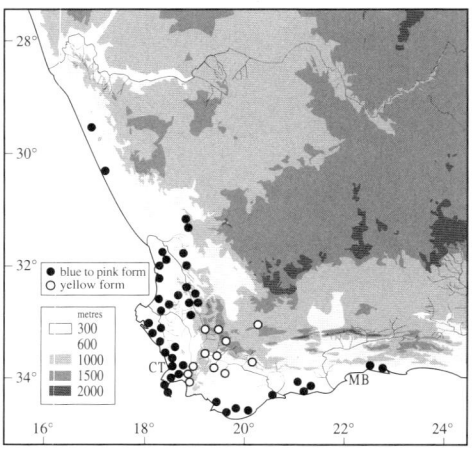

3–4 mm long, extending above the anthers. *Capsules* oblong to ellipsoid, 20–25 mm long, the apices acute; *seeds* oblong, 6–8 x 5–7 mm, broadly and evenly winged, light yellow-brown. *Chromosome number* unknown.

Flowering time June to mid-September.

Distribution and biology

One of the more common of the western southern African species of the genus, *Gladiolus carinatus* extends from the Cape Peninsula northward into central Namaqualand and eastward along the southern coast to the vicinity of Knysna. It is perhaps most frequent in well-drained, coarse sandy soil on coastal flats dominated by tussocks of Restionaceae (typically growing in the clumps), but it may be found in suitable habitats on stony mountain slopes at elevations of up to 1000 m. In Namaqualand the species is poorly recorded, but it apparently occurs there only on the coastal plain in sandy deposits that resemble the deep sandy soils of the western Cape coast. The blue-flowered form of the species is the one most often seen, but yellow-flowered plants are also common, always found on lower mountain slopes or in alluvial washes in the south-western Cape interior.

Like other relatively short-tubed and blue- (or yellow-)flowered southern African species of *Gladiolus*, *G. carinatus* is pollinated by a range of long-tongued bees of the family Anthophoridae and by the honey bee, *Apis mellifera*. The bees visit the flowers to feed on the nectar in the lower part of the perianth tubes. Although plants are rarely visited with any frequency, they normally have full complements of capsules densely packed with seeds.

This particularly charming species deserves more horticulural attention, as it is a good pot and open-garden plant, and an excellent cut-flower. The unpollinated blooms last for several days and produce a scent rivalling that of the most fragrant garden freesia.

Diagnosis and relationships

Gladiolus carinatus is readily recognized by the three more or less plane foliage leaves on the stems which bear the large, extremely sweetly scented blue, yellow, or occasionally pink flowers, and by the distinctively purple-mottled upper cataphyll. The rather soft-textured fibres of the corm tunics are also fairly characteristic, and this together with the constant leaf number and mottled cataphylls makes the species easy to identify. It is closely related to *G. griseus*, a species of

the Cape west coast that has similar leaves and mottled cataphylls, but smaller, grey-green flowers, typically 22–32 mm long. The marshland species, *G. quadrangulus*, also shares these vegetative features and we assume that it is closely related to *G. carinatus* and *G. griseus*. However, its relatively small, actinomorphic flower leaves no possibility of confusing this plant with either. The firm-textured leaf with raised and somewhat thickened margins and midrib places *G. carinatus* with the small western Cape series *Carinatus* that includes *G. comptonii*, with yellow flowers, and *G. violaceolineatus*, both of which have four leaves, always much narrower than those of *G. carinatus*.

Variation in floral colour is marked in *Gladiolus carinatus* and quite constant locally. Yellow-flowered plants occur from the foot of the Hottentots Holland Mountains near Sir Lowry's Pass and the Stellenbosch Flats through the Breede River valley to Ceres and the Warm Bokkeveld. Vegetatively these plants are typically smaller than the blue-flowered plants and usually have narrower leaves. We assume that they represent a separate race of the species that has not diverged sufficiently to merit separate taxonomic recognition. Blue-flowered populations extend along the south coast and along the west coast and interior, including the Cedarberg and Bokkeveld escarpment.

History

Although *Gladiolus carinatus* is a common plant in the southwestern Cape and was recorded by all the early botanists and collectors, it took surprisingly long for the species to be formally named. C.P. Thunberg collected the species on several occasions and named the yellow-flowered form *Gladiolus tristis* var. *inodorus* in 1784. We believe that Thunberg's *G. tristis* var. *humilis* and var. *luteus* also represent the species, but the specimens on which these names are based are in poor condition and some authorities have associated both with *G. trichonemifolius* (Lewis et al., 1972). *Gladiolus carinatus* was formally named at species rank in 1789 when it was described by William Aiton in *Hortus Kewensis*, a botanical account of the plants grown at Kew Gardens, London. The type specimen, now preserved at the British Museum of Natural History, was originally collected by Francis Masson some years earlier at the Cape and was grown and flowered at Kew Gardens. In 1790 plants flowered in England from a second source were described as *G. ringens* by Henry Andrews in the *Botanists' Repository*. At about the same

time, Nicholas Jacquin described the controversial *G. tenellus*, actually published in 1791, although the title page of volume 4 of his *Collecteana* is dated 1789. The illustration that serves as the type of *G. tenellus* was only published in 1795. It is a poor representation of the dull yellow-flowered form of *G. carinatus*, but some authorities, including G.J. Lewis, associated the species with *G. trichonemifolius* and in the revision of *Gladiolus* in South Africa (Lewis et al., 1972) that species is called *G. tenellus*.

In 1800 Thunberg published his *Prodromus Plantae Capensium* in which *Gladiolus carinatus* is again described, there under the name *G. laevis*. Despite this sudden accumulation of names for the reasonably distinctive *G. carinatus*, a good illustration published in *Curtis's Botanical Magazine* was identified there by John Ker Gawler as *G. recurvus* of Linnaeus. The following year, 1803, plants collected by James Niven, a paid plant collector for the British magnate, George Hibbert, were flowered in England and provided the material used for a painting reproduced later that year in the *Botanists' Repository* where it was described as *G. ringens* var. *undulatus*. Nevertheless, the name used by Ker Gawler for the species was the one that was most often applied, and *G. carinatus* remained known as *G. recurvus* in *Flora Capensis*. It was also treated there under a second epithet, *G. nivenii*, the name J.G. Baker gave *G. ringens* var. *undulatus* at species rank (the name *G. undulatus* was already in use for another species). Only in 1933 did N.E. Brown point out that *G. carinatus* was the correct name for the plant then known as *G. recurvus*.

In their revision of *Gladiolus* in South Africa, Lewis et al. (1972) recognized two subspecies of *G. carinatus*, subsp. *carinatus* and subsp. *parviflorus*. The latter is treated here as a separate species under the name *G. griseus*. This treatment parallels that of Thunberg (1800), who recogized both *G. carinatus* (as *G. laevis*) and *G. griseus* (as the illegitimate homonym, *G. punctatus*).

In past times *Gladiolus carinatus* was a favourite cut-flower in the western Cape. Plants gathered in the wild were sold to home owners and many houses in the southwestern Cape boasted a vase of this *Gladiolus* in their parlours. A few stems would fill a home with a delightful fragrance. Even today, wild-collected *G. carinatus* is occasionally seen for sale, although the practice of picking wild flowers for sale is now illegal.

118. GLADIOLUS GRISEUS Goldblatt & Manning

PLATE 101

Gladiolus griseus Goldblatt & Manning, new species. Type: South Africa, Western Cape, Melkbos, 21 June 1959, *Horrocks 9* (NBG, holotype and isotype).

griseus = grey-coloured, referring to the slate-grey flowers.

Synonymy

Gladiolus tristis var. *permeabilis* Thunberg, *Dissertatio de Gladiolo* 167 (1784). *G. carinatus* subsp. *parviflorus* Lewis, *J. S. African Bot.*, Suppl. 10: 216 (1972), not *G. parviflorus* Jacquin (= *Tritoniopsis parviflorus* (Jacquin) G. Lewis). Duncan, *Fl. Pl. Africa* 48: pl. 1906 (1985). Type: South Africa, Western Cape, Paarden Island, Cape Peninsula, without date, *Thunberg s.n.* (UPS–Herb. Thunberg 1058, holotype, BOL, photo).

Gladiolus punctatus Thunberg, *Prodr. Pl. Capensium* 185 (1800), an illegitimate homonym for *G. punctatus* Jacquin and a superfluous name for *G. permeabilis* Delaroche (1766). Baker, *Fl. Capensis* 6: 149 (1896). Based on *Thunberg s.n.* (UPS–Herb. Thunberg 1058), from Paarden Island, Cape Town.

Latin diagnosis

Plantae 20–80 cm altae, cormo globoso 8–15 mm in diametro, tunicis papyraceis, foliis 3, infimo longioribus usitate spica breviter excedentibus lamina brevi vel vestigiali lineari, 2–7.5 mm lati costis leviter incrassatis marginibus non vel parum incrassatis, caule eramoso, spica flexuosa (3–)6–15 florum, bracteis 25–35(–40) mm longis, floribus malvinis ad griseus tepalis inferioribus flavo- et rubronotatis, tubo perianthii infundibuliformi 6–10 mm longo, tepalis inaequalibus dorsali 16–26 x 11–17 mm, inferioribus 13–16 x c. 6 mm, filamentis 5–9 mm longis, antheris 7–9.5 mm longis.

Description

Plants 20–80 cm high. *Corm* globose, 8–15 mm in diameter, the tunics of papery layers, decaying with age into soft-textured fibres. *Cataphylls* pale and membranous, the upper usually reaching 5–12 cm above the ground and then purple mottled with white. *Leaves* three, the lowermost longest, usually shortly exceeding the spike, the blade linear, 2–7.5 mm wide, the midrib lightly to moderately thickened, the margins not or very little thickened, the second leaf sheathing the lower half of the stem, the blade like the basal leaf but fairly short, uppermost leaf shortest and inserted on the upper third of the stem, the margins united in the lower half of the sheathing part. *Stem* erect below, flexed outward above the sheaths of the two upper leaves, unbranched, 1–2 mm in diameter below the spike.

Spike inclined, lightly flexuose, occasionally 3-, usually 6- to 15-flowered; *bracts* pale green or greyish green or flushed with purple, translucent on the veins, the outer 25–35(–40) mm long, the margins and veins transparent, the inner c. 5 mm shorter than the outer, minutely notched apically. *Flowers* shades of mauve to grey, the lower tepals each with a yellow band across the upper third edged in green and maroon, usually with a light, sweet scent; *perianth tube* obliquely funnel-shaped, 6–10 mm long, the lower cylindrical part 3–5 mm long; *tepals* unequal, the dorsal largest, arching forward, recurved near the apex, 16–26 x 11–17 mm, the upper laterals equal or slightly smaller, 15–21 x 10–14 mm, directed forward and scarcely curving outward near the apices, the lower tepals joined to the upper laterals for 2–3 mm and to one another for up to 3.5 mm, 13–16 x c. 6 mm, widest in the upper third, the lower laterals deeply channelled below with the margins more or less vertical, the lower median curving downward in the upper half, in profile the lower tepals exceeding the dorsal by 3–4 mm. *Filaments* 5–9 mm long, exserted 3–6 mm from the tube; *anthers* 7–9.5 mm long, cream, the pollen cream. *Ovary* ovoid, c. 5 mm long; *style* arching over the stamens, dividing between the middle and upper third of the anthers, the branches c. 2 mm long, extending above the anthers. *Capsules* oblong to ellipsoid, 15–20(–25) mm long, the apices acute; *seeds* ovate, c. 7 x 4–5 mm, broadly and evenly winged, light golden brown. *Chromosome number* unknown.

Flowering time late May to mid-July.

Distribution and biology

An endemic of the west coast of Western Cape Province, *Gladiolus griseus* is found almost exclusively in calcareous sands and limestone gravel. Nearly all populations occur within sight of the ocean or only a short distance inland where there are calcareous soils. Plants mostly grow in low, coastal scrub, or occasionally in clumps of Restionaceae, an association also favoured by its close relative, *G. carinatus*. Plants flower surprisingly early in the season, in May and June after early autumn rains, or in late June or July after late rains. By the end of July seeds have usually ripened and been dispersed.

The fairly small, lightly scented flowers are adapted for pollination by bees, and the only insect visitors we have recorded are honey bees *Apis mellifera*, one of the few bee species active in the cool winter months.

Diagnosis and relationships

Morphologically similar to *Gladiolus carinatus* and sometimes indistinguishable from it

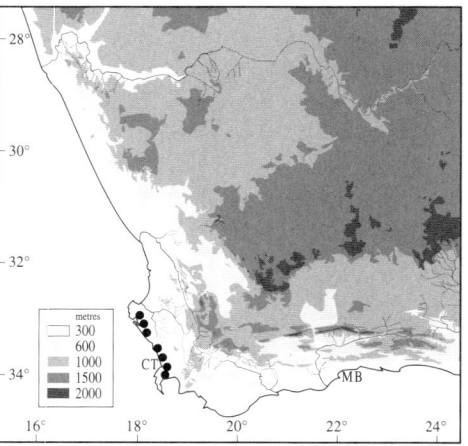

in the vegetative state, *G. griseus* always has an unbranched stem and just three super-posed foliage leaves, the blades sword-shaped and with lightly thickened margins but a prominent midrib. The flowers of *G. griseus* are slate-grey and 22–36 mm long with a tube 6–10 mm and dorsal tepal 16–26 mm long. Although similar in structure, the flowers are thus consistently smaller than those of *G. carinatus* which has flowers 37–54 mm long (perianth tube 13–16 mm long and dorsal tepal 24–38 mm long) and upper lateral tepals typically directed forward for most of their length. Spikes of *G. griseus* usually have 6–12 flowers, whereas those of *G. carinatus* normally have fewer flowers, mostly 4–6 per stalk. *Gladiolus carinatus* is known for the strong, delightful scent of its flowers, a few of which will fill a room with

their freesia-like odour. Flowers of *G. griseus* are, in contrast, just faintly scented. The flowering time as well as the habitat of the two also differ. *Gladiolus griseus* flowers in winter and favours lime-enriched substrates such as calcareous sand, while *G. carinatus* flowers in spring and grows in well-drained, nutrient-poor acidic sandstone soils, either in rocky sites or deep sands.

History

The winter-flowering *Gladiolus griseus* appears to have first been collected by Carl Thunberg, probably in 1772. Thunberg initially described his collection as *G. tristis* var. *permeabilis* (Thunberg, 1784). Later, in 1800 when he had refined his understanding of *G. tristis* (in which he recognized 16 varieties, most of which are now regarded

as separate species), he renamed the species *G. punctatus*. In doing so, Thunberg obviously regarded *G. punctatus* as distinct from the morphologically similar *G. carinatus* which he called *G. laevis*. Unfortunately, Thunberg's name is a homonym, for Nicholas Jacquin had already used the name in the genus for a species that is now *Tritoniopsis parviflora*. Moreover, Thunberg cited *G. permeabilis* Delaroche in the proto-logue of *G. punctatus*, rendering the name superfluous and illegitimate. *Gladiolus punctatus* was maintained by J.G. Baker (1896) in *Flora Capensis* but was reduced to sub-species rank in *G. carinatus* by Lewis et al. (1972), who replaced the specific epithet, calling the plant subsp. *parviflorus*.

119. GLADIOLUS QUADRANGULUS (Delaroche) Barnard

PLATE 102

Gladiolus quadrangulus (Delaroche) Barnard, *J. S. African Bot.* 36: 300 (1970). Lewis et al., *J. S. African Bot.*, Suppl. 10: 172 (1972). *Ixia quadrangula* Delaroche, *Plantae Aliquot Novarum* 16 (1766). *Geissorhiza quadrangula* (Delaroche) Ker Gawler, *Genera Irid.* 88 (1827). Baker *Fl. Capensis* 6: 73 (1896). Type: South Africa, Western Cape, Cape Flats, Isoetes Vlei, waterlogged sandy soil, 22 Sept. 1970, *Goldblatt 525* (BOL, neo-type; BOL, K, isoneotype).

quadrangulus = quadrangular, four-sided, for the shape of the leaf in transverse section, with the midrib strongly thickened and the margins unthickened, thus imparting a more or less four-sided shape to the narrow leaves.

Synonymy

Ixia linearis Linnaeus fil., *Suppl. Pl.* 92 (1781). *Gladiolus linearis* (Linnaeus fil.) N.E. Brown, *J. Linn. Soc.* 48: 48 (1928). Type: South Africa, Western Cape, flats between Cape Town and Stellenbosch, *Thunberg s.n.* (UPS–Herb. Thunberg 964, holotype).
Gladiolus biflorus Klatt, *Trans. S. African Phil. Soc.* 3: 197 (1885). Baker, *Fl. Capensis* 6: 145 (1896), an illegitimate homonym, not *G. biflorus* Thunberg (= *Olsynium biflorum* (Thunberg) Goldblatt, from South America). Type: South Africa, Western Cape, Cape Flats near Wynberg, seasonally flooded sandy ground, Sept. 1884, *MacOwan 2279* (*Herb. Norm. Austr. Afr.* 279) (S, possible holotype; B [not seen], BOL, G, K, P [not seen], SAM, UPS [not seen], isotypes).

Description

Plants 14–35 cm high. **Corm** globose, 10–15 mm in diameter, the tunics of firm-papery to lightly coriaceous layers, fragmenting with age into narrow segments from the base. **Cataphylls** pale and membranous, the uppermost reaching 2–5 cm above the ground and then purplish to dry and brown with white spots. **Leaves** three, the lower two basal, the lowermost longest, as long as or slightly longer than the spike, the blade linear, thickest in the midline, the margins not thickened or raised, 1–2 mm wide, the second leaf sheathing the lower half of the stem, reaching to between the upper third of the stem and the base of the spike, the upper-most leaf inserted on the upper third of the stem and shortest, the margins of the sheathing part united around the stem. **Stem** erect, unbranched, straight or lightly flexed above, the sheaths of the two upper leaves united around the stem, c. 1 mm in diameter below the spike.

Spike straight, erect, 2- to 5-, occasionally to 8-flowered; **bracts** pale green, translucent on the veins, the outer 27–33(–45) mm long, the inner half to two-thirds as long as the outer. **Flowers** actinomorphic, stellate, pale pink, mauve or white, the tepals with darker veins and often with a dark median streak toward the base, lightly scented of narcissus, closing tightly in the mid-after-noon; **perianth tube** narrowly funnel-shaped, 6–10 mm long, the lower cylindrical part filiform, 4–6 mm long, tightly enclosing the style; **tepals** subequal, ovate to broadly

lanceolate, 18–25 x 7–12 mm, the outer slightly larger than the inner. **Filaments** erect and symmetrically disposed, c. 5 mm long; **anthers** 7–8 mm long, yellow, pollen yellow.

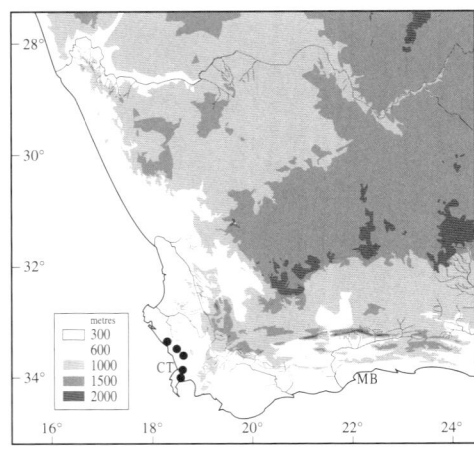

Ovary ovoid, c. 3 mm long; *style* erect and central, dividing just below the anther apices, the branches recurved, 3–4 mm long. *Capsules* ellipsoid, 13–16 mm long; *seeds* ovate, 4–6 x c. 4 mm, broadly and evenly winged, translucent brown, the seed body darker than the wing. *Chromosome number* $2n = 30$.

Flowering time mid-August to mid-September, occasionally in early October.

Distribution and biology

Gladiolus quadrangulus is a local endemic of the southwestern Cape coastal forelands. Its original range extended from the Cape Flats near Wynberg northward to the Malmesbury District. Farming activity and urban development has eliminated it from most of its range, but it still persists in the Edith Stephens Reserve on the Cape Flats and at a few undisturbed sites along the west coast. Its long-term survival in the wild is unlikely given the rapid urban expansion and intensified farming activity in the Cape Town area. The species grows in sandy soil in seasonally wet, poorly drained and often slightly saline or brackish habitats and it is basically a wetland plant. It does, however, respond well to cultivation, and for many years has been successfully grown at Kirstenbosch Gardens where it is not given special treatment.

Unlike nearly all the species of section *Homoglossum,* flowers of *Gladiolus quadrangulus* do not produce nectar and the filiform perianth tube is too narrow to allow access to an insect's tongue. This actinomorphic-flowered species is also unusual in the section in having prominently displayed anthers and only a light floral fragrance. Thus, we assume that the flowers of *G. quadrangulus* offer pollen as a reward for pollinators, a rare strategy in the genus, as far as we know shared only by *G. stellatus* (section *Hebea*) and *G. aureus* (section *Linearifolius*).

Diagnosis and relationships

Gladiolus quadrangulus is one of just a handful of unrelated species of *Gladiolus* that have an actinomorphic flower. The pale pink, mauve or white flowers have a short, filiform tube and erect stamens surrounding a central style. The species is thus easily recognized by the flower and, in addition, has a fairly distinctive vegetative morphology. A member of section *Homoglossum, G. quadrangulus* has superposed leaves, and consistently three leaves with more or less plane blades and margins which are only lightly raised. The three leaves with plane blades, together with softly textured, more or less fibrous corm tunics and distinctive mottled purple and cream cataphylls suggest that *G. quadrangulus* is most closely related to *G. carinatus* and its immediate ally, *G. griseus.* We presume the three species comprise a single lineage within the primitive series *Carinatus* of section *Homoglossum.*

History

There is no record of who first collected *Gladiolus quadrangulus*, but plants were in cultivation in Holland at least by the 1760s. There, at the University of Leiden, Daniel Delaroche, a student of the Dutch botanist, David van Royen, described the plant among several new species of southern African Iridaceae in his doctoral dissertation. Published in 1766, the dissertation, *Plantarum Aliquot Novarum*, included *G. quadrangulus* in the genus *Ixia*. For many years the identity of *Ixia quadrangula* was unknown and it was only discovered to belong to *Gladiolus* after the type specimen was found in the Leiden Herbarium in 1970 (Goldblatt & Barnard, 1970).

The species was also collected by Thunberg in the 1770s and the younger Linnaeus described *Ixia linearis* in 1781 on the basis of Thunberg's specimens. Only in 1928, when N.E. Brown examined the Iridaceae of Thunberg's herbarium in Uppsala, Sweden, was *Ixia linearis* understood to belong to *Gladiolus* and to be an earlier name for what was until then known as *G. biflorus*, a species described by F.W. Klatt in 1885. Klatt's *G. biflorus* was, incidentally, a homonym for *G. biflorus* Thunberg, a South American species of Iridaceae, now transferred to the genus *Olsynium* (Goldblatt et al., 1990). Because of the poor understanding of *Ixia quadrangula* and *I. linearis*, both these species were recognized in *Flora Capensis*, the former in the genus *Geissorhiza* and the latter in *Ixia*. True *G. quadrangulus* was also recognized, under Klatt's illegitimate epithet, *G. biflorus*. Thus, the three synonyms of what is now recognized as *G. quadrangulus* were included by J.G. Baker in *Flora Capensis* in three separate genera.

Section *Homoglossum*: Series *Mutabilis*

120. GLADIOLUS MUTABILIS G. Lewis

Brownies

PLATE 103

Gladiolus mutabilis G. Lewis in *J. S. African Bot.*, Suppl. 10: 230 (1972), as a new name for *G. muirii*. L. Bolus, *Ann. Bolus Herb.* 3: 181 (1924), not *G. muirii* L. Bolus, 1915 (= *G. involutus* Delaroche). Type: South Africa, Western Cape, Riversdale District, without precise locality or date, *Muir 1412* (BOL, lectotype).

mutabilis = changeable, referring to the variation in the colour of the flowers across the range of the species, from brown to cream to deep blue.

Description

Plants 25–50 cm high. *Corm* elongate-obconic, 8–10 mm in diameter, the tunics of cartilaginous layers, decaying with age into firm vertical strips and ultimately becoming coarsely fibrous. *Cataphylls* pale and membranous, the uppermost reaching up to 5 cm above the ground and then green and coriaceous, accumulating with age into a fibrous neck around the base of the stem. *Leaves* four, the lowermost basal and longest, sheathing the lower third of the stem, the blade more or less linear, at least 5 cm long or reaching to the top of the spike, fleshy, the margins and midrib neither raised nor thickened, remaining leaves cauline, inserted between the middle and upper third of the stem, progressively shorter above, all with short blades, imbricate and enclosing the stem nearly to the base of the spike. *Stem* erect, flexed outward above the sheaths of the upper two or three leaves, unbranched, 1–1.3 mm in diameter below the spike.

Spike inclined, lightly flexuose, 2- to 5-flowered; *bracts* pale green, glaucous, the outer elongate and attenuate, the apices with the margins inrolled and slightly twisted, 25–40(–50) mm long, the inner about two-thirds as long as the outer. *Flowers* pale or dark blue, mauve or brownish to cream shaded with brown, the lower three tepals yellow across the midline or in the lower two-thirds, lightly and irregularly lined and spotted with purple to brown, sweetly scented of violets; *perianth tube* obliquely funnel-shaped, 13–17 mm long, usually completely enclosed in the bracts; *tepals* unequal, the upper three larger than the lower, obovate, the lower laterals oblanceolate, the margins undulate to crisped, the dorsal arched to hooded over the stamens, the upper edges curling backward, 25–32 x 14–16 mm, the upper laterals directed forward, curving outward in the upper third, the lower tepals united with the upper laterals for 6–10 mm and to one another for 2–3 mm, the lower laterals 16–28 x 7–9 mm, tapering and channelled below but not differentiated into limb and claw, directed forward below, curving toward the ground in the upper half, the lower median 16–25 x 12–14 mm, in profile the lower three tepals longer than the upper. *Filaments* 10–15 mm long, exserted 4–5 mm from the tube; *anthers* 9–11 mm long, lilac to cream, the pollen pale yellow. *Ovary* ovoid, c. 4 mm long; *style* arching over the stamens, dividing just below to shortly above the anther apices, the branches c. 3 mm long. *Capsules* ellipsoid, c. 15 mm long; *seeds* ovate, c. 5 x 3 mm, broadly and evenly winged, translucent reddish brown. *Chromosome number* unknown.

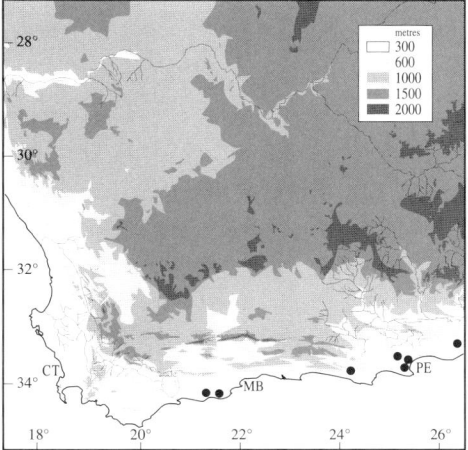

Flowering time May to July, rarely in August.

Distribution and biology

A species of the southern African winter-rainfall region, *Gladiolus mutabilis* occurs in Eastern Cape Province and the adjacent eastern part of Western Cape Province, South Africa. The species is more common in the east of its range in the vicinity of Port Elizabeth and Grahamstown, but populations have been recorded as far west as Albertinia and Riversdale, the latter the type locality. Plants grow in sandstone-derived soils, often in sandy loam and in stony ground.

The flowers of *Gladiolus mutabilis*, which resemble those of better-known species such as *G. carinatus* and *G. gracilis*, are assumed to be similarly adapted for pollination by long-tongued bees.

Diagnosis and relationships

Probably the least specialized member of series *Mutabilis*, *Gladiolus mutabilis* has the slightly fleshy leaf blades without thickened margins or midrib that define the series. The blades are usually fairly short, and that of the lowermost leaf, always the longest, is sometimes only 5 cm long, or it may reach to about the top of the spike. The most characteristic feature of the species is its elongate-conical corm, two to three times as long as it is wide. The firm, cartilaginous tunics that decay with age into vertical strips and ultimately into a fibrous mass are typical of series *Mutabilis*. The flowers are variable in colour, hence the specific epithet. They range from pale to dark blue, mauve, brown or cream, the lower tepals usually cream or yellow below and with dark blue or brown dots and streaks on the yellow background forming the nectar guide. The lower tepals appear longer than the upper and are distinctively broadened in the upper half, a feature useful in recognizing the species. The flowers have a strong, very pleasant odour reminiscent of violets.

Within series *Mutabilis*, *Gladiolus mutabilis* appears to have no particularly close relatives. Like most members of the series, it flowers early in the season, usually in June or July.

History

The earliest record of *Gladiolus mutabilis* that we have traced is a collection from Uitenhage in Eastern Cape Province, without date or exact locality and bearing the name Harvey, presumably William Harvey,

the nineteenth-century Irish botanist who was the major contributor to the early volumes of *Flora Capensis*. Harvey did not visit the eastern Cape Colony, as it was then, and the specimens must have been sent to him by one of his several eastern Cape correspondents. The collection was overlooked by contemporary and later botanists and confused with other species of *Gladiolus*. Thus, when specimens collected by the physician, John Muir, near Riversdale in the southern Cape came to the attention of H.M.L. Bolus in the 1920s, she thought this was a new discovery. She was, of course, correct in realizing that it was an undescribed species and she named it *G. muirii*, after the collector. The name is illegitimate for it is a homonym for *G. muirii* L. Bolus, described ten years earlier, in 1914. The latter is a synonym of another southern and eastern Cape species recorded by Muir, *G. involutus*. A new name, *G. mutabilis*, for the illegitimate *G. muirii* was provided by G.J. Lewis only in 1972.

Gladiolus exilis Lewis in *J. S. African Bot.*, Suppl. 10: 229 (1972). Type: South Africa, Western Cape, Wellington District, near the top of Bain's Kloof Pass on west-facing slopes, 20 May 1962, *Barnard s.n.* (NBG 62038, holotype).

exilis = small, referring to the size of the plants.

Description

Plants 30–60 cm high. *Corm* ovoid, 12–15 mm in diameter, the tunics of firmly textured coriaceous layers, fragmenting with age to form fine vertical fibres. *Cataphylls* pale and membranous, the uppermost reaching 2–12 cm above the ground and then green, often decaying into coarse fibres accumulating in a neck around the underground part of the stem. *Leaves* three, the lowermost basal and longest and sheathing the lower half of the stem, reaching to about the upper third of the stem, the blade relatively short, (50–)80–120 mm long, linear, 1–1.5 mm wide, plane and slightly succulent, neither the margins nor the midrib thickened or raised, the remaining two leaves inserted on the upper third of the stem, much shorter than the basal, sheathing for about half their length, the blades like the basal but slightly narrower, the blade of the uppermost leaf usually reaching just short of the base of the spike. *Stem* slender, usually lightly flexed above the sheaths of the leaves and inclined above, unbranched, c. 1 mm in diameter below the spike.

Spike inclined, weakly flexuose, 1- to 4-flowered; *bracts* grey-green, relatively soft-textured, the outer 20–30 mm long, the apices often attenuate and twisted, the inner slightly shorter than to about as long as the outer, usually minutely forked apically. *Flowers* pale blue-mauve to whitish barely flushed with mauve, the lower lateral tepals with a narrow, pale yellow median streak in the lower half and all three lower tepals irregularly streaked and speckled with dark mauve, usually lightly sweet-scented of violets; *perianth tube* obliquely funnel-shaped, 11–15 mm long, the lower cylindrical part 7–10 mm long; *tepals* unequal, lanceolate, the dorsal 22–35 x 11–18 mm, extending more or less horizontally over the stamens, curved upward in the upper quarter, the upper laterals 22–33 x 10–15 mm, directed forward and subpatent above, the lower three tepals joined to the upper laterals for 5–8 mm and to one another for 3–4 mm, the lower laterals 16–28 x c. 6 mm, narrow below and becoming channelled toward the base, the lower median 18–28 x c. 9 mm, downcurved in the distal half, in profile the lower three tepals extending well beyond the upper three. *Filaments* 11–12 mm long, exserted 5–7 mm from the tube; *anthers* 8–10 mm long, pale mauve, the pollen whitish. *Ovary* ovoid, 3–4 mm long; *style* arching over the stamens, dividing at or just beyond the apices of the anthers, the branches c. 3 mm long. *Capsules* ellipsoid, 12–15 mm long; *seeds* ovate, c. 6 x 4 mm, broadly and evenly winged, dark brown. *Chromosome number* unknown.

Flowering time April and May.

Distribution and biology

Gladiolus exilis is a narrow endemic of the mountains in the southwestern part of Western Cape Province. An inconspicuous plant, its range is not well documented. It is currently known mainly from the mountains surrounding Bain's Kloof Pass and the lower slopes of the Slanghoek Mountains a short distance to the south. It also occurs at a few sites in the Witzenberg range between Michell's Pass and Tulbagh, and on the Groot Winterhoek Mountains to the north. A collection from Gannakraal near Robinson's Pass (*Vlok 601*), some 300 km to the east, appears to match *G. exilis* except for having four leaves instead of three. It is tentatively included here until more information is obtained about the population. *Gladiolus exilis* favours well-drained rocky slopes in loamy, sandstone-derived soil in low fynbos. At the Slanghoek and Bain's Kloof sites the species grows close to the interface of sandstone and clay soils, either near the base of the Cape Sandstone Series, or close to the upper shale band, thus in richer soils than are provided by sandstone alone.

The flowers of *Gladiolus exilis* are very similar to those of the related *G. gracilis* and *G. carinatus* in colour, overall proportion and even fragrance and is, like them, adapted for pollination by long-tongued anthophorid bees, of which the autumn-flying *Amegilla fallax* has been recorded visiting flowers of the species.

Diagnosis and relationships

One of the earliest-flowering members of series *Mutabilis*, *Gladiolus exilis* has the strongly flexuose stem and somewhat fleshy leaves without thickened margins or midrib that are diagnostic for the series. The corm tunics of *G. exilis* are relatively soft-textured and become fibrous with age, its flowers are pale blue to mauve, and the lower three tepals have nectar guides consisting of dark spots and streaks on a pale yellow background. The flowers resemble those of several other species of the series, but especially

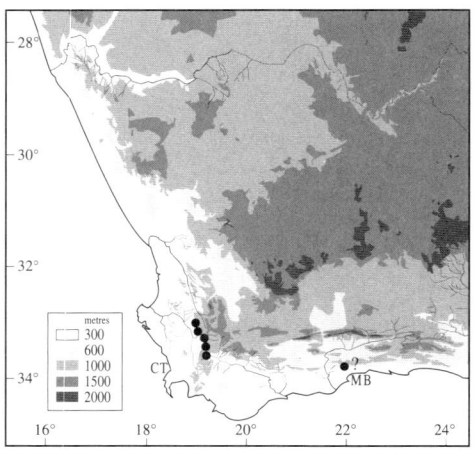

G. mutabilis. The two species are readily distinguished by the leaf number, *G. exilis* having three leaves and *G. mutabilis* four. Also, *G. exilis* has finely fibrous corm tunics and an ovoid corm, whereas *G. mutabilis* has an elongate-obconic corm with firmly coriaceous tunics. *Gladiolus exilis* also resembles fairly closely *G. vaginatus*, another member of series *Mutabilis*, and the two have similar, rather pale-coloured flowers, both blooming in the autumn. Confusion between them is, however, unlikely because *G. vaginatus* has only two leaves, the lower one bladeless and sheathing the stem for most of its length while the second leaf is inconspicuous, being reduced to a short, sheathing scale up to 20 mm long.

History

Relatively rare and with a scattered distribution, *Gladiolus exilis* is poorly collected. The first record we have been able to trace is a collection made in 1935 by the South African botanist, L.E. Taylor at the foot of Bain's Kloof Pass. We think it likely that there are earlier records, for places like Bain's Kloof have been much travelled since the road through the kloof was opened in the nineteenth century. Initial collections of the species were simply assigned to *G. gracilis*, despite the absolute difference in the corm morphology and leaf number and structure in the two species. *Gladiolus exilis* was formally described by G.J. Lewis in the posthumously published *A Revision of the South African Species of Gladiolus* (Lewis et al., 1972). The type of the species, also from Bain's Kloof, was collected by the *Gladiolus* grower and amateur naturalist, T.T. Barnard.

122. GLADIOLUS VAGINATUS F. Bolus
PLATE 105

Gladiolus vaginatus F. Bolus, *Ann. Bolus Herb.* 2: 103 (1917). Lewis et al., *J. S. African Bot.*, Suppl. 10: 272 (1972). Type: South Africa, Western Cape, Albertinia, Apr. 1914, *Muir 1287* (BOL, holotype; K, isotype).

vaginatus = sheathed, referring to the entirely sheathing nature of the basal leaf.

Synonymy

Gladiolus vaginatus var. *subtilis* Obermeyer in *J. S. African Bot.*, Suppl. 10: 273 (1972). Type: South Africa, Western Cape, 16 km west of Elim on road to Gansbaai, 'Gladiolus delicatus,' Lewis ms., 15 Apr. 1962, *Lewis 5956* (NBG, holotype; MO, isotype).

Description

Plants (22–)30–55(–105) cm high. *Corm* globose to slightly depressed-globose, 18–30 mm in diameter, the tunics thin and papery, decaying to become finely fibrous. *Cataphylls* pale and membranous, the uppermost reaching 2–4 cm above the ground and then green or purplish. *Leaves* (of the flowering stem) two, the lowermost basal and entirely sheathing, the blade completely suppressed, sheathing the stem for two-thirds of its length, dark green, the upper leaf reduced to a sheathing scale 1.6–2 mm long, inserted above the apex of the basal leaf, the margins free to the base but overlapping; leaves of non-flowering plants and seedlings single, linear, slightly succulent, 1–1.5 mm wide. *Stem* straight and erect, unbranched, 0.8–1.3 mm in diameter below the spike, continuing to elongate after flowering, remaining green for several months.

Spike flexed at the base and inclined, lightly flexuose, 2- to 6-, occasionally to 12-flowered; *bracts* grey to purplish green, the outer 15–20 mm long, the inner about as long as or slightly shorter than the outer, acute or minutely forked apically. *Flowers* pale blue, the lower tepals streaked and dotted with purple on a cream to yellow ground, the throat yellow and minutely speckled with red-purple, usually with a light, sweet scent (or apparently unscented); *perianth tube* obliquely funnel-shaped, 6.5–20 mm long, the wider part 3–8 mm long; *tepals* unequal, the dorsal ovate to lanceolate, arching over the stamens to nearly horizontal, 18–27 x 8–13 mm, the upper laterals directed forward below, patent in the upper half, 17–23 x 6.5–10 mm, the lower three tepals united with the upper laterals for 2.5–4 mm and to one another for 3–6 mm, the free parts lanceolate, 12–20 x 5–8 mm, narrowed and channelled below, gradually widening above, in profile the lower tepals exceeding the upper by 8–10 mm. *Filaments* 8–14 mm long, exserted 5–8 mm from the tube; *anthers* 5.5–11 mm long, mauve, the pollen pale mauve. *Ovary* oblong, c. 4 mm long; *style* arching over the stamens, dividing opposite the middle of the anthers, the branches 2–4 mm long, tangled in the anthers. *Capsules* broadly ovoid, 10–13 mm long, the apices rounded; *seeds* oblong, 7 x 4 mm, broadly and evenly winged, light yellow-brown. *Chromosome number* unknown.
Flowering time March and April.

Distribution and biology

Restricted to coastal Western Cape Province, *Gladiolus vaginatus* extends from the southern Cape Peninsula in the west to Knysna in the east. Although most records are from coastal calcareous sands or limestone outcrops, there are collections from gravelly clay in the Caledon District, mostly north of the Klein River–Bredasdorp Mountain axis but also near Elim to the south of these ranges. Thus, despite an apparent preference for calcium-enriched soils, *G. vaginatus* also grows on nutrient-intermediate, neutral to acidic soils. Records of the species show a scattered pattern, in part reflecting incomplete sampling across its range, but probably also a rather broken distribution pattern. The species is inconspicuous, and the pale bluish

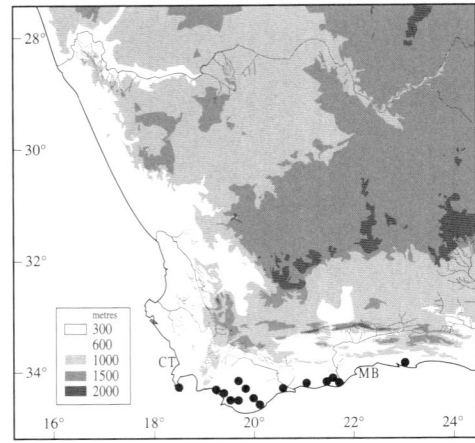

to white flowers are easily overlooked for the few weeks that *G. vaginatus* flowers in the autumn, in March and April.

Like its relatives with similar small, short-tubed flowers, *Gladiolus vaginatus* is assumed to be pollinated by anthophorid bees, but its pollination biology has not been studied.

Diagnosis and relationships

One of the easiest species of *Gladiolus* to recognize, even in the vegetative state, *G. vaginatus* has just two leaves on the flowering stem, both lacking blades. The lower leaf sheaths the stem for most of its length and the second leaf is an inconspicuous, scale-like structure up to 20 mm long, inserted below the base of the spike. The flowers are unremarkable and closely resemble those of related species, including *G. exilis* and *G. mutabilis*. They are pale blue or whitish and the lower tepals have nectar guides consisting of pale yellow in the lower part streaked and dotted with dark blue to purple. The flowers are also sweetly scented.

Variation in the eastern and western populations led Lewis et al. (1972) to recognize two subspecies of *Gladiolus vaginatus*, the eastern subsp. *vaginatus* with spikes of up to nine flowers, the filaments about as long as the anthers, the corm tunics membranous and the flowers with a tube 6.5–9 mm long and a narrow lower lip. The western subsp. *subtilis* was recognized by having spikes of one to four flowers, the filaments longer than the anthers, the corm tunics fibrous and the flowers with a tube 10–20 mm long and a wide lower lip.

These differences are largely correlated with the soils on which the plants grow. Plants with smaller flowers, a shorter tube, narrow lower lip and rather membranous corm tunics grow on limestone or calcareous sands. The larger-flowered plants with a longer perianth tube (and hence longer filaments relative to the anthers) and fairly fibrous corm tunics grow on fairly heavy clay soils. The tougher, more fibrous corm tunics may be directly correlated with the habitat, but other features are not, and presumably indicate that the populations growing on clay represent a distinct genotype or race. The differences between the two forms is, however, quite small and, compared with variation in other species of *Gladiolus*, hardly warrants taxonomic recognition.

Further study of the populations of *Gladiolus vaginatus* would certainly be worthwhile, and may shed more light on what appears to be an early stage in the differentiation of populations into recognizably different races that may ultimately lead to the emergence of a new species.

History

The earliest record of *Gladiolus vaginatus* that we have traced is the collection made by Harry Bolus in 1892. The collection is from the Caledon District, without precise locality, but presumably near the town and thus on the lower slopes of the Swartberg. A second collection made by Rudolf Schlechter in 1897 was also made in the Caledon District, at the eastern end of the Swartberg, thus at the western and inland edge of the range of the species. Curiously, Schlechter's collection is a mixed one, also including the first records of a second autumn-flowering species, *G. subcaeruleus*. These early collections were not at first perceived to be different from *G. brevifolius*, the species with which they were initially associated. The somewhat smaller-flowered and shorter-tubed form of *G. vaginatus* from the southern Cape coast was first collected near Stilbaai by the physician and naturalist, John Muir, in 1909 and a later collection from nearby at Albertinia, made by Muir in 1914, is the type of the species. Surprisingly, *G. vaginatus* was only recorded on the Cape Peninsula in 1979, where it is admittedly rare and restricted to a few sites in the Cape Point Nature Reserve.

Annotations on specimens of the inland form of *Gladiolus vaginatus* indicate that G.J. Lewis intended to recognize it as a separate species, *G. delicatus*. After her death, A.A. Obermeyer decided to recognize this series of populations as subsp. *subtilis*. These plants merit further study, but as presently understood do not appear to warrant taxonomic recognition at even subspecific rank.

123. GLADIOLUS MACULATUS Sweet

Brown Afrikaner, aandblom

PLATE 106

Gladiolus maculatus Sweet, *Hort. Brit.* 397 (as 1827 but published in 1826). Lewis et al., *J. S. African Bot.*, Suppl. 10: 258 (1972). Type: South Africa, Cape, without precise locality, flowered in Great Britain, illustration in *Curtis's Bot. Mag.* 16: pl. 556 (1802), no preserved material known.

maculatus = speckled, referring to the spotted markings on the flower.

Synonymy

Gladiolus versicolor var. *tenuior* Ker Gawler, *Curtis's Bot. Mag.* 16: pl. 556 (1802). *G. suaveolens* Ker Gawler, *Genera Irid.* 136 (1827), an illegitimate name based on the same type as *G. maculatus*. Type: South Africa, without precise locality, illustration in *Curtis's Bot. Mag.* 16: pl. 556 (1802).

Gladiolus breynianus Ker Gawler, *Genera Irid.* 135 (1827). Type: South Africa, without precise locality, *Masson s.n.* (BM, holotype).

Gladiolus vaginatus var. *fergusoniae* L. Bolus, *S. African Gard.* 19: 215 (1929). Type: South Africa, Western Cape, Riversdale District, Still Bay, May 1929, *Ferguson s.n.* (BOL 18922, holotype).

Gladiolus hibernus Ingram, *Gard. Chron.*, Ser. 3, 90: 9 (1931). *Gladiolus maculatus* subsp. *hibernus* (Ingram) Obermeyer in *J. S. African Bot.*, Suppl. 10: 263 (1972). Type: South Africa, without precise locality or date, *Ingram s.n.*, illustration in BM (not seen) (no preserved specimens found by Lewis et al., 1972 or ourselves).

[*Gladiolus recurvus* sensu Baker, *Fl. Capensis* 6: 139 (1896), not *G. recurvus* Linnaeus (1767).]

Description

Plants 30–80 cm high. **Corm** globose, 10–20 mm in diameter, the tunics papery, decaying with age into vertical fibres from below. **Cataphylls** pale and membranous, the uppermost reaching up to 12 cm above the ground and then green or dull purple. **Leaves** three or four, the lower two basal, the lowermost sheathing the lower half of the stem, short, rarely reaching beyond the base of the spike, the blade more or less plane, usually fairly fleshy, the margins and midrib neither raised nor thickened, remaining leaves cauline, inserted between the middle and upper third of the stem, progressively shorter above, all with short blades. *Stem* erect, lightly flexed above the sheaths of the upper two leaves, unbranched, 1–1.5 mm in diameter below the spike.

Spike usually flexed at the base and inclined, flexuose, 1- to 3-, occasionally to 5-flowered; *bracts* green or greyish to dull purple on the upper surface, soft-textured, the outer 30–50(–65) mm long, the inner two-thirds to about as long as the outer. *Flowers* dull yellow to lilac, densely to lightly speckled throughout with small to large, light to dark brown or purplish spots, often darkest toward the tepal apices, the dorsal tepal more or less transparent below the apices on either side of the midline, strongly and sweetly scented both day and night; *perianth tube* obliquely funnel-shaped, 23–35(–48) mm long, the lower cylindrical part 15–25 mm long, with papillose zones in the lower throat opposite the lower tepals; *tepals* lanceolate-attenuate, margins undulate, the dorsal slightly larger, 25–33 x 13–19 mm, extending more or less horizontally over the stamens, the upper laterals directed forward and curving outward in the upper half, 25–33 mm long, the lower three tepals joined to the upper laterals for 5–7 mm and to one another for 1–2 mm, the free parts 23–33 mm long, curving downwards in the upper half. *Filaments* 13–16 mm long, exserted 4–6 mm from the tube; *anthers* 8–11 mm long, brown to cream, the pollen yellow. *Ovary* oblong, c. 4 mm long; *style* arching over the stamens, dividing slightly below to slightly beyond the anther apices, the branches 2–3 mm long. *Capsules* ovoid-ellipsoid, c. 12 mm long; *seeds* ovate, 5–6 x

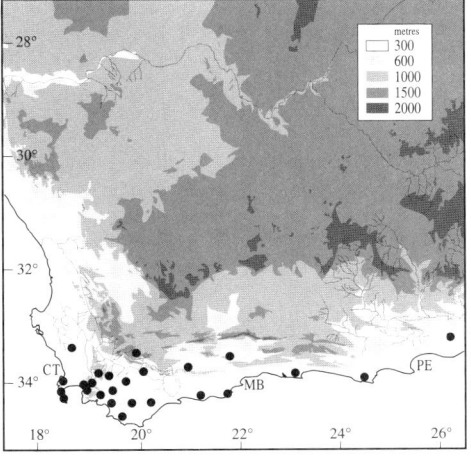

3–4 mm, broadly and evenly winged, golden brown, the seed body tan. *Chromosome number* 2*n* = 30.

Flowering time April to June, occasionally later in the year.

Distribution and biology

Gladiolus maculatus has a relatively wide range across the southern coast and immediate interior of the southern African winter-rainfall zone. Plants extend from the Cape Peninsula in the west along the southern Cape to Grahamstown and Alexandria in the Eastern Cape Province. They favour heavy soils and are most often found on clay soils in renosterveld vegetation or in the lighter soils at the interface of shale and sandstone strata of the Cape System. One of the winter-flowering species of the southern African winter-rainfall region, *G. maculatus* is relatively common in suitable habitats. It is, however, not often seen, both because it flowers in winter when weather discourages outdoor activity and because of the dull-coloured flowers.

The flowers of *Gladiolus maculatus* are relatively long-tubed and always have a strong, sweet fragrance during the day and in the evening. They are evidently adapted for pollination by moths and have the hallmarks of that pollination strategy, notably the production of ample nectar of relatively high sugar concentration, a long, narrow perianth tube, and a strong fragrance. The moth *Cucullia terensis* (Noctuidae) has been noted visiting the flowers and this species at least is presumed to pollinate *G. maculatus*.

Diagnosis and relationships

A moderately large, scented flower, typically with tan or dark brown shading and mottling on a pale background, transparent edges to the upper tepals, a flexuose spike, and somewhat fleshy leaves with short, plane blades characterize *Gladiolus maculatus*. The scented flowers and the pale perianth with dark mottling and transparent dorsal tepal edges combined with three leaves (or sometimes four) per stem might suggest a relationship with *G. hyalinus* and *G. liliaceus* of series *Tristis*. However, the soft-textured, fibrous corm tunics and leaves without thickened margins or midrib are inconsistent with series *Tristis* and we consider *G. maculatus* to belong in series *Mutabilis* close to *G. priorii*, *G. meridionalis* and *G. albens* which have similar leaf blade architecture and corm tunics. The timing of scent production also differs in the two series. Scent is produced only in the evening in *G. liliaceus*

and other species of series *Tristis*, whereas scent is produced both at night and during the day in *G. maculatus*. The flowers of *G. maculatus* are adapted for moth pollination and the floral similarities found in species of series *Tristis* are convergent adaptations for the same pollination strategy.

Gladiolus maculatus was treated by Lewis at al. (1972) as comprising four subspecies. One of these, subsp. *hibernus*, seems to us a minor variant of typical *G. maculatus* and is not recognized as a distinct taxon. Subsp. *eburneus* and subsp. *meridionalis* are, however, very different from *G. maculatus* and we regard them as distinct species, *G. albens* and *G. meridionalis* respectively. The criteria on which we base our decisions are discussed below under the two species.

History

The earliest record of the winter-flowering and fairly inconspicuous *Gladiolus maculatus* dates from the eighteenth century. A rather poor likeness of the species was included in Jakob Breyne's posthumously published book under the polynomial *Gladiolus, floribus e spadiceo et flavo variegatus suprema lacinia brevissima*. Plants were collected later by

Carl Peter Thunberg and Francis Masson in the 1770s. Thunberg associated the species with *G. tristis* and specimens he called *G. tristis* var. *odorus* include both *G. tristis* and *G. maculatus*. Then, in 1802, plants were raised from seed and flowered at the nursery of Grimwood and Wykes in London. A painting was immediately made and was soon published in *Curtis's Botanical Magazine* as *G. versicolor* (now *G. liliaceus*) var. *tenuior*, with a description by John Ker Gawler. Ker Gawler recognized this plant as the same as Thunberg's *G. tristis* var. *odorus*. In this same article Ker Gawler named the plant illustrated by Breyne *G. versicolor* var. *inaequalis*.

Recognition of the plant at species rank took another 24 years when, in 1826, Robert Sweet named the illustration in *Curtis's Botanical Magazine G. maculatus*. The following year Ker Gawler named the plant *G. suaveolens*, citing the same illustration. Ker Gawler described a second species at this time, *G. breynianus*, based on the Breyne illustration and a specimen collected by Francis Masson. The name *G. maculatus* was accepted for some time. J.G. Baker, however, completely misunderstood the species, and in *Flora Capensis* (1896) he included *G. maculatus* in *G. carinatus* (which he called *G. recurvus*), a related but quite separate species. Confusion continued, and

in 1929 H.M.L. Bolus described *G. vaginatus* var. *fergusoniae*, this none other than *G. maculatus* again. The collection is an unusually pale-flowered form of the species, and it is interesting to note that G.J. Lewis annotated the sheets *G. fergusoniae* in 1964, indicating that she intended to recognize the plant as a distinct species. Likewise, *G. hibernus*, named by Collingwood Ingram in 1931, is also *G. maculatus*, albeit a form with a slightly longer perianth tube than is usual. Lewis et al. (1972) restored the use of the name *G. maculatus* for the species, although, as explained above, *G. albens* and *G. meridionalis* were included in that treatment as separate subspecies of *G. maculatus*.

124. GLADIOLUS ALBENS Goldblatt & Manning
PLATE 107

Gladiolus albens Goldblatt & Manning, new species. Type: South Africa, Eastern Cape, Alexandria, 8 May 1931, *Galpin s.n.* (BOL 19568, holotype; K, isotype).

albens = whitish, for the flower colour.

Synonymy
Gladiolus maculatus subsp. *eburneus* Obermeyer in Lewis et al., *J. S. African Bot.*, Suppl. 10: 263 (1972). Type: South Africa, Eastern Cape, Alexandria, 8 May 1931, *Galpin s.n.* (BOL 19568, holotype; K, isotype).

Latin diagnosis
Plantae 25–50(–70) cm altae, cormo globoso 10–20 mm in diametro tunicis papyraceis vel fibrosis, foliis 3 raro 4, laminis linearibus 1–1.5 mm latis usque ad 100 mm longis marginibus costisque non incrassatis, spica 1–3(–5) florum, bracteis 27–40(–45) mm longis, floribus cremeis vel albis, tubo perianthii (35–)45–60 mm longo basem versus cylindricis, tepalis lanceolatis inaequalibus dorsali 23–35 x 12–21 mm, inferioribus 21–31 x 1–13 mm, filamentis 12–15 mm longis, antheris 9–12 mm longis.

Description
Plants 25–50(–70) cm high. *Corm* globose, 10–20 mm in diameter, the tunics papery, decaying with age into vertical fibres from below. *Cataphylls* pale and membranous, the uppermost reaching up to 9 cm above the ground and then green or dull purple. *Leaves* three, rarely four, the lower two basal, the lowermost sheathing the lower third to half of the stem, rarely reaching beyond the middle of the stem, the blade more or less plane, linear, 1–2 mm wide, short, up to 10 cm long, somewhat fleshy, with three equal veins, the margins not raised nor thickened, remaining leaves cauline, inserted between the middle and upper third of the stem, progressively shorter above, all with short blades. *Stem* erect, lightly flexed above

the sheaths of the upper one or two leaves, unbranched, 1–1.5 mm in diameter below the spike.

Spike flexed at the base and inclined, flexuose, 1- to 3-, rarely to 5-flowered; *bracts* green or greyish on the upper surface, soft-textured, the outer 27–40(–45) mm long, the inner two-thirds to about as long as the outer. *Flowers* cream to white, sometimes with small speckles or faint red-brown streaks in the throat or all over the tepals, with a light, somewhat metallic, dusty sweet scent; *perianth tube* narrowly and obliquely funnel-shaped, (35–)45–65 mm long, the lower cylindrical part 25–42 mm long, gradually expanding in the upper part; *tepals* ovate, the margins lightly undulate, subequal, the dorsal slightly larger, 23–35 x 12–21 mm, extending more or less horizontally over the stamens, the upper laterals directed forward and curving outward in the upper third, 23–39 x 9–14 mm, the lower three tepals joined to the upper laterals for 5–6 mm and to one another for 1–2 mm, the free parts 21–31 x 1–13 mm, curving downwards in the upper half. *Filaments* 12–15 mm long, exserted 3–6 mm from the tube; *anthers* 9–12 mm long, cream, the pollen yellow. *Ovary* oblong, c. 5 mm long; *style* arching over the stamens, dividing close to or shortly beyond the anther apices, the branches 3–5 mm long. *Capsules* and *seeds* unknown. *Chromosome number* unknown.

Flowering time March to May, occasionally in February.

Distribution and biology
Gladiolus albens is a narrow endemic of Eastern Cape Province, South Africa. Most

records of the species are from the Albany and Alexandria Districts but there is also a record from Somerset East and another from George in Western Cape Province. The latter is unlikely and in need of confirmation. Although abutting the southern African winter-rainfall zone, the Albany and Alexandria Districts are generally regarded as falling within the southern African summer-rainfall zone. Nevertheless, they do have some winter rain and consequently have a flora that has some species adapted to a winter-rainfall regime. *Gladiolus albens* is one of these, for although it flowers in the autumn, in April and May, its main growth cycle is in the winter months after flowering is complete. *Gladiolus albens* grows on grassy slopes and in low, fynbos-like vegetation on light, well-drained soils.

The flowers have the appearance of being adapted for moth pollination, having a long, slender perianth tube, a pale colour and a light fragrance. The pollination ecology of the species has yet to be investigated.

Diagnosis and relationships

Slightly fleshy leaf blades without thickened margins or midrib place *Gladiolus albens* in series *Mutabilis* of section *Homoglossum* where it can be distinguished solely on floral features. The corm tunics are more or less papery and fairly soft in texture, and they

decay with age into vertical fibres from below, thus being typical of the series. The flowers, however, are unique in series *Mutabilis*, being white to cream and with a fairly long, slender perianth tube, 35–50 mm long, widening gradually in the upper third. No doubt *G. albens* is closely related to the *G. maculatus* complex in which it was recognized as subsp. *eburneus* by Lewis et al. (1972). It differs sufficiently from *G. maculatus*, which has shorter-tubed flowers with larger tepals usually heavily mottled with dark brown, that the two cannot be mistaken for one another. We see no merit in regarding *G. albens* simply as a subspecies of *G. maculatus* and describe it here as a new species.

History

The first collection of *Gladiolus albens* appears to be an undated one from Olifantshoek, now Alexandria, in the eastern Cape. Label information suggests that the Colonial Botanist, Carl Ludwig Pappe, was the collector, but as Pappe never went to the eastern Cape this cannot be so. We suspect that the collection is that of C.L. Zeyher whose collections came into Pappe's possession after Zeyher's death in 1858. Zeyher certainly collected at Olifantshoek and perhaps gathered his specimens of *G. albens* there in 1832 (Gunn & Codd, 1981).

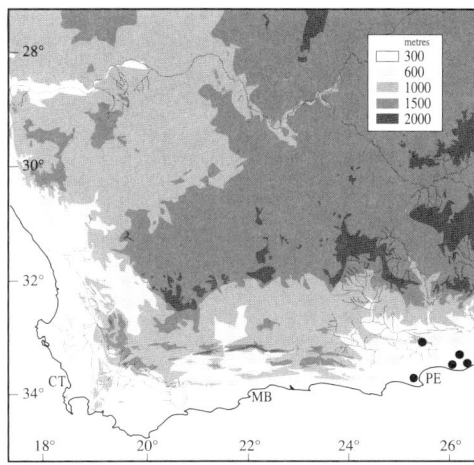

Thomas Cooper, collecting for his patron, W. Wilson Saunders, also collected *G. albens* relatively early, recording plants in the Albany District in 1860. These and later collections of *G. albens* were confused with other species, including *G. longicollis*, and attracted little attention. Specimens at herbaria in Cape Town have annotations in G.J. Lewis's hand indicating that she intended to recognize the species as *G. albus*. *Gladiolus albens* was, however, treated as *G. maculatus* subsp. *eburneus* by A.A. Obermeyer in the revision of *Gladiolus* in South Africa. Evidently Obermeyer decided not to follow Lewis's taxonomy regarding the species.

125. GLADIOLUS MERIDIONALIS G. Lewis

PLATE 108

Gladiolus meridionalis G. Lewis, *Ann. S. African Mus.* 40: 127 (1954). *G. maculatus* subsp. *meridionalis* (G. Lewis) Obermeyer in *J. S. African Bot.*, Suppl. 10: 262 (1972). Type: South Africa, Caledon District, south slopes of mountains near Danger Point, 28 Apr. 1948, *Linley s.n.* (SAM 60214, holotype; BOL, K, isotypes).

meridionalis = southern, referring to the distribution along the southern coast of South Africa.

Description

Plants 30–80 cm high. *Corm* globose, 10–20 mm in diameter, the tunics papery to coriaceous, decaying with age into vertical fibres from below. *Cataphylls* pale and membranous, the uppermost reaching up to 12 cm above the ground and then green or dull purple. *Leaves* three, the lower two basal, the lowermost sheathing the lower half of the stem, short, rarely reaching the base of

the spike, the blade more or less plane, usually fairly fleshy, the margins and midrib neither raised nor thickened, remaining leaves cauline, inserted between the middle and upper third of the stem, progressively shorter above, all with short blades. *Stem* erect, lightly flexed above the sheaths of the upper two leaves, unbranched, 1–1.5 mm in diameter below the spike.

Spike flexed at the base and inclined, flexuose, 1- to 2-, rarely to 3-flowered; *bracts* green or greyish to dull purple on the upper surface, soft-textured, the outer 36–50 mm long, the inner two-thirds to about as long as the outer. *Flowers* pale to deep pink or cream, often with darker spots at the base of the lower tepals and in the lower throat, faintly sweet-scented; *perianth tube* 40–48 mm long, slender and cylindric below for c. 20 mm, the upper part wide and cylindric; *tepals* ovate, subequal, the dorsal 22–28 x 18–23 mm, extending more or less horizontally over the stamens, the upper

laterals 24–32 x 15–17 mm long, the lower three tepals joined to the upper laterals for 6–7 mm and to one another for c. 1 mm, the free parts 20–28 mm long, curving downwards in the upper half. *Filaments* 28–40 mm long, exserted 3–4 mm from the tube; *anthers* 9–10 mm long, light brown, the pollen yellow. *Ovary* globose, c. 5 mm long; *style* extending horizontally over the stamens, dividing up to 5 mm beyond the anther apices, the branches c. 4 mm long. *Capsules* ovoid-ellipsoid, c. 12 mm long; *seeds* ovate, c. 5 x 3 mm, broadly and evenly winged, golden brown, the seed body tan. *Chromosome number* unknown.

Flowering time April to June, occasionally in March or rarely later in the year.

Distribution and biology

Gladiolus meridionalis has a discontinuous distribution along the southern Cape coast. Populations occur between Gansbaai and Elim in the southwestern Cape, and between

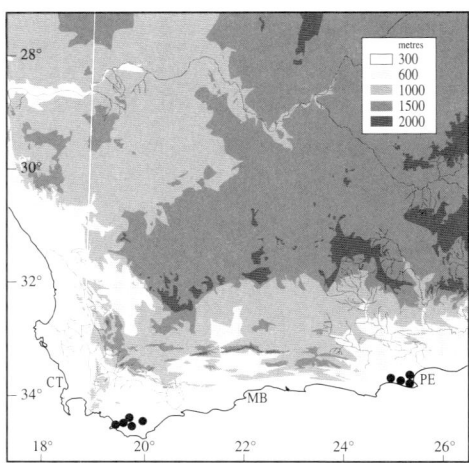

have flowers adapted for pollination by sunbirds. At two sites where we collected *G. meridionalis* it was growing together with *Tritoniopsis pulchra* (also Iridaceae) which has similarly coloured and shaped flowers and is also thought to be pollinated by sunbirds. The small orange-breasted sunbird (*Nectarinia violacea*) was active at the sites and is very likely the pollinator of both species.

Diagnosis and relationships

The outstanding feature of *Gladiolus meridionalis* is its long perianth tube, slender below and wide and cylindric above. The flowers themselves are variable in colour. Populations in the west of its range have dark pink to salmon flowers and the eastern populations have cream to pinkish orange flowers. The leaf blades are slightly fleshy and without thickened margins or visible midrib and thus typical of series *Mutabilis*, as is the strongly flexuose stem and spike. Plants have just three leaves, a feature consistent with *G. maculatus* of the series, to which *G. meridionalis* is most likely closely allied.

Gladiolus meridionalis was included in *G. maculatus* as subsp. *meridionalis* by Lewis et al. (1972), a treatment we cannot accept. Although vegetatively similar, the two have significantly different flowers, *G. maculatus* having cream flowers usually heavily speckled with dark brown, a strong sweet scent, and a perianth tube which widens gradually in the upper third. The floral differences are correlated with the pollination systems of the two species, *G. maculatus* being adapted for pollination by moths and *G. meridionalis* for pollination by sunbirds.

Flowers adapted for sunbird pollination also characterize *Gladiolus priorii* of series *Mutabilis*, but that species is easily distinguished by normally having four leaves whereas *G. meridionalis* has three. The flowers of these two species also differ in tepal shape and slightly in colour, those of *G. priorii* being scarlet with yellow or cream at the base of the lower tepals and in the throat. The dark pink to salmon-coloured form of *G. meridionalis* does not have yellow markings. We think it most likely that the two species of series *Mutabilis* with flowers adapted for pollination by sunbirds evolved this system independently from a common ancestor, most likely *G. maculatus* or its immediate ancestor. Our assumption is based on the differences in leaf number and flower colouring in the two species, as well as on the persistence of scent in only *G. meridionalis*. This is presumably an atavistic trait no longer useful in attracting pollinators.

History

Gladiolus meridionalis was apparently first recorded in 1896 by the energetic collector, Rudolf Schlechter, who gathered plants near Elim in the southern Cape. Initially this early collection was confused with *G. watsonius* and later with *G. priorii*, after that species was described in 1932. Specimens bought from flower-sellers in Cape Town in 1929 were, however, recognized by H.M.L. Bolus as an undescribed species which she intended to name *G. hiemalis*. Both early collections plus several found later formed the basis for *G. meridionalis* which was described by G.J. Lewis in 1954. The species was reduced to subspecies rank in *G. maculatus* by A.A. Obermeyer (in Lewis et al., 1972), a treatment which we do not accept for the reasons outlined above.

Greenbushes and Bethelsdorp in the Port Elizabeth District in Eastern Cape Province. Plants grow in stony sandstone soils in low fynbos on mountain slopes and flats, usually in sight of the ocean and, we assume, in soils of slightly higher nutrient status because of the proximity to the coast.

The flowers of *Gladiolus meridionalis* are adapted for pollination by sunbirds and have the long perianth tube with a wide cylindric upper part and large quantities of nectar associated with this pollination system. Somewhat unusually for sunbird pollination, the nectar has a relatively high sugar concentration and is sucrose-dominant. This pattern is also encountered in *G. watsonius* and its allies in series *Homoglossum*, which also

---

126. GLADIOLUS PRIORII (N.E. Brown) Goldblatt & De Vos

Rooipypie

PLATE 109

Gladiolus priorii (N.E. Brown) Goldblatt & De Vos, *Bull. Mus. Hist. Nat., Paris*, Sér. 4, Sect. B, *Adansonia* 11: 421 (1989). *Antholyza priorii* N.E. Brown, *Kew Bull.* 1929: 244 (1929). *Homoglossum priorii* (N.E. Brown) N.E. Brown, *Trans. Roy. Soc.*

S. Africa 20: 279 (1932). De Vos, *J. S. African Bot.* 42: 329 (1976). Type: South Africa, Western Cape, near Cape Town, open sites in scrub 'in dumetis apricis,' July 1846, *Prior s.n.* (K, holotype).

priorii, named in honour of Richard Prior (born Alexander), British medical doctor and amateur naturalist who collected actively in South Africa in the years 1846 to 1848.

Synonymy

Gladiolus hirsutus var. *tenuiflorus* Klatt, *Linnaea*
32: 713 (1863). Type: South Africa, Western
Cape, flats and hills near Cape Town, without
date, *Ecklon & Zeyher Irid. 152* (MO, isotype).
[*Antholyza revoluta* sensu Baker, *Fl. Capensis* 6:
169 (1896), in part, not *A. revoluta* Burman fil.
= *Tritoniopsis revoluta* (Burman fil.) Goldblatt).]

Description

Plants 30–50(–70) cm high. *Corm* ovoid,
12–20 mm in diameter, the tunics papery,
decaying with age into soft vertical fibres
from below. *Cataphylls* pale and membranous, the uppermost reaching 4–12 cm
above the ground and then green or purplish. *Leaves* usually four, occasionally three,
the lowermost basal and longest, sheathing
the stem in the lower third to half, the blade
relatively short, up to 10 cm long, occasionally reaching the base of the spike, more or
less plane, usually fairly fleshy, the margins
and midrib neither raised nor thickened, the
remaining leaves progressively shorter above.
Stem erect, flexed outward above the sheaths
of the upper three leaves, unbranched,
1–2 mm in diameter below the spike.

Spike inclined, flexuose, 1- to 4-, occasionally up to 5-flowered; *bracts* green, firm-
textured, the outer 25–40(–50) mm long,
the inner two-thirds to almost as long as the
outer. *Flowers* scarlet, the base of the lower
tepals and upper part of the tube pale yellow
or cream, unscented; *perianth tube* 30–
46 mm long, slender and cylindric in the
lower 10–20 mm, abruptly expanded into a
wider cylindric upper part 20–30 mm long;
tepals lanceolate, unequal, the dorsal largest,
extending horizontally over the stamens,
25–32 x 15–19 mm, the upper laterals
spreading from the base, 22–32 x 10–
12 mm, the lower three tepals arching downward from the base, 20–27 x 9–12 mm.
Filaments 25–37 mm long, exserted 6–
10 mm from the tube; *anthers* 7.5–8.5 mm

long, pale yellow, the pollen yellow. *Ovary*
oblong, 5–8 mm long; *style* extending horizontally over the stamens, dividing between
the middle and upper third of the anthers,
2.5–3.5 mm long. *Capsules* ellipsoid, 12–
15 mm long, acute at the apex; *seeds* ovate,
6–7 x c. 5 mm, broadly and evenly winged,
golden-brown, the seed body slightly darker.
Chromosome number $2n = 30$.

Flowering time May and June, occasionally
in April or July.

Distribution and biology

Gladiolus priorii has a fairly narrow distribution in Western Cape Province, South
Africa. It extends from the granite hills north
of Saldanha Bay southward to the Cape
Peninsula, extending only a short distance
inland to the Hottentots Holland Mountains
and eastward along the coast as far as
Hermanus. Although it is common on granite outcrops along the west coast, *G. priorii*
grows equally well on rocky sandstone slopes
on the Cape Peninsula and elsewhere.

Gladiolus priorii is one of the first of the
winter-flowering species of *Gladiolus* to
bloom. Flowering may begin in April, but
it is more often seen in May or June, and
even in July in years when the rains are late.
The flowers are clearly adapted for pollination by sunbirds and they have the red
perianth, elongate flower tube with a wide
cylindrical upper part, and exserted anthers
typical of this pollination strategy. The large
volume of nectar of low sugar concentration
produced by the flowers is also consistent
with sunbird pollination. Observations of
sunbirds visiting the species have yet to be
recorded.

Diagnosis and relationships

Gladiolus priorii has bright red flowers with
a long perianth tube, narrow below and wide
and long above, and elliptic to lanceolate
tepals together with a conspicuously flexuose
stem. These features combined with slightly
fleshy leaves without visible venation when
alive make the species easy to recognize. It
has been confused with a handful of other
western Cape species with similar flowers,
and has even been included (under various
names) in *G. watsonius*. The latter has
straight, erect stems, rigid leaves with the
margins and midribs strongly thickened and
the perianth tube is abruptly expanded into
the wide upper part, unlike *G. priorii* where
the transition to the upper tube is fairly
gradual.

The peculiar leaf morphology of *Gladiolus
priorii* places the species in series *Mutabilis* of

section *Homoglossum* rather than with the
species of series *Homoglossum* which have
flowers adapted for pollination by sunbirds.
Species of the latter series consistently have
three leaves, always with the margins and
midrib thickened, usually elongate capsules,
and somewhat woody corm tunics. Like
some members of series *Mutabilis*, *G. priorii*
may have three or four leaves (always four
in robust individuals), and like all members
of the series the leaf blades always lack
thickened margins and midrib. It is evident
that the affinities of *G. priorii* lie most
closely with *G. maculatus*, which also has
finely fibrous corm tunics and relatively
short capsules. Floral similarity with series
Homoglossum is thus due to convergence
for the same pollinators. It is perhaps
significant in this regard that although
the flowers of *G. priorii* closely resemble
several species of series *Homoglossum*, it
differs in nectar characteristics. While
species of series *Homoglossum* have nectar
of high sugar concentration, usually above
30%, nectar of *G. priorii* has a concentration
of about 21%, supporting our hypothesis
of an independent origin of the pollination
system.

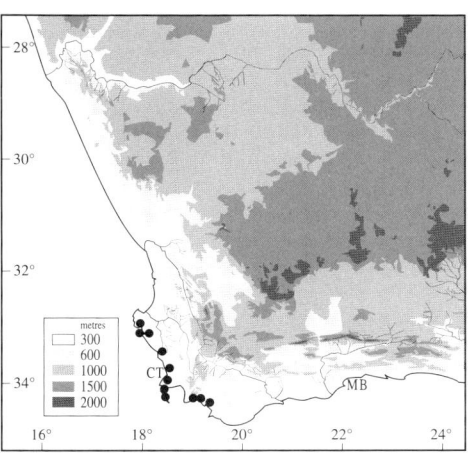

History

Poorly known until the twentieth century, *Gladiolus priorii* was only recognized as a distinct species in 1929 when the English botanist, N.E. Brown, named it *Antholyza priorii*. Brown based the species on a collection made the previous century in 1846 by Richard C. Prior (born Richard Alexander), who Brown assumed had discovered the species. This is not the case, for there is a much earlier record in the Kew Herbarium, collected in 1815 on Devil's Peak on the Cape Peninsula by the explorer-naturalist,

William Burchell. The species was also collected by C.F. Ecklon and C.L. Zeyher near Cape Town in the late 1820s. Their collection formed the basis for F.W. Klatt's *G. hirsutus* var. *tenuiflorus* (in which he included *G. merianellum* sensu Thunberg, now *G. bonaspei*). Mostly, however, these few early collections of *G. priorii* were assigned either to *Gladiolus* (or *Antholyza* or *Homoglossum*) *watsonius* which it closely resembles in its flowers, or to *Antholyza* (or *Homoglossum*) *revolutum*. This last species was described by Burman fil. in 1768 and its

identity remained unknown until the discovery of what is presumed to be the type illustration (Wijnands & Goldblatt, 1992). The type plant is a species of *Tritoniopsis* and not a species of *Gladiolus* as was often thought. As generic concepts in the Iridaceae changed in the twentieth century, *Gladiolus priorii* was placed in *Homoglossum* (Brown, 1932; De Vos, 1976) where it remained until 1989 when that genus was formally included in *Gladiolus* (Goldblatt & De Vos, 1989).

Section *Homoglossum*: Series *Brevitubus*

127. GLADIOLUS BREVITUBUS G. Lewis

PLATE 110

Gladiolus brevitubus G. Lewis in *J. S. African Bot.*, Suppl. 10: 171 (1972), as a new name for *Tritonia ventricosa* Baker, *Handbook Irideae* 193 (1892), not *Gladiolus ventricosus* Lamarck (= *G. carneus* Delaroche). Type: South Africa, Western Cape, Appelskraal near Riviersonderend, Sept., *Zeyher 3793* (K, lectotype designated by

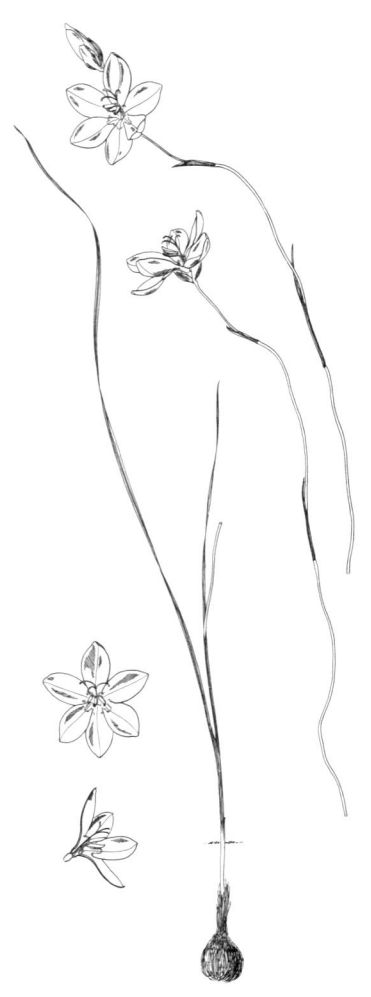

Lewis, *J. S. African Bot.*, Suppl. 10: 171, (1972); SAM, isolectotype).

brevitubus = short-tubed, so named for the extremely short perianth tube, a most unusual feature in *Gladiolus*.

Description

Plants 12–40 cm high. *Corm* globose, 8–12 mm in diameter, the tunics of fine to medium-textured reticulate fibres. *Cataphylls* pale and membranous, accumulating with age in a short fibrous neck around the underground part of the stem. *Leaves* three or four, the lower one or two basal and longest, the lowermost usually somewhat longer than the spike, the sheath scabridulous, the blade linear, 1.5–3.5 mm wide, the midrib and margins lightly thickened, remaining leaves progressively shorter above, the upper two leaves largely to entirely sheathing, the margins of the uppermost free to the base. *Stem* erect or inclined, more or less straight, unbranched, c. 1 mm wide below the spike.

Spike inclined, the flowers borne on the upper side, lightly flexuose, 2- to 8-flowered; *bracts* green, broadly ovate and obtuse, the margins often reddish, the outer 10–13 mm long, the inner slightly shorter to slightly longer than the outer. *Flowers* nearly actinomorphic, the stamens unilateral, orange to salmon or brick-red, the lower lateral tepals yellow near the base, with a light, sweet scent; *perianth tube* funnel-shaped, 2.5–4 mm long, papillate at the throat; *tepals* subequal, the outer whorl slightly larger than the inner, ovate, 11–18 x 5.5–10 mm. *Filaments* 2–3 mm long, reaching to about the mouth of the tube, the bases papillate;

anthers 4–5 mm long, yellow, the pollen yellow. *Ovary* ovoid, c. 2 mm long; *style* straight, 4–5 mm long, dividing opposite the base of the anthers, the branches 3–4 mm long. *Capsules* obovoid, 8–11 mm long, three-lobed apically and retuse; *seeds* oblong, 4.5–6 x 3.5–4 mm, broadly and evenly winged, light reddish brown, the seed body darker brown. *Chromosome number* unknown.

Flowering time late October and November, occasionally in early December.

Distribution and biology

Gladiolus brevitubus is a local endemic of the southwestern Cape, South Africa. Its range extends from the Hottentots Holland Mountains in the west to Riviersonderend in the east and to the Klein River Mountains in the south. Plants grow on rocky mountain slopes in sandy soil at elevations of between 200 and 1500 m. Although the species flowers well after fire, as do so many geophytes of similar habitats, *G. brevitubus* can usually be found flowering in unburnt low fynbos as long as the vegetation is less than 40 cm tall.

The bright orange flowers have a light sweet scent, but lack nectar and appear to be pollinated by fairly small, pollen-collecting bees of the family Halictidae. These bees visit the flowers with some frequency and appear especially attracted to the papillae at the base of the filaments and the throat of the perianth tube. They often also gather pollen from the anthers in a purposeful manner.

Diagnosis and relationships

Few species of *Gladiolus* look less like the genus. The orange to brick-red flowers are almost actinomorphic, with the lower tepals nearly equal in size to the upper, and the perianth tube is extremely short, 2.5–4 mm in length, but the stamens are unilateral. The plants are usually low in stature and have small corms with fairly soft-textured fibrous tunics and three or four linear, superposed leaves more than 1.5 mm wide with lightly thickened margins and midrib. The species undoubtedly does belong in *Gladiolus*, for it has the broadly winged seeds that characterize the genus and are unique in the Iridaceae.

The relationships of *Gladiolus brevitubus* are puzzling. We tentatively include the species in section *Homoglossum* in a separate series *Brevitubus*. This placement is somewhat doubtful because the small, three-lobed and apically retuse capsules are discordant in section *Homoglossum*, as are the fibrous corm tunics. Species of the section typically have ellipsoid capsules and woody to papery corm

tunics. We considered an alternative placement in section *Hebea*, especially because of the fibrous corm tunics and orange to brick flower colour, but not only are the capsules inconsistent with that section, but the leaves are more or less superposed and have thickened margins, and the stem is unbranched, features which make inclusion in section *Hebea* even less likely than in section *Homoglossum*.

History

Although first recorded in the 1820s by C.L. Zeyher, probably together with his frequent companion on plant-collecting expeditions, C.F. Ecklon, *Gladiolus brevitubus* was described only in 1892 by J.G. Baker. The short perianth tube and orange flower prompted Baker to assign the species to *Tritonia*, naming it *T. ventricosus*. It lacks the membranous to scarious bracts of *Tritonia* and is now known to have winged

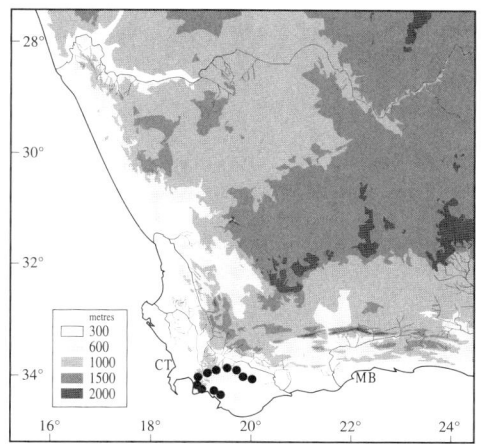

seeds typical of *Gladiolus* as well as green, soft-textured floral bracts characteristic of the genus. Because of the existence of the name *G. ventricosus* Lamarck, a later synonym of *G. carneus*, *T. ventricosus* was renamed *G. brevitubus* when it was transferred to *Gladiolus* by Lewis et al. (1972).

Section *Homoglossum:* Series *Gracilis*

128. GLADIOLUS ROGERSII Baker
Outeniqua or George bluebell, Riversdale bluebell
PLATE 111

Gladiolus rogersii Baker, *Handbook Irideae* 208 (1892); *Fl. Capensis* 6: 146 (1896). Lewis et al., *J. S. African Bot.*, Suppl. 10: 242 (1972). Type: South Africa, Western Cape, without precise locality (as Cape Flats), 1859, *Moyle Rogers s.n.* (K, holotype).

rogersii, named in honour of the early collector of southern African plants, Rev. Moyle Rogers, who made the type collection.

Synonymy

Gladiolus bolusii var. *burchellii* F. Bolus, *Ann. Bolus Herb.* 2: 101 (1917). *G. burchellii* (F. Bolus) Ingram, *Gard. Chron.*, Ser. 3, 92: 482 (1932). Type: South Africa, Western Cape, Knysna District, west end of Groenvlei, 9 Aug. 1814, *Burchell 5631* (K, holotype).
Gladiolus rogersii var. *graminifolius* G. Lewis in *J. S. African Bot.*, Suppl. 10: 246 (1972). Type: South Africa, Western Cape, Montagu District, Scheepersrus, 5 Aug. 1949, *Barker 5427* (NBG, holotype; BOL, isotype).
Gladiolus rogersii var. *vlokii* Goldblatt, *J. S. African Bot.* 50: 453 (1984). Type: South Africa, Western Cape, Ruitersbos Forest Reserve in deep sand, 1 Mar. 1948, *Vlok 348* (NBG, holotype; MO, PRE, isotypes).
[*Gladiolus involutus* sensu Baker, *Fl. Capensis* 6: 147 (1896), not sensu Delaroche (1766)].

Description

Plants (20–)30–50(–65) cm high. *Corm* globose, 8–14 mm in diameter, the tunics usually of soft-textured layers of fine fibres, occasionally the fibres of medium texture, rarely coarse and claw-like below. *Cataphylls* pale and membranous, the uppermost usually extending well above the ground and then uniformly purple, mottled purple and white, or dry and dark brown, sometimes minutely puberulous. *Leaves* three or four, rarely five, the lowermost or lower two longest, usually at least reaching the base of the spike or shortly exceeding it, sometimes half as long again as the spike, the blades linear, rarely narrowly sword-shaped, 1.5–4(–6.5) mm wide, straight, soft-textured to somewhat wiry, the margins and midrib lightly to heavily thickened, rarely more or less solidly terete, remaining leaves inserted on the stem well above the ground and progressively shorter above, sheathing for about half their length, the blades like the basal, when leaves four the uppermost sheathing entirely, the margins of the sheathing parts of the upper two leaves united below, or sometimes free to the base. *Stem* erect below, flexed outward above the sheaths of the two upper leaves, thus inclined above, unbranched, c. 1.5 mm in diameter below the spike.

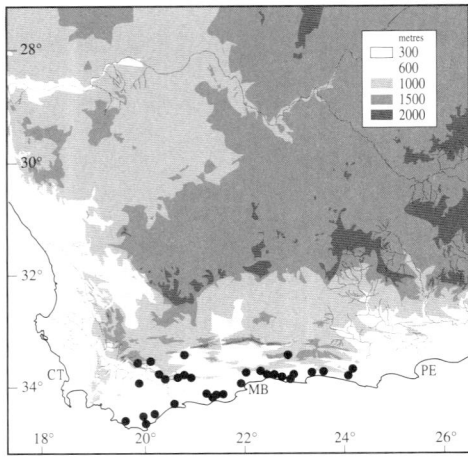

Spike inclined, occasionally 1-, usually 3- to 6-flowered; *bracts* green or lightly flushed with grey to purple above, diverging markedly from the axis, the outer 15–22 mm long, with transparent margins c. 2 mm wide, the inner similar but slightly smaller than the outer, minutely forked apically. *Flowers* inflated and bell-like, pale to dark blue or purple, darker on the reverse of the upper tepals, the lower tepals lightly spotted with purple in the lower half and with a transverse or median yellow to white mark outlined or surrounded with dark blue to purple in the upper third, usually lightly sweet-scented; *perianth tube* 12–19 mm long, obliquely funnel-shaped, the lower half erect and cylindric, the upper half bent at right angles and flaring outward; *tepals* unequal, the upper obovate, the lower spathulate, the dorsal largest, extending horizontally over the stamens, 20–30 x 16– 26 mm, upper laterals slightly shorter and extending forward, only slightly curving outward apically, the lower three initially curving downward, then arched upward, the apices directed forward or downward, united for 5–9 mm with the upper laterals, the free parts 16–22 x 8–16 mm, not clearly divided into a limb and claw, raised distally in the midline, margins curved upward, the apices broadly obtuse or emarginate, in profile exceeding the upper by c. 5 mm. *Filaments* 10– 12(–15) mm long, exserted 5–8 mm from the tube; *anthers* 5.5–9 mm long, lying more or less horizontally, blue, the pollen cream. *Ovary* oblong, c. 3 mm long; *style* arching over the stamens, dividing close to the anther apices, the branches c. 4 mm long, extending beyond the anthers. *Capsules* and *seeds* unknown. *Chromosome number* unknown.

Flowering time mainly late August to mid-October, occasionally to mid-November, exceptionally March to July.

Distribution and biology

Gladiolus rogersii is restricted to the cooler and wetter parts of the southern African winter-rainfall zone. Its range extends from the Agulhas Peninsula near Pearly Beach in the west along the coastal plain and through the Langeberg and Outeniqua Mountains of the southern Cape almost to Humansdorp. A notable variant, called var. *graminifolius* by Lewis et al. (1972), occurs interior to the main Langeberg axis in a few sites in the Little Karoo between Barrydale and Montagu and on the isolated Touwsberg a short distance to the north. These plants only occur in rocky sites on south-facing slopes, whereas in the mountains and coastal plain to the south plants may grow in less sheltered places. The range of habitats in which *G. rogersii* occurs is unusual, although not without precedent in the genus. Plants seem to be as well adapted to well-drained sandstone slopes at elevations of up to 1000 m as to limestone flats near the coast where they thrive under a completely different nutrient and soil pH regime. Plants from limestone habitats typically have more fibrotic leaves, often solidly oval or terete, and they often have the cataphylls and leaf sheaths mottled reddish or purple, features that may be related to the habitat.

The short-tubed blue to mauve flowers of *Gladiolus rogersii* are assumed to be pollinated primarily by long-tongued anthophorid bees for they have the same basic structure, sweet scent and colouring of many species of the genus for which the pollination system is known.

Diagnosis and relationships

Although *Gladiolus rogersii* is a typical member of section *Homoglossum*, its immediate relationships are uncertain. This is largely because of the variation in its corm tunics, which range from coarsely fibrous and almost woody to fairly softly fibrous. We have placed the species in series *Gracilis* and assume that the coarsely fibrous to woody tunics are the ancestral condition and that the softer, more fibrous tunics are derived and associated with some of the more mesic habitats in which the species grows. Plants usually have four narrow leaves, the margins and midrib of which are lightly raised, but plants from populations growing in exposed habitats on limestone have solidly oval or even terete leaves. Whatever the substrate on which *G. rogersii* grows, the flowers are always similar, having a perianth tube 12–19 mm long and tepals forming an inflated bell. The perianth is usually deep blue, but mauve and

pale blue flowers are also common. The lower tepals are typically marked with white to yellow V-shaped transverse bands outlined in dark purple, and the lower throat is speckled with mauve on a pale ground. Flowers are typically lightly fragrant, but some populations apparently have unscented flowers.

The flowers of *Gladiolus rogersii* most closely resemble those of *G. bullatus* and we assume the two are immediately related. The two species are, however, readily distinguished for *G. bullatus* has attenuate floral bracts, 30–70 mm long, which are prominently striated above the veins; large flowers, 35–50 mm long; and spikes of only one or two flowers. *Gladiolus bullatus* also has corms with coarse fibres of a woody texture, very much like the corms of other members of series *Gracilis*. The smaller flowers of *G. rogersii* are 32–48 mm long, the bracts are shorter, 15–22 mm long, and they have a smooth surface. Although individual plants of *G. rogersii* may have only one or two flowers per spike, robust plants rarely have less than four flowers and occasionally up to six per stem.

Variation

There are three main series of populations of *Gladiolus rogersii*. The typical form of the species occurs on the Langeberg and Outeniqua Mountains and has relatively soft-textured corm tunics and green or purplish, but not mottled, cataphylls. The leaves are usually more or less plane and have lightly to moderately thickened margins and midrib.

Plants from coastal limestone outcrops along the southern Cape coast have flowers exactly like those of other populations, but they often have coarsely fibrous corm tunics, mottled purple and green or white cataphylls and leaf sheaths, and the leaves tend to have a coarse texture. They are also often oval or terete in transverse section, and without evident midrib or margins.

Plants from drier habitats, especially in the Little Karoo, often have a weakly developed neck of fibres, derived from decaying cataphylls, around the base of the stem. These plants also have unusually broad leaves, at least 3–4 mm wide, and in the type of var. *graminifolius*, 4–6 mm wide. Despite the wider leaves and weakly developed neck in these populations, we prefer not to recognize a formal infraspecific taxon. There seems to be a gradient in leaf width and in the texture of the corm tunics depending on whether the habitat is drier or wetter. More information is, however, needed to better understand the plants from the Little Karoo.

A last group of populations consists of early-flowering plants with unscented flowers that were formally recognized by Goldblatt (1984) as var. *vlokii*, based not only on the flowering time, March to May, but on the coarsely fibrous corm tunics, at least in the type (other specimens cited lack corm tunics). Although the corm tunics of *Gladiolus rogersii* are often composed of fairly fine, softly textured fibres, coarse fibres are known in some populations that flower in the spring, on both sandstone and limestone soils. It now seems likely that the coarse corm tunics in *G. rogersii* may be related to habitat conditions, as are leaves with particularly strongly thickened midrib and margins, such as found in plants from the Kammanassie Mountains.

History

Described by J.G. Baker in 1892 based on plants collected in about 1860 by the clergyman and botanist, Rev. William Moyle Rogers, *Gladiolus rogersii* has a confused history. The locality indicated on the type collection, Cape Flats, is obviously wrong and is largely responsible for the ensuing confusion about the identity of the species.

Rogers' collection was not, however, the first record of *Gladiolus rogersii*, for it had already been recorded in the early 1770s by the Swedish botanists, C.P. Thunberg in 1772 or 1773 and Anders Sparrmann in 1775. Thunberg understandably confused these collections with the superficially similar *G. inflatus*. Plants were later collected by the naturalist-explorer, William Burchell, in 1814 at Groenvlei near Knysna and elsewhere in the southern Cape, and *G. rogersii* was also found by C.L. Zeyher and J.F. Drège in the 1830s. In *Flora Capensis* Baker (1896) cited one Burchell collection of the species under *G. hastatus*, and the Drège and Zeyher collections under *G. inflatus*, the latter otherwise including mostly *G. ornatus*. Baker also included two more Burchell collections under a third name, *G. involutus*, a quite different and unrelated species. The only specimen included in *G. rogersii* by Baker was the type of the species. No wonder then, that plants from Groenvlei actually collected by Burchell were segregated by Frank Bolus as *G. bolusii* var. *burchellii* when he was working on the genus in the years before 1917. *Gladiolus bolusii* is a later synonym of *G. inflatus* which is a Western Cape montane species with inflated flowers resembling those of *G. rogersii*. The variety was raised to species rank by the British *Gladiolus* grower, Collingwood Ingram, in 1932. Both Bolus and Ingram were clearly quite unaware of the identity of *G. rogersii* at this time, and with its wrong locality information had no reason to compare it with specimens from the southern Cape. Other botanists, including N.E. Brown in 1928 and R.A. Dyer in 1943, did however, associate the type of *G. rogersii* with collections from the southern Cape. Thus by the end of World War II, the Outeniqua or Riversdale bluebell was correctly known as *G. rogersii*.

129. GLADIOLUS BULLATUS Thunberg ex G. Lewis
Caledon bluebell
PLATE 112

Gladiolus bullatus G. Lewis, *Bot. Not.* 119: 286 (1966) (as *G. bullatus* Thunberg ex N.E. Brown), as a new name for *G. spathaceus* Pappe ex Baker, *Handbook Irideae* 208 (1892); *Fl. Capensis* 6: 147 (1896), an illegitimate homonym, not *G. spathaceus* Linnaeus fil. (1782) (= *Babiana spathacea* (Linnaeus fil.) Ker Gawler). Lewis et al., *J. S. African Bot.*, Suppl. 10: 247 (1972). Type: South Africa, Western Cape, Caledon Swartberg, Sept. 1884, *MacOwan 2167* (Herb. Norm. Austr. Afr. 282) (K, lectotype designated here; G, BOL, isolectotypes).

bullatus = bubble-like, alluding to the inflated or rounded shape of the flowers.

Description

Plants 45–100 cm high. **Corm** ovoid, 10–20 mm in diameter, the tunics of hard woody layers, usually divided from the base into thick, claw-like ribs. **Cataphylls** pale and membranous, the uppermost reaching up to 15 cm above the ground and then firm and green or often purplish, sometimes minutely puberulous. **Leaves** usually four, occasionally three, the lowermost longest, reaching to about the middle of the stem, usually largely sheathing, the blade linear, c. 2 mm wide, the midrib heavily thickened, margins lightly so, thus oval in transverse section and with two narrow grooves on each surface, the upper leaves sheathing for at least two-thirds of their length, decreasing in size above, the upper two channelled throughout, the sheaths of all the leaves overlapping. **Stem** erect and more or less straight, lightly flexed outward above the sheaths of the three upper leaves and at the base of the spike, unbranched, 1–1.4 mm in diameter below the spike.

Spike inclined, 1- to 2-flowered; **bracts** brownish purple to grey, lightly ridged above the veins, drawn into attenuate apices, the outer 30–37(–70) mm long, the inner similar but about two-thirds as long as the outer. **Flowers** forming an inflated bell, blue to violet, darker on the reverse of the upper tepals, whitish within, the lower tepals lightly spotted with purple in the lower half, the lower laterals with yellow marginal marks or a transverse band in the upper third, the lowermost with a yellow median streak, unscented; **perianth tube** 10–16 mm long, obliquely funnel-shaped, erect and cylindric below for 5–8 mm long, curved and expanding abruptly in the middle, the upper half bent at right angles and flaring outward; **tepals** unequal, the upper broadly obovate, the lower oblanceolate, the dorsal largest,

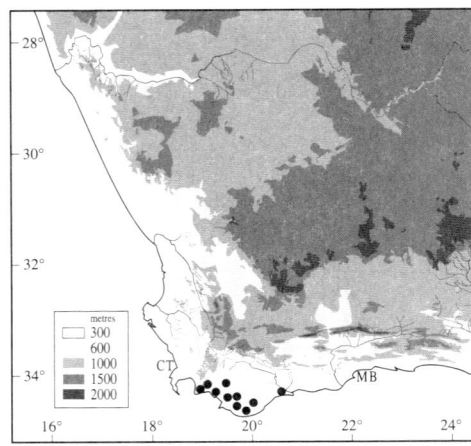

extending horizontally over the stamens, 25–35 x 20–28 mm, the upper laterals slightly shorter and extending forward, only slightly curving outward apically, the lower three united with the upper laterals for 4–10 mm and to one another for c. 2 mm, the lower laterals 20–32 x 10–18 mm, initially curving downward, then arching upward and the apices directed forward, the lower median slightly shorter, 18–22 x 12–16 mm, in profile the lower tepals exceeding the upper by c. 10 mm. *Filaments* 12–14 mm long, exserted 3–6 mm from the tube; *anthers* 7–10 mm long, reaching only to the middle of the dorsal tepal, white to grey-blue, the pollen cream. *Ovary* oblong, 5–6 mm long; *style* arching over the stamens, dividing near the anther apices, the branches c. 3 mm long, extending beyond the anthers, very broad in the upper half. *Capsules* narrowly ovoid, 18–20 mm long, the apex obtuse, concealed in the bracts; *seeds* ovate, pointed at one end, 4–5 x 3 mm, broadly and evenly winged, light translucent brown.
Chromosome number unknown.

Flowering time mid-August to mid-September.

Distribution and biology

Gladiolus bullatus is a narrow endemic of the southern African winter-rainfall zone. It is centred in the mountains of the Caledon District of Western Cape Province, South Africa, but extends from Houw Hoek and the Kogelberg in the west to Bredasdorp and the Potberg near Cape Infanta in the east. Plants grow in stony sandstone-derived soils, or occasionally on limestone outcrops, in low fynbos vegetation. The handsome

flowers stand 20 or more centimetres above the surrounding bush and make a striking display. It is a fairly successful species and quite common in low fynbos vegetation in the mountains of the Caledon District. It is, however, difficult to grow and flower successfully, and *G. bullatus* is one of very few common Cape species of *Gladiolus* that was not figured in *Curtis's Botanical Magazine*, presumably because it simply never flowered in cultivation.

We assume that the flowers are adapted for pollination by long-tongued bees, as are those of other species of *Gladiolus* with bell-shaped, short-tubed flowers.

Diagnosis and relationships

Gladiolus bullatus is unmistakable in its large, blue-purple flower with a short perianth tube and inflated, bell-like perianth, the flower nodding on slender stems up to 1 m high. The whole flower is 35–50 mm long and the dorsal tepal 25–35 mm long. The nectar guides on the lower tepals are variable, as they are in most bell-flowered species of the genus, but there are always bright yellow marks on the upper halves of the lower three tepals, and the lower part of the tepals and throat are streaked and speckled with dark blue. The species is distinctive not only in the relatively large flowers, but also in having only one- or two-flowered spikes, long floral bracts with striate ridges over the veins, and narrow leaf blades, about 2 mm wide, with a thickened midrib but only lightly raised margins.

Gladiolus bullatus appears to be closely related to *G. rogersii*, but is readily distinguished from that species which has short, smooth floral bracts, 15–22 mm long and up to six flowers per spike. The cataphylls of *G. rogersii* are often mottled, another useful feature in distinguishing the species, even though the character is not consistent. The striate floral bracts of *G. bullatus* suggest a possible relationship with *G. blommesteinii* and *G. debilis* which also have these unusual bracts. The rather different flowers of the species, with their striate bracts, make confusion with either unlikely.

History

Although *Gladiolus bullatus* was first collected in the later eighteenth century, it was

only validly described in 1892 by J.G. Baker. He adopted a manuscript name, *G. spathaceus*, used for duplicate specimens of the species sent to the Kew Herbarium by the Cape Colonial Botanist of the time, C.L. Pappe.

The existence of *Gladiolus bullatus* was established at least by the early 1770s by the Swedish botanists, C.P. Thunberg and Anders Sparrman, and it was described as *G. bullatus* in an unpublished manuscript in the Thunberg Herbarium in Uppsala, Sweden. It is not clear why Thunberg or his successors never published the name, but we assume that *G. bullatus* became confused with another species. Collections of the species in European herbaria were often associated with a second bell-flowered species, *G. inflatus*, to which F.W. Klatt, for example, referred specimens. The name proposed for the species by Baker is a homonym for *G. spathaceus* Linnaeus fil., now *Babiana spathacea* (Linnaeus fil.) Ker Gawler (Lewis, 1959).

G.J. Lewis realized that a new name was needed (1966) and she proposed substituting Thunberg's manuscript epithet, *Gladiolus bullatus*. She attributed the substitution to N.E. Brown (1928) who, however, merely pointed out the existence of Thunberg's manuscript name for the plant known as *G. spathaceus*. Brown was unaware at the time that the name was a later homonym. Lewis also mistakenly cited the type of the substitute name as the Thunberg collection associated with his manuscript name. In the posthumously published revision of the South African species the name *G. bullatus* is attributed to Lewis (1966), again with the Thunberg collection indicated as the type.

The type of *Gladiolus bullatus*, when regarded as a substitute name for *G. spathaceus*, is the type of the latter name. Baker cited just two specimens in the protologue and we choose one of these, *Macowan 2167*, as the lectotype. Lewis regarded a Pappe collection at the Kew Herbarium as the holotype of *G. spathaceus*, but since Baker did not mention this specimen with the original description we feel this specimen is inappropriately regarded as a holotype and it should probably not even be considered a type of any kind.

130. GLADIOLUS BLOMMESTEINII L. Bolus

PLATE 113

Gladiolus blommesteinii L. Bolus, *Ann. Bolus Herb.* 3: 183 (1924). Lewis et al., *J. S. African Bot.*, Suppl. 10: 218 (1972). Type: South Africa, Western Cape, without precise locality or collector, obtained from the Caledon Wild Flower Show in 1915 (BOL 14786, lectotype designated by Lewis et al. 1972: 218).

blommesteinii, named in honour of G. van Blommestein, who made one of the first wild collections of the species.

Synonymy

Gladiolus blommesteinii var. *major* L. Bolus, *Ann. Bolus Herb.* 3: 183 (1924). Type: South Africa, Western Cape, without precise locality or collector, obtained from the Caledon Wild Flower Show in 1915 (BOL 14791, holotype).

Description

Plants 30–60 cm high. *Corm* ovoid, 12–20 mm in diameter, the tunics of hard woody layers, fragmenting from the base into claw-like vertical segments. *Cataphylls* pale and membranous, the uppermost reaching to 8 cm above the ground and then green or purple. *Leaves* four, the lower one basal and longest, sheathing the lower half of the stem, reaching to between the middle of the stem and just above the top of the spike, the blade linear, 1–1.5 mm wide, straight, the margins and midrib lightly raised, the upper leaves decreasing in size above, usually sheathing for half their length, the blades similar to the basal but often slightly narrower, the apex of the uppermost leaf usually reaching to the middle of the first flower. *Stem* erect, flexed outward above the sheaths of the three upper leaves and weakly inclined in the upper third, unbranched, c. 1 mm in diameter below the spike.

Spike inclined c. 30°, occasionally 1-, usually 2- to 4-flowered; *bracts* green to dark grey-green, lightly ridged above the veins, the outer 20–36 mm long, the inner two-thirds as long as to slightly shorter than the outer, minutely forked apically. *Flowers* pale mauve-pink to pale blue-mauve, the lower tepals yellowish below with streaks or lines of dark red, blue or purple along the veins, and with red spots and streaks in the throat, lightly sweet-scented; *perianth tube* 13–24 mm long, obliquely and narrowly funnel-shaped, the lower cylindrical part 7–13 mm long, the lower part enclosed in the bracts and the upper part reaching to about the apex of the outer bract; *tepals* unequal, lanceolate, the dorsal largest, ascending, 28–44 x 14–20 mm, upper laterals about as long or somewhat shorter and slightly narrower, directed forward and held more closely to the dorsal than to the lower tepals, the lower three tepals joined to the upper laterals for 4–6 mm and to one another for 2–3 mm, usually with a short tooth at the point of union, narrowly lanceolate, not notably clawed below, the free parts 20–35 x 8–10 mm, ascending and straight below, flexed in the upper half and distally more or less horizontal, in profile the lower tepals exceeding the upper by 10–15 mm. *Filaments* 12–14 mm long, exserted 2–6 mm from the tube but included in the flower; *anthers* 8–11 mm long, extended horizontally, whitish, the pollen light grey. *Ovary* oblong, 3.5–5 mm long; *style* arching over the stamens, dividing opposite the upper third of the anthers, the branches 3–4 mm long, extending past the anther apices. *Capsules* and *seeds* unknown. *Chromosome number* unknown.

Flowering time August to mid-October.

Distribution and biology

Restricted to the southwestern part of the southern African winter-rainfall region, *Gladiolus blommesteinii* is a narrow endemic of rocky mountain soils in the Caledon and adjacent districts of Western Cape Province. The species is not often seen, although its range extends from the Hottentots Holland–Kogelberg mountain axis eastward to Riviersonderend, and encompasses both Sir Lowry's Pass and Franschhoek Pass where they can sometimes be seen from the road. Like so many geophytes of the rocky sandstone-derived soils of the Cape

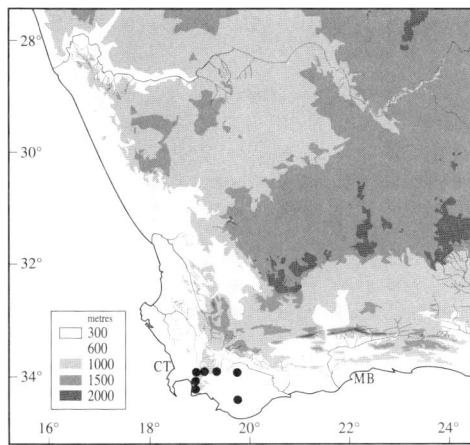

mountains, *G. blommesteinii* flowers well only in the first two seasons after fire.

The species is pollinated by the large, rapid-flying bee *Anthophora diversipes* (Anthophoridae). The size and shape of the flowers and their sweet fragrance are typical of species in the section that are pollinated by long-tongued anthophorid bees.

Diagnosis and relationships

Rare and poorly known, *Gladiolus blommesteinii* can nevertheless be readily recognized by its short-tubed, but relatively large, pale blue or pale pink flower with purple-streaked nectar guides and lanceolate-attenuate tepals. The four superposed linear leaves with moderately thickened margins and midrib and the woody corm tunics place the species in series *Gracilis* of section *Homoglossum*, and its prominently striate floral bracts suggest a close relationship with the few other species of the series with this odd feature. These species include *G. bullatus*, *G. debilis* and *G. virgatus*, and as this last species has flowers most closely resembling those of *G. blommesteinii* we assume them to be immediately related. Vegetatively identical, *G. virgatus* also has lanceolate-attenuate tepals, but the pink flowers are unscented and have a longer perianth tube, 22–27 mm long, while the lower tepals each have a spear-shaped whitish median streak edged with dark pink. This series of attributes constitute adaptations for pollination by long-tongued flies, whereas *G. blommesteinii* has flowers adapted for pollination by bees. The two species actually grow near one another in the Hottentots Holland Mountains and in similar habitats, but whereas *G. blommesteinii* flowers early in the spring, *G. virgatus* flowers in late spring, occasionally in September but more often in late October or early November.

History

Although a fairly rare plant, *Gladiolus blommesteinii* is certainly one that should have caught the attention of botanists and naturalists travelling east from Cape Town toward Caledon in spring. It seems to have avoided scientific attention until 1924 when it was described by H.M.L. Bolus. The earliest record of the species that we have found was made in 1894, and even this is surprisingly late. The collection was made by Charles Grisbrook, who made a number of others in the Caledon District at this time. No doubt *G. blommesteinii* was seen earlier than this, but it may not have seemed worth collecting. Not particularly distinctive, it may have been confused with any of several other pale blue-flowered species of series *Gracilis*. The variety *major* distinguished by Bolus represents no more than larger-flowered plants of *G. blommesteinii* and does not warrant taxonomic recognition.

131. GLADIOLUS VIRGATUS Goldblatt & Manning
PLATE 114

Gladiolus virgatus Goldblatt & Manning, new species. Type: South Africa, Western Cape, west end of Dutoit's Peak, 16 Oct. 1949, *Esterhuysen 16061* (BOL, holotype; K, NBG, SAM, isotypes).

virgatus = willowy, for the slender, arching stems.

Latin diagnosis

Plantae 40–50 cm altae, cormo globoso 12–14 mm in diametro, tunicis fibris lignosis, foliis 4, infimo longiore lamina lineari, 1.5–2.5 mm lata marginibus costaque incrassatis, caule eramoso, spica 2–4 florum, bracteis attenuatis striatis 30–45(–50) mm longis, floribus roseis ad malvinis tepalis inferioribus albis infra rubro notatis, tubo perianthii cylindrico 22–27 mm longo, tepalis inaequalibus dorsale 30–37 x 18–19 mm, inferioribus 20–28 x 10–12 mm, filamentis 12–16 mm longis, antheris 8–9 mm longis.

Description

Plants 40–50 cm high. *Corm* globose, c. 12–14 mm in diameter, the tunics of hard, cartilaginous to woody layers, breaking into claw-like vertical segments. *Cataphylls* membranous and pale, the uppermost reaching 7–12 cm above the ground and then firm-textured and green. *Leaves* four, the lower two basal and longest, the lowermost reaching to between the base and slightly beyond the apex of the spike, sometimes sheathing the stem up to the middle, the remaining leaves progressively shorter and sheathing for half their length, the blades linear, plane with the midrib and margins lightly raised, 1.5–2.5 mm wide. *Stem* erect below or inclined from the base, forming a gentle arc, flexed laterally above the sheaths of the two upper leaves, unbranched, c. 1.3 mm in diameter below the spike.

Spike inclined, 2- to 4-, or occasionally 6-flowered; *bracts* green or flushed purple above, the outer 30–45(–50) mm long, the surface ridged above the veins, the inner slightly shorter than the outer, notched apically for c. 1 mm. *Flowers* pale to dark pink or nearly white, the lower tepals each with a spear-shaped whitish median streak in the lower half, distally edged with a dark pink V-shaped mark, speckled pink in the throat, unscented; *perianth tube* 22–27 mm long, slender below, abruptly expanded in the upper 4–7 mm, slightly exceeding the bracts; *tepals* lanceolate, unequal, the margins usually undulate, the dorsal largest, inclined over the stamens, c. 30–37 x 18–19 mm, the upper laterals about as long, 14–16 mm wide, arching outward in the upper third, the lower three tepals joined to the upper laterals for 2–3 mm and to one another for c. 3 mm, more or less horizontal below, gently to sharply flexed downward in the midline, the lower laterals 20–28 x 10–12 mm, the lower median 25–28 x 10–12 mm, in profile the lower tepals exceeding the upper by 10–15 mm. *Filaments* 12–16 mm long, exserted c. 8 mm from the tube; *anthers* 8–9 mm long, light mauve, the pollen whitish. *Ovary* oblong, c. 6 mm long; *style* arching over the stamens, dividing between the upper third and slightly beyond the apices of the anthers, the branches spreading, c. 6 mm long. *Capsules* and *seeds* unknown. *Chromosome number* unknown.

Flowering time mid-September to early November, rarely as early as the end of August.

Distribution and biology

Gladiolus virgatus is restricted to cool, well-watered slopes of the coastal ranges of the southwestern Cape. The species is poorly collected and is so far known from the Helderberg at Somerset West in the south to Dutoit's Kloof in the north, a distance of 50 km. All collections are from mountain slopes where the soil is composed of clay derived either from shales of the Malmesbury System or from bands of shale in strata of the Cape System. Not uncommonly, the ground is covered with scattered boulders of Cape Sandstone, fallen from the slopes and cliffs higher up. On north-trending slopes plants tend to flower earliest, usually in September, but at high elevations and on southern or southeastern slopes they flower later, lasting into early November. The plants illustrated on Plate 114 are from the Helderberg near Somerset West. There they bloom very late and have unusually large flowers for the species. Smaller-flowered plants from Dutoit's Kloof (see line illustration, opposite page) are less striking, but show the same series of features that characterize the species.

The pink to mauve flowers of *Gladiolus virgatus* are unscented and have a relatively long cylindrical perianth tube. This plus the reddish markings on the lower tepals constitute adaptations for pollination by long-tongued flies. Occasional visits by the horse-fly *Philoliche rostrata* have been noted, and we assume this insect is the main or sole pollinator of *G. virgatus*.

Diagnosis and relationships

Distinguished by having fairly large flowers and a relatively long cylindrical perianth tube, *Gladiolus virgatus* is otherwise like other species of series *Gracilis* in having four

narrow leaves with moderately thickened margins and midrib, and hard, woody corm tunics that split into sections from below. The two to four salmon or occasionally mauve-pink flowers are borne on slender, willowy spikes and have a straight, cylindrical perianth tube 22–27 mm long. The tepal markings consist of white median streaks edged distally with reddish chevrons. One unusual feature of the plant is the striate floral bracts and this, as well as the lanceolate-attenuate tepals, indicates a relationship with *G. blommesteinii*. The latter has flowers with a campanulate perianth tube and pale blue or mauve flowers, and its flowers are pollinated by anthophorid bees whereas those of *G. virgatus* are pollinated by long-tongued flies. Although the two have adjacent geographic ranges, they have different habitat preferences, *G. virgatus* favouring clay soils and *G. blommesteinii* stony sandstone soils.

History

A new species described for the first time in this monograph, *Gladiolus virgatus* was apparently completely unknown until 1907 when specimens bought from flower-sellers in Cape Town were brought to the Bolus Herbarium. Flowers obtained at the 1914 Stellenbosch Wild Flower Show were also preserved, but the species remained unknown in the wild until 1943, when it was gathered independently by the South

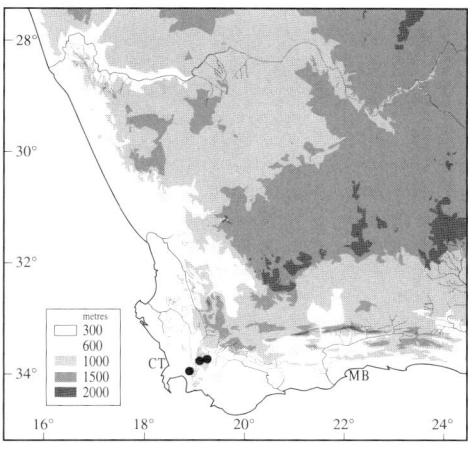

African botanist, Elsie Esterhuysen, and by retired forester, R.N. Parker. Their collections, as well as several more made in the following decades, were often confused with the shorter-tubed *G. blommesteinii* until our own research on *Gladiolus* began. Esterhuysen did, however, point out that the markings on flowers she collected in 1949 on Dutoit's Peak differed from those of *G. blommesteinii*. In 1993 we were alerted to the presence of an unusual pink-flowered *Gladiolus* species in the Helderberg Nature Reserve at Somerset West by P.A. Runnals who was making collections of the flora of the reserve. The plants were flowering conspicuously after a fire a few seasons before and were thought by Runnals to be unusual. Our investigations proved her to be correct.

132. GLADIOLUS DEBILIS Sims

Painted lady

PLATE 115

Gladiolus debilis Sims, *Curtis's Bot. Mag.* 52: pl. 2585 (1825). Baker, *Fl. Capensis* 6: 145 (1896). Lewis et al., *J. S. African Bot.*, Suppl. 10: 184 (1972). Type: South Africa, Cape, without precise locality or collector, cultivated in Great Britain, illustration in *Curtis's Bot. Mag.* 52: pl. 2585 (1825).

debilis = dainty, for the delicate, porcelain-like appearance of the flowers.

Synonymy

Gladiolus cochleatus Sweet, *British Fl. Gard.*, Ser. 2: pl. 140 (1832). Baker, *Fl. Capensis* 6: 146 (1896). Type: South Africa, Cape, without precise locality or collector, cultivated in Great Britain, illustration in *British Fl. Gard.*, Ser. 2: pl. 140 (1832).
Gladiolus lambda Klatt, *Linnaea* 32: 708 (1863). Baker, *Fl. Capensis* 6: 143 (1896). Type: South Africa, Cape, without precise locality or date,

Reynaud s.n. (S, lectotype designated by Lewis et al., 1972: 186).
[*Geissorhiza albens* E. Meyer, *Zweipflanzen Documente, Flora* 26: 187 (1843), a name without description, based on *Drège 1568*, 1839, and *Ecklon & Zeyher 3988*.]

Description

Plants (12–)25–45 cm high. *Corm* 8–15 mm in diameter, the tunics of coarse wiry to more or less woody, claw-like vertical strips. *Cataphylls* membranous and pale, reaching 2–6 cm above the ground and then green or purplish. *Leaves* four, rarely three, the lowermost basal and longest, reaching to between the upper third of the stem and the apex of the spike, the blade plane, 1–2 mm wide, the midvein moderately thickened, the margins usually also thickened and raised, upper leaves sheathing for about half their length, decreasing in size above, the blades like the basal. *Stem* erect

below, flexed outward above the sheaths of the upper leaves, inclined in the upper half, unbranched, 1–1.3 mm in diameter below the spike.

Spike inclined, flexuose, 1- to 3-flowered; *bracts* light green to greyish, sometimes flushed grey-purple above or entirely, the outer 15–28 mm long, the surface lightly ridged above the veins, the margins hyaline, the inner slightly shorter than the outer, minutely forked apically. *Flowers* white to pale pink, the lower three tepals each variously marked with red, often each with a pair of large spots near the base and a V-shaped or spade-shaped distal mark, the shaft of the spade sometimes extending back along the midline, the lower part of the throat speckled and streaked with red and the filaments red, unscented; *perianth tube* cylindric, very slightly curved and expanded in the upper 2 mm, (10–)15–22 mm long,

slightly shorter than to as long as the bracts; **tepals** unequal, ovate to nearly orbicular, the dorsal largest, either inclined or erect (occasionally reflexed), 17–27 x 10–18(–22) mm, upper laterals arching outward from the base and spreading at right angles to the tube, 15–27 x (9–)13–18 mm, the lower tepals sometimes united basally for up to 3 mm, nearly straight or weakly flexed downward in the upper half, weakly differentiated into claws below, 15–24 mm long, the limbs 8–12 mm long, subequal or the lower median slightly larger. **Filaments** 4–10 mm long, exserted 2–7 mm from the tube, often reddish, or red abaxially; **anthers** 5–9 mm long, white above, purple below, the pollen cream. **Ovary** oblong, c. 4.5 mm long; **style** dividing opposite to slightly beyond the anther apices, the branches 2–5 mm long, extending beyond the anthers. **Capsules** ellipsoid, subacute, (12–)14–18 mm long; **seeds** ovate, 5–6 x c. 4 mm, broadly and evenly winged. **Chromosome number** $2n = 30$.

Flowering time September to early October.

Distribution and biology
Restricted to the southern African winter-

rainfall zone, *Gladiolus debilis* is a fairly narrow endemic of the mountains of the southwestern Cape. Its range extends from Bain's Kloof in the north to the Cape Peninsula and eastward through the Caledon District to Bredasdorp. Plants grow on rocky sandstone slopes in mountain fynbos vegetation. In these nutrient-poor, acidic soils, they tend to flower well in the two or three years after fire when the surrounding vegetation is removed and the soil is enriched with additional nutrients, and competition for water and light is reduced. As the perennial and shrubby vegetation regrows, fewer plants flower and eventually a once large colony becomes invisible until the next fire.

The white, unscented flowers of *Gladiolus debilis*, with their strongly contrasting dark red markings and long, straight perianth tubes, are thought to be adapted for pollination by long-tongued flies.

Diagnosis and relationships
One of the loveliest species of *Gladiolus*, and one of the best known of the many in the southwestern Cape, *G. debilis* is readily recognized by its fairly large white flowers, the lower tepals with distinctive red markings. These include chevrons, diamonds or streaks and lines in very regular patterns, and the base of the throat is also marked with a circle of red. Except for the ascending or inclined dorsal tepal, the tepals spread widely, giving the flower a distinctive appearance. The tepals are pinkish on the outside and this imparts a somewhat pinkish colour to the whole flower when viewed against back lighting. The flowers have a relatively long, cylindrical perianth tube, usually 15–22 mm long, which reaches to the top of the floral bracts. *Gladiolus debilis* otherwise has the typical attributes of series *Gracilis*: hard, woody corm tunics that fragment into vertical segments, and four narrow, linear leaves with moderately thickened margins and midrib. The species also has conspicuously striate floral bracts, a feature shared with just three other species in the genus – *G. blommesteinii*, *G. bullatus* and *G. virgatus*.

Gladiolus debilis was regarded by Lewis et al. (1972) as comprising three subspecies. One of these, subsp. *cochleatus*, does not seem to us to differ significantly from typical *G. debilis* and we do not recognize this taxon. Subsp. *variegatus*, however, differs from it in several ways, not least by lacking striate floral bracts. It also has only three leaves in contrast to four in most populations of *G. debilis*. We regard *G. variegatus* as a separate species and are not convinced that

it is even immediately related to *G. debilis*. In any event, the two species can readily be distinguished from one another, as detailed below under the discussion of *G. variegatus*.

Tepal markings are variable across the range of *Gladiolus debilis* but show clear correlation with geography. The smallest flowers encountered in the species occur in the populations in the Hottentots Holland Mountains where the markings are reduced to a small red diamond on each of the lower tepals. On the Cape Peninsula the markings consist of V-shaped chevrons, and in the southeast, from Hermanus to Bredasdorp, the markings are paired spots and streaks in the lower halves of the lower tepals.

History
It is puzzling that such an attractive and relatively common plant as *Gladiolus debilis* should not have been discovered during the earliest years of botanical exploration at the Cape. There is no record of who made the first gathering of the species, but plants were grown in Great Britain in the 1820s. The bulb specialist and early *Gladiolus* hybridizer, William Herbert, provided the plants that formed the basis for a description and painting published in *Curtis's Botanical Magazine* in 1825. The protologue is attributed to John Sims, then editor of the series, and not to John Ker Gawler, who described so many species of Cape Iridaceae in that publication. The brief description does not conclude with Ker Gawler's initials as was the normal practice, and the description is unusually terse for Ker Gawler. Nevertheless, J.G. Baker and later authors often attributed authorship of *G. debilis* to Ker Gawler, for example in *Flora Capensis* (Baker, 1896).

A minor variant of *Gladiolus debilis* was described in 1832 by Robert Sweet under the name *G. cochleatus* in the illustrated magazine, *British Flower Garden*. The flowers are drawn from an unusual perspective and

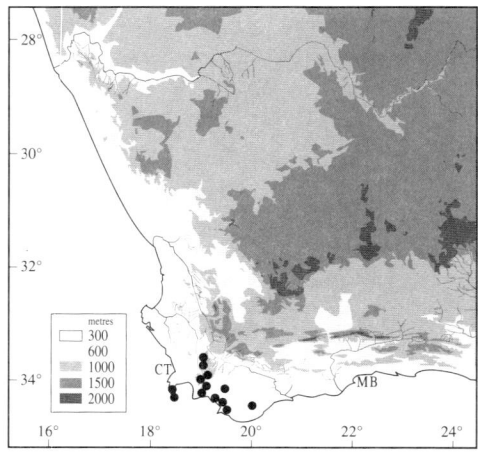

so appear rather different from typical *G. debilis* as illustrated in *Curtis's Botanical Magazine*. Both *G. cochleatus* and *G. debilis* were recognized in *Flora Capensis*, as was a third synonym, *G. lambda*, a species described by F.W. Klatt in 1863. This last species closely matches typical *G. debilis* and it is difficult to see why it should ever have been thought distinct.

133. GLADIOLUS VARIEGATUS (G. Lewis) Goldblatt & Manning
PLATE 116

Gladiolus variegatus (G. Lewis) Goldblatt & Manning, *Novon* 6: 178 (1996). *Gladiolus debilis* var. *variegatus* G. Lewis in *J. S. African Bot.*, Suppl. 10: 188 (1972). Type: South Africa, Cape, Bredasdorp District, Brandfontein, 14 Oct. 1951, *Esterhuysen 19087* (BOL, holotype; BOL, NBG, PRE, isotypes).

variegatus = variegated, marked with two colours, referring to the nectar guides.

Description
Plants 20–40(–60) cm high. *Corm* conical, 10–15 mm in diameter, the tunics of hard, woody, claw-like segments. *Cataphylls* membranous and pale, the uppermost reaching a short distance above the ground and then green, sometimes flushed with purple. *Leaves* three, rarely four, the lower two basal and longest, reaching at least to the base of the spike and sometimes shortly exceeding it, the lowermost sheathing the stem for a short distance (about one-fifth the length of the stem), the blade straight and linear or falcate, 1.5–2.5 mm wide, the midrib lightly raised, the margins not thickened or raised, the second leaf sheathing the lower two-thirds of the stem, the blade like the basal but relatively short, the uppermost leaf sheathing for half its length, the margins of the sheathing part free to the base. *Stem* erect below, flexed outward above the sheaths of the two upper leaves, inclined and flexuose in the upper half, unbranched, 0.8–1.5 mm in diameter below the spike.
   Spike lightly inclined, flexuose, 1- or 2-, occasionally to 4-flowered; *bracts* dark green, sometimes flushed grey-purple above, the outer 25–40(–50) mm long, the surface smooth, the margins narrowly hyaline, the inner slightly shorter than the outer, minutely forked apically. *Flowers* white to palest pink, the lower part of the throat and lower three tepals each marked variously with red spots irregularly distributed on the lower two-thirds, unscented; *perianth tube* cylindric for most of its length, very slightly curved and expanded in the upper 3–4 mm, 14–20 mm long, the cylindric lower part 11–15 mm long, shorter than the bracts; *tepals* unequal, the dorsal largest, inclined,

25–30 x 16–22 mm, upper laterals arching outward from the base and ultimately spreading at right angles to the tube, 23–28 x 11–15 mm, the lower tepals united with the upper laterals for c. 2 mm and to one another for c. 2 mm, nearly straight and horizontal below, weakly arching downward in the upper half, not differentiated into claws below, 24–27 x 10–13 mm, the lower median slightly smaller than the laterals, in profile the lower much exceeding the upper. *Filaments* 12–15 mm long, exserted 5–6 mm from the tube, often reddish, or red abaxially; *anthers* (7–)8–10 mm long, white above, dark purple, the pollen cream. *Ovary* oblong, c. 6 mm long; *style* dividing 1–2 mm below to c. 2 mm beyond the anther apices, the branches c. 4 mm long, extending over the anthers. *Capsules* and *seeds* unknown. *Chromosome number* unknown.
   Flowering time mid-September to mid-October.

Distribution and biology
A narrow edaphic endemic of the southern African winter-rainfall region, *Gladiolus variegatus* is restricted to the southern Cape coast between Stanford and the western side of Cape Agulhas in Western Cape Province, South Africa. Plants only occur short distances from the coast, usually within sight of the sea, at elevations between 10 and 250 m and they are found only in cracks in limestone outcrops or in stony calcareous sands.
   Gladiolus variegatus flowers fairly late in the spring, in September and October, and flowering is strongly enhanced by fire. After a burn the previous summer, plants flower profusely. Both the number of plants that flower and the number of flowers produced decrease in the following seasons, until the surrounding vegetation has reached a height at which the plants are completely shaded and cease to bloom.
   The white to pink flowers coloured with red markings and the relatively long, cylindrical perianth tube suggest an adaptation for pollination by long-tongued flies. We have observed populations of *Gladiolus variegatus* in full flower for many hours and have not yet seen any insect visitors.

Diagnosis and relationships
Gladiolus variegatus can be recognized immediately by its white or pale pink flowers with unusual nectar guides which consist of unevenly sized, deep red spots distributed across the lower two-thirds of the lower tepals and the lower part of the throat. The markings differ slightly on each tepal and on every flower, a condition without parallel in the Iridaceae, let alone in *Gladiolus*. Apart from the peculiar nectar guides, *G. variegatus* resembles fairly closely the southwestern Cape mountain species, *G. debilis*, and it was treated as a variety of that species by Lewis et al. (1972). As pointed out by Lewis et al., the two taxa differ in several respects. The differences include the nectar guides which are variously shaped in *G. debilis* but most often consist of diamond-, spade-, or chevron-shaped markings on the lower

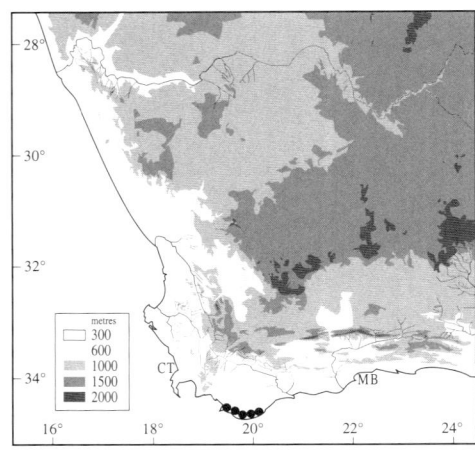

more significantly, *G. variegatus* normally has three leaves per stem and the leaf blades are 1.5–2.5 mm wide and plane with lightly raised midribs and virtually unthickened margins. In contrast, *G. debilis* has four leaves, 1–2 mm wide, and the blades have moderately to strongly thickened midrib and margins. In addition, the flowers of *G. variegatus* appear to be consistently larger than those of *G. debilis*, and have a dorsal tepal 25–30 x 16–22 mm, the lower tepals 24–27 mm long, the filaments 12–15 mm long, and the anthers 7–10 mm long. Flowers of *G. debilis* have a dorsal tepal 17–27 x 10–22 mm, the lower tepals 15–24 mm long, the filaments 4–10 mm long, and the anthers 5–9 mm long. The combined qualitative and quantitative differences between *G. variegatus* and *G. debilis* make it easy to distinguish the two, and are of sufficient magnitude that they warrant

tepals, bract length and texture, and leaf number and structure. The floral bracts are 15–28 mm long in *G. debilis* and have a distinctively striate or ridged surface, whereas those of *G. variegatus* are (25–)30–45 mm long and completely smooth. Perhaps even

recognition as separate species. The differences in leaf number and bract texture make a direct relationship between the two somewhat equivocal, although they are certainly fairly closely allied.

History

Although known at least since 1951 when it was collected by the South African botanist, Elsie Esterhuysen, this beautiful plant has attracted little attention until recently. It was long thought to be a local form of the fairly widespread *G. debilis*, but careful examination of the leaves and flowers shows it to be consistently different from that species in several features. It was recognized as var. *variegatus* of *G. debilis* by Lewis et al. (1972), but the differences between the two taxa are of such magnitude that we consider them separate species, and possibly not even immediately related to one another.

134. GLADIOLUS VIGILANS Barnard

PLATE 117

Gladiolus vigilans Barnard in Lewis et al., *J. S. African Bot.*, Suppl. 10: 82 (1972).

Type: South Africa, Western Cape, Cape Point Nature Reserve, Vasco da Gama Ridge, Nov. 1939, *Lewis s.n.* (SAM 54296, holotype; PRE, isotype).

vigilans = watchful, vigilant, so named for its habitat, hills on the southern Cape Peninsula which have wide views across False Bay and the southern Atlantic Ocean.

Synonymy

[*Gladiolus prismatosiphon* sensu G. Lewis in Adamson & Salter, *Fl. Cape Peninsula* 259 (1950); *Fl. Pl. Africa* 29: pl. 1160 (1953), not *G. prismatosiphon* Schlechter (1900) = *G. carneus*.]

Description

Plants 30–45 cm high. *Corm* 12–15 mm in diameter, the tunics of hard, woody, claw-like segments. *Cataphylls* membranous and pale, reaching 5–8 cm above the ground and then dark green or flushed with purple. *Leaves* four, the lower two basal and longest, usually shortly exceeding the stem, the blades linear, plane, 1.2–1.5 mm wide, the margins and midrib moderately thickened, the two upper leaves inserted in the middle and upper third of the stem respectively, progressively smaller in size above, sheathing for about half their length, the uppermost channelled throughout, the margins of the sheathing part free to the base. *Stem* erect below, flexed outward above the sheaths of

the two upper leaves, inclined in the upper two-thirds, unbranched, c. 1 mm in diameter below the spike.

Spike inclined, flexuose, 1- to 3-flowered; *bracts* pale green, not ridged, the outer 20–25 mm long, the margins hyaline, the inner slightly shorter than the outer, minutely forked apically. *Flowers* pale pink, the lower three tepals each with a median spade-shaped white mark edged in dark red, the shaft of the spade extending onto the tube, the base of the upper part of the tube with fine red lines running from the tepal sutures, unscented; *perianth tube* narrowly funnel-shaped, 35–40 mm long, straight and cylindric below for 30–35 mm, curved outward and flared in the upper 5 mm, about twice as long as the bracts; *tepals* unequal, the dorsal largest, inclined over the stamens, c. 28 x 18 mm, the upper laterals arching outward from the base and ultimately spreading at right angles to the tube, c. 25 x 17 mm, the lower tepals united with the upper laterals for c. 3 mm and to one another for c. 3 mm, weakly differentiated into claws below, straight and horizontal in the lower half, lightly flexed in the middle and then curving downward, the lower laterals c. 18 x 8 mm, the lower median c. 22 x 6 mm. *Filaments* 12 mm long, exserted 6 mm from the tube, white below, pink toward the apices; *anthers* c. 6.5 mm long, dark purple, the pollen cream. *Ovary* oblong, c. 4 mm long; *style* dividing between the upper third and apices of the anthers, the branches c. 4 mm long, broadly expanded in the upper half. *Capsules* ellipsoid, 12–14 mm long; *seeds* oblong, c. 5.5 x 3 mm, light brown, broadly and evenly winged. *Chromosome number* unknown.

Flowering time late October to early November.

Distribution and biology

Gladiolus vigilans is an extremely local endemic of the southern African winter-rainfall zone. It occurs only on a few low hills on the southern Cape Peninsula and possibly also on the slopes of Kogelberg which lies opposite the Peninsula across False Bay. Only one collection, made by T.P. Stokoe in 1939, is known from there and the record has never been confirmed. Stokoe's lack of careful record-keeping makes the report highly suspect. Plants grow in stony sandstone soil on well-drained slopes in low fynbos.

The pale pink flower colour, long perianth tube and lack of fragrance conform to the floral characteristics of Iridaceae pollinated by long-tongued flies, and we have recorded the horsefly *Philoliche rostrata* foraging for nectar on the flowers of *Gladiolus vigilans.* This fly, almost certainly the sole pollinator of the species, is very active on the southern Peninsula when *G. vigilans* flowers in late October and November.

Diagnosis and relationships

Gladiolus vigilans has the woody corm

tunics, superposed leaves and linear leaf blades with raised margins and midrib that characterize series *Gracilis* of section *Homoglossum*. Within the series the pink flower with red markings and the elongate perianth tube, 35–40 mm long, suggest a relationship with other species with similar floral features. These include *G. debilis* and *G. virgatus*, both of which have conspicuously striated floral bracts, and *G. ornatus* and *G. variegatus* which have smooth bracts as does *G. vigilans*. We suspect its immediate relationships lie with the latter two species. They are unlikely to be confused with *G. vigilans* as both have flowers with much shorter tubes, only 14–20 mm long.

Lewis et al. (1972) associated *Gladiolus vigilans* with species of section *Blandus* in their revision of *Gladiolus* in southern Africa, in particular with the *G. carneus* group. This was no doubt due to floral similarities which we now understand to be due to convergence for long-tongued fly pollination. The coarsely fibrous corm tunics and linear leaves, up to 1.5 mm wide and lacking evident secondary veins, are unknown in that section, but characterize most species of series *Gracilis* of section *Homoglossum*.

History

Although *Gladiolus vigilans* occurs less than 40 km from the centre of Cape Town, around which the flora has been studied for more than 250 years, its existence was evi-

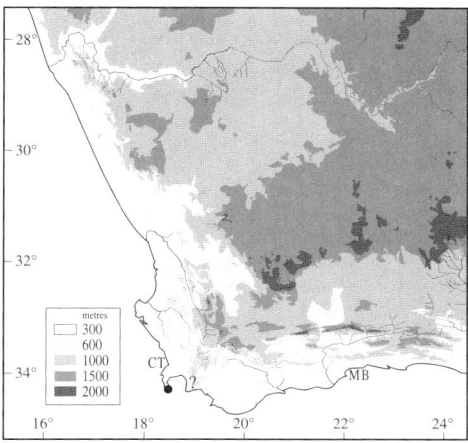

dently unknown until 1932 when plants were found near Cape Point by T.M. Salter and W.F. Barker. The species was for many years confused with *G. prismatosiphon* (Lewis, 1950), a later synonym of *G. carneus* (section *Blandus*), owing to the elongate perianth tube and pink colouring of the flowers. Only after G.J. Lewis's death, when A.A. Obermeyer and T.T. Barnard completed studies for her revision of South African *Gladiolus* species, was it understood that this was an undescribed species. These authors, however, continued to associate *G. vigilans* with *G. carneus*, owing to a dependence on flowers to determine its relationships, and ignoring the corm tunics and leaves which are quite unlike those of any member of the *G. carneus* group or, in fact, the entire section *Blandus*.

135. GLADIOLUS ORNATUS Klatt
Pienk klokkie, marsh bell
PLATE 118

Gladiolus ornatus Klatt, *Trans. S. African Phil. Soc.* 3: 198 (1885). Lewis et al., *J. S. African Bot.*, Suppl. 10: 222 (1972). Type: South Africa, Western Cape, flats near Rondebosch, Oct. 1884, *MacOwan 2553* (G, S, SAM, isotypes).

ornatus = adorned, for the boldly marked flowers.

Synonymy

[*Gladiolus inflatus* sensu Baker, *Fl. Capensis* 6: 147 (1896), in part, not *G. inflatus* Thunberg.]
[*Gladiolus thunbergii* Ecklon, *Topographisches Verzeichniss* 37 (1827), name without description, based on *Ludwig & Beil s.n.* (S), from Stellenbosch, near Eerste River.]

Description

Plants 25–35 cm high. *Corm* globose, 9–

12 mm in diameter, the tunics of soft-textured, more or less membranous layers, fragmenting irregularly, not accumulating with age. *Cataphylls* pale and membranous or reddish, the uppermost reaching up to 8 cm above the ground and then dark green or flushed with purple. *Leaves* three, the lowermost basal and longest, sheathing the lower third of the stem, the blade linear, 1–2 mm wide, the midrib lightly raised, the margins lightly to heavily thickened, the remaining two leaves inserted on the upper half of the stem, progressively smaller in size, the uppermost sheathing for half its length, the blade channelled throughout, the margins of the sheathing part overlapping, free to the base. *Stem* erect below, flexed outward above the sheathing parts of the upper three leaves and inclined, unbranched, c. 1.2 mm in diameter below the spike.

Spike inclined, flexuose, occasionally 1-, usually 2- to 3-flowered; *bracts* dull green or purplish grey, the outer 27–32 mm long, the inner about as long as to slightly shorter than the outer. *Flowers* pale to deep pink, the lower tepals each with a spear- to spade-shaped white median streak edged in dark pink, the lower part of the throat white, irregularly streaked with dark pink, unscented; *perianth tube* obliquely funnel-shaped, 18–20 mm long, the lower cylindrical part c. 12 mm long, enclosed in the bracts; *tepals* unequal, the dorsal largest, inclined, 30–33 x 18–21 mm, the upper laterals directed forward below, curving outward in the upper halves, 28–31 x 15–17 mm, the lower tepals united for c. 5 mm, straight below and tilted c. 20° toward the ground, curving downward in the upper third, 25–28 x 10–14 mm, in profile the lower

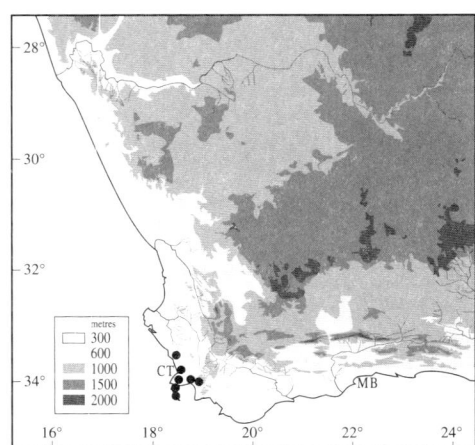

in the southern African winter-rainfall region. Most common on the Cape Peninsula, it extends north to the Mamre hills some 40 km north of Cape Town and eastward to Jonkershoek near Stellenbosch. There is also an isolated record from Hermanus ascribed to Harry Bolus that we assume is an error. A collection cited in the literature from Piketberg is, likewise, incorrect. Over the past 50 years the range of *G. ornatus* has decreased substantially. It once occurred on the lower slopes of Table Mountain and on the Cape Flats from Rondebosch to Stellenbosch. Urban development and increased land cultivation have destroyed all suitable habitats there. The species does, however, seem secure in nature reserves at Silvermine and Cape Point on the Cape Peninsula, and at Jonkershoek, a water-catchment and conservation area.

The pink perianth with bold white nectar guides edged in darker pink and the funnel-shaped tube of the flowers of *Gladiolus ornatus* suggest that they are adapted for pollination by long-tongued bees, as are the similarly proportioned, though usually blue or mauve, flowers of many species of section *Homoglossum*. However, the pollination ecology of *G. ornatus* remains to be investigated.

Diagnosis and relationships

In most respects a typical member of section *Homoglossum*, *Gladiolus ornatus* has the narrow, superposed leaves and unbranched, flexuose stem that define the alliance. However, its corm tunics, important in assessing relationships within the section, are puzzling. They are soft-textured and more or less membranous, and do not accumulate with age. The soft tunics are probably correlated with the habitat – marshes and seeps – and may not be an indication of relationship. The leaf blades, which are very narrow and have thickened margins and midrib, suggest a relationship with species of series *Gracilis*, and we assume that *G. ornatus* is

allied to members of the series with similarly coloured pink flowers and red nectar guides. We thus place *G. ornatus* in series *Gracilis* close to species with similar long-tubed flowers with unspecialized, smooth floral bracts. These include *G. vigilans* and *G. variegatus*, but we remain uncertain that this is where its immediate relationships lie.

History

Before the middle of the twentieth century *Gladiolus ornatus* was relatively common in wet sites on the Cape Peninsula and the surrounding coastal flats, and it was probably recorded near Cape Town by C.P. Thunberg in the 1770s. Thunberg's collection went unnamed and this and a few later records, among them specimens collected by J.F. Drège and by C.L. Zeyher and Nathaniel Wallich in the early nineteenth century, attracted no botanical attention. Thus, when duplicates of a collection made by Peter MacOwan in 1884 came into the hands of F.W. Klatt, it was thought to be a new discovery. Klatt named the species *G. ornatus* in 1885. J.G. Baker, however, associated the species with *G. inflatus*, and both the name *G. ornatus* and collections of the species are cited under *G. inflatus* in *Flora Capensis*.

three tepals exceeding the upper. *Filaments* 12–14 mm long, exserted 5–6 mm from the tube; *anthers* c. 10 mm long, mauve to violet, the pollen cream to mauve. *Ovary* oblong, c. 6 mm long; *style* arching over the stamens, dividing opposite the upper third of the anthers, the branches 4–5.5 mm long. *Capsules* ellipsoid, 16–20 mm long; *seeds* unknown. *Chromosome number* unknown.

Flowering time late September to November.

Distribution and biology

Gladiolus ornatus has a narrow distribution

136. GLADIOLUS INFLEXUS Goldblatt & Manning

PLATE 119

Gladiolus inflexus Goldblatt & Manning, new species. Type: South Africa, Cape, Worcester District, c. 8 km west of the town in rocky alluvial ground with sandy loam, 27 July 1994, *Goldblatt & Manning 9888* (NBG, holotype; K, MO, PRE, isotypes).

inflexus = bent abruptly, describing the upper part of the stem and spike.

Latin diagnosis

Plantae (15–)20–35 cm altae, cormo 10–15 mm in diametro, tunicis lignosis, foliis 3, laminis planis marginibus costisque incrassatis, spica 1–3 florum flexuosa, floribus malvinis ad purpureis tepalis inferioribus cremeis purpureo-maculatis odoratis, tubo perianthii infundibuliformi 11–20 mm longo, tepalis inaequalibus lanceolatis dorsali

majore 21–25 x c. 14.5 mm, inferioribus 15–25 x 10 mm, filamentis 12–14 mm longis, antheris 6–7 mm longis, capsulis oblongo-ellipsoideis 15–22 mm longis.

Description

Plants (15–)20–35 cm high. *Corm* conic, 10–15 mm in diameter, the tunics of hard concentric layers, fragmenting below or completely with age into flat vertical segments, these sometimes acute basally and apically, or truncate. *Cataphylls* pale and membranous, the uppermost reaching 2–4 cm above the ground and dark green or purple near the apex. *Leaves* three, the lowermost basal and longest, sheathing the lower half to two-thirds of the stem, reaching to about the apex of the spike or slightly exceeding it, the blade linear, 1–2 mm wide, grey-green, plane, the midrib paler and slightly raised, the margins not raised or hyaline, the upper two leaves short, reaching to about the middle of the spike, sheathing in the lower third to half, the blades similar to the basal but narrower. *Stem* erect below, sharply flexed outward above the sheathing parts of each leaf, thus flexuose in the upper third, c. 0.8–1.2 mm in diameter below the spike.

Spike flexuose, 1- to 3-flowered; *bracts* grey-green, the margins transparent, the outer 20–38 mm long, the apices often attenuate and twisted, the inner slightly longer or slightly shorter than the outer, the margins broadly hyaline below. *Flowers* shades of purple with the bases of the tepals and the throat cream, the lower tepals speckled to blotched with purple in the lower two-thirds, sweetly scented; *perianth tube* obliquely funnel-shaped, 11–20 mm long; *tepals* unequal, broadly lanceolate, the dorsal largest, 21–25 x 14–15 mm, extended forward and spreading in the upper 7–10 mm, upper laterals as long as but slightly narrower than the dorsal, the lower three tepals joined to the upper laterals for 3–5 mm and to one another for 3–4 mm, 15–25 x 10 mm, with weakly defined limbs 10–12 mm long and curving outward and downward. *Filaments* 12–14 mm long, exserted 3–8 mm from the tube but included in the flower; *anthers* 6–7 mm long, light purple-grey, the pollen cream. *Ovary* oblong, 3–5 mm long; *style* arched over the stamens, dividing opposite

or just below the anther apices, the branches 2–3 mm long, arching beyond the anther apices. *Capsules* oblong-ellipsoid, 15–22 mm long; *seeds* ovate, 6–7 x 4–5 mm, yellowish brown, broadly and evenly winged. *Chromosome number* unknown.

Flowering time mid- to late July, rarely in early August.

Distribution and biology

Gladiolus inflexus is a narrow endemic of the southwestern Cape, South Africa. Known from just a handful of collections, it has a range extending from the alluvial plain of the Breede River west of Worcester to the Potberg near the mouth of the same river. Collections of the coastal form of the species from near Arniston on the Agulhas Peninsula, Bredasdorp Poort and De Hoop Nature Reserve suggest that the species favours lime-enriched soils, for these are areas where limestone substrates are common. This is, however, not the case for at most, if not all, of the sites where it has been recorded, *G. inflexus* grows in sandstone-derived gravel. The soil where it grows near Worcester consists of sandy loam on flats where the ground is covered to considerable depth by small rounded rocks. The corms are buried to a depth of 10 cm and tightly wedged between rocks, making them secure from predation by molerats, porcupines and baboons. As in so many other species of plants that grow on sandstone-derived soils, flowering is stimulated by fire or clearing of the surrounding vegetation.

Like other Cape species of *Gladiolus* with relatively short, funnel-shaped perianth tubes and predominantly blue, sweetly scented flowers, the flowers of *G. inflexus* are adapted for pollination by long-tongued bees, of which only the honey bee *Apis mellifera* and *Anthophora krugeri* (Anthophoridae) have been observed visiting the flowers.

Diagnosis and relationships

Gladiolus inflexus is reminiscent of *G. gracilis* in the wiry, flexouse stem, narrow leaves and flowers with a strongly arched dorsal tepal and large lower median tepal that exceeds the lower laterals. The resemblance also extends to the hard, woody corm tunics that fragment into vertical segments from the base, a feature shared with most species of series *Gracilis*. It differs from *G. gracilis* and other species of the series in the flower colouring, which is bluish to creamy mauve liberally mottled with dark blue to mauve spots. It is also unusual in having only three leaves, the blades of which are plane with

lightly thickened margins and midrib. In both the plane leaf blades and leaf number but not in the flower patterning, *G. inflexus* particularly recalls the Cedarberg species, *G. taubertianus*, and we suggest that the two species are closely allied. A close relationship with *G. gracilis* seems unlikely because that species has unusual leaves in which the margins are raised into broad wings extending at right angles to the blade, rendering them more or less H-shaped in transverse section.

Floral variation in *Gladiolus inflexus* is correlated with its geographic range. Plants from the Breede River valley have creamy mauve flowers with dark mauve spots, while those from the Bredasdorp coast have pale flowers with bright violet streaks rather than discrete spotting. The latter is evidently a distinct race of the species and requires further investigation before its taxonomic status can be satisfactorily established.

History

The earliest record of *Gladiolus inflexus* that we have traced is a collection of two plants reported to have been found near Bredasdorp in 1933 and later grown at

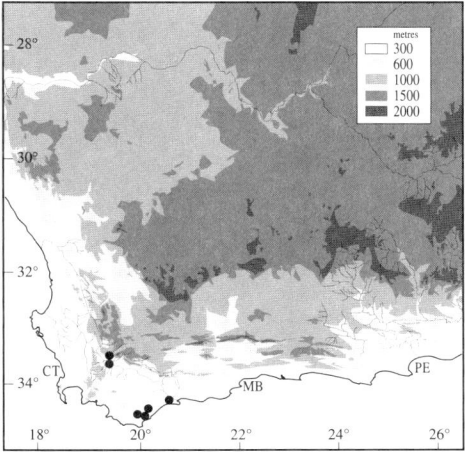

Kirstenbosch Gardens. No collector is associated with these specimens. The species was subsequently collected in 1947 by the late W.F. Barker, who devoted much of her time to studies of southern African geophytic plants. This and the few other collections that have accumulated since then were either regarded as variants of *G. gracilis*, or were thought to be hybrids between this species and *G. maculatus*. An extensive population was discovered by bulb enthusiasts, Margie and Cameron Taswell-Yates, during highway construction near Worcester in the early 1990s and it was through their plant rescue efforts at this site that *G. inflexus* became known to us. The study of living plants made it possible for us to assess its relationships. It then became apparent how different *G. inflexus* was from *G. gracilis* and its allies.

137. GLADIOLUS TAUBERTIANUS Schlechter
PLATE 120

Gladiolus taubertianus Schlechter, *Bot. Jahrb. Syst.* 27: 103 (1900). Type: South Africa, Cape, Cold Bokkeveld, 4000 ft, 5 Sept. 1896, *Schlechter 8860* (BOL, lectotype here designated, PRE, isolectotype; note *Schlechter 8860* at BM, G, and K are *G. inflatus*, those at B and Z, not seen).

taubertianus, named in honour of Paul Taubert, German botanist at Berlin, a contemporary of Rudolf Schlechter.

Description

Plants 18–50(–90) cm high. *Corm* globose, c. 12 mm in diameter, the tunics of hard concentric layers, fragmenting below or

completely with age into flat vertical segments, these sometimes acute basally and apically, or truncate. *Cataphylls* pale and membranous, the uppermost reaching 3–5 cm above the ground and then purple, minutely puberulous. *Leaves* three, the lower two basal, the lowermost longest, sheathing the lower half of the stem, reaching to about the top of the spike, sometimes much exceeding it, the blade linear, (1–)2–3 mm wide, more or less plane, the midrib lightly raised and often a second vein also prominent and the blade then flexed and L-shaped in section, the margins barely thickened, the second leaf sheathing the lower two-thirds of the stem, the blade short, the upper leaf inserted in the upper third of the stem, sheathing in the lower half to one-third, the margins of the sheathing part united around the stem. *Stem* erect, flexed outward above the sheathing parts of the two upper or all three leaves, unbranched, c. 1 mm in diameter below the spike.

Spike lightly inclined and weakly flexuose, 1- to 3-, occasionally to 5-flowered; *bracts* grey-green, soft-textured, the outer 20–35(–40) mm long, the inner about two-thirds as long as the outer, forked apically, twisted around to lie under the outer. *Flowers* pale mauve to slate blue, the upper lateral and lower tepals lightly streaked with purple inside, the lower laterals each with a pale yellow transverse band across the middle, with a light sweet scent; *perianth tube* obliquely funnel-shaped, 9–14 mm long, the lower cylindrical part 5–7 mm long; *tepals* unequal, ovate, the dorsal largest, 20–26 x 11–16 mm, hooded over the stamens and horizontal to inclined slightly toward the ground, upper laterals directed forward and curving outward in the upper third, 16–21 x 10–12 mm, the lower three tepals joined to the upper laterals for 2–5 mm, the free parts 15–19 x 7–9 mm, more or less straight and horizontal, in profile the lower laterals exceeding the upper by c. 5 mm. *Filaments* c. 12 mm long, exserted c. 7 mm from the tube, adpressed to the dorsal tepal; *anthers* 6–9 mm long, pale lilac, the pollen cream. *Ovary* ovoid, c. 3 mm long; *style* arching over the stamens, dividing opposite the upper third of the anthers, the branches 4–5 mm long, extending beyond the anthers. *Capsules* and *seeds* unknown. *Chromosome number* unknown.

Flowering time mid-August to mid-September.

Distribution and biology

A narrow endemic of the southern African winter-rainfall zone, *Gladiolus taubertianus* is restricted to the eastern Cedarberg and the northern edge of the Cold Bokkeveld, in the north of Western Cape Province, South Africa. The only record from the Cold Bokkeveld is the type collection which is from the farm Rosendal at the southern end of the Cedarberg. Plants grow in coarse, stony, sandstone-derived ground in rather dry mountain fynbos in well-drained sites. Relatively rare, *G. taubertianus* has been collected on only a handful of occasions, and never have more than a few plants been seen at any site.

Pollination biology is unknown, but the short-tubed, dark blue-mauve and sweetly scented flowers of *Gladiolus taubertianus* are assumed to be adapted for pollination by long-tongued bees.

Diagnosis and relationships

Although the flowers of *Gladiolus taubertianus* broadly resemble those of the so-called bluebells, particularly *G. inflatus* and *G. patersoniae* of series *Teretifolius*, they are actually not or only very slightly inflated, and the markings consist of longitudinal streaks or spots on the lower and upper lateral tepals and in the throat. This is rather different from the spear-shaped or transverse yellow marks with dark edges characteristic of *G. inflatus* or *G. patersoniae*. Moreover, the leaves of *G. taubertianus* are plane with lightly to moderately thickened margins and midrib, quite different from the terete, four-grooved leaves of species of series *Teretifolius*.

Thus, although it was treated as a synonym of *G. inflatus* by Lewis et al. (1972), we do not regard the two species as at all closely related. We prefer to place *G. taubertianus* in series *Gracilis*, species of which usually have more or less plane leaf blades with moderately thickened margins and midrib, and flowers with nectar guides consisting of dark lines and streaks rather than discrete pale markings outlined in darker pigment. We suggest that *G. taubertianus* is most closely related to *G. inflexus*, which is one of the few other members of the section with three plane leaves. The flowers are similar in shape and size, but quite different in their markings. Those of *G. inflexus* have dark mottling on a pale bluish or creamy mauve background, whereas those of *G. taubertianus* are blue-mauve with dark streaks on the lower and upper lateral tepals and the lower tepals have a fairly well-defined pale yellow transverse band across the middle.

History

Gladiolus taubertianus was first recorded by the German botanist, Rudolf Schlechter, at Rosendal in the northern Cold Bokkeveld in 1896. A year later, A.E. Bodkin, who often collected with Harry Bolus, found the species in the northern Cedarberg at Ezelsbank. Schlechter correctly regarded it as a new species which he described in 1900, naming it in honour of the contemporary German botanist, Paul Taubert. *Gladiolus taubertianus* remained poorly collected and misunderstood well into the twentieth century, and in their revision of *Gladiolus* in South Africa, Lewis et al. (1972) treated the species as a synonym of *G. inflatus* subsp. *inflatus*. Sheets bearing Schlechter's type number, at least in some herbaria, consist of a mixture of plane-leafed *G. taubertianus* and terete-leafed *G. inflatus*, and some sheets consist only of the latter. Very likely the confusion over the status of *G. taubertianus* is

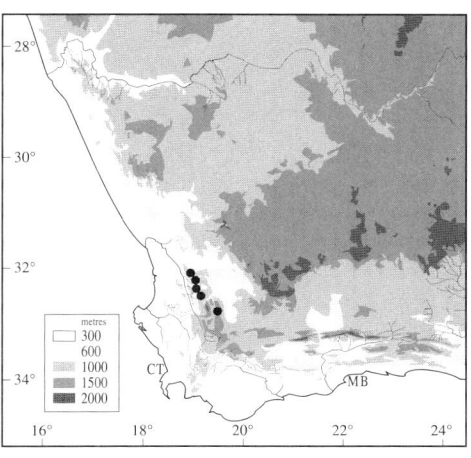

partly the result of this mixed collection, but no doubt its rarity is also partly to blame. We have chosen as lectotype a sheet in the Bolus Herbarium which only has plants that have flat leaves, thus corresponding exactly to Schlechter's description.

138. GLADIOLUS GRACILIS Jacquin

Bloupypie

PLATE 121

Gladiolus gracilis Jacquin, *Collecteana* 4: 159 (1792); *Icones Pl. Rar.* 2: pl. 246 (1795). Baker, *Fl. Capensis* 6: 141 (1896). Lewis et al., *J. S. African Bot.*, Suppl. 10: 224 (1972). Type: South Africa, Western Cape, Cape Peninsula, without precise locality or collector, illustration in Jacquin, *Icones Pl. Rar.* 2: pl. 246 (1795).

gracilis = graceful, alluding to the slender habit.

Synonymy

Gladiolus tristis var. *punctatus* Thunberg, *Dissertatio de Gladiolo* no. 8g (1784). Type: South Africa, Western Cape, without precise locality, *Thunberg s.n.* (UPS–Herb. Thunberg 1028, holotype).
Gladiolus pterophyllus Persoon, *Synopsis Pl.* 1: 43 (1805). Type: South Africa, Western Cape, without precise locality or collector, cultivated in Austria, illustration in Jacquin, *Icones Pl. Rar.* 2: pl. 244 (1795), as '*G. tristis* var.'
Gladiolus scaber Sprengel & Link, *Jahrb. 1*, 3: 70 (1820). Type: South Africa, Western Cape, without precise locality, without collector (Herb. Willdenow, B, holotype [not seen]).
[*Gladiolus setifolius* Ecklon, *Topographisches Verzeichniss* 37 (1827), name without description, based on *Ecklon & Zeyher s.n.* from Caledon Swartberg near the Baths.]

Description

Plants 25–50(–70) cm high. **Corm** 12–20 mm in diameter, the tunics of hard concentric layers, fragmenting below or completely with age into flat vertical segments, these sometimes acute basally and apically, or truncate. **Cataphylls** membranous, the upper reaching 3–5 cm above the ground and then green or flushed purple. **Leaves** usually four, occasionally three, the lowermost basal and longest, reaching to between the upper third of the stem and the middle of the spike, the lower half sheathing the stem and often lightly scabrid below, the blade linear, 1.5–2.5 mm wide, the margins raised into wings extending at right angles to the blade and arching over the laminar surface, sometimes the edges contiguous over the midrib, the margin edges usually scabrid to ciliate, the remaining three leaves cauline, progressively shorter above and sheathing for at least half their length, the sheaths closed, the blades channelled or terminally isobilateral and more or less plane. **Stem** erect or inclined, the upper portion inclined at c. 45°, flexed outward above the sheaths of the three upper leaves, unbranched, c. 1 mm in diameter below the spike.

Spike inclined at c. 45°, flexuose, 2- to 5-, rarely to 7-flowered; **bracts** grey-green or becoming dull purple, the margins membranous, the outer 17–35(–43) mm long, the

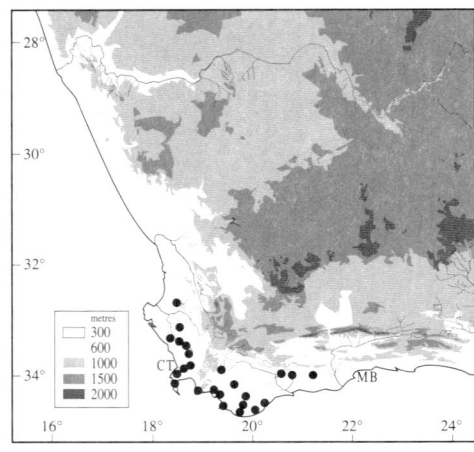

apices often attenuate and somewhat twisted, the inner usually about three-quarters as long as the outer. *Flowers* pale blue to grey, or occasionally pink or dull yellow, the lower three tepals and the abaxial side of the throat pale yellow to cream and marked irregularly with dark purple streaks and spots, usually delicately and sweetly scented, sometimes apparently unscented; *perianth tube* obliquely funnel-shaped, 12–18 mm long; *tepals* unequal, obovate, the margins undulate, the dorsal largest, 24–33 x 14–19 mm, nearly horizontal below, curving outward in the upper half, upper laterals similar but slightly smaller, the lower three tepals joined to the upper laterals for 3–8 mm and to one another for 2–3 mm, narrowed below and shallowly channelled, sometimes more or less claw-like, the lower laterals 20–25 x 7–9 mm, arching downward in the upper halves, the lower median 5–10 mm longer than the lower laterals, up to 12 mm wide, extending downward distally, in profile the lower tepals much exceeding the upper. *Filaments* 11–14 mm long, exserted 4–8 mm from the tube; *anthers* 7–11 mm long, pale to dark blue, the pollen cream or pale blue. *Ovary* oblong, 5–6 mm long; *style* arching over the stamens, dividing shortly below to 2 mm beyond the anther apices, the branches 3–4 mm long, arching outward well beyond the anthers. *Capsules* oblong-ellipsoid, 12–18 mm long, acute; *seeds* ovate, 4.5–5 x c. 3 mm, broadly and evenly winged, translucent yellow-brown. *Chromosome number* 2n = 30.

Flowering time mainly July and early August, sometimes in June.

Distribution and biology

A well-known species of the southern African winter-rainfall zone, *Gladiolus gracilis* extends from Albertinia in the east along the southern Cape coastal belt to the Cape Peninsula. Northward *G. gracilis* extends along the west coast to Darling and Hopefield, and the most northerly population recorded is near Aurora in the western foothills of the Piketberg. Plants are mostly found on heavy soils in renosterveld, thus on clay or granite substrates, but at its northern-most station it grows on sandy soil in fynbos, an odd ecological shift.

Gladiolus gracilis flowers quite early in the season, often in July, and can rarely be found in flower after the end of August. The sweetly scented and short-tubed flowers are adapted for pollination by long-tongued bees. Early-flowering plants are most often pollinated by the honey bee *Apis mellifera* as well as by *Anthophora diversipes*, which is the more important pollinator later in the season when more plant species are in flower and honey bees seem to favour other plants.

Diagnosis and relationships

Gladiolus gracilis typically has short-tubed, clear light blue to dark slate-coloured flowers distinctive in having elongate lower tepals, the lower median exceeding the lower laterals, and nectar guides consisting of irregular longitudinal lines and streaks of dark blue to purple on a pale yellow to cream background. The corm tunics are hard and woody, breaking into vertical segments from below and thus typical of series *Gracilis*. The leaves of *G. gracilis* are its most distinctive feature. The narrow blades appear terete and hollow at first glance, but the blades are plane with a lightly thickened midrib and the margins are raised into wings extending outward at right angles to the surface, then arching inward. The wing edges are usually minutely ciliate to scabrid. The leaf blades are thus more or less H-shaped in transverse section. This peculiar leaf type is shared with only two other species of the genus, *G. caeruleus* and *G. recurvus*. In *G. recurvus* the marginal wings also curve toward one another, giving the blades a rounded appearance, and *G. gracilis* and *G. recurvus* can be distinguished from one another only by their flowers. Those of *G. recurvus* are usually pearl to cream, have a cylindrical perianth tube, 27–36 mm long, and attenuate tepals with the apices somewhat twisted. The flowers of *G. gracilis* have a funnel-shaped tube 12–18 mm long, and the upper tepals are more or less obtuse. The floral scent, pronounced in both species, also differs. The two species have different pollination strategies, *G. gracilis* being pollinated by large, long-tongued bees, and *G. recurvus* by moths.

More closely resembling *Gladiolus gracilis*, at least in its flowers, is *G. caeruleus* which typically has spikes of 8–14 pale blue flowers of more or less similar shape and proportion. The lower tepals are somewhat broader, spotted rather than streaked, and the lower median is usually about as long as the lower laterals. More importantly, the leaves of the two species are rather different. The marginal wings of *G. caeruleus* are held at right angles to the blade surface and the blades are usually 3–6 mm wide, thus appearing quite different to those of *G. gracilis*. *Gladiolus caeruleus* was treated as subsp. *latifolius* of *G. gracilis* by Lewis et al., but this endemic of limestone and calcareous sand is best regarded as a separate species (Goldblatt & Manning, 1996c).

Variation in floral colour in *Gladiolus gracilis* is marked. Most populations from the Cape west coast have pale blue or occasionally pale mauve to pink flowers, with the nectar guides dark blue to violet on a cream ground. In the east, flowers tend to be darker blue or slate-coloured, or occasionally cream and then with brown shading and streaking. Some isolated populations growing on shale in the limestone belt of the Bredasdorp District have light brown flowers with the lower tepals yellow in the lower halves and the typical dark streaking hardly developed. Floral scent is also variable in its intensity and is occasionally absent, or not evident to the human nose. Normally, however, *G. gracilis* has one of the strongest and sweetest fragrances in the genus.

History

This common Western Cape species must have been known long before it was formally named in 1792 by Nicholas Jacquin. It was certainly collected by C.P. Thunberg in the early 1770s, and then by Franz Boos and Georg Schol, who gathered plants for Jacquin in the 1780s. Thunberg was the first to recognize *G. gracilis* taxonomically, calling it *G. tristis* var. *punctatus* and thus treating it as one of 16 varieties of that species. By the time Thunberg revised his ideas about the taxonomy of *G. tristis*, Jacquin had described *G. gracilis* in 1792 and the illustration that serves as the type was published shortly thereafter in 1795. Thunberg immediately recognized *G. gracilis*, and cited his var. *punctatus* in the synonymy of that species. Most authorities accepted and correctly understood *G. gracilis*, but Hendrik Persoon, always idiosyncratic, provided the name *G. pterophyllus* in 1805 for a second, rather poor illustration of *G. gracilis* in Jacquin's *Icones Plantarum Rariorum*, there identified as 'G. tristis var.'. There is no reason at all for the recognition of either *G. scaber* Sprengel & Link, described in 1820, or *G. setifolius* Ecklon, the latter a name without description. Both are quite typical *G. gracilis*.

The account of *Gladiolus gracilis* by Lewis et al. (1972) in *A Revision of the South African Species of* Gladiolus recognized for the first time the Cape west coast relative of *G. gracilis* which they called subsp. *latifolius*.

That plant appears to have first been collected only in 1933, and is now recognized as *G. caeruleus*, a limestone endemic closely related to *G. gracilis*.

139. GLADIOLUS CAERULEUS Goldblatt & Manning
PLATE 122

Gladiolus caeruleus Goldblatt & Manning, *Novon* 6: 179 (1996), as a new name for *G. gracilis* var. *latifolius* G. Lewis in *J. S. African Bot.*, Suppl. 10: 228 (1972), not *G. latifolius* Lamarck (1791) (= *Babiana villosa* (Aiton) Ker Gawler). Type: South Africa, Western Cape, road to Donkergat north of Churchhaven, 15 Aug. 1966, *Barker 10395* (NBG, holotype; K, isotype).

caeruleus = clear blue, for the flower colour.

Description
Plants 40–60 cm high. *Corm* 10–14 mm in diameter, the tunics of hard concentric layers, fragmenting below or completely with age into flat vertical segments, these sometimes acute basally and apically, or truncate. *Cataphylls* pale and membranous, the upper reaching 3–5 cm above the ground and then green. *Leaves* four, the lowermost basal and longest, reaching to between the base and apex of the spike, the lower half sheathing the stem and often lightly scabrid below, the blade linear, 3–7 mm wide, the margins extended at right angles to the laminar surface, thus H-shaped in section, the margin edges smooth, the midrib lightly raised, the remaining three leaves cauline, progressively shorter above and sheathing for at least half their length, the sheaths closed, the blades channelled or terminally isobilateral, linear and more or less plane. *Stem* erect below, the upper portion inclined at c. 20°, flexed outward above the sheaths of the two upper leaves, unbranched, 1.7–2 mm in diameter below the spike.

Spike occasionally 4-, usually 8- to 10-, rarely to 14-flowered, flexed at the base and inclined c. 45°, weakly flexuose; *bracts* grey-green to dull purple-grey, the outer with narrowly membranous margins, 25–35(–40) mm long, the inner c. two-thirds as long as the outer, the margins broadly membranous, barely notched at the apex. *Flowers* pale blue, darker along the midline of the tepals, the abaxial side of the throat and the lower two-thirds of the lower three tepals irregularly speckled with dark purple on a cream background, lightly scented; *perianth tube* obliquely funnel-shaped, c. 15 mm long; *tepals* obovate and obtuse, unequal, the dorsal largest, 30–32 x c. 20 mm, arching forward and nearly horizontal below, curving outward in the upper third, upper laterals directed forward, patent in the upper third, c. 25 x 17 mm, the lower three tepals joined to the upper laterals for c. 6 mm and to one another for 2–3 mm, tapering toward the bases but not forming claws below, subequal, c. 20 x 8–10 mm, arching gently outward, widest just below the apices and obtuse, in profile the lower tepals exceeding the upper by c. 5 mm. *Filaments* c. 12 mm long, exserted c. 6 mm from the tube; *anthers* 9–10 mm long, pale blue, the pollen cream or pale blue. *Ovary* oblong, 5–6 mm long; *style* arching over the stamens, dividing opposite or shortly beyond the anther apices, the branches c. 4 mm long, arching outward well beyond the anthers. *Capsules* ellipsoid, c. 15 mm long; *seeds* ovate, 5–7 x c. 4 mm, broadly and evenly winged, translucent light brown. *Chromosome number* unknown.

Flowering time August to mid-September.

Distribution and biology
Gladiolus caeruleus is a local endemic of the southwestern coast of Western Cape Province, South Africa, where it occurs from Yzerfontein in the south to the hills above Cape Columbine north of Saldanha Bay. Plants are known from only a few sites, always growing in sandy soils in limestone outcrops or on calcareous sands close to the coast.

The short-tubed and sweetly scented flowers are believed to be adapted for pollination by long-tongued bees.

Diagnosis and relationships
The most striking feature of *Gladiolus caeruleus* is its fairly broad leaf blades, 3–7 mm wide, which diverge sharply from the stem and have strongly winged margins that extend upward at right angles to the blade surface. The spikes bear up to 14 flowers, although more often there are only eight to ten, and the flowers are pale blue. Leaves with winged margins are known in only two other species of *Gladiolus* – *G. gracilis* and its long-tubed relative, *G. recurvus* – and it is to these two species that *G. caeruleus* is most closely allied. *Gladiolus caeruleus* was treated by Lewis et al. (1972) as var. *latifolius* of the more widespread *G. gracilis*, but the two taxa are best regarded as separate species. No doubt they are closely related, sharing at least the specialized winged leaf margins and similar flowers, but they differ in several notable features. *Gladiolus caeruleus* has weakly flexuose spikes with up to 14 flowers, and the floral bracts are 25–40 mm long but without the attenuate, twisted apices characteristic of *G. gracilis*. The margins of the leaf blades also differ in their right-angled orientation to the blade surface. In *G. gracilis*

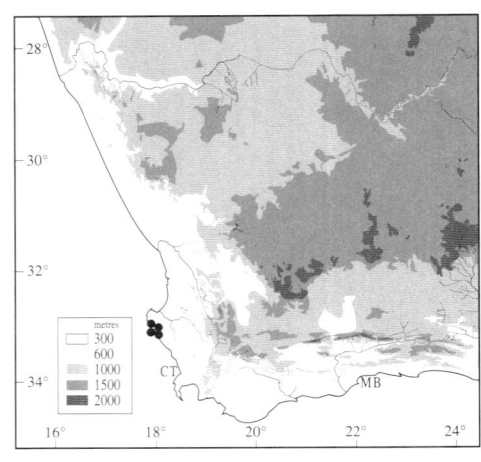

the spikes are strongly flexuose, normally with two to five flowers, and the leaf blades are 1.5–2.5 mm wide, with the marginal

wings arching inward over the blade surface and almost meeting over the midrib. In addition, the flowers of *G. caeruleus* are fairly large, 45–47 mm long, and the lower tepals are more or less equal and obtuse to nearly acute. In contrast, *G. gracilis* usually has smaller flowers, 36–44 mm long, floral bracts 17–35 mm, occasionally to 43 mm long, and the lower tepals are acute or almost so and unequal, with the lowermost being notably longer than the lower laterals.

History

Despite occurring relatively near Cape Town and close to places that have been well botanized, records of *Gladiolus caeruleus* are scanty. The first collection seems to have been made in 1933 on the farm Doorn-

fontein near Darling by Miss M. Mansergh, who brought specimens to H.M.L. Bolus at the Bolus Herbarium. Later, in 1946, the South African botanist, Frances Leighton, made a gathering at Danger Bay. A collection made in 1966 by the Kirstenbosch botanist, W.F. Barker, on the Donkergat Peninsula near Churchhaven formed the basis for *G. gracilis* subsp. *latifolius* which was described by Lewis et al. in 1972. The existence of the name *G. latifolius* Lamarck, dating from 1791, a later synonym of *Babiana villosa*, made it necessary to rename this plant when it was recognized at species rank as recommended by Goldblatt & Manning (1996c). They called the species *G. caeruleus*.

140. GLADIOLUS RECURVUS Linnaeus
Voorloopertjie
PLATE 123

Gladiolus recurvus Linnaeus, *Mantissa* 1: 28 (1767). Baker, *Fl. Capensis* 6: 139 (1896), misapplied to *G. maculatus* and including

G. carinatus. Lewis et al., *J. S. African Bot.*, Suppl. 10: 205 (1972). *Watsonia recurva* (Linnaeus) Persoon, *Synopsis Pl.* 1: 43 (1805). *Gladiolus ramosus* Burman fil., *Prodromus Fl. Capensis* 2 (1768), an illegitimate and superfluous name, based on the same type as *G. recurvus*. Type: South Africa, Western Cape, without precise locality, illustration in Miller, *Figures of Plants* 2: 157, pl. 235, fig. 2 (1758).

recurvus = recurving, referring to the attenuate and recurving tips of the tepals.

Synonymy

Gladiolus modestus Ingram, *Gard. Chron.*, Ser. 3, 90: 9 (1931). Type: South Africa, Western Cape, Breede River Valley, without date, *Ingram s.n.* (BM, colour illustration only).

Description

Plants 25–35 cm high. *Corm* globose, 10–14 mm in diameter, the tunics of hard concentric layers, fragmenting below with age into vertical claw-like segments, acute below. *Cataphylls* membranous, the uppermost reaching 3–8 cm above the ground and then green. *Leaves* four, the lowermost basal and longest, sheathing the lower third of the stem, the blade fairly short or reaching to just below the base of the spike, 30–100 mm long, linear, c. 2 mm wide, the margins raised into wings extended at right angles to the blade and arching inward over the laminar surface, the edges lightly scabrid, the midrib lightly raised but not thickened.

Stem erect below, inclined above and flexed outward above each leaf sheath, unbranched, c. 1 mm in diameter below the spike.

Spike inclined, flexuose, 1- to 2-, occasionally to 4-flowered; *bracts* dull green to grey-brown, the outer 30–40 mm long, the apices somewhat twisted and with the margins inrolled, the inner about two-thirds as long as the outer. *Flowers* pearl-grey to yellowish, the lower lateral tepals irregularly lined and spotted with dull purple, the upper tepals also sometimes lightly lined or spotted toward the midline, the lower margins of the upper tepals transparent, with a strong, sweet, cloying scent; *perianth tube* more or less cylindric, curving outward and widening in the throat, 27–36 mm long; *tepals* lanceolate-attenuate, the upper margins undulate and the apices twisted outward or downward, unequal, the dorsal largest, ascending, 22–30 x 14–18 mm, the upper laterals 22–26 x 8–11 mm, directed forward, patent in the upper third, the lower three tepals joined to the upper laterals for 4–7 mm and to one another for 2–3 mm, the free parts c. 20 x 8–9 mm, patent and directed toward the ground in the upper halves. *Filaments* 14–15 mm long, exserted 4–5 mm from the tube; *anthers* 8–10 mm long, dull grey to blackish, the pollen pale yellow. *Ovary* oblong, c. 5 mm long; *style* arching over the stamens, dividing 2 mm below to 2 mm beyond the anther apices, the branches fairly short, 2.5–4 mm long, expanded to 1.5 mm wide in the upper halves. *Capsules* ovoid-ellipsoid, 14–16 mm long; *seeds* ovate,

c. 5 x 3 mm, broadly and evenly winged, golden brown. *Chromosome number* 2*n* = 30.

Flowering time mainly June to early August at low elevations, September to early October at higher elevations.

Distribution and biology

Endemic to the southwestern Cape, South Africa, *Gladiolus recurvus* has a fairly narrow range extending from the Witzenberg above Tulbagh and the Warm Bokkeveld in the north through the upper Breede River valley to Somerset West in the south. There is also a single record from Caledon (*Zeyher s.n.*) that requires confirmation before it can be accepted. Plants said by Rudolf Schlechter to have been collected on Table Mountain on the Cape Peninsula flowering in March must be mislabelled (Lewis et al., 1972).

Gladiolus recurvus flowers in mid- or late winter, June and July at low elevations, rarely in early August, and from late August to early October at higher elevations. It appears to be restricted to lower mountain slopes and flats where it grows in renosterveld on stony, clay loam soil.

The sweetly and heavily scented, long-tubed and pale creamy grey flowers of *Gladiolus recurvus* are clearly adapted for pollination by moths. The flowers resemble in form and colouring – although not in size – those of the distantly related *G. liliaceus*. The latter blooms one to two months later than *G. recurvus* and is also pollinated by moths, as is the winter-flowering *G. maculatus* with which *G. recurvus* is often confused.

Diagnosis and relationships

The inconspicuous flowers of *Gladiolus recurvus* are pale cream to pearl-grey or occasionally pale lilac, and are readily identified by their elongate perianth tube, 27–36 mm long, and tapering attenuate tepals, the tips twisted, undulate and recurved. In its vegetative features *G. recurvus* is virtually identical to the more common and well-known *G. gracilis* and shares with it the woody corm tunics and specialized leaves with winged margins that arch inward over the leaf blades, which are thus more or less H-shaped in transverse section. The morphological similarity between the two species is so close that it is little doubt that they are immediately related. They thus differ only in their floral adaptations to different pollinators, which are moths for *G. recurvus* and long-tongued bees for *G. gracilis*. Over most of their ranges the two species are allopatric, but in the south at Stellenbosh and Somerset West they grow sympatrically, but then flower at different times, *G. gracilis* a week or more later than *G. recurvus*.

History

Although not a common plant and absent from the Cape Peninsula, *Gladiolus recurvus* was one of the first species of southern African *Gladiolus* to be described. It first appeared in the literature in 1758 in Philip Miller's *Figures of Plants*, that lavishly illustrated collection of watercolour reproductions of rare plants cultivated at the Physick Garden in Chelsea, London. In that work plants were not given binomials, but in 1767 Carl Linnaeus named the species based on Miller's illustration. At this period the Cape was still maintained by the Dutch East India Company and Miller would have obtained his plants from Holland, so it may be assumed that *G. recurvus* was grown there some years before 1758. Just a year after Linnaeus named the species, the younger Burman renamed Miller's plate *G. ramosus* in

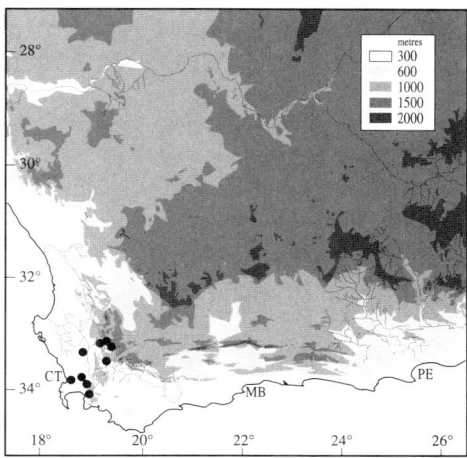

1768. Slow communication may be blamed for this unfortunate circumstance. Almost immediately *G. recurvus* became the subject of botanical confusion, despite the existence of the fine watercolour in *Figures of Plants*.

The Dutch naturalist, Maarten Houttuyn, identified *Gladiolus watsonius* as *G. recurvus* in 1780 and John Ker Gawler used the name for *G. carinatus* in all his treatments of *Gladiolus*. Even worse, C.P. Thunberg associated *Hesperantha radiata* with *G. recurvus* in his accounts of southern African Iridaceae. J.G. Baker treated *G. maculatus* as conspecific with *G. recurvus* in *Flora Capensis* (1896), also citing specimens of *G. carinatus* and *G. longicollis* in his account of the species (see Lewis et al., 1972).

The confusion over the correct identity of *G. recurvus* is presumably the reason for Collingwood Ingram, the British *Gladiolus* grower, describing *G. modestus* in 1931, based on plants he collected in the Breede River valley. The series of misunderstandings surrounding *G. recurvus* was only unravelled by G.J. Lewis in the 1960s when she was completing her studies for her revision of *Gladiolus* in South Africa.

Section *Homoglossum:* Series *Teretifolius*

141. GLADIOLUS INFLATUS Thunberg

Tulbagh klokkie, Tulbagh bell
PLATE 124

Gladiolus inflatus Thunberg, *Prodromus Pl. Capensium* 185 (1800). Baker, *Fl. Capensis* 6: 147, name applied partly to *G. ornatus*. Lewis et al., *J. S. African Bot.*, Suppl. 10: 236 (1972), in part. Type: South Africa, Western Cape, '*Gladiolus inflatus* a,' without precise locality or date, *Thunberg s.n.* (UPS–Herb. Thunberg 1033, holotype; BOL, photo).

inflatus = inflated, referring to the inflated, and thus bell-like, flower.

Synonymy

Gladiolus tristis var. *violaceus* Thunberg, *Dissertatio de Gladiolo* 166 (1784). Type: South Africa, Western Cape, without precise locality, *Thunberg s.n.* (UPS–Herb. Thunberg 1034, holotype).
Gladiolus tristis var. *hastatus* Thunberg, *Dissertatio de Gladiolo* 167 (1784). *G. hastatus* Thunberg, *Prodromus Pl. Capensium* 185 (1800). Baker, *Fl. Capensis* 6: 140 (1896), name applied partly to *G. rogersii*. Type: South Africa, Western Cape, without precise locality, *Thunberg s.n.* (UPS–Herb. Thunberg 1032, holotype).
Gladiolus bolusii Baker, *Handbook Irideae* 208 (1892); *Fl. Capensis* 6: 145 (1896). Type: South Africa, Western Cape, Winterhoek Mountains near Tulbagh, Nov. 1879, *H. Bolus 5244* (K, holotype; BM, BOL, isotypes).

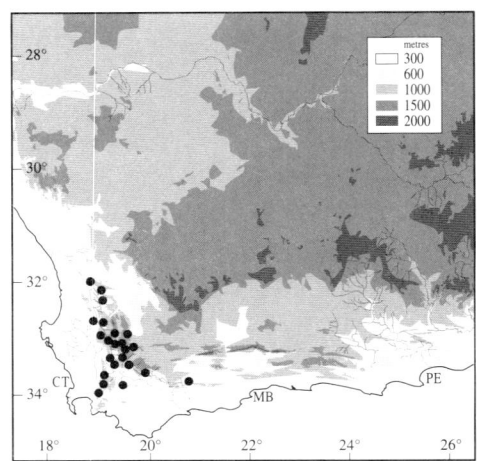

Gladiolus louiseae L. Bolus, *Ann. Bolus Herb.*
3: 182 (1924). *Gladiolus inflatus* var. *louiseae*
(L. Bolus) Obermeyer, *J. S. African Bot.*, Suppl.
10: 241 (1972). Type: South Africa, Cape,
without precise locality, from the Ceres exhibit
at the Cape Town Wild Flower Show, Oct.
1921, *L. Guthrie s.n.* (BOL 14195, holotype;
BOL, K, isotypes).

Description

Plants (12–)20–45 cm high. *Corm* globose,
9–20 mm in diameter, the tunics usually of
dark brown to blackish more or less woody
layers, splitting below into thickened claw-
like ridges, sometimes the layers cartilaginous and becoming fibrous with age.
Cataphylls membranous, entirely pale or
purplish above the ground, sometimes
scabrid, sometimes accumulating with age to
form a slender fibrous neck around the base
of the stem. *Leaves* three, the uppermost
sometimes reduced to a sheathing, bract-like
scale 20–30 mm long, the lower two leaves
basal, the lowermost reaching to about the
base of the spike, sheathing the stem for a
short distance or only near the ground,
usually the sheath and sometimes the blade
lightly pubescent or scabrid, the hairs

pointed downward, the blade oval to terete
with four hairline longitudinal grooves
(rarely centric and cross-shaped in section
with wide longitudinal grooves), straight or
curving outward, c. 1 mm in diameter, the
second leaf sheathing the lower two-thirds of
the stem, the upper portion fairly short,
reaching or occasionally exceeding the spike,
the sheath and blade sometimes puberulous.
Stem erect, flexed outward above the sheath
of the second leaf, unbranched, 1.2–1.5 mm
in diameter below the spike.

Spike flexed at the base, lightly inclined,
lightly flexuose, 2- to 3-flowered; *bracts* pale
green to dull greenish grey, the veins transparent, the outer 18–26(–30) mm long, the
inner two-thirds to nearly as long as the
outer, the apices more or less entire, acute.
Flowers slightly bell-shaped or tepals not
belled, deep pink to mauve, purple or violet,
rarely white, the lower tepals each with dark
spear- or more or less spade-shaped purple to
red marks in the middle third, the mark on
the lower median tepal usually yellow in the
centre, sometimes the throat also ringed with
red to purple, unscented; *perianth tube*
obliquely funnel-shaped, 12–25(–30) mm
long, the lower cylindrical part 5–12(–
20) mm long; *tepals* unequal, the upper
ovate, the lower lanceolate to spathulate, the
dorsal largest, 22–27 x 12–17 mm, inclined
to horizontal, the upper laterals 20–24 x
12–14 mm, directed forward, straight or
recurved in the upper third, the lower three
tepals united with the upper laterals for
4–7 mm and sometimes to one another for
c. 1 mm, 17–20 x 9–11 mm, straight or
lightly curving downward in the upper third,
in profile about as long as or slightly longer
than the upper. *Filaments* 12–18 mm long,
exserted 5–11 mm from the tube; *anthers*
7–9 mm long, grey-purple, the pollen yellow.
Ovary oblong, 4–5 mm long; *style* arching
over the stamens, dividing between the middle and apex of the anthers, the branches
3–4 mm long. *Capsules* ellipsoid, subacute,
12–18 mm long; *seeds* broadly ovate, 5–6 x
c. 4.5 mm, broadly and evenly winged, light
brown. *Chromosome number* unknown.

Flowering time late September and
October, sometimes to late November at
higher elevations.

Distribution and biology

Gladiolus inflatus is a relatively common
species of the northern and western moun-
tains of Western Cape Province, South
Africa. Its range extends from the
Franschhoek Pass northward through the
Groot Winterhoek and Cold Bokkeveld

Mountains into the Cedarberg. To the east,
populations occur on the Matroosberg and
Langeberg, and in the south on Jonaskop in
the Riviersonderend Mountains. A population from the central Langeberg near
Riversdale is somewhat atypical and is discussed below. Plants generally grow in fairly
open habitats on sandstone substrates, often
in rocky outcrops or in stony ground among
scattered shrubs and restios. Although the
species can be found in flower in healthy
mature vegetation, clearing or fire promotes
flowering and the most striking displays are
often seen the first and second seasons after a
late summer veld fire.

The short-tubed flowers conform to the
pattern for bee-pollinated species of
Gladiolus and the anthophorid bee *Amegilla
spilostoma* has been found to pollinate plants
on Jonaskop.

Diagnosis and relationships

The presence of three leaves with narrow,
terete leaf blades and corm tunics of thick
woody fibres places *Gladiolus inflatus* in
series *Teretifolius* of section *Homoglossum*.
Within the series, it has flowers that most
closely resemble those of *G. patersoniae* and
the two are evidently closely allied, possibly
representing descendants of a common
ancestor. They can usually be distinguished
by perianth tube length, tepal markings and
the presence or absence of fragrance. In general, *G. inflatus* has unscented, violet, purple
or pink, rarely white, flowers, often with
somewhat inflated tepals, the lower three, or
at least the lowermost, with longitudinal,
darkly pigmented markings. The perianth
tube is usually gently curved and 15–25 mm
long, exceptionally only 12 mm long or up
to 30 mm. Flowers of *G. patersoniae* have a
tube 10–12 mm long with a sharp geniculate bend, have transverse yellow markings
and are, as far as we know, always sweetly
scented. Careful attention to floral differences

makes it relatively easy to distinguish the two species. *Gladiolus patersoniae* was treated as subsp. *intermedius* of *G. inflatus* by Lewis et al. (1972).

Both flower colour and tube length are unusually variable in *Gladiolus inflatus*. Plants with pink flowers and longer tubes, 20–30 mm long, frequently have the uppermost leaf reduced to a short scale completely sheathing the stem. These plants, corresponding to the type of *G. louiseae* L. Bolus, were regarded as var. *louiseae* of *G. inflatus* by Lewis et al. (1972). Variation in tube length in *G. inflatus* is continuous and, moreover, plants with longer-tubed flowers do not consistently have the reduced scale-like upper leaf. We prefer not to recognize infraspecific taxa within the species. Particularly short-tubed plants occur in an apparently erratic pattern across the range of the species. They have been recorded on the Olifants River and Porterville mountains, the northern Cedarberg, and on Jonaskop. In this last site the flowers have a tube no more than 12 mm long. Late-flowering plants from the higher peaks in the mountains around Ceres have the longest floral tubes in the species, sometimes up to 30 mm long, and these plants also have pale pink flowers with reddish markings. Both the tube length and flower colouring are similar to that in the related *G. cylindraceus* which grows in the same places. The possibility that the long-tubed form of *G. inflatus* occasionally hybridizes with *G. cylindraceus* seems likely. The two can, however, be distinguished by flower shape. *Gladiolus cylindraceus* has a slender perianth tube and tepals spreading outward from the base, quite different from the bell-like flower of *G. inflatus*.

We include in *Gladiolus inflatus* unusual plants from the Langeberg near Riversdale (*Stirton 10208*), the most easterly record of the species. These have bright pink flowers with a tube c. 25 mm long that correspond fairly well with those of *G. inflatus*. The corms, however, have finely fibrous tunics and the old corms accumulate in a chain below the current corm, features not known in other populations of *G. inflatus*. More material of this variant may resolve questions about its status.

Historically, *Gladiolus inflatus* has often been confused with the southern Cape *G. rogersii* for the inflated, bell-like flowers of the two species are very similar. The similarity may well be due to convergence, floral shape and colour being particularly labile in the genus. In any event, the vegetative differences between the two species make it impossible to mistake them. Normal and robust individuals of *G. rogersii* have four foliage leaves and the leaf blades are either plane with moderately thickened margins and midrib, or solid and oval to terete in transverse section. Both *G. inflatus* and closely related *G. patersoniae* have three foliage leaves, the blades of which are terete with four hairline grooves, the result of the enlargement and thickening of the margins and midrib. The terete leaf in *G. rogersii* thus has a quite different origin and is not homologous. The differences in leaf number and blade structure are the reasons for the separation of *G. rogersii* and *G. inflatus* in different series of section *Homoglossum*.

History

The species we now know as *Gladiolus inflatus* was first described in 1784 by Carl Peter Thunberg under three different epithets, *G. tristis* var. *inflatus*, var. *hastatus* and var. *violaceus*, all based on specimens he had collected some ten years earlier. Then in 1800 he raised two of the varieties to species rank as *G. inflatus* and *G. hastatus* respectively. Because the two were known only from description and herbarium specimens in Sweden, they were much confused by botanists in the nineteenth century. Early interpretation of the two species was incorrect largely due to ignorance of the identity of the type material rather than to an informed decision about their status. J.G. Baker (1892, 1896) associated *G. inflatus* with the largely Cape Peninsula species, *G. ornatus*, in particular, but also cited collections of *G. rogersii* under that name. He likewise confused *G. hastatus* and cited specimens of both *G. rogersii* and *G. priorii* under his account of that species. F.W. Klatt regarded *G. inflatus* as conspecific with *G. spathaceus* (now *G. bullatus*). Baker's uncertainty about the identity of *G. inflatus* led him to describe *G. bolusii*, which is actually conspecific with *G. inflatus*, with the comment, 'May be a variety of *G. inflatus*' (Baker, 1896). After examining the Cape plants in the Thunberg Herbarium, N.E. Brown (1928) reached the conclusion that *G. bolusii* and *G. inflatus* were the same species, and the earlier name was adopted. Plants of *G. inflatus* which had pale pink flowers with a somewhat longer perianth tube were described as the new *G. louiseae* by Louisa Bolus in 1924.

In their revision of *Gladiolus* in South Africa, Lewis et al. (1972) correctly understood Thunberg's concept of *G. inflatus* and included *G. louiseae* within the species as a separate variety. They also recognized subsp. *intermedius* for the plants occurring to the east of the main range of the species. Subsp. *intermedius* is, we believe, a separate species, the earliest name for which is *G. patersoniae* (see discussion of this species, page 280). A third species, *G. taubertianus*, described by Rudolf Schlechter in 1900, was included in *G. inflatus* by Lewis et al., we presume because of a mistaken interpretation of the type. *Gladiolus taubertianus* has plane rather than terete leaves and flowers with nectar guides consisting of dark streaks on a pale yellow background, and its affinities lie with species of series *Gracilis*.

142. GLADIOLUS CYLINDRACEUS G. Lewis

PLATE 125

Gladiolus cylindraceus G. Lewis in *J. S. African Bot.*, Suppl. 10: 183 (1972). Type: South Africa, Western Cape, Hansiesberg, Ceres District, c. 1800–1850 m (5500–6000 ft), 15 Dec. 1963, *Jones s.n.* (NBG, holotype).

cylindraceus = cylindrical, referring to the shape of the perianth tube, which is long and slender.

Description

Plants 30–50 cm high. **Corm** globose, 12–20 mm in diameter, the tunics of light brown, medium-textured to moderately thick, more or less wiry fibres, the corms of previous seasons usually accumulating as flat dry discs below the current corm. **Cataphylls** membranous, entirely pale or purplish above the ground, accumulating with age to form a neck around the base of the stem. **Leaves** three, the lower one or two basal, the lowermost longest, reaching to between the base and apex of the spike, sheathing the stem for a short distance or only near the ground, the blade centric, cross-shaped in section, straight, 2–3 mm in diameter, second leaf sheathing the lower half of the stem, overlapping the uppermost leaf, the blade fairly short, ovate in section, the midrib less prominently thickened than in the basal

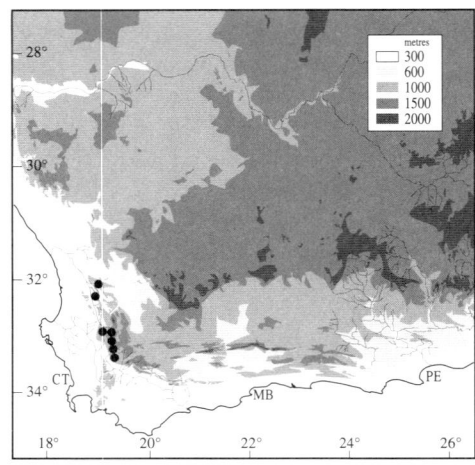

leaf, the uppermost leaf shortest, inserted in the middle to upper third of the stem, sheathing for most of its length, the free portion channelled, the margins of the sheathing part free to the base and overlapping. *Stem* erect, flexed outward above the sheaths of the upper one or two leaves, unbranched, 1.3–1.8 mm in diameter below the spike.

Spike flexed at the base and inclined, lightly flexuose, 5- to 8-flowered; *bracts* pale greenish grey, the outer 18–28 mm long, the inner two-thirds to nearly as long as the outer, the apices more or less entire, acute. *Flowers* hypocrateriform, pale creamy pink to salmon, the lower three tepals each with a reddish diamond-shaped mark sometimes with a white to yellow centre in the lower midline, unscented; *perianth tube* cylindrical, (25–)35–52 mm long, flared only near the apex, half as long again to twice as long as the bracts; *tepals* unequal, broadly lanceolate or the dorsal ovate, the dorsal largest, (18–)20–25 x 11–14 mm, lightly inclined or extended at right angles

to the tube, the upper laterals 18–25 x 9–12 mm, straight and subpatent or patent, lower three tepals sometimes united with the upper laterals for 1–2 mm and to one another for c. 2 mm, 15–21 x 7–8 mm, nearly straight, inclined c. 30°. *Filaments* 7–8 mm long, exserted 3–4 mm from the upper part of the tube; *anthers* 5–8 mm long, white, the pollen cream. *Ovary* oblong, 4–5 mm long; *style* more or less straight, adaxial to the stamens, dividing opposite the upper third of the anthers or shortly beyond them, the branches 3–4 mm long. *Capsules* and *seeds* unknown. *Chromosome number* unknown.

Flowering time mid-December to late January.

Distribution and biology

Gladiolus cylindraceus is a fairly localized endemic of the southwestern Cape. It is virtually confined to the Ceres District and extends from the Mosterthoek Peaks south of Michell's Pass northward along the Witzenberg and Skurweberg ranges to Sneeugat and the Groot Winterhoek Peaks. The most easterly population recorded is from the Waboomsberg, northeast of Ceres, and plants there have the smallest flowers in the species. *Gladiolus cylindraceus* is strictly montane, and has not been recorded below 1500 m. It grows on rocky sandstone slopes and ridges in low fynbos vegetation. Like many other high-elevation species, it flowers in the summer, the first flowers seldom appearing before the middle of December.

The pale pink flowers of *Gladiolus cylindraceus* have reddish to purple nectar guides and a long, slender perianth tube, suggesting that the species is adapted for pollination by the long-tongued fly, *Philoliche rostrata*. No observations have yet been made on the pollination biology of *G. cylindraceus*.

Diagnosis and relationships

Gladiolus cylindraceus is readily recognized by the presence of three foliage leaves, the lowermost leaf with a centric blade cross-shaped in transverse section, and thus with wide grooves between the flanges of leaf tissue. The flowers are pale pink with reddish markings on the lower tepals and, although the tepals are relatively small, the perianth tube is slender and elongate, and up to 52 mm long. The leaf number and general appearance of the flower (not including the long tube) recall in particular *G. inflatus*, although the leaf blades in that species are

thinner and have very narrow grooves. We assume that it is to this species that *G. cylindraceus* is most closely allied. The long perianth tube makes it easy to distinguish it from all other species of series *Teretifolius*, as do the wide grooves of the leaf blades. The shorter-tubed flowers of *G. inflatus* are more or less inflated, with the tepals forming a bell. This contrasts with the more open flower of *G. cylindraceus*, in which the tepals are more or less straight and sometimes spread almost at right angles to the tube.

History

Although described only in 1972 (Lewis et al., 1972), *Gladiolus cylindraceus* has been known since at least 1884, when plants were collected by the German-born botanist and expert on the Cape flora, Rudolf Marloth. He collected plants high up in the Klein Winterhoek Mountains. The specimens attracted no attention and were at first thought to be *G. prismatosiphon*, this actually a synonym of *G. carneus*. A subsequent collection was made by the German chemist and amateur naturalist, H.K. Andreae, in 1923 in the Groot Winterhoek Mountains. This and later gatherings, including ample collections made by Elsie Esterhuysen in 1944 and 1956, also remained unnamed for years or were confused with other species.

The type collection was made in 1963 when living plants were brought to G.J. Lewis from the Hansiesberg, north of Ceres. The plants illustrated here were brought to us from Turret Peak in the Cold Bokkeveld Mountains by Esterhuysen. Her botanical exploration of the high mountains of the Western and Eastern Cape provinces has been invaluable in helping understand the many species restricted to these habitats, including several species of *Gladiolus*.

143. GLADIOLUS NIGROMONTANUS Goldblatt

PLATE 126

Gladiolus nigromontanus Goldblatt, *J. S. African Bot.* 50: 456 (1984). Type: South Africa, Western Cape, Swartberg Pass, Old Toll House, 950 m, 16 Mar. 1981, *Vlok 175* (MO, holotype, K, NBG, PRE–Herb. SAAS, isotypes).

nigromontanus = of the black mountain, for the type locality, a seep in the Swartberg.

Description

Plants 30–40 cm high. *Corm* globose, c. 16 mm in diameter, the tunics of coarse, reticulate fibres. *Cataphylls* membranous, the uppermost reaching shortly above the ground and then green, accumulating with the corm tunics to form a neck around the underground part of the stem. *Leaves* three, the lowermost basal and longest, sheathing the lower half of the stem, reaching the middle or upper third of the stem, the blade relatively short, straight and erect, the midrib and margins heavily thickened and thus terete with four longitudinal grooves, c. 1 mm in diameter, the remaining two leaves entirely sheathing or with short free apices up to 20 mm long, the lower of these fairly long, overlapped by the sheath of the basal, the uppermost short, inserted on the upper fifth of the stem, 40–60 mm long, sometimes overlapped by the second leaf, the margins of the leaf sheaths united around the stem. *Stem* erect below, lightly flexed outward above the sheaths of the upper two leaves or all the leaves, thus inclined above, unbranched, c. 1 mm in diameter below the spike.

Spike inclined, more or less straight, 3- to 6-flowered; *bracts* pale green or flushed with grey-purple, the apices brown, 14–22 mm long, lightly diverging from the stem, the

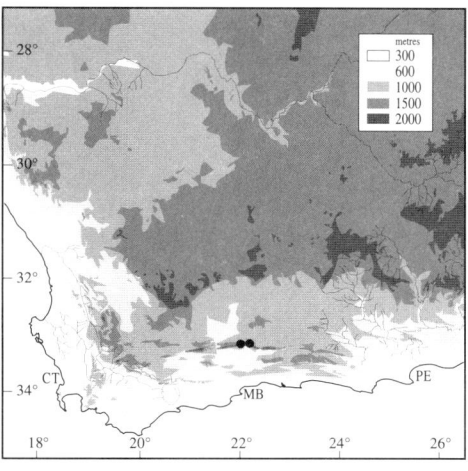

outer 1.2–1.5(–1.8) times as long as the internodes, the inner slightly shorter than the outer, forked apically for 1–1.5 mm. *Flowers* white to pale pink, the lower three tepals each with dark red longitudinal streaks widest at the base of the tepal limbs, the streaks sometimes pale in the centre, red in the base of the tube, unscented; *perianth tube* narrowly and obliquely funnel-shaped, (17–)20–25 mm long, exceeding the bracts and emerging between their apices, curved outward beyond the bracts, the wider upper part c. 8 mm long; *tepals* unequal, the dorsal largest, obovate, inclined to hooded over the stamens 20–24 x 14–16 mm, often curving upwards at the apices, the upper lateral tepals lanceolate, directed forward below, curving outward and becoming reflexed in the upper half, c. 20 x c. 12 mm, the lower three tepals united below for c. 4 mm, the free parts c. 15 mm long, forming narrow horizontal claws below, these lightly channelled, the limbs 4–5 mm at the widest, abruptly curving downward, in profile the dorsal and lower tepals reaching to about the same point. *Filaments* c. 12–14 mm long, exserted 3–4 mm from the tube; *anthers* c. 7 mm long, dark purple, the pollen cream. *Ovary* narrowly ovoid, c. 3.5 mm long; *style* arching over the stamens, dividing 1–2 mm beyond the anther apices, the branches c. 4.5 mm long. *Capsules* obovoid, rounded at the apices, c. 12 mm long (mature capsules not seen); *seeds* unknown. *Chromosome number* unknown.

Flowering time February to early May.

Distribution and biology

Gladiolus nigromontanus is a narrow endemic restricted to the Swartberg in Western Cape Province, South Africa. It is known from only two sites in the mountains north of Oudtshoorn, both perennially moist in this semi-arid area, and at least one of the sites is on a cooler, south-facing slope. Plants flower poorly in the first season after a burn, but well in the following two or three years. After that flowering is poor as the fynbos vegetation regrows, shading out the plants.

The pollination biology of this species is unknown. The white flower with red markings, the absence of floral fragrance and the relatively long perianth tube all suggest the possibility that long-tongued flies are responsible for the pollination of *Gladiolus nigromontanus*.

Diagnosis and relationships

When first described (Goldblatt, 1984a), *Gladiolus nigromontanus* was compared with the *G. gracilis*–*G. mutabilis* alliance, largely because of the reduced leaf blades on the stems of flowering plants. The leaves of *G. nigromontanus* are, however, slender and terete with four narrow, longitudinal grooves. This is quite different from the H-shaped leaves with lightly thickened midrib of *G. gracilis* or the plane, firm, somewhat fleshy leaves of *G. mutabilis*. We consider leaf type to be an important indication of species relationships within section *Homoglossum* and place *G. nigromontanus* in series *Teretifolius*, among other species of the section with such leaves. The white flowers with bright red longitudinal streaks on the lower three tepals and relatively long perianth tube, 20–25 mm long, suggest a relationship with the Langeberg endemic, *G. engysiphon*. Like *G. nigromontanus*, that species flowers in the late summer and autumn, but it can be readily distinguished by the considerably longer perianth tube of most populations, (22–)40–60 mm long, and by the leaves of the flowering stem having the blades completely reduced. In *G. nigromontanus* the lowermost leaf has a fully developed, although short blade, but the upper two leaves are entirely sheathing. Neither of these two species produces foliage leaves during the wet winter and spring months, as was sometimes thought. Plants that flower in a particular year obtain all their carbohydates for seed production and new corm development from the photosynthetic tissue in the green stems, leaf sheaths and short leaf blades of the flowering stems which remain green long after the capsules have ripened. Only plants that do not flower produce a single leaf, the blade of which is slender and terete, quite typical of series *Teretifolius.*

History

When it was described in 1984, *Gladiolus nigromontanus* was thought to have only recently been discovered, by M.J. Viviers in 1981. However, a careful search through the larger southern African herbaria has brought to light two earlier collections. One of these was made in 1976 by the Stellenbosch botanist, Mary Thompson, at the type locality. An earlier collection, evidently the first record of the species, was made by the pioneer South African ecologist, J.H.P. Acocks, in 1966 northeast of Oudtshoorn, also in the Swartberg.

144. GLADIOLUS ENGYSIPHON G. Lewis
PLATE 127

Gladiolus engysiphon G. Lewis in *J. S. African Bot.*, Suppl. 10: 266 (1972). Type: South Africa, Western Cape, Riversdale District, foothills of the Langeberg at the Vette River, cultivated at Kirstenbosch, 22 Mar. 1934, *Ferguson s.n.* (SAM as National Botanic Gardens 481/34, holotype; BOL, PRE, isotypes).

engysiphon = elongate tube, describing the shape of the flower.

Description

Plants 35–45 cm high. *Corm* ovoid, 9–12 mm in diameter, the tunics of light brown papery layers, splitting below into vertical fibres, the layers often unbroken above. *Cataphylls* membranous, entirely pale or purplish above the ground, minutely pubescent, sometimes accumulating with age to form a neck around the base of the stem. *Leaves* (of the flowering stem) three, all largely or entirely sheathing, the lowermost basal, longest, sheathing the lower two-thirds of the stem, overlapping the second leaf, the remaining two leaves inserted in the upper third of the stem, uppermost of these a bract-like scale 20–30 mm long; non-flowering plants producing a solitary terete leaf, the margins and midrib heavily raised and thickened, thus the blade more or less terete in section with four longitudinal grooves, c. 1 mm in diameter, sometimes lightly scabrid. *Stem* erect, lightly flexed outward above the sheath of the uppermost leaf, unbranched, 0.8–1 mm in diameter below the spike.

Spike flexed at the base, inclined, straight, 2- to 6-flowered; *bracts* dull greenish grey to brown, the outer 15–20(–27) mm long, the inner slightly shorter than to about as long as the outer, the apices forked apically for c. 1 mm. *Flowers* predominantly white or cream, occasionally lightly flushed with mauve, the tube pinkish on the reverse, the upper lateral and lower three tepals each with a red streak in the lower midline, sometimes the dark streak cream to yellowish in the centre, the upper part of the tube lightly streaked with red, the filaments pink to red or violet, unscented; *perianth tube* nearly cylindrical or narrowly funnel-shaped, (22–)40–60 mm long, widening in the upper 5–8 mm, sometimes rather abruptly; *tepals* unequal, narrowly lanceolate, the dorsal largest, 16–20 x 9–10 mm, inclined over the stamens, the upper laterals 14–17 x 5–7 mm, directed forward below, spreading in the upper third, the lower three tepals united with the upper laterals for 3–4 mm, the free parts 12–15 x 3–4 mm, in profile slightly exceeding the upper. *Filaments* 10–12 mm long, exserted 3–4 mm from the tube; *anthers* 6–8 mm long, purple, the pollen cream. *Ovary* oblong, c. 2.5 mm long, *style* arching over the stamens, dividing opposite or up to 6 mm beyond the anther apices, the branches c. 3 mm long. *Capsules* obovoid, 10–14 mm long, the apices 3-lobed and retuse; *seeds* ovate, 4.5–5 x 3.5–4 mm, broadly and evenly winged, light brown, the seed body darker brown. *Chromosome number* unknown.

Flowering time March to early April, occasionally as early as mid-February.

Distribution and biology

One of the many endemic *Gladiolus* species of the southern African winter-rainfall zone, *G. engysiphon* occurs along a narrow strip of the southern Cape coastal plain at the foot of the Langeberg. The plants grow on clay soils or clay loam along the interface of shale and sandstone strata of the Table Mountain System. *Gladiolus engysiphon* is one of just a few species of Iridaceae that flower in the autumn, before the winter rains have begun. After flowering in March and April, the plants continue growing as the wet season begins. Immature individuals and those that did not flower produce a single leaf that remains green until the end of spring, in late September or October, and only then become dormant.

Flowering so late in the season allows *Gladiolus engysiphon* to join a limited number of species that share the long-tongued fly *Prosoeca longipennis* (Nemestrinidae) as their pollinator. This fly, which has a tongue about 30 mm long, is the only insect on the wing at that time that is adapted to forage for nectar in the long-tubed flowers of species like *G. engysiphon* (Manning & Goldblatt, 1995). The guild of plant species that use this fly as their pollinator includes several species of long-tubed *Pelargonium* (Geraniaceae), *G. bilineatus*, probably *Tritoniopsis revoluta* (also Iridaceae), a species of *Cyrtanthus* (Amaryllidaceae), and a very few others. Populations of *G. engysiphon* with shorter-tubed flowers are presumably pollinated by long-tongued bees or flies with somewhat shorter tongues than is the rule for *Prosoeca longipennis*.

Diagnosis and relationships

One of several species of *Gladiolus* that have leaves with reduced blades, *G. engysiphon* can be recognized by this characteristic combined with the pale creamy white or sometimes pale mauve-flushed flowers streaked red on the lower tepals and an elongate perianth tube that is usually 40–60 mm long, but occasionally only 22–35 mm long. Leaf blades of non-flowering plants are terete with four narrow longitudinal grooves, and this feature places the species in series *Teretifolius* of section *Homoglossum*. We sus-

pect that *G. engysiphon* is most closely related to *G. nigromontanus* which has flowers with a perianth tube of intermediate length, 17–25 mm long versus (22–)40–60 mm in *G. engysiphon*, and both have a similar autumn-flowering habit. The leaf blades of *G. nigromontanus* are also reduced but not to the same extent as in *G. engysiphon*, and this, as well as the frequent difference in tube length, make it easy to distinguish the two. Populations of *G. engysiphon* from the northern foothills of the Langeberg near Tradouw Pass and on the slopes below Grootvadersbos have reduced leaf blades and relatively soft corm tunics typical of the species, but the flowers have a tube 22–35 mm long, which overlaps the range of tube length in *G. nigromontanus*.

History

Gladiolus engysiphon was first collected by the Riversdale collector, Emily Ferguson, in the early 1930s. She sent corms collected in the Langeberg foothills along the Vette River to Kirstenbosch Botanic Gardens where they flowered in March 1934. Specimens were pressed and apparently forgotten, although new species of *Gladiolus* were being

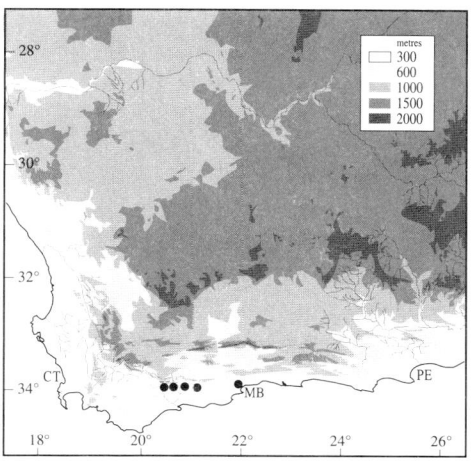

described by H.M.L. Bolus and others at this time. A description was drawn up by G.J. Lewis many years later, and the species was only formally named in 1972 in her posthumously published revision of the South African species of the genus. At that time *G. engysiphon* was known from only two collections, the original, and one other made in 1970. Subsequently, we have found *G. engysiphon* to be relatively common locally in the Langeberg foothills between Swellendam and Riversdale.

145. GLADIOLUS PATERSONIAE F. Bolus

Swartberg klokkie

PLATE 128

Gladiolus patersoniae F. Bolus, *J. Bot.* 66: 14 (1928). Type: South Africa, Eastern Cape, Steytlerville [District], Sept. 1910, *Paterson 17* (GRA, holotype).

patersoniae, named for Florence Paterson, Eastern Cape naturalist, who made the type collection in the Baviaanskloof Mountains of the Eastern Cape in 1910.

Synonymy

Gladiolus inflatus subsp. *intermedius* G. Lewis in *J. S. African Bot.*, Suppl. 10: 241 (1972). Type: South Africa, Cape, Laingsburg District, Witteberg Mountains, 14 Oct. 1928, *Compton 3326* (BOL, holotype; K, isotype).

Description

Plants slender, 30–60(–100) cm high. *Corm* globose, 15–25 mm in diameter or more if the tunics thickly accumulated, the tunics of hard, coarse fibres, usually thickened below into claw-like ridges. *Cataphylls* membranous and pale below the ground, the upper reaching above the ground and coriaceous, purple or dry and brown, sometimes

puberulous, often concealed by a thick mass of coarse fibres accumulating around the base of the stem. *Leaves* three, the lower two basal, the lowermost leaf longest, reaching or slightly exceeding the spike, usually sheathing only near the base of the stem, the blade terete with four narrow longitudinal groves, straight and more or less erect, c. 2 mm in diameter, second leaf sheathing for most or all its length, about two-thirds as long as the stem, the blade vestigial or like the basal, 30–100 mm long, the uppermost leaf largely to entirely sheathing and sometimes reduced to a grey-green scale 10–20 mm long, the margins of the sheathing part united in the lower half. *Stem* erect, flexed slightly in the upper third above the sheath of the second leaf, unbranched, usually dull grey-green, 1.2–1.5 mm in diameter below the spike.

Spike inclined, flexuose, 2- to 4-, occasionally to 5-flowered; *bracts* grey-green, often flushed light purple, the margins hyaline, the outer 18–37 mm long, the inner about three-quarters as long and enclosed by the outer, forked apically. *Flowers* inflated and bell-like, pale to deep blue, slate, greyish or cream, the dorsal tepal usually more intensely coloured, the lower three tepals each with a flexuose transverse yellow band across the distal third, the bands usually outlined in dark blue or purple, strongly scented of apple and carnation; *perianth tube* obliquely funnel-shaped, geniculate, 10–12(–15) mm long, the lower half erect and cylindrical, the upper half bent at right angles and flaring outward; *tepals* unequal, the upper broadly obovate, the dorsal largest, hooded over the stamens, 20–30 x 11–28 mm, the upper laterals slightly shorter and extended forward, only slightly curving outward apically, the lower three tepals united with the upper laterals for 3–6(–10) mm and to one another for 2–3 mm, straight, the upper margins curving upward, more or less plane, oblanceolate-spathulate, not differentiated into distinct limb and claw, 12–22 x 6–10(–14) mm, the lower laterals slightly exceeding the lower median, in profile the lower tepals usually exceeding the upper by 5–8 mm. *Filaments* 10–16 mm long, exserted 6–8 mm from the tube but enclosed in the flower; *anthers* 6–8(–11) mm long, reaching to about the middle of the upper tepal, whitish, pollen whitish. *Ovary* oblong, c. 3 mm long; *style* arching over the stamens, dividing between the middle and apices of the anthers, the branches c. 3.5 mm long, ultimately extending well beyond the anthers. *Capsules* ovoid-ellipsoid, (14–)20–24 mm long; *seeds* ovate, c. 7 x 5–6 mm, broadly and evenly winged, translucent yellowish brown. *Chromosome number* unknown.

Flowering time mid-August to mid-September at lower elevations, October and November at higher elevations.

Distribution and biology

A relatively widespread species of the southern African winter-rainfall zone, *Gladiolus patersoniae* is distributed from the Worcester District of the Western Cape through the southern Cape mountains and eastward to the Baviaanskloof Mountains in the eastern Cape. Although largely a species of exposed, rocky habitats in the interior ranges of the Cape Flora Region, plants also occur near the coast at Cape Infanta and on stony alluvial flats in the Breede River valley near Worcester. The wide range is unusual. Few other species of Iridaceae occur both near the coast and on the most arid of the Cape

mountain ranges, including the Witteberg, Bonteberg and Swartberg that lie along the margin of the Great Karoo. Like most of the species with bell-like flowers, *G. patersoniae* is always associated with soils derived from sandstones of the Cape System.

The short-tubed and sweetly scented flowers of *Gladiolus patersoniae* are adapted for pollination by long-tongued bees. The reward offered the bees is nectar, small quantities of which are stored in the lower part of the perianth tube. Features including flower form, fragrance, and nectar volume and concentration are consistent with many species of *Gladiolus* that have been confirmed to be pollinated by long-tongued anthophorid bees. We have recorded visits to *G. patersoniae* by honey bees and the anthophorid, *Anthophora diversipes*, one of the most common pollinators of Cape species of *Gladiolus*.

Diagnosis and relationships

Gladiolus patersoniae is one of several species loosely called the bluebells, and it can be distinguished from most other species with inflated, bell-like flowers and blue, mauve or pink tepals marked with bright yellow nectar guides by several features. Most importantly, plants have three leaves, the lower one of which is longest and has a terete, four-grooved blade. Secondly, the corm tunics consist of coarse reticulate fibres which accumulate with age and, together with the cataphylls and leaf bases, form a thick neck around the base of the stem.

Of the several other species of *Gladiolus* with bell-like flowers, only *G. inflatus* has terete leaf blades and the tunics and cataphylls also sometimes accumulate in a neck around the stem base, although the neck is less well developed in that species. The differences between the two species lie mainly in the flowers. Flowers of *G. patersoniae* are pale blue to grey, or even cream, and have yellow nectar guides in the form of transverse bands, while the flowers of *G. inflatus* are sometimes white, pink, or purplish as well as blue to violet and the nectar guides are always longitudinal and usually spear-shaped. The flowers of *G. patersoniae* are typically lightly scented, the plants we examined with a pleasant apple-like fragrance. The perianth tube is usually 10–12 mm in length and has a geniculate bend in the midline. The perianth tube in *G. inflatus*

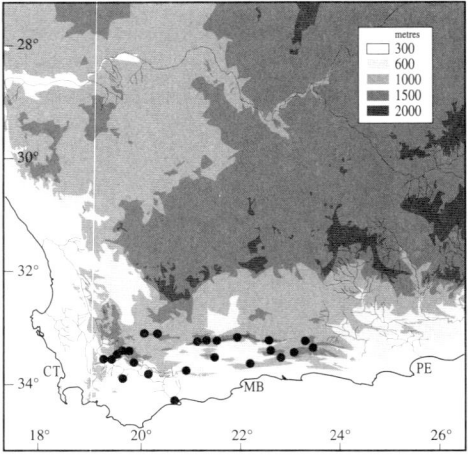

ranges from 12–30 mm long and is usually gently curved, thus without a markedly geniculate bend. Flowers of *G. inflatus* that we have examined alive are unscented and there is no mention of fragrance on the many herbarium specimens of the plant.

History

Gladiolus patersoniae was described by Frank Bolus in 1928, based solely on a rather poor collection of the species from the Baviaanskloof Mountains made by the enthusiastic amateur collector, Florence Paterson, in 1910. This gathering is one of the earliest on record. *Gladiolus patersoniae* seems to have first been collected by Harry Bolus in the Swartberg in 1905, and was also collected by E.P. Phillips in Seweweekspoort in the Swartberg in 1912. These collections were confused with *G. rogersii* (or its synonym *G. spathaceus* var. *burchellii*) and were, in any event, not available to Frank Bolus who was working in Grahamstown at the time.

Gladiolus patersoniae remained poorly understood for many years, and specimens collected by Emily Ferguson near Riversdale in 1932 were considered to be a new species by H.M.L. Bolus and annotated *G. fibrosus*. Mrs Bolus did not, however, ever publish the name, and the same specimens were later annotated *G. fibrosus* L. Bolus ex G. Lewis by Lewis, indicating that she, too, at one time regarded the plant as a distinct species. *Gladiolus patersoniae* was excluded from the genus by Lewis et al. (1972) as being based on material too poor to identify. The plant that they called *G. inflatus* subsp. *intermedius* is, however, identical in all critical taxonomic features with *G. patersoniae*.

146. GLADIOLUS SUBCAERULEUS G. Lewis

PLATE 129

Gladiolus subcaeruleus G. Lewis, *Fl. Pl. Africa* 29: pl. 1158 (1953); Lewis et al., *J. S. African Bot.*, Suppl. 10: 285 (1972). Type: South Africa, Western Cape, southern foothills of the Riviersonderend Mountains, Apr.-May 1950, *Lewis & Davis 2316* (SAM, holotype; K, SAM, isotypes).

subcaeruleus = nearly blue, describing the very pale blue flower.

Description

Plants 15–30 cm high. *Corm* conic, 8–12 mm in diameter, the tunics of fairly firm papery layers, decaying into irregularly shaped fragments and becoming fibrous from the base, extending upward with the fibrous remains of the cataphylls in a neck around the underground part of the stem. *Cataphylls* pale and membranous, the uppermost reaching 2–5 cm above the ground and then purple or dry and brown, minutely pubescent, sometimes only near the base. *Leaves* of the flowering stem three, rarely four, all entirely sheathing, the lowermost longest, sheathing the lower third to half of the stem, usually 60–80 mm long, the second leaf inserted at about the middle of the stem, slightly shorter than the basal, the apex sometimes just overlapping the next leaf, the uppermost leaf 20–40 mm long; foliage leaf produced after flowering, solitary, filiform, oval in transverse section with four longitudinal grooves, with fine scattered deflexed hairs along the edges of the grooves. *Stem* erect below, flexed lightly above the sheaths of the two upper leaves, unbranched, 0.8–1.3 mm in diameter below the spike.

Spike lightly inclined, nearly straight, 3- to 5-flowered; *bracts* grey-green, the outer 11–17 mm long, the inner acute, slightly shorter to slightly longer than the outer. *Flowers* pale blue, the lower lateral tepals with yellow transverse or spear-shaped median markings outlined in dark purple, the lower median less prominently marked or unmarked, unscented or with a light metallic odour; *perianth tube* obliquely funnel-shaped, c. 15–17 mm long, the lower cylindrical part 8–10 mm long; *tepals* lanceolate to ovate, unequal, the upper three largest, ovate, the dorsal inclined over the stamens, 20–24 x 10–12 mm, the upper laterals directed forward, becoming patent in the upper third, c. 20 x 9–12 mm, the lower three tepals united with the upper laterals for

c. 4 mm and to one another for c. 2 mm, the lower laterals narrowed below into channelled claws c. 9 mm long, the limbs c. 7 x 4.5 mm, lower median tepal lanceolate, not distinctly clawed, c. 15 x 7 mm, in profile the lower tepals exceeding the upper by c. 10 mm. *Filaments* 12–15 mm long, exserted 7–8 mm from the tube; *anthers* 4.5–5 mm long, pale mauve, the pollen cream. *Ovary* ovoid, c. 3 mm long; *style* arching over the stamens, dividing opposite the anther apices, the branches 2–3.5 mm long. *Capsules* obovoid, rounded apically, 10–12 mm long, mature capsules not seen; *seeds* unknown. *Chromosome number* unknown.

Flowering time March and April.

Distribution and biology

Gladiolus subcaeruleus is a fairly narrow endemic of the southern African winter-rainfall region. It is confined to the western Cape, where it extends from Bot River in the Caledon District eastward along the southern slopes of the Caledon Swartberg and Riviersonderend Mountains to the town of Riviersonderend. Populations also occur at sites on the Agulhas Peninsula near Elim and at Fairfield Farm at Napier. The plants are usually found in heavy clay soils, or less often on clay loam near the interface between shale and sandstone strata of the Cape System.

An autumn-flowering species, *Gladiolus subcaeruleus* produces vestigial leaf blades on the flowering stem, the cauline leaves being represented by fairly short sheaths. Plants produce solitary foliage leaves from a separate shoot after flowering and these remain green in the winter months, providing nutrition to the new corm.

The short-tubed flowers of *G. subcaeruleus* are assumed to be pollinated by bees, but the pollination biology of the species has not been investigated.

Diagnosis and relationships

At first examination *Gladiolus subcaeruleus* appears unremarkable among the autumn-flowering species of Cape *Gladiolus* that have leafless flowering stems. Only the very pale blue flowers with transverse to broadly spear-shaped yellow and violet nectar guides make it stand out among this heterogeneous group of species. *Gladiolus martleyi* (also series *Teretifolius*) and *G. brevifolius* (section

Linearifolius) which, like *G. subcaeruleus*, flower in the autumn and have the leaves of the flowering stem reduced to short sheaths, usually have flowers in shades of pink or mauve. The nature of the corm tunics and especially of the foliage leaves, produced later in the season, are the key to distinguishing these species. *Gladiolus subcaeruleus* has a single terete, four-grooved foliage leaf, unusual in having a few scattered long hairs along its length. *Gladiolus martleyi*, in contrast, has a smooth but also terete foliage leaf (occasionally two leaves), and its corm tunics consist of softly membranous layers, unlike the firm, almost cartilaginous tunics of *G. subcaeruleus* which decay into medium-textured fibres and, with the cataphylls, tend to accumulate around the stem base to form a neck. *Gladiolus brevifolius*, which belongs in a different section, has plane, linear to lanceolate foliage leaves usually covered in a dense long or short pubescence. The short-tubed flowers of *G. subcaeruleus* suggest a fairly close relationship with *G. patersoniae* and *G. inflatus*, in particular.

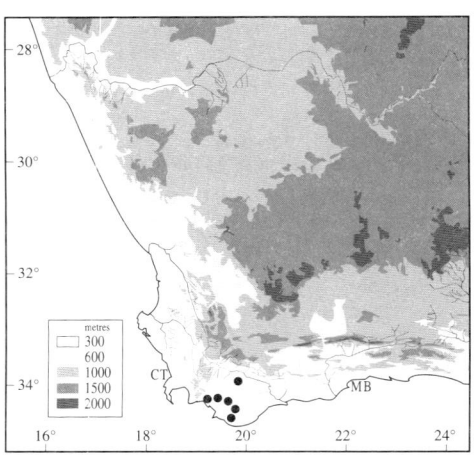

Gladiolus subcaeruleus has also been confused at times with *G. vaginatus* (series *Mutabilis* of section *Homoglossum*), but such confusion can be easily avoided. Although the flowers of the two species are similar in colour, those of *G. vaginatus* are marked with a few dark blue streaks on the lower tepals, in contrast to the transverse yellow bands outlined in dark blue which characterize *G. subcaeruleus*. They are also quite different vegetatively. The flowering stems of *G. vaginatus* have two foliage leaves, the basal one sheathing the lower two-thirds of the stem, and the second leaf relatively short and inserted on the upper third of the stem. The short leaf blades of *G. vaginatus* plane, without a thickened midrib or margins, hence its assignment to series *Mutabilis*. Moreover, corms of the flowering plants of *G. vaginatus*

do not produce foliage leaves later in the season.

History

The first recorded collection of *Gladiolus subcaeruleus* is that made by Rudolf Schlechter in April 1897 near Sandfontein at the eastern end of the Caledon Swartberg. Curiously, the gathering made by Schlechter is a mixed one, consisting of a few specimens of *G. subcaeruleus* and large numbers of *G. vaginatus*. The collection appears also to be the first record of that species. The two grow together at other sites near Caledon and can easily be confused unless attention is paid to the foliage. A second collection, made at the foot of the Riviersonderend Mountains in 1950 by G.J. Lewis and D.K. Davis, both then of the South African Museum, is the type of *G. subcaeruleus*. Obviously Lewis

realized this plant was novel and drew up a description immediately. She also completed a painting of the species and this, together with the protologue, was published in the series *Flowering Plants of Africa* in 1953.

147. GLADIOLUS MARTLEYI L. Bolus

Basterherfspypie

PLATE 130

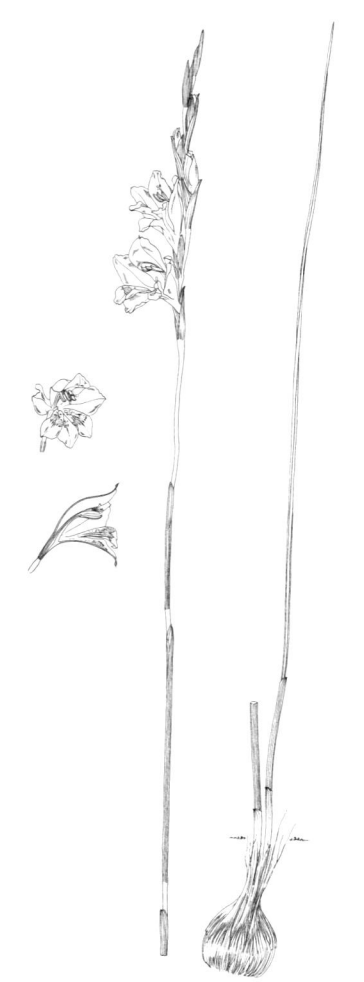

Gladiolus martleyi L. Bolus, *S. African Gard.* 23: 47 (1933). Lewis et al., *J. S. African Bot.*, Suppl. 10: 284 (1972). South Africa, Western Cape, without precise locality, Banhoek, Stellenbosch, cultivated at Kirstenbosch Gardens, 29 Mar. 1932, *Martley s.n.* (BOL 371/32, holotype; K, isotype).

martleyi, named in honour of J.F. Martley, a local bulb-grower, who collected the species and brought it to the attention of botanists at Kirstenbosch Gardens in the early 1930s.

Synonymy

Gladiolus pillansii G. Lewis, *S. African Gard.* 25: 57 (1935). Lewis et al., *J. S. African Bot.*, Suppl. 10: 280 (1972). Type: South Africa, Western Cape, Melkbosch road, 3 Apr. 1932, *Barnard s.n.* (BOL 20518, holotype; K, isotype). *Gladiolus pillansii* var. *roseus* G. Lewis, *Fl. Pl. Africa* 29: pl. 1159 (1953). Type: South Africa, Western Cape, Cape Peninsula, Scarborough, damp sandy ground near the vlei, 20 Feb. 1953, *Minicki s.n.* (SAM 60953).

Description

Plants (20–)30–60 cm high. *Corm* globose, 15–30 mm in diameter, the tunics of papery layers, with age fragmenting irregularly or becoming somewhat fibrous. *Cataphylls* pale and membranous, the uppermost extending up to 5 cm above the ground and then green to dull purple, closely resembling the leaves of the flowering stem. *Leaves* (of the flowering stem) two or three, sheathing entirely, evenly spaced along the stem, sometimes the

lowermost very shortly overlapping the second leaf, 50–150 mm long, decreasing in size above, the uppermost sheathing leaf with the margins free to the base, sometimes the margins overlapping; foliage leaves produced after flowering, one or two, the blades terete with four narrow longitudinal grooves, c. 1 mm in diameter. *Stem* erect below, inclined above, unbranched, 0.8–1.5 mm in diameter below the spike.

Spike lightly inclined, weakly flexuose, occasionally 3-, usually 5- to 11-flowered; *bracts* greyish to purplish green, the outer 14–22 mm long, the inner about as long as to slightly shorter than the outer, acute, evidently not forked apically. *Flowers* pale to deep pink or lilac to mauve, the lower lateral tepals each with a transverse band or a spear-shaped to trilobed-spathulate yellow mark edged with darker pink or purple in the upper half, often with a strong sweet fragrance, sometimes unscented; *perianth tube* obliquely funnel-shaped, 11–12 mm long, the cylindrical lower part c. 7.5 mm long; *tepals* unequal, narrowly to broadly lanceolate, the dorsal largest, inclined over the stamens, curving upward toward the apex, 22–27 x 11–12 mm, the upper laterals directed forward below, becoming recurved toward the apices, 19–22 x 8.5–10 mm, the lower three tepals united with the upper laterals for 2–5 mm and to one another for 3–4 mm, the free parts 15–19 mm long, gradually narrowed below into channelled claws, the limbs arching downward, the lower laterals c. 6 mm wide, the lower median c. 8 mm wide, in profile the lower tepals exceeding the upper. *Filaments* 12–13 mm long, exserted c. 8 mm from the tube; *anthers* 6–8.8 mm long, dark purple-blue, the pollen pale bluish. *Ovary* ovoid, c. 3 mm long; *style* arching over the stamens, dividing close to or up to 2 mm beyond the anther apices, the branches c. 3 mm long. *Capsules* narrowly ellipsoid, c. 12–20 mm long; *seeds* ovate, c. 7 x 4–5 mm, broadly and evenly winged, light brown, the seed body dark brown. *Chromosome number* $2n = 30$.

Flowering time mainly March to April, occasionally as early as mid-February and as late as May.

Distribution and biology

A widespread, but never common species of the southern African winter-rainfall zone, *Gladiolus martleyi* occurs almost throughout the southwestern and northern Cape. It extends from the Bokkeveld Escarpment in the north along the western coastal plain and coastal mountains of the Western Cape to the Cape Peninsula and eastward to Albertinia. Plants favour deep sandy soils, sometimes also fairly rocky sites, but they may occasionally be found in heavier soils along the interface of sandstone and shale strata of the Cape System.

The short-tubed, pink to mauve flowers of *Gladiolus martleyi* have the same shape and structure as many bee-pollinated species of the genus and are thus assumed to be pollinated by long-tongued bees, probably of the genus *Amegilla* which is common and active in the autumn when *G. martleyi* blooms. The flowering stems bear short sheathing leaves and the long-bladed foliage leaves are produced from a separate shoot in the late autumn after the flowers have faded.

Diagnosis and relationships

Gladiolus martleyi is one of several small-flowered, autumn-flowering species of *Gladiolus* that occur in the southern African winter-rainfall zone and, as in most of this group, the leaves of the flowering stem are reduced and lack leaf blades. The flowering stems thus bear two or three short sheathing

leaves. The flowers are usually pale to deep pink or occasionally lilac to mauve, and have strongly marked nectar guides on the lower tepals consisting of transverse bands or spear-shaped yellow marks edged with dark pink or purple. They are also usually sweetly fragrant, although several populations with unscented flowers are known. The general appearance of *G. martleyi* when in flower is very like that of the well-known *G. brevifolius*, although that species seldom has such strongly differentiated nectar guides and rarely has scented flowers. The differences between the two species lie in the corms and in the structure of the foliage leaves. The corms of *G. martleyi* are fleshy and moist, with dry and membranous to papery tunics, and the single or occasionally two foliage leaves produced by corms after flowering are terete with four fine, longitudinal grooves. *Gladiolus brevifolius*, in contrast, has fairly hard, more or less leathery corm tunics that break up into coarse vertical fibres with age, and the foliage leaf produced after flowering has a plane and usually hairy blade, a character associated with species of section *Linearifolius*. The terete and hairless leaves of *G. martleyi* place that species unambiguously in series *Teretifolius* of section *Homoglossum*.

The corms of *Gladiolus martleyi* are unusual in the section and are shared with only one other species, the southwestern Cape endemic, *G. jonquilliodorus*. The latter also has the flowering stems lacking leaf blades, and the flowers closely resemble those of *G. martleyi* except for their cream, yellow or lilac colouring. Apart from flower colour, the main differences between the two species are that *G. martleyi* flowers in the autumn, usually in March and April, and produces only one, or sometimes two, foliage leaves. *Gladiolus jonquilliodorus*, however, flowers in the summer, in December and early January, and has two, or more often three, foliage leaves.

Both *Gladiolus martleyi* and the very similar *G. pillansii* were recognized by Lewis et al. (1972). They distinguished the two on the basis of tepal shape and fragrance, *G. pillansii* described as having lanceolate and acute tepals and a strong scent, in contrast to *G. martleyi* with ovate and obtuse

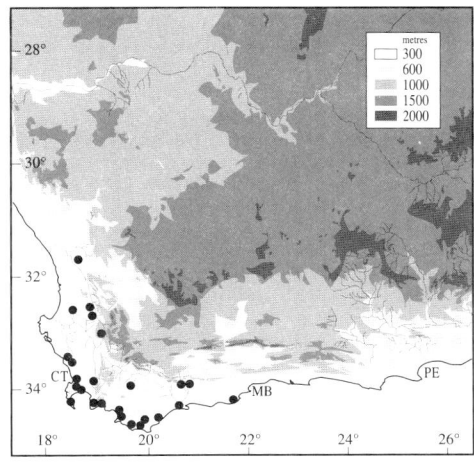

tepals and flowers lacking scent. We see no significant difference between them and unite the two species. *Gladiolus pillansii* var. *roseus* is no more than a pink-flowered and less robust *G. pillansii* with darker tepals.

History

Although *Gladiolus martleyi* has a wide range across the southern African winter-rainfall region and should thus have been encountered by at least some of the early botanical explorers in southern Africa, we have found few early records of the species. The traveller and naturalist, William Burchell, may have been the first to record it in 1815, but his collection was understandably confused with *G. brevifolius*. A collection made by C.W. Pappe at Zeekoeivlei on the southern Cape Flats, probably in the 1850s, may also be this species. Without a corm or foliage leaf identification is uncertain, but the flowering time, February, and the fact that it was scented make this seem likely. The specimen is annotated, '*G. fragrans* Mihi', that is, Pappe himself. The name was never published. Only in 1932 did *G. martleyi* finally attract botanical attention when plants collected by J.F. Martley were grown and flowered at Kirstenbosch Gardens. The terete foliage leaf and absence of pubescence dispelled any possible association with *G. brevifolius* and it was described in 1934 by H.M.L. Bolus. The conspecific *G. pillansii* was described in 1935 by G.J. Lewis and is here reduced to synonymy.

---

148. GLADIOLUS JONQUILLIODORUS Ecklon ex G. Lewis

PLATE 131

Gladiolus jonquilliodorus G. Lewis, *Ann. S. African Mus.* 40: 124 (1954), as *G. jonquilodorus*. Lewis et al., *J. S. African*

Bot., Suppl. 10: 287 (1972). Type: South Africa, Western Cape, Cape Flats, damp places, Feb., *Zeyher 450* (SAM, lectotype).

jonquilliodorus = scented like a jonquil, describing the strong, sweet, narcissus-like scent.

Description

Plants (25–)40–70 cm high. *Corm* globose, 20–40 mm in diameter, the tunics membranous to soft-papery, decaying into fine netted fibres, these not accumulating to any extent. *Cataphylls* membranous, the uppermost reaching up to 10 cm above the ground, often dry and brown. *Leaves* (of the flowering stem) two or three, entirely sheathing, the lowermost usually imbricate, the uppermost with the margins free to the base, sometimes the margins overlapping; foliage leaves produced after flowering, two or three, 300–500 mm long, terete with four hairline longitudinal grooves, c. 1 mm in diameter, dry and dead at flowering time and usually lacking. *Stem* straight and erect, unbranched, 1.3–1.8 mm in diameter below the spike.

Spike lightly flexed at the base and inclined, occasionally 4-, usually 7- to 14-flowered; *bracts* soft-textured, yellow-green to grey, the outer 14–18 mm long, the inner slightly shorter than the outer. *Flowers* cream, pale yellow or pearl grey, sometimes flushed with pink or light mauve, the lower three tepals (or only the lower laterals) each with a transverse yellow band and often streaked with narrow purple lines below,

sweetly scented during the day; *perianth tube* obliquely funnel-shaped, 8–9 mm long, largely enclosed in the bracts; *tepals* narrowly oblanceolate, unequal, the dorsal c. 20 x 9 mm, extending horizontally over the stamens, the upper laterals slightly shorter, directed forward and twisted obliquely to overlap the dorsal tepal, when fully open the upper third lightly curving outward, the lower tepals united for c. 5 mm, the free parts 8–10 mm long, narrowed below into claws, abruptly expanded into limbs c. 5 mm long, arching downward, in profile the lower tepals exceeding the upper. *Filaments* c. 10 mm long, exserted c. 5 mm from the tube; *anthers* c. 6 mm long, mauve, the pollen pale yellow. *Ovary* oblong, c. 3 mm long; *style* arching over the stamens, dividing at or just beyond the anther apices, the branches 1–1.5 mm long. *Capsules* ellipsoid, 16–19 mm long; *seeds* ovate, c. 6 x 3.5 mm, broadly and evenly winged, light translucent brown, the seed body darker brown. *Chromosome number* unknown.

Flowering time mainly December to mid-January, occasionally in late November or early February.

Distribution and biology

A rare and fairly localized endemic of the southwestern Cape, South Africa, *Gladiolus jonquilliodorus* occurs on the Cape Peninsula and along the Cape west coast northward as far as Yzerfontein, south of Saldanha Bay. For many years the species was believed to be restricted to the southern Cape Peninsula and nearby Cape Flats (Lewis in Adamson & Salter, 1950) and it was thought to be seriously endangered through habitat loss at the few known sites where it grew. As a result of botanical surveys conducted north of the Peninsula in the 1980s by the Cape Town biologist, Pixie Littlewort, the range of *G. jonquilliodorus* has been extended well to the north along the western Cape coast and additional sites have also been located on undisturbed land on the Peninsula. Although it can no longer be considered threatened with extinction over its entire range, *G. jonquilliodorus* remains rare on the Cape Peninsula and is threatened elsewhere by uncontrolled development along the Cape west coast.

Plants usually grow in deep alluvial sand among clumps of Restionaceae, at some sites in low-lying areas that remain moist well into the dry summer. The preference for moist habitats is not, however, consistent and plants may occur on well-drained, stony hill slopes. Flowering occurs in December

and January, sometimes even in February, by which time the slender leaves are usually dry and have often decayed or become detached from the flowering stalks. Plants thus lack foliage leaves at flowering, and the stem bears a few short, entirely sheathing leaves along its length. After the onset of cool weather, the corms produce new leaves from a separate shoot. These grow during winter and spring, nourishing the new corm.

The short-tubed and intensely sweet-scented flowers of *Gladiolus jonquilliodorus* are clearly adapted for pollination by long-tongued bees, and to date we have recorded the honey bee *Apis mellifera*, the anthophorid *Amegilla spilostoma*, and small halictid bees visiting the flowers and emerging from them covered with pollen. These bees, and perhaps other species as well, must be assumed to be the pollinators of *G. jonquilliodorus*.

Diagnosis and relationships

One of several summer- and autumn-flowering species of *Gladiolus* that have the leaves of the flowering stem reduced to sheaths, *G. jonquilliodorus* particularly resembles *G. martleyi* and, to a lesser extent, *G. brevifolius*. All three species have a similar general appearance, with the stem bearing only sheathing leaves and relatively small, short-tubed flowers. Only *G. jonquilliodorus* and *G. martleyi*, however, have terete, four-grooved foliage leaves and both consistently have sweetly scented flowers. In *G. brevifolius*, a member of section *Linearifolius*, the foliage leaves are plane and usually hairy and the flowers are seldom scented. The relationships of *G. jonquilliodorus* obviously lie with *G. martleyi* and the two belong in series *Teretifolius* of section *Homoglossum* where they are linked not only by their reduced leaves, but also by similar, somewhat fleshy corms with very soft, membranous tunics. *Gladiolus martleyi*, a much more widespread

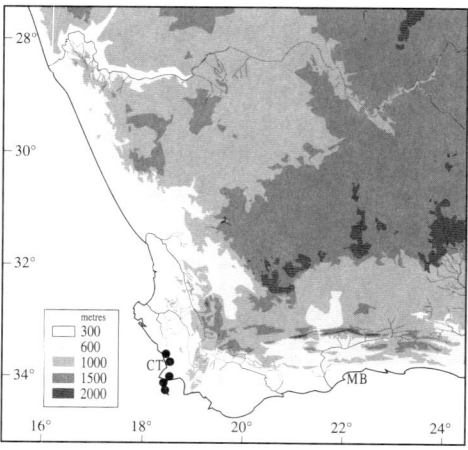

and common species, has one, or sometimes two foliage leaves and flowers in the autumn, usually from late February to April, whereas *G. jonquilliodorus* flowers in the summer, in December and January, and plants have two or three foliage leaves.

History

Gladiolus jonquilliodorus was recorded at least as early as the 1830s, when C.L. Zeyher discovered the species on the Cape Flats outside Cape Town. Specimens in the Geneva, Stockholm and South African Museum herbaria are labelled *G. jonquilliodorus* Ecklon. That name was, however, never published and does not appear in the list of *Gladiolus* species in C.F. Ecklon's 1827 publication, *Topographisches Verzeichniss*. This volume included miscellaneous information about known southern African bulbous plants and some new ones, occasionally with diagnoses, that were grown in Cape Town or had been collected elsewhere in the Cape Colony. The species was collected again by Zeyher, together with Nathaniel Wallich, near Manenberg on the Cape Flats, circa 1843. Plants collected years later in 1894 by Francis Guthrie and in 1896 by A.H. Wolley Dod, on the Cape Flats at Vygekraal farm, were subsequently matched with the Zeyher specimen in the South African Museum collection in Cape Town by Harry Bolus and Wolley Dod. They recognized the species as a native of the Cape Peninsula flora and attributed the name *G. jonquilliodorus* to Ecklon, evidently not realizing that neither the name nor a description had ever been published (Bolus & Wolley Dod, 1903). G.J. Lewis validly published the name in 1954, also attributing the specific epithet to Ecklon but spelling it *jonquilodorus*. In the revision of *Gladiolus* published in 1972, A.A. Obermeyer assumed that by recognizing the existence of *G. jonquilliodorus*, Bolus and Wolley Dod had validly described the species. They clearly had no intention of doing so and Obermeyer's attribution of the name *G. jonquilliodorus* to Bolus and Wolley Dod is unwarranted. We follow Ecklon's spelling of the name of the species on the type in the South African Museum herbarium. Lewis's change to *jonquilodorus*, a spelling used by Ecklon on collections of the species in other herbaria, seems unnecessary.

149. GLADIOLUS TRICHONEMIFOLIUS Ker Gawler

Geelpypie, ruikpypie

PLATE 132

Gladiolus trichonemifolius Ker Gawler, *Curtis's Bot. Mag.* 35: pl. 1483 (1812). Baker, *Fl. Capensis* 6: 141 (1896). Type: South Africa, Western Cape, without precise locality, cultivated in Great Britain at Kew Gardens, *Masson s.n.* (BM, holotype).

trichonemifolius = with leaves like *Trichonema* (a synonym of *Romulea*), referring to the leaves for which the genus *Trichonema* (hair-like thread) was named.

Synonymy

Gladiolus citrinus Klatt, *Abh. Naturforsch. Ges. Halle* 12: 340 (Ergänz. 6) (1882). Lewis et al., *J. S. African Bot.* 72: 175 (1972). Type: South Africa, Cape, between Paarl Mountain and Paardeberg, 1840, *Drège 8457* (B [not seen], ?holotype; BM, G [not seen], K, L [not seen], S, isotypes).
Gladiolus erectiflorus Baker, *Fl. Capensis* 6: 146 (1896), an illegitimate homonym, not *G. erectiflorus* Baker (1895) from tropical Africa. Type: South Africa, Western Cape, sandy places near Malmesbury, Oct. 1892, *Macowan 1548* (K, holotype; BOL, G, SAM, isotypes).
Gladiolus symmetranthus G. Lewis, *Ann. S. African Mus.* 40: 122 & fig. 4 (1954). *Gladiolus trichonemifolius* forma *symmetranthus* (G. Lewis) Bullock, *Curtis's Bot. Mag.* 173: tab. 357 (1960). Type: South Africa, Cape, near Koelenhof, between Stellenbosch and Muldersvlei, 14 Sept. 1950, *Lewis 2237* (SAM, holotype; K, PRE, isotypes).
[*Gladiolus tenellus* sensu G. Lewis et al., *J. S. African Bot.*, Suppl. 10: 177 (1972), not *G. tenellus* Jacquin, 1791.]

Description

Plants 12–25 cm high. **Corm** globose, 7–12 mm in diameter, the tunics of woody to coriaceous, concentric layers, fragmenting from the base into vertical strips, dark brown or blackish. **Cataphylls** pale and membranous, the uppermost barely reaching above the ground, then usually dark purple. **Leaves** three, the lower two basal, the lowermost longest, usually shortly exceeding the spike, sheathing in the lower third, the blade oval to terete and four-grooved in transverse section, the margins and midrib heavily thickened, 1–2 mm in diameter, the second leaf sheathing for half its length, reaching to about the base of the spike, the uppermost leaf inserted in the upper third of the stem, sheathing for at least half its length and channelled throughout. **Stem** erect below, flexed outward above the sheaths of the two upper leaves, unbranched, c. 1 mm in diameter below the spike.

Spike inclined or erect, flexuose, often strongly so, 1- to 3-, rarely 4-flowered; **bracts** dark green, the veins nearly transparent, attenuate, the outer 30–40(–50) mm long, the inner half to two-thirds as long as the outer. **Flowers** zygomorphic or the tepals similarly disposed, rarely completely actinomorphic, cream to yellow, the outer tepals often flushed purple on the reverse, the lower three tepals each with paired brown lines either side of the midline in the lower half and deeper yellow across the middle, sometimes purplish toward the apices, the mouth of the tube edged with a star-shaped purple band of colour or the base of the tepals and the upper part of the tube with a blackish star-shaped mark, often with a strong, sweet, freesia- or violet-like scent, or unscented; **perianth tube** straight or obliquely and narrowly funnel-shaped, 16–20 mm long; **tepals** unequal, the upper three usually slightly larger than the lower three, or subequal, the outer three slightly larger than the inner, lanceolate-elliptic,

24–35(–40) mm long, the dorsal 10–13(–18) mm wide, the lower three usually united for 1–4 mm, 22–35(–44) x 7–9 mm, arching outward in the upper half to two-thirds. *Filaments* 8–15 mm long, exserted c. 2 mm from the tube; *anthers* 8–10 mm long, pale yellow, the pollen cream. *Ovary* oblong, c. 5 mm long; *style* arching over the stamens or occasionally nearly straight, dividing just below the anther apices, the branches 4–5 mm long. *Capsules* ovoid-ellipsoid, 12–15 mm long; *seeds* ovate, 4–5 x c. 3–3.5 mm, rich glossy brown, broadly and evenly winged. *Chromosome number* $2n = 30$.

Flowering time August to mid-September, occasionally in late July or late September.

Distribution and biology

Gladiolus trichonemifolius, today a fairly rare species of the southwestern corner of Western Cape Province, South Africa, was a common wetland plant until the 1930s. Its original range extended from Bredasdorp in the south to Hopefield in the north, although it did not occur on the Cape Peninsula. Largely a low-altitude species, it also occurs inland in the Cold Bokkeveld and Agterwitzenberg Vlakte. It may still be found in relatively undisturbed vlei land, but agriculture, with its attendant development of dams, pasture enrichment and wetland drainage, has left the species little of its once plentiful habitat. Marshy ground that is left relatively undisturbed, although intensively grazed after seeding of the spring flora, still boasts abundant colonies of *G. trichonemi-folius* at farms near Darling on the western Cape coast. These wetlands are managed by landowners so that the native flora is maintained and plants are proudly displayed at the Darling Wild Flower Show year after year.

Gladiolus trichonemifolius is pollinated by long-tongued bees foraging for nectar. The only consistent visitor to the flowers that we have seen is the honey bee *Apis mellifera* (family Apidae).

Diagnosis and relationships

A fairly typical member of section *Homoglossum* series *Teretifolius*, *Gladiolus trichonemifolius* has the hard, woody corm tunics and three foliage leaves with terete, four-grooved blades that define the series. Within the series it can be recognized by its pale yellow or sometimes white flowers with the tepals subequal or sometimes perfectly equal. In most populations the stamens are unilateral and arcuate, but occasionally even

the stamens may be symmetrically arranged. We assume its relationships lie with the other species of the series with yellow flowers. These include *G. pritzelii*, *G. sufflavus* and the rare, high-altitude species, *G. delpierrei*. *Gladiolus sufflavus* has bell-like, greenish yellow flowers and distinctive leaves with villous sheaths and blades with broad grooves along their length. The flowers of *G. pritzelii* are also bell-like, with a hooded dorsal tepal and lower tepals with reddish markings on the nectar guides. Equally important, the leaf blades are plane, but with the midrib consisting of a pair of closely set veins on one surface only, a feature shared with *G. delpierrei*.

We follow A.A. Bullock's contention (1960) that *Gladiolus citrinus* is conspecific with *G. trichonemifolius*. The two species are identical vegetatively and in their ecology, and similar in their flowers. The only substantive difference between them is that the tepals of *G. citrinus* are subequal and the stamens nearly, or rarely fully symmetrically disposed, and have a dark central mark around the base of the tube. There appears to be a grade from typical *G. trichonemifolius* with an ascending dorsal tepal and arcuate, unilateral stamens, to plants with the tepals subequal in size and symmetrically arranged and with unilateral stamens, to rare popula-tions in which the stamens are fully erect and obscurely unilateral or perfectly symmet-rical. Plants matching *G. citrinus* are thought to be consistently scentless, but both strong-ly scented and evidently scentless forms of more typical *G. trichonemifolius* are known.

History

Although a common wetland species in past times, and one encountered along all the routes from Cape Town to the interior, *Gladiolus trichonemifolius* seems to have been missed by most of the early botanists who collected at the Cape. The species first appears unambiguously in the literature when it was formally described by John Ker Gawler in 1812. The type plants were culti-vated in Great Britain at Lee and Kennedy's Nursery at Hammersmith, London. Earlier collections made by C.P. Thunberg in the early 1770s have been attributed to *G. trichonemifolius* and both his *G. tristis* var. *humilis* and var. *luteus* have been identified as that species. We question these determinations and suggest instead that the specimens are yellow-flowered *G. carinatus*. Thunberg seems to have been of the same opinion, for in his later writings (Thunberg, 1800), he cited both varieties under Jacquin's

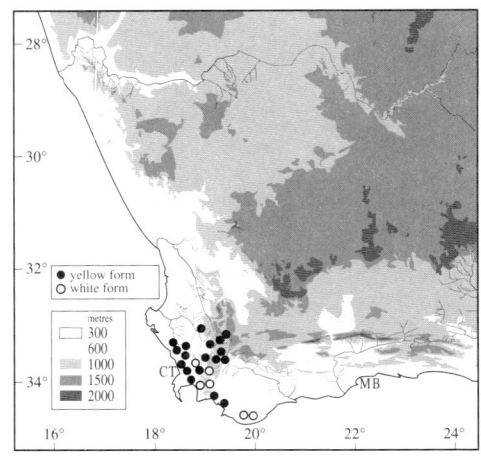

G. tenellus, which represents the yellow-flowered form of *G. carinatus* and is a later synonym of that species. In their revision of *Gladiolus*, however, Lewis et al. (1972) regarded *G. tenellus* as the earliest name for *G. trichonemifolius* and treated it under this earlier name. We discuss our reasons for regarding *G. tenellus* as a synonym of *G. carinatus* under that species.

Despite the excellent watercolour painting that serves to identify *Gladiolus trichonemi-folius*, specimens fairly closely resembling the illustration, collected by Peter MacOwan near Malmesbury, were named *G. erectiflorus* by J.G. Baker in 1896 in *Flora Capensis*. In his account of the genus there, Baker actual-ly recognized *G. trichonemifolius* as well as *G. erectiflorus*, incidentally an illegitimate homonym, the two distinguished on erro-neous grounds.

The rather different form of *Gladiolus trichonemifolius* with an actinomorphic, cupped perianth and a dark purple blotch in the centre was described by F.W. Klatt in 1882 as *G. citrinus*. The species was based on a collection made by J.F. Drège in about 1830 near Paarl and was not known to Klatt from living plants. Not realizing that this plant had already been described at species rank, G.J. Lewis described exactly the same form of *G. trichonemifolius* as *G. symmetranthus* in 1954. The British botanist, A.A. Bullock, questioned the validity of Lewis's species and in 1960 he reduced the scentless, actinomorphic-flowered *G. symmetranthus* with its dark central mark to the rank of forma in *G. trichonemifolius*. Lewis did not accept Bullock's arguments for the reduction of *G. symmetranthus* and the species was recognized by Lewis et al. (1972) in their revision of the South African species of *Gladiolus* under its earlier name, *G. citrinus*.

150. GLADIOLUS SUFFLAVUS (G. Lewis) Goldblatt & Manning

Groenklokkie

PLATE 133

Gladiolus sufflavus (G. Lewis) Goldblatt & Manning, new combination and rank. *Gladiolus pritzelii* var. *sufflavus* G. Lewis in *J. S. African Bot.*, Suppl. 10: 250 (1972). Type: South Africa, Cape, Calvinia District, Glenridge Farm, Nieuwoudtville, 20 Aug. 1960, *Lewis 5730* (NBG, holotype; K, holotype).

sufflavus = pale yellow, for the colour of the flowers.

Description

Plants 45–70 cm high. *Corm* 18–25 mm in diameter, the tunics of coriaceous layers, decaying with age to become more or less fibrous. *Cataphylls* membranous and pale below the ground, the upper two reaching above the ground and purple, puberulous to villous, often accumulating in a fibrous neck around the base of the stem. *Leaves* usually four, occasionally three, the lower two basal, the lowermost longest, reaching to between the base and middle of the spike, sheathing only near the base, the blade more or less terete or cross-shaped in section with four narrow longitudinal grooves, 2–3 mm in diameter, glabrous on both sheath and blade, remaining three leaves villous on the sheaths, the second leaf sheathing the stem for most of its length, free for a short distance and with a short blade like the basal leaf, the upper two leaves inserted on the stem above the ground, largely sheathing, channelled to the apices, although free of the stem for their upper 20–40 mm, sheaths of the two uppermost leaves with the margins free to the base, open or overlapping. *Stem* erect, flexed outward above the sheathing part of the two upper leaves, unbranched, c. 1.2 mm in diameter below the spike.

Spike inclined, lightly flexuose, 4- to 6-flowered; *bracts* olive-green, the outer 20–30 mm long, the inner nearly as long as the outer. *Flowers* pale yellow to greenish, the lower tepals with obscure greenish, dark yellow or brown markings, the tepals together forming a somewhat inflated bell, with a sweet, lily-like fragrance, especially reminiscent of apple and the orchid *Pterygodium*; *perianth tube* obliquely funnel-shaped, 14–15 mm long, the lower half erect and cylindrical, the upper half bent at right angles, flaring outward; *tepals* unequal, the upper broadly ovate, the lower spathulate,

the dorsal largest, arched to hooded over the stamens, 18–20 x 13–16 mm, the upper laterals slightly smaller, directed forward and only curving outward near the margins, the lower laterals united with the upper for 4–5 mm, forming a shallow spoon, c. 11 x 8 mm, not differentiated into claws below, curving upward below, the distal quarter deflexed, in profile the lower tepals slightly exceeding the upper tepals. *Filaments* 13–14 mm long, exserted 7–8 mm from the tube, but enclosed; *anthers* 6–7 mm long, purple, pollen cream. *Ovary* oblong, 4–5 mm long; *style* arched over the stamens, dividing just below the anther apices, the branches c. 3.5 mm long, usually ultimately exceeding the anthers by 1–2 mm and sometimes just exserted from the flower. *Capsules* ovoid-ellipsoid, 17–21 mm long; *seeds* ovate, 5–6 x 4 mm, broadly and evenly winged, golden-brown, the seed body darker. *Chromosome number* $2n = 30$.

Flowering time late August and September.

Distribution and biology

Gladiolus sufflavus is restricted to the Bokkeveld Escarpment in Northern Cape Province, South Africa. There, at the northernmost limit of the sandstone strata of the Table Mountain Series and true fynbos vegetation, it occurs in seasonally waterlogged sandy soils and at the edges of streams, marshes and ponds. Most commonly seen after fires or on cleared land, it does nevertheless flower, although poorly, in old growth, especially in very wet conditions.

The flowers of *Gladiolus sufflavus* are typical of members of the genus adapted for pollination by long-tongued bees, having a relatively short perianth tube and a strong, sweet scent. We have recorded repeated visits to flowers by the large anthophorid bee *Anthophora diversipes* which we assume to be the main pollinator. Nectar is the main reward for bees, which ignore the pollen adhering to the anthers concealed under the arched dorsal tepal.

Diagnosis and relationships

Gladiolus sufflavus has some of the features associated with members of series *Teretifolius* of section *Homoglossum*, including a centric leaf blade, flexuose stem and rather inflated perianth. The occasional presence of four leaves, rather than three which are consis-

tently present in other species of the series, is however, puzzling, as is the fact that the leaf sheaths are villous whereas the blades are not. This conspicuous pubescence is unknown elsewhere in section *Homoglossum*, although some species of series *Teretifolius* have scabrid cataphylls or leaf bases, and *G. pritzelii* occasionally has lightly hairy leaf blades. The greenish yellow, somewhat inflated, nodding flowers of *G. sufflavus* are very much like those of several species of series *Teretifolius*, notably *G. inflatus* and *G. patersoniae*, and we provisionally assume that it belongs in the series and is most closely related to *G. pritzelii* and *G. patersoniae* of this group. Although regarded as no more than a variety of *G. pritzelii* by Lewis et al. (1972), *G. sufflavus* is morphologically quite distinct from that species. It can readily be distinguished from all short-tubed species of series *Teretifolius* by the densely villous leaf sheaths and the broadly four-grooved leaf blades, thus cross-shaped in transverse section. By contrast, the basal leaf of *G. pritzelii*

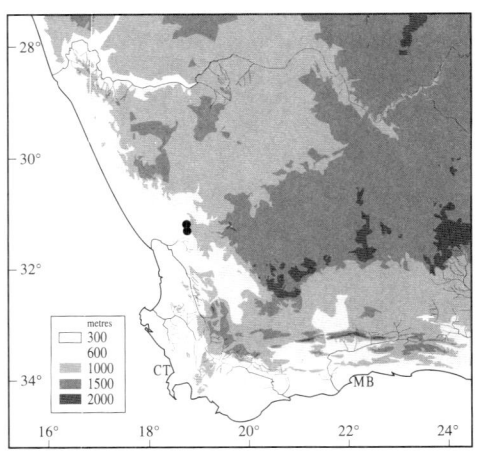

is almost plane with a double midrib on one surface only, and the two upper leaves are usually glabrous. The flowers of *G. pritzelii* are larger than those of *G. sufflavus*, are borne on spikes of only two or three flowers, and the tepals are clear yellow, usually with contrasting transverse red markings across the upper ends of the lower tepals. These several differences make confusion between the two unlikely.

The widely grooved, cruciform leaf of *Gladiolus sufflavus* is not unique in series *Teretifolius*. It also occurs in the long-tubed *G. cylindraceus*, a species closely allied to *G. inflatus*. The leaves of *G. sufflavus* also recall those of *G. marlothii* (section *Heterocolon*), in which the blades as well as the sheaths are villous. Other features of *G. marlothii*, including absence of scent,

rounded capsules and a fairly straight spike, make it seem unlikely that the relationships of *G. sufflavus* lie with that species and, hence, that it is misplaced in section *Homoglossum*.

History

Gladiolus sufflavus was first recorded by the Cape Town botanist, H.M.L. Bolus in 1930, south of the village of Nieuwoudtville on the Bokkeveld Escarpment between Calvinia and Vanrhynsdorp. The species did not attract botanical attention then or later when it was occasionally re-collected. G.J. Lewis, however, saw live plants in the course of her research on *Gladiolus* in South Africa and intended to recognize it as a distinct species. After her death in 1968 A.A. Obermeyer, who brought the incomplete manuscript to

publication, decided to treat *G. sufflavus* as a subspecies of *G. pritzelii*. We concur with Lewis's earlier decision to regard *G. sufflavus* as a distinct species.

151. GLADIOLUS PRITZELII Diels

Geelklokkie

PLATE 134

Gladiolus pritzelii Diels, *Bot. Jahrb. Syst.* 44: 118 (1909). Pole Evans, *Fl. Pl. S. Africa*, 2: pl. 68 (1922). Type: South Africa,

Northern Cape, Hantamsberg, Sept. 1900, *Diels & Pritzel 738* (B [not seen], holotype, NBG, PRE, photo.; Z, isotype).

pritzelii, named in honour of G.A. Pritzel, nineteenth-century German botanist.

Description

Plants 20–60 cm high. *Corm* globose, 15–20 mm in diameter, the tunics of woody to coriaceous layers, fragmenting from the base into coarse teeth or becoming fibrous. *Cataphylls* membranous below ground, the upper reaching above the ground and then green or purple, often becoming dry toward the apex, often puberulous. *Leaves* three, the lowermost basal and reaching to about the base of the spike, glabrous or puberulous to villous, sheathing the lower third of the stem, the blade linear, 1–3.5 mm wide, usually with two raised central veins, the laminar surface broad or vestigial, the margins not noticeably thickened, the second leaf sheathing the stem to shortly below the spike, the blade similar to the basal, short but exceeding the third leaf and sometimes the spike, the third leaf inserted shortly below the spike. *Stem* more or less erect throughout, flexed above the sheaths of the upper two leaves, unbranched, c. 1.3–2 mm in diameter below the spike.

Spike lightly inclined and flexuose, usually 1- to 3-, occasionally to 5-flowered; *bracts* pale to olive-green, the margins membranous, the outer 20–40 mm long, the inner

two-thirds to nearly as long as the outer, minutely forked apically. *Flowers* with the tepals forming an inflated bell, clear to dull yellow, the lower lateral tepals each with a bright yellow transverse band outlined in red to purple or brown, the lower median with a spade-shaped yellow mark outlined in red or purple (upper lateral tepals sometimes also with reddish markings), the lower part of the throat also streaked with dark colour, sweetly fragrant; *perianth tube* 11–12 mm long, obliquely funnel-shaped, the lower half erect and cylindrical for c. 6 mm, the upper half bent sharply at right angles and flaring outward; *tepals* unequal, the upper ovate, the dorsal largest, arched over the stamens, (16–)22–24 x 13–20 mm, upper laterals slightly smaller, directed forward, only the apices recurving, the lower three tepals united with the upper laterals for 4–6 mm and to one another for 2–3 mm, not differentiated into claws below, curving downward in the upper third, scabridulous in the lower midline, the lower laterals spathulate, 12–16 x c. 7 mm, the lower median about as long, lanceolate, in profile the lower tepals exceeding the upper by 3–5 mm. *Filaments* 10–12 mm long, exserted 5–6 mm from the tube, but included in the flower; *anthers* c. 8.5 mm long, reddish to brown, the pollen yellow. *Ovary* obovoid, c. 3 mm long; *style* arching over the stamens, dividing shortly below the anther apices, the branches 3–4 mm long, the apices ultimately exceeding the anthers by 1–2 mm. *Capsules* and *seeds* unknown. *Chromosome number* unknown.

Flowering time late August and September.

Distribution and biology

Centred in the northern mountain ranges of Western Cape Province, South Africa, *Gladiolus pritzelii* extends from the Cold Bokkeveld near Ceres northward to the southern Cedarberg. Rather surprisingly for a *Gladiolus* of the Cape Flora Region, few species of which range outside the area, *G. pritzelii* also occurs on the Hantamsberg near Calvinia as well as in wetter portions of the Roggeveld Escarpment in the western Karoo.

In the Cape mountains plants grow in rocky habitats in thin, nutrient-poor sandstone soils. There they flower poorly, if at all, except in seasons following summer fires. In the western Karoo *Gladiolus pritzelii* grows on heavy clay soils derived from dolerite and flowering there is not dependent on fire, but is rather related to rainfall, which is often too scanty to permit the plants to grow to flowering in the short growing season. This ecological shift is perhaps even more remarkable than the odd distribution pattern. Despite its unusual distribution and diverse habitats, *G. pritzelii* plants from the Cape mountains and the western Karoo appear virtually identical and show no sign of local differentiation.

The short-tubed, bell-like flowers are assumed to be pollinated by anthophorid bees, but the pollination biology of *Gladiolus pritzelii* has not yet been studied.

Diagnosis and relationships

Although apparently nested in series *Teretifolius* of section *Homoglossum*, *Gladiolus pritzelii* does not have the terete, four-grooved leaf that characterizes the alliance. In other important features, however – including the few-flowered, flexuose spike, the short-tubed, inflated flower, the rather coarsely fibrous corm tunics and the presence of three leaves – it accords well with other species of the section. Its leaf blades are more or less plane and distinctive in having a prominent pair of raised central veins on one surface and obscure venation on the other. The leaf blades and sheaths are also sometimes lightly villous or scabrid, but in some populations are entirely smooth. The flowers of *G. pritzelii* are somewhat inflated and a bright clear yellow, often with pale markings on the lower tepals, these occasionally outlined in red or brown. Like those of other species of section *Homoglossum*, the flowers are pleasantly fragrant.

The only other species of the section that has a leaf blade like that of *Gladiolus pritzelii* is the local Cedarberg endemic, *G. delpierrei*, and we believe the two species are immediately related. *Gladiolus delpierrei* has pale yellow to cream flowers and the perianth is not inflated, but otherwise the two species are very alike. We assume the peculiar leaf venation of the two species arose in their common ancestor and we speculate that the plane blade is a reversal to the condition in species ancestral to series *Teretifolius*. The conspicuous paired veins on one leaf surface may represent the vestiges of

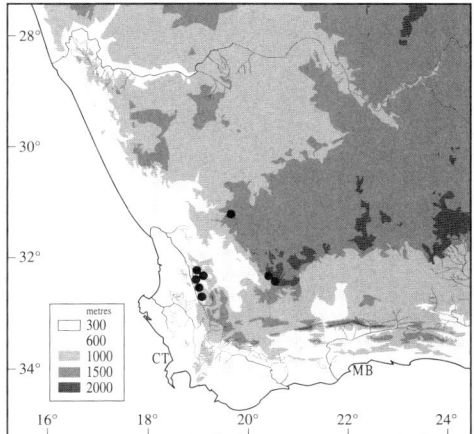

two of the flanges that constituted a centric, cross-shaped, broadly four-grooved blade.

History

Native to an area that did not receive much botanical attention until the late nineteenth century, *Gladiolus pritzelii* was not discovered until 1900 when the German botanist, Friedrich Diels, collected plants on the Hantamsberg at Calvinia during his short visit to South Africa in October of that year. The species was described just a few years later in 1909 by Diels himself. The presence of *G. pritzelii* in the mountains of the Ceres District, much closer to Cape Town, was not discovered until the 1920s when plants were routinely exhibited at the Ceres Wild Flower Show. *Gladiolus pritzelii* was only found in the Cedarberg in 1930. As explained above, the related *G. sufflavus* was regarded as a subspecies of *G. pritzelii* by A.A. Obermeyer (in Lewis et al., 1972).

152. GLADIOLUS DELPIERREI Goldblatt

Gladiolus delpierrei Goldblatt, *J. S. African Bot.* 45: 84 (1979). Type: South Africa, Western Cape, Cedarberg Mountains, Sneeuberg, 5 Jan. 1975, *Delpierre 456* (NBG, holotype).

delpierrei, named in honour of its discoverer, mathematician and bulb enthusiast, Georges Delpierre.

Description

Plants 55–65 cm high. *Corm* globose, c. 20 mm in diameter, the tunics of fine reticulate fibres, the old corms persisting as dry discs below the current corm. *Cataphylls* pale and membranous, the uppermost reaching to 8 cm above the ground, lightly mottled but dry and light by flowering time, decaying with age to become fibrous and

persisting as a fibrous neck around the base of the stem. *Leaves* three, the sheaths of the two lower leaves lightly scabrid on the veins, the lowermost basal and longest, reaching to about the base of the spike, the blade linear, 3–4 mm wide, the midrib and one other vein moderately thickened and prominent, the margins lightly thickened, the remaining two leaves inserted well above the ground, widely spaced, entirely sheathing, the margins united below around the stem. *Stem* erect, unbranched, c. 2 mm in diameter below the spike.

Spike lightly inclined, 5- to 7-flowered; *bracts* pale grey-green or flushed light purple on the upper sides, the outer 18–20 mm long, the inner slightly shorter than the outer. *Flowers* creamy yellow, the lower tepals dark yellow in the lower third and

each with a pair of fine red lines at the base, the upper laterals with a narrow red streak in the midline, the mouth of the tube dark red, probably lightly scented;

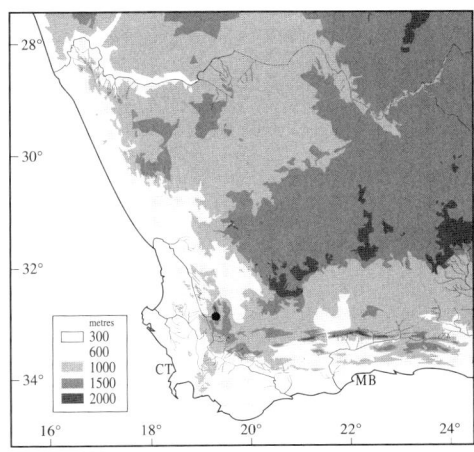

perianth tube narrowly funnel-shaped, c. 8 mm long; *tepals* unequal, the upper ovate, the lower lanceolate, the dorsal largest, ascending to inclined over the stamens, c. 21 x 11 mm, the upper laterals c. 17 x 8.5 mm, the lower three tepals joined to the upper laterals for c. 3 mm, 12–13 mm long, narrowed below into channelled claws, the limbs minutely puberulous in the lower midline, the lowermost slightly longer than the lower laterals, in profile windowed between the dorsal and upper lateral tepals. *Filaments* c. 12.5 mm long, exserted c. 9 mm from the tube; *anthers* c. 8 mm long, light mauve, the pollen cream. *Ovary* obovoid, c. 4 mm long; *style* arching over the stamens, dividing just short of the anther apices, the branches 2–3 mm long. *Capsules* and *seeds* unknown. *Chromosome number* unknown.

Flowering time December and early January.

Distribution and biology

Gladiolus delpierrei is a narrow endemic of the Cedarberg range in Western Cape Province, South Africa. The only known populations occur at an elevation of 1300 m on Cedarberg Sneeuberg and adjacent peaks. Plants grow in leached, stony sandstone soil on a seepage zone that is waterlogged in winter and remains moist at least until mid-summer when the plants flower.

The pollination biology is unknown, but the general form of the flower suggests that long-tongued bees are the pollinators of *Gladiolus delpierrei*.

Diagnosis and relationships

As explained under the previous species, *Gladiolus pritzelii* and the presumably closely related *G. delpierrei* fit somewhat uncomfortably in series *Teretifolius* because they have plane rather than centric, usually terete leaf blades. *Gladiolus delpierrei* is also discordant in the series in having soft-textured, finely fibrous corm tunics, unlike the characteristically hard woody or coarsely fibrous tunics of most other species of the series. In fact, without reference to *G. pritzelii* with which it shares the odd, double central vein on one surface of the plane leaf blades, it would not have been possible to place *G. delpierrei* in any of the series of section *Homoglossum*.

Finely fibrous corm tunics are often associated with permanently wet, montane habitats and as *Gladiolus delpierrei* grows in seepage zones that remain wet well into the summer, we assume that its corm tunics reflect its habitat rather than phylogenetic relationship. The species has short-tubed creamy yellow flowers with a short perianth tube, c. 8 mm long, and darker yellow nectar guides on the lower tepals, marked with a network of fine red lines near the base. We assume the flowers are fragrant, for although this feature was not recorded at the time of collection, the lower tepals have zones of papillae in the lower midlines, which are thought to represent osmophores, the sites of scent production.

History

Gladiolus delpierrei was discovered in the summer of 1975 by Georges Delpierre while climbing Cedarberg Sneeuberg. At the time Delpierre was actively growing indigenous bulbous plants and was also preparing a wild flower guide to the genus *Gladiolus* in winter-rainfall South Africa. Thus, he immediately realized that he had found an unusual and possibly new species. The latter proved to be the case and *G. delpierrei* was described a few years later (Goldblatt, 1979), being named in honour of its discoverer. The comparatively recent discovery of rare and extremely localized species such as this one, and others such as *G. atropictus* and *G. roseovenosus* which were first recorded in 1978 and 1982 respectively, leaves us with the feeling that several more unknown species of *Gladiolus* remain to be found on isolated peaks in the Cape mountains.

Section *Homoglossum:* Series *Tristis*

153. GLADIOLUS HYALINUS Jacquin
Small brown Afrikaner, klein bruin Afrikaner, bowties
PLATE 135

Gladiolus hyalinus Jacquin, *Icones Pl. Rar.* 2: conspectus of illustrations (1793), as a new name for *Gladiolus strictus* Jacquin, Collecteana 4: 170 (1792); *Icones Pl. Rar.* 2: pl. 242 (1795), an illegitimate homonym, not *G. strictus* Aiton (1789) (= *Babiana stricta* (Aiton) Ker Gawler). Baker, *Fl. Capensis* 6: 148 (1896). Lewis et al., *J. S. African Bot.*, Suppl. 10: 202 (1972). Type: South Africa, Western Cape, without precise locality or collector, figure in Jacquin, *Icones Pl. Rar.* pl. 242 (1795).

hyalinus = translucent, alluding to the transparent edges of the upper tepals.

Synonymy

Gladiolus tristis var. *inodorus* Thunberg, *Dissertatio de Gladiolo* 165, sub no. 8 (1784). Type: South Africa, Western Cape, without precise locality, *Thunberg s.n.* (UPS–Herb. Thunberg 1082, 1083, syntypes).
Gladiolus confusus N.E. Brown, *J. Linn. Soc.* 48: 31 (1928). Type: South Africa, Western Cape, Devil's Peak near Mowbray, in rocks at foot of mountain, c. 90 m (300 ft), Aug. 1883, *H. Bolus 4890* (K, holotype; BOL, isotype). [*Gladiolus tenellus* sensu Baker, *Handbook Irideae* 204 (1892), *Fl. Capensis* 6: 141 (1896), not *G. tenellus* Jacquin (1790).]

Description

Plants (12–)30–60 cm high. *Corm* globose-conic, 10–12 mm in diameter, the tunics of more or less woody layers broken below into oblong to linear segments. *Cataphylls* membranous and pale, the uppermost extending 3–5 cm above the ground and then purple. *Leaves* three, imbricate, the lowermost longest, reaching to about the base of the spike, the blade linear, 1.7–2.5 mm wide, plane with the margins and midrib strongly thickened and raised, but with two fairly wide grooves on each surface, second leaf sheathing the lower half of the stem, the blade like the basal, upper leaf sheathing for about half its length, lower margins free to the base and overlapping. *Stem* erect below, flexed outward above the sheaths of the two upper leaves, unbranched, 1–2 mm in diameter below the spike.

Spike inclined, more or less straight, 1- to 3-, rarely 6-flowered; *bracts* pale green to greyish, sometimes long-attenuate, the outer 36–50 mm long, usually exceeding the perianth tube, the inner 5–10 mm shorter than the outer, minutely forked at the apices. *Flowers* shades of light to reddish brown on a pale cream background, the tepals usually darker along the midlines, the lower half of the dorsal tepals transparent on the edges, the abaxial part of the throat transparent with dark spots and streaks, becoming translucent in the evenings, completely odourless or rarely sweetly scented, especially in the evenings; *perianth tube* slightly shorter to slightly longer than the bracts, 25–36 mm long, erect and cylindrical below for 20–24 mm, abruptly bent at 30–45° and flared outward for 10–12 mm; *tepals* unequal, lanceolate to ovate, the dorsal largest, arching over the stamens, 23–30 x 12–16 mm, the upper laterals 23–30 x 7–8 mm, the lower three tepals united for 2–4 mm, the lower laterals 14–20 mm long, in the upper half curving obliquely away from the lower median, the lower median more or less straight, 16–25 x 5–6 mm. *Filaments* 12–16 mm long, exserted 2–5 mm from the tube; *anthers* 9–11 mm long, cream, the pollen yellow. *Ovary* oblong, c. 5 mm long; *style* arching over the stamens, dividing at or shortly beyond the anther apices, the branches 2–3 mm long, broadly expanded above. *Capsules* elongate-ellipsoid and three-sided, (20–)25–30 mm long, enclosed in the bracts; *seeds* more or less ovate, 4–7 x 3–4 mm, broadly and evenly winged, light translucent brown, the seed body darker brown. *Chromosome number* 2*n* = 30.

Flowering time June to mid-August, at higher elevations until late September, rarely in October.

Distribution and biology

Gladiolus hyalinus is one of the most widespread species of *Gladiolus* in the southern African winter-rainfall region. Although most common in the southwestern Cape, it extends northward to the Bokkeveld Escarpment near Nieuwoudtville, and isolated populations occur in the Kamiesberg in central Namaqualand, west of Springbok, and near Steinkopf at the southern edge of the Richtersveld. The range eastward of the Cape Peninsula is poorly documented beyond the Riviersonderend Mountains.

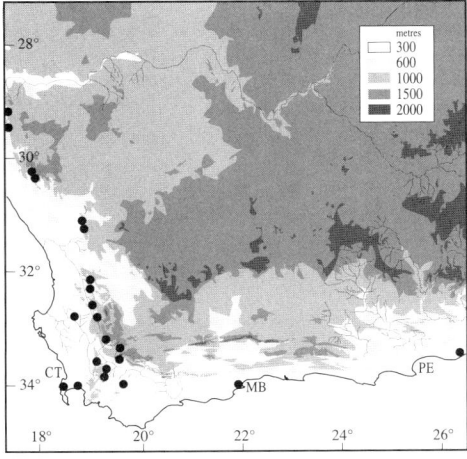

There are, however, indisputable records of the species from two isolated localities, one at Plettenberg Bay near Knysna, and the other near Kenton-on-Sea in the Eastern Cape, east of Port Elizabeth. Within the Cape Flora Region plants most often occur in transitional fynbos–renosterveld on heavy soils or along the transition zone between sandstone and shale, but populations also occur on granite or on sandstone in fynbos communities. In Namaqualand, plants occur either on granite-derived sand or on clay, as at Spektakel Pass west of Springbok, in renosterveld-like communities. The species typically flowers early in the season, sometimes as early as May, but more often in July and early August. At higher elevations plants flower as late as mid-September, or occasionally in October.

Like other species of series *Tristis*, *Gladiolus hyalinus* is probably pollinated by moths. The partly translucent perianth with a dull brown sheen and brown to purple mottling certainly suggests this pollination strategy, as do the long, straight perianth tube and the flower that opens fully only toward sunset. The unusual tendency for the tepals to become translucent or somewhat mauve at sunset is a very odd feature, and no doubt comparable to the pronounced colour change from brown or reddish to purple that occurs in the related *G. liliaceus*.

Diagnosis and relationships

The comparatively small, dull-coloured flower, pale yellow to cream with brown to purple mottling on the midlines of the upper tepals and on the limbs of the lower tepals, transparent lower edges of the dorsal tepal, combined with the presence of three leaves and an unbranched stem, place *Gladiolus hyalinus* firmly in series *Tristis*. It resembles most closely the common southwestern and eastern Cape species, *G. liliaceus*, and like that species *G. hyalinus* has linear leaves with heavily thickened and raised margins and a prominent midrib. The two differ most conspicuously in the size of the flowers, the upper tepals of *G. liliaceus* being 40–45 mm long and the bracts 70–100 mm long, in contrast to upper tepals c. 25 mm long and bracts 30–50 mm long in *G. hyalinus*. The two can also readily be distinguished by the lower lateral tepals, which are characteristically twisted obliquely in *G. hyalinus* but straight and attenuate with undulate margins in *G. liliaceus*. The flowers of *G. maculatus*, here referred to series *Mutabilis*, also resemble fairly closely those of *G. hyalinus*,

but the former has leaves without thickened margins and midrib, and the flowers have straight lower lateral tepals and are always strongly scented. Scent production is at best occasional in *G. hyalinus*. A few collections indicate the flowers are scented, but several populations that we have examined on the Cape Peninsula, near Gydo Pass, in the Riviersonderend Mountains and on the Bokkeveld Escarpment consistently have unscented flowers both during the day and in the evenings. Evidently, the species reproduces successfully without the production of floral odour to attract insect visitors.

History

The confusion surrounding the name of the plant that is now *Gladiolus hyalinus* was resolved by Lewis et al. (1972), but is worth repeating briefly as an aid to readers of the literature on *Gladiolus* prior to that time. The species was first described as *G. strictus* by Jacquin in 1792, but by the time the volume of the *Icones Plantarum Rariorum* was published with the illustration of the plant so entitled, Jacquin had learned that his name was a homonym for *G. strictus* Aiton, dating from 1789, now *Babiana stricta*. The name was changed to *G. hyalinus* in the conspectus of plates at the beginning of volume 2 of that work. That name was intended to replace the homonym and is the earliest valid name for the species. Baker, however, seems to have misunderstood the situation for he recognized a *G. hyalinus*, known to him only from the description, in *Flora Capensis* (1896) and elsewhere (e.g., Baker, 1892) for Jacquin's plant. At the same time he referred specimens of what we now call *G. hyalinus* to a second species, *G. tenellus*

Jacquin (actually a poor representation of *G. carinatus*). Then in 1928, N.E. Brown, realizing that the specimens that Baker cited as *G. tenellus* were not that species, decided to publish a new name for *G. hyalinus* sensu Jacquin, calling the plant *G. confusus*. Thus, for the last decade of the nineteenth century and until 1928 *G. hyalinus* was known as *G. tenellus*, and from 1928 until 1972 it was called *G. confusus*. In the literature prior to Baker's dealing with the plant, however, the name *G. hyalinus* was applied correctly, that is, in the sense intended by Jacquin.

Other early collections of *Gladiolus hyalinus* include several made by C.P. Thunberg in the early 1770s, and in fact they appear to be the earliest on record. Thunberg associated these specimens with *Gladiolus tristis* var. *inodorus* (b) (as opposed to var. *inodorus* (d) which is *G. carinatus*). Among the sheets in the Thunberg collection at Uppsala, Sweden, that are annotated *G. tristis* var. b, there is one of *G. maculatus*. The confusion is not surprising, for the two look reasonably alike, although the flowers of *G. maculatus* are always sweetly scented.

154. GLADIOLUS LILIACEUS Houttuyn

Large brown Afrikaner, groot bruinaandblom, kaneelaandblom

PLATE 136

Gladiolus liliaceus Houttuyn, *Naturl. Hist. 2*, 12: 55, pl. 79, fig. 2 (1780). Lewis, *Bot. Not.* 119: 228 (1966); Lewis et al., *J. S. African Bot.*, Suppl. 10: 196 (1972). Type: South Africa, Cape, without precise locality or

collector, cultivated in Holland, *Houttuyn s.n.* (G–Herb. Burman, holotype).

liliaceus = resembling a lily, referring to the large flower with long flaring tepals.

Synonymy

Gladiolus tristis var. *grandis* Thunberg, *Dissertatio de Gladiolo* no. 8c (1784). *Gladiolus grandis* (Thunberg) Thunberg, *Prodromus Pl. Capensium* 185 (1800). Baker, *Fl. Capensis* 6: 138 (1896). Type: South Africa, Western Cape, without precise locality or date, *Thunberg s.n.* (UPS–Herb. Thunberg 1031, holotype; BOL, K, photo).
Gladiolus versicolor Andrews, *Bot. Repository* 1: pl. 19 (1798). *G. versicolor* var. *major* Ker Gawler, *Curtis's Bot. Mag.* 16: sub. pl. 556 (1802). Type: South Africa, Western Cape, without precise locality, *Pringle s.n.* cultivated in Great Britain at Hammersmith, London, illustration in Andrews, *Bot. Repository* 1: pl. 19 (1798).

Description

Plants 30–80 cm high. *Corm* globose, 12–18 mm in diameter, the tunics of more or less woody to coriaceous layers broken below into triangular to linear segments. *Cataphylls* pale and membranous, green and coriaceous above the ground. *Leaves* three, overlapping, the two lowermost basal, the lowermost longest, reaching to between the base of the spike and shortly beyond the spike apex, sometimes sheathing the lower part of the stem or diverging close to ground level, the second leaf sheathing the lower half of the stem, the blades linear, (1.5–)2–4(–6) mm wide, plane with the margins and midribs strongly thickened and raised, the margin and midrib edges minutely scabrid, the laminar surface between the midrib and margins narrow or fairly broad, the upper-

most leaf inserted between the middle and upper quarter of the stem, channelled for its entire length. *Stem* erect or inclined below, sharply flexed outward above the sheathing part of the uppermost leaf and strongly inclined above, unbranched, 1.5–2 mm in diameter below the spike.

Spike inclined, lightly flexuose, 1- to 4-, sometimes to 6-flowered; *bracts* pale green or suffused with dull grey, the outer 55–80(–115) mm long, long-attenuate and often twisted above, especially the lowermost, the lower margins united around the spike axis, the inner slightly more than half as long as the outer, forked in the upper 1–2 mm, twisted around to lie against the outer. *Flowers* light brown, dull pinkish red, tan, purplish or cream to greenish yellow, the tepals darker along the midlines, the lower tepals pale yellow to cream in the lower third, the abaxial half of the throat speckled with dark brown, the lower edges of the dorsal tepal and the sutures between the tepals transparent, changing colour after nightfall, becoming bluish to mauve and intensely sweetly clove-scented; *perianth tube* narrowly and obliquely funnel-shaped, 40–45(–53) mm long, the wider upper part 12–20 mm long and exserted between the bracts, sometimes lightly pubescent in the lower part of the tube; *tepals* lanceolate-attenuate, slightly twisted in the upper halves, the margins undulate, sometimes crisped toward the apices, the dorsal largest, strongly inclined over the stamens, curving upward in the upper quarter, 38–45 x 14–20 mm, the upper laterals 37–44 x 10–15 mm, arching outward from the base, the lower three tepals joined to the upper laterals for 3–6 mm and sometimes to one another for up to 2 mm, directed forward

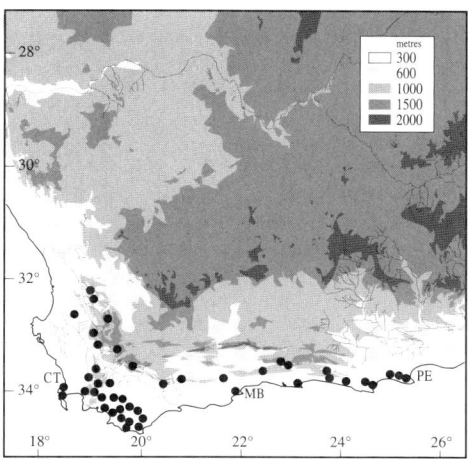

below, curving outward in the upper halves, 31–40 x 10–15 mm. *Filaments* 15–25 mm long, usually exserted 1.5–5 mm from the tube, occasionally only reaching the top of the tube, lightly pubescent below; *anthers* nearly horizontal, 11–16 mm long, brown or dull purple, the pollen pale yellow or cream. *Ovary* oblong, c. 9 mm long; *style* arching over the stamens, lightly pubescent in the lower half, usually dividing opposite the anther apices, the branches c. 3 mm long, very broad in the upper halves, arching well beyond the anthers. *Capsules* oblong-ellipsoid, triangular in transverse section, 30–50 mm long; *seeds* ovate 4–5 x 3–3.5 mm, broadly and evenly winged, the wing light brown and semi-transparent, the seed body red-brown. *Chromosome number* $2n = 30$.

Flowering time late August to mid-September at lower elevations, occasionally later, especially at higher elevations; a late-blooming form flowers in November and December.

Distribution and biology

Restricted to the southern African winter-rainfall region, *Gladiolus liliaceus* is a relatively common species of the southwestern and southern Cape. Its range extends from the Cedarberg in the north to Port Elizabeth in the east, thus encompassing almost the entire Cape Flora Region. Plants favour heavy soils and are most common on clay soils and in lowland habitats. Thus, the species is a common sight in early September on the rolling plains of the southern Cape between Caledon and Bredasdorp, and to Riversdale. It also occurs in the interior Cape mountains and has been recorded in the Cold Bokkeveld and occasionally in the Cedarberg.

Gladiolus liliaceus has a particularly interesting flowering physiology. The tepals of the large flowers are partly closed during the day and coloured a translucent rusty red, cream or brown. At sunset the tepals undergo a

rapid change to a light, translucent mauve. At the same time the tepals open more widely and the flowers release a strong, heady scent redolent of carnation with cloves. The behaviour is clearly an adaptation to pollination by moths, two species of which have been recorded visiting the flowers. Large, long-tongued moths such as those belonging to the Sphingidae, and some Noctuidae, for example *Cucullia extricata*, are rewarded by large quantities of sweet nectar which is held in the lower half of the perianth tube.

Diagnosis and relationships

A particularly distinctive species, *Gladiolus liliaceus* is easily recognized by its large translucent rust-coloured to cream flowers, 78–95 mm long with a slender tube 40–53 mm long, and lanceolate tepals with attenuate apices and undulate margins. The species is also unusual in section *Homoglossum* in having pubescent filaments. This odd feature is otherwise found in a few species of section *Hebea*. *Gladiolus liliaceus* is a member of series *Tristis* and plants have just three leaves, the lowermost by far the largest, and the leaf blades with strongly thickened margins and midrib. Like those of the majority of species of section *Homoglossum*, the corm tunics of *G. liliaceus* are hard and woody. Among the species of series *Tristis*, *G. liliaceus* is most closely related to the smaller-flowered *G. hyalinus* (see above). Difference in flower size alone makes the two species easy to distinguish. They are the only species of the series with leaves linear rather than cross-shaped to terete and four-grooved in transverse section.

History

A well-known plant, affectionately called the large brown Afrikaner, *Gladiolus liliaceus* was known by a later synonym, *G. grandis*, from the beginning of the nineteenth century until 1965. It was collected by several of the early plant collectors at the Cape and was

the subject of much confusion with other species that have long-tubed, brown-speckled flowers with a strong sweet scent. The Swedish botanist C.P. Thunberg collected the species and initially called it *G. tristis* var. *grandis* in 1784. He subsequently realized that it was quite separate from *G. tristis* and so named it *G. grandis* in 1800. By this time, however, the species had acquired another synonym, *G. versicolor*, provided by Henry Andrews in 1798. The unique feature that the flowers of *G. liliaceus* have of changing colour to pale blue in the early evening was first noted by Andrews, who carefully described the colour change in his description of *G. versicolor*.

Neither Thunberg nor Andrews was aware that the plant described as *Gladiolus liliaceus* in 1780 by the Dutch nurseryman and natural historian, Maarten Houttuyn, was this same species. It was only when G.J. Lewis had completed her nomenclatural research for her revision of *Gladiolus* in South Africa in the 1960s that the identity of *G. liliaceus* became clear. The illustration published by Houttuyn was too poor to be associated with *G. versicolor* and *G. grandis*, but the type specimen, preserved in the Burman Collection at Geneva, leaves no doubt about the identity of *G. liliaceus*.

155. GLADIOLUS TRISTIS Linnaeus
Vlei-aandblom, marsh Afrikaner, rheebokblom, trompetters
PLATE 137

Gladiolus tristis Linnaeus, *Sp. Pl.* ed. 2: 53 (1762). Lewis et al., *J. S. African Bot.*, Suppl. 10: 191 (1972). Type: South Africa, Western Cape, without precise locality or collector (Herb. Linn. 59.9, holotype).

tristis = sad, alluding to the pale and often dull colour of the flowers.

Synonymy
Gladiolus tristis var. *odorus* Thunberg, *Dissertatio Gladiolo* 8 (1788). Type: South Africa, Western Cape, without precise locality, *Thunberg s.n.* (UPS–Herb. Thunberg 1080, holotype).
Gladiolus concolor Salisbury, *Paradisus Londonensis* pl. 8 (1806). *G. tristis* var. *concolor* (Salisbury) Baker, *J. Linn. Soc. Bot.* 16: 172 (1877);

Fl. Capensis 6: 139 (1896). Lewis et al., *J. S. African Bot.*, Suppl. 10: 191 (1972). Type: South Africa, Western Cape, without precise locality, cultivated in Britain, figure in *Paradisus Londonensis* pl. 8 (1806).
Gladiolus spiralis Persoon, *Synopsis Pl.* 1: 43 (1805). Type: South Africa, Western Cape, without precise locality or collector, illustration

in Jacquin, *Icones Pl. Rar.* 2: pl. 245 (1795). *Gladiolus aestivalis* Ingram, *Gard. Chron.* Ser. 3, 88: 301 (1930). *G. tristis* var. *aestivalis* (Ingram) Lewis in *J. S. African Bot.*, Suppl. 10: 195 (1972). Type: South Africa, Western Cape, Drakenstein Mountains, flowered in Britain, 1930, *Ingram s.n.* (not seen and not found at BM or K).
Gladiolus flavidus Ingram, *Gard. Chron.* Ser. 3, 88, with fig.: 301 (1930). Type: South Africa, Eastern Cape, Tsitsikamma, estuary of the Groot River, 13 Nov. 1927, *Ingram s.n.* (not located at BM or K).
Gladiolus fulvescens Ingram, *Gard. Chron.* Ser. 3, 88: 301 (1930). Type: South Africa, Western Cape, dry ground near Elgin, 11 Oct. 1927, *Ingram s.n.* (not located at BM or K).

Description

Plants (17–)40–120 cm high. *Corm* 12–18 mm in diameter, the tunics of firm papery to hard, nearly woody layers, becoming notched into sections from the base, or more or less broken into coarse vertical fibres. *Cataphylls* pale and membranous, reaching up to 4 cm above the ground and then purple or more or less dry and brown.

Leaves three, rarely a fourth entirely sheathing bract-like leaf present just below the spike, the lower two basal, the lowermost longest, usually reaching at least to the base of the spike and often exceeding it, sheathing the stem below, sometimes to the middle, the blade usually centric, 2–4 mm in diameter, cruciform in section with the midrib raised at right angles to the blade, usually the laminar surface c. 1.5 mm wide and exposed, but sometimes the thickened edges of the margins and midrib almost meeting and the leaf appearing terete, rarely oval in section with two narrow grooves on each surface, the two upper leaves largely to entirely sheathing, or the second leaf with a short blade, the margins of the third leaf free to the base and imbricate below. *Stem* erect, usually flexed outward above the sheaths of the two upper leaves, unbranched, 1.3–3 mm in diameter below the spike.

Spike lightly inclined, more or less straight, rarely 1-, usually 2- to 4-, occasionally to 11-flowered; *bracts* pale to dark green, sometimes flushed greyish above, the outer (25–)40–50 mm long, the inner slightly shorter than to about as long as the outer, forked apically for 1–2 mm or entire and acute, twisted to lie against the outer. *Flowers* pale yellow to greenish yellow or cream, the tepals darker on the midlines, or sometimes with purple to reddish median streaks, the reverse of the tepals usually dark grey-purple to reddish on the midlines and sometimes on the tips, the sutures between the tepals transparent for c. 10 mm, most conspicuously so between the dorsal and upper laterals, weakly scented during the day, strongly scented of carnation and cloves in the early evening; *perianth tube* 40–63 mm long, narrowly and obliquely funnel-shaped, the narrow cylindrical part 20–30 mm long, lightly papillose in the lower throat; *tepals* broadly lanceolate, unequal, the dorsal largest, inclined to almost horizontal, 22–28 x 16–20 mm, the upper laterals lightly arching outward from the base, 22–28 x 9–16 mm, the lower tepals united with the upper laterals for c. 2 mm and sometimes to one another for up to 2 mm, more or less straight and directed forward, or gently curving outward distally, 16–22 x 9–12 mm. *Filaments* (15–)18–25 mm long, included in the upper part of the tube or barely exserted for up to 2 mm; *anthers* 10–17 mm long, the lower 4–5 mm included in the tube, pale yellow to light purple, the pollen pale yellow. *Ovary* oblong, 8–10 mm long; *style* arching over the stamens, minutely scabridulous below,

dividing between the upper third and apex of the anthers, the branches (2–)4–6 mm long, the apices usually reaching beyond the anthers, very broad in the distal third. *Capsules* more or less oblong-ellipsoid, 34–36 mm long, about as long as or slightly longer than the outer bract; *seeds* ovate, 5–7 x 4–5 mm, broadly and evenly winged, light reddish brown, the seed body slightly darker brown. *Chromosome number* 2*n* = 30.

Flowering time September to November, rarely in August, occasionally in December or early January at high elevations.

Distribution and biology

One of the more widespread of the *Gladiolus* species of the southern African winter-rainfall region, *G. tristis* extends from Port Elizabeth in the east to the Bokkeveld Escarpment near Nieuwoudtville in the northwest. It is particularly common in the southern Cape between Bredasdorp and Riversdale where it can sometimes be seen in dense colonies on damp flats. Across this wide range the species occurs from near sea level to fairly high elevations in montane habitats. Soil conditions appear to be relatively unimportant and plants may be found in clay or granite- or sandstone-derived soils. However, almost wherever found, *G. tristis* is associated with seasonally or perennially wet habitats, hence its common name, marsh Afrikaner. To say that it always grows in marshes is, however, not true, for plants may be found on banks above streams or on cool south-facing slopes as well as in poorly drained seeps, wet bottoms and marshes. Flowering time is also variable. In lowland habitats plants may be found flowering in the southern Cape in August, or even in July, while in the mountains plants may not flower until December or January. The range of flowering times and habitats is reflected to some extent in morphological variability, discussed below.

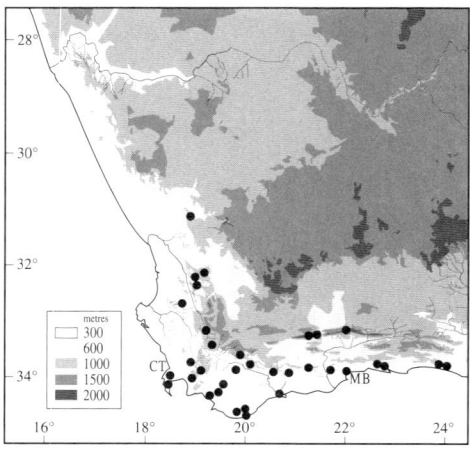

Gladiolus tristis has pale-coloured flowers that only open completely in the late afternoon and evening and are then wonderfully scented. This flowering pattern, as well as the long perianth tube, are adaptations for pollination by large moths, including species of Noctuidae and most likely sphinx moths (Sphingidae). The flowers produce large amounts of nectar with a high sugar concentration.

Diagnosis and relationships

Gladiolus tristis is recognized by the combination of a tall, erect flowering stem, presence of just three leaves per plant, and large, pale-coloured flowers with a well-developed perianth tube. The leaf blades are also distinctive, always being radially symmetric in transverse section and either cross-shaped (with wide sinuses between the flanges of the blade) or terete (with narrow sinuses). Like all the members of series *Tristis*, *G. tristis* has three leaves with thickened margins and midrib but, unlike the others, it has fairly soft, more or less fibrous corm tunics. This apparently derived condition may be an adaptation to its moist habitat. Flower colour is somewhat variable and may be uniformly cream to palest yellow, or the midlines of the lower three tepals, or the upper laterals as well, may have light brown to reddish shading. The reverse of the tepals may also be flushed with brown or dull purple. Flower colouring is not always constant and in experimentally grown plants has been found to vary depending on moisture and soil regimes. Colour variants have, however, been accorded taxonomic recognition, the more or less uniformly coloured forms being called var. *concolor* by Lewis et al. (1972), and the strongly discolorous forms var. *tristis*. There seems little merit in recognizing these trivial variants, at least some of which are not genetically fixed for flower colour. Later-flowering plants have been accorded taxonomic recognition too. The amateur *Gladiolus* enthusiast and grower, Collingwood Ingram, named a summer-flowering population *G. aestivalis* in 1930 and Lewis et al. continued to recognize the variant as var. *aestivalis* in their 1972 revision. They point out that the summer-flowering plants have spikes of eight to 20 flowers, implying a second genetic character in summer-flowering plants. Plants with this many flowers are, however, exceptional. Most summer-flowering plants that we have seen have stems of no more than four flowers, and often only one or two. There is no reason to believe that the so-called summer-flowering form of *G. tristis* is even a single race. We suspect that variation in flowering time is a trend that has developed repeatedly in the species.

Gladiolus tristis can easily be crossed with other species of the genus and hybrids between it and *G. watsonius* were once thought quite remarkable, given the very different flowers of the two species. These are the so-called *Homoglad* hybrids, named thus because *G. watsonius* was considered to belong in a different genus, *Homoglossum*, when the hybrids were made. A natural hybrid between *G. tristis* and *G. caryophyllaceus* was described as *G. lewisiae* by A.A. Obermeyer in 1970 and the hybrid origin of this plant was only understood much later when field studies were made at Seweweekspoort, where the plants occurred (Goldblatt & Vlok, 1989).

History

Common around the sites of the earliest European settlement in southern Africa, *Gladiolus tristis* makes its appearance in the literature in 1750 in the German botanist, C.J. Trew's *Plantae Selectae*. In this selection of rare and new plants it was called *Lilio-Gladiolus bifolius biflorus foliis quadrangularis* – note the reference to the distinctive leaf blades. In 1762 the species was given the binomial, *G. tristis*, by Carl Linneaus. The name was immediately adopted by Linnaeus' contemporaries and *G. tristis* has little of the nomenclatural confusion that surrounds so many other Cape plants known at this time. The understanding of variation in plant species was rudimentary in the eighteenth century, and it is not especially surprising that more than one species was included under the name *G. tristis*. Thunberg, for example, who had extensive field experience in southern Africa, nevertheless recognized some 16 varieties of *G. tristis* in 1784, including *G. liliaceus* as var. *grandis*, *G. maculatus* as var. *odorus*, and *G. carinatus* as var. *inodorus*. As more collections reached Europe, it became clear that most of these varieties represented quite separate species and in the following 20 years most were raised to species rank by Thunberg or were described independently by other authors.

The English botanist, R.A. Salisbury, recognized *Gladiolus concolor* as separate from *G. tristis* because of the obovate and acuminate tepals which he considered different from true *G. tristis*. There seems to be no basis for the recognition of this species, even at varietal rank in *G. tristis*, as was proposed by Lewis et al. (1972). As outlined above, the colour variation is trivial and sometimes not even genetically based, but a result of environmental conditions including soil and water availability. Likewise, the variation in tepal shape is trivial at best. Plants with uniformly coloured flowers of pale yellow were described as *G. flavidus* by Ingram in 1930, and he named plants with deeper yellow flowers *G. fulvescens*. These plants represent minor colour variants of *G. tristis* and should never have been admitted as botanical species. The status of a third so-called species named by Ingram, *G. aestivalis*, is discussed above.

---

156. GLADIOLUS LONGICOLLIS Baker

Graspypie

PLATE 138

Gladiolus longicollis Baker, *J. Bot.* 14: 182 (June, 1876). Lewis et al., *J. S. African Bot.*, Suppl. 10: 188 (1972). Type: South Africa, Eastern Cape, Baziya Mountain, c. 900–1500 m (3000-4000 ft), Oct.–Nov., *Baur 505* (K, holotype; BOL, NU, isotype).

longicollis = long-necked, actually referring to the elongate perianth tube.

Synonymy

For synonyms see under the subspecies.

Description

Plants (20–)30–50(–60) cm high. **Corm** conic, (7–)10–14 mm long, the tunics of fairly fine to moderately coarse light brown fibres, sometimes more or less claw-like below. **Cataphylls** membranous and pale, the uppermost reaching shortly above the ground, and then green and coriaceous, sometimes becoming dry and brown above. **Leaves** three, only the lowermost with a well-developed blade about two-thirds as long as the stem, cross-shaped in section to terete with four narrow longitudinal grooves, the second leaf sheathing the stem for 100–150 mm, the apex often free but channelled

throughout, or with a short terete apex, the third leaf inserted in the upper quarter of the stem and reaching to the first flower, also largely sheathing with a short free channelled portion. *Stem* erect, unbranched, usually flexed outward just above the sheathing part of the second leaf, 1.2–2 mm in diameter below the spike.

Spike 1- to 3-, occasionally to 7-flowered; *bracts* grey-green, the outer 35–50(–65) mm long, with a short acuminate apex, the inner three-quarters as long as to slightly longer than the outer, entire or notched apically for up to 3 mm. *Flowers* white to pale yellowish, uniformly coloured or lightly or occasionally heavily mottled with brown, more densely so along the midlines, the outer tepals flushed and veined purplish to brown or green on the reverse, the tube obscurely lined with green or purple on the veins, lower margins of the upper tepals and tepal sutures usually transparent, opening in the evening (or during the day in misty weather) and then sweetly scented of carnation and cloves, closing the next morning after 07h30; *perianth tube* slender and cylindrical, more or less horizontal and slightly curved above or erect below and curving outward above, (45–)50–110 mm long,

2–3 mm in diameter in the middle, papillose in the lower throat; *tepals* subequal or unequal and those of the outer whorl equal and larger than those of the inner whorl, acute to acuminate or attenuate, the dorsal nearly horizontal for its entire length or curving upward in the distal half, 25–32 x 12–17 mm, the upper laterals 25–42 x 10–19 mm, the lower tepals sometimes joined to the upper laterals for up to 2 mm and to one another for up to 2 mm, the lower lateral tepals 22–32 x 9–14 mm, the lower median 25–32 x 12–14 mm. *Filaments* 5–13 mm long, usually included in the tube, occasionally exserted 1–2 mm; *anthers* 9–15 mm long, included in the tube below, sometimes for half their length, or the bases exserted 1–2 mm, sometimes lying on the lower tepals, green or purplish, the pollen cream to yellow. *Ovary* cylindrical, 7–9 mm long; *style* arching over the stamens, puberulous in the lower half, dividing opposite the upper third of the anthers, the branches usually short, 3–6 mm long, slender throughout or broad above. *Capsules* elongate-ellipsoid, 20–25 x c. 6 mm; *seeds* ovate, 5–6 x 3.5–4 mm, broadly and evenly winged, transparent light brown, the seed body darker brown. *Chromosome number* 2n = 30.

Flowering time October to mid-February.

Distribution and biology

One of the most widespread species in southern Africa, *Gladiolus longicollis* extends from the Swartberg and Kammanassie Mountains in the Oudtshoorn District of the southern Cape northeastward through the eastern Cape and Free State to Northern Province. Often surprisingly common, plants are usually seen in low grassland in late spring and early summer, but flowering in some populations may be as late as January, especially in Mpumalanga where plants with ripe capsules may be found only a short distance from others with buds and flowers.

The long to extremely long perianth tube and the pattern of flowers opening in the evening and then producing a strong, sweet, clove-rich fragrance show the species to be adapted for pollination by night-flying moths. We have demonstrated one pollinator, the convolvulus hawkmoth *Agrius convolvuli* in Mpumalanga. This moth has a tongue c. 100 mm long, complementing the perianth tube of the species which is

80–100 mm long in this part of its range. The Natal botanist, John Medley Wood, also reported hawkmoth pollination in plants from KwaZulu-Natal (Scott Elliot, 1891). He identified the insect as the spurge hawkmoth, actually a European species, but the observation of hawkmoth pollination is nevertheless useful. Plants produce ample amounts of nectar of relatively high sugar concentration.

Diagnosis and relationships

Most readily recognized by it tubular flower, *Gladiolus longicollis* has a perianth tube at least 45 mm long, and usually 60–110 mm. The tube is also narrow for almost its entire length, widening only in the upper 10–15 mm. The flowers are usually shades of cream to pale yellow, often with brown shading in the midline of the tepals on both the inside and the reverse, and in many populations the flowers are also irregularly speckled with light to dark brown. As noted above, the flowers usually open fully in the late afternoon and then produce a strong sweet scent. Like all species of series *Tristis*, *G. longicollis* has an unbranched stem and produces only three leaves per plant, the lowermost longest and the two upper much smaller and largely sheathing. The leaf blades are radially symmetric in transverse section, and cross-shaped with wide or narrow sinuses between the four ridges of the blade.

The leaf blade and the flowers of *Gladiolus longicollis* resemble those of *G. tristis* and we assume that the two species are immediately related. Shorter-tubed populations of *G. longicollis* are easily confused with *G. tristis*, and only a close examination of the flowers makes it possible to distinguish such plants. *Gladiolus tristis* has flowers with the upper part of the perianth tube fairly wide and tubular, and the filaments which are inserted at the base of the upper part of the tube are usually 18–25 mm long. The anthers are either included in the mouth of the tube or are exserted for up to 3 mm.

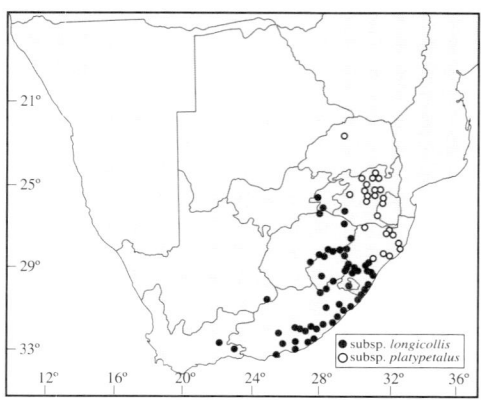

In *G. longicollis* the upper part of the tube is shorter, rarely more than 15 mm long, the filaments are 5–13 mm long, and the anthers are usually partly included in the tube. Both the length of the upper part of the tube and the length of the filaments can thus be used to distinguish these two species when perianth tube length is similar.

Perianth tube length in *Gladiolus longicollis* shows a clear correlation with geographic distribution. Tube length is shortest in the west and plants from the southern and eastern Cape have tubes 45–65 mm long. Flowers of these plants are also typically more strongly marked with brown speckles and shading. In the north of its range, in coastal and northern KwaZulu-Natal, Swaziland and north of the Vaal River, plants usually have a tube 65–110 mm long and little or no speckling on the perianth, which ranges from almost white to pale yellow. The two forms, treated as varieties by Lewis et al. (1972), are recognized as separate subspecies here.

History

Gladiolus longicollis was apparently first collected by C.F. Ecklon, probably in November 1829, or by Ecklon together with C.L. Zeyher in 1832, in the eastern Cape near the mouth of the Great Fish River. The collection, distributed under the name *G. versicolor* Ecklon (a synonym of *G. maculatus*), elicited no botanical interest. The western part of the range of *G. longicollis*, including East London, Grahamstown and the Swartberg, was subsequently visited by several botanists in the following years, yet the next record of the species that we have traced is actually the type specimen of *G. longicollis* subsp. *platypetala*. This was collected by John Sanderson, a British journalist, trader and plant collector who arrived in Natal Colony in 1851.

It was only in 1872 that the species was again collected, by the South African botanist, Peter MacOwan, at Somerset East. The type collection of *Gladiolus longicollis* was, however, made a few years later by the Rev. L.R. Baur. After establishing a mission in the Transkei at Baziya in the 1860s, Baur began collecting there in 1873 (Gunn & Codd, 1981). The several duplicates of the type collection bear no year, stating only the month, October. *Gladiolus longicollis* was described by J.G. Baker in June 1876. In November of the same year Baker described *Acidanthera platypetala* based on Sanderson's collection. The species was no doubt placed in *Acidanthera* because of its very long perianth tube.

Over the following two decades several collections from across almost the entire range of *Gladiolus longicollis* came to Baker's attention. These included plants with tubes of intermediate length from southern KwaZulu-Natal, collected by William Tyson in 1883, short-tubed plants from Grahamstown, collected by Ernest Galpin in 1888, and long-tubed plants from what was then the eastern Transvaal, collected by Christopher Mudd in 1877 and Galpin in 1889. These and other collections persuaded Baker that *Gladiolus longicollis* and *Acidanthera platypetala* were a single species, and in his *Handbook of the Irideae* (1892) and *Flora Capensis* (1896) Baker united the two under the name *Acidanthera platypetala*.

Although time has shown that Baker's decision about the species was correct, his choice of genus is no longer accepted. The genus *Acidanthera*, based on a distantly related tropical African species, is included in *Gladiolus*. Baker's choice of species name was also nomenclaturally inadmissible, for he included the earlier synonym under the later one. This situation was corrected by Lewis et al. (1972) in their revision of the South African species of *Gladiolus*. These authors also recognized the long- and short-tubed forms of *G. longicollis* as separate varieties.

The extremely long-tubed form of *Gladiolus longicollis* from Swaziland and Mpumalanga was described as a separate species, *G. praelongitubus*, by G.J. Lewis in 1941. Although perianth tube length exceeds 100 mm in Swaziland and Mpumalanga, tube length grades across KwaZulu-Natal and it has proven impossible to maintain this taxon, even at infraspecific rank. These plants are, however, recognizably different from the southern KwaZulu-Natal form of subsp. *platypetalus* when seen alive and they apparently represent a distinct race of the subspecies.

Key to the subspecies

1. Perianth tube 45–60(–65) mm long; tepals narrowly lanceolate and tapered distally; anthers exserted or the bases included in the tube; spike with 2–7 flowers .... subsp. *longicollis*
1'. Perianth tube 65–110 mm long; tepals broadly lanceolate to ovate, not markedly tapered distally; lower half of the anthers usually included in the tube; spike with 1–2(–3) flowers .... subsp. *platypetalus*

GLADIOLUS LONGICOLLIS subsp. LONGICOLLIS

Description

Spike occasionally 1-, usually 2- to 4-, rarely to 7-flowered; *bracts* grey-green, the outer 35–40 mm long, the inner three-quarters as long as the outer, entire or minutely notched apically. *Flowers* cream to yellow with brown speckling variously developed, upper tepals and tepal sutures transparent; *perianth tube* usually erect below, curving outward above, 45–65 mm long, sometimes only shortly exserted from the bracts; *tepals* lanceolate, distally attenuate, subequal, 30–35 x 8–12 mm. *Filaments* 10–12 mm long, reaching to the mouth of the tube; *anthers* c. 9 mm long, exserted or sometimes the bases included in the tube. *Style* dividing opposite the anther apices, the branches c. 3 mm long.

Flowering time October to November.

Distribution

The southern subspecies of *Gladiolus longicollis*, subsp. *longicollis* extends from the Oudtshoorn District in Western Cape Province through the Eastern Cape and Free State, South Africa, into Lesotho. In the south of its range it grows in low grassy fynbos vegetation, mostly on moist, south-facing slopes, but elsewhere it occurs in rocky grassland.

GLADIOLUS LONGICOLLIS subsp. PLATYPETALUS (Baker) Goldblatt & Manning

Subsp. *platypetalus* (Baker) Goldblatt & Manning, new combination and rank. *Acidanthera platypetala* Baker, *J. Bot.* 14: 339 (November, 1876); *Fl. Capensis* 6: 13 (1896). *Gladiolus longicollis* var. *platypetalus* (Baker) Obermeyer in Lewis et al., *J. S. African Bot.*, Suppl. 10: 190 (1972). Type: Natal, Attercliffe, 30 Sept. 1860, *Sanderson 265* (K, holotype; BOL, probable isotype).

platypetalus = broad-petalled, for the comparatively broad petals, at least compared to its immediate relatives.

Synonymy

Gladiolus praelongitubus G. Lewis, *J. S. African Bot.* 7: 27 (1941). Type: South Africa, Mpumalanga (Transvaal), Barberton District, Saddleback Range, 3000-5000 ft, Sept.-Oct. 1889, *Galpin 530* (SAM, holotype; BOL, K, PRE, isotypes).
Gladiolus cygneus Ingram, *Gard. Chron.* Ser. 3, 90: 347, fig. 144 (1931). Type: South Africa, Natal,

Mont aux Sources, 9,500–10,000 ft, cultivated in Britain, 1930, *Ingram s.n.* (BM, holotype).

Description

Spike 1- to 2-, occasionally 3-flowered; *bracts* green or flushed with purple, the outer 45–55 mm long, with a short acuminate apex, the inner slightly shorter to slightly longer than the outer, entire or notched apically for up to 1 mm. *Flowers* white or cream, the outer tepals flushed and veined purplish to brown or green on the reverse, the tube obscurely lined with green or purple, the tepal sutures sometimes transparent;

perianth tube slender and cylindrical, more or less horizontal and slightly curved, 65–110 mm long, well exserted from the bracts; *tepals* lanceolate, those of the outer whorl subequal and larger than those of the inner whorl, these also subequal, the dorsal 30–32 x 14–17 mm, the upper lateral and lower median tepals 40–42 x 15–19 mm. *Filaments* 5–9 mm long, included in the tube; *anthers* 11-13 mm long, the lower halves included in the tube, often lying on the lower tepals. *Style* dividing between the middle and apex of the anthers, the branches 5–6 mm long.

Flowering time October to December, occasionally until mid-February.

Distribution

Subspecies *platypetalus* extends from the southern Drakensberg near Barkly East in Eastern Cape Province through the midlands of KwaZulu-Natal to Mpumalanga and Northern Province of South Africa and to neighbouring Swaziland. It is common at elevations above 1000 m in low or rocky grassland, but may also occur along the eastern Cape and KwaZulu-Natal coast close to sea level in suitable grassy habitats.

157. GLADIOLUS SYMONSII F. Bolus

Gladiolus symonsii F. Bolus, *Ann. Bolus Herb.* 2: 99 (1917). Lewis et al., *J. S. African Bot.,* Suppl. 10: 206 (1972). Type: KwaZulu-Natal, Underberg District, Bushman's Peak, c. 3000 m (9500 ft), 8 Jan. 1915, *Symons 336* (BOL, holotype).

symonsii, in honour of R.E. Symons, the South African forester and naturalist who made the first collection of the species in January 1915.

Description

Plants 25–45 cm high. *Corm* globose, c. 10 mm in diameter, the tunics of fine to medium-textured (to wiry) vertical fibres. *Cataphylls* membranous, barely reaching

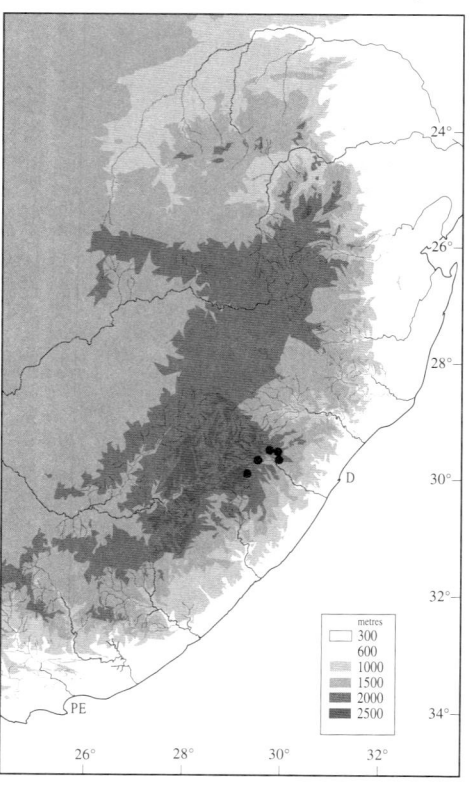

above the ground and then purplish. *Leaves* three, the lowermost longest, sheathing the lower third of the stem, reaching to about the middle to upper third of the stem, the blade 60–120 mm long, centric with four broad longitudinal grooves (cross-shaped in section and with the midrib raised and thickened), second leaf sheathing the stem for 80–150 mm, the apex often free but channelled, third leaf inserted near the stem apex and sometimes reaching to the first flower, also largely sheathing with a short free apex. *Stem* erect, unbranched, usually flexed outward just above the sheathing part of the second leaf, up to 1 mm in diameter below the spike.

Spike inclined, 2- to 4-flowered; *bracts* soft-textured, dull green, 20–30 mm long, the inner about two-thirds to three-quarters as long as the outer. *Flowers* almost actinomorphic but the stamens unilateral and the style arcuate, bright pink to pale rose, whitish in the throat, the lower three tepals feathered or streaked with fine red lines or spots at the base, apparently unscented; *perianth tube* obliquely funnel-shaped, 8–12 mm long; *tepals* subequal, lanceolate, the dorsal slightly larger, 20–25 x 10–13 mm, inclined over the stamens, the upper laterals directed forward, slightly shorter than the dorsal, the lower three tepals 19–23 x 10 mm, directed forward. *Filaments* c. 7 mm long, included or exserted up to 1 mm from the tube; *anthers* 8–10 mm long, yellow, the pollen also yellow. *Ovary* globose, c. 3 mm long; *style* arching over the stamens, dividing opposite the upper quarter of the anthers, the branches c. 3.5 mm long. *Capsules* and *seeds* unknown. *Chromosome number* unknown.

Flowering time December to late January.

Distribution and biology

A species of montane habitats, *Gladiolus symonsii* has a relatively narrow distribution in the main Drakensberg range of KwaZulu-Natal and eastern Lesotho. Records of the species are from sites extending from the Bushman's Nek area in the south to Giant's Castle in the north. Most collections are from above 2000 m, sometimes 2500 m, but a population on the isolated Kamberg occurs at c. 1800 m. Plants flower in early to mid-summer, December and January, and grow in short grassland or on rocky pavement either on sandstone or basalt. We have, unfortunately, not seen living plants ourselves and owe information about its habitat to the collections of others, notably O.M. Hilliard and B.L. Burtt, who have extensive knowledge of the Drakensberg flora and have recorded the species on several occasions, extending its known range considerably in doing so.

The pollination biology of *Gladiolus symonsii* is unknown. The short perianth tube, 8–12 mm long, suggests an unspecialized pollination syndrome. The relatively small, pink flowers are probably pollinated by anthophorid and mellitid bees, as are the similar flowers of species like *Hesperantha* and *Dierama* that also occur in the Drakensberg and flower at the same time.

Diagnosis and relationships

Gladiolus symonsii is easy to recognize by its relatively small flowers with a short tube, 8–12 mm long, and nearly equal, lanceolate tepals. The stems are always unbranched and the spikes bear only two to four flowers. Among the *Gladiolus* species of the southern African summer-rainfall zone, *G. symonsii* stands out in having only three leaves, the lowermost longest, and the leaf blades slender and terete with four narrow, longitudinal grooves. The vegetative features match those of section *Homoglossum* series *Tristis*, and we tentatively refer *G. symonsii* to this alliance. The only other summer-rainfall species of the series is *G. longicollis* which has larger flowers with long to extremely long

perianth tubes, and there is no possibility that the two can be confused. Additional collections, especially capsules and seeds, will be useful in better evaluating the relationships of the plant.

History

Gladiolus symonsii was discovered by the

South African forester and game warden, R.E. Symons, in January 1915 in the southern high Drakensberg on Bushman's Peak. Just a handful of collections have been made since then, and the species is far from being fully understood. It is, nevertheless, well founded and apparently taxonomically isolated.

Section *Homoglossum:* Series *Homoglossum*

158. GLADIOLUS WATSONIUS Thunberg

Red Afrikaner, suikerkannetjie
PLATE 139

Gladiolus watsonius Thunberg, *Dissertatio de Gladiolo* no. 10 (1784). *Antholyza watsonia* (as *watsoniana*) (Thunberg) Pax in Engler & Prantl, *Nat. Pflanzenf.* 2(5): 156 (1888). *Homoglossum watsonium* (Thunberg) N.E. Brown, *Trans. Roy. Soc. S. Africa* 20: 227 (1932). *Watsonia revoluta* Persoon, *Synopsis Pl.* 42 (1805), as a new name for *Gladiolus watsonius* Thunberg in *Watsonia*, the combination *W. watsonia* being inadmissable. Type: South Africa, Western Cape, Lion's Rump (and elsewhere on the Cape Peninsula), *Thunberg s.n* (UPS–Herb. Thunberg 1092, holotype).

watsonius = watsonia-like, referring to the similarity of the bright red, long-tubed flowers to those of many species of *Watsonia*.

Synonymy

Gladiolus praecox Andrews, *Bot. Repository* 1: pl. 38 (1797), as *G.praecox* var. *flore rubro.* *Watsonia praecox* (Andrews) Persoon, *Synopsis Pl.* 1: 42 (1805). *Homoglossum praecox* (Andrews) Salisbury, *Trans. Hort. Soc.* 1: 325 (1812). Type: South Africa, Western Cape, without precise locality or collector, illustration in Andrews, *Bot. Repository* 1: pl. 38 (1797).
Homoglossum revolutum var. *gawleri* Baker, *J. Linn. Soc. Bot.* 16: 161 (1877). *Gladiolus watsonius* var. *gawleri* (Baker) Baker, *Handbook Irideae* 227 (1892). *Antholyza revoluta* var. *gawleri* (Baker) Baker, *Fl. Capensis* 6: 169 (1896). *Gladiolus gawleri* (Baker) Klatt, *Abh. Naturforsch. Ges. Halle* 15: 341 (Ergänzungen 7) (1882). *Antholyza gawleri* (Baker) L. Bolus, *Ann. Bolus Herb.* 3: 11 (1920). *Homoglossum gawleri* (Baker) N.E. Brown, *Trans. Roy. Soc. S. Africa* 20: 278 (1932). Type: South Africa, Western Cape, without precise locality or collector, illustration in *Curtis's Bot. Mag.* 16: pl. 569 (1802) as *G. watsonius* var. ß, *corollis majoribus luteo-variegatus.*
Antholyza acuminata N.E. Brown, *Kew Bull.* 1929: 133 (1929). *Homoglossum acuminatum* (N.E. Brown) N.E. Brown, *Trans. Roy. Soc.*

S. Africa 20: 278 (1932). Type: South Africa, Western Cape, without precise locality or collector, *Burman s.n.* (Herb. Burman 'Gladiolus recurvus,' G, holotype [not seen]).
Homoglossum flexicaule N.E. Brown, *Trans. Roy. Soc. S. Africa* 20: 278 (1932). Type: South Africa, Western Cape, without precise locality or date, *Villet s.n.* (K, holotype).
Gladiolus watsonius var. *maculosus* De Vos & Goldblatt, *Bull. Mus. Nat. Hist. Nat., Paris* Sér. 4, sect. B, *Adansonia* 11: 421 (1989). Type: South Africa, Western Cape, Kapteinskloof, Farm Banghoek, 22 Aug. 1989, *De Vos 2711* (NBG, holotype and isotype).
[*Homoglossum revolutum* sensu Baker, *J. Linn. Soc. Bot.* 16: 161 (1877). *Antholyza revoluta* sensu Baker, *Fl. Capensis* 6: 169 (1896), not *A. revoluta* Burman fil.]

Description

Plants 30–70 cm high. *Corm* globose, 12–20 mm in diameter, the tunics of more or less woody to coriaceous layers broken below into triangular to linear segments, often with numerous small cormlets around the base. *Cataphylls* pale and membranous, the uppermost reaching up to 18 cm above the ground and then greenish to purple. *Leaves* three, imbricate, the lower two basal, the lowermost longest, sheathing the lower third of the stem, reaching at least to the base or sometimes exceeding the spike, the blade linear, (1.5–)4–6 mm wide, the margins heavily thickened and forming wings extended at right angles to the blade, the midrib moderately thickened and raised, the second leaf sheathing the stem for two-thirds of its length, with a short plane blade, the uppermost leaf inserted on the upper third of the stem, sheathing for half its length, channelled throughout, the lower margins united around the stem. *Stem* erect, barely flexed outward above the sheath of the terminal leaf, unbranched, 2.2–3 mm in diameter below the spike.

Spike erect and more or less straight, rarely 2-, usually 4- to 6-, occasionally to 8-flowered; *bracts* pale green or flushed light brown, the outer (35–)40–55(–70) mm long, the lower margins closed around the stem, the inner about two-thirds to nearly as long as the outer. *Flowers* orange-red to scarlet, the lower tepals often translucent, occasionally yellowish below and speckled dark red near the bases, the dorsal tepal with the edges translucent in the lower half,

unscented; *perianth tube* 43–55 mm long, narrow, cylindrical in the lower 15–22 mm, abruptly expanded and curved outward into a wider cylindrical and horizontal upper part 30–35 mm long, 6–10 x 4–6 mm wide; *tepals* unequal, broadly lanceolate, the dorsal largest, extended horizontally over the stamens, 26–33 x 12–18 mm, upper laterals 25–32 x 7–12 mm, more or less patent, the lower three tepals arching outward from the base, the lower laterals 20–26 x 5–11 mm, the lower median 23–32 x 6–12 mm. *Filaments* 32–40 mm long, exserted 8–13 mm from the tube; *anthers* 7–10 mm long, dark purple, the pollen cream to pale yellow. *Ovary* cylindrical, 6–8 mm long; *style* arching over the stamens, dividing between the upper third and apex of the anthers, the branches 4–6 mm long, broadly expanded in the upper halves. *Capsules* elongate-ellipsoid, 40–45 x c. 14 mm; *seeds* ovate, 6–9 x 5–6 mm, broadly and evenly winged, light brown. *Chromosome number* $2n = 30$.

Flowering time August to mid-September, occasionally in July.

Distribution and biology

Restricted to the southwestern portion of Western Cape Province, *Gladiolus watsonius* is found on lower mountain slopes and flats from Piketberg in the north to the Cape Peninsula, and no further east than the upper Breede River valley west of Worcester. Plants favour heavier soils and are typically found growing on clay soils derived from Malmesbury or Bokkeveld Shales. Along the west coast *G. watsonius* grows on sandy soils derived from decomposed granite. Throughout its range *G. watsonius* has become rare as a result of urban expansion and intensive agriculture on the fertile coastal forelands and low hills that originally formed its habitat.

The long-tubed and bright red flowers of *Gladiolus watsonius* are widely assumed to be adapted for pollination by sunbirds, *Nectarinia* species (Rebelo, 1987), but critical observations remain to be recorded. A flower produces up to 36 microlitres of nectar, with a mean sugar concentration of 31%. Surprisingly, the nectar sugar in two populations sampled is largely sucrose, with just traces of glucose and fructose. While the nectar quantity is typical of many bird-pollinated species of Iridaceae, the high sugar concentration and sucrose dominance is unusual. More often the nectar of bird-pollinated flowers has high ratios of glucose and fructose and only traces of sucrose; the sugar concentrations are typically less than 20% (Goldblatt, 1989).

Diagnosis and relationships

Type species of *Homoglossum*, a genus now included in *Gladiolus*, *G. watsonius* has the classic flower of southern African Iridaceae pollinated by sunbirds: bright red in colour, with an elongated and tubular upper part of the perianth tube, outspread tepals and well-exserted anthers. Other taxonomically important features of *G. watsonius* are corm tunics of woody texture and three linear foliage leaves. Among the several species with this type of flower and vegetative organization *G. watsonius* is distinguished by its robust habit, relatively large flower with subequal tepals, and leaf blade which is linear with enlarged and raised margins and midrib. The species is most closely related to *G. teretifolius* which has similar, though usually slightly smaller flowers, but different leaves and capsules. The leaves of *G. teretifolius* are much narrower than those of *G. watsonius*, rarely more than 2 mm wide, sometimes so narrow as to appear terete, but the margins are usually more strongly raised and thickened than the midrib. The capsules of *G. teretifolius* are ellipsoid, up to 20 mm long, and contain unusually small seeds c. 4.5 mm long. This is quite different from the remarkably long capsules, 40–45 mm, of *G. watsonius* which have seeds 6–9 mm long.

History

A common lowland species and, until the 1930s, conspicuous on the lower slopes of the hills and mountains around Cape Town, *Gladiolus watsonius* was first recorded in the early eighteenth century and was in cultivation in Holland by the 1760s, if not before. It was not formally named until 1784 when C.P. Thunberg described plants that he had collected on Signal Hill (Lion's Rump) in Cape Town and elsewhere on the Cape Peninsula (Thunberg, 1784). He called it *G. watsonius*, recognizing the resemblance of its flowers to those of species of *Watsonia*, particularly *W. meriana*, another common Cape plant. An early collection of the species grown in Holland and preserved in the Burman Herbarium, now in Geneva, was described by N.E. Brown in 1929 as *Antholyza acuminata*, but as far as we can tell this differs in no significant way from the type of *G. watsonius*.

Plants grown in Great Britain in the 1790s, obtained from a commercial source in Holland, were illustrated by Henry Andrews in 1797 under the name *Gladiolus praecox*, so named for its early flowering habit. Hendrik Persoon, frequently a taxonomic splitter, transferred both *G. praecox*

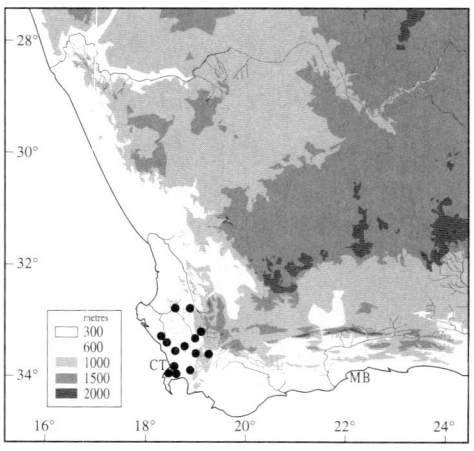

and *G. watsonius* to *Watsonia* in 1805, renaming the latter *W. revoluta* (the combination *W. watsonia* would certainly have been considered unacceptable).

The appropriate genus for this and other species of *Gladiolus* with red, long-tubed flowers like those of *G. watsonius* remained a problem for botanists for many years, and J.G. Baker first placed the species in *Homoglossum* and later in *Antholyza*. Baker also made the mistake of associating *G. watsonius* with the earlier *Antholyza revoluta* Burman fil. (1768). Thus, in *Flora Capensis* and elsewhere, Baker (1896) placed *G. watsonius* Thunberg in synonymy under *A. revoluta*. Although the type of Burman's species was not discovered until the 1990s, N.E. Brown (1932) realized that Burman's species was not the same as *G. watsonius* which he recognized, although he placed the species in *Homoglossum*. *Antholyza revoluta* is now known to be a species of *Tritoniopsis* (Wijnands & Goldblatt, 1992).

The west coast variant of *Gladiolus watsonius* which has pale yellow streaks of colour usually covered with small reddish speckles on the lower tepals has had its own complex history. The plant was first illustrated in *Curtis's Botanical Magazine* in 1802 under the name *G. watsonius* var. ß (dwarf Watson's corn-flag) by John Ker Gawler. Baker recognized the plant illustrated in *Curtis's Botanical Magazine* as *Homoglossum revolutum* var. *gawleri* in 1877, but F.W. Klatt raised the variety to species rank in 1882 as *Gladiolus gawleri*. Klatt, like Ker Gawler, preferred to place species like *G. watsonius* in *Gladiolus*. N.E. Brown recognized the species in 1932, but placed it in the genus *Homoglossum*. The plants from Kapteinskloof in the Piketberg, called *G. watsonius* var. *maculosus* by M.P. de Vos (Goldblatt & De Vos, 1989) appear to represent the same race of the species as var. *gawleri*.

159. GLADIOLUS TERETIFOLIUS Goldblatt & De Vos

Overberg or Swellendam suikerkannetjie, rooipypie

PLATE 140

Gladiolus teretifolius Goldblatt & De Vos, *Bull. Mus. Hist. Nat., Paris* Sér. 4, Sect. B, *Adansonia* 11: 422 (1989), as a new name for *Antholyza muirii* L. Bolus, *Contrib. Bolus Herb.* 3: 12 (1920), not *G. muirii* L. Bolus, (= *G. involutus* Delaroche). *Homoglossum muirii* (L. Bolus) N.E. Brown, *Trans. Roy. Soc. S. Africa* 20: 279 (1932). De Vos, *J. S. African Bot.* 42: 344 (1976). Type: South Africa, Western Cape, south slopes of the hills behind Albertinia, 20 May 1914, *Muir 1348* (BOL, holotype; SAM, isotype).

teretifolius = terete-leafed, referring to the leaf blade which is rounded (actually oval) in transverse section.

Description

Plants 40–70 cm high. *Corm* globose, 12–20 mm in diameter, the tunics of more or less woody to coriaceous layers broken below into triangular to linear segments, often with numerous small cormlets around the base. *Cataphylls* pale and membranous, reaching 5–7 cm above the ground and then bright green. *Leaves* three, the lower two basal, the lowermost longest, about as long as the spike, the blade 1–2 mm wide, oval to nearly terete in transverse section with the margins and midrib thickened, each surface with two narrow longitudinal grooves, the second leaf sheathing the stem for most of its length, the blade 10–25 mm long, the uppermost leaf inserted in the upper quarter of the stem, almost entirely sheathing, overlapped by the second leaf. *Stem* erect or slightly inclined above the sheath of the uppermost leaf, unbranched, 1.2–2 mm in diameter below the spike.

Spike nearly straight, more or less erect,

1- to 5-, occasionally up to 7-flowered; *bracts* grey-green, sometimes flushed with purple, the outer 35–45(–50) mm long, the inner about two-thirds as long as the outer. *Flowers* scarlet (rarely cream), the dorsal tepal translucent along the lower edges, the tube pale with red speckles and with paired dark red lines running from the base of the upper lateral tepals, unscented; *perianth tube* 35–45 mm long, straight and narrowly cylindrical in the lower 12–17 mm, abruptly expanded and curved outward into a wider cylindrical upper part, 22–27 mm long, 6–7 mm in diameter; *tepals* unequal, obovate, dorsal largest, 20–27 x 12–15 mm, extended horizontally, upper laterals patent, 16–25 x 10–12 mm, the lower tepals arching outward and downward, the lower laterals 12–18 x 8–12 mm, the lower median 15–22 mm long. *Filaments* 32–38 mm long, exserted 9–12 mm from the tube, more or less horizontal; *anthers* 7–9 mm long, dark violet, the pollen yellow. *Ovary* oblong, c. 6–8 mm long; *style* arching over the stamens, dividing opposite the anther apices, the branches c. 3.5 mm long, broadly expanded in the upper half. *Capsules* ovoid-ellipsoid, acute, c. 20 mm long; *seeds* broadly ovate, c. 4.5 x c. 3 mm, broadly and evenly winged, brown and hardly transparent. *Chromosome number* 2*n* = 30.

Flowering time mid-July to the end of August.

Distribution and biology

A local endemic of Western Cape Province, South Africa, *Gladiolus teretifolius* is restricted to the Caledon, Montagu, Swellendam and Riversdale Districts of the southern Cape. Plants occur in renosterveld at low elevations in heavy soils, either in clay or clay loam. Despite a range extending from near Bot River in the west to Albertinia in the east, a distance of more than 200 km, *G. teretifolius* is poorly collected. Today it is rare, being mostly confined to small islands of undisturbed vegetation among pastures and grain fields.

The plants flower relatively early in the season, mostly in late July and August. Like its relatives, *Gladiolus watsonius* and *G. quadrangularis*, *G. teretifolius* has flowers adapted for pollination by sunbirds. However, its pollination biology has not yet been investigated.

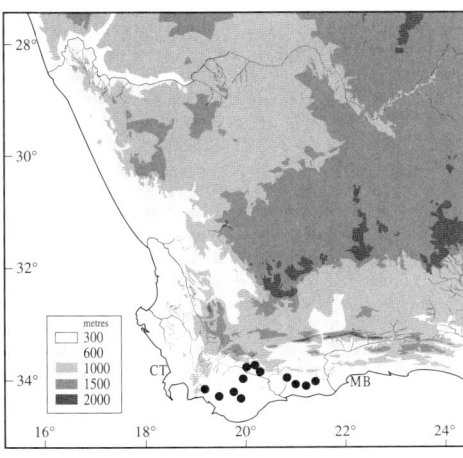

Diagnosis and relationships

With the classic flower form of southern African Iridaceae adapted for bird pollination, *Gladiolus teretifolius* has a bright red perianth and an elongate perianth tube with a slender basal and wide tubular upper part, and widely spreading tepals. The corm has hard, woody to coriaceous tunics and the three foliage leaves have narrow blades 1–2 mm wide with two narrow grooves on each surface. *Gladiolus teretifolius* is most closely related to *G. watsonius* but can be distinguished from that species by its more slender habit, somewhat smaller flowers with broader, ovate tepals and especially by the leaf blades which are no more than 2 mm wide and either oval or nearly terete in transverse section, the margins and midrib being so thickened that the leaves have two narrow grooves on each surface. Altogether more robust, *G. watsonius* has flowers with longer, lanceolate tepals more than twice as long as wide, and broader leaves that have raised

301

margins but the midrib only lightly thickened. The capsules of *G. teretifolius* are ellipsoid and up to 20 mm long, and contain unusually small seeds, c. 4.5 mm long. This is quite different from the capsules of *G. watsonius* which are 40–45 mm long and contain seeds 6–9 mm long.

History

Long included in, or confused with *Gladiolus watsonius*, *G. teretifolius* was only recognized as a distinct species in 1920 when H.M.L. Bolus described it as *Antholyza muirii*. Bolus assigned the species to the genus *Antholyza* following the contemporary generic convention, naming it in honour of the Riversdale physician and naturalist, John Muir, who she believed had made the first collection. In fact, *G. teretifolius* had been discovered at least 90 years earlier, as evinced by collections made by C.L. Zeyher near Caledon. The British botanist, N.E. Brown

transferred *A. muirii* to *Homoglossum* in 1932 when he proposed restricting *Antholyza* to include one species, *A. ringens*, now *Babiana ringens*. When Goldblatt & De Vos (1989) decided to include *Homoglossum* in *Gladiolus* they proposed the new specific epithet, *teretifolius*, for the plant because the name *G. muirii* had already been used in *Gladiolus* (*G. muirii* is a later synonym of *G. involutus*).

160. GLADIOLUS QUADRANGULARIS (Burman fil.) Ker Gawler

Rooi Afrikaner

PLATE 141

Gladiolus quadrangularis (Burman fil.) Ker Gawler, *Curtis's Bot. Mag.* 16: pl. 567 (1802), misapplied to *G. abbreviatus*. Baker, *Fl. Capensis* 6: 166 (1896), misapplied to *G. abbreviatus*. Goldblatt & De Vos, *Bull. Mus. Hist. Nat., Paris* Sér. 4, Sect. B, *Adansonia* 11: 422 (1989). *Antholyza quadrangularis* Burman fil., *Fl. Capensis Prodromus* 1 (1768). *Petamenes quadrangularis* (Burman fil.) J.W. Loudon, *Ladies' Flower-Garden Ornamental Bulbous Plants* 42, pl. 8 (1841),

misapplied to *G. abbreviatus*. *Homoglossum quadrangulare* (Burman fil.) N.E. Brown, *Trans. Roy. Soc. S. Africa* 20: 279 (1932). De Vos, *J. S. African Bot.* 42: 347 (1976). Type: South Africa, Western Cape, without precise locality or collector, *Burman s.n.* (Herb. Burman, G [not seen], holotype).

quadrangularis = four-angled, referring to the leaf which has four radiating wings and is thus cross-shaped in transverse section.

Description

Plants 35–70 mm high. *Corm* globose-conic, 12–20 mm in diameter, the tunics of papery to coriaceous layers, decaying with age into fine or medium-textured vertical fibres, often remaining unbroken above, but fraying into short fibres apically. *Cataphylls* membranous, pale throughout or the upper reaching up to 5 mm above the ground and then greenish purple. *Leaves* three, the lowermost longest, reaching to between the base and apex of the spike, sheathing the lower part of the stem for 50–80 mm, the blade erect or trailing above, centric, 2.5–4 mm wide, cross-shaped in section, the midrib raised on each surface into a wing-like ridge, the edges of the midrib wings and the margins lightly thickened, the second leaf sheathing the stem for about two-thirds of its length, with a short free apex, the blade like the basal, the third leaf inserted shortly below the spike, sheathing below, 60–80 mm long, channelled throughout. *Stem* erect, flexed outward above the sheaths of the two upper leaves, unbranched, 1.5–2 mm in diameter below the spike.

Spike erect, nearly straight, 4- to 8-, occasionally up to 11-flowered; *bracts* green, the outer 30–45 mm long, the inner two-thirds as long as to only slightly shorter than the outer, forked apically for c. 2 mm. *Flowers*

scarlet to brick-red, rarely pink, the dorsal and lower lateral tepals becoming transparent along the lower edges, occasionally the lower tepals orange toward the bases, unscented; *perianth tube* (35–)43–55 mm long, the lower part slender and cylindrical (10–)12–22 mm, abruptly expanded and curved outward into a wider cylindrical upper part (15–)24–32 mm long, 5–7 x 4–6 mm wide; *tepals* unequal, lanceolate, the dorsal largest, extended horizontally over the stamens, (20–)25–29 x (8–)12–15 mm, the upper laterals spreading outward almost from the base, (16–)20–28 x (6–)9–11 mm, the lower tepals curving outward from the base, the lower laterals 10–15 x 4.5–6.5 mm, the lower median (12–)15–20 x (3–)5–6 mm. *Filaments* 30–40 mm long, exserted (4–)6–10 mm from the tube, more or less horizontal; *anthers* 9–11(–14) mm long, yellow, the pollen yellow. *Ovary* cylindrical, 7–10 mm long; *style* extending horizontally over the stamens, dividing between the middle and apices of the anthers, the branches 3.5–4.5 mm long. *Capsules* ellipsoid, c. 25 mm long, mature capsules not seen; *seeds* unknown. *Chromosome number* $2n = 30$.

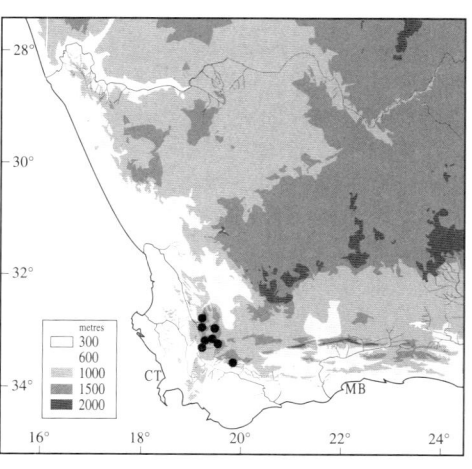

Flowering time mainly in late September, occasionally until late October at high elevations.

Distribution and biology

A narrow endemic of the southern African winter-rainfall zone, *Gladiolus quadrangularis* is restricted to the Ceres District of Western Cape Province. Plants occur from the fertile plains around Ceres, the Warm Bokkeveld, extending eastward to Karoopoort and Verlorenvlei and northward into the high Cold Bokkeveld.

Like most species of series *Homoglossum*, *Gladiolus quadrangularis* favours heavier soils and it is thus found mostly on clay and loamy soils which are now largely farmed for grain or deciduous fruit. Its range has been much reduced and *G. quadrangularis* is not often encountered. We have no reason to believe, however, that its existence is threatened. Its range includes areas of fairly low rainfall that are used only for grazing.

The flower, which closely resembles that of *Gladiolus watsonius*, is adapted for pollination by sunbirds but observations of floral visits are lacking.

Diagnosis and relationships

The long perianth tube with an abruptly widened cylindrical upper half and almost pouched at its base, the brick-red flower, the erect, unbranched stem and spike, and the three-leafed habit place *Gladiolus quadrangularis* squarely in series *Homoglossum* of section *Homoglossum*. The lanceolate to elliptic tepals seem to place the species close to *G. watsonius* and *G. teretifolius*, and it can be distinguished from these two species largely by its softer corm tunics and different leaf blades, the midribs of which are raised into wings so that the leaf is cross-shaped in transverse section. In *G. watsonius* and *G. teretifolius* the leaf blades have midribs strongly thickened but not winged, so the leaves are quite different in section.

De Vos (1976) associated *G. quadrangularis* with *G. huttonii* and *G. fourcadei* in her section *Quadrangulares*, thus emphasizing leaf morphology over tepal shape. Because the flowers of *G. quadrangularis* more closely resemble those of *G. watsonius* and *G. teretifolius* in their lanceolate to elliptic tepals, we believe its relationships lie more closely with these two species.

History

Although it occurs a fair distance from Cape Town and the centres of early settlement, *Gladiolus quadrangularis* was collected quite early and was grown and flowered in Holland at least by 1760. A specimen in the Linnaean Herbarium was sent to Linnaeus by the Dutch botanist, David van Royen, in 1763. Linnaeus initially drafted a description of the species with the name *Antholyza capensis*, but later changed his mind about its status (De Vos, 1976) and did not publish the name. A specimen in the younger Burman's collection, which may have come from the same source, is the type of *Antholyza quadrangularis* which Burman fil. described in 1768. The leaf architecture was understood, hence the name chosen for the species. Some 45 years later, when Ker Gawler saw what we now know as *G. abbreviatus*, he recognized the same four-flanged leaf and assumed this was Burman's *A. quadrangularis*. Accordingly, he transferred the species to *Gladiolus*, believing it to be the earlier name for *G. abbreviatus* (the transfer of *A. quadrangularis* to *Gladiolus* was mistakenly attributed to William Aiton by De Vos, 1976). Until 1932, when N.E. Brown examined the Iridaceae in the Burman Herbarium, the two species were thought to be one and the same. Brown realized they were different once he saw the type and so separated the two species, referring true *A. quadrangularis* to *Homoglossum* and *G. abbreviatus* to a second genus, *Petamenes* (see discussion of *G. abbreviatus*, page 307).

161. GLADIOLUS HUTTONII (N.E. Brown) Goldblatt & De Vos

PLATE 142

Gladiolus huttonii (N.E. Brown) Goldblatt & De Vos, *Bull. Mus. Hist. Nat., Paris* Sér. 4, Sect. B, *Adansonia* 11: 422 (1989). *Homoglossum huttonii* N.E. Brown, *Trans. Roy. Soc. S. Africa* 20: 278 (1932). Type: South Africa, Eastern Cape, Albany District, without date, *Hutton s.n.* (K, holotype).

huttonii, named in honour of Henry Hutton, soldier and administrator, who collected plants in the Eastern Cape in the latter half of the nineteenth century.

Synonymy

Homoglossum hollandii L. Bolus, *S. African Gard.* 23: 47 (1933). Type: hills at Bethelsdorp, 4 miles from Port Elizabeth (bought from a flower seller), July 1931, *Bolus s.n.* (BOL 20077, lectotype designated here; BOL, K, SAM, isolectotypes).
Homoglossum hollandii var. *zitzikammense* L. Bolus, *S. African Gard.* 23: 47 (1933). Type: South Africa, Eastern Cape, flats at Ratelsbosch, Aug. 1905, *Fourcade 59* (BOL, holotype and isotype).

Description

Plants 20–45(–80) cm high. *Corm* globose, 8–12 mm in diameter, the tunics of firm cartilaginous layers, with age decaying from the base into coarse vertical fibres or woody segments. *Cataphylls* pale and membranous, the upper green above the ground. *Leaves* three, the lower two basal, the lowermost longest, sheathing only near the ground, about as long as or slightly longer than the spike, the blade centric, the midribs raised into flanges as wide as half the blade width, thus cross-shaped in section, 2–3 mm wide, the second leaf sheathing the lower two-thirds of the stem, the blade short, more or less like the basal or the midribs not raised as much, the uppermost leaf inserted in the upper third of the stem, reaching to about the base of the spike, channelled for its entire length, the margins open to the base. *Stem* erect, very lightly flexed above the sheathing part of the second leaf, unbranched, c. 2 mm in diameter below the spike.

Spike erect or very slightly inclined, 3- to 4-, occasionally up to 8-flowered; *bracts* green or flushed purple, the outer 30–55 mm long, clasping the stem below, seldom exceeding the lower part of the tube, the inner slightly shorter than the outer, notched apically for c. 1 mm. *Flowers* orange-red, the dorsal tepal with broad translucent edges in the lower half, the lower tepals light orange to yellow in the lower half, sometimes yellow on the reverse, the reverse of the tube with deep red along the tepal sutures, unscented; *perianth tube* 50–53 mm long, narrow and cylindric below for 20–32 mm, abruptly expanded above into a tubular upper part 22–33 mm long, 5 x 6.5 mm wide, ascending; *tepals* unequal, broadly lanceolate, the dorsal largest, ascending, 25–40 x 15–25 mm, the upper laterals 22–32 x 12–18 mm, directed forward below, patent in the upper half, the lower three tepals much smaller, the lower laterals 12–20 x 10–14 mm, the lower median 13–22 x 7–10 mm. *Filaments*

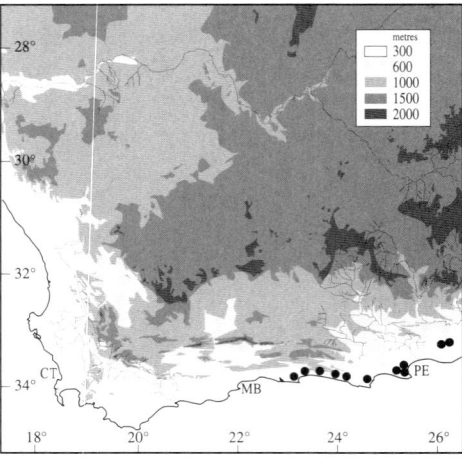

32–45 mm long, exserted 13–15 mm from the tube; *anthers* 8–10 mm long, orange, the pollen yellow. *Ovary* cylindrical, 8–10 mm long; *style* arching over the stamens, dividing at or up to 3 mm beyond the anther apices, the branches c. 4 mm long, very broad in the upper half. *Capsules* elongate-ellipsoid, 38–40 mm long, as long as or slightly longer than the outer bracts; *seeds* ovate, 5–6 x 3.5–4.5 mm, usually broadly and evenly winged, transparent light brown, the seed body darker brown. *Chromosome number* $2n = 30$.

Flowering time August to mid-September.

Distribution and biology

The only member of series *Homoglossum* that occurs in Eastern Cape Province, South Africa, *Gladiolus huttonii* extends along the southern coastal plain from Plettenberg Bay eastward to East London. It ranges inland a short distance and is found in the Albany District between Highlands and Coldstream near Grahamstown. Plants grow in fertile soils, often in sandy loam, light clay or on the moderately fertile soils derived from sandstones of the Witteberg Series.

The floral morphology, including the long perianth tube with a wide cylindrical upper part, prominent dorsal tepal and reduced lower tepals suggest that *Gladiolus huttonii* is adapted for pollination by sunbirds. The floral ecology remains to be studied.

Diagnosis and relationships

Gladiolus huttonii is usually readily distinguished by its large orange-red flowers, the lower tepals marked with yellow, and the wide upper part of the perianth tube with characteristic deep red lines on the outer surface. The tube is 50–53 mm long with the wide cylindric upper part 20–33 mm long, and the tepals are broadly lanceolate. Plants have only three foliage leaves, the blades of which are cross-shaped in transverse section, the midribs being raised above the surface into broad flanges. This character is shared with three other species of series *Homoglossum*, among which *G. huttonii* can be recognized by having the upper lateral tepals only slightly shorter than the dorsal, and the lower three tepals reduced in size and slightly more than half as long as the dorsal. The deep red lines on the reverse of the tube and the yellow markings on the lower tepals are also unique in series *Homoglossum*.

Within the series, *Gladiolus huttonii* appears to be most closely related to *G. fourcadei* and may sometimes be confused with that species. *Gladiolus fourcadei* has smaller flowers, the dorsal tepal 15–20 mm long and the perianth tube 35–46 mm long. More importantly, the lower tepals are only half as long as the dorsal, or even smaller, 7–9 mm long compared with the dorsal tepal, 15–20 mm long. In *G. huttonii* the dorsal tepal is 25–40 mm long and the lower tepals are 13–22 mm long.

History

Apparently first collected by C.F. Ecklon and C.L. Zeyher near the Swartkops River on their expedition to the eastern Cape Colony in 1831–1832, *Gladiolus huttonii* was not at first recognized as a distinct species. A subsequent collection from the Albany District, made by the soldier and administrator, Henry Hutton, in the mid-nineteenth century likewise attracted no immediate attention and specimens of the species were included by J.G. Baker in *Antholyza revoluta*, the name he used for *G. watsonius*.

Only in 1932, when N.E. Brown was trying to bring some order to the species then included in the heterogeneous genus *Antholyza*, did it become evident that this was indeed a distinct species. Brown placed the species in the genus *Homoglossum* as *H. huttonii*. Only a year later, in 1933, H.M.L. Bolus described *H. hollandii* based on specimens collected at St Albans near Port Elizabeth by the businessman and enthusiastic naturalist, F.H. Holland, and on additional specimens from Bethelsdorp, bought from a flower seller. Bolus thought her *H. hollandii* was separate from *H. huttonii* by virtue of a somewhat longer and nearly straight perianth tube. However, additional collections have shown that there is no basis for *H. hollandii* and that it is the same species as Brown's *H. huttonii*. The two species were united by M.P. de Vos (1976) in her revision of *Homoglossum* together with *H. hollandii* var. *zitzikammense* L. Bolus, a robust variant with 3–7 flowers per spike. With the reduction of *Homoglossum* by Goldblatt & De Vos in 1989, *H. huttonii* was transferred to *Gladiolus*.

Gladiolus fourcadei (L. Bolus) Goldblatt & De Vos, *Bull. Mus. Hist. Nat., Paris* Sér. 4, Sect. B, *Adansonia* 11: 422 (1989). *Antholyza fourcadei* L. Bolus, *Ann. Bolus Herb.* 4: 118 (1927). *Homoglossum fourcadei* (L. Bolus) N.E. Brown, *Trans. Roy. Soc. S. Africa* 20: 279 (1932). De Vos, *J. S. African Bot.* 42: 352 (1976). Type: South Africa, Eastern Cape, near Uniondale, headwaters of Wagenboom River, c. 750 m (1900 ft), Sept. 1922, *Fourcade 2337* (BOL, lectotype here designated; K, NBG, isolectotypes).

fourcadei, named for H.G. Fourcade, French-born botanist who surveyed the forests of the southern Cape and later the entire flora of the region.

Description

Plants 40–60 cm high. *Corm* globose-obconic, 10–12 mm in diameter, the tunics of firm cartilaginous layers, with age decaying from the base into soft, vertical fibres. *Cataphylls* pale and membranous, the uppermost reaching 5–7 cm above the ground and then green or flushed with purple. *Leaves* three (rarely two), the lower two (or one) basal, the lowermost longest, diverging from the stem a short distance above the ground, reaching to about the middle of the spike, the blade centric, the midrib raised into flanges as wide as half the leaf width, the blade thus cross-shaped in transverse section, the margins and the edges of the midrib wings thickened, the second leaf sheathing the lower half of the stem, usually with a short non-sheathing channelled portion, the uppermost leaf inserted on the upper third of the stem, largely or entirely sheathing, the margins fused below. *Stem* erect,

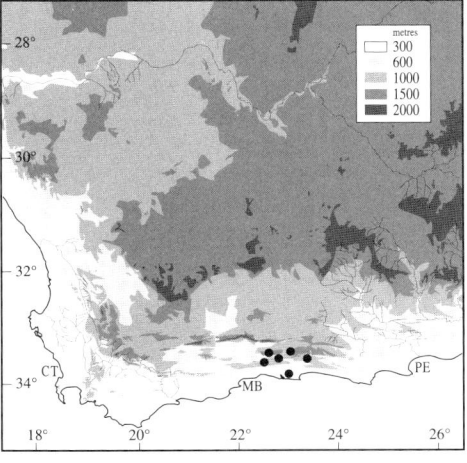

flexed outward above the sheath of the second leaf, becoming erect again, unbranched, c. 2.3 mm wide below the spike.

Spike erect and more or less straight, 3- to 5-flowered; *bracts* pale brownish green, sometimes lightly flushed with purple, the outer 42–60 mm long, enclosing the lower portion of the upper part of the tube, the inner about two-thirds as long as the outer, minutely forked apically, twisted to lie against the outer bract. *Flowers* yellowish green on the tepals, the upper three tepals flushed and veined with dusky red, the reverse of the tepals and the tube red, somewhat streaked on the tepals, the lower three tepals unicoloured or speckled with minute dark red spots in the lower half, unscented; *perianth tube* 35–46 mm long, slender and cylindric below for 15–18 mm, curving abruptly into a wider cylindric part 20–28 mm long, c. 5 mm in diameter, ascending to almost horizontal; *tepals* unequal, the dorsal ovate, ascending, 15–20 x c. 12 mm, the upper laterals patent, 14–15 x c. 10 mm, the lower three tepals much smaller, patent, subequal, c. 9 x 7 mm, directed downward. *Filaments* 26–32 mm long, exserted 4–8 mm from the tube; *anthers* 8–10 mm long, light purple, the pollen yellow. *Ovary* cylindric, c. 8 mm long; *style* extending horizontally over the stamens, dividing opposite the upper third of the anthers, the branches 4 mm long, gradually and weakly expanded above, the apices just exceeding the anthers. *Capsules* and *seeds* unknown. *Chromosome number* unknown.

Flowering time occasionally in September, mostly October to November.

Distribution and biology

A rare and poorly collected species of the southern African winter-rainfall region, *Gladiolus fourcadei* is centred in the Uniondale District of the southern Cape. Most collections are from the upper Longkloof but there is also a single record of the species from the coast near Knysna and one more from the southern slopes of the Kammanassie Mountains which lie between Uniondale and Oudtshoorn. Plants appear to favour heavy soils, either clay or clay loam, at the sandstone-shale transition and grow in renosterveld or transitional fynbos.

The relatively small, greenish yellow flowers of *Gladiolus fourcadei*, with their reddish veins and spots, are most likely adapted for

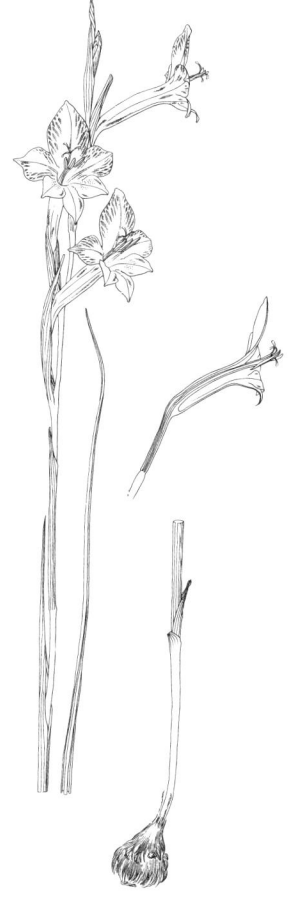

pollination by sunbirds. Their comparatively small size, delicate appearance and predominantly greenish yellow colouring are, however, somewhat unusual with this mode of pollination and its floral ecology needs to be investigated.

Diagnosis and relationships

A rather atypical member of series *Homoglossum*, *Gladiolus fourcadei* does not have the predominantly scarlet to dark red flowers of other members of the alliance. Instead, the upper three tepals are yellowish green, flushed and veined with dusky red, and the lower tepals are either uniformly yellow-green or are minutely dotted with dark red in the lower half. The reverse of the tepals and the tube are red and, as in related species, the flowers are unscented. In other respects *G. fourcadei* conforms closely to series *Homoglossum* in having three foliage leaves, the blades with thickened margins and midrib. The leaf blades are cross-shaped in transverse section, and the leaf architecture plus the ovate tepals suggest a close relationship with the larger-flowered *G. huttonii* in particular. This species is altogether more robust and has red flowers with the lower tepals more or less orange inside and yellow on the reverse, and it is also distinctive in having dark red lines on the outside of the tube.

History

Described in 1927 by H.M.L. Bolus, *Gladiolus fourcadei* was named in honour of the French-born botanist, G.H. Fourcade, who appears to have made the first collection of this rare species. There are just a handful of later collections and the species is now presumably threatened by the farming activity that has encroached on much of its original range.

Following the generic conventions established in *Flora Capensis*, Bolus assigned *Gladiolus fourcadei* to *Antholyza*. In 1931 N.E. Brown transferred this and several of its close relatives to *Homoglossum* where it remained until that genus was merged with *Gladiolus* in 1989.

163. GLADIOLUS ABBREVIATUS Andrews

Slangkop, suikerkannetjie

PLATE 144

Gladiolus abbreviatus Andrews, *Bot. Repository* 3: pl. 166 (1801). Goldblatt & De Vos, *Bull. Mus. Hist. Nat., Paris* Sér. 4, Sect. B, *Adansonia* 11: 422 (1989). *Antholyza abbreviata* (Andrews) Persoon, *Synopsis Plantarum* 1: 42 (1805). *Petamenes abbreviatus* (Andrews) N.E. Brown, *Trans. Roy. Soc. S. Africa* 20: 276 (1932). *Homoglossum abbreviatum* (Andrews) Goldblatt, *J. S. African Bot.* 37: 443 (1971). De Vos, *J. S. African Bot.* 42: 353 (1976). Type: South Africa, Western Cape, without precise locality or collector, cultivated in Great Britain at Hammersmith, illustration in Andrews, *Bot. Repository* 3: pl. 166 (1801).

abbreviatus = abbreviated, referring to the very short lower tepals, only 3–6 mm long in comparison to the long dorsal which is up to 30 mm long.

Synonymy

[*Gladiolus quadrangularis* sensu Ker Gawler, *Curtis's Bot. Mag.* 16: pl. 567 (1802). Baker, *Fl. Capensis* 6: 166 (1896), not *G. quadrangularis* (Burman fil.) Ker Gawler.]

[*Petamenes quadrangularis* sensu J.W. Loudon, *Ladies' Flower-Garden Ornamental Bulbous Plants* 42, pl. 8 (1841), but not according to the basionym, *Antholyza quadrangularis* Burman fil.]

Description

Plants 35–55 cm high. *Corm* globose-obconic, 10–12 mm in diameter, the tunics of firm cartilaginous layers, with age decaying from the base into vertical fibres. *Cataphylls* pale and membranous below the ground, reaching up to 12 cm above the ground and then dark purple. *Leaves* three, the lower two basal, the lowermost longest, reaching or shortly exceeding the top of the spike, the blade centric, 3–4 mm wide, the midrib raised into ridges as wide as half the leaf width, the margins lightly thickened, the blade thus cross-shaped in transverse section, the second leaf sheathing the stem almost to the base of the spike and overlapping the third leaf, with a short filiform blade up to 50 mm long, the uppermost leaf inserted on the upper quarter of the stem, entirely sheathing and bract-like, often dull purple, the lower margins usually free almost to the base. *Stem* straight and erect, unbranched, c. 3 mm in diameter below the spike.

Spike erect, nearly straight, 4- to 6-, occasionally to 9-flowered; *bracts* greenish grey flushed with dark red, or olive green flushed with brown or orange, the outer 45–60(–70) mm long, the inner half to two-thirds as long as the outer, acute or barely notched at the apices. *Flowers* predominantly red to orange, the dorsal tepal flushed brownish and pale to almost transparent on the lower edges, upper lateral and lower tepals either darker red, brown or green, unscented; *perianth tube* 40–52 mm long, the lower half narrow and cylindric, ascending, 17–25 mm long, abruptly expanded into a wide cylindric upper part, 24–30 mm long, ascending, 5–6 mm in diameter; *tepals* unequal, the dorsal longest, elliptic, 16–31 x 11–15 mm, extended horizontally, upper laterals joined to the dorsal for c. 3 mm, ovate and apiculate, 7–11 mm long, directed forward, the lower laterals deltoid, 5–6 mm long, lower median 3–4 mm long. *Filaments* 28–44 mm long, exserted 8–17 mm from the tube; *anthers* 8–12 mm long, dull reddish, the pollen yellow. *Ovary* oblong, 6–7 mm long; *style* arching over the stamens, dividing opposite the upper third of the anthers, the branches 4–6 mm long, weakly expanded above, the apices just overtopping the anthers. *Capsules* elongate-elliptic, (20–)30–38 mm long; *seeds* ovate, usually tapering at one end, 6–7 x c. 4 mm, broadly and evenly winged, yellowish brown. *Chromosome number* 2n = 30.

Flowering time mid-June to mid-August, occasionally in September.

Distribution and biology

Gladiolus abbreviatus is confined to clay and shale banks and slopes and occurs in a fairly restricted area of the southwestern Cape. The range extends along the rolling plains south of the Riviersonderend Mountains from a short distance west of Caledon eastward to Stormsvlei. Plants are most common on wetter, more sheltered south-trending slopes. The species grows in renosterveld among low shrubs, less commonly among grass when most of the native flora has been eliminated. Plants are most conspicuous

when the surrounding bush is heavily grazed or partly cleared.

The flowers of *Gladiolus abbreviatus* are adapted for pollination by sunbirds, and produce large quantities of sweet nectar typical of bird-pollinated species. Visits have been reported by the malachite sunbird, *Nectarinia famosa*, presumably the pollinator. *Gladiolus abbreviatus* frequently co-occurs with other red-flowered species also pollinated by sunbirds, including *Lessertia* (= *Sutherlandia*) *frutescens* (Fabaceae) and *Watsonia aletroides* (Iridaceae), and we have even encountered *G. teretifolius* growing close to *G. abbreviatus* at one site.

Diagnosis and relationships

One of the oddest species of *Gladiolus*, *G. abbreviatus* has flowers which look very little like those of any other species of the genus and it is no wonder that the generic disposition of this species should have caused so much confusion. The tepals are grossly unequal and the flowers are predominantly dark red to scarlet with the lower tepals tipped with dark red, brown or dull green. The dorsal tepal, usually 18–30 mm long, is elliptic and concave, the upper laterals are 7–11 mm long and the lower tepals are a mere 3–6 mm long, tiny by comparison with the dorsal.

Gladiolus abbreviatus is the only species of the genus in southern Africa in which the floral bracts may be coloured. They are pale to olive green flushed with red, brown or orange, and we assume they are part of the signal for pollinators. A few species of *Gladiolus* in tropical Africa also have coloured bracts, notably *G. dichrous* and *G. abyssinicus*, but these species are only

distantly related to *G. abbreviatus*.

The peculiar flowers are contrasted by a vegetative and fruit morphology not at all unusual and which places the species squarely in series *Homoglossum*. Plants have three slender foliage leaves, and the blades are cross-shaped in transverse section. Like several species of series *Homoglossum*, *Gladiolus abbreviatus* has rather large capsules, usually 30–38 mm long, occasionally shorter. The broadly winged seeds of *G. abbreviatus* leave no doubt about its generic position. The very reduced lower tepals suggest that it is most closely related to *G. fourcadei* which has the lower tepals 7–9 mm long.

History

The striking and highly specialized *Gladiolus abbreviatus* makes its appearance in botanical literature in 1801 in Henry Andrews' *Botanists' Repository*, an eclectic collection of articles accompanied by hand-coloured reproductions of plant species cultivated in Britain, many of them new to science. Andrews recorded neither the collector nor the wild origin of the plants. A year later John Ker Gawler provided the text for another painting of the species, this one in *Curtis's Botanical Magazine*. There, he identified the plant as *G. quadrangularis*, a combination based on N.L. Burman's *Antholyza quadrangularis*. Ker Gawler was mistaken in his identification of Burman's species as the same as the one illustrated in *Curtis's Botanical Magazine*. The extreme reduction of the lower tepals and relatively short upper laterals combine to give the species such a distinctive appearance that the eccentric botanist, R.A. Salisbury, was moved to pro-

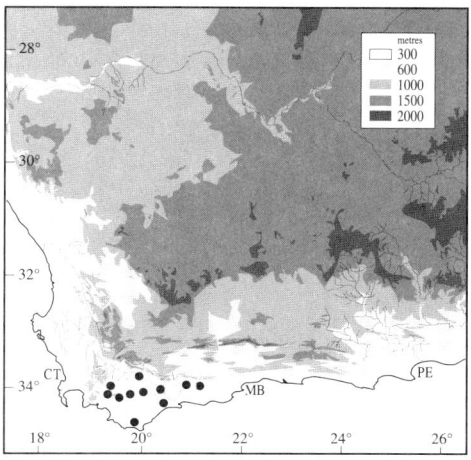

pose a new genus, *Petamenes*, for the species. The generic name was actually validated by Jane Loudon in her *Ladies' Flower Garden of Ornamental Bulbous Plants* (Loudon, 1841). Loudon called the species *P. quadrangularis*, thus perpetuating Ker Gawler's misapplication of the younger Burman's *Antholyza quadrangularis*.

The misapplication of the specific epithet *quadrangularis* was continued by Baker and *Gladiolus abbreviatus* was treated as *Antholyza quadrangularis* in *Flora Capensis* (Baker, 1896). N.E. Brown realized that *G. abbreviatus* and *G. quadrangularis* were different species after examining the type specimen of the latter in the Burman Herbarium (Brown, 1929) and in 1932 he transferred *G. abbreviatus* to *Petamenes*, one of the genera to which he referred most species we now regard as belonging to series *Homoglossum*. Evidently ignorant of Loudon's valid publication of the genus, Brown actually redescribed *Petamenes* at this time and authorship of the genus is often attributed to him.

EXCLUDED SPECIES

For the identity of species in southern Africa referred to *Gladiolus* since 1753 we refer readers to the list provided by Lewis et al. (1972: 299–303). We include two additions to this list as follows:

Gladiolus bakeri

Klatt in Durand & Schinz, *Conspectus Florae Africae* 5: 213 (1895), a new name for *G. micranthus* Baker, *Handbook Irideae* 212 (1892), an illegitimate homonym, not

G. micranthus Stapf, from Eurasia. Type: Botswana, Leshumo valley, *Holub s.n.* (location of the type unknown).

A small-flowered plant, almost certainly a species of *Gladiolus*. The type is not at the Kew or Bolus herbaria where other collections made by Emil Holub and seen by J.G. Baker are preserved, and it has not been located elsewhere. In the absence of a specimen, the species cannot be identified.

Gladiolus lewisiae

Obermeyer, *Fl. Pl. Africa* 40: pl. 1596 (1970); Lewis et al., *J. S. African Bot.*, Suppl. 10: 117 (1972).

Believed to be a hybrid between *G. tristis* and *G. caryophyllaceus* (Goldblatt & Vlok, 1989).

Glossary

abaxial: describing the side of an organ furthest from the vertical axis

actinomorphic: describing a regular flower, i.e., radially symmetric

acuminate: having an acumen, a short, comma-like structure at the apex of an organ

acute: having a sharply angled apex

adaxial: describing the side of an organ nearest the vertical axis

allopatric: growing apart, i.e., geographically isolated

anatropous: describing an ovule curved back on itself

androecium: the male part of the flower (filaments and anthers)

angiosperm: an informal term for a flowering plant

anther: part of the male organ, consisting of two lobes (thecae), each with two chambers that contain the pollen

anthesis: the time of opening of the flower

aperture: in a pollen grain, the opening or pore in the wall through which the pollen tube emerges

apiculus: a small, usually sharply pointed structure borne at the apex of an organ, e.g., leaf, tepal, anther; *adj.* apiculate

apomorphic: describing a specialized or derived characteristic; *see also* **synapomorphic**

arcuate: arching, usually upward

attenuate: with an elongated and tapering apex

auriculate: with ear-like lobes

autogamous: self-pollinating, without an external pollinating agent

axil: the angle between the leaf and stem axis; also lateral as opposed to terminal; *adj.* axillary

basifixed: attached basally; *see also* **subbasifixed**

bifid: divided into two for a short distance

bilabiate: two-lipped, usually of a vertically oriented flower with clearly defined upper and lower tepals

blade: the expanded part of the leaf or lamina, as opposed to the petiole; in *Gladiolus*, the sheath

bract: the modified leaf associated with a flower or inflorescence, sometimes much reduced in size, with a leafy or non-leafy texture

campanulate: bell-shaped

capsule: a dry, dehiscent fruit, as found in *Gladiolus* and most other Iridaceae, comprising three or more fused carpels, and containing two or more seeds per chamber

carinate: keeled, i.e., with a fold or ridge along the midline

cataphyll: modified leaf without a blade, produced at the base of a plant, often underground

cauline: relating to the stem

centric: describing a leaf which is radially symmetric in transverse section

chalaza: anatomical part of the ovule and seed, the point at which the vasculature enters the ovule, often prominent in the mature seed

chlorenchyma: plant tissue, usually in the leaf, containing chloroplasts, the main site of photosynthesis

chromosome: cell organelle located in the nucleus that contains genetic information

clade: evolutionary lineage defined by derived characters

columellate: in pollen exine, with small pillars, the pillars supporting the tectum (roof layer)

connective: sterile tissue between the pollen-bearing chambers of the anther, sometimes extended apically into a distinctive appendage

conspecific: the same species

cordate: heart-shaped

coriaceous: having a leathery texture

corm: underground storage organ consisting of tissue derived from the base of the flowering stem

corm tunic: fibrous, papery or leathery covering of the corm, derived from the bases of cataphylls and sometimes leaves

cucullate: hooded or spooned

cuticle: thin impermeable layer external to the epidermal cells

cytology: study of the cells, often specifically of the chromosomes

decurrent: continuing downward (onto another structure)

dehiscence: method of splitting, e.g., of capsules, anthers; *adj.* dehiscent

deltoid: D-shaped or more or less triangular

distal: distant from an originating point

distichous: arranged in two opposed ranks

dorsal: upper

dysploid: chromosome number differing from the norm or basic number for a group

edaphic: relating to the soil

ellipsoid: having a solid elliptical shape, rounded but tapering symmetrically at opposite ends

emarginate: with the apex notched or indented

endemic: restricted to a particular geographic area (sometimes used as a noun)

endothecium: tissue in the anther with distinctive wall thickenings

epidermis: distinct plant tissue layer, usually one cell thick

exine: outer wall of the pollen grain

falcate: shaped like a sickle

fasciated: flattened, often in an abnormal or pathological way

filament: part of the male organ of the flower, the stalk bearing the anther, usually thread-like

filiform: thread-like, i.e., very slender

flavone: class of flavonol compounds of simple structure

flavonol: group of chemical compounds in plants

flexuose: bent abruptly, zigzag

genome: entire genetic constitution of an organism

girder: strengthening structure of the leaf in which thickened cells extend from the vascular trace to the epidermis

glabrous: smooth, without hairs

glaucous: with a waxy or whitish bloom on the surface

globose: having a solid round shape (like a globe)

glycoside: primitive type of flavonol

gynoecium: the female part of the flower, composed of one or more units called carpels (ovary, style, and stigma)

herbarium: museum collection of preserved plant specimens

herkogamous: referring to the spatial separation of the receptive part of the stigma from pollen-bearing structures

hispidulous: with minute sturdy hairs

holotype: the single designated specimen or element on which a species name is based

homologous: derived from the same structure, used in the evolutionary or phylogenetic sense

homonym: specific name matching exactly one already given to a plant in the same genus

hyaline: translucent, clear

hypocrateriform: describing a flower with a long narrow tube and tepals spread at right-angles

imbricate: overlapping

inaperturate: describing pollen grains without pores or openings of any kind

inflorescence: a cluster of flowers on a stem

involute: with the margins curving inward

isobilateral: both surfaces of a leaf identical

isolectotype: duplicate of a lectotype

isotype: duplicate of a holotype

karyotype: appearance of the chromosome complement

lanceolate: shaped like a spear; *see also* **oblanceolate**

lectotype: type designated after the original publication from among several elements available

linear: narrow and with sides parallel

locule: chamber of the ovary

meristem: area of active cell division, giving rise to cells that form tissues and organs

mesophyll: green, chloroplast-filled tissue

monophyletic: taxonomic group sharing a common ancestor

monosulcate: with one sulcus, the narrow groove which is often the germination pore of a monocot pollen grain

morphology: external appearance, usually at macroscopic level

muri: walls, often of the pollen wall sculpturing

obconic: a solid cone shape, widest at the top

oblanceolate: shaped like a spear, with the broadest end at the base

obovate: egg-shaped in outline, with the widest end at the top

obovoid: egg-shaped of a solid body, with the widest end at the top

obtuse: with a rounded or bluntly angled apex

ontogeny: the developmental process of an organ

operculum: literally, a lid; the discrete covering of the pollen grain aperture

ovary: part of the gynoecium that contains the placentas and ovules

ovate: egg-shaped in outline

ovoid: egg-shaped of a solid body

ovule: the organ within the ovary that bears the embryo sac and egg cell

pandurate: fiddle-shaped

papilla: a small, round protuberance extending shortly above the cell surface; *adj.* papillate

papyraceous: with a papery texture

paraphyletic: a natural (monophyletic) group which does not include all of the descendents of the common ancestor, e.g., a genus with one or more genera nested within it

parenchyma: tissue consisting of large, rounded, thin-walled cells

patent: spreading, outspread; *see also* **subpatent**

pedicel: the stalk of an individual flower; *adj.* pedicellate

peduncle: the stalk of an inflorescence

perforate: describing a pollen grain with the exine having small, widely spaced pores (perforations)

perianth: the petal-like parts of a flower, used when the calyx (green rather than petaloid) and the corolla are identical in texture and often colour

periclinal: parallel to the external surface

pericycle: tissue between the epidermis and vasculature, usually in stems

petaloid: with a texture like a petal

phenology: relating to timing, e.g., the open and closing time of a flower; *adj.* phenological

phylogeny: relating to evolutionary relationships; *adj.* phylogenetic

pilose: hairy, specifically with short hairs

plesiomorphic: describing an unspecialized or primitive characteristic; *see also* **symplesiomorphic**

polyphyletic: describing a group of organisms of any rank that do not share an immediate common ancestor, i.e., having multiple origins

polyploid: having more than two sets of the basic chromosome complement

proboscis: the elongated mouth part of some insects, enclosing the tongue; *adj.* proboscid

protologue: the formal description of a species

proximal: close to an originating point

pseudomidrib: the central and main vein of the unifacial leaf of many Iridaceae, so named to indicate that it is not homologous with the midrib of bifacial leaves of other plants

puberulous: with minute hairs

pubescent: with hairs; *noun* pubescence

reticulate: netted

retuse: describing a flat or solid structure with a broad end, the apex of which is sunken

rhizome: a creeping stem, usually underground, filled with stored food, capable of undergoing a dormant period and producing shoots for a new growth cycle

sagittate: shaped like an arrow

scabrae: small raised surface sculpturing, usually in a regular pattern; *adj.* scabrid

scarious: dry and papery

sclerenchyma: tissue consisting of cells with heavily thickened walls

secund: facing to one side, often relating to flowers on a spike

sensu: in the sense of, or as interpreted by

septa: walls, as those separating the locules of the ovary

septal nectary: a nectar-producing gland located within the septa of the ovary

sessile: without a stalk, stemless

spathulate: like a spatula, markedly broad and rounded at one end

stamen: male organ within the flower

stigma: apical part of the style and receptive surface for pollen grains; *adj.* stigmatic

stolon: a slender process produced from the base of the plant, arising from the stem and terminating in a bud, bulb or cormlet

style: portion of the gynoecium linking the stigma(s) to the ovary. It is hollow, allowing the passage of pollen tubes

sub-: prefix denoting below, nearly or somewhat less than, e.g., subacute, almost acute

subbasifixed: attached shortly above the base; *see also **basifixed***

subpatent: partly or incompletely spreading

sulcus: a narrow groove, often specifically the germination pore of a monocot pollen grain; *adj.* sulcate

superposed: describing organs placed apart, well above one another

sympatric: growing together, sharing the same habitat

symplesiomorphic: describing a shared primitive characteristic; *see also **plesiomorphic***

synapomorphic: describing a shared derived characteristic; *see also **apomorphic***

syntype: one of several elements on which a species name is based in the absence of a designated holotype

tannin: plant chemical, located in specific dark-coloured cells

taxonomy: study of nomenclature, relating to classification; *adj.* taxonomic

tectate: describing a particular pollen exine structure with a roof layer (tectum) supported by pillars

tepal: term for units of the petaloid part of the flower when the calyx (green outer segments of some flowers) is not distinguishable, as in the lilioid monocots

terete: round in cross-section

theca(e): the lobe(s) of the anther containing the pollen

trichome: surface outgrowth, including hairs of various types

trigonous: three-lobed to broadly three-angled

truncate: abruptly terminated at one end

unifacial: describing leaves in which the two surfaces have the same anatomical origin

unilateral: twisted to lie on one side

velutinous: velvety

verrucose: describing a rough surface with small, wart-like outgrowths

vicariant: an immediately related species

villous: hairy, specifically with long hairs

vlei: term used in southern tropical Africa and South Africa for a seasonally or permanently marshy area

windowed: describing a *Gladiolus* flower with a wide gap between the bases of the dorsal and upper lateral tepals so that in profile one can see through the flower

xeric: describing arid or semi-arid habitats

xeromorphic: describing a feature related to adaptation for dry habitats

xylem: plant tissue responsible for conducting water

xylem vessel: cell in the xylem with perforations in the walls which allow unrestricted water passage

zygomorphic describing a non-symmetric flower, e.g., bilaterally symmetric

BIBLIOGRAPHY

Baker, J.G. 1877. Systema Iridacearum. *J. Linn. Soc. Bot.* 16: 61–180.

Baker, J.G. 1892. *Handbook of the Irideae.* George Bell & Sons, London.

Baker, J.G. 1896. Iridaceae. Pp. 7–171 *in* W.T. Thiselton-Dyer (editor), *Flora Capensis*, Volume 6. Reeve & Co., Ashford.

Baker, J.G. 1898. Irideae. Pp. 337–376 and 573–578 *in* W.T. Thiselton-Dyer, *Flora of Tropical Africa*, Volume 7. Reeve & Co., London.

Baker, J.G. 1904. Iridaceae. Pp. 1003–1007 *in* H. Schinz, Beiträge zur Kenntniss der Afrikanischen Flora (neue Folge). *Bull. Herb. Boissier*, Sér. 2, 4.

Bamford, R. 1935. The chromosome number in *Gladiolus*. *J. Agric. Res.* 51: 945–950.

Bamford, R. 1941. Chromosome number and hybridisation in *Gladiolus*. *J. Hered.* 32: 418–422.

Barnard, T.T. 1972. On hybrids and hybridization. Pp. 304–310 *in* G.J. Lewis et al., A revision of the South African species of *Gladiolus*. *J. S. African Bot.*, Suppl. 10.

Bentham, G. & J.D. Hooker. 1883. *Genera Plantarum*, Volume 3(2). Reeve & Co., London.

Bolus, F. 1917. *In* Novitates Africanae. *Contrib. Bolus Herb.* 2: 95–111.

Bolus, H. & A.H. Wolley-Dod. 1903. A list of the flowering plants and ferns of the Cape Peninsula, with notes on some critical species. *Trans. S. African Phil. Soc.* 14: 207–373.

Brown, N.E. 1928. The South African Iridaceae of Thunberg's herbarium. *J. Linn. Soc., Bot.* 48: 15–55.

Brown, N.E. 1929. The Iridaceae of Burmann's Florae Capensis Prodromus. *Kew Bull. Misc. Inform.* 1929: 29–139.

Brown, N.E. 1932. Contributions to a knowledge of the Transvaal Iridaceae. *Trans. Roy. Soc. S. Africa.* 20: 261–280.

Bullock, A.A. 1930. *Oenostachys*, a new genus of Iridaceae from East Africa. *Kew Bull. Misc. Inform.* 1930: 465–466.

Bullock, A.A. 1960. *Gladiolus trichonemifolius* forma *symmetranthus*. *Curtis's Bot. Mag.* 173: pl. 357.

Burman, N.L. 1768. *Prodromus Florae Capensis*. Cornelius Haak, Leiden.

Cowling, R.M. 1992. *The Ecology of Fynbos*. Oxford University Press, Cape Town.

Dalziel, J.M. 1937. *The Useful Plants of West Tropical Africa*. Crown Agents for the Colonies, London.

Delaroche, D. 1768. *Plantae Aliquot Novarum*. Leiden.

Delpierre, G.R. & N.M. du Plessis. 1973. *The Winter-growing Gladioli of South Africa*. Tafelberg, Cape Town.

De Vos, M.P. 1972. The genus *Romulea* in South Africa. *J. S. African Bot.*, Suppl. 9.

De Vos, M.P. 1976. Die Suid-Afrikaanse species van *Homoglossum*. *J. S. African Bot.* 42: 301–359.

De Vilmorin, R. & M. Simonet. 1927. Nombre des chromosomes dans les genres *Lobelia*, *Linum* et chez quelques autres espèces végétales. *Compt.-Rend. Hebd. Séances Mém. Soc. Biol.* 96: 166–168.

De Wet, G.C. & R.H. Pheiffer. 1979. *Simon van der Stel's Journey to Namaqualand in 1685*. Human & Rousseau, Cape Town.

Duncan, G.D. 1985. *Gladiolus carinatus* subsp. *parviflorus*. *Fl. Pl. Africa.* 48: pl. 1906.

Ecklon, C.F. 1827. *Topographisches Verzeichniss der Pflanzensammlung der C.F. Ecklon*. Reise Verein, Esslingen.

Ernst-Schwarzenbach, M. 1931. Contribution à l'étude des chromosomes chez le genre *Gladiolus* L. *Ann. Sci. Nat. Bot.*, Sér. 10, 13: 345–351.

Forbes, V.S. & J.P. Rourke. 1980. *Paterson's Cape Travels*. Brenthurst Press, Johannesburg.

Geerinck, D. 1972. Révision du genre *Gladiolus* L. (Iridaceae) au Zaïre, au Rwanda et au Burundi. *Bull. Jard. Bot. Nat. Belgique.* 42: 269–287.

Goldblatt. P. 1971. Cytological and anatomical studies on the southern African Iridaceae. *J. S. African Bot.* 37: 317–460.

Goldblatt, P. 1979. New species of Cape Iridaceae. *J. S. African Bot.* 45: 81–89.

Goldblatt, P. 1984a. New taxa and notes on southern African *Gladiolus* (Iridaceae). *J. S. African Bot.* 50: 449–459.

Goldblatt, P. 1984b. A revision of *Hesperantha* (Iridaceae) in the winter-rainfall area of southern Africa. *J. S. African Bot.* 50: l5–l4l.

Goldblatt, P. 1985. Systematics of the southern African genus *Geissorhiza* (Iridaceae–Ixioideae). *Ann. Missouri Bot. Gard.* 72: 277–447.

Goldblatt, P. 1989. Systematics of *Gladiolus* (Iridaceae) in Madagascar. *Bull. Mus. Hist. Nat., Paris*, Sér. 4, Sect. B, *Adansonia*. 11: 235–255.

Goldblatt, P. 1989. The Genus *Watsonia*. A Systematic Monograph. *Ann. Kirstenbosch Bot. Gard.* 17.

Goldblatt, P. 1990a. Phylogeny and classification of Iridaceae. *Ann. Missouri Bot. Gard.* 77: 607–627.

Goldblatt, P. 1990b. Status of the southern African *Anapalina* and *Antholyza* (Iridaceae), genera based solely on characters for bird pollination, and a new species of *Tritoniopsis*. *S. African J. Bot.* 56: 577–582.

Goldblatt, P. 1991. An overview of the systematics, phylogeny and biology of the southern African Iridaceae. *Contrib. Bolus Herb.* 13: 1–74.

Goldblatt, P. 1993. Iridaceae. Pp. 1–106 *in* G.V. Pope (editor), *Flora Zambesiaca*, Volume 12(4). Flora Zambesiaca Managing Committee, London.

Goldblatt, P. 1996. Gladiolus *in Tropical Africa*. Timber Press, Portland, Oregon.

Goldblatt, P. & T.T. Barnard. 1970. The Iridaceae of Daniel de la Roche. *J. S. African Bot.* 36: 291–318.

Goldblatt, P. & M.P. de Vos. 1989. The reduction of *Oenostachys*, *Homoglossum* and *Anomalesia*, putative sunbird-pollinated genera in *Gladiolus* L. (Iridaceae–Ixioideae). *Bull. Mus. Hist. Nat., Paris*, Sér. 4, Sect. B, *Adansonia*. 11: 417–428.

Goldblatt, P. & J.C. Manning. 1990. *Devia xeromorpha*, a new genus and species of Iridaceae–Ixioideae from the Cape Province, South Africa. *Ann. Missouri Bot. Gard.* 77: 359–364.

Goldblatt, P. & J.C. Manning. 1995. Phylogeny of the African genera *Anomatheca* and *Freesia* (Iridaceae–Ixioideae), and a new genus *Xenoscapa*. *Syst. Bot.* 20: 161–178.

Goldblatt, P. & J.C. Manning. 1996a. Reduction of *Schizostylis* (Iridaceae: Ixioideae) in *Hesperantha*. *Novon*. 6: 262–264.

Goldblatt, P. & J.C. Manning. 1996b. Phylogeny and speciation in *Lapeirousia* subgenus *Lapeirousia* (Iridaceae subfamily Ixioideae). *Ann. Missouri Bot. Gard.* 83: 346–361.

Goldblatt, P. & J.C. Manning. 1996c. Two new edaphic endemic species and taxonomic changes in *Gladiolus* (Iridaceae) of southern Africa, and notes on Iridaceae restricted to unusual substrates. *Novon*. 6: 172–180.

Goldblatt, P. & M. Takei. 1997. Chromosome cytology of Iridaceae base numbers, patterns of variation and modes of karyotype change. *Ann. Missouri. Bot. Gard*. In press.

Goldblatt, P. & J.H.J. Vlok. 1989. New species of *Gladiolus* (Iridaceae) from the southern Cape and the status of *G. lewisiae*. *S. African J. Bot.* 55: 259–264.

Goldblatt, P., P. Bernhardt & J.C. Manning. 1989. Notes on the pollination mechanisms of *Moraea inclinata* and *M. brevistyla* (Iridaceae). *Pl. Syst. Evol.* 163: 201–209.

Goldblatt, P., P. Rudall & J.E. Henrich. 1990. The genera of the *Sisyrinchium* alliance (Iridaceae–Iridoideae): phylogeny and relationships. *Syst. Bot.* 15: 497–510.

Goldblatt, P., A. Bari & J.C. Manning. 1991. Sulcus variability in the pollen grains of Iridaceae subfamily Ixioideae. *Ann. Missouri Bot. Gard*. 78: 950–961.

Goldblatt, P., M. Takei & Z.A. Razzaq. 1993. Chromosome cytology in tropical African *Gladiolus* (Iridaceae–Ixioideae). *Ann. Missouri Bot. Gard*. 80: 461–470.

Goldblatt, P., J.C. Manning & P. Bernhardt. 1995. Pollination biology of *Lapeirousia* subgenus *Lapeirousia* (Iridaceae) in southern Africa: floral divergence and adaptation for long-tongued fly pollination. *Ann. Missouri Bot. Gard*. 82: 517–534.

Gunn, M. & L.E. Codd. 1981. *Botanical Exploration of Southern Africa*. Balkema, Cape Town.

Hennig, W. 1966. *Phylogenetic Systematics*. University of Illinois Press, Urbana (translated by D.D. Davis & R. Zangerl).

Hepper, F.N. 1968. *Flora of West Tropical Africa*. 2nd edition, 3(1). Crown Agents for Oversea Governments, London.

Herbert, W. 1837. *Amaryllidaceae*. James Ridgway & Sons, London.

Herbert, W. 1842. *Gladiolus aequinoctialis*. *Bot. Reg*. 28, Misc: 85, no. 97.

Herbert, W. 1843. *Gladiolus splendens. Edwards' Bot. Reg*. 29 (new ser. 6), Misc: 46.

Hilliard, O.M. & B.L. Burtt. 1979. Notes on some plants of southern Africa, chiefly from Natal (VIII). *Notes Roy. Bot. Gard. Edinburgh*. 37: 284–325.

Hilliard, O.M. & B.L. Burtt. 1983. Notes on some plants of southern Africa, chiefly from Natal (X). *Notes Roy. Bot. Gard. Edinburgh*. 41: 299–319.

Hilliard, O.M. & B.L. Burtt. 1986. Notes on some plants of southern Africa, chiefly from Natal (XIII). *Notes Roy. Bot. Gard. Edinburgh*. 43: 189–228.

Hilliard, O.M., B.L. Burtt & A. Batten. 1991. *Dieramas, Hairbells of Africa*. Acorn Press, Johannesburg.

Irvine, F.R. 1930. *Plants of the Gold Coast*. Oxford University Press, London.

Jacot-Guillarmod, A.F.M. 1971. *Flora of Lesotho*. J. Cramer, Lehre, Germany.

Johnson, S.D. & W.A. Bond. 1994. Red flowers and butterfly pollination in the fynbos of South Africa. Pp. 137–148 *in* M. Arianoutsou & R. Grooves (editors), *Plant–Animal Interactions in Mediterranean-type Ecosystems*. Kluwer Academic Press, Dordrecht.

Ker Gawler (as Bellenden Ker), J. 1827. *Iridearum Genera*. P.J. de Mat, Brussels.

Klatt, F.W. 1863. Revisio Iridearum. *Linnaea*. 32: 689–784.

Klatt, F.W. 1882. Ergänzungen und Berichtigungen zu Baker's Systema Iridacearum. *Abh. Naturforsch. Ges. Halle*. 15: 44–404.

Klatt, F.W. 1895. Iridaceae. Pp. 143–230 *in* T. Durand & H. Schinz (editors), *Conspectus Florae Africae*, Volume 5. Jardin Botanique de l'État, Brussels.

Kuntze, O. 1898. *Revisio Generum Plantarum*, Volume 3. Arthur Felix, Leipzig.

Lewis, G.J. 1950. Iridaceae. Pp. 217–264 *in* R.A. Adamson & T.M. Salter, *Flora of the Cape Peninsula*. Juta, Cape Town.

Lewis, G.J. 1959. The genus *Babiana*. *J. S. African Bot*., Suppl. 3.

Lewis, G.J. 1966. Thunberg's South African species of *Gladiolus* – four name changes. *Bot. Notiser*. 1966: 286–296.

Lewis, G.J., A.A. Obermeyer & T.T. Barnard. 1972. A revision of the South African species of *Gladiolus*. *J. S. African Bot.*, Suppl. 10: 1–316.

Linder, H.P. 1985. Gene flow, speciation and species diversity patterns in a species-rich area: the Cape Flora. Pp. 53-57 *in* E.S. Vrba (editor), Species and Speciation. *Transvaal Mus. Monogr*. 4. Transvaal Museum, Pretoria.

Linnaeus, C. 1753. *Species Plantarum*. Salvius, Stockholm.

Loudon, J. 1840. *The Ladies' Flower-Garden of Ornamental Bulbous Plants*. William Smith, London.

Malaisse, F. & G. Parent. 1985. Edible wild vegetable products in the Zambezian woodland area: a nutritional and ecological approach. *Ecol. Food & Nutrition*. 18: 43–82.

Manning, J.C. & P. Goldblatt. 1995. Cupid comes in many guises: the not-so-humble fly and a pollination guild in the Overberg. *Veld & Flora*. 81(2): 50–52.

Manning, J.C. & P. Goldblatt. 1996. The *Prosoeca peringueyi* (Diptera: Nemestrinidae) pollination syndrome in southern Africa: long-tongued flies and their tubular flowers. *Ann. Missouri Bot. Gard.* 83: 67–86.

Manning, J.C. & P. Goldblatt. 1997. The *Moegistorhynchus longirostris* (Diptera: Nemestrinidae) pollination guild: long-tubed flowers and a specialized long-tongued fly-pollination system in southern Africa. *Pl. Syst. Evol.* In press.

Marais, W. 1973. Notes on African Iridaceae. *Kew Bull.* 28: 311–317.

Matthews, W.S., A.E. van Wyk & G.J. Bredenkamp. 1993. Endemic flora of the north-eastern Transvaal escarpment, South Africa. *Biol. Conserv.* 63: 83–94.

Mayr, G. 1942. *Systematics and the Origin of Species*. Columbia University Press, New York.

Miller, P. 1756–1759. *Figures of the Most Beautiful Plants, etc*. P. Miller, London.

Morrey, D.R., K. Balkwill & M.-J. Balkwill. 1989. Studies on serpentine flora: preliminary analyses of soils and vegetation associated with serpentinite rock formations in the southeastern Transvaal. *S. African J. Bot.* 55: 171–177.

Obermeyer, A.A. 1979. A new species of *Gladiolus* from the Transvaal. *Bothalia*. 12: 636.

Obermeyer, A.A. 1982. A new species of *Gladiolus*. *Bothalia*. 14: 78.

Persoon, H. 1805. *Synopsis Plantarum*, Volume 1. Cramer, Paris.

Plukenet, L. 1691. *Phytogeographia*. London.

Pole Evans, I.B. 1925. *Gladiolus cruentus*. *Fl. Pl. S. Africa*. 5: pl. 182.

Rebelo, A. 1987. Bird pollination in the Cape flora. Pp. 83-108 *in* A. Rebelo (editor), *A Preliminary Synthesis of Pollination Biology in the Cape Flora*. CSIR, Pretoria.

Rendle, A.B. 1938. *Gladiolus crispulatus*. *Curtis's Bot. Mag.* 147: pl. 8923.

Rudall, P. & P. Goldblatt. 1991. Leaf anatomy and phylogeny of Ixioideae (Iridaceae). *Bot. J. Linn. Soc.* 106: 329–345.

Salisbury, R.A. 1866. *The Genera of Plants*. Van Voorst, London.

Scott Elliot, G.E. 1891. Notes of the fertilisation of South African and Madagascar flowering plants. *Ann. Bot.* 5: 333–405.

Sealy, J.R. 1950. *Gladiolus ceresianus*. *Curtis's Bot. Mag.* 167: pl. 104, new series.

Simpson, G.G. 1953. *The Major Features of Evolution*. Columbia University Press, New York.

Thunberg, C.P. 1784. *Dissertatio de Gladiolo*. J. Edman, Uppsala.

Thunberg, C.P. 1800. *Prodromus Plantarum Capensium*. J. Edman, Uppsala.

Thunberg, C.P. 1823. *Flora Capensis*. J.A. Schultes, Stuttgart.

Van Raamsdonk, L.W.D. & T. de Vries. 1989. Biosystematic studies in European species of *Gladiolus* (Iridaceae). *Pl. Syst. Evol.* 165: 189–198.

Vogel, S. 1954. Blütenbiologische Types als Elemente der Sippengliederung. *Bot. Stud.* 1: 1–338.

Watt, J.M. & M.G. Breyer-Brandwijk. 1962. *Medicinal and Poisonous Plants of Southern and Eastern Africa*, 2nd edition. Livingstone Ltd., Edinburgh.

Wijnands, O. & P. Goldblatt. 1992. A volume of South African plant drawings for Johannes Burman (1707–1779). *Candollea*. 47: 357–366.

Wood, J.M. 1904. *Natal Plants*. 4: pl. 342.

LIST OF SUBSCRIBERS

COLLECTORS' EDITION

STEVE BALES
ALEX A. BARRELL
VIV & LORI BARTLETT
THE BRENTHURST LIBRARY
JOHN E. BRYAN, FELLOW OF THE INSTITUTE OF
HORTICULTURE
THE COMPTON HERBARIUM
CHARLES CRAIB
JALAL & KULSUM DHANSAY
DR H.R. DINGLE
MR & MRS J.M.R. DOWER
J.P. DU PLESSIS
FRIEDA DUCKITT
FRITZ ECKL
THE ERNEST OPPENHEIMER MEMORIAL TRUST
FERNWOOD PRESS (PTY) LIMITED
EUGENE & LALIE FOURIE
PETER FRANKS
E.S.C. GARNER
LYDIA GORVY
ALEC & CATHY GRANT
CHARLES EUGENE HARDMAN
BASIL HERSOV
YOSHITO IWASA
R.G. JEFFERY
G.E. JEWELL, NEW ZEALAND

CHRIS F. KLEISS
G.G. LESLIE
ANN C. MACMICHAEL
PETER MARTENS
ALLEN & CAROL MILLER
MISSOURI BOTANICAL GARDEN
ROBIN MOSER
DESMOND NIELSEN
HIROKI NUMATA
RAINER PAMPEL
PETERSFIELD NURSERIES
BERYL & LAWRENCE POSNIAK
RICHARD J. PRICE
PIETER & PAM STRUIK
AKIRA SUZUKI
DENNIS TSANG
JEAN TURCK
J.A. JANSE VAN RENSBURG
N.P. JANSE VAN RENSBURG
W. VAN RŸSWŸCK
M. & A. VAN SANDWYK
ILSE & STEPHAN WENTZEL
BEVERLEY WILSON
DERRIC H. WILSON
J.A. WINDELL

STANDARD EDITION

AFROFLORA NURSERY
F. ALBERS
IVONNE & OSWALD ALBERS
M. ALBERTS
ALBERTUS DELPORT-BIBLIOTEEK
C.D. ANDERSON
C. & R.H. ARCHER
MARION ARNOLD
AUCKLAND REGIONAL COUNCIL,
    NEW ZEALAND
DAVID & LINDA BALDIE
KEITH MITCHELL BALES
MARIJKE & KEN BALL
NIGEL BARKER
MALCOLM & PETA ANNE BASFORD
AURIOL BATTEN
CHRISTOPHER BATTEN
CLAIRE MARIANNE BATTEN
NICHOLAS ASHLEY BATTEN
THOMAS BATTEN
ANDRE VANDEN BAVIERE
P. ANNE BEAN
YVONNE BECKER
IAN & JOCELYN BELL
JEAN-LUC BESTEL
PROFESSOR GERHARD & ISOLDE
    BEUKES
DR ERICH BIGALKE

PROFESSOR JAMES W. BIZZELL
IAN R.M. BLACK
M. BLERSCH
BERT BONGERS
KATHY & ROB BORTHWICK
THE BOTANICAL SOCIETY,
    KIRSTENBOSCH BRANCH
C. BOUCHER
MAURICE BOUSSARD
RUPERT BOWLBY
PAUL D. BRINK
GILBERT BRISCOE
THE BRITISH GLADIOLUS SOCIETY
MARLEEN BRUWER
ANTON BRYANT
DR H.J. BRYNARD
BULB ARGENCE, DE JAGER
NINO & KARIN BURELLI
JOHN & SANDIE BURROWS
CHRIS, ANNELIZE, ANDRÉ & DEON
    BUYS
MRS AUDREY CAIN
R. CAMPBELL
DIANA CHAPMAN
M.J. & T.A. CHAPMAN
P.J. CILLIÉ
NIGEL COE
PAUL G. COETZEE

TERTIUS COETZEE
DES & NAUREEN COLE
ANN E. COLEMAN
GAEL & NEVILLE CONSTABLE
ALAN CORTIE
DR MARJORIE COURTENAY-LATIMER
C.L. COUSINS
GRAHAM & JILLIAN COX
PIERRE & FRANÇOISE CUÉNOD
FAMILLE E. DANDRIEUX
H.G. DANTU
DR JOHN DAVID
SANDRA FAY DE CONING
LINDA DE LUCA
NICHOLAS DE ROTHSCHILD
D.L.L. (DEE) DE SOUZA
DR & MRS C. DEACON
PROFESSOR & MRS G.R. DELPIERRE
ANNE DEW
JALAL & KULSUM DHANSAY
TONY DOLD
CLIFFORD DORSE
BARRY DU PLESSIS
E.P. DU PLESSIS
N.M. DU PLESSIS
PROFESSOR JAN G.H. DU PREEZ
JOHN & PAMELA DUFF
JIM DUGGAN

G.D. DUNCAN
LOUISE & PIETER DUYS
ANNE EARHART
URS EGGLI
MR W.A.D. ELLIOTT
R. ERNSTZEN
JOHN FANNIN
LIZ & ALEX FICK
ISTVAN FIGURA
FIRST NATIONAL BANK
D.J. FÖLSCHER
KIRBY W. FONG
FOREST FERNS CC
MALCOLM FOSTER
EUGENE & LALIE FOURIE
JOHAN & HENNY FOURIE
J.E. FRASER-JONES
WALTER W.B. FRIEDRICH
THINNES GABY
IAN GARLAND
MR GERRIT GERMISHUIZEN
FRED & SARAH GESS
R. GETTLIFFE
HOWARD & MONICA GIE
JAN GILIOMEE
DEANNE VOIGTS & MARK GILROY
ANDREW & MOIRAGH GIRDWOOD
TOSHINARI GODO

PETER & TESSA GOEMANS
PETER GOLDSCHMIDT
PAT & KARIN GOSS
JILL GOUGH-PALMER
DR J.E. GRANGER
ADIN GREAVES
DAVE GREEN
MICHAEL & MARIETTE GREGOR
GWYN GRIFFITHS
GROW WILD
GERALD A.H. GUY
RICHARD HALL
ADRIAAN HANEKOM, CALEDON
    FYNBOS KWEKERY
CHARLES EUGENE HARDMAN
E.M. HARTLEY
C.A.S. HAYNE
GEOFFREY A. HEDGECOCK
GAVIN & CATHY HELLSTRÖM
NICK HELME
DONALD HENLEY
BERT HENNIPMAN
HERMANUS BOTANICAL SOCIETY
INGRID SOPHIE HOERSCH, GERMANY
J.L. HOLMES
HORTUS BOTANICUS
    JOHANNESBURGENSIS
JAN & PROTEA HUBREGTSE
MR EINION HUGHES
DAVID HUMAN
C.R. HUNTING
L.R. HUNTING
M.S. HUNTING
IAN B. HUNTLEY
ERIKA HUNTLY
DR KATHLEEN IMMELMAN
MARY & GARY IRISH
MARY SUE ITTNER
NIELS H.G. JACOBSEN
ELODIE JANOVSKY
A.E. JEANES
MRS D.E. JEFFERIES
G.E. JEWELL, NEW ZEALAND
DR M.H. JOHANNSMEIER
PETER JOHNSTON
JEAN JORDAAN
GERALD JOSMAN
PAUL A.M. JOUBERT
SIDNEY H. KAHN
M. KIEN
MICHAEL W. KING
MIRIAM C. KIRK
COLIN & LESLEY KITCHENER
PETER & BARBARA KNOX-SHAW
AIKICHI KOBAYASHI
MORIJI KOBAYASHI
JOHAN KRIEL
NOELLINE KROON
DEON EN NELIE KÜHN
MRS U.E. KUUN
ERIC & ISOBYL LA CROIX
GREG & JEAN LABUSCHAGNE
NINA LANDMAN

MEV ANNETTE OTILLIA LEMMER
JEAN-MARIE & ASTRI LEROY
LONGWOOD GARDENS, INC.
PAT STUART LORBER
JOHAN W. LOUBSER
BETTY LOUW
CHRISTOPHER R. LOVELL
NICOLE LUDEKE
HEINER LUTZEYER
JAMES R. MABEE, TILLSONBURG,
    ONT.
ROGER M. MACFARLANE
O.J. MACKENZIE
ME. M. MALAN
MIMMIE MALHERBE
NEELS MARITZ
WILLIAM MASSYN
BRIAN MATHEW
THE MATLOCK FAMILY
DIANA MATLOCK
MR GORDON C. MATTHEWS
VICTORIA MATTHEWS
RON MCDONALD
DR C.R. MCDOWELL
DERICK MCKENZIE
CAMERON & RHODA MCMASTER
C. MCNEILL
ADVOCATE ABRI MEIRING
ROBERT & MARIJKE MIDDELMANN
WALTER MIDDELMANN
WILLIAM B. MILLER
JOHN D. MITCHELL
N.J. MITCHLEY
HIROKI MIYAZAKI
BUDDY & JENNY MOCKFORD
BARBARA ROSHAN MOORE
PATRICK MORANT
ANDRÉ MORENCY
J.W. MORRIS
TAKESHI MOTOZU
NIGEL JOHN MUDGE
NIGEL WAYNE MUDGE
RENÉ EN ELMARIÉ MULDER
R.M. MURRAY
NATIONAL BOTANICAL INSTITUTE,
    PRETORIA
PROFESSOR W. DU T. NAUDÉ
J.H. NEETHLING
DAVID NEWTON
DOUGLAS NEWTON
SARAH NEWTON
L-M. NICHOLLS
GEOFF, LYNNE & DOUGLAS
    NICHOLS
DESMOND NIELSEN
GERRIT NIEUWOUDT
MRS G.A. NIKSCHTAT
K. NIXON
DR BARBARA NORTON
HIROKI NUMATA
NERINE & JOHN OAKES
JO ONDERSTALL
O. OPPENHEIMER

NEVA Y. PAHLER
P.A. PALMER
DALE & ELIZABETH PARKER
DEAN PENDRIGH, NEW ZEALAND
ENG. MANUEL PITA, FUNCHAL,
    MADEIRA, PORTUGAL
DARREL C.H. PLOWES
DR THOMAS R. PRAY
N.W. PRIMICH
DR NEVIL QUINN
LORRAINE & ROGER RAAB
JOHN RAE
R.F. RAYMOND
ARDA & KOTIE RETIEF
P. RETIEF
D.D. RIACH
DAVE, CORLIA & SEAN RICHARDSON
STEPHEN & MARION RICHARDSON
G.A. RISSIK
DOUGLAS STUART ROBERTS
MR BASIL ROBINSON
CEDRIC ROCHÉ
H.P. ROHLANDT, SANDBAAI,
    HERMANUS
J.A. ROHLANDT, HEILFONTEIN,
    CALEDON
J.P. ROURKE
SAFARI TUINSENTRUM
K. SAHIN
ROBIN J. SANDY
SANTA BARBARA BOTANIC GARDEN
ROD & RACHEL SAUNDERS
H.L. SCHAARY
GORDON SCHACHAT
A.& P. SCHAEFER
ERNST SCHMIDT
R.J. SCHOLES
LIONEL SCHRODER
MICHAEL H.P. SCHURR
FREDERICK R. SELLMAN, LONDON
EDUARD SEVENSTER
DR R. SEVIN, LAUSANNE,
    SWITZERLAND
G.M. SIMMONS
GENARD SIZER
D. & H. SKAWRAN
ANSIE, DENNIS & JENNIFER SLOTOW
KEITH & DOROTHY SMITH
ROSEMARY SMUTS
THE SOUTH AFRICAN FLOWER
    GROWERS ASSOCIATION
SOUTHERN AFRICAN SOCIETY FOR
    HORTICULTURAL SCIENCES
HUMPHREY & ELAINE STAINTHORPE
MARY & ROBERT STOBIE
SANDRA JACQUELINE STOWELL
SANDY STRETTON
SUMMERFIELD'S INDIGENOUS BULBS
    & SEED
SUNBURST FLOWER BULBS
QUENTIN SUSSMAN
SUTHERLAND SEEDLINGS
ANDREW C. SUTHERLAND

AKIRA SUZUKI
YASUMASA TAKATSU
SUMITRA TALUKDAR
HIROYUKI TANAKA
COLIN TEDDER
MRS M.L. THOMAS
JENNETTA ANNE TILNEY
PETER & SUE TIMM
DAVID L. TRAYLOR
DAVE & JO TRICKETT
TERRY H. TRINDER-SMITH
ROD & FRANCHESCA TRITTON
DENNIS TSANG
DAVID & EILEEN TUFENKIAN
ROBERT TURLEY
DI & BILL TURNER
JAMES W.D. TURNER
MRS C.J. TURNLEY-JONES
PRIMROSE & MICHAEL UPWARD
A.L. VAN COLLER
DR ERIKA VAN DEN BERG
DICK & LIZ VAN DER JAGT
ANDRÉ & BIBI VAN DER MERWE
TROOS VAN DER MERWE
KIM VAN DER RIET
JOHAN & ANNATJIE VAN DER SANDT
ANNE & ERROL VAN GREUNEN
H. VAN KERKEN
M. & M.A. VAN RIJSWIJCK
HANS VAN ROOYEN
BRIAN & JANE VAN WILGEN
MISS H.J. VANDERPLANK
MICHAEL G. VASSAR
VEGMOFLORA (PTY) LIMITED
SUSAN EN VELLIES VELDSMAN
NATASCHA VISSER
WILNAND VLOK
VOGELGAT NATURE RESERVE
M. VON FINTEL
JAMES R. VOORTMAN
DR & MRS P. VORSTER
JOAN WALKER
ERIC WALTON
C.J. WARD
WILLIAM WATERFIELD
RON WEDDERBURN
DEZ WEEKS
ROLAND A. WELSBIE
W. WETSCHNIG
DR J. DE V. WICHT
DR ANDREW WILSON
B.C. WINKLER
B. NIGEL WOLSTENHOLME
ROB & JANET WOOD
G.W. WOODLAND
SONIA WORTHINGTON-SMITH
ALBINA WORTLEY
GERALD & HELEN WRIGHT
SATOSHI YOKOYAMA
DR G.J. YOUTHED
TOMOHISA YUKAWA